Handbook of
# NONWOVEN FILTER MEDIA

*Front cover image:* Nanowave® Filter Media by Hollingsworth and Vose Co (H&V). East Walpole, Massachusetts USA. Courtesy of Hollingsworth and Vose (H&V)

Handbook of
# NONWOVEN FILTER MEDIA

IRWIN M. HUTTEN

ELSEVIER

AMSTERDAM • BOSTON • HEIDELBERG • LONDON
NEW YORK • OXFORD • PARIS • SAN DIEGO
SAN FRANCISCO • SINGAPORE • SYDNEY • TOKYO
Butterworth-Heinemann is an imprint of Elsevier

B H

Butterworth-Heinemann is an imprint of Elsevier
The Boulevard, Langford Lane, Kidlington, Oxford OX5 1GB, UK
225 Wyman Street, Waltham, MA 02451, USA



ISBN: 978-0-08-098301-1

**British Library Cataloguing-in-Publication Data**
A catalogue record for this book is available from the British Library

**Library of Congress Cataloging-in-Publication Data**
A catalog record for this book is available from the Library of Congress

For information on all Butterworth-Heinemann publications
visit our website at http://store.elsevier.com/

# DEDICATION

Soon my lovely wife, Susan, and I will start preparing for our fiftieth wedding anniversary.

During all this time, Susan amazed me with her love, support, and patience. She understood the time it took to write the first edition of this book, and once again she understood as I wrote this second edition. She provided encouragement, support, and assistance for this endeavor. I am very fortunate to have a loving wife like Susan. It is with great pride that I dedicate this book to her.

# CONTENTS

# PREFACE

When I wrote the preface of the first edition of the *Handbook of Nonwoven Filter Media*, I reported that the writing was a labor of love and there was an awful lot of labor to love. Four years ago I undertook the writing of the second edition and learned that upgrading and revising the first edition required as much love.

The intent of this book is to provide the reader with a fundamental understanding of nonwoven filter media and to upgrade the technology from the 8 years since the first edition was published in 2007. This edition provides the following information on nonwoven filter media: how it is used, how it is made, its raw materials, how it is processed, and how it is managed and regulated. Several new features were added in this second edition. This includes a section on Computational Flow Dynamics (CFD) prepared by Prof. George C. Chase, Ph.D., University of Akron, Akron, Ohio, USA, and short sections on Global and North American marketing statistics prepared by Brad Kalil, INDA (Association of the Nonwovens Fabrics Industry), Cary, North Carolina, USA.

The definition of nonwoven filter media was modified to include membrane filtration technology. Membrane filtration has grown so much and so inter-wrapped with nonwoven filter media that to write about the former without including the latter is like trying to write about a large family without including the in-laws. It can be done but it is awkward. Accordingly, the second edition's discussion of membrane filtration and separation is greatly expanded.

Another expansion is the discussion of nanofiber technology. Nanotechnology is one of the emerging technologies of the twenty-first century. Nanofibers and nanofiber webs, as used in filter media, are an important part of that emerging technology. The filtration and separation of nanoparticle sized contaminants is a major concern of modern day nanotechnology.

The accelerating rate of global technology is demanding extensive legislation, regulation, standardization, and control. The emphasis is on safety and environmental concerns. Sustainability has become an important issue in the marketplace and in manufacturing. The demand includes the technology of nonwoven filter media. For this reason, a whole new chapter (Chapter 10) on federal law and regulatory activity has been added to the second edition. The new legislative and regulatory activities place new burdens and hardships on the manufacturers of nonwoven filter media. However, they create abundant opportunities for nonwoven filter media to solve the problems of the environment and safety. The opportunities far outweigh the burdens.

The manufacturers of nonwoven filter media should be prepared to take advantage of these opportunities.

Other new items that have been added to the second edition include color technology and new separation processes. For example, barrier fabrics that separate oil from water as a result of oil spills are described. The second edition includes membrane separation processes such as ion exchange, electrodialysis, and pervaporation.

Much information in the first edition had to be updated. The products of many companies had become obsolete and had to be updated. Standards and regulations have been revised. The author has attempted to present the most recent ones at the time of writing. Companies have been bought out and organizational names have been changed. Again, the author has attempted to present the company and its most recent form.

A guided tour of the second edition starts off with introductory remarks in Chapter 1 followed by the marketing statistics prepared by Mr. Kalil. The Introduction includes a description of what nonwoven filter media is all about and how it is used. Definitions of filtration and separation, nonwovens, and nonwoven filter media are included in Chapter 1. These definitions are necessary to set the scope of the book. In addition to the filtration of particulate matter from a fluid stream other forms of separation phenomena are included. Among these are adsorption, electrostatic and electrokinetic effects, coalescence, diffusion, membrane separation phenomena, and others.

The definition of nonwoven filter media includes paper or wet lay media, felts, and membrane media. These items are not necessarily included in the ISO definition of nonwoven filter media. The defining factor is that they be used as a filter medium.

Chapter 1 classifies nonwoven filter media based on forming process. The major processes come under the major headings of dry formed, wet laid, and composite structures. The dry formed processes are further divided into polymer sourced and fiber sourced. Polymer sourced are those processes in which a molten or solution polymer is spun through spinnerets to form a random web. Fiber sourced are those processes such as carding or air laid in which the raw feed to the process is in a fiber form.

The mechanisms and theory of nonwoven filter media are presented in Chapter 2. The theoretical presentation is classical. It includes the Monte Carlo techniques of Piekars and Clarenburg (87) to demonstrate random web structure and how pores are defined. It is in Chapter 2 that we find Dr. Chase's introduction to Computational Flow Dynamics (CFD).

Chapter 3 is an attempt to explain the broad range and variety of the properties of nonwoven filter media. The intent is to relate the properties to the use of the medium and provide guidance and its selection. The properties not only include classical engineering properties such as strength and rigidity, but also such as permeability and pore size that relate to filtration performance.

To truly understand nonwoven filter media, one must understand how it is made. Chapter 4 discusses the raw materials of nonwoven filter media and Chapter 5 describes the processes by which they are made. The raw materials of Chapter 4 are subdivided into polymers, fibers, resins, and additives (finishes). The processes of Chapter 5 not only include the forming processes by which nonwoven filter media are classified, but also include the downstream converting and finishing operations. For example, needle punching and/or hydroentanglement of carded webs are well described. Since so much of nonwoven filter media is used in a pleated cartridge or pleated panel form, the various pleating processes are detailed. The second edition additions to Chapters 4 and 5 include the materials and processes for making filter membranes, both organic and inorganic.

Test methods pertinent to nonwoven filter media are discussed in Chapter 6. The methods are presented in a way to help the reader understand the property being tested. For example, in the section on strength properties, the stress–strain curve is presented to help the reader identify and understand the relationship of each of the strength properties to the medium. While many of the tests described are medium sheet tests, also described are filtration tests in which the medium is tested as part of a filter.

Applications of nonwoven filter media are broken down into Liquid Filtration (Chapter 7), Air Filtration (Chapter 8), and Engine Filtration (Chapter 9). Engine filtration is concerned with filters for automobiles, heavy duty vehicles, off-the-road equipment, and rotating machinery. It is of sufficient market size to justify its own chapter. Gas turbine air filtration, which was part of Chapter 9 in the first edition, is reviewed in Chapter 7 of the second edition.

Chapter 10 has already been mentioned above as discussing federal law and regulatory activity that may relate to nonwoven filter media. Included is a review of major federal laws that may affect nonwoven filter media. Among the major laws included are the Clean Air Act, the Clean Water Act, OSHA, etc. Chapter 10 also provides a review of the "Globally Harmonized System of Classification and Labeling of Chemicals (GHS)." This is a system administered by the United Nations and adopted by the USA. Chapter 10 also includes an attempt to explain the concept of "sustainability." I found this difficult to do. This attempt in Chapter 10 to introduce the reader to federal and regulatory law that may affect his or her job and his or her business is only a beginning. As this book was being written, there was so much regulatory activity in progress, that probably, by the time this book is published, many changes in the business of filtration media will already be in place.

It is important to identify those organizations that are involved in developing, coordinating, promulgating, revising, amending, reissuing, interpreting, certifying, regulating, or otherwise maintaining standards that are related to filtration and filter media. This was done in Chapter 11. The list is by no means complete, but does present a representative spectrum of those organizations involved. In Chapter 11, many of the worldwide standards applicable to nonwoven filter media are tabulated. Included is the table of contents for the most recent (2012) INDA – EDANA Harmonized Test Methods (347).

In the world of nonwovens, words are everything. A number of words are bandied about that have different meanings to different people. Sometimes they are used in the wrong context. For example, many people use the word porosity when they mean permeability. These two words are related to each other; however, they are two different properties of nonwoven filter media. Another example is the variety of words used to describe the distribution of particles in a fluid. Included are words such as sol, aerosol, emulsion, latex, dispersion, suspension, and solution. In many cases they are used interchangeably and incorrectly. This book includes a glossary of filtration terms and terms related to filtration. The primary purpose of the glossary is to provide the reader with meanings of words related to filtration and separation media that he may not be familiar with, and to provide specific definitions to confusing words or terms.

There is a bibliography with approximately 350 references. They are numbered in the order in which they appear in the book. In the text they are identified by a number in parentheses. When I wrote the first edition, much of my references came from the archives of support organizations such as INDA, TAPPI, and the American Filtration Society. In writing this second edition, I was able to take great advantage of Google. I could Google just about every topic that I was researching and find all kinds of information concerning that topic. Some of the best information came from the websites of companies involved in the filtration industry. Some very good information came from Wikipedia, although, I always had to check it out to make sure it was not erroneous or misleading.

I found the following books very helpful:

*Advances in Aerosol Filtration* Editor: Spurny, Lewis Publishers, 1998
*Air Filtration* Brown, Pergamon Press, 1993
*Air Filtration* Davies, Academic Press, 1973
*Filters and Filtration Handbook* 5th Edition, Dickensen, Elsevier, 2008
*Filtration – Equipment Selection Modeling and Process Simulation* Wakeman and Tarleton, Elsevier, 1999
*Filtration Technology Handbook* Butler, INDA, 2000
*Handbook for Pulp and Paper Technologists* 2nd Edition, Smook, Angus Wilde Publications, 1997
*Handbook of Filter Media* Purchas and Sutherland, Elsevier, 2002
*Introduction to Nonwovens Technology* Batra and Pourdeyhimi, DES*tech* Publications, Inc., 2012
*Membrane Technology* Cui and Muralidhara, Elsevier, 2010
*NAFA Guide to Air Filtration* 4th Edition, National Air Filtration Association, 2007
*Nonwovens: Theory, Process, Performance, and Testing* Editor: Turbak, TAPPI Press, 1993

*Perry's Chemical Engineers' Handbook* 7th Edition, Perry and Green, McGraw Hill, 1997
*Transport Phenomena* Bird, Stewart, and Lightfoot, John Wiley and Sons, 1960

A "Handbook" like this cannot be written without the assistance of organizations and people in the filtration industry. The following organizations and their people were particularly supportive of this work and should be acknowledged as such:

*INDA (Association of the Nonwovens Fabrics Industry)* was particularly supportive. Brad Kalil has already been acknowledged above for writing the sections on Marketing Statistics. In addition, President David Rousse and his predecessor Rory Holmes provided additional support by allowing me to visit their facilities, consult with their personnel, review their research material, and provide me with contacts and items of information that were of great value in writing the book. Personnel who were extremely helpful included Jessica Franken, Steve Ogle, Phil Pitt (no longer with INDA), Joan Izzo, Ian Butler (now retired), and Cindy Garcia.

*AFS (American Filtration and Separations Society)* was very important to me in providing reference material, and the names of contacts. Whenever I needed help, Executive Manager Lyn Shol and her predecessor Suzanne Sower went out of their way to provide it.

*H&V (Hollingsworth and Vose Company)*, my former employer, was one of the biggest providers of information used in this book. Much of this required copyright permission. Cathy Aikman and Angelika Mayman were very thorough in coordinating my permission requests and making sure that I got the proper response. Other personnel who made themselves available were Bob Murphey, Andrew Shepard, Per Lindblom, Eric Westgate, and Dr. John Fitzgerald.

*Ahlstrom* was another big contributor of information for this book. Many times I consulted with Rod Komlenic who either provided the information I needed or referred me to the person who could. I also received help from Kent Williamson, Stephanie Mulligan, and Tomi Tekala.

*Donaldson Company* provided a large amount of information in this book. Laura Russell was my prime contact and did an outstanding job in coordinating the large number of permission requests that I had submitted. Joe Israel and Kristine Graham were other personnel who I had consulted with. Donaldson Tetratex was another part of the Donaldson organization that helped me. Sara Meyer did a lot of work in helping to prepare the Tetratex® product tables. Ed Ricketts was also involved.

*Cerex Advanced Fabrics* provided powerful support to me in the writing of the book. President Jim Walker and his management team of John Hancock, Jim Bostick, Erin Carter, Albert Ortega, and Warren Whitfield invited me to their facilities and provided me with a wealth of information about nylon: its fiber, its fabric, and its spunbond nonwovens. John Hancock was excellent in helping me coordinate with the organization.

*Elmarco* President Fred Lybrand and Lee Baker were kind enough to invite me to their facility and observe the operation of their pilot electrospun web making machine. They provided me with much of the information on the production of electrospun webs discussed in this book.

*Hills Inc.* is an important manufacturer of melt-blown web making equipment. They have also pioneered the development of bicomponent and multicomponent fibers, much of which is discussed in this book. President Arnold Wilkie and Tim Robson allowed me to visit and tour their facilities. They provided me with important information about the production of melt-blown webs.

In addition to the above there were a large number of people who took the time from their work to make some form of contribution to the "second edition." This may have been an advisory discussion, help with copyright permission approval, providing me with contacts and/or providing me with technical information from their organization. For this help I acknowledge the following:

Michelle Ecken, Pam Henley, and Hugo Hemel, AAF International, Debby Leglu, Adventec, Don Largent, Air Techniques International, Gary Smook, and Hilde Wiebe, Angus Wilde Publications, Kathe Hooper, ASTM International, Dr. Roy Broughton, Auburn University, Becky Shaver and Paul Hume, Bekeart Corporation, Vincie Albritton, Clemson University, Frank Baker, Consultant, Ed Gregor, Consultant, Ed Homonoff, Consultant, Dr. Norman Lifshutz, Consultant, Devika Patel, Dr. Simon Frisk, and Alan Eaton, DuPont, Mark Antilla (now with Lydall), Eastman Chemicals, Dave Merrill, Engineered Fibers Technology, David Bowers, Ertel Alsop, Jeff Dugan, and Mike Hodge, Fiber Innovation Technology, Kial Gramley, Roger Lipton, and Eduardo Martinez, FibeRio, Carrie Jones, Christine Sun, Filtration Technologies International, John Scrivener, Frazier Instrument Company, Chris Murray, Freudenberg Filtration Technologies, Amy Cotton, GE Healthcare Life Sciences (formerly Whatman), Frank Ludwick, Georgia-Pacific, Jeff Riggi, Glens Falls Interweb (GFI), Pat Paquet (no longer with Gravure) and Harriet Kurtz, Graver Industries (Formerly KX), Robert McCurdy, GTI Graphic Technology, Alexander Koukoulas, and David White (now retired), Herty Foundation, Steve Donker, Industrial Filter Corporation, Klaas De Waal, International Filtration News, Chris Lyons, JCEM GmbH, Brian Collins, John Wiley, Tanya Bradby, and Jared Cox, Johns Manville, Jim Mash, Kaydon, Rob Martin, Ron Cox, Jane Thiel, Nancy Zimmerman, Tony Fedel, and Chris Creagon, Kimberly-Clark, Isaac Goan, KmX Corporation, Traute Nieuwoudt, Koch-Glitsch, Susan Havey, Labthink International, Camden Arthur, Lauscha, Andrew Slater, Lenzing, Mark Wollinger, Lubrizol, Rob Bender (now with Eastman Chemicals) and Bill Cambo, Lydall, Dr. Pius Trautmann, Mann + Hummel, Ron Sunderland, Manufacturing Machine Corporation, Jack Duffie, Matrix Membranes, Chris Noe, Midwest Filtration, Dana Hubbard, Millipore, Ira Stone, MiniFIBERS, Inc., Ken Rubow, Mott Corporation, Jim

Hanson, MTS Publications, Terry Driscoll and Alan Veeck, NAFA (National Air Filter Association), Dr. Benham Pourdeyhimi, Nonwovens Institute, Martin Hubbe, North Carolina State University, Ted Oberlin, Oberlin Filter Company, Ted Donno and Pablo Escobar, Omnova, Jeff Brinks, Onyx Specialty Papers, Mary Harkins, Pall Corporation, John Neely and Bryce Davis, Polymer Group Inc. (PGI), Dr. Krishna Gupta, Porous Materials, Inc. (PMI), Jeff Frierderichs, Powder Technology, Inc. (PTI), Jan Lawson and Stefan Hans, Karl Rabofsky GmbH, Lavon Harper and Dr. Norman Marsolan, Renewable Bioproducts Institute (RBI) (formerly IPST), Bethany Donovan and Anthony Lotto, Schroeder, Dr. Valery Antipov, Simpex, Gary Bessee and Maria Martinez, Southwest Research Institute (SwRI), Lisa Stephens and Jeff Wells, TAPPI, Steve Buck, Bryan Xu, and André Grundahl, TDC Filter, Dr. Larry Wadsworth, Dr. Peter Tsai, and Dr. Randall Bresee (retired), University of Tennessee, Steve Berg, Thwing Albert, Frank Mariani, Tiger Vac, Tim Warriner, Valmet, Dr. Gajanan Bhat, Nonwovens Research Laboratory at the University of Tennessee (UTNRL), Tim Johnson, TSI, Dr. Garth Wilkes, Virginia Tech, Coby Leber and Simone Rampmeier, Voith, Dr. Wilson Poon and Bob Rodriguez, W.L. Gore & Associates, Miles Wright, Xanofi.

I could not have put this book together without the assistance of the personnel at Elsevier. Jonathan Simpson is the publisher who got the second edition rolling. Fiona Geraghty was my Editorial Project Manager contact until she was promoted. Afterwards, Cari Owen took over the role and remains my prime contact. Both Fiona and Cari were excellent. Anytime I needed help, advice, or assistance they were there to provide it. As this book went into the publishing stage, Debbie Clark became involved as Project Manager for the Publishing Services Department. Debbie had previously worked with me during the publishing of the first edition.

Any endeavor to write a book like this needs assistance, patience, and encouragement from the family. I have already acknowledged the contributions of my wife, Susan Hutten, in the dedication section. I also acknowledge my sons Scott and Ethan Hutten, Scott's wife Laura Hutten, grandson Lucas Hutten, Ethan's wife Trish Hutten, and granddaughter Lilah Hutten.

# CHAPTER 1

# Introduction to Nonwoven Filter Media

## 1.1 INTRODUCTION

### 1.1.1 The challenges of filtration and separation

A filter medium is that part of a filter that does the filtering or that part of a separator that does the separating. Two important goals of filtration and separation are to protect humankind from the environment and the environment from humankind. People need fresh water to drink, clean air to breathe, and healthy foods for nutrition. Often, the only way to achieve these requirements for filtration and separation to remove the toxic and dangerous contaminants that are inherent in these sources. To illustrate, most of the water on earth is undrinkable ocean and seawater that can only be made drinkable by using separation techniques such as desalinization to remove the over-abundance of salts and minerals. The filters in heat, ventilation, air conditioning (HVAC) systems help remove allergens, microbes, toxic compounds, and other chemical irritants from the fresh air we breathe and in so doing makes the air more breathable, healthy, and comfortable. Grown and processed foods often contain chemicals, biocides, and preservatives that by themselves are unhealthy for human consumption. The technologies of filtration and separation are often necessary to clean up these foods and make them safer, healthier, and more tasteful.

Yes, much of the contamination in the environment is natural; however, humankind does its fair share in contributing to this contamination. Human activities that are a serious source of the contaminants in our environment include waste discharges into our streams, rivers, and lakes; air contamination from the burning of fossil fuels; and disposal of solid waste. It is a big challenge of the Filtration and Separation Industry to develop the filtration and separation technologies that prevent the discharge of dangerous and undesirable contaminants into the environment. It is also a goal to protect people from contaminant dangers in the environment, whether they are annoying chemicals, toxic drugs, disease-causing microorganisms, and/or corrosive fumes.

Environmental and safety concerns and the issues of sustainability are major factors driving the use of filtration and separation technology and application. However, non-environmental problems are also a challenge to the Filtration and Separation Industry. Lube oil filters help automobile engines run longer and more efficiently. Tea bags and coffee filters provide the consumer with better tasting tea and coffee. In industry many chemical and pharmaceutical recoveries, separations, and purifications can only be accomplished with the right type of separation medium.

*Handbook of Nonwoven Filter Media*
ISBN: 978-0-08-098301-1

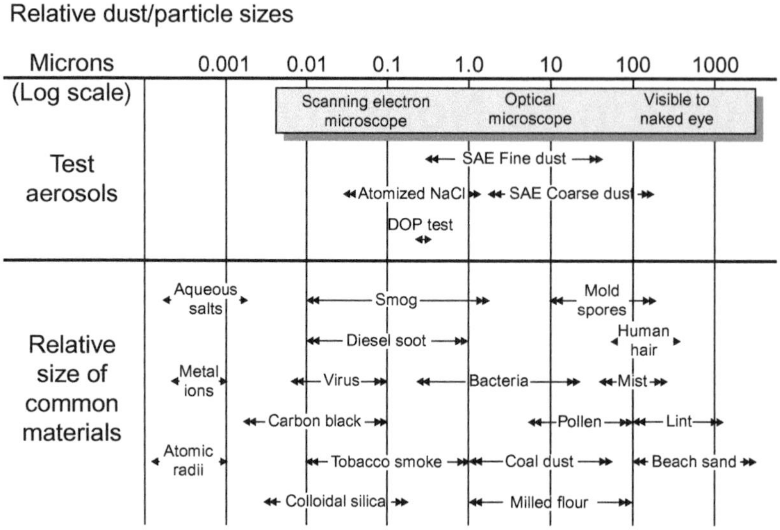


**Figure 1.1** The Sizes of Contaminants in Air. *Reproduced with permission of Donaldson Company, Inc. All rights reserved.*

Fortunately, the Filtration and Separation Industry is rising to the challenge. New designs in filters and separation devices are dealing with the ever-increasing complexities of separating contaminants from a fluid stream. More and more filters are being made from recyclable polymers and biodegradable biomaterials. Filter media being introduced into the market place demonstrate new technologies to remove a variety of particles from fluid streams. This includes separation of particles of all shapes, sizes, and forms. As an illustration, Figure 1.1 shows the sizes of some of the various contaminants in air. Note the range of particle sizes vary from eye-visible beach sand particles that are as large as 1 mm (1000 μm) to particles of molecular and element dimensions that are less than 1 nm (0.001 μm). Some test dusts and aerosols are also shown in Figure 1.1 for comparison purposes[1].

Technologies for removing these particles extend beyond simple mechanical filtration, that is, the separation is something more than just trapping random particles in the matrix fibers of a nonwoven mat. Other separation technologies such as adsorption, electrostatic effects, coalescence, reverse osmosis, ion exchange, antimicrobial effects are playing more and more important roles in meeting the challenges of modern-day separations. New materials and filter designs are greatly affecting the performance of modern-day filtration and separation media. The most significant of these are membrane technology, nanotechnology, adsorption media, coalescing media, and composite and impregnated medium structures. It should be noted that the terms

[1] Test dusts and aerosols for filtration testing are discussed in Chapter 6, Section 6.9.5, of this book.

filter and filter medium have become generic for many forms of separation that are in addition to the simple concept of mechanical separation of contaminant particles by a porous medium.

Like any other industry, the Filtration and Separation Industry is buried in a massive bureaucracy of laws, regulation, environmental and safety matters, and quality management systems that makes operations exasperating and expensive. Although the new regulatory activity provides increased demand on the manufacture and performance of filter/separators and their media, the opportunities are vast. New and emerging technologies, such as nanotechnology, electric vehicles, electronics and computer development, advances in medicine and medical practice, all require new kinds of thinking in the development of filters and filter media and it well may be worth it.

The marketplace for filters and filter media are growing at a rapid pace. This is evidenced by the marketing statistics reports in Sections 1.1.2 and 1.1.3. These reports were provided by Brad Kalil, Director of Market Research and Statistics at INDA. Reported in Section 1.1.2 are excerpts from the "Worldwide Outlook for the Nonwovens Industry" published by INDA/EDANA in 2012 (1).

It indicates that the global consumption of nonwoven filter media was 465,000 tons in 2011 and is projected to grow to 744,000 tons in 2016. Although filtration media are projected to be only 8.7% of the total global nonwoven market in 2016, it is the fastest growing segment. From 2011 through 2016 the growth rate for nonwoven filter media are projected to be 9.8% per year. The largest market growth will be in Asia, which in 2016 is projected to consume 48% of all nonwoven filter media.

Section 1.1.3 is the filtration section of the "North American Nonwovens Industry, 2013–2018" report that will be published by INDA in 2015 (2) roughly at the same time this "Handbook of Nonwoven Filter Media" are being published. It indicated the filtration market will grow from $5.4 billion in 2013 to $19.1 billion in 2018, with a growth rate of 4.4% per year. In tons, the growth rate is projected to be 3.5% per year. The report also indicates that for air and liquid filtration, air filtration will be 70% of the marketplace and liquid filtration will be 30%. In terms of unit sales volume ($), transportation, which includes filters for engine air intake and for the cabin air of automobiles, heavy duty vehicles, trains, aircraft, etc., is by far the largest segment of the air filtration market.

### 1.1.2 Worldwide outlook for the nonwovens industry—leading applications by major region (1) Written by Brad Kalil, Director of Market Research and Statistics, INDA.

The spectrum of nonwovens' end uses is very broad and each market segment has its own drivers. The split between "durables/disposables" is far from being sufficient when you want to analyze the evolution of the nonwovens market. We can say that absorbent hygiene and wipes personal care products have some common drivers (such as demography, incomes), but they are still not at the same stage of their market developments at a global level. Within the industrial applications, as seen during the economic crisis (at least in the period 2008–2011), the construction and the automotive sectors did not react in the same way.

This chapter proposes an analysis of the consumption of nonwovens in each of these main applications. Obviously, various types of nonwovens can meet the same requirements for a particular end use. Hereafter, though, all types of fabrics are considered without any differentiation by web making process. The different ways to substitute nonwovens with other nonwovens to fulfill the same demand in a particular end use will not be explained. Moreover, the presentation of data in tonnage does not allow us to catch the products' weight reductions recorded during the last decade. This decline in grammage has been particularly important in hygiene and wipes for instance.

Table 1.1 provides the outlook for nonwovens by leading applications for the combined three main regions of the world: NAFTA, Europe, and Asia-Pacific. In 2011 they represented together nearly 87% of the nonwoven roll goods produced worldwide. These figures by leading applications give the total deliveries coming from the three regions, including not only the local consumption but also the nonwovens exported outside the region.

The sales of nonwovens coming from these three regions will continue to increase in tonnage in the upcoming five years. During the next five years, the average growth

**Table 1.1** Outlook for Nonwovens by Leading Applications in Asia-Pacific, Europe, and NAFTA (in Tons)

| | 2001 | 2006 | 2011 | 2016 | Growth 2001-2011 (%/Year) | Growth 2011-2016 (%/Year) |
|---|---|---|---|---|---|---|
| **Hygiene** | 960 | 1,224 | 1,799 | 2,416 | 6.5% | 6.1% |
| **Medical** | 149 | 222 | 334 | 459 | 8.4% | 6.5% |
| **Wipes** | 383 | 658 | 907 | 1,335 | 8.9% | 8.0% |
| **Apparel** | 159 | 242 | 245 | 304 | 4.4% | 4.5% |
| **Home** | 363 | 466 | 547 | 705 | 4.2% | 5.2% |
| **Filtration** | 199 | 256 | 465 | 744 | 8.9% | 9.8% |
| **Automotive** | 194 | 250 | 342 | 505 | 5.8% | 8.1% |
| **Other Industrials** | 99 | 172 | 173 | 254 | 5.7% | 8.0% |
| **Construction** | 407 | 696 | 872 | 1,241 | 7.9% | 7.3% |
| **Others** | 378 | 624 | 666 | 866 | 5.8% | 5.9% |
| **Total** | **3,294** | **4,809** | **6,351** | **8,849** | **6.8%** | **6.9%** |

Source: INDA, EDANA estimates.

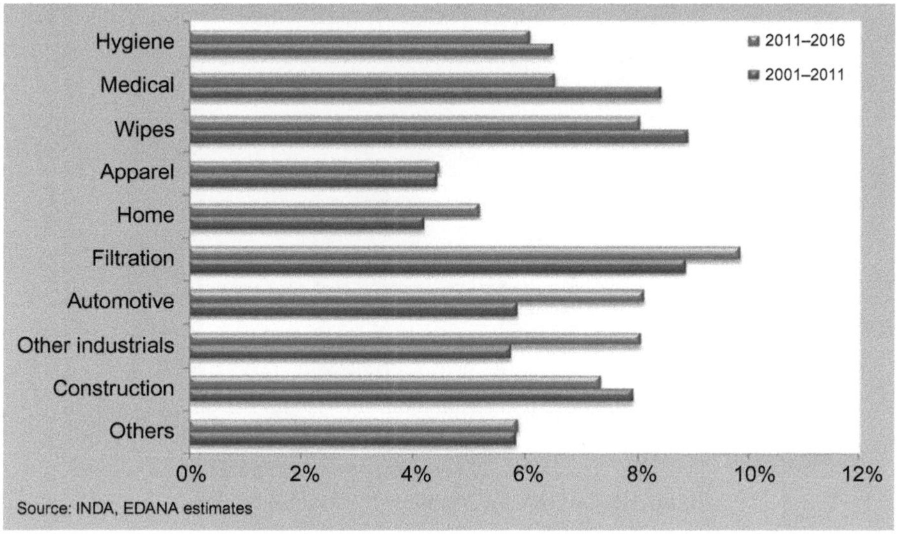


**Figure 1.2** Growth Rate by Application.

rate will be around 6.9% per year. The same calculation for the period 2001–2011 gives a 6.8% growth rate on average.

Figure 1.2 compares the growth rates of the various markets using the growth data in Table 1.1. The figure compares the annual average growth for the last ten years (2001–2011) and the evolution foreseen for the next five years. All segments demonstrate a sustained growth. Wiping applications will grow at a lower rate in the 8% range, while the filtration market will be the fastest, with growth forecast at 9.8% per year. Hygiene at 6.1%, automotive at 8.1%, and construction at 7.3% are significant markets and will continue to grow in the coming five years. The developments expected in Asia in general, and in China in particular, are the principal reasons for these forecast growths, as they should compensate slower rates expected in the two other regions.

Air and liquid nonwoven *Filtration* medium items[2] are presented together in Figure 1.3. The average annual growth rate for the next five years is forecast to be about 10%, while the market growth rate over the period 2001–2011 was 9%. According to such a trend, the quantity of nonwovens used in filtration should reach the level of 744,000 tons in 2016. We estimate that 358,000 tons, equivalent to 48% of this total volume will be produced in Asia-Pacific and 213,000 tons in the NAFTA region. Europe will produce about 173,000 tons.

[2] Air filtration regroups any fabric for gas (air or dust) filtration. This also includes filters for the car industry. Filters for hospital facemasks are registered under "medical." Liquid filtration products are any fabric for liquid (food or nonfood) filtration as well as fabrics for tea or coffee (filtration bags). This also includes oil filters for the car industry.

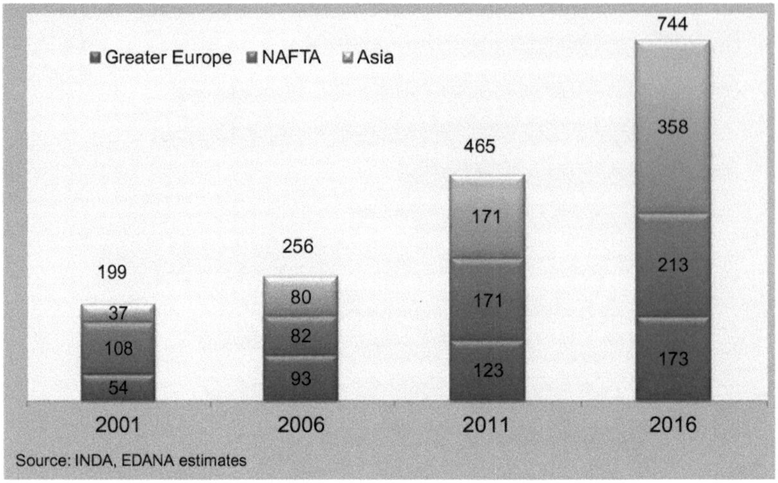


**Figure 1.3** Outlook for Nonwoven Filtration Market.

Figure 1.3 shows the development of the filtration market over the past decade with a five-year forecast. The fastest growing region will be Asia-Pacific with a growth expected to average 15.9% per year throughout the 2011–2016 time period—it was 16.5% over the last decade. The NAFTA region is forecast to increase about 4.5% per year through 2016. Even though Europe recorded the lowest volume compared to Asia and NAFTA, this segment has grown on average, on the "Old Continent," by 8.6% every year since 2011. European growth is forecast at 7.0% per year through 2016.

### 1.1.3 North American nonwoven industry outlook 2013–2018: 1. filtration (3) Written by Brad Kalil, Director of Market Research and Statistics, INDA.

**End Use Comprises**

- Air Filtration
  - Transportation
    - Induction/Air Intake
    - Cabin Air (Standard and Combi)
  - Commercial HVAC
    - High Efficiency (HEPA, ULPA)
    - Mid-Range Efficiency (MERV 9-15)
    - Low Efficiency (MERV 1-8)

- Residential HVAC/Furnace
    - Low End
    - High End
- Personal Protection
    - Face Masks (Molded/Pleated)
    - Respirator
- Industrial
    - Baghouse Dust Removal (Low Temperature and High Temperature)
    - Air Intake
- Vacuum Cleaner Bags

- Liquid Filtration
  - Food
    - Water
    - Beverage (Tea, Coffee, Milk, Bottled Beverage, Others)
    - Cooking Oil
  - Non-Food
    - Transportation (Oil/Hydraulic/Lubrication/Fuel)
    - Industrial/Manufacturing
    - Swimming Pool/Spa
    - Blood

**Main Drivers**

- Consumer Awareness and Sensitivity
- Legislation and Regulatory Controls with Enforcement
- Transportation Market
- Residential Home Starts
- Nonresidential Construction, Such As Office and Manufacturing

**Secondary Drivers**

- Filter Efficiency Improvements
- Use of Nanofibers
- Increasing Share of Filter Media
- Sustainability

Filtration has become so important that you can find it almost everywhere in your life. Filtration is a separation process with a filter medium—this section attempts to quantify the nonwoven filter medium used in the many filtration processes.

Mechanical filtration refers to a device used for separating one substance from another. The mechanism is to pass the contaminated fluid through a filter medium and to trap and contain the particles. Filters are constructed with filter media that can be paper, woven (including woven metal), nonwoven (including glass), microporous membranes, or a combination/hybrid of different media. Nonwoven fabrics typically add backup support and/or mechanical strength to the comparatively thin and fragile polymer-based membranes that are produced by directly coating onto the carrier material's surface.

Nonwovens are well suited to the filtration market, because the fabric can be specifically engineered to provide the precise porosity and flow rate needed for the particular filtering application. Filter media are manufactured from nonwovens produced using the dry laid, wet laid, and spunmelt (spunbond and melt-blown) processes from a variety of materials. The melt-blown technology is important to the filtration market as melt-blown material significantly improves a filter's efficiency and dust-holding capabilities. Because of these properties, the filtration market is one of the largest end markets for the melt-blown technology.

Nonwoven fabrics and membrane filtration are the predominant media, with approximately two-thirds of market share combined, in terms of volumes of filtration medium roll goods in comparison to the other material forms.

The filtration market customer base is the most diversified of all the nonwoven end use categories with more than 30 major market segments.

Transportation is the biggest market for nonwoven filter media, as it uses both air and liquid filtration, the two segments of the filtration market. The average passenger car or truck contains more than 50 filters which perform various functions, including improvements in the performance of the engine, removing contaminants in the oil, fuel, and transmission systems, and improvements in the quality of air in the cabin. In the first five years, an average passenger car will have the filters changed 3–5 times, that is a lot of filters in a five-year span.

The sales to end users of filtration products in North America were estimated at $15.4 billion in 2013, a 4.0% annual increase from 2008. The filtration market is forecast to rise to $19.1 billion by 2018, a 4.4% annual growth (refer Table 1.2 and Figure 1.4).

The most important factors affecting market growth is the consumer demand for purer air, water and cleaner products, and government environmental regulations, which have become progressively more rigorous. The market for nonwovens in filtration media are also growing—due to nonwoven media being more economical and better performing—by taking share from woven and paper-based filter media.

The general health of the economy is also a driver, as many filtration products are in residential and commercial buildings, in vehicles, and used in industrial processes.

**Table 1.2** North American Filtration Market Sales to End Users by End Use

| | 2008 Sales ($MM) | 2013 Sales ($MM) | 2018f Sales ($MM) | 2008–2013 Sales% (CAGR) | 2013–2018f Sales% (CAGR) |
|---|---|---|---|---|---|
| **Air** | 9,011 | 11,028 | 13,703 | 4.1 | 4.4 |
| **Liquid** | 3,695 | 4,411 | 5,412 | 3.6 | 4.2 |
| **Total** | **12,706** | **15,439** | **19,115** | **4.0** | **4.4** |

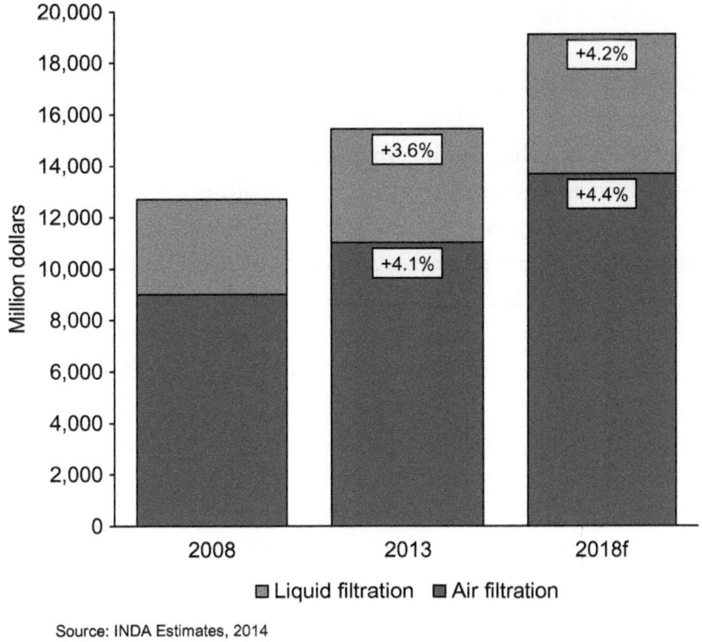


**Figure 1.4** ***North American Filtration Market.*** Dollar Sales to End Users by Principal End Use with Annual Percentage Change (CAGR) from Preceding Five Year Plan.

The filtration market, therefore, grows when new buildings are built, new cars are sold, and products are consumed. Capital investments in equipment that were delayed during the financial downturn should also provide a boost for nonwoven filter medium demand as business spending is poised to accelerate over the forecast period to upgrade and add new capacity, in addition to the potential for "reshoring" of manufacturing (refer "North American Nonwoven Industry Outlook, 2013–2018," (3)). Much of the filtration market demand is based on replacement, that is, filters are exchanged after a useful service life. This feature is a factor across the consumer, commercial, and industrial filtration markets, though it is in the consumer end uses that, given tighter economic times, consumers may have used filtration products past their useful life in order to save money. As the economy strengthens, consumers should return to replacing their filtration products closer to the recommended replacement periods.

Consumers are increasingly aware of the importance of indoor air quality and drinking water quality. We all want to breathe clean air whether it is at work, at home, or inside our vehicles. Concerns over the air quality inside buildings is now receiving the attention than the outdoor air received in the 1970s and this will be a driver of nonwoven filter media through the forecast period. The desire for companies to have zero-energy buildings will also drive nonwoven usage as it is necessary to have ventilation systems that supply high indoor air quality that require high quality air filters and need to be frequently changed.

New legislation and regulatory controls are increasing at local, state, and national levels. Enforcement of these regulations by the EPA and other agencies has increased the demand for products that limit emissions as well as an increased demand for products that protect people from the effect of emissions.

The Congressional Clean Air Act of 1990—with enforcement by the EPA—was the first of many pieces of broad legislation with real teeth. It provided new regulations for particulate and other contaminant control. Other specific examples have been the enforcement of coal-based power generating facilities including mercury, $SO_x$, $NO_x$, and $CO_2$ exhaust. Beginning in 2007, new diesel car and truck soot and $NO_x$ exhaust were regulated in the United States.

To meet these consumer demands and the demands of the commercial and industrial markets, some of which were driven by governmental regulations, nonwoven producers continually work to *improve the efficiency* of filter media while enhancing the removal capability of both particulate and chemical contaminant. These have resulted in filter manufacturers increasing the range of spunmelt nonwoven materials, incorporating *nanofiber layers* or coatings, and developing *hybrid composite materials* all while optimizing the medium structure and element configuration by computational modeling. This is in addition to the ever-present demand of sustainability.

Higher efficiency is one of the key requirements for the filter media market and further improving the efficiency is a continuous development. Continued improvements in air medium efficiency, for example, to provide better pressure drop characteristics (meaning it is much easier for the ventilation system to push air through the filter), allow significant energy savings throughout the life time of the filter.

The ability to reliably manufacture materials at the nanoscale level has created product development opportunities that are not previously possible. The launch of new nonwoven products enabled by nanotechnology is increasing in pace as the understanding of the properties and benefits continues to rise. Applications that will benefit from nanofiber integration are those where high surface area combined with little additional volume or weight is desired. Air filtration end uses range from residential to industrial HVAC systems, as material with nanofiber achieves higher levels of filtration efficiency at reduced pressure drops. Automotive fluids such as oil and fuel are benefitting from the unique combination of a very thin nanofiber nonwoven and its inherent high surface area.

The U.S. National Nanotechnology Initiative defines nanotechnology as "the science, engineering and technology conducted at the nanoscale level—about 1–100 nanometers." Scientists have been conducting research at the nano level for decades; in fact, wet laid glass medium has been using nanofibers for more than 70 years. Until recently, scientists had difficulty measuring and manipulating parameters at the nanoscale level.

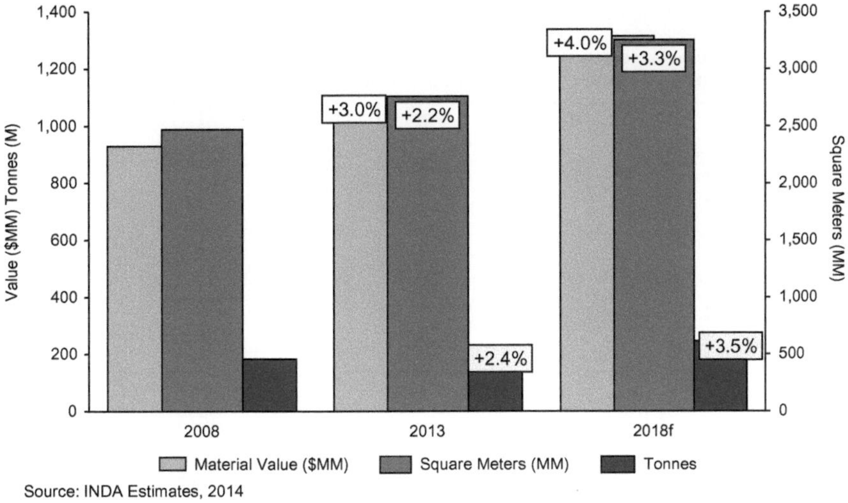


**Figure 1.5** ***Filtration Nonwoven Material Consumption.*** 2008, 2013, 2018 with Annual Percent Change (CAGR) from Preceding Five-Year Period.

Nanotechnology is likely to boost the filter market substantially, as new research shows that particles just 20 nm in diameter are contributing to autoimmune and other diseases. To capture these particles, it is necessary to use ultra-high efficiency (ULPA) particulate filters.

Nanofibers do not exist independently in filtration, but are produced onto a substrate, typically a nonwoven fabric.

These demands and developments result in the North American filtration market consuming an ever greater amount of nonwoven materials, 2.8 million square meters, weighing 206,400 tons in 2013. This is an increase of 2.2 % annually in square meters and 2.4% in tons from 2008. The filtration market is forecast to consume 3.2 square meters, weighing 245,100 tons in 2018, an annual increase of 3.3% and 3.5%, respectively (Figure 1.5).

### *1.1.3.1 Air filtration*

Filtration and purification of air is required in homes, hospitals, commercial buildings, and in hundreds of industries. Air filtration products include filters for vehicle air intake/induction, vehicle cabin air, commercial and residential heating and air-conditioning (HVAC) systems, clean rooms, laboratory hoods, personal protection such as face masks and respirators, large baghouse filters, flue gas scrubbers, industrial dust, vacuum cleaner bags filters, and so on. The most common fibers in air filtration include polypropylene (PP), polyester (PET), glass, and cellulose.

The transportation category is the largest in value of the air filtration market, accounting for more than half of the sales to end users (52.3%), while personal

**Table 1.3** North American Air Filtration Market Sales to End Users and Units by End Use

| | 2008 | | 2013 | | 2018f | | 2008–2013 | | 2013–2018f | |
|---|---|---|---|---|---|---|---|---|---|---|
| | Sales ($MM) | Units (MM) | Sales ($MM) | Units (MM) | Sales ($MM) | Units (MM) | Sales% (CAGR) | Units% (CAGR) | Sales% (CAGR) | Units% (CAGR) |
| Vacuum Cleaner Bags | 300 | 277 | 241 | 200 | 227 | 173 | −4.3 | −6.3 | −1.2 | −2.9 |
| Industrial | 640 | 25 | 924 | 32 | 1,361 | 42 | 7.6 | 4.6 | 8.0 | 5.5 |
| Personal protection | 802 | 3,648 | 1,099 | 4,218 | 1,505 | 5,257 | 4.1 | 2.3 | 4.4 | 3.9 |
| Residential HVAC | 1,014 | 317 | 1,641 | 352 | 2,056 | 396 | 10.1 | 2.1 | 4.6 | 2.4 |
| Commercial HVAC | 1,466 | 330 | 1,739 | 362 | 2,167 | 409 | 3.5 | 1.9 | 2.5 | 4.5 |
| Transportation | 4,791 | 256 | 5,382 | 284 | 6,386 | 333 | 2.4 | 2.1 | 3.5 | 3.2 |
| **Total** | **9,013** | **4,853** | **11,028** | **5,449** | **13,703** | **6,610** | **4.1** | **2.3** | **3.9** | **4.4** |

Source: INDA Estimates, 2014

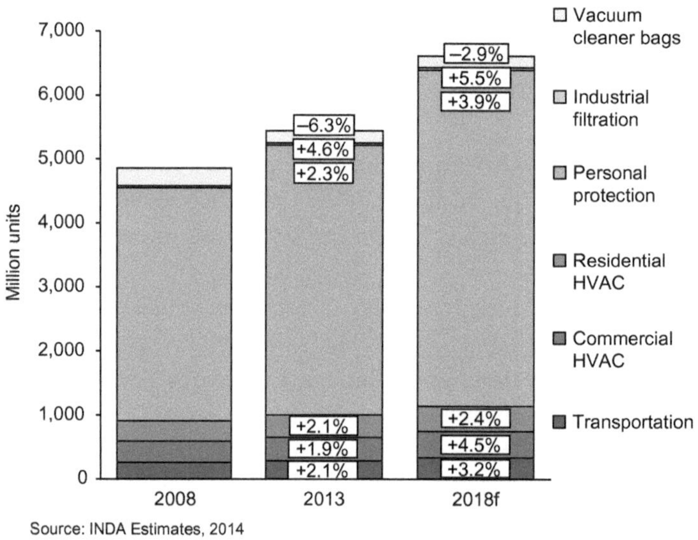


**Figure 1.6** ***North American Air Filtration Market.*** **Unit Sales to End Users by Main End Use, with Annual Percent Change (CAGR) from Preceding Five-Year Period.**

protection, due to the small size of face masks and respirators, accounts for the majority of individual units (75.2%) (Table 1.3 and Figure 1.6).

Much of the measurement for air filtration is based on the size of contaminants and particles, which are usually described in microns (μm), a metric unit of measure where one micron is one-millionth of a meter.

Most large particles—those above 10 μm—settle out in the air. The smaller particles under 10 μm are those that get entrained in the moving air and can be filtered out by air filters. Some really small particles stay suspended in air forever, and never make it through a filter.

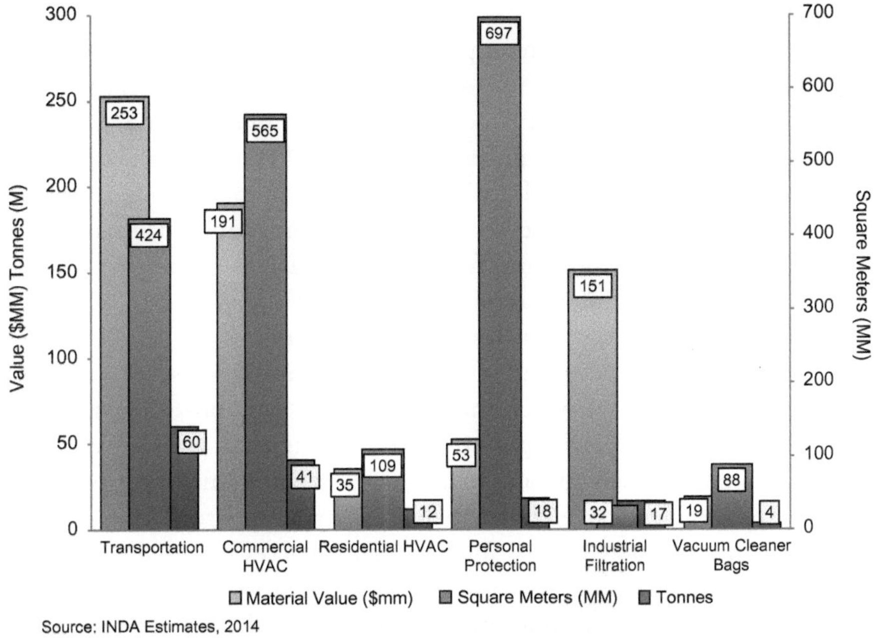


**Figure 1.7** Air Filtration Nonwoven Material Consumption, 2013.

The Minimum Efficiency Reporting Value (MERV) is a method that shows removal by a filter based on particle size based on the minimum efficiency of the filter. Mold and fungi spores tend to be large and reproduce by spores somewhat like a dandelion, and can be filtered with most MERV 6 air filters. Bacteria reproduce by cell division and are sized in a broader spectrum and can only be filtered with more efficient filters—MERV 11 and higher.

Viruses are the smallest of the contaminants and cannot survive for long periods of time outside the host. They normally "ride" on dust or droplet nuclei and can only effectively be removed with high-efficiency particulate absorption (HEPA) filtration. Minimum efficiency of a filter that can be called HEPA is 99.97% retention of 0.3-μm particles. Filters can have efficiencies of 99.999% and are called ULPA (ultra-low particulate air), while filters with 99.9999% are called SULPA (super ultra-low penetration).

The strengthening economy will drive growth in industrial, commercial HVAC, and transportation markets, in addition to increases in the replacement rate of transportation and consumer HVAC filters (Table 1.3 and Figure 1.6).

In terms of North American nonwoven consumption for air filtration, personal protection is the largest square meter category, accounting for a third (36.3%), followed closely by commercial HVAC (29.5%). In terms of tons used, transportation is the largest category (40.0%), followed by commercial HVAC (26.9%) (Figure 1.7).

### *1.1.3.2 Liquid filtration*

As with the air filtration industry, the liquid filtration market captures a myriad of end uses, though they can be separated into two categories: liquid food filtration and liquid nonfood filtration. The liquid food filtration end uses include water filtration (fresh water and wastewater including reverse osmosis filters for desalination and water purification projects); tea bags and coffee filters; cooking oil; and other food and beverage filtration, such as milk, beer, and wine. Liquid nonfood filtration end uses include oil, hydraulic, lubrication, and fuel filters for various modes of transportation, industrial/manufacturing liquid filtration (including cutting oil filters), swimming pool and spa filters, and biochemical filters for blood purification.

In addition to a myriad of end uses, liquid filtration makes use of numerous filter types, including membranes of which the nonwoven is only providing support. From the point of view of equipment types employed in filtration, a much wider range is used for liquid filtration than for air. In fact, almost all air filtration is achieved by some kind of replaceable filter element, held in an appropriate housing; there being no equivalent to the complex process filters used for harvesting solids from a concentrated liquid suspension. Because of this it is very difficult to estimate some of the end use markets by units, for this section, only the sales to end users are presented in the aggregate (Table 1.4).

For liquid filtration, wet laid nonwovens are important media and are found mainly in products such as tea bags, coffee filters, swimming pool filters, and coolant oil filters. The consumption of melt-blown media, in addition to nanofibers, is rising in liquid filtration applications where it is used in bag filters for filtering large quantities of fluids (drinking water, for example), blood filters, prefilters, and postfilters for other filtration media. A large end use for melt-blown media are depth cartridges. The high performance in this product has resulted in the capture of a sizable portion of the depth cartridge market replacing string-wound cartridges. Spunbond polypropylene medium is used widely in liquid filtration systems, often as a support material, particularly in pleated cartridge filters.

The most common fibers used in liquid filtration are polypropylene (PP), polyethylene (PE), polyester (PET), polybutylene terephthalate (PBT), nylon, glass, and cellulose.

**Table 1.4** North American Liquid Filtration Market Sales to End Users by End Use

| | 2008 Sales ($MM) | 2013 Sales ($MM) | 2018f Sales ($MM) | 2008–2013 Sales% (CAGR) | 2013–2018f Sales% (CAGR) |
|---|---|---|---|---|---|
| **Total** | **3,695** | **4,411** | **5,412** | **3.6** | **4.2** |

Source: INDA Estimates, 2014.

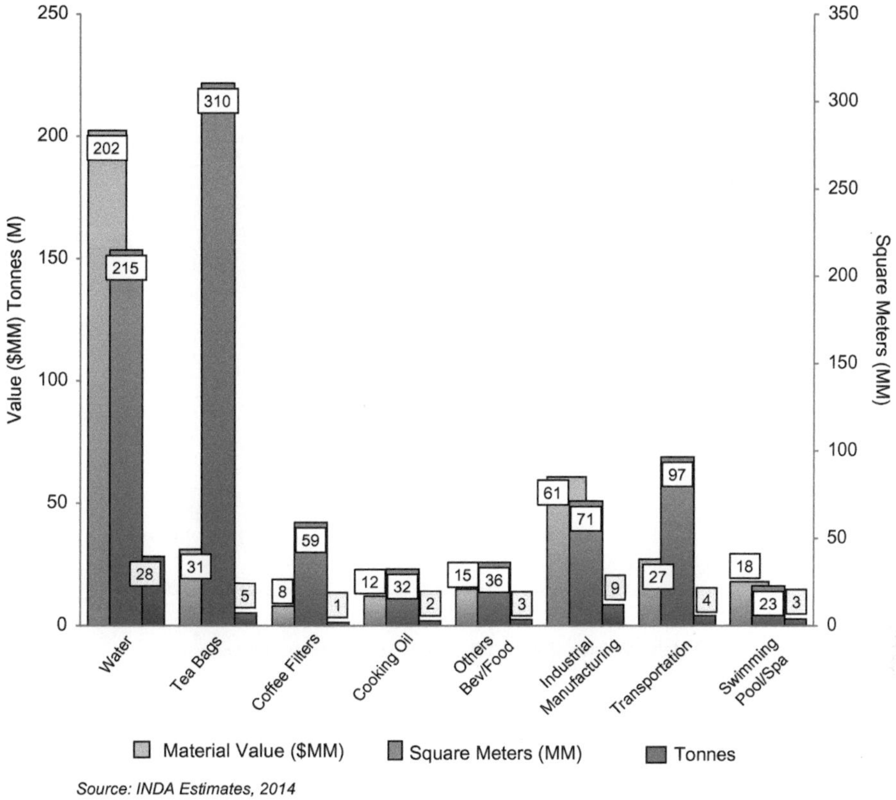


**Figure 1.8** Liquid Filtration Nonwoven Material Consumption, 2013.

Liquid filtration based on nonwoven media are generally not considered an absolute method of purification, because particles of 1 μm or less in size can pass through. Further separation is required in many applications where sub-micron containments are present. Membrane- and nanofiber-based filtration and separation processes employ membranes or nanofibers layered onto a nonwoven support, which increases the strength and improves the particle-retention capacity. Some of these membranes are so thin and fragile that they can only be produced by directly being coated onto the carrier's nonwoven surface.

In 2013, the liquid food filtration end uses consumed three-quarters of the nonwoven material, 77.2% of the square meters and 70.6% of the tonnage. In terms of square meters, tea bags is the largest nonwoven using liquid filtration category, accounting for a third (36.7%) of the volume; while water filtration is the largest in terms of tonnage, consuming half (51.0%) of the tons (Figure 1.8).

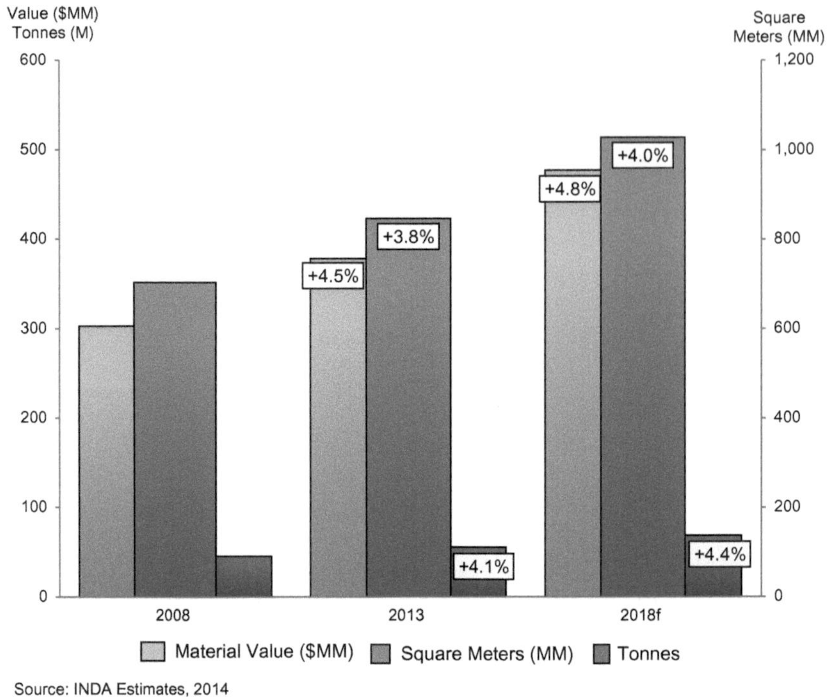


**Figure 1.9** Liquid Filtration Nonwoven Material Consumption for the years 2008, 2013, and 2018 with Annual Percent Change (CAGR) from Preceding Five-Year Period.

Square meter volume is approximate; as we have made allowances for melt-blown media used in depth cartridges. These media are not purchased as roll goods but rather produced on-line as the cartridge is made.

At the forecast growth rate of 4 in the percent range, the liquid filtration market is expected to increase to 1.0 billion square meters, weighing 68,800 tons in 2018, annual increases of 4.0% and 4.4%, respectively (Figure 1.9).

### 1.1.4 Filter/separators—how they work

Filter/separators work in the following manner.

1. They remove undesirable contaminants from a fluid stream.
2. They recover the contaminant because it has value.
3. They do a combination of both steps 1 and 2 above.
4. They separate different contaminants from each other.

These reasons may very well overlap. For example, contaminants may be recovered from a stream not only to clean up the stream, but also to recover the contaminants because they have value. For example, sludge recovered from sewage streams may have value as a recycle material or as a fuel, even though the original intent is to clean up the stream.

Tea bags and coffee filters are good example of a filter used to separate contaminants. In its intended use it allows one contaminant, the tea or coffee flavor to flow through and out of the bag or filter while restraining the tea leaves or coffee grounds, the other contaminant.

Very often protection is the goal of a filter. The purpose may be to protect an inside environment from the contamination of an outside environment or to protect the outside environment from contaminants of the inside environment. To illustrate, HEPA filters were first developed for nuclear processing facilities to provide a barrier against any radioactive dust or contaminants leaking to the outside world. Now they are just as likely to be used in clean room applications where the goal is to prevent even the minutest amount of contaminant from the outside world to enter the workplace.

Sometimes the protection is in two directions. A face mask filter may protect the wearer from undesirable contaminants in the air, but it may also serve to protect the environment and other personnel from germs and moisture emerging from the wearer.

The fluid stream to be filtered may be liquid or gas. For gas filters the fluid is most often air. There are other applications where gases other than air are filtered. Examples are process filters for gaseous fuels such as natural gas and propane, and specialized filters for the manufacture of special gases such as oxygen, nitrogen, helium, hydrogen, etc. However the needs for non-air filters are quite small compared to the demands for air filtration. On the other hand, the markets for liquid filtration can be divided into the filtration of aqueous fluids and the filtration of hydrocarbon fluids. Hydrocarbon fluids include fuels, engine lubricants, hydraulic fluids, cooling fluids for metal working operations, and edible oils.

Eleftherakis (4) defines an "ideal" filter as follows[3]:

- Removes all contaminants, regardless of size.
- Has no restriction, $\Delta P = 0$.
- Has infinite dust holding capacity, lasts forever.
- Is infinitely small, takes up no space.
- Costs nothing, price = $0.00.

Obviously, within all these idealisms there has to be some room for compromise. It becomes apparent that to design a filter and a medium for that filter one has to consider:

1. The properties of the fluid being filtered, including its viscosity, temperature, and its chemical properties such as pH and corrosiveness.

[3] Reproduced by permission of John Elaftherakis (4).

2. The properties of the contaminant including its particle size and concentration.
3. The desired performance of the filter:
   a. The best separation technique for that type of contaminant, be it mechanical filtration, adsorption coalescence, electrostatic effect or otherwise.
   b. Filtration efficiency.
   c. Flow resistance.
   d. Filter life.
   e. Filter size.
   f. Cost.

The most important thing a filter has to do is to separate the contaminants in the fluid. If the contaminant is a particulate material, it will come in a broad range of particle sizes. Figure 1.1 above lists the particle sizes of various contaminants in air. The medium designed for removing larger particles such as pollen, spores, larger bacteria, and dust will be different than the medium for smaller particles such as viruses, tobacco smoke, and fumes. Generally, the smaller the particle, the finer is the fiber in the medium, and the higher will be the cost. It is established that electret media produced from electro-charging techniques can enhance the filtration of fine particles. In liquid filtration cationic charged medium has been established as an effective way of improving liquid filter performance.

The nature of the medium changes even more severely when the particles are in the molecular size range. In this size range are the gas contaminants in air and the soluble components in a liquid. Media to remove molecular contaminants have to be designed with capabilities for ultrafiltration, nanofiltration, reverse osmosis, and/or adsorption.

## 1.2 FILTRATION AND SEPARATION DEFINED

To define the scope of this handbook it is necessary to define filtration and separation, nonwovens, and filter media. INDA, the Association of the Nonwovens Fabrics Industry (5) explains filtration as follows:

> *Filtration is a mechanism or device for separating one substance from another. Filtration may be used to separate contaminants from a fluid or separate value added materials, such as minerals, chemicals, or foodstuffs in a process operation.*

Separations can be broadly classified into six categories:

- Solid—Gas separations
- Solid—Liquid separations
- Solid—Solid separations
- Liquid—Liquid separations
- Gas—Liquid separations
- Gas—Gas separations

One form of solid–solid separations is screening. For example, screening can separate a blend of solid particles into various particle sizes. Solid–solid separation can also be achieved by non-medium methods such as magnetic separation, for example, separating ferrous particles from nonferrous particles. The author is unaware of any significant use of nonwoven media in solid–solid separations; therefore, this form of separation is outside the scope of this book and will not be discussed. The other forms of separation, solid–liquid, solid–gas, liquid–liquid, gas–liquid, and gas–gas, will be the basis of the discussion in this book.

The author will use the "Encyclopedia Britannica" (6) definition of a fluid as follows:

> *"any liquid or gas or generally any material that cannot sustain a tangential, or shearing, force when at rest and that undergoes a continuous change in shape when subjected to such a stress."*

It should be noted that by this definition, both gases and liquids are considered as fluids.

Separations are not limited to the filtration of solid particles from a fluid by entrapment within the matrix structure of the Medium. Other forms of separation are mentioned in the introductory statement above and further discussed in Section 1.4.

Note that in the brief description of nonwoven filter media in the opening paragraph above, the fluid is moving and passes through the medium. This implies a driving force to force the fluid. Usually this force is a pressure differential induced by gravity, a vacuum, capillary action, a pump, a compressor, a fan, or a blower. Other forms of driving forces may also apply. For example, a thermal gradient may cause the fluid to pass through a medium going from hot to cold. The driving force may be a concentration gradient forcing diffusion. A teabag is an example of a "diffusion filter." Hot water extracts soluble taste components from the tea leaves contained within the bag. The soluble taste components then diffuse through the tea bag medium into the greater body of hot water. In some cases, voltage potentials may force the flow of electronic charged particles to the filter surface. In air filtration this may be caused by electrostatic effects. In water filtration this could be the result of electronic or electrokinetic effects involving a zeta potential.

## 1.3 NONWOVENS DEFINED

Table 1.5 lists definitions of nonwovens from various sources. There has been a great deal of difficulty and controversy in the attempt to define nonwovens. A large part of this is due to the desire to differentiate nonwovens from paper and felts. An early definition in "The Manual of Nonwovens" by Krcma (7) in 1971 (Line 1, Table 1.5) failed to make this distinction.

Table 1.5 Definitions of Nonwovens

| Number | Source | Definition |
|---|---|---|
| 1 | **"Manual of Nonwovens"** (3) | "nonwoven fabrics are textile fabrics made of a fibrous layer, which may be a carded web, a fiber web, or any system of randomly laid or orientated fibers or threads, possibly combined with textile or nontextile materials such as conventional [woven] fabrics, plastic films, foam layers, metal foils, etc., and forming them with a mechanically bound or chemically bonded textile product." |
| 2 | "**Annual Book of Standards,**" **1989, ASTM International** (8) | "Nonwoven fabrics," a structure produced by bonding or interlocking of fibers, or both, accomplished by mechanical, chemical, thermal, or solvent means and the combination thereof. The term does not include paper or fabrics that are woven, knitted, tufted, or those made by wool or other felting processes. |
| 3 | **"Man Made Fibers and Textiles Dictionary"** (9) | **Nonwoven fabric:** An assembly of textile fibers held together by mechanical interlocking in a random web or mat, by fusing of the fibers (in the case of thermoplastic fibers), or by bonding with cementing medium such as starch, glue, casein, rubber, latex, or one of the cellulose derivatives or synthetic resins. Initially, the fibers may be oriented in one direction or may be deposited in a random manner. This web or sheet of fibers is bonded together by one of the methods described above. Normally, crimped fibers are used which range in length from 0.75 to 4.5 inches.<br>Nonwoven fabrics are used for expendable items, such as hospital sheets, napkins, diapers, and wiping |

(*Continued*)

**Table 1.5** (Continued)

| Number | Source | Definition |
|---|---|---|
| | | cloths, as the base material for coated fabrics, and in a variety of other applications. They have also been used for semi disposable items and for permanent items such as in interlinings and similar end uses. |
| 4 | **"Nonwovens: Theory, Process, Performance, and Testing," Chapter 2: Nonwoven Terminology** (10) **p. 14** | **Nonwoven:** A manufactured sheet, web, or batt of directionally oriented fibers, made by bonding or entangling fibers through mechanical, thermal, or chemical means. They exclude paper and products which are woven, knitted, tufted, or felted by wet milling. The fibers may be of natural or man-made origin. |
| 5 | **The Dictionary of Paper, Fourth Edition p. 285** (11) | **NONWOVEN FABRIC.** A cloth-like material made of fibers longer than those normally used in papermaking which, instead of being woven on a loom, is formed by felting on a fine mesh screen from an air or water suspension with or without binders. |
| 6 | **Nonwoven Fabrics Handbook (INDA)** (5) | Nonwoven: A sheet, web, or batt of natural and/or man-made fibers or filaments, excluding paper, that have not been converted into yarns, and that are bonded to each other by any of several means.<br>"NOTES<br>To distinguish nonwovens from papers, a material shall be defined as a nonwoven if:<br>More than 50% by mass of its fibrous content is made up of fibers (excluding chemically digested vegetable fibers) with a length to diameter ratio greater than 300: or |

(*Continued*)

**Table 1.5** (Continued)

| Number | Source | Definition |
|---|---|---|
| | | More than 30% by mass of its fibrous content is made up of fibers as in "a" above and meeting one or both of the following criteria:<br>Length to diameter ratio more than 600.<br>The density of the fabric is less than 0.4 g/cc.<br>Bonding methods may include any of the following means or any combination thereof, including but not limited to:<br>Adding an adhesive.<br>Thermally fusing the fibers or filaments to each other or to other meltable fibers of powders.<br>Fusing fibers by first dissolving then resolidifying their surfaces.<br>Creating physical entanglements or tufts among the fibers.<br>Stitching the fibers or filaments in place.<br>A nonwoven may be a structural component of a composite.<br>Nonwoven structures may incorporate monofilaments or yarns.<br>Reproduced by permission of INDA |
| 7 | **EDANA Website** (12) | A nonwoven is a sheet of fibers, continuous filaments, or chopped yarns of any nature or origin, that have been formed into a web by any means, and bonded together by any means, with the exception of weaving or knitting.<br>Felts obtained by wet milling are not nonwovens.<br>Wet laid webs are nonwovens provided they contain a minimum of 50% of man-made fibers or other fibers of non-vegetable origin with a length to diameter ratio equals or superior |

(*Continued*)

**Table 1.5** (Continued)

| Number | Source | Definition |
|---|---|---|
| | | to 300, or a minimum of 30% of man-made fibers with a length to diameter ratio equals or superior to 600, and a maximum apparent density of 0.40 g/cm<br>Composite structures are considered nonwovens provided their mass is constituted of at least 50% of nonwoven as per the above definitions, or if the nonwoven component plays a prevalent role. |
| 8 | **"A New System for Classifying Nonwovens," TAPPI Press** (4) | **Structure-based definition of nonwoven:**<br>• Fiber web structures: Includes all textile sheet structures made from fibrous webs, bonded by frictional/mechanical restraints and/or covalent bonds obtained through the use of resins, thermal fusion, or formation of chemical complexes. Here, fibers or filaments are bonded so that the average bond-to-bond distances are greater than 50–100 times the fiber diameter giving textile-like qualities of low bending and low in-plane stiffness.<br>• Net-like structures: Structures formed by extruding one or more fiber-forming polymers in the form of a network or film. The film may then be uniaxially or biaxially oriented to fibrillate into a net-like structure.<br>• Multiplex structures: This class of fabrics combine and utilize characteristics of several primary and/or secondary structures, at least one of which is a recognized nonwoven textile structure, into a single unitized structure. |

*(Continued)*

**Table 1.5** (Continued)

| Number | Source | Definition |
|---|---|---|
| 9 | **ISO 9092:2011 Nonwoven Definition** (13) | A Nonwoven material is a web containing fibers of man-made, naturally derived fibers or combination of fiber origins where the fibers are randomly formed from staple, continuous filament, or formed in situ by mechanical, chemical, thermal, or other processes or combinations from which textiles products are made.<br>Note 1: Felt, film, and paper are not considered nonwoven materials. |

A revised version of ISO 9092 (ISO 9092:2011 (13)) is listed in line 9 of Table 1.5. This definition excludes felt, film, and paper, however, provides no guidance for distinguishing the excluded materials.

Later definitions excluded paper and felt from the definition of nonwovens and tried to set up criteria for differentiating nonwovens. Batra et al. (14) developed a structure-based definition as listed in line 8 of Table 1.5. The definition listed is as that reported by Hansen (15). Batra et al. excluded paper by specifying the fiber length between bond points. Batra et al. also defined "multiplex structures" which in this book is identified as composite structures (see Section 1.7.3).

Both INDA ((5), p. vii (Association of the Nonwoven Fabrics Industry) (Line 6, Table 1.5)) and EDANA ((12) (European Disposables and Nonwovens Association) (Line 7, Table 1.5)) proposed definitions for ISO 9092:2011 (13). Both attempted to differentiate a nonwoven from paper based on fiber L/D (length/diameter) ratio and bulk. Nonwovens are generally perceived as being formed by a dry form process where aerodynamics, mechanical devices, and spinnerets are used to form the web. On the other hand, paper is formed from a water medium using a wet lay or paper machine process. Paper is generally perceived as being denser than nonwovens and containing shorter wood and/or vegetable fibers.

Both EDANA and INDA differentiate a nonwoven from paper by specifying at least 50% of its mass is composed of fibers whose L/D is 300 or greater. EDANA specifies that the 50% be made of man-made fibers or of fibers of non-vegetable origin, whereas INDA excludes chemically digested vegetable fibers. Both definitions provide for an alternative composition whereby 30% of the mass is composed of fibers whose L/D is 600 or greater, and the density is less than 0.4 g/cm$^3$. The EDANA alternative again requires the 30% to be man-made fibers. The EDANA definition excludes felts obtained by wet milling. INDA works around this by specifying the bonding methods for nonwovens. There are wet lay products that satisfy the INDA and/or EDANA definitions and are recognized as nonwovens. This includes glass fiber paper media (including glass microfiber paper), and wet laid substrates with high percentages of synthetic long fibers such as those of nylon and polyester. Wet laid media made from inorganic fibers such as carbon, ceramic, and metal might also satisfy the definition provided that they meet the bulk requirement of less than 0.4 g/cm$^3$.

The INDA and EDANA definitions allow for the use of natural fibers; however, if they are chemically digested, they cannot be counted as long fibers. This exclusion eliminates certain types of wet laid long fiber media that might be perceived as being nonwoven. An example would be teabag paper made from chemically digested abaca (Manila hemp) and/or sisal.

There are long natural fibers that are not chemically digested and used to make random fiber webs, mats, and batts by the wet mill process (see Chapter 5, Section 5.1.2.3.5, for a description of the wet mill process). These may not satisfy the EDANA definition because they are considered as felts. Examples are wool, cotton, sisal, hemp, abaca, and certain forms of kenaf. Indeed these fibers have the historical significance of being the origins of nonwoven materials[4]. There are some forms of wood pulp used in paper that are not chemically digested (e.g., groundwood), however, they are very short fibers and do not meet the L/D ratio requirements for nonwovens. The wood pulp fibers used for making paper normally have an L/D ratio of less than 200.

There is a reviving interest in natural fibers for nonwoven webs, because they are biodegradable and are friendlier to sustainability issues. The development of nonwovens from natural fiber sources is also motivated by the economics of synthetics derived from petrochemical sources.

Some dry formed products do not necessarily satisfy the definitions of nonwoven. Examples are the absorbent webs and other products produced on an air laid machine. These webs are formed from hammer milled wood pulps and adsorbent materials that have very low L/D ratios. They generally have a minimal amount of long-fibered synthetics depending on product design and application.

The author understands the importance of having a distinction between nonwovens and paper. He less understands why there has to be a distinction between nonwovens and felts. There is a great deal of gray area in making the distinctions. In addition, there is a great deal of overlay in the market place and technology of filter medium products. Nonwoven, felt, and paper products compete with and/or complement each other in many filter and separation applications. Many composite structures contain some combination of nonwoven, felt, and/or paper components.

[4] A more advanced discussion on the history of nonwovens can be found in "Introduction to Nonwovens Technology" (16) Batra, and Pourdeyhimi.

Several filter medium manufacturers produce both nonwoven and paper media and do not differentiate their products by the nonwoven definition. Almost every nonwoven conference that includes sessions on filtration will have discussions and presentations concerning paper media and felt media. Although, the title of this book is the "Handbook of Nonwoven Filter Media," the author finds it awkward and difficult to present an adequate discussion of the topic without including paper and felt structures. Therefore for purposes of this book the authors accept the definition of nonwoven as stated by ISO 9092:2011 (line 9 in Table 1.5), however, with the following modification:

Any wet laid structure, including paper, whose specific purpose is to be a medium in a filtration and/or separation application and any felt structure whose specific purpose is to be a medium in a filtration and/or separation application.

The ISO definition also excludes film. Understandably, a nonfiber fabric such as a film is not considered as a nonwoven. However, the discussion of nonwoven filter media cannot be complete without including the growing technology of membrane filtration (see Section 1.4.10), even though many membrane materials are sourced from film. In deed, membrane filters can be tied to nonwovens because their support structure is often a nonwoven. In addition, materials identified as nanofiber membranes fit the definition of nonwovens and have membrane-like filtration properties.

## 1.4 DEFINITION OF FILTER MEDIUM

The "Filtration Dictionary and Glossary" (Wakeman (17)) defines filter medium as "any permeable material used in filtration and upon which, or within which, the solids are deposited." Sutherland and Purchas (18) argue that this definition is not broad enough. The glossary in their handbook defines filter medium as "The porous material in a filter that does the actual filtering." In Chapter 1 of their Handbook they broaden the definition as follows:

> *A filter medium is any material that, under the operating conditions of the filter, is permeable to one or more components of a mixture, solution, or suspension, and is impermeable to the remaining components.*

The Sutherland and Purchas' definition is consistent with the purposes of this handbook, because it allows for possible separation processes other than simple particle entrapment. Examples of other types of separation media are as follows.

### 1.4.1 Sorption

Wikipedia (19) defines sorption as a "physical and chemical process by which one substance becomes attached to another." The Wikipedia lists three forms of sorption: adsorption, absorption, and ion exchange.

#### 1.4.1.1 Adsorbent media

Filter media containing adsorbents remove liquid and soluble contaminants from fluid streams by surface adsorption, i.e. the contaminants are attracted to the surfaces of the medium and held there by physical chemical effects. Activated carbon is a common adsorbent used in filter media. In liquid filtration it is used to remove hydrocarbons and bad taste contaminants from drinking water. It is also used to remove toxic and obnoxious chemicals from air such as in gas masks and respirators.

In military affairs, adsorbents such as activated carbon are used to protect both military personnel and civilians from the toxic effects of chemical agents. Impregnation of the activated carbon with metallic compounds, such as silver or copper, creates a chemically reactive surface for trapping many of these toxic agents. Other adsorbent materials used in filter media include activated alumina, zeolites, and baking soda (sodium bicarbonate). Adsorbent media are gaining widespread use in automotive cabin air filters where they are effective in removing vile odors and emission fumes. In recent years adsorbents have become a significant medium in HVAC systems in the workplace and in the home. They also play a role in recovering volatile organic compound (VOC) solvent material in waste air streams.

#### 1.4.1.2 Absorbent media

Whereas an adsorbent medium removes contaminants from a fluid stream by surface effects, an absorbent medium removes the contaminants by absorbing them into the structure of the medium, much as a sponge soaks up water from a wet surface. Nonwovens are often used with absorbent structures to contain the absorbent material and to act as a transport layer for fluid to flow to the absorbent material. This would be the case for diaper and incontinent pads. Polymer fibers such as polyethylene, polypropylene, and nylon are used in oil booms to soak up oil spills at sea and in fresh water lakes, streams, and rivers. A more detailed discussion on absorbent materials can be found in Chapter 3, Section 3.10.2.

#### 1.4.1.3 Ion exchange

Ion exchange is widely used in the following industries: food and beverage; hydrometallurgical; metals finishing; chemical and petrochemical; pharmaceutical; sugar and sweeteners; ground and potable water; nuclear; softening and industrial water; semiconductor; power; and a host of other industries. Ion exchange is a reversible interchange of ions. The ions in an insoluble solid exchange with ions of the same charge that are in the solution surrounding the solid. Filters based on this reaction are used for softening or demineralizing water, purifying chemicals, or separating substances in an aqueous medium (including aqueous foods and drinks).The separation system usually consists of a tank containing small beads of synthetic resin. In some processes the beads are used as a granular bed in a separation column. The beads are most often made of polystyrene,

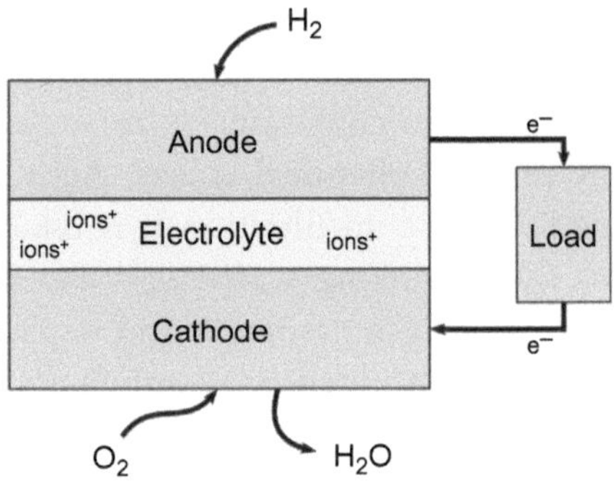


**Figure 1.10** Fuel Cell Ion Exchange Process.

specially treated to selectively adsorb either cations (positive) or anions (negative) and exchange certain ions based on their relative activity compared to the resin. One notable manufacturer of ion exchange resins is Dow Chemical Company[5].

Membrane filters with ion exchange capability are available in the commercial marketplace. For example, PCA GmbH, Lebach, Germany; Hangzhou Iontech Environmental Technology Co., Ltd, Hangzhou, China; FuMA-Tech GmbH, Baden-Württemberg, Germany all manufacture ion exchange membranes and ion exchange membrane systems.

Among the more notable applications for membrane ion exchangers are water desalinization and electrodialysis. Other applications include electrodialysis reversal and high purity water for power engineering, electronic and nuclear industries. Polymeric or mineral insoluble ion exchangers are widely used for water softening, water purification, water decontamination, etc.

Fuel cells are a way of converting fuels into electricity. A notable use of fuel cells is in space vehicles such as the Apollo spacecraft and the space shuttles. There are many types of fuel cells, but they all consist of an anode, a cathode, and an electrolyte that allows charges to move between the two sides of the fuel cell. Electrons are drawn from the anode to the cathode through an external circuit, producing direct current electricity. The process is diagrammed in Figure 1.10. In membrane fuel cells, the membrane is the electrolyte. Referred to as a PEM (polymer electrolyte membrane), it allows positively charged ions (protons) generated from the oxidation of hydrogen (the fuel) to pass through to the cathode; negatively charged electrons do not pass through the membrane, but must take an alternate path to the cathode. The separation of protons and electrons generate a voltage and a flow of electricity.

[5] Dow obtained the ion exchange resin business in 2009 as part of its purchase of Rohm and Haas.

## 1.4.2 Coalescence

Coalescence is a process by which a filter medium can remove immiscible liquid particles from a liquid or gas stream. This includes dispersed hydrocarbon oils from water, oil mists from air, moisture and vapor from air, and water contamination from aviation fuel. The phenomena involve collecting the minute droplets, either at the surface of a medium or on the fibers of the medium. The droplets coalesce with each other to form larger droplets that eventually become large enough to either fall away or be driven away from the medium. The surface collection phenomena require a highly oleophobic or hydrophobic surface on the medium that rejects the liquid contaminant particle so that it cannot penetrate the surface. The liquid particles are either rejected from the surface or collected and coalesced on the surface until they are large enough to fall away from the surface, usually by gravity forces.

The fiber coalescing phenomenon is a particulate filtration, in that it requires a matrix of fine fibers to trap the liquid particles and hold them on the fibers. Depending on the surface properties of the fibers and the liquid particles, the particles coalesce into larger droplets that become large enough that they can no longer cling to the fiber surfaces because either they are too heavy or they are blown away by the "wind velocity" of the fluid flowing through the fine fiber structure.

In coalescing processes, provision has to be provided for the transport of the released liquid particles so that they can be removed from the filter structure as a separate stream, either by a flotation mechanism or by a gravity settling mechanism.

## 1.4.3 Barrier properties

There are certain types of media that exhibit barrier properties, that is, they can hold back one component of a liquid–liquid mixture and allow the other component(s) to pass though. The phenomenon appears to be based on a combination of absorbency and coalescence. The use of barrier media, such as OIL SHARK®,[6] for oil spill control is discussed in Chapter 3, Section 3.14.

## 1.4.4 Electro-filtration

There are two principal types of filtration separation that employ an electric charge: electrostatic filtration for air streams and electrokinetic separation for liquids/streams.

### *1.4.4.1 Electrostatic filtration*

Filter media and/or fibers for filter media can be electrostatically charged to enhance their filtration properties. Electrostatic fibers so charged are referred to as electrets.

[6] OIL SHARK® is a registered trademark of Cerex Advanced Fabrics, Pensacola, Florida USA.

Although it can be argued that electrets remove particles by filtration, the author considers this a different type of separation because a different technology is introduced. Electrets are dielectric materials that exhibit an external electric field in the absence of an applied field. In this way they act like a magnet in attracting particles. When used in air filtration applications, they can greatly increase initial filtration efficiency and reduce pressure drop. Electrostatic media have become an important entry into air filtration markets. However, they do have a problem in that they lose their charge in a relatively short period of use (3–6 months).

An electrostatic charge on a filter medium is not always desirable. For example, in flammable fuel filtration there is a fear that an electrostatic discharge might ignite the fuel. Another major concern is baghouse filtration where there may be large concentrations of airborne dust. A simple spark from the medium could cause a serious explosion. Savage et al. (20) have written about this. The solution to this problem is a conductive material such as one containing epitropic fibers that dissipates the electrostatic charge. Savage et al. mention the development, by W.L. Gore, of a membrane-felt composite whereby the membrane contains carbon particles to give it conductive properties. Cerex has developed a nylon spunbonded material, trade named Nystat 50® that has conductive properties.

#### *1.4.4.2 Electrokinetic separation*

Electrokinetic filtration is used in liquid applications. The medium possess an electronic charge that is more positive than the mostly anionic charges on the contaminant particles in a polar fluid stream (water is one example of a polar fluid). The difference in electric voltage potential is called the "zeta potential." A high zeta potential enables the medium to attract the anionic particles and separate them from the polar fluid stream. Cuno Inc.[7] (21) markets an electrokinetic medium identified as "Zeta Plus®. More recently Ahlstrom Corporation (22) introduced a cationic medium which they market as their Disruptor™ medium. Another producer of electrokinetic media are ErtelAlsop.

### 1.4.5 Viscous impingement

Viscous impingement is the case where the medium is impregnated with a sticky mineral oil or other sticky substance that causes particles to stick to fibers. It is used in air filtration applications. It is effective for filtering large particles, but not so effective for filtering fine particles. AAF International, Louisville, Kentucky, USA markets a substance called Viscosine, a mineral type oil that is applied to filter media for viscous impingement properties.

[7] Cuno is now part of 3M Corp. and is identified as 3M Purification Inc.

### 1.4.6 Antimicrobial media

Antimicrobial filter media contain agents to protect filters by inhibiting the growth of bacteria, fungi, and yeast. In some cases the antimicrobial action is also intended to prevent the migration of biological microorganisms into the filtrate or filtered product. The antimicrobial agent can be applied to the medium as a finishing step or it can be incorporated into the fiber. Antimicrobial agents containing silver have proven to be effective and are widespread. There are, of course, regulatory concerns related to antimicrobial agents. These are regulated both by the United States Environmental Protection Agency (USEPE) and by the United States Food and Drug Administration (USFDA), depending on the application.

Antimicrobial media and agents are discussed in Sections 3.1 and 4.5.4 respectively.

### 1.4.7 Catalytic media

Catalytic media contain a catalytic agent or substance to remove or break down undesirable contaminants in a fluid. Potassium permanganate or manganese dioxide added to a water filter medium helps remove manganese, iron, arsenic, and hydrogen sulfide. In air, undesirable contaminants such as nitrogen oxides and carbon monoxide may be catalyzed by a sintered metal medium or a fibrous medium containing metal fibers. Palladium is a frequently used metal for this purpose.

Another catalyst system receiving high level attention is photo-catalytic oxidation (PCO). PCO utilizes $TiO_2$-coated substrate combined with ultraviolet radiation to remove airborne contaminants. It is particularly effective against microorganisms and VOCs. Trane, a subsidiary of Ingersoll Rand, Dublin, Ireland, markets an air cleaning system (TCACS) containing three separation stages:

1. MERV 13-rated filtration
2. Ultraviolet light to kill microbiological activity
3. PCO to remove VOCs and microorganisms.

### 1.4.8 Extraction

Extraction is a phenomenon that is not built into the filter medium, but rather the medium serves as a structural support for the extraction process. A coffee filter is a good example. Hot water extracts the soluble coffee taste components from the coffee grinds supported in the coffee filter medium. The coffee rich water passes through the medium, however the grinds are retained. Teabag and tea filters act in a similar manner.

### 1.4.9 Filter support

There are applications where a medium does not do the filtering but acts as a support structure for the medium that does. Certain types of nonwoven media such as melt-blown and electrospun media are generally too thin and weak to provide for their own structural integrity. They require a stronger material such as a woven fabric, spun-bonded fabric, needle punch felt or cellulose web to provide mechanical strength for the application. Very often metal and plastic mesh materials are used for this purpose.

In many cases, a medium used to provide mechanical support to a filter medium also participates in the filtration process. Often the support layer is a prefilter layer for the more efficient filter medium it is supporting, resulting in a gradient density filter medium—a filter medium composed of two or more layers; each layer being successively more efficient than the previous layer.

### 1.4.10 Membrane filtration

Membrane filter media are generally not considered as nonwoven media, however, they are considered within the scope of this handbook, because:

- Nonwoven support structures are often used with the membrane.
- Nanofiber membrane filters (often from the electrospun process) are consistent with the definition of nonwovens.
- The growing and emerging markets for membrane filter technology are too important to be ignored in any discussion of filter media technology.
- Membrane filter technology introduces new separation technology not achievable with more conventional filter media. This includes microfiltration, ultrafiltration, reverse osmosis, dialysis, and electro dialysis.
- Although not produced from a fiber or fiber-producing source as is other nonwoven forms of filter media, some membrane media have the appearance of a fibrous structure. This is evident in the SEM microphotograph of Figure 1.11 (Advantec MFS Inc.).

Membrane filter media can be subdivided into the following filtration and separation categories:

- Microfiltration
- Ultrafiltration
- Nanofiltration
- Reverse osmosis

These categories are discussed in more detail in Section 2.1.4.

### 1.4.11 Composite structure

The nonwoven supports for nonwoven filter media and membrane filter media as discussed in Sections 1.4.6 and 1.4.7 are examples of composite structures. Composites provide the opportunity to bring more than one separation process into one medium. They are further discussed in Section 1.7.3. Robert Murphey (23) of Hollingsworth and Vose Co. defined composites as “Two or more different web technologies that are

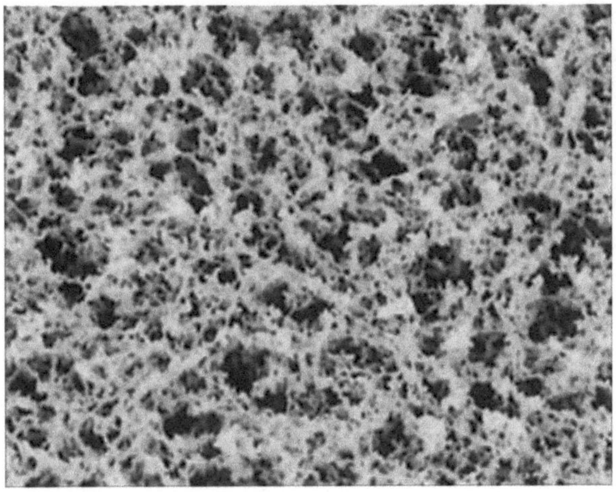

**Figure 1.11** Nitrocellulose Filter Membrane Looks Like a Fibrous Structure. *Reproduced with permission of Advantec MFS Inc. Dublin, California USA.*

combined to optimize the characteristics of a filter medium." Batra et al. (9) refer to composites as multiplex structures as defined in Table 1.5, line 8 above. For purposes of this book a composite filter media structure will be considered a nonwoven filter media if at least one layer is a nonwoven.

## 1.5 NONWOVEN FILTER MEDIA DEFINED

Combining the discussion points of Sections 1.2–1.4, the following definition of a nonwoven filter medium emerges:

A nonwoven filter medium is a porous fabric composed of a random array of fibers or filaments and whose specific function is to filter and/or separate phases and components of a fluid being transported through the medium or to support the medium that does the separation. The fabric is a sheet-like structure produced in lengths long enough to be wound into rolls. Although the random fiber structure is the backbone of the web it may contain other components that are part of the forming process including (but not limited to) particulate fillers (clays, calcium, adsorptive powders etc.), sizing agents, wet strength agents, anti-microbial additives, plasticizers, flame retardants, dyes and pigments, softening agents, and wetting agents.

It is implicit in the definition that the fibers and filaments be bonded in some form or fashion. The formed fabric may be bonded as it is formed. An example is the hydrogen bonding of cellulose fibers in webs formed by the wet lay process. Nonwoven webs may also be bonded by a process downstream of the fabric former that is either in-line or off-line with the forming process. The downstream bonding process may include any of the following: fiber entanglement either by needle punching and hydroentanglement or by stitch-bonding; water based latex treatment; solvent-based resin treatment; and thermal bonding. As long as the web is a random fiber structure to be used for filtration and/or separation, the specific bonding process is not critical to the definition.

The fabric may be exposed to other chemical and mechanical treatments to enhance its properties. Examples are coatings and finishes, flame retardants, antimicrobial agents, water repellents, dyes, and softening agents. It may also be exposed to downstream mechanical conversions such as creping, corrugating, embossing, slitting and rewinding, pleating, bag making, sheeting, dye cutting, and stamping as necessary for the end-use filter and/or separator application.

## 1.6 WHAT IS NOT A NONWOVEN FILTER MEDIUM

The definition of nonwoven filter media above defines the scope of this handbook. Among the many types of filter media that are not within the scope of this book are:

- Woven filter media
- Screens (metal, plastic, etc.)
- Sintered metals (unless formed from metal fibers).
- Sand bed filters
- Cake filters
- Ceramic media (unless formed from ceramic fibers).
- Fluidized bed filters

Film-formed membrane filter media are not considered as nonwovens, however, they are often supported by a nonwoven substrate and/or preceded by a nonwoven prefilter medium. This type of media will be included in the scope of this book (see Section 1.4.10).

## 1.7 CLASSIFICATION OF NONWOVENS BY PROCESS

Nonwovens are best classified by process, in the following list. Each process produces a unique class of nonwovens utilizing different raw materials and/or different forms of the same raw materials. There are two major process groups for forming nonwovens: dry formed and wet laid. A simple distinction between the two groups is that in dry formed processes the webs are formed in an air medium, whereas in the wet lay process the webs are formed in water. In addition there are composite structures whereby webs of different technologies are combined. Membrane structures are not formed by standard nonwoven forming processes, but rather, by several different unique processes particular to membrane formation.

Classification of nonwoven filter media

- Dry Formed
  - Fiber sourced
    - Air laid
    - Dry Laid
  - Polymer sourced
    - Meltspun
      - Spunbond
      - Melt-blown
    - Flash spun
    - Nanofiber spun webs
      - Electrospun
      - Centrifugal spinning
  - Wet laid
  - Filter membranes
  - Composite structures

Air laid and dry laid involve the dispersion of individualized fibers to form a random fiber web. Spunbonded and melt-blown are identified as meltspun processes. These utilize melted polymers that are spun into fibers and laid on some form of collector. Flash spun is the case where the web or fiber is spun from a polymer solution rather than a polymer melt. Electrospun nanofiber webs are a form of flash spun web in that they are spun from a solvent dissolved polymer rather than a melted polymer. Centrifugal processes for producing nanofiber webs may go either way. The DuPont® Melt Spun Nanofiber Process is strictly meltspun whereas FibeRio's Forcespinning® process can produce nanofiber webs from either meltspun polymers or solution dissolved polymers.

The wet laid processes are similar to the processes for making paper, and the wet lay machines for producing wet laid webs, particularly filter paper, are often referred to as paper machines. Highly porous filter papers and webs formed from long fibers require specially designed paper machines.

Following are short descriptions of each process. More detailed descriptions will be found in Chapter 5.

## 1.7.1 Dry formed processes

### *1.7.1.1 Fiber sourced*

#### 1.7.1.1.1 Air laid

Air laid structures, mostly formed from cellulose fluff pulp, are generally used for absorbent materials such as diaper fluff or personal hygienic pads. The fluff pulp is hammer milled into individualized fibers, which are then air conveyed to a moving belt or forming wire to be formed into a fabric. Short cut or staple synthetic fibers

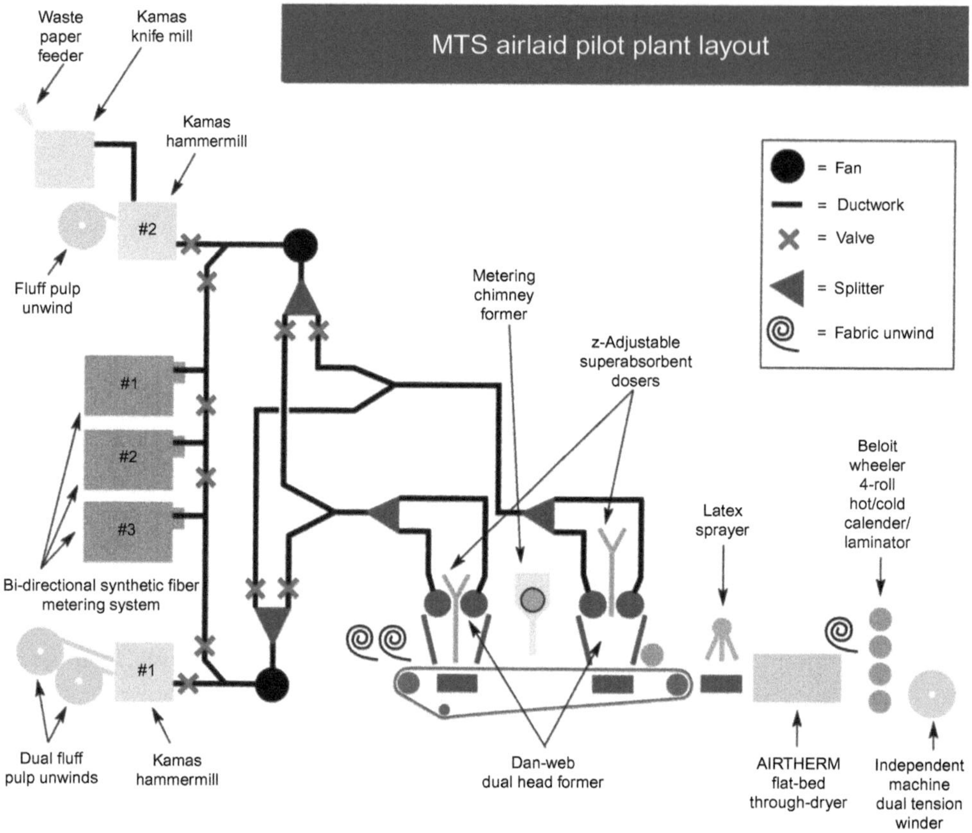


**Figure 1.12** MTS Air Laid Pilot Plant Layout. *Printed by permission of MTS Publications, Kalamazoo, Michigan.*

such as polyester, polypropylene, Nylon, Rayon, etc. can also be fed into the forming area. Super absorbent polymers (SAP), such as carboxy methyl cellulose (CMC) are often incorporated. Composite structures can also be formed. The air laid former often has an in-line bonder for applying latex binder. If the web produced contains a considerable amount of long synthetic fibers, the web can be needle punched or Spunlaced (hydroentangled). Figure 1.12 is a flow diagram of the air laid pilot machine at MTS Testing Laboratories in Kalamazoo, Michigan, USA. The machine utilizes a Dan-Webb former produced by Dan-Webb in Aarhus, Denmark.

Although air laid webs are bulky structures that would appear to be likely candidates for filter media, the author is unaware of any major filtration market area that utilizes this type of media.

#### 1.7.1.1.2 Dry laid webs and felts

**1.7.1.1.2.1 Forming of dry laid webs** Dry laid forming is the process for forming many of the carded webs and felt materials used in filtration. The process has been well described by Crook (24). A detailed discussion is also found in Chapter 5, Section 5.1.2.1. Both carded webs and felts are produced on similar types of equipment. High loft webs used in HVAC systems are usually thermo-bonded. In some cases they are chemically bonded. Felts produced by the layering of carded webs are usually much heavier in grammage. Needle punched felts are a major medium source in filtration markets. Woven fabrics and needle punched felts are the largest source of filter media for bag house filter operations. In recent times some applications have replaced needle punched felts with spunlace (hydroentangled) versions. In recent times, some baghouse operations have started reverting to cartridge filters in place of felt and woven medium bag filters.

Dry laid webs are differentiated from air laid webs by the nature of raw materials and the way they are handled by process. Air laid webs are primarily composed of cellulose wood pulp fibers and possibly a partial content (up to 20%) of staple fibers. Wipes and absorbent pads for hygienic products are important production items of the air laid process. On the other hand, longer staple fibers are the majority fiber source for dry laid structures. Very often, these are crimped fibers. In air laid processes hammer mills and aerodynamic action are utilized to separate, individualize, and disperse the fibers and form them into a web. Dry laid processes also utilize aerodynamic action to form webs, however, a considerable amount of mechanical processing is also utilized to individualize, disperse, and orient the fibers and to form them into webs.

Figure 1.13 is a layout of a typical dry laid line. The core of the dry laid process is the card or a related machine called the Garnett. Both consist of a series of rolls containing barbed point surfaces. The barbed points on the surfaces are obtained by coating the rolls with spirally wound sawtooth wire often referred to as metallic clothing. The sawtooth points are engineered to have specified heights, pitches, and angles. Fiber is carded when it passes between two or more metallic coated surfaces moving at different speeds and having their points in opposite directions. A combing action takes place that parallelizes the fiber. One of these surfaces is the main cylinder; the other is a smaller worker roll or a stationary plate with a barbed surface. Other types of rolls perform various other functions in the process. These include stripping rolls that help breakdown fiber tufts, likerin rolls that feed fiber to the main cylinder from the feed bin, and doffer rolls that remove the fiber from the main cylinder. There are also condensation and randomizing rolls that serve to cross-orient and randomize the fiber. A detailed discussion of card rolls and their purposes is presented in Chapter 5, Section 5.1.2.

In addition to cards and garnetts a dry laid line will include bale opening machines, blending machines, and machines to feed the cards (or garnetts). There

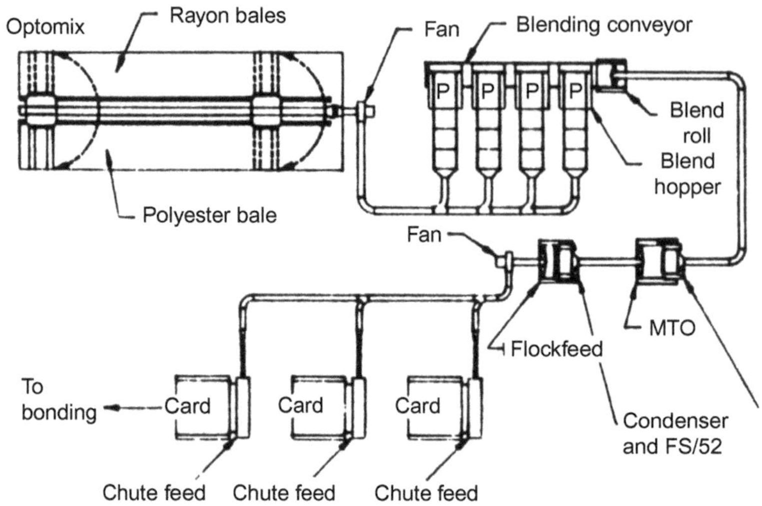


Notes: Fiber bales in this line consist of polyester and rayon
Optomix® is a bale opening machine
MTO is a Micro-Tuft Opener that reduces fiber tuft size
FS/52 is a fiber accumulator.
Flockfeed is used for additional fiber opening
Condenser is a barbed roll that accepts opened fiber and delivers it into the air stream for processing downstream toward the web former.

**Figure 1.13** Dry Laid Nonwovens Line Layout. *Reproduced by permission of TAPPI, Atlanta, Georgia USA (24), p155.*

may be two or more cards, each one forming a separate layer combined with the previously formed layers. The layering allows for different properties between layers, such as fiber orientation. It also allows for different fiber compositions in each layer. Another method for building layers is cross layering. Cross layering is accomplished by folding formed webs that are delivered by a conveyor into a lapper, which continuously transfers the web unto an apron operating at right angles to the cross-layering motion. This is a way to obtain thick webs for heavyweight felted materials.

The process requires a device for forming the final web. This could be the output of the cards or garnetts. There are also air laid formers that form webs from carded fibers coming from the cards. A very popular such former is the Rando Webber® produced by Rando Machine Corporation, Macedon, New York, USA. This machine is further discussed in Section 5.1.2.1.

It should be noted that the Rando Webber® is often used to form dry laid webs directly from bale opening equipment. Although cards are not used, the product is still considered a dry laid fabric, rather than an air laid fabric because the forming action of a Rando Webber® tends to be mechanical rather than aerodynamic. The Rando Webber utilizes barbed surfaces such as rolls and aprons to distribute the fiber unto the moving forming fabric.

**1.7.1.1.2.2 Bonding of dry laid webs** Following is a brief review of the various methods to bond carded webs. More detailed discussions will be found in Chapter 5, Section 5.1.2.3.

*Needle Punch*: Needle punching is a process by which fibers are entangled and mechanically interlocked by puncturing the web with a series of barbed needles. The density and strength of the web can be regulated by the strokes per minute of the web, the advance rate of the web, and the degree of penetration of the needles. Needle-punched webs are one of the more popular filter media for dust collection systems (baghouse and cartridge).

*Hydroentanglement (Spunlace)*: Hydroentanglement bonds webs in a manner similar to needle punching, except that instead of needles very high velocity water jets are used to entangle the fibers. In recent years, spunlaced webs have developed as an alternative to needle-punched webs in air filtration applications.

*Stitch bonding*: Stitch bonding is a mechanical method of consolidating fiber webs by using knitting elements with or without yarn to interlock the fibers. In brief, a yarn is stitched through the fabric in a way that can impart very interesting properties. Quilted fabrics are often made this way. Lycra® will impart stretch properties such as in Ace bandages. Stitch-bonded webs are used in filter applications such as vacuum bags.

*Thermal bonding*: Webs containing low melting thermoplastic material can be thermal bonded to consolidate the web. Usually the low melting material is fibers or the low melting sheaths of bicomponent fibers. Thermoplastic powders are used in some applications. Following are some forms of thermal bonding:

- Calendering or embossing—The heated web is squeezed between two rolls (either smooth or patterned) which forces intimate contact of the molten fibers with other fibers in the web. Often the rolls themselves are heated. If the rolls are smooth (calendering), the web is uniformly bonded. If the rolls are patterned, an embossed web results where bonding is most intense in the dense areas of the embossed structure. Spot bonding is accomplished in this manner. Embossing results in a soft bulky web, because the areas between the bond points are unbonded. Figure 1.14 is an illustration of a pattern-bonded filter medium produced by PGI.
- Through air bonding—The web is thermally bonded by passing the web through air dryer or heater. The results are a bulkier and loftier web. High loft filter media are produced in this manner.
- Ultrasonic bonding—This relatively new technology utilizes high intensity ultrasonic energy to melt and form bonds from thermoplastic fiber. The ultrasonic bonding is like stitch bonding in that it takes place along narrow lines in the web. It is also a way of producing quilted structures.
- Hot melt bonding—A hot melt thermoplastic is extruded unto the web and while still in the molten stage, penetrates and bonds the web.

**Figure 1.14** Thermally Embossed Pattern Bonded Filter Media PGI Synergex® Style 6250. *Reproduced by permission of PGI, Old Hickory, Tennessee USA.*

*Chemical bonding*: Chemical bonding involves the use of a chemical binder or resin to bond the web. Water-based latex binders are the most commonly used resin binders for fiber-sourced dry formed webs. Application of a latex binder to a web requires a drying process to follow the application. Additional heat may be necessary to cure the resin. Following are ways of applying the latex:

- Saturation
- Spray bonding
- Print bonding
- Spot bonding
- Froth binder application
- Curtain coating

The process methods of applying latex binder to a nonwoven filter medium are discussed in more detail in Chapter 5, Section 5.1.2.3.4.

### *1.7.1.2 Polymer sourced*

#### 1.7.1.2.1 Melt spun

There are two major melt spun processes that produce polymer sourced nonwoven webs: spunbond and melt-blown. Both follow the general process flow diagram of Figure 1.15. Polymer in the form of powder granules or pellets is fed from a hopper into an extrusion chamber where the polymer is heated to a molten state. The molten polymer is then extruded through a gear pump and screen (to remove unmelted contaminants or chunks) and fed into a die block or spin pack. The polymer is then forced through a spinnerette block containing thousands of

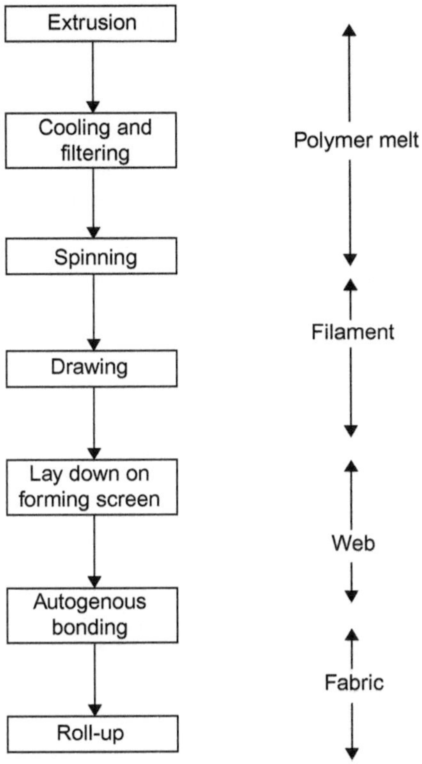


**Figure 1.15** Generic of the Meltspun Processes. *Reproduced by Permission of TAPPI, Atlanta, Georgia USA.*

micro-diameter holes. The fibers, so formed, are attenuated, cooled, and randomized on their way to form a nonwoven web structure on the collector belt. The significant difference between the processes is the manner in which air is introduced to cool and/or attenuate the fibers as they emerge from the spinnerette. This is illustrated in Figure 1.16. Note the quenching air flow for the spunbond process is horizontal to the vertical direction of the fibers and intended to cool the fibers as they emerge from the spinnerette. On the other hand, the melt-blown hot air flow is in the diagonal direction and merges with the fibers as they merge with the hot fibers from the spinnerette. The merging action of the hot air flow attenuates the fiber much more than is done in the spunbond process. As a result, the melt-blown fibers are much finer than the spunbond fibers, and the melt-blown fleece formed is softer and weaker.

**1.7.1.2.1.1 Spunbonded** There are several manufacturers of spunbonded systems. The Reicofil system developed by Reifenhauser GmbH of Germany as described by Malkan and Wadsworth (25) is discussed in detail in Chapter 5,

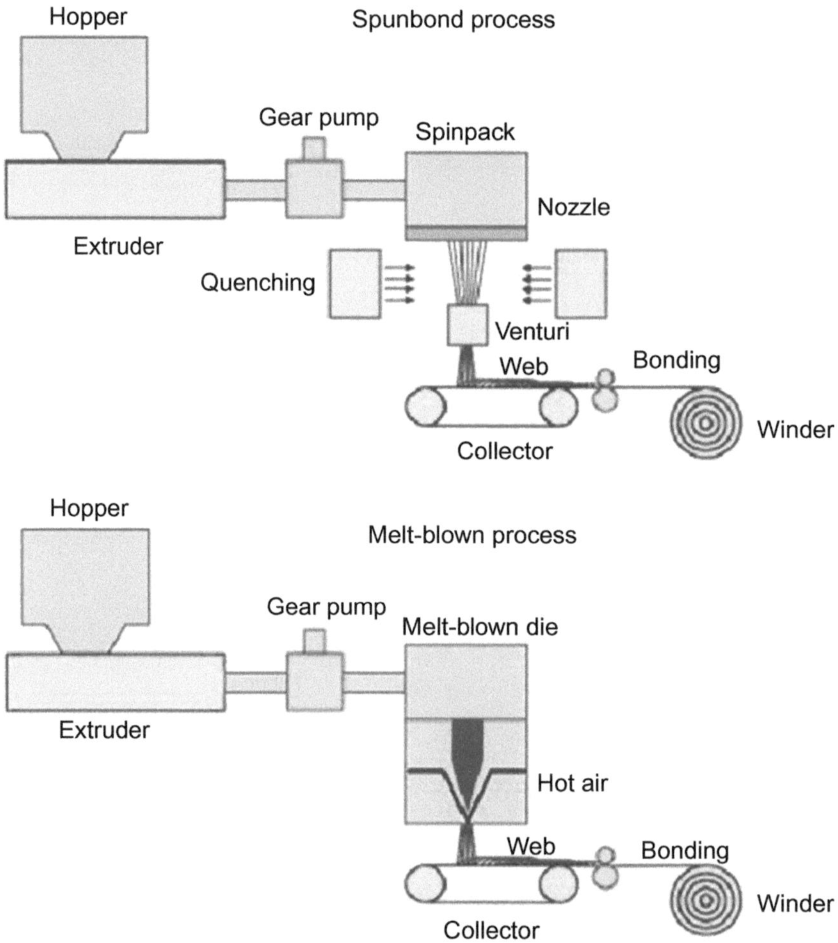


**Figure 1.16** Comparison of Meltspun Processes.

Section 5.2.1. Other manufacturers of spunbond systems are Kohle and Mineraloltechnik GmbH of Germany (Docan System) and Carl Freudenberg Co. of Germany (Lutravil System).

Polymers generally used in the spunbond process are polyester (Reemay®), nylon (Cerex®), and polypropylene (Typar® and Tekton®). Polyurethane and rayon have also been successfully spun into spunbonded webs. More recently Freudenberg Filtration Technologies, Wenheim, Germany has introduced Lutrador® ECO, a spunbonded web produced from exclusively postconsumer recycled (PCR) plastic materials. The trade names mentioned are those that were identified with the materials when originally introduced. Currently, there are at least 80 different trade names for spunbonded filter media.

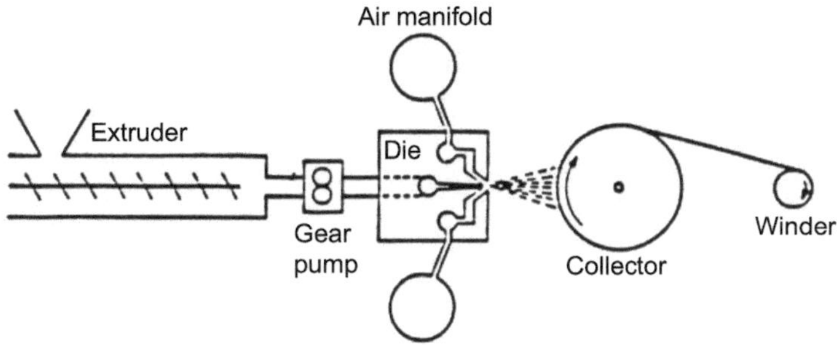


**Figure 1.17** Schematic of Melt-Blown Process. *Reproduced by permission of TAPPI, Atlanta, Georgia USA.*

**1.7.1.2.1.2 Melt-blown** Melt-blown webs are formed directly from a molten resin. The process is similar to spunbonded webs in that thermoplastic polymers are extruded through a spinning die, to form filaments. It is different than spunbonded webs in that high velocity, heated air, injected near the die tips, converges with the filaments to attenuate them to very fine diameters. The attenuated filaments are quenched with cool air and collected on a moving collector screen. The attenuated filaments are generally 1−4 μm in diameter and form a very uniform web at low grammage. Figure 1.17 is the diagram of the typical melt-blown process.

Polypropylene and poly butyl terephthalate are the most frequently used polymers used for producing melt-blown webs.

The principal difference between melt-blown webs and spunbonded webs is that the melt-blown process produces a web composed of finer filaments. Melt-blown webs are generally softer and weaker than spunbonded webs. Melt-blown webs are often used in a composite or laminated form combined with a supporting web or material to provide strength and stability. Butler estimated (5) that two thirds of melt-blown consumption is used in composite form. The largest volume composite structures are spunbonded/melt-blown structures. More will be said about composite structures in Section 1.7.3. The fine fiber structure of melt-blown webs makes them candidates for high efficiency filter media, both in liquid and in air filtration applications.

### 1.7.1.2.2 Flash spinning

In contrast to meltspun products, flash spinning forms nonwoven webs from a solvent solution of a polymer. The solvent solution of polymer is extruded through the spinnerette and the solvent quickly evaporates causing the filaments to form a highly fibrillated form before they are deposited on a screen. The web is thermally bonded with a calender roll to form a strong fabric. It is by this process that the high density polyethylene fabric known as Tyvek®[8] is produced. Tyvek® in itself is generally too

[8] Tyvek® is a registered trademark of DuPont, Wilmington, Delaware, USA.

dense to be used as a filter medium, however a variation to this is a series of filter grades, marketed by DuPont, known as SoloFlo® (see Section 3.4).

#### 1.7.1.2.3 Nanofiber spinning processes

There are two known commercial types of processes for spinning webs made of nanofiber: electospun webs and centrifugal spun webs.

**1.7.1.2.3.1 Electroformed webs** In 1934, Anton Formals (26) was the first inventor to patent an electrospinning phenomenon. Electrospinning is a process that produces nanofiber webs by applying a high-voltage charge to a polymer solution or melt and using the charge to draw the solution from the source to a grounded collector. The source may be a needle, a nozzle, a spinnerette, or a rotating surface.

Voltages range from 5 to 30 kV, sufficient to overcome the surface tension forces of the polymer. The free surface of the charged polymer produces very fine jets of liquid that are rapidly drawn to the grounded collector. The effect causes substantial drawing of the rapidly solidifying fibers as they approach the grounded collector. The highly attenuated fibers collect on a moving surface passing over the grounded collector and form an interconnected web of small filaments. The fiber diameters for filtration purposes are in the range of 0.25 μm. The webs are of very fine thickness (on the order of 1 μm or less) and have limited mechanical properties. The moving surface on which they are collected is usually a substrate that will provide strength and stability for subsequent processing. Very often the substrate will be a nonwoven chosen specifically for the filtration application.

Figure 1.18 is a schematic of the laboratory needle process and Figure 1.19 is a SEM image of a nanofiber web produced in this manner.

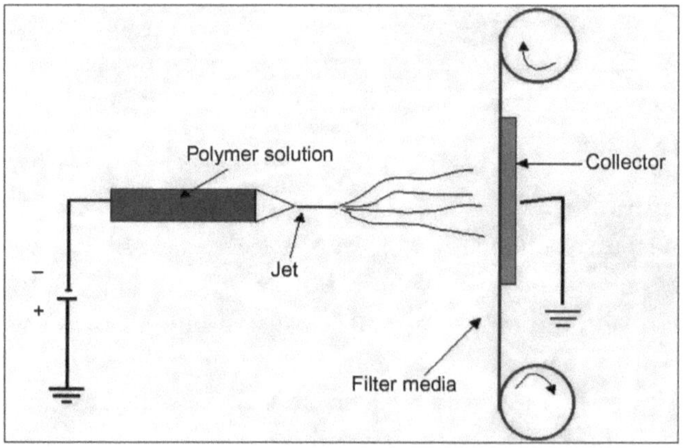


**Figure 1.18** Schematic of the Electrospinning Process. *Reproduced with permission from Donaldson Company, Inc © Donaldson Company, Inc. All rights reserved.*

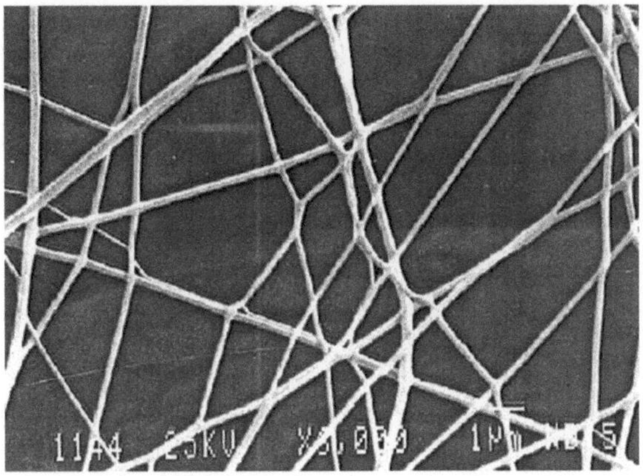

**Figure 1.19** Electrospun Nanofiber Web. *Reproduced with permission from Donaldson Company, Inc* *© Donaldson Company, Inc. All rights reserved.*

DuPont has patented an electro-blowing process (27) that spins a fiber-forming polymer solution from an electro-charged spinnerette. The spinnerette process appears similar in principle to the melt bond process except that the polymer is solution polymer rather than molten polymer and the force driving the polymer solution through the spinnerette is voltage driven as well as pressure driven. In Chapter 5, Section 5.3.4. Figure 1.20 is a schematic of the process (28). Process details are proprietary; however, it appears to be a variation of the melt-blown process in that a high-voltage charge is applied to the spinnertte. Note from Figure 1.20 that the airflows converging around the filament fibers as they emerge from the spinnerette openings, similar to the air flows of the melt-blown spinnerette (see Figure 1.17). The angled air flow around the spinnerette results in a drag motion around the emerging filaments causing them to attenuate. In the electro-blown process, the fiber attenuation is further enhanced by the electric voltage differential between the spinnerette and the collector. The electro-blowing process is further discussed in Chapter 3, Section 3.5.1, and Chapter 5, Section 5.3.4.

Currently, air filtration applications are the significant markets for electrospun nanofiber webs. This includes engine air intake filters, turbine air filters, pulse filters for dust collection systems, and vacuum bag filters. A good discussion on the air filtration applications is provided by Graham et al. (29). Electrospun nanofiber filter media are also finding niches in liquid filtration. For example, they are a source of nanofiber membranes' use in aqueous microfiltration and ultrafiltration applications (see Chapter 7, Section 7.5).

A more detailed discussion of the various electrospun processes is provided in Section 5.3.

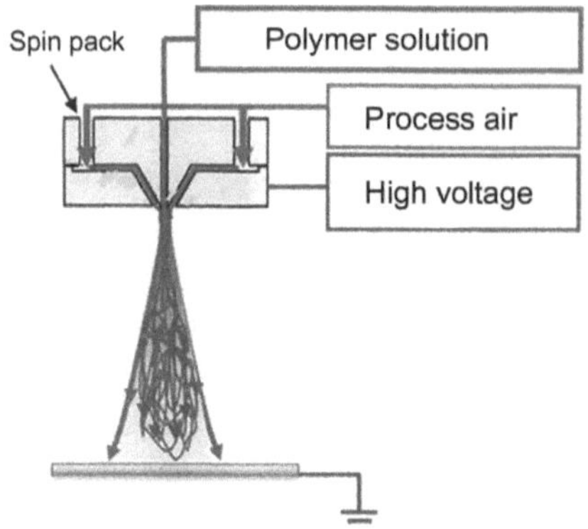


**Figure 1.20** DuPont® Electro-Blowing Process (28). *Reproduced with permission of DuPont, Wilmington, Delaware USA.*

**1.7.1.2.3.2 Centrifugal spinning** There are two known processes for centrifugal spinning of a nanofiber web: Forcespinning®[9] by FibeRio® Technology Corporation of Edinburg, Texas, USA and a Melt Spun Nanofiber Process (MSNP) by DuPont®.

The Forcespinning® process is based on a spinnerette orifice nozzle attached to a rotating disk. Figure 1.21 is a photograph of the FibeRio laboratory process. The disk contains molten or solution polymer that is forced out of attached nozzles and then attenuated to nanofiber dimensions. In the laboratory process, of Figure 1.21, the web is deposited on a set of vertical bars structured around the spinning disk. This is the laboratory process. FibeRio markets commercial machine processes that produce webs in a continuous uniform flat-sheet manner. This is further discussed in Chapter 5, Section 5.4.1.

The DuPont® MSPN process works with melt-based polymers. The centrifugal action of the spinnerette head lays down the fibers in a circle (see Figure 1.22). The gap between circles caused by the advancing collector causes each circle to intersect with the previous circle and with the one after it forms an intersecting set of circular filaments. This action combined with the action of adjacent spinnerettes creates a nanofiber structure that has all the appearance of a random nonwoven web.

## 1.7.2 Wet laid

Practically all filter media that is identified as "paper" filter media, are made by the wet laid process. The wet lay process is essentially a paper machine process. A considerable amount of nonwoven that satisfies the INDA definition is produced by the wet lay process, however, there are a considerable amount of filter media that do not satisfy the INDA definition yet are considered nonwoven filter media by the criteria of

[9] Forcespinning™ is a trademark of FibeRio Technology Corporation in Edinburg, Texas, USA.

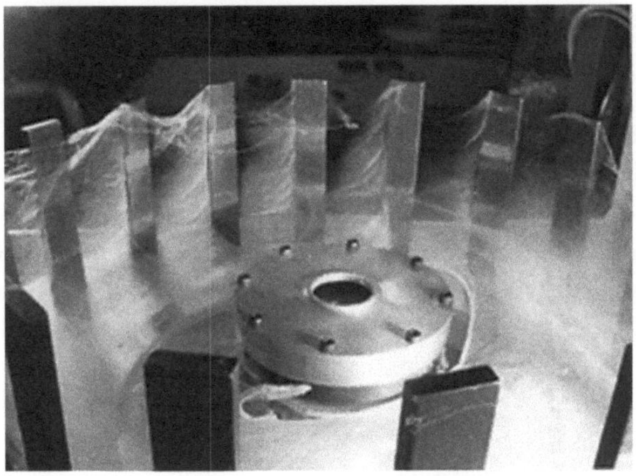

**Figure 1.21** FibeRio Forcespinning™ in Progress. *Reproduced with permission of FibeRio Technology Corporation, Edinburg, Texas USA.*

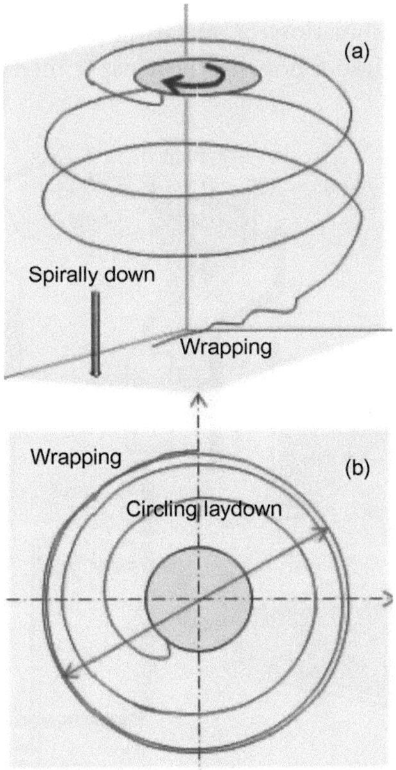


**Figure 1.22** Melt Spun Nanofiber Process (MSNP) by DuPont. *Reproduced with permission of DuPont, Wilmington, Delaware USA.*

this book. The wet lay process produces most glass fiber media used in filtration. Wet lay machines are known to form webs up to 5 m wide and at speeds exceeding 300 m/min. They generally handle short fibers of 6 mm or less, however, commercial products from the wet lay process have been produced with fibers as long as 35–50 mm. In addition to wood pulp and natural fibers, the wet lay process can also handle fibers of polyester, nylon, rayon, glass, Kevlar®, Nomex®, graphite, and any other material whose fiber can be reasonably dispersed in water. How well the fiber can be dispersed will depend on its fiber length and physical properties. The largest volume filter product from the wet lay process are resin bonded filter media. This is the type that is usually pleated into cartridge filter elements.

#### *1.7.2.1 Forming of wet lay webs*

Figure 1.23 is a diagram of the wet end of a typical wet laid process line. The wet end is a process for dispersing fiber into aqueous slurry called stock and feeding it to the headbox of the web former. The headbox is a device to feed the diluted stock to a forming screen where it is formed into a wet sheet. The dry end of the wet lay machine consists of a combination of mechanical and thermal devices to remove moisture and dry the sheet into a nonwoven or paper media.

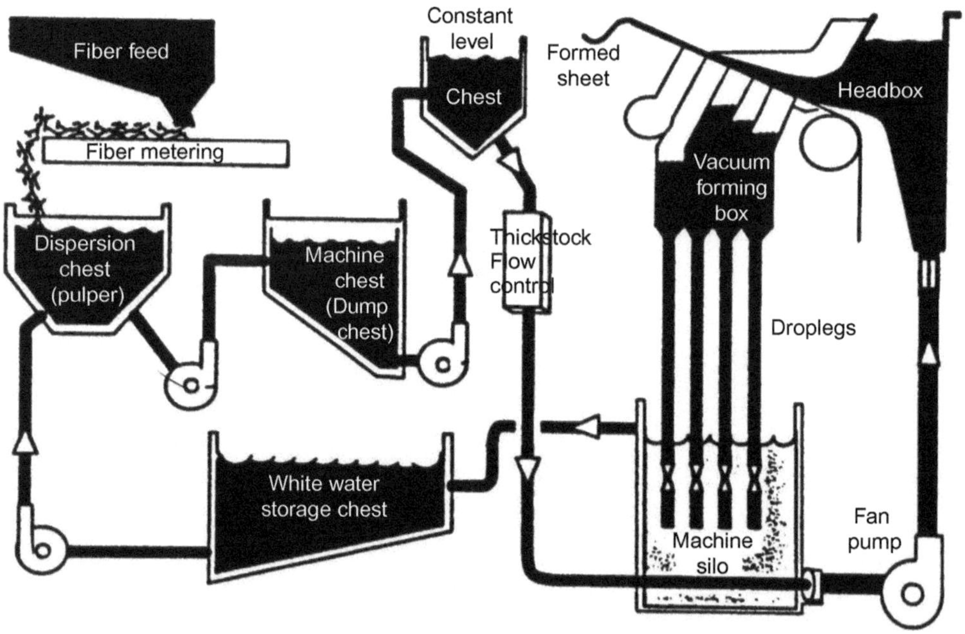


**Figure 1.23** Wet End System for an Inclined Wire Wet Lay Process. *Reproduced by permission of Glens Falls Interweb, Inc. Glens Falls, NY.*

Just as the card is the core of the dry lay process, the headbox is the heart of the wet lay process. Paper, some nonwovens, and microglass fiber media are formed by traditional type of paper machine headbox called the Fourdrinier headbox. Filter media made from long synthetic fiber and difficult-to-disperse furnishes are produced on headboxes specially designed for this purpose. Two of the most common are the Rotoformer® headbox and the inclined wire headbox. Glens Falls Interweb (GFI) in Glens Falls, New York, USA manufactures both. The Rotoformer® forms the sheet on a wire-covered rotating drum. The inclined wire headbox (trade named by GFI as the Delta Former®) forms the sheet on the incline of the wire as it passes through the pond. Voigt-Dorries in Germany also manufactures inclined wire formers under the trade name Hydroformer®. These machines operate at much lower stock consistencies than do traditional paper machines. Chapter 5, Section 5.6.1, describes in greater detail the operation of wet lay headboxes.

After the wet lay sheet is formed, mechanical and thermal devices remove water and dry the sheet. Mechanical devices include wet presses, foils, and suction slots. Most water is removed by thermal drying. The thermal devices are usually hot air dryers, hot air ovens, and steam heated dryer cans. Infrared and microwave drying technologies are also utilized. Usually, wet lay drying systems use a combination of these devices.

#### *1.7.2.2 Bonding of wet laid nonwovens*

Wet laid nonwoven webs can be bonded by adding a bonding agent to the fiber slurry before formation (wet end or beater addition) and/or by passing the web through a resin applicator after formation. Bonding can also be achieved by incorporating thermal bonding fibers in the fiber furnish.

##### 1.7.2.2.1 Beater addition

Beater addition or beater ad is the papermaker's term for adding additives and resins to the fiber stock slurry. Several types of resins or chemical agents can be added to the fiber slurry or stock to enhance the strength of the web after it is dried. Most often, these are cationic agents that use ionic chemistry to be attracted to the negatively charged anionic fibers. If the additive is a latex binder, a complex system of retention aids is used to retain the latex particles on the fiber surfaces. Similar types of chemistry can also be used to retain solid fillers such as clays and activated carbon powder. Cationic wet strength agents are frequently used to give the medium enough strength to be transported through the wet lay machine and through subsequent aqueous treatments. They also make the medium strong enough for aqueous filtration applications. Sheeted filter media used in plate and frame industrial filtration applications are often referred to as high wet strength filter paper. Coffee filter paper is another example of a filter medium bonded with a wet strength agent.

##### 1.7.2.2.2 Resin application to the web

Most wet laid nonwovens used for filter media applications are resin treated after the web is formed. The resin treatment is intended to provide the desired structural properties so that the web can be converted to the end-use filter application and have the integrity to stand up to the hostile challenges of the filtration environment.

Resin can be applied to the web either in-line with the wet lay machine or off-line in a separate operation. Either way, the most common form of resin applicator is the size press. There are many versions of size presses. Two versions, the vertical size press and the horizontal size press are discussed in Chapter 5, Section 5.6.3, and diagrammed in Figure 5.47.

##### 1.7.2.2.3 Thermal bonding fibers

Binder fibers can be included in the fiber furnish. This includes poly vinyl alcohol type fibers such as Kuralon®,[10] vinyon (polyvinyl chloride fiber), and other low melting thermal bonding fibers. In more recent times, bicomponent fibers have come into vogue for this purpose. An established technology is sheath-core fibers with a low melting sheath. Sheath-core fibers are part of the discussion of multicomponent and bicomponent fibers in Chapter 4, Section 4.3.5.2.

#### *1.7.2.3 Converting processes for wet lay webs*

A number of converting processes are applied to wet lay media as necessary for its end-use filtration performance. Included are corrugating, rewinding, slitting, creping, sheet cutting, die cutting, pleating, bag making, and other operations that alter the mechanical structure and geometrical surface of the web. In addition to resin, there are other chemical treatments that add to web properties. Among them are flame retardants, water repellents, extenders, surfactants, adsorbents, and antimicrobial agents. The various converting operations for filter media are discussed in Chapter 5.

### 1.7.3 Composite structures

There are many examples of composite filter media including web-supported membrane and electrospun media. Many needle felts are reinforced with a scrim to provide structural integrity in the filtration application. Kimberly-Clark has pioneered the SMS (spunbond/melt-blown/spunbond) process and variations thereof. Composite structures can combine layers of different density media to form gradient density media. Some structures are designed to provide different separation modes in a single filter medium. For example, automotive cabin air filters often combine a nonwoven particulate filter layer with an activated carbon layer to remove odors and fumes.

[10] Kuralon® is a trade name of Kuraray Co., Inc., Japan.

Following are some of the purposes of composite structures:

1. One or more layers provide mechanical support for the layer or layers doing the filtering.
2. One layer serves as a prefilter for the layer doing the final filtering. In some applications the prefilter layer is referred to as the "capacity layer" which contributes to dirt holding capacity and the final filter layer is referred to as the "efficiency layer" which contributes to filtration efficiency.
3. Several layers are organized in a way to provide gradient density depth filtration. Each layer has a different bulk and filtration efficiency, that is, as the fluid and contaminants progress through the medium, each successive layer provides a higher level of efficiency to remove smaller and smaller particles.
4. Different separation technologies are combined into one filter medium. In the example mentioned above for cabin air filters, an activated carbon layer is combined with one or more nonwoven layers, whereby the nonwoven layers do particulate filtration and the activated carbon layer separates molecular contaminants by adsorption.
5. The downstream layers serve as a containment to inhibit media migration, dusting, and particle fallout from the upstream layers.

It should be noted that for a medium to fit the definition of a composite structure it must be "layerized." For example, there are technologies for combining activated carbon particles into the fibrous structure of a single web. Such a web is often referred to as an impregnated web containing carbon particles and combines the technologies of particulate filtration and adsorption. It is not a composite structure. Likewise, in such cases where additional properties are applied to the web by the use of coatings or impregnants (examples are flame retardants or water repellents), the structure remains a single layer structure and is not a composite.

Not all the layers of a composite are necessarily nonwovens. One or more of the layers could be a membrane material, a woven material, or a plastic or metal mesh material to provide structural support. Figure 1.24 illustrates the cross section of a filter medium that contains a stainless steel mesh for structural support.

Needless to say, the combination possibilities for nonwoven composite structures are endless.

Just as there are numerous combinations for composite structures, there are also numerous ways for them to be combined. Some examples are:

1. Laminating—two or more layers are unwound from unwind stands and combined into a composite structure. The layers are bonded to each other by adhesive application, thermal bonding, or by inherent Van der Waals or electrostatic forces.

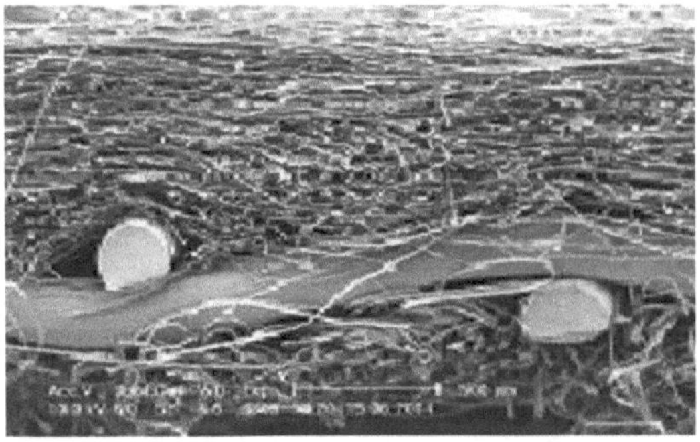

**Figure 1.24** Nonwoven Filter Media Reinforced with a Stainless Steel Scrim. *Reproduced with permission of Sächisches Textilforschungsinstitute e.V.Chemnitz, Germany.*

2. Fiber entanglement between layers by needle-punching or hydroentanglement.
3. Multilayer forming—this can be done in wet lay processes as well as dry formed processes.
4. Combined process forming—Kimberly-Clark's SMS technology is a good example of combining different layers by different processes to form a composite structure.
5. Coating—the fibers or spun web are sprayed or spun unto a substrate that is used as collector and becomes part of the composite structure.

A more detailed discussion of the composite processes will be found in Chapter 5.

# CHAPTER 2

# Filtration Mechanisms and Theory

## 2.1 FILTRATION MECHANISMS

### 2.1.1 Types of filtration mechanisms

Sutherland and Purchas (18) describe four fundamental filtration mechanisms: surface straining, depth straining, depth filtration, and cake filtration. Following is a brief description of each.

Surface Straining—Particles larger than the pore diameters cannot pass through the medium, whereas particles with diameters smaller than the pore diameters pass through the medium and are not filtered. This filtration mechanism collects particles at the medium surface that either affect or are affected by other mechanisms at the filter medium surface. It is particularly important to membrane filtration and separation, particularly cross flow separation where contaminants and particles are continually removed by the cross flow washing of the fluid. This type of separation is also associated with nonwoven fabrics and membranes where dirt and cake release are desired properties. It is also the process by which cake filtration is achieved as described later.

Depth straining—This applies to felts and nonwoven materials that are relatively thick compared to pore diameters, and where the pore diameters are quite variable in their length, the particles penetrate the pores until they reach a necking point where the diameter becomes smaller than the particle and at this point the particle is trapped in the pore.

Depth filtration—Depth filtration is different than depth straining; the former involves mechanisms for removing a particle from a fluid even though the particle may be smaller than the diameter at any point in the pore structure. The mechanisms for depth filtration are discussed in Section 2.1.2.

Cake filtration—Cake (or surface) filtration involves the capture of particles on the surface (or near the surface) of a filter medium so that the buildup of particulate matter into a layer of filter cake participates in the filtration process. Surface-modified needle felts and other media used in baghouse-dust collection systems operate in this manner. Pulsing and/or reverse flow cleaning operations then allow for easy removal of the cake and reuse of the filter medium. In some chemical operations this is important because the cake has value and its recovery is the intended purpose of the filtration. In many operations where a filter aid such as diatomaceous earth is used, the filter cake becomes the filter medium and the nonwoven medium serves as the cake support. Plate and frame filter presses often operate in this manner.

*Handbook of Nonwoven Filter Media*
ISBN: 978-0-08-098301-1

## 2.1.2 Mechanisms of particle capture

Depth filtration is the basis by which mechanisms for filtration and separation of particulate matter are developed. Filtration and separation theory is more concerned with the mechanisms, listed below, for particle separation after the particle penetrates the surface of the medium.

1. Inertial impaction filtration mechanisms—occurs when the particle inertia is so high that it has sufficient momentum to break away from air streamlines and impact the fiber.
2. Interception—occurs when a particle does not have sufficient inertia to break away from the streamline, however, comes close enough to the fiber so that natural forces will attach the particle to the fiber. For mathematical treatment, Lastow and Podgorsk (30) define interception as follows "A particle is intercepted by a particle when the distance from the center of mass of the particle to the fiber surface is equal or less than the radius of the particle."
3. Diffusion—is based on the Brownian (zigzag) motion of very small particles (<0.5 μm). This random and probabilistic motion will cause a particle to vary from the streamline and possibly engage a fiber surface.
4. Electrostatic attraction—is based on an electric or electrostatic charge on the particle and/or fiber that will force the particle to be diverted from the streamline and attracted to the fiber.

Figure 2.1 is a classic illustration of the various mechanisms. It is based on flow past a single fiber, the flow being perpendicular to the axis of the fiber.

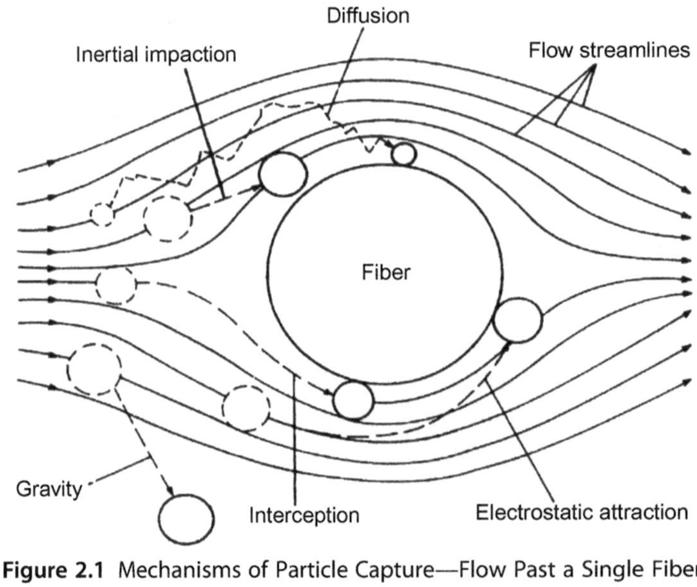


**Figure 2.1** Mechanisms of Particle Capture—Flow Past a Single Fiber.

Hutten (31) introduced similar effects by illustrating, in Figure 2.2, the passage of particles through a cylindrical pore. The illustration is a series showing particles in decreasing particle size order trying to penetrate the pore tunnel. Note that the pore tunnel is bent and not straight. It contains obstructions. It may also be assumed to vary in its diameter.

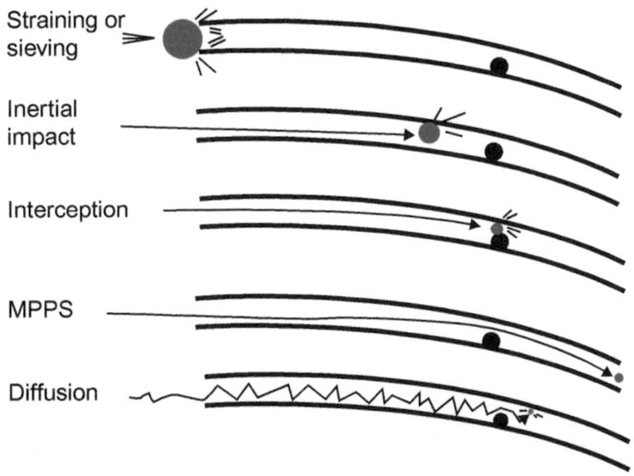


**Figure 2.2** Mechanisms of Particle Capture: Flow Through a Pore.

The surface straining or sieving action is demonstrated when the particle is too big to enter the entrance of the pore. Smaller particles may fit through the entrance, but have sufficient speed and momentum to break loose from the streamline and strike and adhere to the wall of the pore. In other words, the particle wants to go straight even though the streamline wants to follow the curvature of the pore path. This is separation by inertial impact.

Even smaller particles enter the pore and do not have enough inertia to stray from the streamlines. However, the streamlines may carry them too close to the wall surface of the pore or too close to obstructions that the particles are unable to avoid. Obstructions may be necking and narrowing of the pore diameter, fibers bridging across the pore, or protrusions from the wall of the pore. This form of particle capture is interception.

There is even a smaller particle size that is small enough to stay with the streamline, stay away from the pore wall surface and get by or through all the obstacles and emerge unscathed from the exit side of the pore. This is the most penetrating particle size (MPPS). The MPPS is an important property when evaluating filter performance. It is the efficiency value used to rate high-efficiency particulate arrestance (HEPA) filters under EN1822-5:2009 (32). It will be further discussed later.

The term MPPS suggests that not only will bigger particles be less able to penetrate the pore, but so will smaller particles. Very small particles exert Brownian motion and are subject to capture by diffusion. This is because it is the nature of Brownian motion that small particles bounce off other particles and exhibit a random or zig zag behavior in their motion. When the particles travel through the pore passage, their erratic zig zag behavior may cause them to wander too close to the wall surface and be captured. The smaller the particle, the more intense is the Brownian motion and the more likely is the particle to be captured.

Note from the discussion above that the dominant capture mechanism is related to particle size. Very small particles are subject to capture by diffusion. Large particles have more momentum. They are more likely to break loose from the fluid streamlines and be captured in accordance with the inertia mechanism. For small particles, there is a particle size in the range of 0.04 to 0.4 μm, that is too large for substantial diffusion effects and yet too small to have sufficient momentum for inertial effects. Particles of this size are the most difficult to capture by a filter medium. It is referred to as the "most penetrating particle size or MPPS". Efficiency ratings in EN 1822-5:2009[1] (32), the European standard for HEPA and ULPA media, are based on MPPS.

There are numerous examples of MPPS in the literature. Figure 2.3 is an example chosen from a presentation by Goldsmith (33). The high-efficiency filter medium was a synthetic fiber web. The challenge contaminants include three different oil aerosols, latex spheres, and a commonly used solid particulate, potassium chloride (KCl). Figure 2.3 shows that the MPPS is different for each

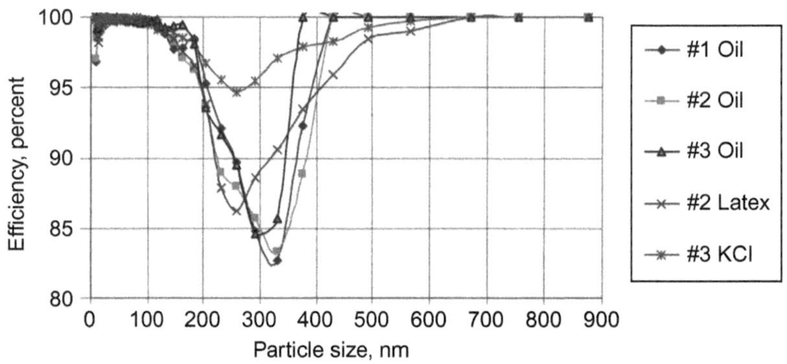


**Figure 2.3** Effect of Challenge on Most Penetrating Particle Size—Synthetic Media. *Reproduced with permission of INDA, Cary, North Carolina USA.*

[1] See Chapter 6, Section 6.14.1.13 and Chapter 8, Section 8.3.5.1 for discussions of EN1822-5:2009.

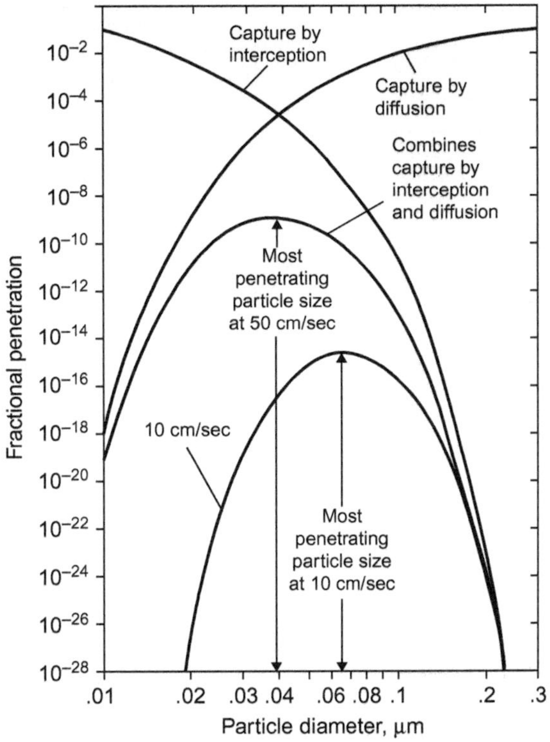


**Figure 2.4** Most Penetrating Particle Size Separation Mechanisms. *Copyright © 1998 from "Advances in Aerosol Filtration" by Editor Kvetoslav R. Spurny (34). Printed with permission of Routledge/Taylor & Frances Group, LLC, Boca Raton, Florida, USA.*

contaminant. The MPPS is also velocity related and will decrease in size as the velocity is increased. This is indicated by the representation of Spurny (34) in Figure 2.4. Note that the MPPS at 50 cm/s is smaller than the MPPS at 10 cm/s. Also note that the range of particle sizes in Spurny's display is of smaller diameters than the range in Goldsmith's presentation.

The suggestion from Spurny's Figure 2.4 is that in addition to an MPPS, there is also for a small particle size a velocity of maximum penetration. Note that Spurny includes the inertial mechanism with the interception mechanism. Ramskill and Anderson (35) reviewed the theories of Irving Langmuir and noted that he had assumed that the inertial effect was not significant. In their paper, the diagram of Figure 2.5 demonstrates the three mechanisms and supports the argument that inertial effects are more significant than Langmuir previously supposed. Ramskill and Anderson provided data to support this contention. Davies (36) supported this contention with his data.

Figure 2.5 supports the argument that for small particles there is a velocity of maximum penetration (i.e., a velocity at which a particle of a given size achieves

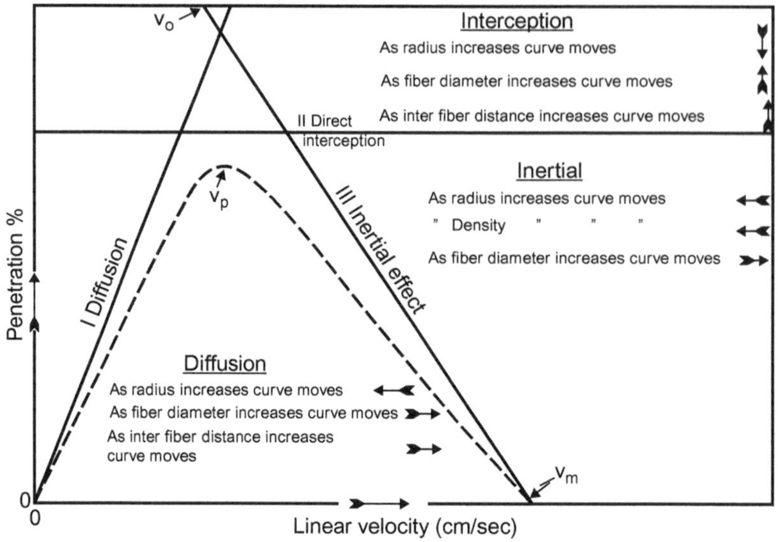


**Figure 2.5** Qualitative Picture of Filtration Mechanisms. *Reproduced with permission of Elsevier Inc., Oxford England, UK.*

maximum penetration). Above and below this velocity the particle is less penetrating. Below this velocity, diffusion and Brownian motion dominates the separation process. Above this velocity, the competing process of inertia begins to dominate. Note that Ramskill and Anderson have the interception mechanism independent of velocity.

## 2.1.3 Reentrainment and media migration

It is rational to think that as the velocity of a particle increases, its momentum will make it more vulnerable to capture by the fibers of a nonwoven medium. In real-life testing, this effect is confounded by the effects of reentrainment and media migration. Reentrainment is the result of particles that originally attach to the fiber surfaces being blown free by the "wind" forces of the fluid flowing through it. Media migration is the release of "fines" or small fibers and particles in the medium structure that were there as residue from the manufacturing process. They confound efficiency measurements and the cleanliness of the exiting stream. The tendencies of reentrainment and media migration tend to increase with increasing fluid velocity.

## 2.1.4 Membrane separation processes

The mechanisms discussed in Section 2.1.2 are generally applied to air and aerosol filtration. Processes for fine filtering and/or separation of contaminants in liquid fluids are microfiltration, ultrafiltration, nanofiltration, and reverse osmosis. These are generally membrane processes, although very high-efficiency nonwovens composed of fibers in the nanofiber category are often used for microfiltration and ultrafiltration purposes. Following is a description of the each membrane separation processes.

#### 2.1.4.1 Microfiltration

Microfiltration is used for the filtration of particles and finely suspended solids down to less than 0.1 μm (100 nm). The process operates at relatively low pressures 1–4 bars (100–400 kPa). Liquid clarification and sterile filtration are two applications of microfiltration.

#### 2.1.4.2 Ultrafiltration

Ultrafiltration starts to bridge the gap between particulate filtration and molecular filtration. The particle size filtration range is from roughly 0.004 to 0.1 μm (4–100 nm). Ultrafiltration rejects finer particulates such as viruses and molecules of molecular weight greater than 10,000. Process pressures are in the range of 5–10 bar (500–1,000 kPa). Ultrafiltration applications include separation of macromolecular solutions.

#### 2.1.4.3 Nanofiltration

Nanofiltration is a form of reverse osmosis (see Section 2.1.4.4) for the separation of molecules in the 300–1000 molecular weight range and for larger ions such as $Ca^{2+}$ and $Mg^{2+}$. The size range for particle and molecule separation is approximately 0.0012–012 μm (1.2–12 nm). The pressure range is 20–40 bar (2,000–4,000 kPa). An application of nanofiltration is the separation of small organic compounds and selected salts from solutions.

#### 2.1.4.4 Reverse osmosis

Osmosis is the passage of a solvent (e.g., water) through a semipermeable membrane from a solution of lower concentration to a solution of higher concentration so as to equalize the concentrations on either side of the membrane. Reverse osmosis reverses this flow. It utilizes a membrane that is under very high pressure, 30–60 bar (3,000–6,000 kPa) and impermeable to most molecular species in the solution. The back pressure must be sufficient to overcome the natural osmotic pressure of the solution, so that water will flow from the high-salt concentration side of the membrane to the dilute side. The size range of the molecules to be separated is approximately from 0.0015 μm (1.5 nm) to less than 0.0005 μm (0.5 nm). A major application of reverse osmosis is the desalinization of brackish water.

### 2.1.4.5 *Electrodialysis and pervaporation*

#### 2.1.4.5.1 Electrodialysis

Electrodialysis (ED) is used to transport salt ions from one solution through ion-exchange membranes to another solution under the influence of an applied electric current. It is an alternative to reverse osmosis (RO) as a desalinization mechanism. Whereas RO uses pressure drop to force the separation, ED utilizes a voltage potential. The separation is accomplished in what is called an ED cell. Multiple cells form an ED stack as depicted in Figure 2.6[2]. Both cationic and anionic membranes are used to attract and then migrate charged ions from a feed stream into an exit concentrate stream.

The ED process in Figure 2.6 shows two feed streams, C and D, approaching the ED stack. The C stream will become the concentrate exit stream and the D stream will become the purified product stream. In this case, both streams contain

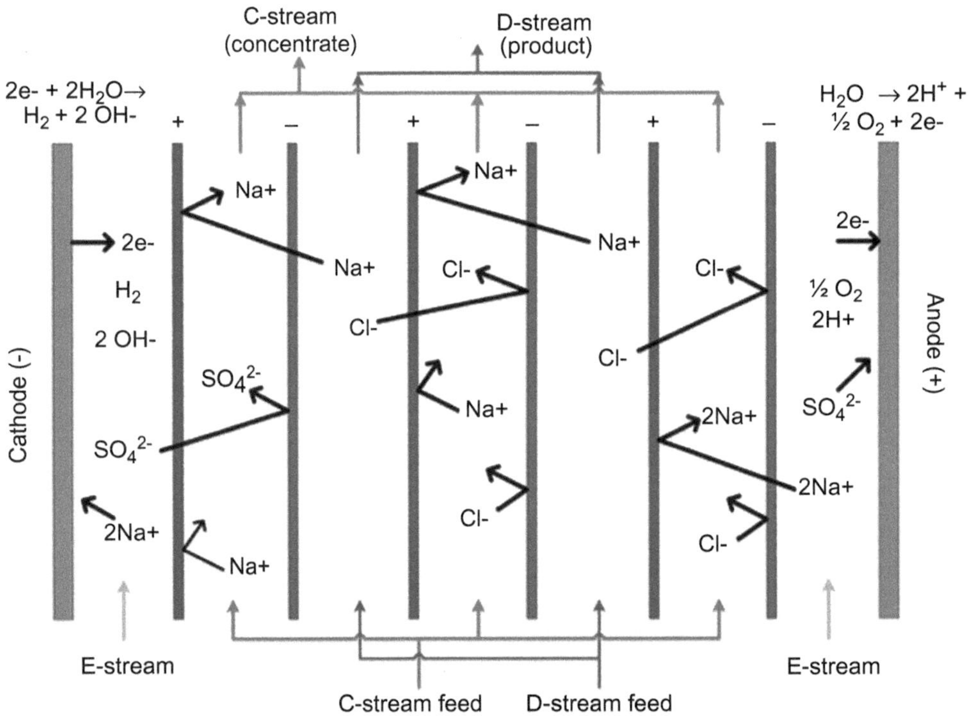


**Figure 2.6** Schematic of an Electrodialysis Stack for Water Desalinization. *Source: Wikepedia.*

[2] Refer Wikepedia (37).

dissolved sodium chloride which ionizes into positive sodium ($Na+$) ion and a negative chloride ($Cl-$) ion. The vertical lines in the stack diagram represent charged membranes. Note there are alternating positively charged (anionic) membranes and negatively charged (cationic) membranes. On the left side of the stack is a negatively (−) charged cathode and on the right side is a positively (+) charged anode. By applying a voltage charge from one cathode to the other an electrical current across the stack is created. This will cause ions in the feed streams to migrate; positive ions toward the cathode and negative ions toward the anode. There is also an E stream that flows past each electrode. This stream may be of the same composition as the feed streams or it may contain a different salt species such as sodium sulfate which ionizes to $Na+$ and $SO_4^{2-}$.

When the current is applied, $Na+$ ion in the D stream (to be purified) will migrate toward the cathode. It will pass through the cationic membrane into the C stream. It will be rejected by the anionic membrane on the other side of the C stream and thereby trapped in the C stream. Likewise the $Cl-$ ion in the D stream will be attracted to the anode and pass through an anionic membrane into the C stream. There it will be trapped, because it will be rejected by the cationic membrane on the other side of the C stream. It is necessary to maintain an ionic balance across the stack. That is why the E streams are there, to provide the necessary ions ($Na+$ and $SO_4^{2-}$) to maintain this balance.

#### 2.1.4.5.2 Pervaporation

KmX Corporation of Ontario, Canada (38) describes pervaporation as a membrane process used to purify chemicals. It is a selective membrane operation which can dehydrate organic solutions. It can remove water from organic liquids and volatile organic contaminants from water. It can also separate components of organic liquid mixtures. It is an alternative to distillation in cases where one or more of the components are unable to stand up to the heat process. The membrane operation is driven by a chemical potential difference across the membrane in which the major component of the feed stream (the retentate) remains in the liquid phase. The permeate is preferentially absorbed into the structure of the membrane where it diffuses to the exit or downstream side of the membrane. The permeate is then desorbed or removed from the membrane as a vapor. Very often a vacuum is introduced to create this effect. The membrane is designed to be preferentially absorbed or wetted out by the more-volatile component. If the permeate is water, then the membrane is hydrophilic. If it is a volatile organic compound being separated from water it would be hydrophobic. Organophyllic membranes are used to separate liquid mixtures of organic compounds. The three aforementioned pervaporation process mechanisms are diagrammed in Figure 2.7.

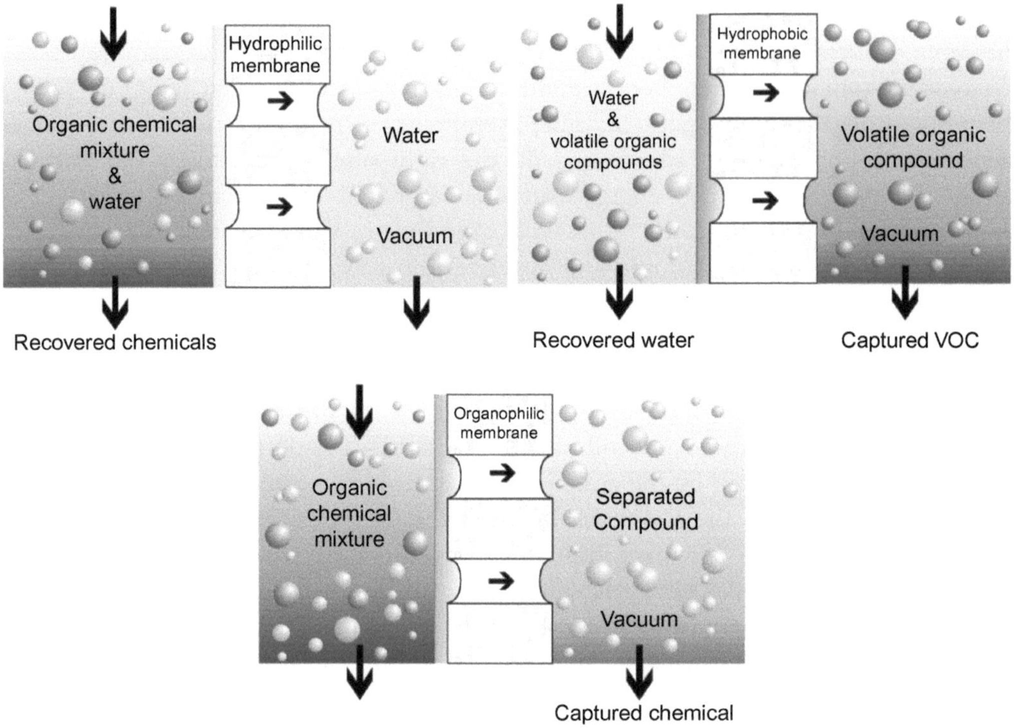


**Figure 2.7** Pervaporation Mechanisms. *Reproduced with permission of KmX Membrane Technologies Corp., Oakville, Ontario, Canada.*

## 2.2 FILTRATION THEORY

### 2.2.1 Filtration theory flow dynamics

A nonwoven filter medium is a porous medium and the theories concerning flow-through porous medium apply. There are two major avenues of theory concerning nonwoven filter media: channel theory and cell model theory. A variation of cell model theory is drag model theory. The original channel theory was based on filtration through nonfibrous packed beds such as sand. Often referred to as the capillary tube model, it assumes that the medium is a bundle of cylindrical tubes passing from one surface of the medium to the other surface, and not necessarily perpendicular to the surfaces. Channel theory can be applied to nonwoven filter medium used in liquid filtration, particularly if the medium is a tight structure with a high packing density (high solidity).

Much of cell model and drag model theories were developed for fibrous air filter medium. They are based on flow past a single fiber and the organization of fibers that compose the medium. They are best applied to open structures of low packing density. For nonwoven medium, the fibers are represented as cylinders. Cell model theory assumes an array of circular cylinders, each cylinder contained in a cell of fluid surrounded by cylindrical envelopes. Each cylinder with its fluid and envelope is treated as a cell. Drag model theory, the variation of cell model theory, analyzes the drag on each envelope.

Cell model theory is applicable to liquid filtration. Wakeman and Tarleton (39) discuss the cell model work of Happel (40) and Happel and Brenner (41) in regards to liquid filtration, where the medium is an assemblage of spheres[3]. They report that the Happel model closely agrees with channel theory and is superior at low packing densities.

The criterion is the packing density or solidity of the medium, which is the volume of solids in the medium per unit volume of medium. Schaeffer (42), summarizing the comments of several investigators, indicated that channel theory would apply to solidities greater than 0.2. Cell model theory and/or drag model theory is preferred for solidities less than 0.2. Cell model theory is the most commonly used theory for the theoretical development of air filtration.

## 2.2.2 The equations of motion and continuity

The equations of motion and continuity are the fundamental equations from which filtration theory is derived. The simplified and abbreviated explanations below are based on Bird, Lightfoot, and Stewart (43). Chapter 3 of this classic textbook contains a comprehensive teaching of the two equations for isothermal systems including the equation of mechanical energy.

### 2.2.2.1 The equation of continuity

The equation of continuity is simply a mass balance of a fluid flowing through a stationary volume element. It states that the rate of mass accumulation in this volume element equals the rate of mass in minus mass out. In vector form the balance is as follows:

$$\frac{\partial \rho}{\partial t} = -(\nabla \cdot \rho \boldsymbol{v}) \tag{2.1}$$

$\rho$ is the density, kg/m$^3$.
$t$ is the time variable, s.
$v$ is the velocity vector, m/s.
$\nabla \cdot \rho \boldsymbol{v}$ is the vector operator indicating the divergence of the mass flux $\rho v$.
Note that $\nabla$ has the units of reciprocal length, 1/m.

### 2.2.2.2 The equation of motion

In the case of an incompressible fluid, Equation 2.1 simplifies to the equation of motion:

$$(\nabla \cdot \boldsymbol{v}) = 0 \tag{2.2}$$

Analogous to the equation of continuity, the equation of motion is a momentum balance around a unit volume of fluid. It states that rate of momentum accumulation

[3] The work of Happel in developing cell model theory for fiber or cylinder structures is presented in Section 2.2.3.3.

equals the rate of momentum in minus momentum out plus the sum of all the other forces acting on the system. Its vector form is

$$\rho \frac{D\mathbf{v}}{Dt} = -\nabla p - [\nabla \cdot \boldsymbol{\tau}] + \rho \mathbf{g} \tag{2.3}$$

$\rho$ is the density, kg/m$^3$.
$t$ is the time variable, m/s.
$\rho \frac{D\mathbf{v}}{Dt}$ is the rate of momentum accumulation per unit volume, kg/m$^2 \cdot$ s$^2$.
$\nabla p$ is the pressure force on the element per unit volume, Pa (Pa = kg/m $\cdot$ s$^{-2}$).
$\nabla \cdot \boldsymbol{\tau}$ is the viscous force on the element per unit volume, kg/m$^2 \cdot$ s$^2$.
$\tau$ is the shear stress tensor, kg/m$^1 \cdot$ s$^{-2}$.
$\rho g$ is the gravitational force on the element per unit volume, kg/m$^2 \cdot$ s$^2$.
$\mathbf{g}$ is the acceleration of gravity (9.807 m/s$^2$).
$\nabla \cdot \tau$ can be revised in terms of the fluid viscosity μ, assuming μ and ρ are constant.

$$[\nabla \cdot \boldsymbol{\tau}] = \mu \nabla^2 \cdot v \tag{2.4}$$

For the case of constant $\mu$ and constant $\rho$, Equation 2.4 becomes

$$\rho \frac{D\mathbf{v}}{Dt} = -\nabla p - \left[\mu \nabla^2 \cdot v\right] + \rho \mathbf{g} \tag{2.5}$$

Equation 2.5 is known as the Navier–Stokes equation. For the case of $\nabla \cdot \boldsymbol{\tau} = 0$, Equation (2.5) reduces to:

$$\rho \frac{D\mathbf{v}}{Dt} = -\nabla p + \rho \mathbf{g} \tag{2.6}$$

Equation 2.6 is known as the Euler equation.

### 2.2.3 Flow-through porous media—channel theory

#### *2.2.3.1 Darcy's law*

For flow-through porous media, the equations of continuity and motion may be replaced by

$$\varepsilon \frac{\partial \rho}{\partial t} = -(\nabla \cdot \rho \mathbf{v}_0) \tag{2.7}$$

$$\mathbf{v}_0 = -\frac{k}{\mu} \nabla p - \rho \mathbf{g} \tag{2.8}$$

where

$\varepsilon$ the is the porosity or the void volume ratio, cm$^3$/cm$^3$ or m$^3$/m$^3$.
$\mathbf{v}_0$ is the flow rate or superficial fluid velocity through the media, m/s.
$\mu$ is the fluid viscosity, Pa $\cdot$ s.
$\nabla p$ is the pressure gradient across the media, Pa/m.
$\rho$ is the fluid density, kg/m$^3$.

**g** is the gravitational constant, $m/s^2$.

$k$ is a property of the porous medium called the permeability constant[4], $m^2$.

Equation 2.8 is known as Darcy's law. It is one of the fundamental equations concerning flow through a porous medium. In the case of a flat porous medium where the fluid flow is perpendicular to the surface of the medium, Darcy's law reduces to:

$$v_0 = -\frac{k}{\mu}\frac{dp}{dz} \tag{2.9}$$

where $dp/dz$ (Pa/m) is the differential pressure gradient across the thickness or $z$-direction of the medium. If the medium is a nonwoven fabric, Equation 2.9 can be further simplified to:

$$v_0 = \frac{k}{\mu}\frac{\Delta p}{L} \tag{2.10}$$

where $v_0$ is the face velocity of the medium expressed as volume flow per unit area of medium ($m^3/m^2 \cdot s$) and $\Delta p$ is the pressure drop (Pa) across the thickness, $L$ (m), of the fabric.

It should be noted that $v_0$ is the superficial velocity. Bird, Lightfoot, and Stewart (43) describe it as the volume rate of flow through a unit cross-sectional area of the solid plus fluid. It is averaged over a small region of space—small with respect to macroscopic dimensions in the flow system but large with respect to the pore size.

Darcy's law was originally developed for flow through packed beds. It is limited to viscous flow and for Reynolds numbers less than 1. Wakeman and Tarleton (39) state that the dimensionless Reynolds number, Re, as calculated for packed beds is based on the particle size $x$(m) and usually determined from the mean volume to surface size ratio.

$$\mathrm{Re} = \frac{\rho v x}{\mu} \tag{2.11}$$

In the case of nonwoven filter media, the pore structure is of sufficiently small size that the concerns of high Reynolds number and related turbulence generally do not apply. Darcy's law can be applied to nonwoven media. The air permeability test that is fundamental to describing nonwoven filter medium is based on Darcy's law. The test measures the air flow rate through a specified area of filter media at a specified pressure drop. The air flow rate so measured is called the air permeability of the medium.

[4] The permeability constant as defined here is often referred to as the permeability. It is not quite the same as the air permeability test property (frequently referred to as the Frazier permeability) of nonwoven filter media as discussed in Chapter 6, Paragraph 6.5. The air permeability is actually the $v_0$ of Darcy's Law. The air permeability is directly related and proportional to Darcy's permeability constant because all the other variables in the test are fixed. The air permeability and the permeability constant are directly related to the properties of the medium or porous structure.

#### 2.2.3.2 The Hagen–Poiseuille equation

The Hagen–Poiseuille equation describes laminar flow through a cylinder:

$$Q = \frac{\pi(P_0 - P_L)r_c^4}{8\mu L} = \tag{2.12}$$

where Q is the volume flow rate through the cylinder ($m^3/s$).

$r_c$ is the radius of the cylinder (m).

$P_0$ is the pressure at the inlet side of the cylinder (Pa).

$P_L$ is the pressure at the exit side of the cylinder (Pa).

$L$ is the length of the cylinder (m).

#### 2.2.3.3 The Kozeny–Carman equation

The permeability constant, $k$, of Darcy's law (Equation 2.10) is defined from the Kozeny–Carman equation.

$$k = \frac{1}{KS_0^2}\frac{\varepsilon_0^3}{(1-\varepsilon_0)^2} \tag{2.13}$$

$K$ is the Kozeny constant that accounts for the tortuosity of the pore structure through the media.

$\varepsilon_0$ is the porosity or void volume expressed as a decimal fraction of the total media volume ($m^3/m^3$).

$S_0$ is the surface area per unit volume of solid material ($m^2/m^3$).

The Kozeny–Carman equation (Equation 2.13) is derived from the velocity form of the Hagen–Poiseuille equation (Equation 2.12).

$$\mu' = \frac{\pi(P_0 - P_L)r_c^2}{8\mu L} \tag{2.14}$$

where $\mu'$ is the average velocity through the cylinder (m/s).

The following explanation of Kozeny's (44) and Carman's (45) work is based on Wakeman and Tarleton ((39) pp, 19–21). The assumptions are based on a packed bed that is a bundle of capillary tubes whose orientation is at 45° to the surface of the medium and 45° to the flow approaching the medium. It is assumed that the flow energy lost by the fluid when it contacts the capillary surface is similar to the energy lost by the flowing fluid when it is in contact with the actual internal surfaces of the media. A hydraulic radius, $D_h$ (m), is defined as

$$\begin{aligned} D_h = 2r_h &= 4 \times \frac{\text{flow cross section area}}{\text{wetted perimeter}} \times \frac{\text{length of flow path}}{\text{length of flow path}} \\ &= 4 \times \frac{\text{volume of voids in bed}}{\text{total surface of particles forming bed}} \\ &= 4 \times \text{voids ratio} \times \frac{\text{volume of the particles}}{\text{surface of particles}} \end{aligned} \tag{2.15}$$

$$D_h = 4\frac{\varepsilon_0}{S_0(1-\varepsilon_0)} \tag{2.16}$$

For a straight circular pipe $D_h$ and $D_c$ ($D_c = 2r_c$) are identical, however, in real media, pores are not perfect cylinders. Their lengths are not straight lines, but rather follow tortuous paths. The length of a tortuous path through a pore is $L_e$ (m). $L_e$ divided by the thickness of the media, $L$, is the tortuosity factor.

$$\frac{L_e}{L} = \text{tortuosity factor} > 1 \tag{2.17}$$

If $\mu'$ from Equation 2.14 represents the velocity through a cylindrical pore, then $v$ the velocity through a tortuous pore (m/s) is related to $\mu'$ by

$$v = \mu'\frac{L_e}{L} \tag{2.18}$$

If we let $L = L_e$ and $v = \mu'$, Equation 2.14 becomes

$$v = \frac{\pi(P_0 - P_L)D_h^2}{32\,\mu L} \tag{2.19}$$

$v$ is related to $\upsilon_0$, the superficial velocity of Darcy's law is as follows:

$$v = \mu'\frac{L_e}{L} = \frac{\upsilon_0}{\varepsilon_0}\frac{L_e}{L} \tag{2.20}$$

Combining Equations 2.16 and 2.20 into Equation 2.19 gives:

$$\upsilon_0 = \frac{\varepsilon_0^3(P_0 - P_L)}{\kappa_0\mu S_0^2(1-\varepsilon_0)^2 L}\left(\frac{L}{L_e}\right)^2 \tag{2.21}$$

$k_0$ has the value of 2 if the pores are perfect circles; however, for rectangular, annular, and elliptical shapes $k_0$ has values ranging from 2 to 2.5. One of the assumptions was that the capillary flow through the medium is at 45° to the medium surface. Then $(L_e/L)^2 = 2\ K$, and the Kozeny constant of Equation 2.13 above becomes:

$$K = k_0\left(\frac{L_e}{L}\right)^2 = \text{Kozeny Constant} \tag{2.22}$$

Based on $k_0 = 2.5$ and $(L_e/L)^2 = 2$, the value for $K$ will be 5. Substituting into Equation 2.21 gives the Kozeny–Carman Equation.

$$\upsilon_0 = \frac{\varepsilon_0^3(P_0 - P_L)D_h^2}{5\mu S_0^2(1-\varepsilon_0)L} \tag{2.23}$$

The Kozeny–Carman equation applies to substrates whose solidity ($\chi_0 = (1 - \varepsilon_0)$) is greater than 0.2 or $\varepsilon_0 < 0.8$. It should be noted that the author has chosen to use $\varepsilon_0$, the available or extra-fiber porosity rather than $\varepsilon$ the true porosity and $\chi_0$, the apparent solidity rather than $\chi$, the solidity based on nonporous solid materials in the web. This is because much of the void space in a nonwoven medium may not be available to permeable airflow. For example, cellulose fibers such as cotton, wood pulp, and vegetable fibers have internal cellular structures. That is why they are called cellulose. In a filter medium, these internal cells contained within the walls of the fiber may contribute to the void volume of the medium, but may not be part of the permeable void space where significant flow occurs. $\varepsilon_0$ is the void volume outside the fiber structure. For solid and nonporous fibers such as glass and many synthetic polymer fibers it would appear that $\varepsilon_0 \cong \varepsilon$. However, Jena and Gupta (46) also identified blind pores and closed pores that may exist outside the fiber walls of the filter media. Jena and Gupta (46) noted that although blind pores will not permit fluid flow, they may adsorb fluids, capture small particles, and participate in reactions.

Lindsay (47) discussed this problem for water in pulp mats. He refers to the fiber cell structure that is not available to fluid flow as being swollen fiber. He defined $\varepsilon_{rel}$ as the fraction of total pore space open to flow. Lindsey's measurements indicated that $\varepsilon_{rel}$ ranged from as low as 45% up to 70%.[5]

Lindsay defined $\varepsilon_0$ in terms of $\alpha$ and $c$.

$$\varepsilon_0 = 1 - \alpha c \tag{2.24}$$

$\alpha$ is the volume of swollen fiber per unit mass of fiber in the medium ($m^3/kg$).

$c$ is the mass concentration of fiber per unit volume of medium ($kg/m^3$).

Substitution of Equation 2.24 into the permeability constant of the Kozeny–Carman, Equation 2.13 above, and rearrangement yields:

$$(kc^2)^{1/3} = \left(\frac{1}{5.55S^2}\right)^{1/3}(1 - \alpha c) \tag{2.25}$$

$S$ is the flow exposed surface of the fibers per unit mass ($m^2/kg$), commonly called the specific surface area.

$$S = \alpha S_0 \tag{2.26}$$

Equation 2.25 is a classic equation used for pulp mats. Note 5.55 is used here as the value of the Kozeny constant $K$. One feature of this equation is that it provides for an experimental way of determining $\alpha$ and $S$ in fibrous webs under various degrees of compression. This is accomplished by using experimental techniques to develop the plot of $(kc^2)^{1/3}$ vs $c$. $\alpha$ is determined from the slope and $S$ from the intercept.

## 2.2.4 Flow-through porous media—cell model theory

### *2.2.4.1 Davies equation*

The Kozeny–Carman equation (Equation 2.23) was based on liquid flow through a flat bed of filter medium such as sand. It can also applied to fibrous webs. Generally,

[5] See Chapter 6, Paragraph 6.7 for further discussion of Lindsay's work.

the limitation is that the solidity be greater than 20%. Unfortunately most air filtration media require porosities greater than 80% (or solidities less than 20%). For air filtration a different approach was required. Davies (48) developed the following relationship for the flow of air through fibrous pads.

$$\frac{\Delta P d_f^2 A}{\mu Q L} = 64(1-\varepsilon)^{1.5}[1+56(1-\varepsilon)]^3 \tag{2.27}$$

$v_0$ is the Darcy's Law velocity from Equation 2.10 above (m/s).
$L$ is the thickness of the fibrous pad (m).
$A$ is the filter area ($m^2$).
$d_f$ is the mean fiber diameter (m).
$\mu$ is the fluid viscosity (Pa · s).
$1 - \varepsilon$ is the solidity or packing density ($\chi = (1 - \varepsilon)$) ($m^3/m^3$).

Davies' development was based on flow past a single fiber and Darcy's Law. He asserted that a "unique relationship must exist between $\chi$, the packing density and the dimensionless group":

$$\frac{\Delta p d_f^2 A}{\mu Q L} \tag{2.28}$$

Davies (48) gathered data on a number of filter pads (Figure 2.8) and found their scatter best fit the model as proposed in Equation 2.27[6]. He noted that the scatter is related to clumps, fiber orientation, and type of fiber, but not to the fiber length. He also noted that the equation is only valid to Reynolds numbers (Re) up to 1 where $Re = \nu d_f/\mu$ and $\nu$ is a mean internal velocity within the depth of the path. Note that Davies theory is based on flow past a single fiber and therefore is dependent on fiber diameter $d_f$. This is in contrast to the Kozeny–Carman Equation, which is developed from pore size considerations of cylindrical openings through the medium. This is not to say that fiber diameter is not important to the Kozeny–Carman Equation. Fiber diameter is important to determine the effective surface area term, $S_0$.

#### 2.2.4.2 Langmuir's theory

Prior to Davies' work several pioneering investigators (Albrecht (49), Sell (50), and Kaufman (51)) investigated flow patterns around parallel arrays of cylinders. Albrecht was the first to employ a single cylinder model. His work is considered as

[6] For solidities where $\chi = (1 - \varepsilon) > 0.02$, Davies (48) determined that

$$\frac{\Delta P d_{fe}^2 A}{\mu \nu \varepsilon L} = 70(1-\varepsilon)^{1.5}[1+52(1-\varepsilon)]^3$$

gives better agreement with experimental results. $\nu$ is the mean velocity inside the filter and $d_{fe}$ is the effective fiber diameter in contrast to $d_f$, the mean fiber diameter.

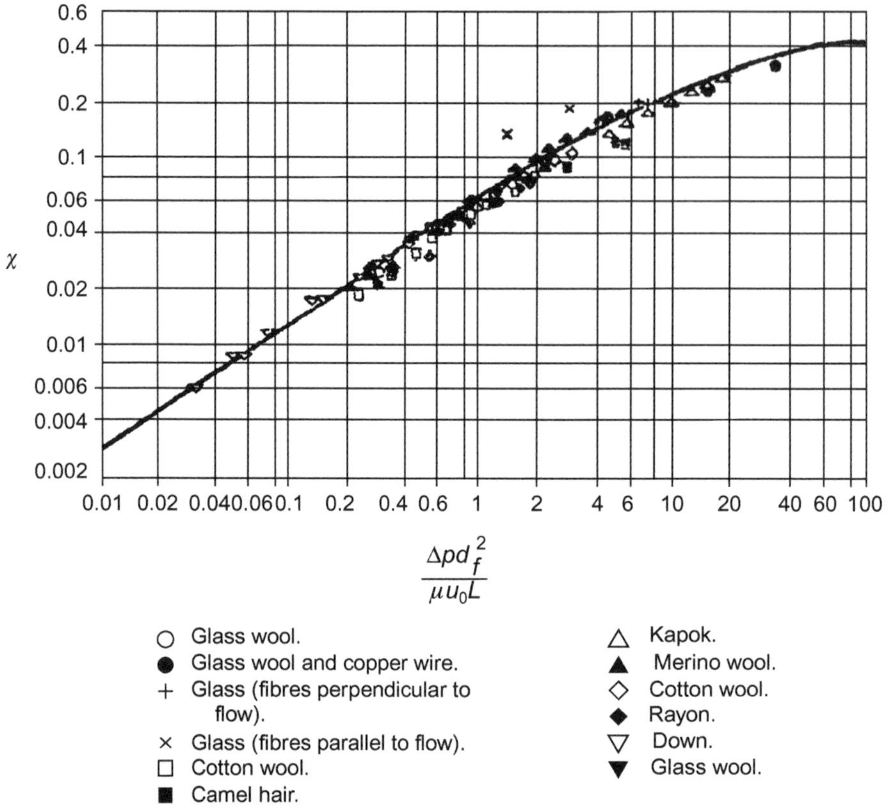


**Figure 2.8** Fluid Flow Through Fibrous Media. *Reproduced with permission of the Council of the Institution of Mechanical Engineers, London, England, UK.*

pioneering. Further progress had to be made in order to consider the interference effects of adjacent fibers. Langmuir (52) in an ingenuous approach assumed an array of parallel cylinders parallel to the direction of flow. Each cylinder, representing a fiber was surrounded by a hexagon shaped void space that bordered with the similar void space of the adjacent fibers. The honeycomb arrangement is depicted in Figure 2.9.

Langmuir (52) then assumed that each hexagon could be approximated by a circle whose area was equal to the hexagon. This reduced the problem to flow through a coaxial cylinder: the inner cylinder being the fiber perimeter and the outer cylinder being the fluid boundaries to the adjacent cylinders. In the classic solution to this problem (see Bird, Lightfoot and Stewart (41)) the velocity profile from inner cylinder to outer cylinder is parabolic. The boundary layer velocities at both the inner and outer cylinder walls are zero. Langmuir changed this by assuming a maximum velocity

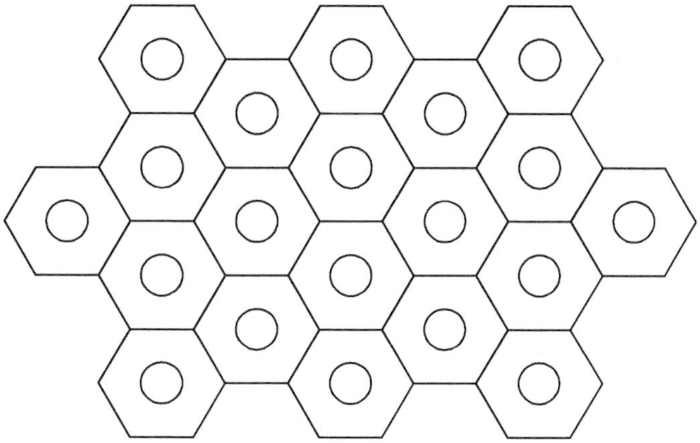

**Figure 2.9** Langmuir's Cylinder Model.

at the outer cylinder wall, which in reality was the assumed fluid boundary of the flow around each cylinder. Langmuir also assumed that the pressure gradient was only along the axis.

($x$) of each cylinder, and that of the velocity was a function of the radial distance ($r$) from the axes. Langmuir also assumed there were no inertial effects. The resulting equation was:

$$\frac{dp}{dx} = \mu\left(\frac{1}{r}\frac{dv}{dr} + \frac{d^2v}{dr^2}\right) \tag{2.29}$$

Note that in Equation 2.29 the shearing force equals the pressure force, so that no acceleration or deceleration takes place in the direction of the axis ($x$). Double integration of Equation 2.29 gives the velocity profile from the fiber surface to the outer cylinder wall.

$$v = \frac{1}{2\mu}\frac{\Delta P}{L}\left[b^2 \ln\left(\frac{r}{a}\right) - \frac{1}{2}(r^2 - a^2)\right] \tag{2.30}$$

where

$L$ is the length of the cylinder model (m).
$r$ is a radial position from the fiber axis (m).
$b$ is the radius of the outer cylinder (m).
$a$ is the radius of the fiber (m).

The volumetric flow rate $Q$ ($m^3/s$) is obtained by the integration of Equation 2.30 as follows:

$$Q = \int_0^{2\pi}\int_a^b v r dr d\theta \tag{2.31}$$

This gives:

$$Q = \frac{\pi \Delta P a^4}{4\mu L \chi^2}\left(-\ln \chi + 2\chi - \frac{\chi^2}{2} - \frac{3}{2}\right) \tag{2.32}$$

$\chi$ the solidity or the "fraction of the filter volume occupied by the fibers ($m^3/m^3$) is

$$\chi = \left(\frac{a}{b}\right)^2 \tag{2.33}$$

Langmuir continued with his analysis by defining a function as being the reciprocal of the portion of Equation 2.32 that is bracketed.

$$\varphi = \frac{1}{\left(-\ln \chi + 2\chi - \frac{\chi^2}{2} - \frac{3}{2}\right)} \tag{2.34}$$

Langmuir (52) drew an analogy between the Hagen–Poiseuille equation for flow through a cylinder and Ohm's law for electrical resistance and developed the following expression.

$$Qp = \frac{p_a(P_0 - P_L)}{R} \tag{2.35}$$

Q is the cylinder flow volume from Equation 2.32.
$p$ is the pressure at any point in the cylinder (Pa).
$P_0 - P_L$ is the pressure drop (Pa).
$p_a$ is the average pressure in the cylinder, $(P_0 - P_L)/2$, (Pa).
$R$ is the resistance to flow and has the units of $\text{Pa} \cdot \text{s/m}^3$

$$R = R_0(1 + (\Delta L/L)) \tag{2.36}$$

$L$ is the length of the cylinder (m).
$\Delta L$ is an end correction for the fluid entering the cylinder (m).
$R_0$ is the Hagen–Poiseuille resistance of the flow within the cylinder

$$R_0 = 8\mu L/\pi r^4 \tag{2.37}$$

Combining Equations 2.32, 2.34, 2.35, and 2.37, Langmuir developed the resistance of his model as:

$$R = 4\mu L \chi \varphi / A a^2 \tag{2.38}$$

$A$ is the total area of the filter ($m^2$).

Langmuir (52) recognized that his model did not represent real filter media because:

1. The fibers of an actual filter have their axis approximately parallel to the surface of the media.

2. The fibers are not parallel to one another but cross at all angles and their distribution in space is far less than in the model.

A correction factor was applied based on the flow of ellipsoids through fluids. The model was a prolate ellipsoid. A prolate ellipsoid is obtained by rotating an ellipse around its long axis so that the three-dimensional shape is something like an egg. If 'a' is the long axis of the ellipsoid and 'b' is the diameter at its center or thickest part, the ratio a/b approximates the L/D (length/diameter) ratio of a fiber. Langmuir calculated that the force $F_T$ to move an ellipsoid with its long axis transverse to the flow was greater than the force $F_L$ to move the ellipsoid when its long axis was parallel to the direction of flow. The force ratio $F_T/F_L$ increased as a/b increased, however, it was asymptotic to 2 as a/b approached infinity. By relating this rational to fibers of a filter media, Langmuir argued that the flow resistance of transverse fibers should never be more than two times the resistance of parallel fibers as calculated by his model and more likely to be in the neighborhood of 1.4[7]. Accordingly, the correction factor $B$ was introduced in Equation 2.39.

$$R = 4B\mu\chi\varphi/Aa^2 \tag{2.39}$$

#### 2.2.4.3 Happel and Kuwabara models

It should be noted that Langmuir (52) was studying aerosol filtration and based his model for a gas fluid. He developed the model for flow-through filter media based on fibers parallel to the flow, and then justified a correction factor $B$ for fibers transverse to the flow. Happel (53) and Kuwabara (54) both solved the problem for the transverse flow situation. Both solutions were based on two concentric cylinders being used to represent a model of a fluid flowing through an assemblage of cylinders. Note in Figure 2.10, that the starting point for this model is very similar to Langmuir's model.

The solidity of the cell in Figure 2.10 is

$$\chi = \left(\frac{a}{b}\right)^2 \tag{2.40}$$

Note that Equation 2.40 is identical to Langmuir's Equation 2.33 above. If there are $N$ such cells in the filter domain, then Equation 2.40 is the solidity of the entire model.

For this discussion, we will follow Happel's (53) development and then note the differences between Happel's model and Kuwabara's model.

[7] Using Happel's theory as discussed in Section 2.2.4.3, Davies (36) showed that Langmuir's estimate of 1.4 for the transverse to parallel resistance ratio is low. Davies' calculations showed the number to range from 1.43 to 2.0.

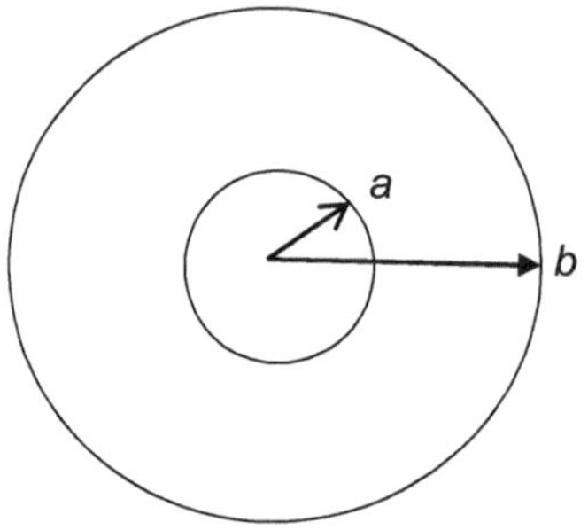


**Figure 2.10** Happel–Kuwabara Cell Model.

Happel made the following assumptions:

1. There is no slippage at the cell boundaries, $a$ and $b$.
2. The mean free path of gas molecules is very small compared to the dimension $a$ of the inner cylinder.
3. The cross-sectional shape of the cell is represented by a circle whose area is the same as the cross section of the actual cell geometry. In effect, the outer boundary of the cell at $b$ is a cylinder.

Happel started with the equations of motion and utilized the stream function $\psi$.

$$v_r = \frac{1}{r}\frac{\partial \psi}{\partial \theta} \quad \text{and} \quad v_\theta = -\frac{\partial \psi}{\partial r} \tag{2.41}$$

$v_r$ and $v_\theta$ are the velocity components (m/s) at the radial coordinates $\theta$ (•) and $r$ (m), based on the axis of the inner cylinder being $r = 0$.

He then assumed that the system could be represented by the biharmonic equation:

$$\nabla^4 \psi = 0 \tag{2.42}$$

The general solution to Equation 2.42 is

$$\psi = \sin\theta\left[\frac{1}{8}Cr^3 + \frac{1}{2}Dr\left(\ln r - \frac{1}{2}\right) + E'r + \frac{F}{r}\right] \tag{2.43}$$

$C$, $D$, $E'$, and $F$ are constants obtained from boundary conditions.

In order to develop the boundary condition constants, Happel assumed that the cylinder in Figure 2.12 was moving and the fluid envelope was standing still. This would give the same, but negative, result as if the fluid envelope were moving and the cylinder standing still. This allowed for the following boundary assumptions at the surface of the cylinder, where $r = a$:

$$u = v_f \tag{2.44}$$

$$v_r = v_f \cos\theta \tag{2.45}$$

$$v_\theta = v_f \sin\theta \tag{2.46}$$

$u$ and $v_f$ are the velocity and superficial velocity of the cylinder, respectively (refer Figure 2.11).

At the outer cylinder of Figure 2.11, where $r = b$, Happel (53) assumed there is no shearing stress, $\sigma_{r\theta} = 0$, and there is no radial velocity component, $v_r = 0$. Yuan (55) derived the equations of motion in cylindrical coordinates and gave the following relationships for normal ($\sigma_{rr}$) and shearing stress ($\sigma_{r\theta}$):

$$\sigma_{rr} = -p + 2\mu\varepsilon_{rr} - 2/3\mu(\nabla \cdot v) \tag{2.47}$$

$$\sigma_{r\theta} = \mu\gamma_{r\theta} \tag{2.48}$$

$\varepsilon_{rr}$ is a normal strain.

$\gamma_{r\theta}$ is a shearing strain.

It has already been noted in Equation 2.2 above that for a viscous incompressible fluid, the equation of continuity is equal to zero.

$$\nabla \cdot \boldsymbol{v} = 0 \tag{2.49}$$

Happel (53) assumed that if $v_r = 0$, then

$$\varepsilon_{rr} = \frac{\partial v_r}{\partial r} = 0 \tag{2.50}$$

Accounting for Equations 2.49 and 2.50 in Equation 2.47 reduces it to

$$\sigma_{rr} = -p \tag{2.51}$$

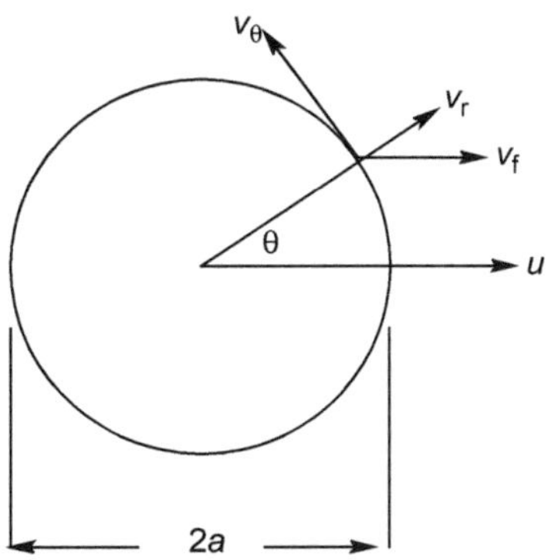


**Figure 2.11** Happel Model Boundary Conditions at the Inner Cylinder.

The shearing strain $\gamma_{r\theta}$ is

$$\gamma_{r\theta} = \frac{\partial v_\theta}{\partial r} - \frac{v_\theta}{r} + \frac{1}{r}\frac{\partial v_r}{\partial \theta} \tag{2.52}$$

If Happel's assumption that $\sigma_{r\theta} = 0$ is accepted, then combining Equation 2.52 into Equation 2.48 gives

$$\frac{\partial v_\theta}{\partial r} - \frac{v_\theta}{r} + \frac{1}{r}\frac{\partial v_r}{\partial \theta} = 0 \tag{2.53}$$

Equation 2.53 along with $v_r = 0$ becomes the boundary conditions for Happel's development.

At this point, it would be useful to compare Kuwabara's (54) boundary condition. Like Happel, Kuwabara assumed $v_r = 0$ at the outer cylinder where $r = b$. In contrast to Happel, Kuwabara assumed no vorticity at the outer cylinder. As a result, his boundary equation is

$$\frac{v_\theta}{r} + \frac{\partial v_\theta}{\partial r} - \frac{1}{r}\frac{\partial v_r}{\partial \theta} = 0 \tag{2.54}$$

By comparing Equation 2.54 to Equation 2.53, it is noted that the only difference is a sign change for two of the terms, however, this creates significant differences in the resultant equations. Both Happel and Kuwabara were identical to each other and to Langmuir in their solutions of flow parallel to the cylinders (Equation 2.39). Their models for flow transverse to the cylinders are as follows:

Happel (53):

$$\frac{\Delta p d_f^2}{\mu v_0 L} = \frac{32\chi}{-\ln\chi - (1-\chi^2)/(1+\chi^2)} \tag{2.55}$$

Kuwabara (54):

$$\frac{\Delta p d_f^2}{\mu v_0 L} = \frac{32\chi}{-\ln\chi + 2\chi - \chi^2/2 - 3/2} \tag{2.56}$$

Note that Equations 2.55 and 2.56 are presented in the dimensionless arrangement (Equation 2.29) of Davies (36). The difference between the Happel and Kuwabara models raises the question as to which is the most accurate. Kirsch and Fuchs (56), using spheres, 20–30 μm in diameter and equal in density to the surrounding fluid, determined velocities that were more in agreement with Kuwabara (54) than Happel. Brown (57) states that "Of the two theories, Kuwabara's is found to give better agreement with experiment and is probably even now the most popular model of air flow through a filter." The denominator in the Kuwabara equation (Equation 2.56) divided by two gives the Kuwabara hydrodynamic factor *Ku*. Several investigators in subsequent theory development have used this factor.

$$Ku = -\frac{1}{2}\ln\chi - 3/4 + \chi - \frac{\chi^2}{4} \tag{2.57}$$

### 2.2.4.4 Advances in cell model theory

There have been several updates on the theories offered by Kuwabara and Happel. Pich ((58) and (59)) modified Kuwabara's model by assuming there was some slippage at the surface of the inner cylinder. The resulting equation in Davies' dimensionless terms is

$$\frac{\Delta p d_f^2}{\mu v_0 L} = \frac{32\,\chi(1 + 1.996\,Kn)}{-\ln\chi + 2\chi - \chi^2/2 - 3/2 + 1.996\,Kn(-\ln\chi + \chi^2/2 - 1/2)} \tag{2.58}$$

where *Kn* is the Knudsen number

$$\mathrm{Kn} = \frac{2\lambda}{d_f} \tag{2.59}$$

$\lambda$ is the mean free path of the gas molecules, m.

$d_f$ is the fiber diameter, m.

Pich claims his equation to be valid for conditions where $\lambda$ is very small compared to the diameter of the fiber, $d_f$. The applicable region is $\mathrm{Kn} < 0.25$. Grafe and Graham (60) have noted that at low fiber diameters (0.5 μm) slip flow should be considered. Slip flow has become an important consideration in nanofiber filtration such as with electrospun webs. Chase (61) argues that "One of the advantages of nanofibers for gas phase depth filtration is the slip flow phenomena."

Brown (57) mentions that the strength of cell model theory is its simplicity. He assumes that finding the airflow around a single circular fiber is typical of the filter as a whole. Brown notes that, in a real media, not all fibers are perpendicular to the flow nor are they of uniform diameter. In addition, the fiber cross section is not necessarily circular. He also notes that the packing density varies through the medium. Brown acknowledges that there is no existing theory that deals with all of these problems, however, a number of refinements exist. Miyagi (62) used a complex variable approach to treat the filter as a single infinite row of uniformly spaced parallel fibers. According to Brown, this approach gave results comparable to cell theory, provided that the packing distance between successive layers of the rows was equal to the inter-fiber distance within a row.

Kirsch and Fuchs ((63) and (64)) explored a fan model approach. This approach consists of successive planes of parallel fibers; however, the fibers in each plane are rotated at some arbitrary angle with respect to the parallel fibers in the previous plane.

The fan model is based on a perfect homogeneous dispersion of fibers. In real filters the fibers are not so homogeneously dispersed. The fan model predicted higher resistance than real filters, therefore an inhomogeneity factor was introduced to bring the real filter in alignment with the fan model (Fuchs et al. (65)).

The assumption that the distance between layers of a filter medium is the same as the interfiber distance within a layer is not necessarily true. In a depth filter, successive arrays of parallel fibers are the simplest approach for developing a theoretical model. Two such arrays are depicted in Figure 2.12. The cylinders represent segments of parallel fibers perpendicular to the direction of flow. Each vertical row of cylinders represents a layer of filter media. The distance between parallel fibers within a layer is 2L. The distance between layers is 2e. The channel structure assumes that the fibers in each layer line up directly behind the fibers of the layer in front of it. The staggered model represents the case where the fiber alignment from layer to layer is staggered. Brown (57) illustrated four structures: the channeled and three versions of the staggered structure.

The parameters are the scale, the packing fraction, and the ratio between interfiber spacing and interlayer spacing. If there is stagger between successive layers then a fourth parameter is required describing the extant of stagger. Brown (57) mentions four approaches that are listed below.

1. Extension of cell model to account for adjacent fibers—In this approach the solution of the biharmonic equation (Equation 2.42) requires a series solution of higher order than the simple cell model solution of Equation 2.43. The radial components vary continuously. They cannot be simply defined by the cell radius

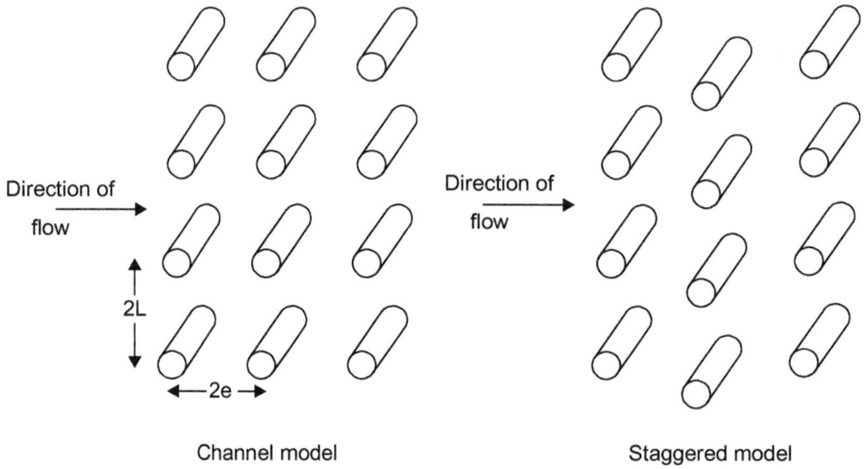


**Figure 2.12** Arrays of Parallel Fibers.

and fiber radius of the cell model. Sangani and Acrivos (66) are one set of investigators who used this approach. They used their methodology to obtain an exact numerical solution on the creeping flow field in arrays of high and low porosities.

Referring to a personal communication with Majid Zia of the University of Rhode Island, Kahn (67) reports that Zia had compared Sangani and Acrivos' solution to the streamlines and radial velocities of the Kuwabara solution and determined that disagreement between the two solutions increases with increasing solidities, particularly near the upstream cell boundary and at increasing radial distances from the fiber surface. Zia did note that calculated drag for filter solidities up to 0.4 were near identical. Above 0.4 the difference increased rapidly. Kahn also notes that for $\chi \leq 0.2$ the pressure drop associated with inline and staggered arrangements of parallel cylinders is almost identical, however, for increased solidities the drag for inline arrays is considerably higher than for staggered structures.

2. Numerical methods—Finite difference methods are a way of breaking down complex differential equations into meshes of simultaneous equations. Each point in the mesh is solved one by one to approximate the analytical solution. Fardi and Liu (68) used this approach to solve the simplified components of the Navier–Stokes equations along with the equation of continuity for the fluid velocity. Numerical analysis will give very good results provided appropriate boundary conditions are applied at the surface. An expansion of numerical analysis for problem solving is computational fluid dynamics (CFD). CFD is discussed by Dr. George G. Chase of Akron University in Section 2.5.
3. Variational method—Helmholtz's principle is used to find the flow pattern that gives rise to the lowest rate of dissipation of energy by viscous drag. The periodic symmetry or the fiber array is exploited by the use of a Fourier series. Brown ((69) and (70)) conducted calculations using the variational model and determined that although the pressure drop is not sensitive to fiber arrangement when the interfiber and interlayer distances are equal, the structure behaves very differently when they are not. If the distance between layers is less than the distance between fibers, then the effect of compression may be simulated. In addition, the structure behaves differently depending on whether the successive layers are channeled or staggered.
4. Boundary element method—Brown (57) mentions Hildyard et al. (71) as one set of researchers who used the boundary element method. In this method, the biharmonic equation is split into two equations by way of vorticity resulting in two integral equations. The problem is then reduced to the solution of simultaneous equations. Brown reports that flow patterns and pressure drops obtained by this method are similar to that obtained by the variational method.

#### 2.2.4.5 Drag model theory

Drag model theory is based on the drag on an object moving through a fluid. The drag equation is

$$F = \frac{1}{2}\rho u^2 A C_d \tag{2.60}$$

where

$F$ is the drag force (N).
$\rho$ is the density of the fluid (kg/m$^3$).
$u$ is the velocity of the object moving through the fluid (m/s).
$A$ is the reference area of the object (m$^2$).
$C_d$ is the drag coefficient (dimensionless).

If the object is a fiber represented by the unit length of a long cylinder transverse to the flow, then

$$F = \frac{1}{2}\rho u^2 d_f C_d \tag{2.61}$$

$d_f$ is the fiber diameter (m).

Note that in Equation 2.61 the reference area $A$ of Equation 2.60 is the cross section of the unit length cylinder ($d_f$ × unit length).

Applying Lamb's (72) solution for an isolated fiber

$$F = \frac{4\pi\mu v}{2.0022 - \ln \mathrm{Re}} \tag{2.62}$$

where, Re is the Reynolds number

$$\mathrm{Re} = \frac{v d_f \rho}{\mu} \tag{2.63}$$

The Lamb drag coefficient for Equation 2.62 is

$$C_d = \frac{8\pi}{\mathrm{Re}(2.022 - \ln \mathrm{Re})} \tag{2.64}$$

The problem with Lamb's equation is that it is for an isolated fiber and does not consider the interference effect of other nearby fibers in the filter media network. Some investigators such as Langmuir, Irving, Happel, and Kuwabara, discussed above, dealt with this by organizing fibers into parallel arrays and placing a fluid cell envelope around each fiber. Boundary conditions were established at each cell border to allow for the interaction of the adjacent cell.

Other investigators such as Chen (73) established drag coefficients by using neighboring fibers as boundaries. Chen assumed that at low Reynolds number, the ratio of interfiber distance ($d_b$) to fiber diameter ($d_f$) is inversely proportional to the square root of the solidity ($\chi$). White (74), a previous investigator, had determined that the drag coefficient of a cylinder moving in a tank could be correlated by the following equation:

$$\frac{C_{d\chi}}{2} \cdot \mathrm{Re} = \frac{k'}{\ln k'' \times d_b/d_f} \tag{2.65}$$

where, k′ and k″ are constants of the equation.

Chen used White's Equation 2.65 for his own model. Chen's model was a screen formed from a square matrix of fibers. The in-plane distance between fibers in both the x and y directions was $d_b$. Each screen formed a separate layer in the filter; $d_b$ was also the distance between layers.

The drag equations developed from Chen's model were as follows:

$$\frac{C_{d\chi}}{2} \cdot \mathrm{Re} = \frac{k_4}{\ln k_5 \times \chi^{-0.5}} \tag{2.66}$$

$$\Delta p = \frac{4}{\pi} \frac{k_4}{\ln k_5 \chi^{-0.5}} \frac{\chi}{1-\chi} \frac{\mu v_s L}{(d_f)_s^2} \tag{2.67}$$

$k_4$ and $k_5$ are constants of the equation.
$\Delta p$ is the pressure drop across the filter (Pa).
$\mu$ is the fluid viscosity (Pa · s).
$v_s$ is the superficial velocity through the filter (m/s).
$L$ is the thickness of the filter (m).
$(d_f)_s$ is the surface average fiber diameter (m).

Experiments by Chen (73) indicated that the drag coefficient term, $\frac{C_{d\chi}}{2} \cdot \mathrm{Re}$, is constant over a broad range of Reynolds numbers ($\mathrm{Re} = 10^{-3}$ to $10^{-1}$). Chen also determined the best-fit data for the constants $k_4$ and $k_5$ as 6.1 and 0.64, respectively.

Kahn ((67) and (75)) developed a screen model approach which he termed an "offset screen model." It consisted of two layers of parallel fibers. The parallel fibers in the first layer were perpendicular to the fibers in the second layer. The layout is pictured in Figure 2.13. Each layer of area $\overline{L} \times \overline{L}$ is composed of $\overline{N}$ equally spaced fibers in the first layer and $\overline{M}$ equally spaced fibers in the second layer and perpendicular to the fibers in the first layer. Kahn defined an anisotropy parameter as being

$$R_{pa} = \frac{\overline{M}}{\overline{N}} \tag{2.68}$$

Kahn related the anisotropy parameter to the orientation of fibers in real media. He claimed that his model accounted for structural properties: fiber size, solidity, and

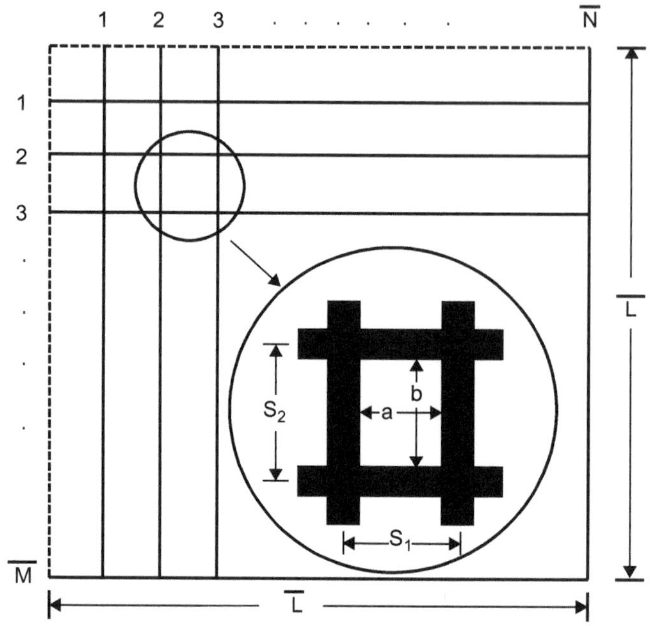


**Figure 2.13** Kahn Offset Screen Model. *Reproduced with permission of the American Filtration and Separation Society, Nashville, Tennessee, USA.*

anisotropy. It also calculated a three dimensional flow field. Kahn argued that his model predicted pressure drop very well and that particle capture efficiencies compared well with experimental data, except in domains where diffusion, interception, and inertial mechanisms were of equal importance. He also showed that anisotropy affects pressure drop only for solidities greater than 0.2.

## 2.3 PARTICLE FILTRATION

The theoretical discussion to this point has been mostly concerned with the theories of flow-through porous media. It is necessary to understand the flow behavior through a nonwoven filter medium in order to develop theories for the separation of particles in that flow. Most of the following discussion will be based on the separation of particles in gas streams, particularly airborne particles.

### 2.3.1 Single fiber theory

Single fiber theory was developed for filtration of aerosol particles from an air stream. The following explanation will be mostly based on that offered by Davies (36).

Consider a filter element of length $\delta x$ (m) in a fibrous filter of unit cross section and thickness $h$ (m) in the x-direction. The filter is at right angles to the airflow as shown in Figure 2.14.

If $d_f$ is the fiber diameter (m) and $L_f$ is the length of fiber (m) in the unit thickness $\delta x$ of the filter pad, then the solidity or packing density of the element is

$$\chi = \frac{\pi d_f^2 L_f}{4} = 1 - \varepsilon \tag{2.69}$$

where $\varepsilon$ is the porosity or fractional void volume of the web.

Now consider one fiber in that filter whose axis is transverse to the flow as indicated in Figure 2.15. $d_f$ (m) is the diameter of the cross section of this fiber. At some distance upstream of this fiber is a stream of width $y$ (m), where all the particles in this stream are the ones that will strike the fiber. The single fiber efficiency is then:

$$E = \frac{y}{d_f} \tag{2.70}$$

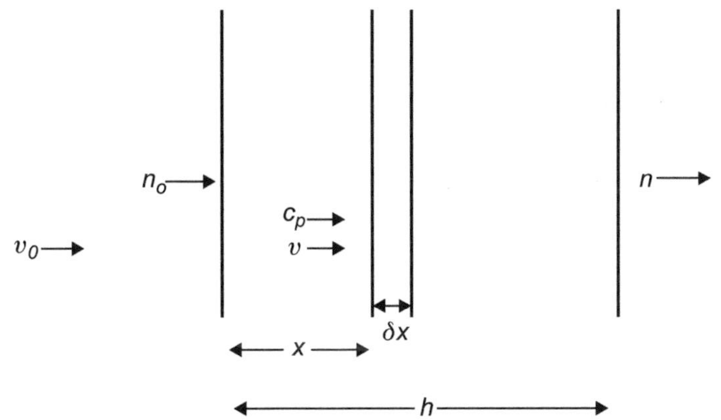


**Figure 2.14** Element $\delta x$ of a Filter of Thickness $h$.

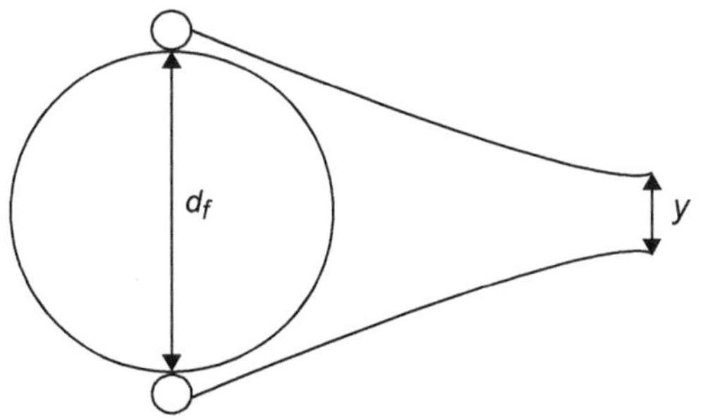


**Figure 2.15** Single Fiber Efficiency Model.

Let $\nu$ (m/s) be the velocity of the flow inside the filter

$$\nu = \frac{\upsilon_0}{1 - \chi} \tag{2.71}$$

where $\upsilon_0$ is the face velocity of the filter (m/s).

Let $c_p$ be the particle concentration (particles/m$^3$) of the air stream approaching the element. Then $\upsilon \times c_p$ is the particle rate (particles/m$^2 \cdot$s) approaching the filter. If the fiber cross section to remove particles is $L_f \times d_f$ (m$^2$), then

$$\delta c_p = -\, c_p E L_f d_f \delta x \tag{2.72}$$

and the rate of particulate removal is

$$\frac{dc_p}{dx} = -\, c_p E L_f d_f \tag{2.73}$$

Integration of Equation 2.73 across the thickness $h$ yields

$$\frac{n}{n_0} = \exp(-E L_f d_f h) \tag{2.74}$$

$n_0$ is the concentration of aerosol particles entering the filter (particles/cm$^3$).

$n$ is the concentration of aerosol particles leaving the filter (particles/cm$^3$).

A filter is often characterized by its penetration, $P$ where $P = 1 - E$. Note that $n/n_0$ is not the efficiency of the filter but rather the fraction of upstream particles penetrating the filter (the penetration). In terms of efficiency, filter performance is generally defined as

$$E = \frac{n_0 - n}{n_0} = 1 - \exp(-\, E L_f d_f h) \tag{2.75}$$

The value $\gamma = E L_f d_f$ is referred to as the filtration index and has the units of 1/m. Expressing the penetration as a percentage, Equation 2.75 becomes

$$P = 100 \frac{n}{n_0} = 100 \exp(-\, \gamma h) \tag{2.76}$$

If the fibers are cylindrical and of constant radius, combination of Equations 2.69 and 2.76 yields

$$\gamma = \frac{4E\chi}{\pi d_f} \tag{2.77}$$

Davies (36) organized Table 2.1 to demonstrate the calculated relationship between $\gamma h$ and $P$.

**Table 2.1** Relationship Between Thickness of Filter, Filtration Index, $\gamma$, and Percentage Penetration

| Filtration index $-\gamma h$ | $P$, percentage penetration (%) |
|---|---|
| 13.9 | 0.0001 |
| 12.2 | 0.0005 |
| 11.5 | 0.001 |
| 9.9 | 0.005 |
| 9.2 | 0.01 |
| 7.6 | 0.05 |
| 6.9 | 0.1 |
| 5.3 | 0.5 |
| 4.6 | 1.0 |
| 3.0 | 5.0 |
| 2.3 | 10 |
| 1.61 | 20 |
| 0.694 | 50 |
| 0.511 | 70 |
| 0.223 | 80 |
| 0.105 | 90 |

Reproduced with permission of Elsevier Ltd, London, England, UK.

## 2.3.2 Particle capture based on the Kuwabara flow model

### *2.3.2.1 Justification for Kuwabara flow model*

The Davies presentation in Section 2.3.1 is based on an isolated fiber. Davies (36) acknowledged that the single fiber efficiency found this way is smaller than what would be calculated from an accurate theory, because the dispersion of fibers is imperfect. An accurate theory must allow for the proximity of other fibers.

Much of the current theory is developed from the Kuwabara and Happel cell models discussed above. These flow models consider the effects of adjacent fibers. The Kuwabara model has been preferred because several investigators have argued that "it is more representative of the flow around the fiber in the case of low Reynolds number" (Bouland and Renoux (76) p. 55).

### *2.3.2.2 Mechanisms of particle capture*

In Section 2.1.2, the mechanisms of particle capture were discussed. These were inertia $E_{\mathrm{I}}$, interception $E_{\mathrm{R}}$, and diffusion $E_{\mathrm{D}}$. Electrostatic attraction is not included for this discussion. Many investigators either ignore inertial efficiency $E_{\mathrm{I}}$ or account for it with the interception efficiency $E_{\mathrm{R}}$ in the term $E_{\mathrm{RI}}$. For fine particle filtration, Kowalski et al. (77) argues that the inertia process "is not significant for normal filter velocities and microbial sizes, and is neglected in most filter models since interception satisfactorily accounts for it. Davies (36) (p. 72) introduced a combination term $E_{DR}$ for diffusion, and so that the total overall single fiber efficiency, $E_,$ when calculating filter efficiency is

$$E = E_D + E_{DR} + E_R + E_I \tag{2.78}$$

If the inertia-interception term ($E_{RI} = E_R + E_I$) is used then Equation 2.78 reduces to

$$E = E_D + E_{DR} + E_{RI} \tag{2.79}$$

Figure 2.16 by Davies (36) indicates the particle size domains for the various mechanisms of single fiber efficiency. The medium is a paper filter containing fibers 2 μm in diameter. The curves are based on Kuwabara flow theory as discussed in the following paragraphs. For this particular filtration condition, diffusion $E_D$ is the dominating mechanism for particles less than 0.1 μm in radius. Note that for particles in this range of <0.1 μm, the single fiber efficiency decreases with increasing particle size. Interception, $E_R$, first becomes notable at particle radii of 0.075 μm. At this point, the contribution of interception efficiency starts causing an increase in efficiency with increasing particle size as indicated by the turnaround in the curve of

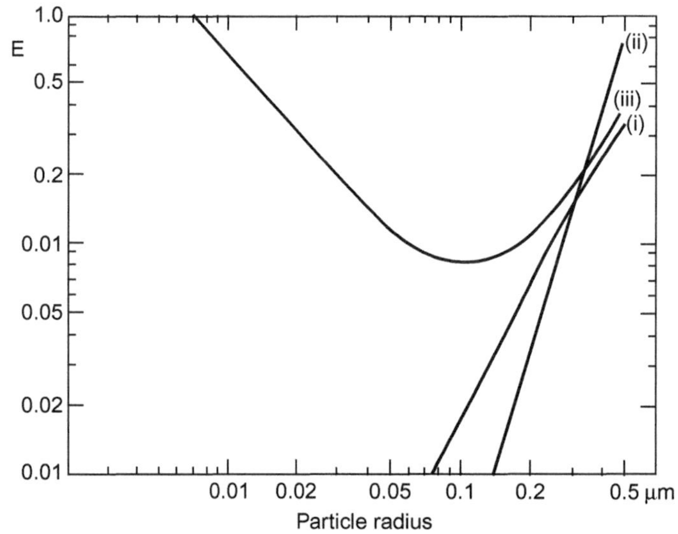


**Figure 2.16** Filtration Mechanisms. *Reproduced by permission of Academic Press (36), pg 76.*

$E_D + E_{DR} + E_R$. The effects of inertia, $E_I$, do not show until particle size radius is in the range of 0.1−0.2 μm. Figure 2.16 suggests that at particles size radii of 0.5 μm and greater, particle inertia becomes the dominating efficiency mechanism.

### 2.3.2.3 Dimensionless numbers

Before we proceed further, several numbers are identified.

1. The Knudsen number *Kn*: See Equation 2.59 above. Note there are two forms of the Knudsen number.

   $Kn_f = 2\lambda/d_f$ is based on fiber diameter whereas, $Kn_p = 2\lambda/d_p$ is based on particle diameter.

2. The Peclet Number *Pe*: This number accounts for the diffusivity of the particles in the flow field. It is defined as:

$$Pe = \frac{v_0 d_f}{\mathscr{D}_{AB}} \tag{2.80}$$

$v_0$ is the face velocity of the filter, m/s.
$d_f$ is the fiber diameter, m.
$\mathscr{D}_{AB}$ is the particle diffusion coefficient, m/s$^2$.

$\mathscr{D}_{AB}$ is defined by the Einstein equation.

$$\mathscr{D}_{AB} = \mu_p k_b T \tag{2.81}$$

$\mu_p$ is the particle mobility, N-s/m.
$k_b$ is Boltzman's constant[8], $1.3805 \times 10^{(-23)}$ J/K.
$T$ is the Temperature, K.

The particle mobility, $\mu_p$, is determined from the Cunningham slip factor $C_k$

$$\mu_p = \frac{C_k}{\pi \eta d_p} \tag{2.82}$$

$C_k$ is the Cunnungham slip factor, dimensionless.
$\eta$ is the absolute viscosity, N-s/m$^2$.
$d_p$ is the particle diameter, m.

[8] The author has noted some slight inconsistencies in various sources reporting the value of the Boltzman constant. Kowalski et al. (77) report the value as $1.3708 \times (10)^{(-23)}$ J/K. "Perry's Chemical Engineers' Handbook (78) reports it as $1.38048 \times (10)^{-23}$ J/K. Bird, Lightfoot, and Stewart (Bird et al, Transport Phenomena, 2nd Edition, 2002) report it as $1.3805 \times (10)^{-16}$ erg/K (One $J = 1 \times (10)^7$ erg). The Encyclopedia Brittanica (6) reports two values; $1.38062 \times (10)^{-23}$ J/K and $1.38622 \times (10)^{-46}$ erg/K. The author can only assume that the latter Brittanica value is the result of a typographical error. The author chose to list the values reported by Perry and by Bird et al., since these were the only two sources that were consistent with each other.

The Cunningham slip factor accounts for the aerodynamic slip that occurs at the particle surface. It is defined as

$$C_k = 1 + \left(\frac{\lambda}{d_p}\right)\left(2.492 + 0.84e^{\left(\frac{-0.435d_p}{\lambda}\right)}\right) \tag{2.83}$$

$\lambda$ is the gas molecule free path, N-s/m$^2$. It is the same $\lambda$ as used in the Knudsen number of Equation (2.59).

3. The Stokes Number St: The Stokes number "is a dimensionless group expressing the ratio of kinetic energy of a particle, traveling with the velocity $v_p$, through air or another fluid, to the work done against viscous drag over a distance $R$ (m) (Davies (36)). It is defined by:

$$St = \frac{d_p \rho_p v}{9\mu R} \tag{2.84}$$

$d_p$ is the particle diameter, m
$\rho_p$ is the particle density, kg/m$^3$
$v$ is the particle velocity, cm/s
$\mu$ is the air or fluid viscosity, N-s/m = kg/m-s
$R$ is the viscous drag distance, m

4. The Kuwabara hydrodynamic factor: This factor has been previously discussed in Section 2.2.3.3 above and defined by Equation 2.57

$$Ku = -\frac{1}{2}\ln\chi - 3/4 + \chi - \frac{\chi^2}{4} \tag{2.85}$$

### 2.3.2.4 *Results of various investigators*

#### 2.3.2.4.1 Diffusion and interception

Stechkina and Fuchs (79) were the first to use the Kuwabara flow approach for diffusion ($E_D$). Their result was

$$E_D = 2.6\ Ku^{-1/3}Pe^{-2/3} \tag{2.86}$$

Referring to Equation 2.78, Davies (36) determined the following for diffusion and interception in the Kuwabara flow field.

$$E_D = 2.9\ Ku^{-1/3}Pe^{-2/3} + 0.62Pe^{-1} \tag{2.87}$$

$$E_R = \frac{1}{2Ku}\left\{\begin{array}{l} 2\left(1+\frac{d_p}{d_f}\right)\ln\left(1+\frac{d_p}{d_f}\right) - \left(1+\frac{d_p}{d_f}\right) + \left(1+\frac{d_p}{d_f}\right)^{-1} + \\ \chi\left(-2\frac{d_p^2}{d_f^2} - \frac{d_p^4}{2d_f^4} + \frac{d_p^5}{2d_f} + \ldots\right) \end{array}\right\} \tag{2.88}$$

$$E_{\mathrm{DR}} = 1.24\ \mathrm{Ku}^{-1/2}\mathrm{Pe}^{-1/2}\left(\frac{d_{\mathrm{p}}}{d_{\mathrm{f}}}\right)^{2/3} \tag{2.89}$$

Lee and Liu (80) used a boundary layer theory along with the Kuwabara flow field to combine the effects of interception and diffusion. They obtained the following equations:

$$E = E_{\mathrm{D}} + E_{\mathrm{R}} \tag{2.90}$$

$$E_{\mathrm{D}} = 2.6\left[\frac{1-\chi}{Ku}\right]^{1/3} Pe^{-2/3} \tag{2.91}$$

$$E_{\mathrm{D}} = \left[\frac{1-\chi}{Ku}\right]^{1/3} \frac{R^2}{(1+R)} \tag{2.92}$$

$R$ is the interception parameter $d_p/d_f$.

Lee and Liu (81) then amended the equations to better fit with their experimental data.

$$E_{\mathrm{D}} = 1.6\left[\frac{1-\chi}{Ku}\right]^{1/3} Pe^{-2/3} \tag{2.93}$$

$$E_{\mathrm{R}} = 0.6\left[\frac{1-\chi}{Ku}\right]^{1/3} \frac{R^2}{(1+R)} \tag{2.94}$$

Liu and Rubow (82) took into account Pich's ((58) and (59)) slip flow correction (Equation 2.58 above) and applied two correction terms ($C_{\mathrm{D}}$ and $C_{\mathrm{R}}$) to the equations of Lee and Liu.

$$E_{\mathrm{D}} = 1.6\left[\frac{1-\chi}{Ku}\right]^{1/3} Pe^{-2/3} C_{\mathrm{D}} \tag{2.95}$$

$$E_{\mathrm{R}} = 0.6\left[\frac{1-\chi}{Ku}\right]^{1/3} \frac{R^2}{(1+R)} C_{\mathrm{R}} \tag{2.96}$$

where

$$C_{\mathrm{D}} = 1 + 0.388 Kn_f\left[\frac{(1-\chi)Pe^{1/3}}{Ku}\right] \tag{2.97}$$

$$C_{\mathrm{R}} = 1 + \frac{1.996 Kn_f}{R} \tag{2.98}$$

Bouland and Renoux (76) reported on the work of Payet (83) and Payet et al. (84). Their work introduced a new correction of Equation 2.95 related solely to the mechanism of diffusion.

$$E'_{\mathrm{D}} = 1.6\left(\frac{1-\chi}{Ku}\right)^{1/3} Pe^{-2/3} C_{\mathrm{D}} C'_{\mathrm{D}} \tag{2.99}$$

where

$$C'_{\mathrm{D}} = \frac{1}{1+E_{\mathrm{D}}} \tag{2.100}$$

#### 2.3.2.4.2 Inertial impaction

Bouland and Renoux (76) describe three flow regimes for inertial impaction based on the Reynolds number (Equation 2.63 above)

1. Low Reynolds number ($\mathrm{Re} < 0.2$)—Inertial forces are very low and the flow around the fiber is purely viscous. Disturbance of the streamlines due to the fiber starts well upstream.
2. High Reynolds number ($\mathrm{Re} > 1000$)—Inertial forces are dominant and the flow around the cylinder is considered as ideal, nonviscous, and nonrotational. This is a potential flow regime where streamlines come very close to the fiber before deviating strongly in order to get around it. Generally there are very few filter situations that operate at high Reynolds numbers of this order.
3. Intermediate range ($1 < \mathrm{Re} < 1000$)—The flow is considered as transitory.

Most theory has been concerned with the low Re domain where inertia effects are minimal. The Navier–Stokes equation from which most theory derives can be simplified in this region. Transition flow (intermediate range) requires a broader development of the equations.

Davies (36) reported on the work of Stechkina, Kirsch, and Fuchs (85) for calculating collision efficiency. They determined that for Stokes numbers less than one the following model applies:

$$E_{\mathrm{RI}} = E_{\mathrm{R}} + \frac{J}{2Ku} St = E_{\mathrm{R}} + E_{\mathrm{I}} \tag{2.101}$$

$E_{\mathrm{RI}}$ is the combined single fiber efficiency term from $E_{\mathrm{R}}$ and $E_{\mathrm{I}}$.

$St$ and $Ku$ are the Stokes number and the Kuwabara hydrodynamic factor from Equations 2.83 and 2.84, respectively.

$J$ derives from a complicated integral.

Davies extracted from Stechnila et al. (85) the following tabulation scheme (Table 2.2) for determining $J/2\mathrm{Ku}$.

**Table 2.2** Values of $J/2Ku$ for Equation 2.102

| $\chi$ | $\frac{d_p}{d_f} = 0.01$ | 0.02 | 0.05 | 0.10 | 0.20 | – 0.50 |
|---|---|---|---|---|---|---|
| 0.005 | 0.00076 | 0.00294 | 0.0176 | 0.0638 | 0.217 | — |
| 0.01 | 0.00090 | 0.00339 | 0.0208 | 0.0755 | 0.257 | — |
| 0.02 | 0.00110 | 0.00408 | 0.0255 | 0.0926 | 0.313 | 1.21 |
| 0.05 | 0.00157 | 0.00590 | 0.0365 | 0.134 | 0.440 | 1.78 |
| 0.1 | 0.00230 | 0.00852 | 0.0528 | 0.1923 | 0.624 | 2.36 |

Reproduced with permission of Elsevier Ltd. London, Engalnd, U.K.

## 2.4 STRUCTURE OF FIBROUS FILTERS

Several investigators have attempted to model the structure of a fibrous filter and thereby gain insight into the medium properties. A frequent approach is to use Monte Carlo techniques to simulate the random fiber structure of a nonwoven filter medium. One of the more classical presentations is the line model work of Piekaar and Clarenburg (86) in 1967. Piekaar and Clarenburg simulated a nonwoven filter medium with a number of straight lines of equal length *l*, each line representing a fiber. All the lines were perpendicular to the direction of flow and multilayered in the plane of the filter medium. Each layer has a thickness of twice the mean fiber diameter, $\overline{d}$. A random number technique was used to assign a random orientation to each fiber. Figure 2.17 is a typical array containing a line density of 25 per unit area. Note there are both an inner square and an outer square. The inner square whose side is *l*

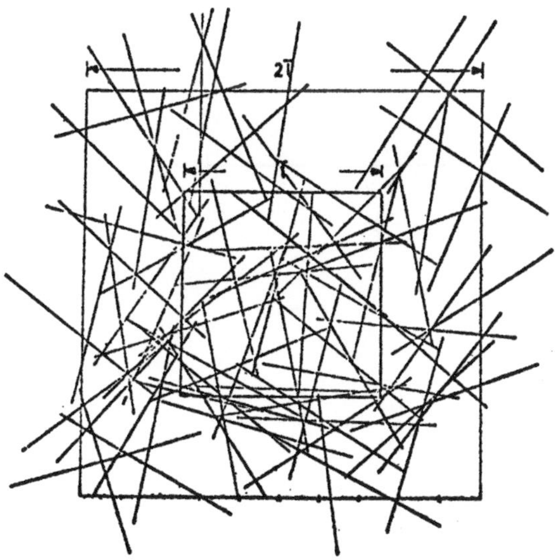


**Figure 2.17** Monte Carlo Method for a Filter Model. *Reproduced by Permission of Elsevier Ltd. London, England, UK.*

(the same as the fiber or line length) represents the area of consideration. The outer square has a side of length $2l$ and accounts for those fibers that extend beyond the inner square.

The random intersection of the lines in Figure 2.17 forms a number of randomly shaped polygons, each polygon simulating a pore. The number of pores per unit surface area of filter is a parameter of interest. Piekaar and Clarenburg (86) expressed this as

$$n_\mathrm{p} = \frac{n \cdot \check{n}_v}{n_v} \tag{2.102}$$

where

$n_p$ is the number of polygons per unit of surface area.
$n$ is the mean number of vertices of a polygon.
$\check{n}_v$ is the number of polygons sharing one vertex per unit area of surface area.

Three types of intersections were identified as illustrated in Figure 2.18.

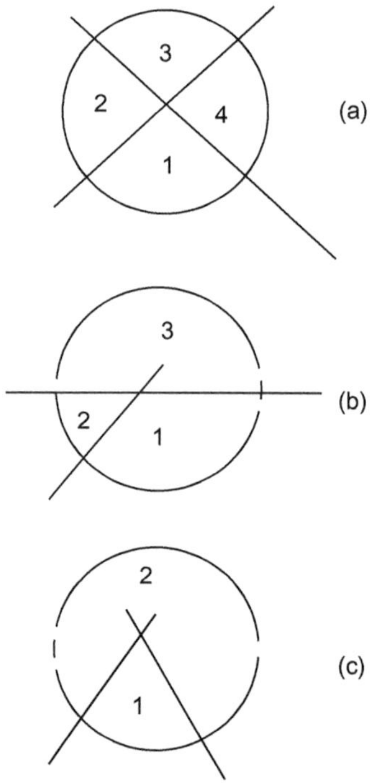


**Figure 2.18** Three Types of Line Intersections. (a) "Normal" Case—One Vertex Shared by Four Polygons; (b) "Normal-Extreme"—One Vertex Shared by Three Polygons; (c) "Extreme-Extreme"—One Vertex Shared by Two Polygons. *Reproduced by Permission of Elsevier Ltd., London, England, UK.*

Based on probabilities, the total number of interceptions per unit of surface area is

$$n = N(N-1)P/2 \tag{2.103}$$

where

$N$ is number of lines of unit surface area (25 in Figure 2.17).

$P$ is the probability that two lines intersect.

By analyzing the three types of line intersections of Figure 2.18 and their probabilities, Piekaar and Clarenburg derived the following expression for $n_p$.

$$n_p = \frac{2N(N-1)P - 2N}{n_v} \tag{2.104}$$

The value for $P$ was determined to be

$$P = 2/\pi \tag{2.105}$$

In a previous assessment, Miles (87) determined that $\check{n}_v$, the number of polygons sharing one vertex is

$$\overline{n}_v = 4 \tag{2.106}$$

This was confirmed by the Monte Carlo technique of Piekaar and Clarenburg as indicated in Table 2.3. The probabilistic outcomes were based on $N = 25$, 37, and 49. The values of $\overline{n}_v$ are 3.98, 3.96, and 3.93, respectively, close enough to justify the four value of Equation 2.106.

The number of polygons per unit of surface area then simplifies to

$$n_p = N(N-1)/\pi - N/2 \tag{2.107}$$

It was mentioned above that Piekaar and Clarenburg conducted line model studies at $N = 25$, 37, and 49. They determined that the "polygon's surface areas form a logarithmic-normal distributon, of which the geometric standard deviation is independent of the specific line density." They also calculated hydraulic radii based on the

**Table 2.3** Percentage of Polygons as a Function of the Number of Lines

| | **Number of lines per unit area, *N*** | | |
|---|---|---|---|
| **Number of vertices** | **25** | **37** | **49** |
| 3 | 38.9 | 39.7 | 39.3 |
| 4 | 35.1 | 34.4 | 36.9 |
| 5 | 16.9 | 16.9 | 15.9 |
| 6 | 7.6 | 7.9 | 7.0 |
| 7 | 1.6 | 1.0 | 0.8 |
| Mean number of vertices per polygon $\overline{n}_v$ | 3.98 | 3.96 | 3.93 |

ratio of the polygon surface area divided by the perimeter. Their sequel conclusion is that "the hydraulic radii form a logarithmic-normal distribution, of which the geometric standard deviation is independent of the specific line density."

Piekaar and Clarenburg related their results to the pore properties of actual filters. The mean surface area of a polygon in the line model is

$$\bar{s} = l^2/n_p \tag{2.108}$$

Allowing for line thickness (fiber diameter), Equation 2.107 is corrected to

$$\bar{s}' = (l^2 - N\bar{d}l)/n'_p \tag{2.109}$$

where

$\bar{s}'$ is the "corrected" mean surface area of a polygon.
$\bar{d}$ is the mean line width or simulated fiber diameter.
$n'_p$ is the total number of pores in a slice of thickness $2\bar{d}$ and a unit square $l^2$ after correction for line width $\bar{d}$.
Miles (87) proved that for isotropic line arrays

$$\bar{s}' = \bar{s} \tag{2.110}$$

Hence, the important result that the mean pore surface area can be calculated from

$$\bar{s}' = l^2/n_p \tag{2.111}$$

and

$$n'_p = (1 - N\bar{d}/l)[N(N-1)/\pi - N/2] \tag{2.112}$$

Piekaar and Clarenburg used a complicated integral to define a parameter $\bar{s}_h$ as being the mean hydraulic surface area of the polygons of the line model.

$$\bar{s}_h = 4\pi m_{hg}^2\, exp(2ln^2\sigma_{hg})\cdot\frac{1}{\sqrt{2\pi}}\int_{-\infty}^{t_2} exp\left[-\frac{s^2}{2}\right]ds \tag{2.113}$$

where

$$s = \frac{\ln m_h - \ln m_{hg} - 2\ln\sigma_{hg}}{\ln\sigma_{hg}}$$

$$t_2 = \frac{\ln M_h - \ln m_{hg} - 2\ln\sigma_{hg}}{\ln\sigma_{hg}}$$

$m_h$ is the hydraulic radius.
$m_{hg}$ is the geometric mean radius based on a log-normal hydraulic radius distribution.

$\sigma_{hg}$ is the standard deviation radius based on a log-normal hydraulic radius distribution.

$M_h$ is the maximal hydraulic radius.

Piekaar and Clarenburg determined that $\sigma_{hg}$ is independent of line density and is to be considered as a constant. They determined the value of $\sigma_{hg}$ to be 1.9.

A parameter $\beta$ was defined as being the conversion factor from effective polygon area to real polygon area. It represents the deviation of the actual pore perimeter from the ideal pore perimeter, which is the perimeter of a circle.

$$\beta \bar{s}_h = l^2/n_p \tag{2.114}$$

From their Monte Carlo techniques Piekaar and Clarenburg determined the value of $\beta$ to be 1.28 and confirmed it with a series of calculations based on a range of Johns Manville glass microfibers.

Piekaar and Clarenburgalso identified $\bar{s}'_h$, a term which was corrected for the mean line width or fiber diameter. They determined from their Monte Carlo methods that $\bar{s}'_h = \bar{s}_h$. Then

$$\bar{s}'_h = \frac{l^2}{\beta n_p} \tag{2.115}$$

The argument made from Equation 2.115 is that it demonstrates that the filter property is equated to the properties of the mathematical line model.

Substituting Equation 2.113 into Equation 2.115, recognizing that $\sigma_{hg}$ is independent of line density, and that $\beta = 1.28$, gives the following expression for the mean hydraulic radius of the pores in a fibrous filter.

$$m_{hg}^2 = 0.0295 l^2/n_p \tag{2.116}$$

In summary, Piekaar and Clarenberg arrived at the following conclusions based on their line model work:

1. Pore surface area and hydraulic radii form logarithmic-normal distributions of which the geometric standard deviation is independent of line density.
2. The mean pore surface area can be calculated from the fiber length and the number of pores per unit area of filter medium (see Equation 2.114).
3. The mean hydraulic radius can also be calculated from fiber length and the number of pores per unit area of filter medium per Equation 2.115.
4. The number of pores in a unit area of filter, $n_p$, can be determined from the line density, $N$, or number of fibers per unit area of filter medium in accordance with Equation 2.106[9].
5. The conversion factor from effective to real pore area is 1.28.

[9] The fiber number per unit area for a filter medium is calculated from the mean fiber diameter $\bar{d}$ and solidity $\chi N = \frac{8\chi}{\pi} \cdot \frac{ld}{\bar{d}^2}$.

## 2.5 INTRODUCTION TO CFD BY GEORGE G. CHASE, PH.D.[10]

The theories and models described in the prior sections are useful to help us to understand filter mechanisms and predict performance. In most of those applications the theories and models simplifying assumptions were used to make the set of equations tractable. Typical assumptions include arrangement of monodispersed fibers in arrays with uniform and periodic spacing of the fibers as well as uniformity of the fiber structure. Real nonwoven media do not have such regular fiber spacing and uniformity of structures. In real media, localized variations in porosity and permeability commonly occur and the fiber diameters are polydispersed. The spacing between fibers in real media may be irregular with some regions having such short distances between fibers that assumptions used in single fiber efficiency models do not apply (Figures 2.19 and 2.20).

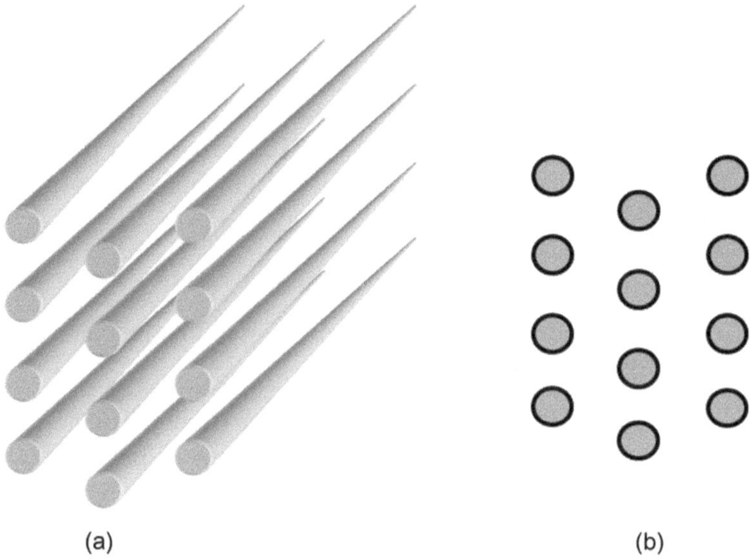


**Figure 2.19** Array of Monodispersed Fibers. (a) Perspective View, (b) 2D View.

[10] Dr. Chase is a professor in the Department of Chemical Engineering, University of Akron, Akron, Ohio, USA.

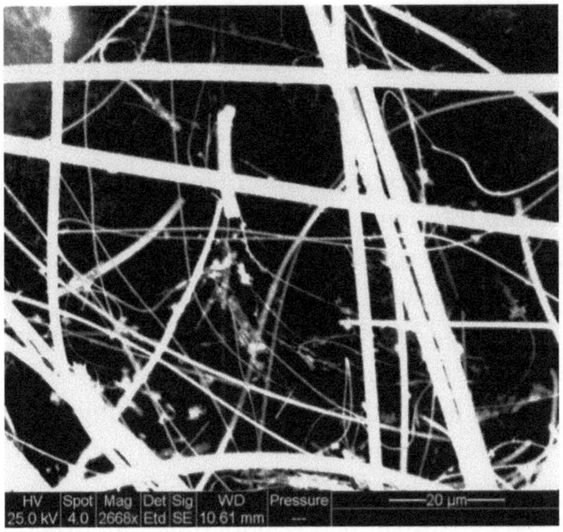


**Figure 2.20** Irregular Fiber Structures in a Filter Medium, Kulkani (88). *Reproduced with Permission of Dr. Prashant Kulkani, University of Akron, Akron, Ohio, USA.*

The prior models in some cases rely on empirically fitted parameters to account for deviations from the assumed conditions. As long as the filter structure and operating conditions do not deviate too far from the assumed conditions and the conditions used for fitting of parameters, then the prior models can be used to make useful predictions. However, to improve prediction without reliance on empirically fitted parameters and to predict performances of filters that do not fit the assumptions used to simplify the models, a more general approach is needed.

Computational fluid dynamics (CFD) refers to the application of numerical methods to solve the general field equations (mass, momentum, energy, species balances) and can include multiphysics mechanisms (electromagnetics, radiant heat transfer, photonics, and others). It is used in many industries such as automotive and aerospace to predict performance of designs of vehicles and vehicle components. It can also be used to predict the performance of filter medium. It is a powerful tool but it comes with a trade-off. It is computationally intensive requiring large computer memories and CPU time; It requires accurate description of the filter medium fiber structure, usually in 3D; and application of CFD requires specialized training of users because the systems and codes require user interactions to properly mesh the geometry and to interpret the calculation results.

A number of commercial and free-ware CFD codes are available. Some examples include ANSYS, FLUENT™, STAR CD, CFX, FlexPDE, FIDAP, CD-adapco, and many more (see for example, CFD Review (www.cfdreview.com (89)). The CFD software, GEODICT (90) is specifically designed for modeling performance of fiber medium.

### 2.5.1 Advantages and disadvantages of CFD

There are many advantages to using CFD. Some of these advantages are

- CFD allows numerical simulation of fluid flows, for which results are available for study even after the analysis is over. This is a big advantage over an experimental approach for which detection and measurement of flow fields and particle concentrations are difficult and transient.
- CFD allows observation of flow properties without disturbing the flow itself, which is not always possible with conventional measuring instruments.
- CFD allows non-intrusive observation of flow properties at locations which may not be accessible to measuring instruments.
- CFD can be used as a qualitative tool for discarding (or narrowing down the choices between) various designs. Designers and analysts can study prototypes numerically and then test by experimentation only those which show promise.
- In general CFD is less expensive than conducting experimental studies.

CFD is not the preferred approach to all filter design questions. Some of the disadvantages of CFD are

- CFD is not a turn-key technology where it can be applied as a black box by users without a high level of working knowledge of the numeric methods applied by the solver.
- Despite the increasing speed of computation available, CFD has not yet matured to a level where it can be used for real time computations. CFD requires significant setup time to prepare the geometry and grid. The numerical solution to CFD requires significant computational time on the computer.
- CFD requires experimental validation to prove that the solution generated by the CFD is the correct solution and the results are reasonable.
- There are many CFD codes and software packages. Switching between packages is not trivial because the user interfaces are not standardized. This is further complicated by CFD code suppliers frequently change or update their software and some retraining of users may be required.
- CFD encompasses a range of physical mechanisms and different numerical tools to solve the expressions describing the physics. The appropriate mechanisms for the problem at hand must be properly selected. The appropriate equations must be solved for accounting for particle motion and particle capture. In some applications the CFD code must be modified by the user to incorporate physics specific to the filtration problem at hand.
- Some commercial CFD codes licenses are expensive and have annual maintenance fees. The employment of in-house trained CFD programmers adds to the cost of using CFD by companies. Industrial users typically weigh the costs between conducting experiments and the flexibility provided by CFD to spread the costs over multiple projects. If an industrial user has only one specific application it may be more practical to conduct experiments instead of CFD or to hire CFD services on a per-project basis.

## 2.5.2 How CFD is applied

In simplified terms, CFD is applied in three steps: (1) set up the geometry, (2) run the solver, and (3) visualize and evaluate the results.

The geometry setup requires information on the dimensions of the system components such as the filter, filter housing, and inlet/outlet ports. If the CFD goal is to evaluate the filter medium structure itself then the geometry setup requires information about the locations and orientations of the individual fibers in the filter. In some cases simplifications can be applied to reduce the size of the system geometry, such as periodic boundary conditions and symmetry.

The geometry setup includes meshing to setup a grid space for controlling the calculations and solving the differential equations that describe the physics. A number of different meshing techniques and mesh geometries are available. The choice of which mesh approach to use depends on the complexity of the system being modeled and the specific CFD software being used. Figure 2.21 shows the pyramid mesh structure generated for flow in 3D box with circular inlet and outlet.

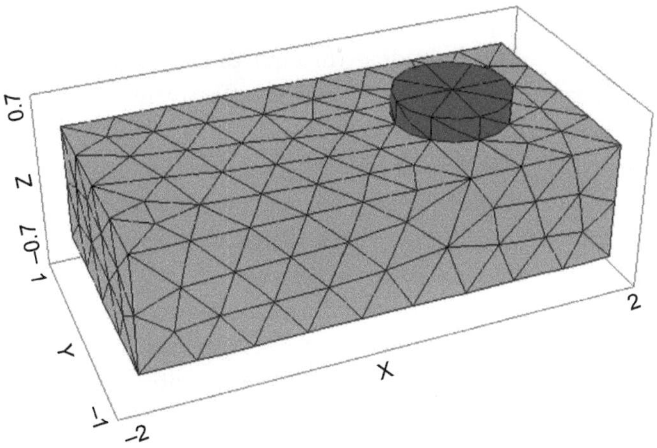


**Figure 2.21** Pyramid Mesh for 3-D Flow Box Generated by FlexPDE.

Before running, a typical solver, information is needed on the fluid material properties, the properties of the solution space such as whether it is to be modeled as a porous medium, the solid surface properties and boundary conditions, and information regarding laminar versus turbulent flow and the type model used to solve turbulent flow conditions. The program also needs information about monitoring the progression of the calculations toward convergence to the final solution.

In some applications the program requires modifications by the user. For example, many CFD programs are designed with a no-slip boundary condition as its default. The user can override this default but may have to write a "user code" to tell the program how to calculate the boundary conditions.

When the code is run most software packages display plots showing the progress of the code toward convergence. The plotted information is updated in real time as the computer code is run and the user can observe how long the calculation takes to reach convergence and whether or not it appears that convergence will be reached.

The final step is to visualize and evaluate the results. The visualization is a very useful aid to help the user to understand and interpret the results. Usually the visualization is a graphical mapping of the calculated values over position or time. The more realistic the simulation appears in comparison to experimental observations the more convincing the model is calculating correctly.

### 2.5.3 CFD approaches to filtration

Depending on the level of detail required of the CFD simulations, the CFD may be applied to predict performance of the filter medium structure or the performance of the filter, filter holder, and how the filter interacts with the environment. In the latter case, treating the filter as a porous medium may be sufficient. The porous medium approach does not consider the internal structure of the filter medium but uses parameters that account for the properties of the medium at a scale larger than the pores or fibers of the medium. The fluid drag through the medium in the porous medium approach is modeled through a resistance tensor (effectively, the inverse of the permeability constant in Darcy's law (Holdich (91)) and correlations may be used to account for particle capture. The predictions in the latter case are limited to the accuracy of the correlations for the resistance tensor and the particle capture, but the calculations are faster than in simulations of the medium structure.

The simulation of the medium structure requires an accurate geometric mapping of the structure, such as the locations of the fibers and pores. This can require significant effort but it can result in accurate prediction of filter performance in calculating the pressure drop and particle capture efficiency. The complexity of the structural geometry requires a very fine mesh for the calculations and makes the CFD simulation challenging.

In the porous medium simulations, Eulerian equations may be used for each phase. The particles in the fluid phase may be described in terms of their concentration, as a species in the species mass balance in continuum mechanics. The fluid phase can be modeled using an Eulerian approach while the particle movement can be modeled using a Lagrangian approach (Crowe et al. (92)).

### 2.5.4 Example CFD applications to filtration

#### *2.5.4.1 Modeling the many scales of a diesel particulate filter*

Konstandopoulos et al. (93) apply CFD to predict performance of diesel particulate filters (DPFs). They consider the DPF at different length scales, depending on the goals of the computational models (Figure 2.22). The scale of the filter wall structure is considered at the smallest scale on the order microns, the pores within the filter

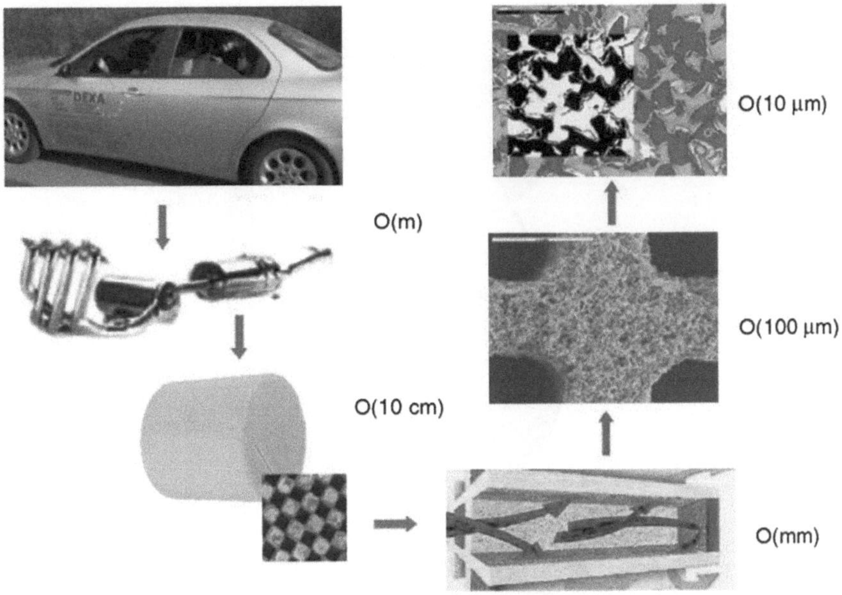


**Figure 2.22** The Multiple Scales That Must Be Simulated for Diesel Particulate Filters (Konstandopoulos et al. (93)).

structure are considered at the scale of tens of microns, the flow channels within the monolithic filter are considered in the millimeter scale while the overall dimensions of the filter are at the centimeter scale. The DPF is part of the exhaust system of the vehicle at the scale of meters. This hierarchy of scales allows the filter model effort to focus on the relevant physics at each scale and provides a comprehensive and systematic approach to design optimization of a DPF.

At each scale level in the hierarchical approach, to set up a tractable set of model equations information of physical phenomena at the next lower scale must be applied to the model. At the lowest scale, the micron scale, the next lower scale is the molecular scale. Molecular scale phenomena are accounted for in the material parameters such as viscosity, thermal conductivity, and molecular diffusivity. As the scales progressively increase, the lower scale phenomena are applied to the model through semi-empirical correlations. This is the strength of the multiscale approach because the lower scale correlations greatly speed up the calculations. However, this is also a weakness of the approach because correlations tend to be limited to the conditions for which the correlation was derived and the model results may only be as accurate as the correlations themselves.

#### *2.5.4.2 Visualization of velocity profiles to improve filter performance*

Cortés-Juan et al. (94) modeled laminar fluid flow in a flat circular filtration cell for cross-flow filtration of nanoparticles (Figure 2.23). The goal was to enhance the

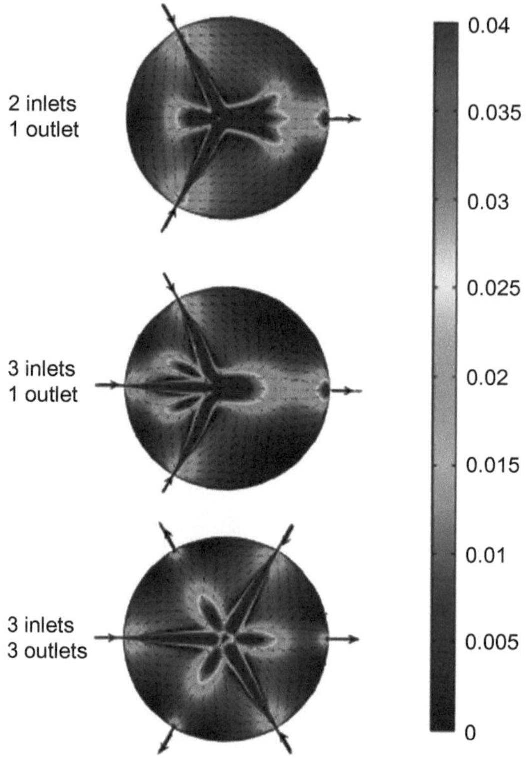


**Figure 2.23** Contour Plot of Velocity Magnitude (m/s) and Vector Directions in the Middle-Height Plane of the Feed Chamber for Different Number and Location of Inlet and Outlet Points (Cortés-Juan et al. (94)). *Reproduced with permission of Elsevier (Pergamon Press), Oxford, England.*

design performance. With the help of CFD simulations geometric design parameters were varied to decrease the size of the recirculation areas. High velocity and wall shear stress fluctuations hindered development of a polarization layer and fouling of the membrane surface. The CFD allowed modeling of grooves on the filter surface that create velocity fluctuations. This is an example of how CFD can directly reduce experimental effort to improve the design of filter housing.

#### *2.5.4.3 Comparison of 2D- and 3D-based correlations for prediction of pressure drop in filter media*

Over the past 70 or so years, since the development of synthetic fibrous filter media, correlations have been developed to predict properties such as pressure drop and particle capture efficiencies (Spurny (33)). Most of these correlations apply to coarse sized fibers with dense gas flow conditions and the no-slip boundary between the fibers and the gas phase (Brown, (95)). The no-slip boundary models the gas flow as having the same velocity as the solid surface of the fibers). Usually the solid surface is stationary hence the velocity is zero (Bird et al. (43)).

However, in rarified gas flow conditions the mean-free path between molecules is relatively large compared to the boundary structures, and slip flow can occur. For filtration applications this occurs when the fiber diameter is of the same order of magnitude as the mean-free path of the gas molecules. This can occur at very low gas pressure conditions with microfiber filter medium, or with fibers in the submicron diameter range at atmospheric pressure. In slip flow the gas velocity at the boundary wall is nonzero (Knudsen (96), Kennard (97)). While some slip flow correlations are available, this is a topic of current research.

Hosseini and Tafreshi (98) describe the application of CFD to model 3D filter structures in slip flow conditions. They use the CFD to compare 3D simulations with slip flow to predictions based on 2D fiber arrangements predictions based on the 2D cell model of the Kuwabara flow field with slip boundary condition presented by Brown (95)

$$\Delta P_s = \frac{4\mu chU(1 + 1.996Kn)}{r^2[Ku + 1.996Kn(-0.5\ln(c) - 0.25 + 0.25c^2)]} \tag{2.117}$$

where $c$ is the solid (fiber) volume fraction and Ku is the Kuwabara constant given by

$$Ku = -0.5\ln(c) - 0.75 + c - 0.25c^2 \tag{2.118}$$

The Knudsen number is defined by

$$Kn = \frac{2\lambda}{d_f} \tag{2.119}$$

where $\lambda$ is the mean free path of the gas molecules and $d_f$ is the fiber diameter.

Hosseini and Tafreshi (98) determined the permeability of the media increased by almost a factor of 2 for 100 nm diameter fibers accounting for the slip flow condition as compared to the no-slip condition. They compared their predictions with those in literature for 3D simulations and found good agreement with Jackson and James (99), Spielman and Goren (100), and Ogorodnikov (101) (for fiber diameters of about 200 nm). Using the permeability constant expression of Jackson and James, they determined the correlation for permeability based on 3D CFD modeling is

$$k = \frac{3r_f^2}{20c}[-\ln(c) - 0.931]\frac{\Delta P_s}{\Delta P_{ns}} \tag{2.120}$$

where $\Delta P_{ns}$, the pressure drop for no-slip conditions is given by

$$\Delta P_{ns} = \frac{4\mu chU}{r_f^2 Ku} \tag{2.121}$$

Hence, the improved correlation for permeability becomes

$$k = \frac{3r_f^2}{20c}[-\ln(c) - 0.931]\frac{Ku(1 + 1.996Kn)}{[Ku + 1.996Kn(-0.5\ln(c) - 0.25 + 0.25c^2)]} \qquad (2.122)$$

The pressure drop over the medium is related to the permeability in Darcy's law as

$$\Delta P = \frac{\mu h U}{k} \qquad (2.123)$$

where $\mu$ is the gas viscosity, $h$ is the filter thickness, and $U$ is the face velocity of the gas at the filter surface.

#### 2.5.4.4 Use of CFD to define average fiber diameter in a bi-modal fiber distribution medium

Numerous publications report analytical expressions for predicting pressure drop or permeability of filter media consisting of unimodal fiber distributions, there are very few expressions available for media with bimodal fiber diameter distributions. Tafreshi et al. (102) used CFD to solve the Stokes flow governing equations, with no-slip boundary conditions, in a series of 3D virtual media geometries with a bimodal fiber distribution (Figure 2.24). He compared the CFD results with unimodal analytical expressions.

They compared eight different methods of defining an average fiber diameter that would allow them to use the unimodal analytical expressions to predict permeability. The use of CFD to do this analysis eliminated experimental uncertainties and

**Figure 2.24** An Example of the Virtual Bimodal Media Modeled in the Work by Tafreshi et al. (102). *Reproduced with permission of Elsevier (Pergamon Press), Oxford, England.*

difficulties in making fair and relevant comparison between the methods. Tafreshi et al. concluded that the area-weighted average diameter of Brown and Thorpe (103) and the volume-weighted resistivity model of Clague and Phillips (104), a cube root relation, offer the best predictions for fiber diameter ratios, 1 to 5, and solidities, 0.05 to 0.15.

### 2.5.4.5 Effect of filter geometry on filter performance

Filter media are often designed with pleats to increase surface area. Filters normally scale with the surface area, but in pleated media the fluid flow may not be uniform across the pleat. The non-uniform flow can cause some regions of the pleats to load with particles more rapidly than others. Also, if surface filtration occurs, caking can form at the apex of the pleats, further causing the non-uniformity in the flow.

Rabaï et al. (105) describe an application of CFD to model flows within pleat channels to validate a model of the pleated filter geometry. The model is easier and faster to compute than CFD and allows rapid comparison of different designs, but the model in general is not as precise as can be obtained with CFD. Comparison of the model to CFD and experiments shows the model gives accurate enough results for pleat designs. They conclude that for high filtration velocities the flow profiles within pleats are nonparabolic and have a great impact on filter loading.

The model can be used to determine the optimal pleat density, i.e., the pleat density minimizing the overall pressure drop across the filter for given flow rate, pleat length, and given filter medium properties. The model significantly reduces the computational effort compared to the standard CFD approach in particular by not requiring re-meshing across the pleat channels when a filter cake forms on the channel walls. The CFD application described in this paper was useful as a tool to verify the accuracy of the simplified model approach.

### 2.5.4.6 CFD model of heterogeneous properties in filter media

Ceramic filter media are often used for removal of dust particles from hot gas streams. The dust particles form dust cakes on the surface of the filters. Periodically the filters are regenerated by a pulse of reversed gas flow to dislodge the cake. Occasionally the dust cake separation is incomplete and the regeneration leaves a pattern of patches of cake on the filter medium surface.

There are several approaches one can use to model these phenomena. The method of particle tracking in CFD can provide exact data on the dust cake structure and the deposition phenomena can be investigated in detail. However, such this approach requires high computational times and computer resources. To reduce some of the time and resources some simplifications are commonly applied, such as modeling only the cake structure while neglecting the medium structure and by modeling only a very small region of the filter medium. These latter simplifications force the model to neglect interaction effects between a dust cake and the surface of the filter medium over distances at which patchy regeneration occurs.

Dittler and Kasper (106) modeled the formation of the dust cake at a coarser scale with a two-dimensional cell model. The coarser scale has less resolution of the dust cake with respect to the computational grid. The advantages of this higher scale approach are shorter computational times, the media structure in depth can be included in the computations, and larger sections of the medium surface could be modeled so that the patchy phenomena could be better represented. The authors showed that the structure of the filter medium as well as the regeneration efficiency and the patchy regeneration of the medium surface have a significant effect on the flow field and the pressure drop of the filter medium. The reduced computational times made practical multiple simulations to observe the effects of regeneration efficiency and cycle times on pressure drop. The pressure curve plot shown in Figure 2.25 was generated by running the CFD code multiple time to generate the data to create the plot.

Extreme operating conditions in temperature, pressure, and flow rate, and extreme conditions in fluid properties such as toxicity, radioactivity, and chemical reactivity can be very challenging to study experimentally. CFD models of such conditions can greatly reduce the experimental effort and the experimental risks. Tanthapanichakoon et al. (107) use CFD to model filtration of high-temperature gas. In their work they modeled two ceramic candle filters near each other. Figure 2.26 shows the grid

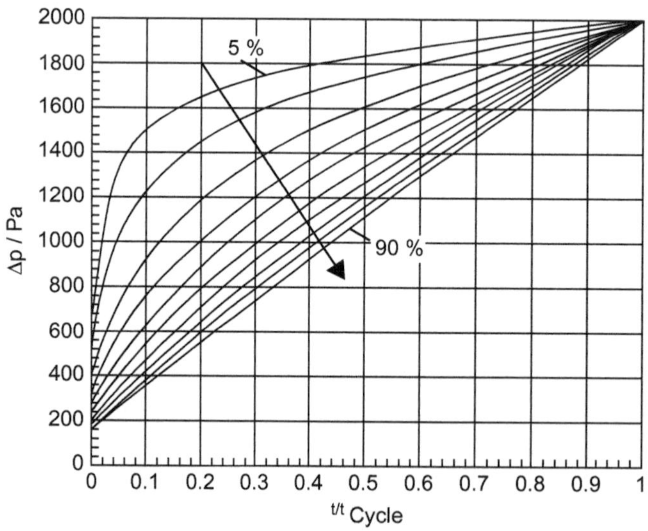


**Figure 2.25** Calculated Pressure Curves for Regeneration Efficiencies Ranging from 5% to 90% (106). *Reproduced with permission of Elsevier BV, Oxford, England.*

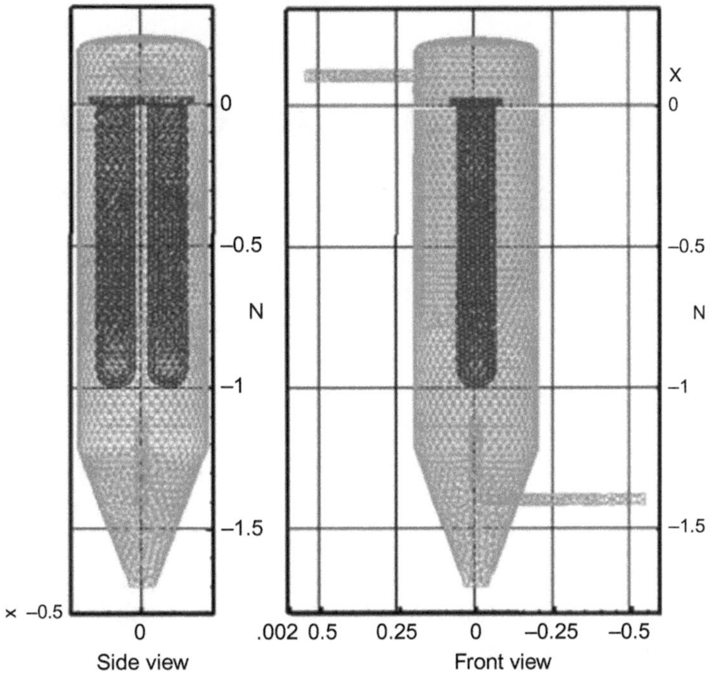


**Figure 2.26** Illustration of the Computational Grid Around the Twin-Candle System (107). *Reproduced with permission of Elsevier S.A., Oxford, England.*

structure used to model the filters. The CFD results showed an unanticipated non-uniform cake formation on the surface of the filters due to the presence of the adjacent filter in the gas flow path.

## 2.5.5 CFD summary

This section provides a brief introduction to the application of CFD modeling to filters and filter media. Some of the strengths and weaknesses of the CFD approach are discussed. Some example applications are presented. Many more applications are reported in the general literature. The interested reader is encouraged to obtain more information on applying CFD to a specific application from the software suppliers.

# CHAPTER 3

# Properties of Nonwoven Filter Media

## 3.1 CHARACTERISTICS AND PROPERTIES OF AIR LAID WEBS

Air laid webs tend to be bulky structures designed for absorbency. The major markets are diaper padding, feminine hygiene pads, incontinence pads, wipes, and other applications where absorbency plays a major role. The absorbency properties are achieved by the use of specially produced high bulk wood pulp fibers known as fluff pulps and by the use of superabsorbent polymers. Superabsorbent polymers can include superabsorbent powders, such as carboxy methyl cellulose, and super absorbent fibers. Air laid webs can incorporate synthetic fibers for improved strength and resiliency. Bicomponent sheath core fibers are often used to provide thermobonded structures. The products used for wipes are often latex and/or foam bonded. Spunlace (hydroentangled) forms of the web are becoming popular for the wipes market. Skyt (108) has identified possibilities for combining the air laid process with other technologies to produce advanced composite structures. This includes multilayer combinations of air laid webs with dry laid (carded), melt blown, spunbonded, and wet laid webs. In addition, composite structures with membrane and nanofiber coatings are finding important applications in the marketplace.

## 3.2 CHARACTERISTICS AND PROPERTIES OF DRY LAID WEBS

There are several forms of dry laid carded webs for filter media. Included are needle punch felts, hydroentangled felts, and high loft webs. PGI (Polymer Group, Inc.), Mooresville, North Carolina produces a variety of dry laid webs as indicated in Table 3.1. The table indicates the extent of filtration technologies, polymers, products, and applications that can be achieved with dry laid webs and does not even include needle punched felts.

### 3.2.1 Needle punched felts

Needle punched felts, produced from a wide variety of fiber materials, have a broad range of properties, depending on the application. The biggest uses of needle punch filter felts are in baghouse-type dust collection and fume systems. The traditional fibers for this type of fabric are wool; however, the industry has converted much of this production to cotton polyester and 100% polyester. Other more specialized fibers are used to provide needle punch felts with resistance to a broad spectrum of hostile



*Handbook of Nonwoven Filter Media*
ISBN: 978-0-08-098301-1

**Table 3.1** Carded Web Filter Media of PGI

| Technology | Raw materials | Products | Applications |
|---|---|---|---|
| APEX®Spunlace | Polyester (PET), polyolefin (PE and PP), bicomponent, acetates (Rayon), polyaramid, polyphenylene sulfide (PPS), natural fibers (cotton, wool, wood pulp), and blends | Spunlace Durapex Aquapex™ | Dust filtration, liquid filtration, membrane support pool and spa, dust bag |
| Thermal Bond | Polyester (PET), polyolefin (PE and PP), bicomponent, acetates (Rayon), cellulose, and blends | Kiara® Ultraloft™ | Potable water, coolant filtration, membrane backing, HVAC filtration, liquid filtration, air filtration, |
| Adhesive bond | Polyester (PET), polyolefin (PE and PP), bicomponent, acetates (Rayon), cellulose, and blends | Adhesive bond | Fuel, oil and gas filtration, turbine in-take filtration, air filtration, specialty applications |
| Resin bond | Polyester (PET), polyolefin (PE and PP), bicomponent, cellulose, and blends | Resin bond | Coalescing filters, liquid filtration, air filtration, specialty applications |

Reproduced with permission of Polymer Group, Inc., Charlotte, North Carolina, USA.

environments. Often the requirement is for very-high temperature resistance and resistance to some very corrosive chemicals. Several types of staple fibers may be used to provide the necessary heat resistance for high temperature applications. These include glass microfiber, polyphenylene sulfide (PPS), polyaramid fibers such as Kevlar® and Nomex®, and polyimide fiber (P84®). Ceramic, metal and carbon fibers are used in applications where a very high degree of temperature and corrosion resistance is required.

In pulse jet filter bags and cartridges, needle punch filter media have to provide surface filtration and release properties so that the filter cake will readily fall off the medium in the pulsing operation. Several possible finish treatments are used to enhance this. These include calendering, glazing, and singeing. Calendering is a treatment that smoothes the web surfaces by passing the web through a set of high-pressure nip rolls. The rolls may be gapped to provide for a uniform thickness. Glazing passes the web through a set of heated, high temperature, nip rolls that are rotating at slightly different surface speeds. The effect is to "iron" onto the web a glazed glass-like surface. Singeing is an operation that burns off protruding surface fibers.

### 3.2.2 Hydroentangled webs

Spunlace and hydroentanglement are the most frequent terms applied to webs consolidated by high-velocity water jets. Other terms are jet entangled, water entangled, and hydraulically needled. The product and process was first developed and patented by DuPont in the 1960s and 1970s. DuPont's product entry into the market place was the product identified as Sontara®. Chicopee Nonwovens (now PGI) and Kendall (now part of Tyco Healthcare) also were the early entries into hydroentangled products.

Hydroentanglement uses high-speed jets of water to entangle the fibers for integrity enhancement. The entanglement is accomplished by applying the jets to the web as it is supported by a foraminous screen. The foraminous screen may be a porous belt, a patterned screen, or a perforated screen. The process of hydroentanglement is discussed in more detail in Chapter 5, Section 5.1.2.3.2.

Carded dry laid webs are common source of spunlace webs. However air-laid, melt-blown, spunbonded webs, and wet lay webs have all been hydroentangled. It provides the possibility of providing strength to a loosely bonded medium without the necessity of a thermobond or chemical binder. In addition to integrity, hydroentanglement can add depth and dimension to a filter medium structure. Hydroentanglement is one way of interbonding the different layers in a composite structure. This is done by a proper choice of pattern or apertures in the support screen.

Kamath et al. (109) indicated that air laid and carded webs are the most frequently used webs for hydroentanglement, however, there has been a trend toward wet laid precursor webs. White (110) reported on hydroentanglement technology applied to wet laid precursor webs. Rupp (111) states that "Basically, every nonwoven web-forming process can be applied in the spunlacing process." Voith Paper GmbH, Düren, Germany and Trützschler Nonwovens, Egelsbach, Germany formed a venture to combine Voith's wet lay HydroFormer® technology (see Section 5.6.1) with Trützschler's AquaJet® hydroentanglement technology to provide the capability to produce spunlace wet laid nonwovens (112).

Spunlace or hydroentangled felts are part of the market place for baghouse filter media. They are also finding applications in other forms of filtration such as turbine intake filters. It will be noted from Table 3.1 above that PGI is a producer of spunlace webs for dust filtration.

### 3.2.3 High loft filter medium

High loft filter medium webs pictured in Figure 3.1 are carded webs that are produced to be bulky and thick. Usually they are composed of polyester fiber, however, other fibers are often utilized. This includes natural fibers such as hog's hair and coconut fiber. High loft webs are most often obtained by thermobonding and less often by chemical or resin bonding. Sheath-core bicomponent fibers with low melting sheaths are often used as the thermal binder. Loft and thickness are accomplished

**Figure 3.1** Samples of High Loft Filter Media.

by through-air oven heating. The loft or thickness can be regulated or controlled to give mats with thicknesses greater than 3 mm (0.12 in). High loft mats are used in residential furnace and air conditioning systems. They are also popular as filters in paint spray booths. In many applications the filters are the mats cut to a specific size. Either the supplier cuts the mats into custom-sized sheets or the user buys the mat in roll or large sheet form and cuts the pads to the dimensions of his filter opening.

## 3.3 CHARACTERISTICS AND PROPERTIES OF MELT-SPUN WEBS

### 3.3.1 Spunbond webs

Malkan and Wadsworth (25) identify the following as being characteristics of spun-bonded webs:

**a.** Random fiber structure.
**b.** Generally the web is white with high opacity per unit area.
**c.** Most spunbonded webs have a layered or shingled structure; the number of layers increases with increasing basis weight.
**d.** Grammage (basis weights) range between 5 and 800 g/m$^2$, typically 10–20 g/m$^2$.
**e.** Fiber diameters range between 1 and 50 μm, but the preferred range is between 15 and 35 μm.
**f.** Web thickness ranges between 0.1 and 4.0 mm, typically 0.2–1.5 mm.
**g.** High strength-to-weight ratios compared to other nonwoven, woven, and knitted structures.
**h.** Planar isotropic properties due to random lay down of the fibers.
**i.** Good fray and crease resistance.
**j.** High liquid retention capacity due to high void content.
**k.** High in-plane shear resistance.
**l.** Low drapeability.

Spunbonded webs can be produced from polymers such as polyester (Reemay®), nylon (Cerex® and PBN-II®), polypropylene (Typar® and Tekton®), and polyethylene (Tyvek® and SoloFlo®). Polymer Group, Inc. (PGI) in Old Hickory, Tennessee, USA is a major manufacturer of spunbond polyester for filtration purposes. PGI manufactures the medium under the trade name of Reemay®. Table 3.2 lists typical properties. Note that Reemay® is produced with straight fibers or crimped fibers and with round or trilobal cross section. The bulking effect of the crimped fibers is evident from the data. For example, straight fiber Grade 2040 has a grammage (basis weight) of 136 gsm and a thickness of 0.51 mm. Crimped fiber Grade 2440 is considerably lighter at 98 gsm and yet has the same or slightly more thickness at 0.53 mm. Also note that the spunbonded products can be spun with either round fibers or trilobal fibers. PGI also manufactures Reemay Elite® as a membrane support medium and a composite filtration product under the trade name of Synergex®

Typar® and Tekton® are the trade names of polypropylene spunbonded filter media marketed by PGI. Typar® is the name used in North America, South America, Israel, and South Africa. Tekton® is the name used elsewhere in the world. Typical properties are listed in Table 3.3. In one grade, T-515LF, the LF stands for ultra-low fuzz. This grade is designed for quick release of filter cake.

Freudenberg Nonwovens, Weinheim, Germany markets spunbond webs made from mostly recycled polyester (bottles). They list three specific styles of polyester spunlaid products made from recycled polyethylene terephthalate (PET) and marketed under the Lutradur® brand. Specialized filtration devices are among the many applications listed for Lutradur® polyester spunlaids. SCS Global Services (SCS), a leader in third-party environmental and sustainability certification, has certified the high-level recycled content claim of Lutradur® products manufactured by the Freudenberg Spunweb Company, a North American subsidiary of Freudenberg Nonwovens (113).

Intrepid™ filtration media, made of continuous polyolefin fibers that are thermally bonded together, are produced by Kimberly Clark Worldwide. The patented process which produces Intrepid® media gives the ability to make a controlled gradient structure within the medium. This fiber structure increases in fiber density from the upstream to the downstream side. The media construction, coupled with a proprietary electrostatic treating technology, allows filters made with Intrepid™ medium to achieve high initial and sustainable filtration efficiencies. Properties of some of the grades of Intrepid™ filter media are listed in Table 3.4.

Midwest Filtration LLC, Cincinnati, Ohio, USA markets a filter medium for the filtration industry. Included are spunbond polyester, spunbond polypropylene, blends of polyester/cellulose and Rayon/polyester, bicomponent (polyester/polyethylene) fiber webs, hydreoentangled polyester and wet laid nonwovens composed of cellulose and polyester. One of their more interesting lines is the Unilayer™ line which consists of multi-ply spunbonded polypropylene. The number of plies range from 3 to 6 as indicated in Table 3.5. The multi-ply approach creates better filtration efficiency because of a more tortuous path through the medium.

**Table 3.2** Reemay® Filtration Grade Spunbonded Polyester

| Style No. | Filament cross section | Basis weight (gsm) | Thickness (mm) | Grab tensile N MD × XD | Trap tear N MD × XD | Mullen burst (kPa) | Frazier air permeablity (l/m²/S) @ 124 Pa | Textest air permeablity (l/m²/S) |
|---|---|---|---|---|---|---|---|---|
| ***Straight fibers*** | | | | | | | | |
| 2004 | T | 14 | 0.13 | 41 × 32 | 14 × 14 | 62 | 7110 | 7750 |
| 2055 | T | 19 | 0.18 | 50 × 36 | 14 × 18 | N/A | 6095 | 7205 |
| 2006 | T | 20 | 0.18 | 45 × 36 | 18 × 23 | 89 | 5995 | 6909 |
| 2011 | T | 25 | 0.23 | 63 × 50 | 23 × 27 | 110 | 5435 | 5649 |
| 2014 | T | 34 | 0.25 | 95 × 77 | 27 × 32 | 151 | 4470 | 4369 |
| 2016 | T | 46 | 0.25 | 144 × 113 | 32 × 41 | 220 | 2745 | 2784 |
| 2024 | T | 71 | 0.30 | 279 × 212 | 41 × 50 | 358 | 1575 | 1641 |
| 2030 | T | 80 | 0.31 | 274 × 209 | 38 × 47 | 135 | 330 | 1625 |
| 2033 | T | 100 | 0.43 | 387 × 351 | 68 × 77 | N/A | 1310 | 1300 |
| 2040 | I | 136 | 0.51 | 563 × 450 | 63 × 86 | 681 | 965 | 885 |
| 2060 | T | 203 | 0.71 | 770 × 583 | 26 × 33 | 1034 | 131 | 610 |
| 2250 | R | 17 | 0.13 | 50 × 32 | 18 × 23 | 76 | 5485 | 6640 |
| 2275 | R | 25 | 0.15 | 68 × 63 | 27 × 32 | 117 | 4410 | 4699 |
| 2200 | R | 36 | 0.20 | 95 × 90 | 32 × 32 | 158 | 3370 | 3378 |
| 2214 | R | 46 | 0.23 | 144 × 135 | 41 × 45 | 193 | 2645 | 2631 |
| 2295 | R | 100 | 0.46 | 338 × 320 | 108 × 108 | 509 | 1270 | 1245 |
| 7061 | T | 46 | 0.32 | 131 × 108 | 32 × 41 | N/A | – | – |

*(Continued)*

Table 3.2 (Continued)

| Style No. | Filament cross section | Basis weight (gsm) | Thickness (mm) | Grab tensile N MD × XD | Trap tear N MD × XD | Mullen burst (kPa) | Frazier air permeablity ($l/m^2/S$) @ 124 Pa | Textest air permeablity ($l/m^2/S$) |
|---|---|---|---|---|---|---|---|---|
| *Crimped fibers* | | | | | | | | |
| 2410 | T | 39 | 0.36 | 68 × 50 | 32 × 36 | N/A | N/A | N/A |
| 2415 | T | 53 | 0.41 | 108 × 81 | 41 × 50 | N/A | N/A | N/A |
| 2420 | T | 63 | 0.43 | 135 × 99 | 50 × 59 | N/A | N/A | N/A |
| 2430 | T | 81 | 0.48 | 198 × 153 | 68 × 81 | N/A | N/A | N/A |
| 2440 | T | 98 | 0.53 | 252 × 189 | 81 × 104 | N/A | N/A | N/A |
| 2470 | T | 203 | 0.81 | 567 × 450 | 167 × 212 | 564 | 845 | 818 |

**The following test methods are used to measure typical properties:**

| | |
|---|---|
| Unit weight | ASTM D-3776 |
| Thickness | ASTM D-1777 |
| Grab tensile | ASTM D-4632 |
| Trap tear | ASTM D-1117 |
| Mullen burst | ASTM D-3786 |
| Air permeability | ASTM D-737 |

NOTES:
2000 Series Trilobal Fiber Diameter = 21 μ
2200 Series Round Fiber Diameter = 16 μ
2400 Series Trilobal Fiber Diameter = 23 μ
PGI makes no warranties, expressed or implied, as to the finished product provided from these fabrics. Physical properties tested are typical of our fabrics at the time of manufacture and are not intended for specification purposes. Reemay® is a registered trademark of Polymer Group, Inc.
Reproduced with permission of PGI, Old Hickory, Tennessee USA.

**Table 3.3** Typar®/Tekton® Filtration Grade Spunbonded Polypropylene

**Typical properties*** ***(Metric Units)***

| Style no. | Basis weight (gsm) | Thickness (mm) | Grab tensile N MD × XD | Trap tear N MD × XD | Mullen burst (kPa) | Frazier air permeablity ($l/m^2/S$) @ 124 Pa | Textest air permeablity ($l/m^2/S$) |
|---|---|---|---|---|---|---|---|
| ***Low denier*** | | | | | | | |
| 3091 L | 31 | 0.20 | 90 × 68 | 27 × 23 | 144 | 2745 | 3302 |
| 3121 L | 42 | 0.23 | 144 × 126 | 54 × 41 | 193 | 1930 | — |
| 3161 L | 54 | 0.25 | 189 × 99 | 54 × 45 | 241 | 1270 | 1219 |
| 3409 L | 136 | 0.38 | 540 × 360 | 158 × 90 | 729 | 147 | — |
| 3609 L | 203 | 0.51 | 765 × 495 | 270 × 158 | 1217 | 80 | 66 |
| ***Standard*** | | | | | | | |
| 3121 (N) | 42 | 0.25 | 239 × 216 | 126 × 108 | 275 | 3302 | 3221 |
| 3141 (N) | 47 | 0.25 | 279 × 239 | 135 × 113 | 303 | 2997 | 2926 |
| 3151 (N) | 54 | 0.30 | 212 × 203 | 104 × 104 | 296 | 2692 | 2921 |
| 3201 (N) | 64 | 0.30 | 356 × 333 | 167 × 153 | 392 | 2030 | 2103 |
| 3251 (N) | 85 | 0.33 | 500 × 477 | 185 × 144 | N/A | 940 | — |
| 3301 (G) | 102 | 0.35 | 599 × 608 | 225 × 225 | 688 | 930 | — |
| 3341 (G) | 115 | 0.41 | 585 × 657 | 216 × 225 | N/A | 680 | — |
| 3351 (G) | 119 | 0.38 | 666 × 648 | 293 × 324 | 540 | 710 | — |
| 3401 (G) | 136 | 0.46 | 666 × 720 | 338 × 351 | N/A | 1075 | — |
| 3601 (G) | 203 | 0.53 | 1215 × 1283 | 482 × 554 | N/A | 195 | — |
| 380 (N) | 271 | 0.64 | 1575 × 1530 | 450 × 392 | N/A | 60 | — |
| T-515 (LF) | 203 | 0.51 | 1035 × 1035 | 198 × 239 | 711 | 147 | 152 |

PGI makes no warranties, expressed or implied, as to the finished product provided from these fabrics. Physical properties tested are typical of our fabrics at the time of manufacture and are not intended for specification purposes. Typar® and Tekton® are registered trademarks of Polymer Group, Inc.

Reproduced with permission of Polymer Group, Inc. Old Hickory, Tennessee, USA.

**Table 3.4** Kimberly Clark Intrepid™ Filter Media

| Grade → / Properties ↓ | 353 H | 355 H | 411 H | 612 L |
|---|---|---|---|---|
| Color | White | White | Gold/White | White |
| Basis weight, $oz/yd^2$ | 1.65 | 2.2 | 3.0 | 3.25 |
| Grammage, $g/m^2$ | 56 | 75 | 102 | 110 |
| Efficiency[a], % | 28 | 39 | 43 | 56 |
| Permeability, $cfm/ft^2$ @ 0.5" $H_2O$ $\Delta P$ | 650 | 495 | 445 | 225 |
| $m^3/s/m^2$ @ 12.7 mm $H_2O$ $\Delta P$ | 3.3 | 2.5 | 2.3 | 1.1 |
| Caliper, in | 0.054 | 0.065 | 0.140 | 0.039 |
| Mm | 1.37 | 1.65 | 3.56 | 0.99 |

[a]Efficiency determined on TSI Model 8130 automated filter tester with 2% NaCl particles at 85 L/min.


**Table 3.5** Midweat Filtration Unilayer™ Multi-ply Grades

| Style | Weight ($oz/yd^2$) | Thickness (mils) | Air permeability ($cfm/ft^2$) | Mullen burst (psi) | Grab tensile (MD/CD, lbf) |
|---|---|---|---|---|---|
| Unilayer 135 (3 ply) | 1.35 | 9 | 379 | 34 | 32/15 |
| Unilayer 180 (4 ply) | 1.80 | 14 | 293 | 44 | 37/27 |
| Unilayer 225 (5 ply) | 2.25 | 18 | 181 | 66 | 48/37 |
| Unilayer 270 (6 ply) | 2.70 | 19 | 169 | 80 | 67/50 |

These are typical values and not intended as specifications.
Unilayer is a unique multi-ply spunbond polypropylene media with these distinguishing characteristics:

- Higher degree of uniformity over single ply
- Provides excellent depth filtration and improved life
- Excellent tramp oil removal, hence, better fungus, algae and bacteria control
- Better efficiency due to a more tortuous path
- Cost effective



Cerex Advanced Fabrics manufactures and markets nylon spunbonded media for several purposes including filter media. Two such products are Cerex® and PBN-II®. Cerex® is autogenously bonded by exposing the web to a chemically activating gas phase that is later removed from the web. It is designed to be a flat, smooth, strong, and crisp fabric. PBN-II® is thermally bonded with a cross-hatch bonding pattern. In contrast to Cerex®, the PBN-II® fabric is drapeable, conformable, and textile-like. Tables 3.6 and 3.7 are tabulations of the Cerex® and PBN-II® products, respectively.

Although not included in the family of spunbonded materials, spun glass is a continuous filament glass fiber spun into a nonwoven structure that is used for residential air filtration.

Table 3.6 Typical Physical Properties of Cerex® Fabrics

**English units**

| Fabric Weight | Thickness | Grab strength | | | | Trap tear | | Burst Strength | Air permeability (Tex test) |
|---|---|---|---|---|---|---|---|---|---|
| | | MD | | CD | | MD | CD | | |
| (osy) | (mils) | (lbs) | (% elongation) | (lbs) | (% elongation) | (lbs) | (lbs) | (psi) | (cfm/ft$^2$) |
| 0.30 | 3.0 | 8.0 | 41 | 4.2 | 63 | 4.4 | 2.3 | 10 | 1974 |
| 0.40 | 3.4 | 10.9 | 44 | 5.9 | 61 | 5.8 | 3.2 | 13 | 1517 |
| 0.45 | 3.6 | 13.2 | 44 | 7.1 | 59 | 6.8 | 3.8 | 15 | 1360 |
| 0.50 | 3.7 | 15.5 | 44 | 8.7 | 59 | 7.1 | 4.4 | 16 | 1179 |
| 0.60 | 3.9 | 19.8 | 47 | 11.5 | 57 | 9.1 | 5.6 | 20 | 981 |
| 0.70 | 4.5 | 23.1 | 49 | 14.1 | 57 | 9.4 | 6.1 | 24 | 785 |
| 0.85 | 5.0 | 30.1 | 48 | 18.6 | 58 | 10.0 | 7.0 | 29 | 625 |
| 1.00 | 5.4 | 36.8 | 50 | 23.5 | 58 | 12.3 | 9.1 | 35 | 511 |
| 1.50 | 6.9 | 58.4 | 54 | 38.2 | 61 | 14.4 | 10.0 | 51 | 277 |
| 2.00 | 8.4 | 73.6 | 60 | 47.8 | 64 | 15.4 | 9.9 | 67 | 170 |
| ASTM D3776 | ASTM D1777 | ASTM D5034 | | | | ASTM D5587 | | ASTM D3786 | ASTM D737 |

*(Continued)*

Table 3.6 (Continued)

**Metric units**

| Fabric Weight | Thickness | Grab strength | | | | Trap tear | | Burst Strength | Air permeability (Tex test) |
|---|---|---|---|---|---|---|---|---|---|
| | | MD | | CD | | MD | CD | | |
| (gsm) | (mm) | (N) | (% elongation) | (N) | (% elongation) | (N) | (N) | (kPa) | ($L/m^2/s$) |
| 10 | 0.08 | 36 | 41 | 19 | 63 | 20 | 10 | 69 | 10030 |
| 14 | 0.09 | 48 | 44 | 26 | 61 | 26 | 14 | 89 | 7706 |
| 15 | 0.09 | 59 | 44 | 32 | 59 | 30 | 17 | 103 | 6911 |
| 17 | 0.09 | 69 | 44 | 39 | 59 | 32 | 20 | 112 | 5992 |
| 20 | 0.10 | 88 | 47 | 51 | 57 | 40 | 25 | 137 | 4985 |
| 24 | 0.11 | 103 | 49 | 63 | 57 | 42 | 27 | 163 | 3989 |
| 29 | 0.13 | 134 | 48 | 83 | 58 | 44 | 31 | 196 | 3176 |
| 34 | 0.14 | 164 | 50 | 105 | 58 | 55 | 40 | 238 | 2596 |
| 51 | 0.18 | 260 | 54 | 170 | 61 | 64 | 44 | 350 | 1409 |
| 68 | 0.21 | 327 | 60 | 213 | 64 | 69 | 44 | 463 | 863 |
| ASTM D3776 | ASTM D1777 | ASTM D5034 | | | | ASTM D5587 | | ASTM D3786 | ASTM D737 |

Cerex® products are nonwoven fabrics made by spinning and autogenously bonding continuous filaments of nylon (PA66) into a flat, strong, and crisp fabric. Autogenous bonding is achieved by exposing the nylon web to a chemically activating gas phase that is later removed from the web. The fabric is available in weights ranging from 0.30 osy (10 $g/m^2$) to 2.0 osy (68 $g/m^2$) in widths up to 120 in (3 m).

Cerex Advanced Fabrics makes no warranties as to the merchantability of fitness for a particular purpose of any products referred to, no guarantee of satisfactory results from reliance upon information on recommendations contained within, and declaims all liability for any resulting loss or damage. Nothing contained herein to be construed as a recommendation to use any product in conflict with any patent.

Reproduced by permission of Cerex Advanced Fabrics, Pensacola, Florida USA.

**Table 3.7** Typical Properties of PBN II® Fabrics

**English units**

| Fabric Weight | Thickness | Grab strength | | | | Trap tear | | Burst Strength | Air permeability (Tex test) |
|---|---|---|---|---|---|---|---|---|---|
| | | MD | | CD | | MD | CD | | |
| (osy) | (mils) | (lbs) | (% elongation) | (lbs) | (% elongation) | (lbs) | (lbs) | (psi) | (cfm/ft$^2$) |
| 0.30 | 4.1 | 6.2 | 56 | 3.6 | 63 | 2.5 | 1.6 | 8 | 2364 |
| 0.45 | 5.5 | 10.0 | 60 | 6.3 | 63 | 3.8 | 2.3 | 11 | 1625 |
| 0.50 | 5.9 | 12.0 | 60 | 7.6 | 63 | 4.4 | 2.8 | 13 | 1516 |
| 0.60 | 6.8 | 14.5 | 63 | 9.9 | 65 | 5.4 | 3.2 | 14 | 1228 |
| 0.70 | 7.5 | 17.3 | 65 | 12.2 | 68 | 6.3 | 3.8 | 17 | 1068 |
| 0.85 | 8.3 | 23.8 | 68 | 16.7 | 71 | 8.1 | 5.0 | 21 | 907 |
| 1.00 | 9.3 | 29.9 | 69 | 20.9 | 75 | 10.5 | 6.5 | 24 | 774 |
| 1.50 | 12.6 | 49.1 | 76 | 35.8 | 85 | 17.2 | 10.9 | 38 | 500 |
| 2.00 | 15.2 | 71.8 | 80 | 52.2 | 89 | 23.6 | 15.6 | 51 | 353 |
| 2.30 | 16.5 | 83.4 | 80 | 64.2 | 92 | 24 | 14.9 | 61 | 263 |
| 3.00 | 19.1 | 115.6 | 84 | 86.0 | 94 | 35.8 | 24.4 | 79 | 197 |
| 4.00 | 22.3 | 156.9 | 91 | 118.8 | 100 | 49.0 | 34.2 | 109 | 125 |
| ASTM D3776 | ASTM D1777 | ASTM D5034 | | | | ASTM D5587 | | ASTM D3786 | ASTM D737 |

*(Continued)*

Table 3.7 (Continued)

**Metric units**

| Fabric Weight | Thickness | Grab strength | | | | Trap tear | | Burst Strength | Air permeability (Tex test) |
|---|---|---|---|---|---|---|---|---|---|
| | | MD | | CD | | MD | CD | | |
| (gsm) | (mm) | (N) | (% elongation) | (N) | (% elongation) | (N) | (N) | (kPa) | (L/m²/s) |
| 10 | 0.10 | 28 | 56 | 16 | 63 | 11 | 7 | 56 | 12010 |
| 15 | 0.14 | 44 | 60 | 28 | 63 | 17 | 10 | 75 | 8256 |
| 17 | 0.15 | 53 | 60 | 34 | 63 | 20 | 12 | 91 | 7703 |
| 20 | 0.17 | 64 | 63 | 44 | 65 | 24 | 14 | 99 | 6240 |
| 24 | 0.19 | 77 | 65 | 54 | 68 | 28 | 17 | 119 | 5427 |
| 29 | 0.21 | 106 | 68 | 74 | 71 | 36 | 22 | 147 | 4610 |
| 34 | 0.24 | 133 | 69 | 93 | 75 | 47 | 29 | 165 | 3930 |
| 51 | 0.32 | 218 | 76 | 159 | 85 | 77 | 48 | 259 | 2542 |
| 68 | 0.39 | 319 | 80 | 232 | 89 | 105 | 69 | 352 | 1791 |
| 78 | 0.42 | 371 | 80 | 286 | 92 | 107 | 66 | 420 | 1336 |
| 102 | 0.49 | 514 | 84 | 383 | 94 | 159 | 109 | 547 | 998 |
| 136 | 0.57 | 698 | 91 | 528 | 100 | 218 | 152 | 754 | 634 |
| ASTM D3776 | ASTM D1777 | ASTM D5034 | | | | ASTM D5587 | | ASTM D3786 | ASTM D737 |

PBN-II® products are nonwoven fabrics made by spinning and thermally bonding continuous filaments of nylon (PA66) into a drapeable, conformable, textile-like fabric, PBN- II® fabrics have a cross-hatch bond pattern. Available weights ranging from 0.30 osy (10 g/m²) to 4.0 osy (136 g/m²) in widths up to 120 in (3 m). Cerex Advanced Fabrics makes no warranties as to the merchantability of fitness for a particular purpose of any products referred to, no guarantee of satisfactory results from reliance upon information on recommendations contained within, and declaims all liability for any resulting loss or damage. Nothing contained herein to be construed as a recommendation to use any product in conflict with any patent.
Reproduced by permission of Cerex Advanced Fabrics, Pensacola, Florida, USA.

### 3.3.2 Melt-blown webs

Malkan and Wadsworth (25) identifies the following as being characteristics of melt-blown webs:

**a.** Random fiber orientation.
**b.** Low to moderate web strength.
**c.** Generally, the web is highly opaque (high cover factor).
**d.** Melt-blown webs derive their strength from mechanical entanglement and frictional forces.
**e.** Most melt-blown webs are layered or shingled structures, the number of layers increase with increasing basis weight.
**f.** Grammages (basis weights) range between 8 and 350 g/m$^2$, typically 20–200 g/m$^2$.
**g.** Fiber diameters range between 0.5 and 30 μm, but the typical range is 2–7 μm.
**h.** Microfibers provide high surface area for good insulator and filter characteristics.
**i.** The fibers have smooth texture and appear to be circular in cross section.
**j.** The fibers vary in diameter along a single fiber, as shown in Figure 3.2.
**k.** Fibers are continuous in length.
**l.** Fibers show thermal branching as indicted in Figure 3.3. It is not certain as to exactly what causes thermal branching, however, it is caused by complexities in the extruded fiber–air stream as it approaches the collector plate. See Chapter 5, Section 5.4.2.

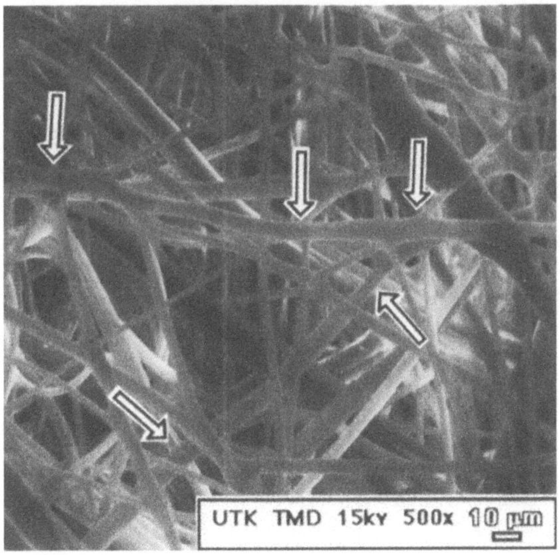


**Figure 3.2** Variation in Fiber Diameter Along a Single Fiber. *Reproduced by permission of TAPPI, Atlanta, Georgia, USA.*

**Figure 3.3** Thermal Branching of Fibers in Melt-blown Webs. *Reproduced by permission of TAPPI, Atlanta, Georgia, USA.*

**Table 3.8** LYPORE®[a] MB Calendered and Single Layer Grades Typical Properties

| Grade | Calendered | Basis weight | Caliper | | Resistance | Air permeability | Mean flow pore |
|---|---|---|---|---|---|---|---|
| Units | Y/N | $g/m^2$ | mils | µm | mm @ 5.3 cm/s | cfm @ 0.5 in $H_2O$ | µm |
| CA9401 | Y | 40 | 6 | 152 | 25 | 8 | 7 |
| CA9603 | Y | 70 | 14 | 356 | 10 | 23 | 13.5 |
| MB9028 | N (single layer) | 52 | – | – | 2.3 | 70 | – |
| MB9031 | N (single layer) | 70 | 27 | | – | 35 | 15 |
| Test Method | – | TAPPI T410 ASTM D-646 | TAPPI T-411 | | MIL STD 282 ASTM D9826-9 | TAPPI T-251 | – |

[a]LYPORE® is a trademark of Lydall, Inc. registered in the United States and other countries.
Reproduced by permission of Lydall, Inc. Manchester, Connecticut, USA.

In most filter applications, melt-blown webs are used as part of a composite structure in combination with spunbond or wet laid webs—but not always. Lydall Filter/Separation Group markets a line of 100% melt-blown polypropylene liquid filter medium under their trade name of LYPORE® MB. The medium is available in calendered and uncalendered versions. Table 3.8 is typical property tabulations of the calendered and uncalendered versions, respectively.

## 3.4 CHARACTERISTICS AND PROPERTIES OF FLASH-SPUN WEBS

According to Batra and Pourdeyhimi (16), most flash-spun fabrics are composed of high density polyethylene. They are permeable to moisture, but not to air, thus making them useful to such applications as house wrap. The fabric can be made to be exceptionally strong and resistant to tearing. They are useful for stationary products such as envelopes and can be designed to readily be printed on. Softer versions of flash-spun products can be achieved by hydroentanglement. A version of Tyvek®, known as Tyvek® SoloFlo® is used for filtration purposes. It is produced with finer fibers and greater distance between fibers (more void space) than in regular Tyvek®.

## 3.5 NANOFIBER-SPUN WEBS

Rankumar and Singh (114) advised that the market for nanotechnology worldwide was estimated to grow at an annual rate of 19% during the years (2011–2013). This was based on a report by RNCOS E-Services Pvt. Ltd, India, an independent market research firm. The report estimated that the value of nanotechnology-based manufactured goods will be $1.6 trillion globally. The US is the leading market in the nanotechnology sector and has an estimated global share of 35%.

Nanofiber-spun webs for nonwoven and filtration purposes are part of the growth. The melt-blown webs discussed in Section 3.3.2 approach the capability of nanofiber webs i.e., fiber diameters of less than 500 μm. Davis and Gramley (115) have suggested that the focus of this technology has shifted "toward finer fiber for advanced air filtration performance." They also report that "research conducted in Germany and the United States has been focused almost solely on polypropylene (PP) and has shown gains in generating fibers in the submicron range."

In addition to melt-blowing, two other prominent web spinning processes for nanofiber webs are electrospinning and centrifugal spinning. The products are characterized as follows.

### 3.5.1 Characteristics of electrospun webs

There are two types of processes where fiber webs of fine diameter are formed from a voltage driven fiber-forming process:

1. Flashspun fibers generated from a voltage charged nozzle or a charged free surface as described in Section 3.5.1.1.
2. Webs formed from electroblown fibers driven through a spinnerette by a combination of pressure drop and/or voltage potential as described in Section 3.5.1.2.

#### *3.5.1.1 Electrospinning*

The Electrospun process for producing nanofiber media has been briefly discussed in Section 1.7.1.2.3.1. Most electrospun fiber webs are spun from solution polymers,

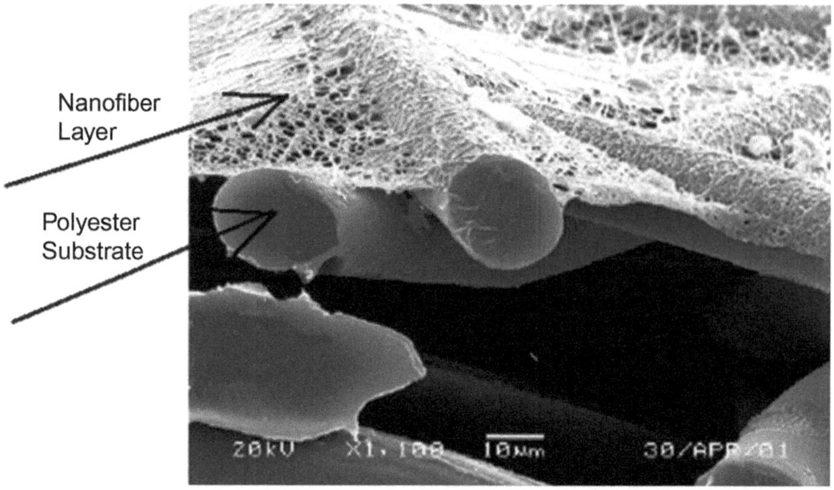


**Figure 3.4** Electrospun Fibers on a Polyester Substrate. *Reproduced with permission from Donaldson Company, Inc. © Donaldson Company, Inc. All rights reserved.*

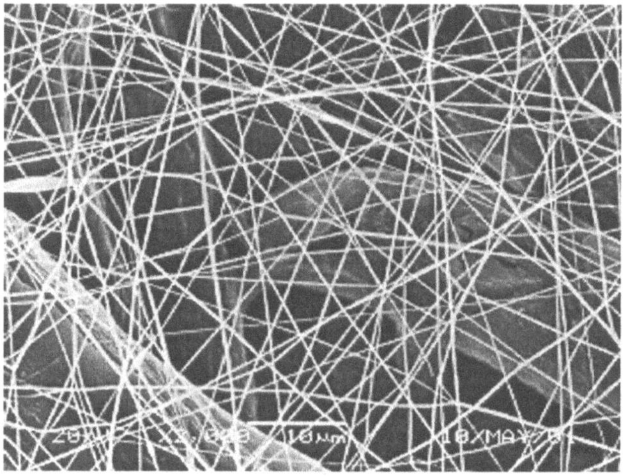

**Figure 3.5** Electrospun Fibers on a Cellulose Substrate. *Reproduced with permission from Donaldson Company, Inc. © Donaldson Company, Inc. All rights reserved.*

however, some have been spun from melt-spun polymers. Figure 1.18 in Chapter 1 above is a simplified diagram of the laboratory procedure. A more detailed discussion of the process is provided in Section 5.3.1 below. The product is a nanofiber web. Usually the web is applied as a coating or laminate to a spunbond or cellulose substrate. Figure 3.4 shows the nanofiber coating as applied to a polyester substrate. Figure 3.5 is a microphotograph of the coating on a cellulose substrate. The electrospun medium achieves high-efficiency filtration performance because of its fine fibered structure. The medium attributes allow for high surface area, small fiber diameter, layer thinness, and low basis weight. The small fiber diameters are significant because they are smaller than available melt-blown fibers. Fiber diameters as low as 0.25 μm (250 nm) have been used in commercial and military filtration operations for more than twenty years (101). The web itself is very thin. A thickness of four 250 nm nanofiber diameters approaches only 1 μm.

The thin electrospun webs have limited mechanical properties, that is why they are used in combination with other substrates. Ultra-Web®, a nanofiber medium combined with a wet laid substrate, is produced by Donaldson Company, Inc. The microphotograph of Figure 3.5 is an example of the Ultra-Web® substrate. It has become available for a broad range of filtration applications (Graham et al. (29)).

Figure 3.6 illustrates the unique air filtration characteristics of nanofiber webs. What is shown is the nanofiber medium on a polyester substrate after a filtration test with nanoparticles of salt crystals. Note that most of the particles have collected on the nanofiber structure and relatively little on the polyester substrate fiber in the upper left hand corner. There is actual dendrite formation because particles are collecting on particles.

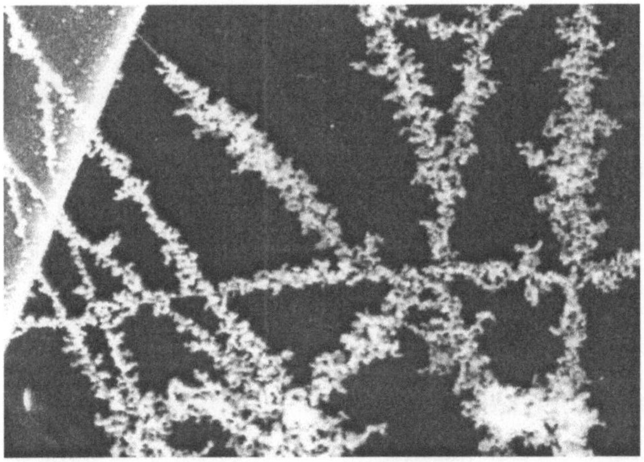

**Figure 3.6** Sub-micro Salt Crystals Preferentially Collect on Nanofibers (Compared to Large Spunbond Fiber in Upper Left) Due to Smaller Fiber Diameter. *Reproduced with permission from Donaldson Company, Inc. © Donaldson Company, Inc. All rights reserved.*

The phenomena are a result of interception and diffusion of the nanoparticles as they try to flow past the nanofiber structure. Graham et al. (29) relate this effect to slip flow and the Knudsen number, previously discussed in Chapter 2 above.

$$Kn = \frac{\lambda}{r_f} \tag{3.1}$$

$Kn$ is the Knudsen number.
$\lambda$ is the gas mean free path.
$r_f$ is the radius of the fiber.

Graham et al. argue that at high $Kn > 0.1$, slip flow will prevail at the fiber surfaces which allows for more air traveling near the fiber surfaces. This translates into more particles traveling near the surface thus increasing their probability for capture.

#### *3.5.1.2 Electroblowing*

Electroblowing has already been discussed in Chapter 1, Section 1.7.1.2.3.1. The process is described in Chapter 5, Section 5.3.4. According to DuPont, (116) electroblown webs can provide thinner webs with lower ionic resistance and high temperature stability, which are not achievable in conventional nonwovens and microporous films. Electroblown polyimide separators have high temperature melt integrity greater than 250°C, low shrinkage, good chemical resistance, etc. DuPont has indicated that polyimide electroblown webs have been shown to have zero shrinkage at 150°C even after exposure for three hours. In Patent US7582247B2 (116) DuPont claims that the fiber spinning process produces fine polymer fibers having average effective diameters of less than about 0.5 μm (less than 500 nm).

### 3.5.2 Centrifugal spinning

In contrast to electrospinning, which uses voltage potential to drive the fiber formation, centrifugal spinning utilizes centrifugal force to drive molten polymer through centrifugal rotating spinerrettes. There are two known processes that produce nanofiber webs in this manner: the Forcespinning®[1] process commercialized by FibeRio® and what Dupont describes as their "Melt Spun Nanofiber Process" (28). The Forcespinning® process has been briefly described already in Chapter 1, Section 1.7.1.2.3.2. It will be described in more detail in Section 5.4.1.

### 3.5.3 Other nanofiber web structures

Hollingsworth and Vose Co. markets two web forms that have nanofiber filtration properties. These are Nanoweb® and NanoWave®.

Nanoweb® is a nanofiber coating technology putting a nanofiber web a substrate. Wertz and Schneiders (117) describes it "as a solvent-free nanofiber coating technology." The coating is formed from fibers ranging from 0.3 to 0.5 μm in size, but can be increased up to 1 μm. The nanofiber layer thickness ranges from 15 to 30 μm. The coating can be applied to a variety of nonwoven substrates such as glass, cellulose, and synthetic fibers. Wertz and Schneiders also suggest that if desired a second nanofiber layer can be applied. Potential applications of the Nanoweb® technology include gas turbine dust collectors, heavy duty air filtration for engine air intake filters, HVAC filtration, fuel filtration, lube oil filtration, and general liquid filtration.

[1] Forcespinning™ is a trademark of FibeRio® Technology Corporation, Edinburg, Texas USA.

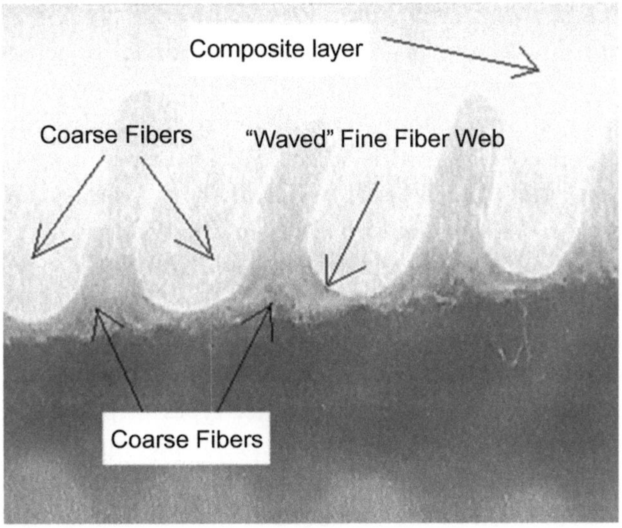


**Figure 3.7** H&V NanoWave® Medium Configuration. *Reproduced by permission of Hollingsworth and Vose Company, Inc., East Walpole, Massachusetts, USA.*

NanoWave® is also a composite structure containing a nanofiber layer that has a wave or sine wave type of configuration. Figure 3.7 shows this configuration. The nanofiber wave layer is embedded or surrounded by a coarse fiber layer. The combination is backed by a support substrate that provides strength and rigidity to the composite structure. The design is for the coarse fiber layer to filter large particles and provide dirt-holding capacity while the nanofiber layer serves as an efficiency layer to filter out fine particles. The wave configuration of the nanofiber layer greatly increases surface area and capacity for the finer particles. Hollingsworth and Vose Co. (H&V) suggests this design provides design efficiency at increased dust holding capacity.

The product is the winner of the prestigious IDEA 13 Achievement Award. The medium is being used in pocket air filters in HVAC systems. It is also being used in aircraft spray booth facilities. It has also potential for vacuum cleaner bag, exhaust filters, respirator filter medium, and auto cabin air filter medium.

## 3.6 CHARACTERISTICS AND PROPERTIES OF WET LAID WEBS

Wet laid filter media have a broad variety of properties and characteristics, depending on application, design, raw materials, and manufacturing process. The two fundamental types of media are base sheet filter medium and resin-treated filter medium.

### 3.6.1 Base sheet

The author defines base sheet filter media as filter paper that is not resin treated or not yet resin treated downstream of the forming process. The term "raw stock" is also applied to this form of wet laid medium. The definition includes paper that may later be resin treated either in-line or off-line. This discussion is concerned with base sheet that is sold as filter media without any post-formation resin treatment. Most base sheets are used in liquid filtration applications and are further discussed in Chapter 7, Section 7.2.5. Examples are the many forms of laboratory filter papers, filter sheets, industrial filter paper for filter presses, coffee filter, and papers for the filtration of edible oils. These papers are produced from cellulose wood and vegetable fiber pulps, blends of synthetic and pulp fibers, and from glass fibers. Base papers may also contain diatomaceous earth, activated carbon, perlite clay, and other powder fillers. Several types of media used in industrial filter applications, such as filter presses, are creped to increase surface area for filtration. Filter sheets are designed for depth filtration and are produced at grammages much higher than other types of paper filter media. Special wet lay machines are used to produce this type of medium. Filter media used for one-shot disposable applications, such as laboratory filter papers, coffee filter, and biological and medical filtration are produced at the low end of the grammage scale. Laboratory filter papers used for analytical purposes have to be composed of extremely pure materials so as not to contaminate the analysis. For example, ash content (noncombustible) is specified for many grades of laboratory filter papers.

Although the author defines base paper as not being resin treated downstream of the forming process, this does not mean to say that the paper does not contain a resin. Most base sheet filter papers contain a resin added by "beater addition" (often referred to as beater ad) i.e., the resin is mixed with the fiber slurry prior to the forming process. Usually the resin is cationic to better adhere to the anionic nature of the fiber surface. Other additives and fillers are added this way to provide special properties such as biological resistance and adsorption. Beater addition resins are intended to provide dry and/or wet strength to the medium. The wet strength is important for the many applications designed for filtration of aqueous fluids.

### 3.6.2 Resin treated filter media

Resin treated filter media are mostly designed for use in pleated filter cartridges and panels. The fiber composition is chosen to provide for filtration properties such as bulk, permeability, and pore size, whereas the resins are chosen to provide processabilty, strength, endurance, and stability. It is important that the paper be stiff and yet flexible enough to go through the pleating process and not degrade in the thermal aspects of this process. In the end use application, the medium has to have the strength and rigidity to maintain its structural integrity against the pressure of the fluid flowing through it. Pleat collapse is to be avoided. It also has to be resistant to the thermochemical environment to which it is exposed. In short the best filter medium is worthless if it cannot be made into a lasting filter.

Resin-treated filter media are categorized by the type of resin: water-based or solvent-based resins (solvent is defined as a non-aqueous fluid such as methanol). The medium is also categorized by whether the resin is a thermoplastic resin or a thermosetting resin. A thermoplastic resin softens when the medium is heated for the pleating operation. After the medium is pleated, the resin is cooled and the pleats are set. A medium treated with a thermosetting resin is usually B-Staged (uncured) when presented to the pleating operation. It is flexible and pliable for pleating. After the pleat folds are formed the medium is subject to an oven-type of heat treatment to cure the resin, stiffen the media, and fix the pleats.

Water-based resins offer a wider choice of polymers than do solvent-based resins. The most common types of polymers are polyvinyl acetate, polyacrylates, styrene acrylates, ethylene vinyl chloride, and water-based phenolics. Most often, they come in the form of a latex emulsion something like latex paint. Most latex resins used by the nonwovens industry are designed to be soft and provide hand to the fabric. In contrast resins used for filter media are intended to provide stiffness and high modulus. $T_g$s (glass transition temperatures) of latex resins used for filter paper are usually in the range of 30–50 °C.

Some polymers are available or prepared as solution polymers. This includes starch, polyvinyl alcohol, and some forms of phenolics. Water-based resins can range from thermoplastic to thermosetting depending on how they are formulated. Catalysts, cross linking agents, and curing agents are often employed to provide some degree of cure or thermoset. Other ingredients can be added to provide for properties such as flame retardancy, water repellency, dimensional stability, formaldehyde control etc.

Almost all solvent-based resins are thermosetting and are based on phenol formaldehyde polymer (phenolic). There are some applications where epoxies are used. At one time solvent-based polyvinyl acetate was used for automotive air intake filter medium. The largest use of phenolic filter medium is for automotive engine oil and fuel filtration. For over fifty years, filter medium made from solvent-based phenol formaldehyde have unique properties that, to date, have not been duplicated by any water-based system. Unlike water, alcohol solvent does not cause fiber swelling and related pinching of pore structure. Better filtration results are obtainable with phenolic media than comparable water-based media. Uncured phenolic media can be B-staged to be pliable for the pleating process and cured to a very rigid structure. The cured media has good moisture resistance, excellent heat and temperature resistance up to over 150 °C, and good resistance to many chemical environments. The biggest concern related to phenolic filter medium is the environmental emissions of volatile organic compounds, which include the solvent, formaldehyde, and phenol. People who produce and process these media have to comply with severe government restrictions on emissions.

An important feature of resin-treated filter medium is that they are often corrugated. The corrugations in filter media are different than the corrugations in corrugated board for containers. The board corrugations are in the cross direction of the board whereas the corrugations in filter media are machine direction oriented. The corrugations are actually embossed into the medium as it passes through a set of rollers that are grooved to have a set of meshing sine wave configurations on their surfaces. The corrugation process is discussed in more detail in Chapter 5, Section 5.6.3.1. Figure 3.8 depicts resin-treated filter paper that is corrugated. The corrugation depth of corrugated filter media are in the range of 0.2–0.5 mm (8–20 mils). Figure 3.9 is a picture of corrugated media from a panel filter pack used for engine air intake filtration. Note that the pleats are tightly packed. The corrugations serve as pleat spacers and provide flow channels to the inside

**Figure 3.8** Corrugated Media.

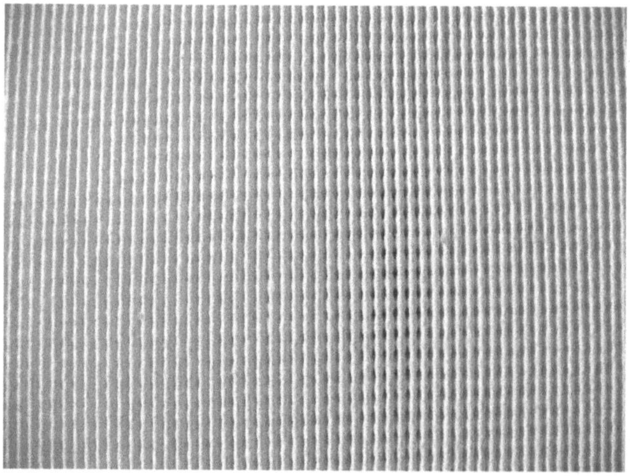

**Figure 3.9** Corrugated Media From an Auto Air Intake Panel Pleat Pack. The Corrugations Serve as Pleat Separators and Provide Flow Channels to the Interior of the Pleat Folds. *Reproduced with permission of Hollingsworth and Vose Co., Inc. East Walpole, Massachusetts, USA.*

pleat folds. The flow channels allow the full area of the media to be utilized in the filtration application. Accordingly, corrugation adds several features to the media:

1. It increases the surface area for filtration.
2. The corrugations act as pleat separators for the pleat folds.
3. In filters, such as panel filters where the pleat folds are close together, the corrugations come together to provide flow channels to the interior of the pleat pack. This is illustrated in Figure 3.9.
4. The corrugations provide additional stiffness for the media in the machine direction, the direction in which the pleat pack is most vulnerable to pressure deformation.

Some pleating lines like to purchase flat (non-corrugated) media and emboss it with carefully designed dimples such as shown in Figure 3.10. The embossments serve as pleat spacers in the pleat pack.

**Figure 3.10** Embossed Filter Media. *Reproduced with permission of Hollingsworth and Vose Co. Inc. East Walpole, Massachusetts, USA.*

### 3.6.3 Glass microfiber filter media

Almost all HEPA and ULPA filter media containing glass microfibers and many ASHRAE (American Society of Heating, Refrigerating, and Air-Conditioning Engineers) grades are produced by the wet lay process. In North America two major producers of glass filtration grades are Lydall Filtration/Separation Group (Lydall) in Rochester, New Hampshire and Hollingsworth and Vose Co., Inc. (H&V) in East Walpole, Massachusetts.

The following tables (Tables 3.9 and 3.10) list some of the glass microfiber filter products of Lydall designated by their LYDAIR®[2] trade name. Table 3.9 lists grades

[2] LYDAIR® is a trademark of Lydall, Inc. registered in the U.S. and other countries.

used in ASHRAE and prefilter applications that will be further discussed in Chapter 8. Table 3.10 lists standard and high-alpha glass microfiber grades used for high-efficiency HEPA applications. Note that properties identified for microglass filter media include water repellency which, in this case, is the height of water that the media will support before penetration occurs and loss on ignition (LOI), which is an indicator of organic binder content (that which is burned off). Dioctyl phthalate percentage is the filter medium penetration by DOP smoke. Resistance is the pressure drop across the medium in mm $H_2O$ at the specified flow rate (1.1 cfm or 32 l/min)[3].

**Table 3.9** Typical Values of Lydair® MG Grades—ASHRAE and Prefilter

| Media type | Grade | Basis weight | Caliper | Tensile strength MD | Tensile strength CD | Stiffness MD | LOI | DOP | Resistance | Water rep. |
|---|---|---|---|---|---|---|---|---|---|---|
| | | Lb/3000 ft² | mil @ 8 psi | g/in | g/in | mg | % | % | mm | In $H_2O$ |
| **High Yield** | 1909 | 40 | 13 | 5000 | 1600 | 1000 | 11.5 | 84 | 1 | 10 |
| **Single Layer** | 1894 | 43 | 13 | 6000 | 1700 | 1000 | 11.5 | 50 | 3.5 | 16 |
| **ASHRAE** | 1923 | 43 | 13 | 6000 | 1600 | 1000 | 11.5 | 37 | 4.8 | 16 |
| **Next** | 1247 | 48 | 17 | 6500 | 2500 | 1800 | 9.5 | 85 | 0.9 | 6 |
| **Generation** | 1296 | 48 | 16 | 6000 | 2000 | 1400 | 9.5 | 48 | 3.1 | 12 |
| **Dual Layer** | 1306 | 48 | 15 | 6000 | 2000 | 1300 | 9.3 | 30 | 5.5 | 12 |
| **Prefilter** | 2400 | 46 | 14 | 5000 | 1600 | 1000 | 6 | 13 | 8.5 | 12 |
| | 2220 | 46 | 14 | 6000 | 1800 | 1000 | 6 | 7 | 11 | 18 |
| **High Yield** | 2991 | 47 | 14 | 6000 | 1700 | 900 | 11 | 6 | 11 | 17 |
| **Prefilter** | 2956 | 48 | 15 | 5000 | 1500 | 900 | 11 | 3 | 14.5 | 17 |

Reproduced with permission of Lydall Inc., Manchester, Connecticut, USA.

**Table 3.10** Typical Values of Lydair® HEPA Grades

| Grade | Basis weight | Caliper @8 psi | Tensile MD | Tensile CD | Stiffness, MD | LOI | DOP | Resistance | Water rep |
|---|---|---|---|---|---|---|---|---|---|
| | lb/3000 ft² | mil | gm/in | gm/in | mg | % | % | mm | In |
| ***Standard HEPA*** | | | | | | | | | |
| 3398 | 55 | 17 | 3800 | 1800 | 1000 | 6 | 0.014 | 35.5 | 30 |
| ***High Alpha HEPA*** | | | | | | | | | |
| 4350 | 48 | 17 | 3000 | 1400 | 1000 | 6 | 0.035 | 27 | 25 |
| 4450 | 48 | 17 | 3000 | 1300 | 1000 | 6 | 0.022 | 28.5 | 25 |
| 4450-HS | 48 | 15 | 4000 | 1400 | 1300 | 6 | 0.018 | 29 | 25 |
| 4470 | 48 | 17 | 3000 | 1300 | 1000 | 6 | 0.010 | 32 | 25 |
| 4470-HS | 48 | 15 | 4000 | 1400 | 1300 | 6 | 0.008 | 33 | 25 |
| 4708 | 42 | 13 | 4000 | 1100 | 1100 | 6 | 0.021 | 30 | 27 |

Reproduced by permission of Lydall Inc., Manchester, Connecticut, USA.

[3] The DOP test and other related tests to measure media filtration efficiency are discussed in Chapter 6, Section 6.14.1.15.

## 3.7 ELECTRET FILTER MEDIA

Electrets are dielectric materials that exhibit an external electric field in the absence of an applied field. When used in air filtration applications, they can greatly increase initial filtration efficiency and reduce pressure drop, because of its electrostatic attraction for air particles. Films, fibers, and nonwoven web structures are among the materials that can be formed into electrets. There are two distinct categories of electrets: space charge electrets and dipolar electrets. Space charge electrets are formed by the deposit or injection of an electric charge directly into the dielectric material. Dipolar electrets are formed or polarized by the application of an electric field. For polarization to occur polymers must be heated sufficiently to a point where dipoles are mobile and then cooled to fix the new orientation of the dipoles. Myers and Arnold (118) report that dipoles can also be formed by charge injection to cause dipole reorientation.

### 3.7.1 Fibrillated electret film

Fibrillated electret film is a unique medium because the fiber source is made differently than the fibers for the other types of filter media discussed in this book. The patent by van Turnhout (119), US Patent 3998916, provides a fundamental description of the process. It describes "A method for the manufacture of an electrically charged fibrous media from a highly molecular nonpolar fiber material wherein a web of the fiber material is continuously fed and stretched. At least one side of the stretched web is homopolarly charged by a plurality of corona charging elements. The charged web material is then fibrillated, collected, and processed into a filter".

The process starts with a film that is stretched in two successive stages, under heat, in the machine direction. While hot, the film is charged by corona discharge. The film is then fibrillated by passing it over a needle roller moving at a different speed than the film. The fibrillation is in the longitudinal direction resulting in a nicely spread layer of ribbon-like fibers which is wound up on a collecting roller. The fiber on this collecting roll is then formed into the desired filter. Myers and Arnold (118) suggest that the collected fibers or fibrils are carded into a filter web. The patent claims that if the charge applied to the top of the film is a negative charge, positive charges are then induced at the bottom side of the medium. This two-sided charging provides a filtration advantage because it will electrostatically collect both positively charged and negatively charged particles in its filtration application. The patent also provides a way to charge both sides of the film.

Fibrillated film filter media are used in home furnace and ventilation filters. 3M Company of St. Paul, Minnesota includes fibrillated film filters under the trade name of Filtrete®[4].

[4] Fibrillated film is not the only medium marketed by 3M under the Filtrete® trade name. Other types of media for ventilation filters marketed by 3M are also covered.

### 3.7.2 Corona-treated melt-blown fibers

Arnold (120) states "An exemplary conventional method of this type is described in detail in US Patent Number 5401446 to Tsai et al. entitled 'Method and Apparatus for the Electrostatic Charging of a Web or Film'" (121). The patent by Tsai and Wadsworth is not limited to melt-blown fiber, but can be used for any sheet material that lends itself to be a web or film material. In contrast to the van Turnhout patent discussed above, this patent allows for the sheet to be "cold charged," that is, the web is charged without heat and stretching. The patent has two embodiments. The first embodiment has the moving web s-wrapping around two charged rollers. While in contact with each roller the web passes a charging bar on the side of the web away from the roll. According to the invention, the charged drums have a negative charge, whereas, the charging bars are assumed to be positively charged. The patent also states that the charging means generates electric fields of 1 to 12 kVDC/cm. The web as it travels around the first drum becomes positively charged on the side against the drum and negatively charged on the side facing the charging bar. The charging reverses itself as the web travels in its s-wrap path around the second drum and past the second charging bar. The web side against the first charging drum now faces the charging bar of the second drum and the web side facing the first charging bar wraps itself against the second charging drum. The web side that was positively charged at the first drum now takes on a negative charge at the second drum and vice versa for the other side of the web.

The second embodiment has the web s-wrapping around two sets of four charge positioning rolls as shown in Figure 3.11. Instead of charging drums there are charging shells that cover the web as it transverses the charge positioning rolls. Charging wires are located on the other side of the web, more or less in the middle of the charge positioning rolls. According to the patent, the charging shells are assumed to be positively charged and the charging wires are assumed to be negatively charged. Just as in the first embodiment of the invention, each side of the web is successively exposed to a negative charge followed by a positive charge or vice versa.

The patent indicates that the alternate switching of polarity of the charges to each side of the web results in an improvement in charge density of the web and an improvement in the lifetime of the charge on the web. The patent notes that the polymer in the web should be nonconductive. Polypropylene, polyethylene, polyester, low density polyethylene, polybutylene terephthalate, polycarbonate, polychlorotrifluoroethylene, and polycyclohexyldimethylene terephthalate all fit in this category. The patent also notes that the method is also suitable for charging composite web structures including those that contain both conductive and nonconductive fibers. Polyolefin mixtures to which a small amount of acrylic acid has been grafted retain their charges for longer than would be expected.

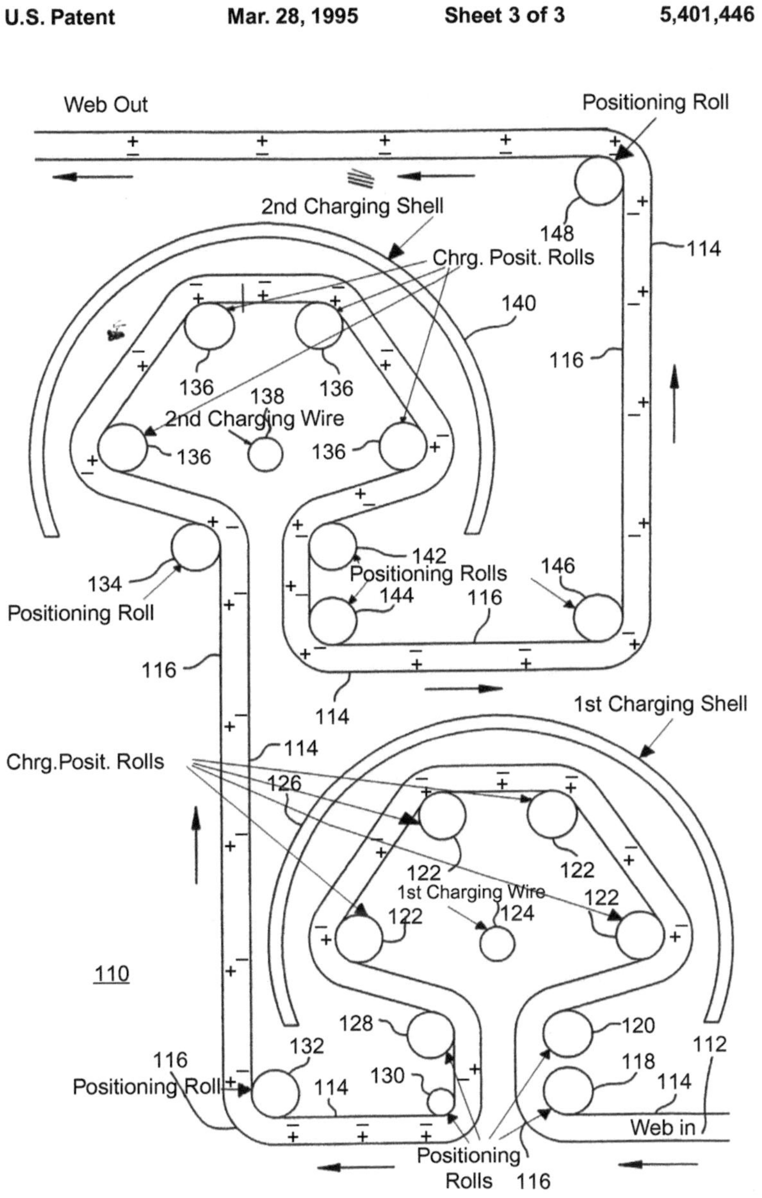


**Figure 3.11** Corona Cold Charge for Films and Webs. Note: Patent diagram has been edited to include text labels. *Reproduced with permission of Peter Tsai and Larry Wadsworth.*

**Table 3.11** Filtration Results From Apparatus of Figure 3.11

| No[a] | Filtration efficiency[b] | Charging voltage[c] | Filtration efficiency[d] | $\Delta p$ | $qF$ |
|---|---|---|---|---|---|
| 15 | 71.65 | +23, −19 (2X) | 99.956 | 6.9 | 1.120 |
| 16 | 71.65 | +21, −19 (1X) | 99.950 | 6.55 | 1.160 |
| 17 | 41.55 | +25, −19 (2X) | 96.278 | 2.3 | 1.431 |
| 18 | 41.55 | +25, −9 (1X) | 95.525 | 2.55 | 1.218 |
| 19 | 24.55 | +25, −19 (2X) | 80.35 | 0.9 | 1.808 |
| 20 | 24.55 | +25, −15 (1X) | 81.90 | 1.0 | 1.709 |

[a]15 and 16 are 3 oz/yd$^2$ PP: 17 and 18 are 1 oz/yd$^2$ PP: 19 and 20 are 1 oz/yd$^2$ recycled PET.
[b]Before charging.
[c]kV for wire, shell (number of chargings).
[d]After charging.
Reproduced with permission of Peter Tsai and Larry Wadsworth.

The patent provides data demonstrating the effectiveness of the corona discharge technique on filtration performance and on charge aging stability. Table 3.11 is an example of filtration results on several materials charged in accordance with the charging shell embodiment. The testing was done with a TSI Model 8110 tester (see Chapter 6, Section 6.9.1.14 for a discussion of air filter testers.) using NaCl aerosol. Filtration efficiency is defined as

$$\text{filtration eficiency} = (100 - P) \tag{3.2}$$

where $P$ is the percent penetration of particles through the medium.

The term $q_F$ in the last column of Table 3.11 is a measure of filtration quality. It relates the filtration penetration and the pressure drop[5].

$$q_F = [\ln(1/P)]/[\Delta p] \tag{3.3}$$

where $\Delta p$ is the pressure drop across the medium.

It should be noted that the higher the value of $q_F$, the better is the quality of the medium. It is apparent from the comparison of filtration efficiencies before and after charging that the charging caused a very significant improvement.

Table 3.12 demonstrates how aging effects the electrostatic filtration performance of several media. The aging was accelerated by exposing the webs to 137°C for 10 minutes. There appears to be very little loss in filtration efficiency and $q_F$ because of the accelerated aging.

[5] The designation $q_F$ is commonly known as the gamma number and is often designated by $\gamma$.

**Table 3.12** Accelerated Aging

| No. | Filtration efficiency[a] | Filtration efficiency[b] | $\Delta p^a$ | $\Delta p^b$ | $q_F{}^a$ | $qF^b$ |
|---|---|---|---|---|---|---|
| 2 | 98.708 | 92.390 | 2.2 | 2.1 | 1.977 | 1.227 |
| 6 | 98.460 | 97.370 | 2.45 | 2.2 | 1.703 | 61.653 |
| 7 | 99.937 | 99.866 | 6.8 | 6.1 | 1.084 | 1.084 |
| 8 | 99.317 | 99.279 | 6.83 | 6.1 | 0.730 | 0.809 |
| 9 | 98.610 | 98.588 | 6.8 | 6.2 | 0.629 | 0.687 |
| 10 | 98.308 | 97.583 | 2.4 | 2.2 | 1.700 | 1.692 |
| 11 | 98.718 | 97.178 | 2.5 | 2.2 | 1.743 | 1.622 |
| 12 | 98.063 | 96.143 | 9.7 | 9.8 | 0.407 | 0.332 |
| 15 | 99.956 | 99.925 | 6.9 | 6.3 | 1.120 | 1.142 |
| 16 | 99.950 | 99.886 | 6.55 | 6.0 | 1.160 | 1.129 |
| 17 | 96.278 | 95.858 | 2.3 | 2.2 | 1.431 | 1.447 |
| 18 | 95.525 | 94.913 | 2.55 | 2.2 | 1.218 | 1.354 |

[a]Before accelerated aging.
[b]After accelerated aging.
Reproduced with permission of Peter Tsai and Larry Wadsworth.

### 3.7.3 Carded blends of triboelectric fibers

Triboelectric charging involves the rubbing or intimate contact of two dissimilar dielectric materials. The kinetic energy dissipated from the rubbing action induces a charge transfer. It is believed that contact electrification is affected by the relative affinities of the two contacting materials. Table 3.12 is the triboelectric series of several types of materials. The series is arranged so that materials that donate electrons are at the top and progressing downward to materials that accept electrons. There are several sources for this series (Brown (122), p.125 and (95), p.222, Sutherland and Purchas (18), p.176, and Lifshutz (123)). They are not necessarily consistent with each other. There are variations in the order of materials listed and the types of materials listed. The series presented in Table 3.13 is a simplified redaction with modifications based on the author's judgment.

It is evident that for a medium to develop a triboelectric charge it must be composed of at least two dissimilar dielectric materials. Lifshutz (123) theorized that the filtration performance of a wet laid medium could be related to the triboelectric difference of the materials. He developed a rating system by assigning numbers to each material, which he defined as triboelectric units (TU). He assigned a TU value to six different fiber materials (glass, polyester, Nylon, acrylic, polypropylene, and polytetraflouroethylene (PTFE)) in accordance with their position in the triboelectric series. The higher the value, the more positive was the triboelectric charge. Lifshutz produced wet lay handsheets from 50:50 volume mixtures of various selected pairs of the six fibers. He hypothesized that the

**Table 3.13** Triboelectric Series of Fiber Materials

| | |
|---|---|
| Most positively charged | ↑ |
| Fiberglass | |
| Wool | |
| Nylon | |
| Viscose | |
| Cotton | |
| Woodpulp | |
| Polyester | ↓ |
| Acetate | |
| Acrylic | |
| Polyethylene | |
| Polypropylene | |
| Teflon | |
| Most negatively charged | |

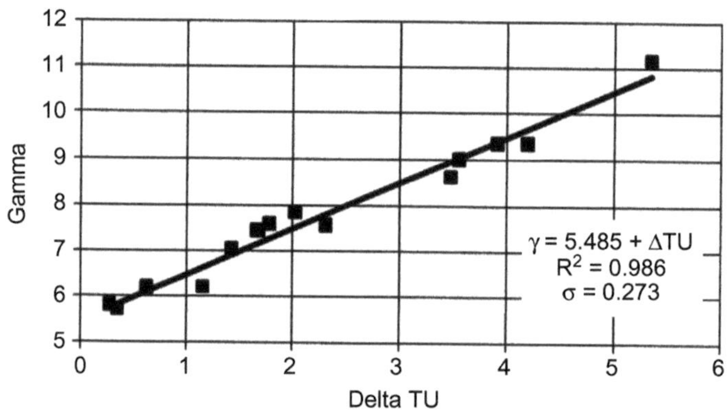


**Figure 3.12** Gamma versus Triboelectric Difference for Wet Laid Fiber Blends. *Reproduced by permission of the American Filtration and Separation Society, Nashville, Tennessee, USA.*

Gamma number (see Footnote 5) or filtration performance from Equation 3.4 would be related to the absolute difference ($\Delta TU$) in the TU values of the two fibers.

$$\gamma = q_F = C + abs\,(\Delta TU) \tag{3.4}$$

Lifshutz tested the 50:50 handsheets in an ATI Q127 DOP penetrometer. The results are presented in Figure 3.12. Based on the $R^2$ value of 0.986, the relationship looks pretty good. Note that the constant, $C$, in Equation 3.4 has the value of 5.485. Presumably, this would be the gamma rating of a filter medium that had no electrostatic charge.

**Table 3.14** Technostat® PLUS—Typical Test Data

| Technostat® grade | Basis weight g/m² | Thickness[a] mm | Resistance Pa | Penetration[b] EN 13274-7 NaCl Aerosol % |
|---|---|---|---|---|
| TPLUS 50/15 | 65 | 1.3 | 5.0 | 16 |
| TPLUS 70/15 | 85 | 1.6 | 8.0 | 10 |
| TPLUS 100/15 | 115 | 2 | 12.0 | 4 |
| TPLUS 150/15 | 165 | 2.6 | 20.0 | 1.8 |
| TPLUS 200/15 | 215 | 3.2 | 29.0 | 0.8 |
| TPLUS 250/15 | 265 | 3.5 | 38.0 | 0.5 |
| TPLUS 300/15 | 315 | 4 | 47.0 | 0.3 |
| TPLUS 400/15 | 415 | 5 | 67.0 | 0.15 |
| TPLUS 500/15 | 515 | 6 | 89.0 | 0.1 |

Technostat® needled to 15 g/m² spunbond polypropylene.
Test Flow: 95 l/min on 100 cm².
Resistance and penetration data were measured in accordance with BS4400 using 0.65 μm NaCl particles.
[a]Test pressure 1.6 kPa.
[b]Aged material.
Reproduced with permission of Hollingsworth and Vose Company, Inc,. East Walpole, Massachusetts, USA.

Lifshutz mentions that the charge on the wet lay sheets is considerably lower than what is seen on needle punched dry laid medium. Gamma values, as high as 200, are encountered in these types of medium. Indeed, manufacturers of dry laid media take advantage of triboelectric effects by composing these webs from fiber blends chosen from appropriate ends of the triboelectric series.

Hollingsworth and Vose Air Filtration Ltd. offers an expanding line of needle punched felt media under the label of Technostat®. Their brochure (124) states that "Technostat® is an electrostatically charged needle-felt comprising two electrically dissimilar synthetic polymers which are especially prepared and processed at the manufacturing stage, creating a charge transfer between the two different types of fiber." Table 3.14 is a tabulation of typical properties of H&V's Technostat®Plus filter medium grades. Note that the structure is a composite since the triboelectric layer is needled to a spunbond polypropylene. The medium is particularly effective in cabin air filtration. In Chapter 9, Table 9.5 tabulates the general data for the H&V Technostat® cabin air filter line of products. Chapter 9, Table 9.6 includes Technostat® in the tabulation of H&V cabin air filter grades that include activated carbon. In addition to automotive cabin air filtration, Technostat® filtration media are used in respirators, industrial air filtration, bacterial filtration for medical filters, residential air filtration, and vacuum cleaner filtration.

### 3.7.4 Undesirable static charge

Electret charge on a filter medium is usually considered as good because it enhances filtration performance. However, it is not always desirable. There are situations where electrostatic discharge (ESD) may generate concerns and undesirable effects. Examples are fuel and lube filtration systems where a static discharge could ignite the volatile organic and cause a fire or explosion. ESD can also cause defects in certain types of products such as electronic components. A filter medium can become electrostatically charged not only from electret materials in the medium, but also by induced charges during operation. For example, in fluid flow an electrostatic charge can be induced as a result of contact between a nonconducting fluid and nonconducting system components in the filter or medium.

Ways of dealing with ESD are by conducting or dissipating the electric charge away from the filter and/or medium. There are several ways of doing this. One way is the inclusion of antistatic additives and coatings as well as added carbon fibers or particles. Metal screens, sleeves, or cores with electric conducting properties can be used to dissipate or conduct static charge away from the medium or filter.

Cerex Advanced Fabrics, Inc. has introduced Nystat™, a nonmetallic static, dissipative fabric. According to Cerex, "Nystat™ is a functionalized nylon 66 with surface resistivity that can be modified to meet a certain desired level" (125).

## 3.8 COMPOSITE STRUCTURES

### 3.8.1 Types of composite structures

Composite structures are multilayer (at least two layers) filter media, each layer serving a different purpose in the filtration/separation phenomena. Examples of composite structures have already been discussed. For example melt-blown and nanofiber structures both have been mentioned as being formed on a support structure such as spunbond polyester or wet laid cellulose. These satisfy the definition of composite structures. Composite structures may serve any of the following purposes:

1. One or more layers provide mechanical support for the other layer or layers doing the filtering. Scrim reinforced needle felts are an example.
2. Provide two or more layers of different filtration efficiency to develop a gradient density medium, i.e., the upstream layers remove the large particle rocks and stones from the fluid whereas the downstream layers remove the fine particles.
3. Combine different separation technologies into one filter medium. For example, an activated carbon layer is combined with one or more nonwoven layers, whereby the nonwoven layers do particulate filtration and the activated carbon layer separates molecular contaminants by adsorption (see Section 3.10).
4. The outer layers serve as a containment to inhibit media migration, dusting, and particle fallout from the inner layers.

There are many forms of composite structures. Prominent are the SMS composites patented by Robert J. Brock and Gary H. Meitner of Kimberly-Clark in 1977 (126). This structure sandwiches a melt-blown web between two spunbond webs that are thermally bonded by passing them through a set of heated rolls. The structure provides a filter fabric with microfiber filtration capability and exceptional strength properties. Variations include SM (spunbond/melt-blown) and SMMS (spunbond/melt-blown/melt-blown/spunbond). Polymer Group International (PGI) markets SMS filter media under the trade name of TINOVA®.

Composite structures are often the way to support nanofiber webs for high efficiency filtration applications. For example, H&V's NanoWave® filter medium is a three layer configuration that includes a capacity layer for dirt holding capacity, a waved, microfiber, efficiency layer for fine particle removal, and a support layer to provide mechanical endurance to the medium in its filtration application. The wave-like construction provides for 2.4 times the surface area of a normal flat sheet media. A primary use of the product is in ASHRAE bag filters. It also can be used in pleated panel air filters, as paint spray booth diffusion media, in respirators and facemasks, vacuum cleaner bags and filters, and in room air cleaners. Other uses for NanoWave® include residential filtration, liquid filtration and gas turbine intake air filtration.

There are a number of ways of forming composite structures in addition to the SMS process. Among them are:

1. Laminating—Two are more different layers of media are sandwiched to achieve the desired composite structure. The different layers may be attached to each other by a glue, latex adhesive, or hot melt adhesive. Sometimes electrostatic or Van der Waals forces are sufficient to hold the layers together for the desired filter making process.
2. Entanglement—Hydroentanglement, needle punching and stitch bonding are ways of combining the individual layers.
3. Multi-ply forming—Some forming processes are capable of forming multilayered media. Carding lines with successive in-line carding machines may be capable doing this. Some wet lay machines are capable of forming dual-ply and triple-ply media. Ahlstrom uses a three-ply machine to produce a three layered filter media composite identified as their Trinitex® technology.
4. Collector—One nonwoven layer serves as the collector for the forming of another type of nonwoven layer. Melt-blown, nanofiber, and microfiber webs are often applied this way as a "coating" to a nonwoven substrate (see Section 3.10.2).

## 3.8.2 Nanofiber and microfiber coatings

There several ways to apply a nanofiber or microfiber coating to a substrate. One way is to use the support substrate as a collector for forming the nanofiber or microfiber web. Most melt-blown, electrospun, and centrifugal spun webs are produced and applied in this manner. The substrate is often a spunbond or wet laid media. In the type of media used for baghouses the substrate may be a needled or hydroentangled felt.

Short fibers, such as glass microfiber or cellulose fibrils, may be mixed with a liquid, most often water, and formed into a slurry. As such it may be applied to a substrate web in a coating operation. In wet lay operations that have dual ply formers, the microfiber or nanofiber layer may be formed as one of the plies and the substrate as the other.

## 3.9 COALESCING MEDIA

Coalescing filters separate liquid particles from a fluid stream. The fluid can be either gas or liquid. Gas phase separations are used to separate moisture and/or oil mists from contaminated air streams.

Examples of gas phase separations are:

1. Separation of moisture from moisture laden air streams
2. Separation of oil mists from air streams
   a. Compressed air operations
   b. Crank case ventilation
   c. Lubricated machining
   d. Cutting processes.

Examples of liquid phase separations are:

1. Separation of liquid moisture particles from organic streams
   a. Moisture removal from fuel pipeline streams
   b. Diesel fuel filtration in engine operated vehicles.
2. Separation of liquid organic particles from water streams.
   a. Environmental cleanup of oily water waste streams
   b. Purification of water for drinking and high purity applications in industry
   c. Removal of unwanted oil in water based coolants used for metal working operations.

At this time it would be advantageous to give a brief course on surface chemistry. Figure 3.13(a) is a cross section diagram of a water droplet resting on solid surface. The point of interest in this diagram (indicated by X) is where the surface of the liquid drop touches the solid surface. This point X is amplified in Figure 3.13(b). There are three forces affecting the drop at point X:

1. The surface tension, $\gamma_L$ of the liquid drop
2. The interfacial tension, $\gamma_{SL}$ of the solid-liquid interface, the face where the bottom surface of the drop rests on the solid surface
3. The interfacial tension, $\gamma_{SG}$ of the solid-gas interface, the face where the solid surface interfaces with gas outside the solid-liquid interface.

The shape of the drop comes to equilibrium when the balance of surface and interfacial tensions = 0[6].

[6] Surface and interfacial tension is measured in terms of force per unit line of surface (N/m) or energy per unit area of surface (J/m$^2$).

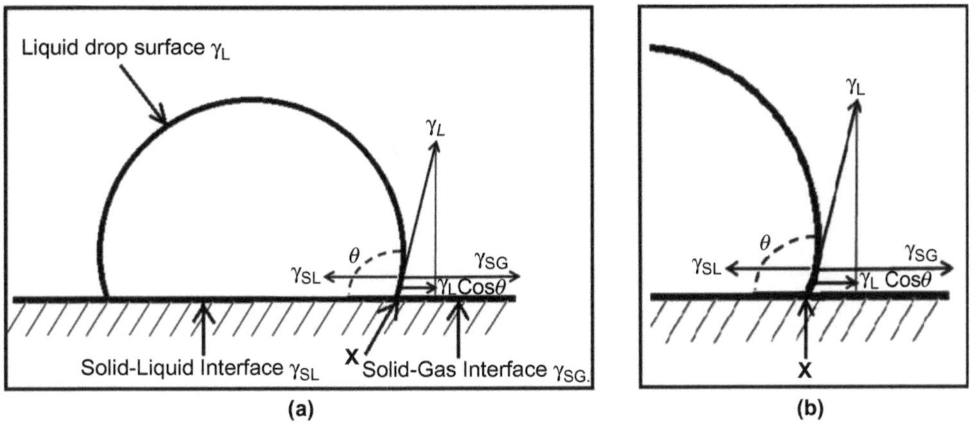


**Figure 3.13** Cross Section of a Water Drop on a Flat Solid Surface.

$$\gamma_{SG} + (\gamma_L \cos\theta - \gamma_{SL}) = 0 \tag{3.5}$$

The forces acting in a surface or interface acts in a direction to reduce the area of that face. $\gamma_{SG}$ pulls on point X so as to spread the drop over the solid surface and reduce the solid surface area exposed. On the other hand, $\gamma_{SL}$ pulls in the opposite direction tries to reduce the interface or contact area between the water drop and the solid surface. The surface tension, $\gamma_L$ tries to reduce the exposed surface area of the drop by pulling in an upward direction as if the drop were trying to lift its skirt. The force exerted at point X by $\gamma_L$ is tangential to the drop surface at point X where it intercepts the solid surface. The horizontal force vector is $\gamma_L \cos\theta$, where $\theta$ is the contact angle between the liquid surface and the solid surface. It can be seen that the three forces are playing tug-of-war on point X: $\gamma_{SG}$ pulling to spread the drop over the solid surface and flatten it, $\gamma_{SL}$ pulling in such a way as to make the drop surface spherical and more cohesive, and the horizontal vector, $\gamma_L \cos\theta$ pulling in a direction depending on the value of $\theta$. If $\theta$ is less than $90°$[7] then $\gamma_L \cos\theta$ pulls inward in the same direction as $\gamma_{SL}$ and if greater than $90°$, $\gamma_L \cos\theta$ pulls outward in the same direction as $\gamma_{SG}$.

The solid surface force $\gamma_{SG}$ is very important. If the force is very strong it is a high energy surface and will pull on the drop in such a way as wet out the surface. If it is very weak, it is a low energy surface (Teflon is an example of a low energy surface material) and the solid–liquid force $\gamma_{SL}$ will dominate and pull the drop surface into a spherical surface. The contact angle will be zero. The spherical drop will have no interfacial area for sticking to the solid surface and will roll off like a ball. It may collide with other drops and coalesce into larger drops.

[7] Note that the value of $\cos\theta = 0$, when $\theta = 90°$. If $\theta$ is less than $90°$, then $\cos\theta$ is positive inwards and if $\theta$ is greater than $90°$, then $\cos\theta$ is negative outwards.

This effect is observed on the hood of a car driving through a rainstorm. If the car and hood had been recently waxed the hood has a low energy surface and the raindrops simply ball up and roll off the hood. If the hood surface has aged since the last waxing, it will have developed a high energy surface. The raindrops will spread and wet out the surface of the hood.

There are two forms of coalescing media: surface coalescing media and depth coalescing media:

1. A surface coalescing medium contains a highly hydrophobic or oleophobic surface to prevent the liquid contaminant particle or droplet from penetrating the surface. This is the effect described by the surface chemistry discussion above. The droplet is either rejected from the surface or collected on the surface, where it coalesces with other droplets to form larger droplets. Eventually the coalesced droplets become large enough to either be washed off or gravity fall from the surface. Generally the surface will consist of a very low energy polymer such as a flouropolymer or a silicone. In fibrous filter media the low surface energy polymer may be present as the fibers themselves, as a binder, or as a surface coating.
2. In depth coalescence the medium filters out the liquid particles just as it would solid particles by entrapment within the fiber structure. Coalescence occurs when the entrapped liquid particles either combine with other nearby droplets and/or are struck by additional droplets approaching and penetrating the medium. For this to happen, the fibers should have a relatively high surface energy. Often depth coalescent media contain a significant percentage of fine fibers such as glass microfiber. The finer fibers better filter out the smaller liquid particles. It should be noted that for best performance, the fibers surface should be neither too hydrophobic (oleophobic) nor too hydrophilic (oleophobic), but somewhere in between. If the fibers are too hydrophobic or oleophobic the liquid particles will not stick and coalescence will not take place. On the other hand, if they are too hydrophilic or oleophylic the liquid particle will spread over and coat the fiber surfaces. The liquid layer will eventually build up and eventually plug the medium.

In water-oil separation, the coalescing filter medium consists of a pleated upstream element and wrapped glass fiber blankets downstream. The pleated elements contain filter papers that contain up to 20% glass fibers and in some cases impregnated with phenolic resin to withstand high pressure and vibration.

Both types of coalescing filters are often used to remove water and moisture contamination in fuel filter/separators systems for marine and aviation fuel. These are two-stage operations. The first stage contains a depth coalescing filter to remove inlet water particles. The second stage is a filter with a hydrophobic, low energy surface that allows hydrocarbon fuel to pass through but repels any entrained water droplets. The system is diagrammed in Figure 9.12 of Chapter 9 and discussed in Section 9.5.2.

The depth coalescing phenomena are illustrated by Chen and Martin (127) in Figures 3.14 and 3.15 below. Figure 3.14 shows a second by second sequence of water

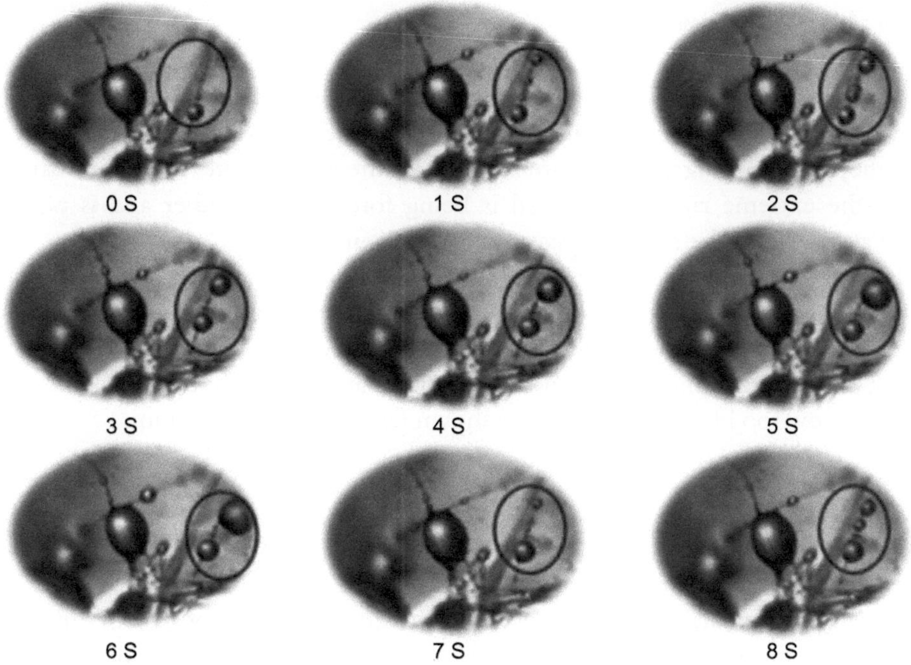


**Figure 3.14** Coalescence on a Glass Fiber. *Reproduced with permission of Kaydon Custom Filtration Corp. Lagrange, Georgia, USA.*

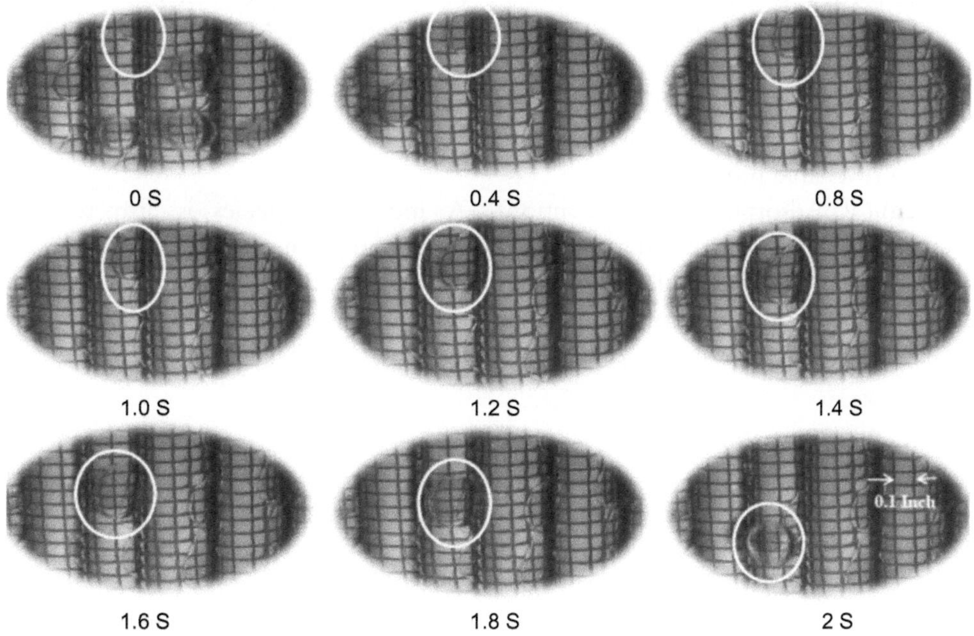


**Figure 3.15** Coalescence—Emerging Droplets. *Reproduced with permission of Kaydon Custom Filtration Corp. Lagrange, Georgia, USA.*

droplets forming on glass fibers in the medium. The oval enclosed portion of the sequences shows droplets forming on the fiber, coalescing to larger drops, and then when they become big enough, are forced off the fiber. Note that at 6S the large drop in the extreme right of the oval is being forced off the fiber and is no longer present at 7S. At 8S the sequence starts all over again.

The oval enclosed section of Figure 3.15 shows a drop emerging from the downstream side of the medium. The screen support structure is visible. In the time sequences the drop emerges from the medium and in the subsequent sequences is shown to be growing as it clings to the combination of medium and screen support filaments. At 2S, the drop is now too big to hang on and is shown falling from the medium.

## 3.10 SORPTION MEDIA

### 3.10.1 Adsorption media

Nonwoven filter media containing adsorbents are used for adsorption of contaminants from both air and water. In both situations, it competes with granular or pellet adsorbents used in bed form.

Adsorption of a contaminant on an adsorbent surface can be by either a physical attraction, or a chemical attraction, or a combination of both. In the physical attraction, the particle is bound to the surface by Van der Waals forces. Chemical attraction causes some form of chemical change in the contaminant. Middlebrooks (128) lists the following mechanisms of adsorption.

**a.** Physical Adsorption (physisorption)—A molecular species is attracted to a surface and held in place by weak forces; similar in the concept of condensation of gases. It is a reversible process.

**b.** Chemical Adsorption (chemsorption)—A molecular species is attracted to a surface and reacts with that surface or something on that surface to form a chemical bond which holds it in place. Typically considered irreversible at constant conditions. Often the surface is impregnated with a chemical that promotes a desired reaction to remove specific contaminants.

#### *3.10.1.1 Physical adsorption*

Adsorbent media are used in a number of application including filters for odor and irritant vapor control in automobile cabins, residences, commercial HVAC systems and workplace environments. It is used for control of VOC emissions in industrial locations that otherwise would source these contaminants to the environment. It is very important in the protective clothing, face masks, and respirators used to protect military personnel and civilians that may be exposed to the toxic and poisonous gases that might be used in warfare and terror incidents. Activated carbon is generally the most common form of adsorbent material used in separation media. There are air filter applications where activated alumina is used.

The factors for choosing an adsorbent medium is as follows:

- Types of contaminant(s)
- Amount of adsorbent
- Type of adsorbent
- Desired removal efficiency of the contaminants
- In air filtration, volatility of the contaminants
- Particle size of the adsorbent
- Impregnation
- Concentration of contaminants
- Humidity
- Filter format.

Following is a list of trace organics with high probability of being adsorbed by activated carbon:

- Aromatic solvents (i.e., benzene, toluene, xylene)
- Polynuclear aromatics(i.e., naphthalene, biphenyls)
- Chlorinated aromatics (i.e., chlorobenzene, PCB's, aldrin, endrin, toxaphene, DDT)
- Phenolics (i.e., phenol, cresol, resorcinol)
- High molecular weight aliphatic amines and aromatic amines (i.e., aniline, toluene diamine)
- High molecular weight ketones, esters, ethers, and alcohols (i.e., dextran, polyethylene glycol)
- Surfactants (i.e., akyl benzene sulfonates).
- Soluble organic dyes (i.e., methylene blue, textile dyes)
- Fuels (I.e., gasoline, kerosene, oil)
- Chlorinated solvents (i.e., carbon tetrachloride, perchloroethylene).

Frequently, the filter contains the media in granular or pelletized form. However, there are many applications where the adsorbent material is part of or combined with a nonwoven separation media. A commercial example is activated carbon filter paper. This wet lay medium is used in home water filters to remove undesirable taste and odor contaminants. The media are produced by impregnating a wet lay fiber slurry with particles of activated carbon and producing a single layer wet lay medium with the carbon content. Activated carbon contents of 50–60% are achievable. A resin binder also added to the fiber-carbon slurry is effective in bonding the carbon particles to the fiber matrix. Onyx Specialty Papers, Inc. in Lee, Massachusetts produces wet laid activated carbon grades which they promote for both water and air filtration. Table 3.15 is a property summary of two of the grades. It is noted that both grades are wet lay media, however Grade745104 is a blend of activated carbon and cellulose fiber whereas Grade 744995 is a blend of activated carbon and synthetic fiber.

**Table 3.15** Target Values of Activated Carbon Grades From Onyx Specialty Papers

| Grade | Composition | Basis weight | | Permeability | | Caliper | | Carbon content |
|---|---|---|---|---|---|---|---|---|
| | | lb/3000 $ft^2$ | g/$m^2$ | $ft^3/ft^2$ ·m @ 0.5″$H_2O$ | $cm^3/cm^2 \cdot s$ @12.7 mm $H_2O$ | mil | mm | % |
| 745104 | Carbon/ cellulose | 130 | 212 | 15 | 7 | 27 | 0.7 | 50 |
| 744995 | Carbon/ synthetic | 130 | 212 | 20 | 10 | 31 | 0.8 | 50 |
| Test Method | | TAPPI T-410 | | TAPPI T-251 | | TAPPI T-241 | | |

Reproduced with permission of Onyx Specialty Papers, Inc., Lee, Massachusetts USA.

Several media manufacturers produce composite structures that contain a carbon layer sandwiched between two nonwoven layers. Figure 3.16 is a photograph showing the layers of PLEKX® adsorption medium by Graver Technologies, LLC, Glasgow, Delaware USA. In this structure the outer layers are polyester nonwovens. Figure 3.17 is a photomicrograph of the structure. The inner layer is activated carbon particles held together with a thermal bonding resin.

**Figure 3.16** PLEKX Adsorption Medium by Graver Technologies, LLC—An Activated Carbon Layer Sandwiched Between Two Nonwoven Layers. *Reproduced with permission of Graver Technologies, LLC, Glasgow, Delaware, USA.*

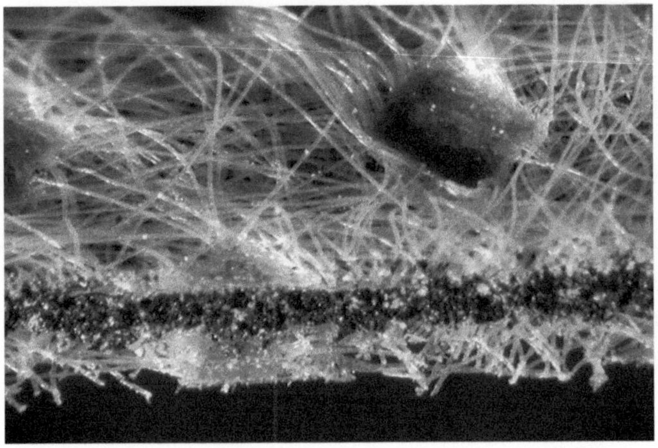

**Figure 3.17** Microstructure of PLEKX Adsorption Medium. *Reproduced with permission of Graver Technologies, LLC, Glasgow, Delaware USA.*

Fiberweb (now PGI) in Charlotte, North Carolina first introduced AQF® technology. H&V, East Walpole, Massachusetts, USA acquired it from Fiberweb in 2010. Production moved to H&V's Floyd, Virginia plant and is now being marketed under the MFM™ (Molecular Filter Media) brand. MFM™ utilizes a patented fiber structure to suspend functional particles, such as activated carbon, for "molecular filtration applications." The technology utilizes bicomponent sheath-core fiber with a low melting sheath to bond the structure into a three dimensional network as illustrated in Figure 3.18.

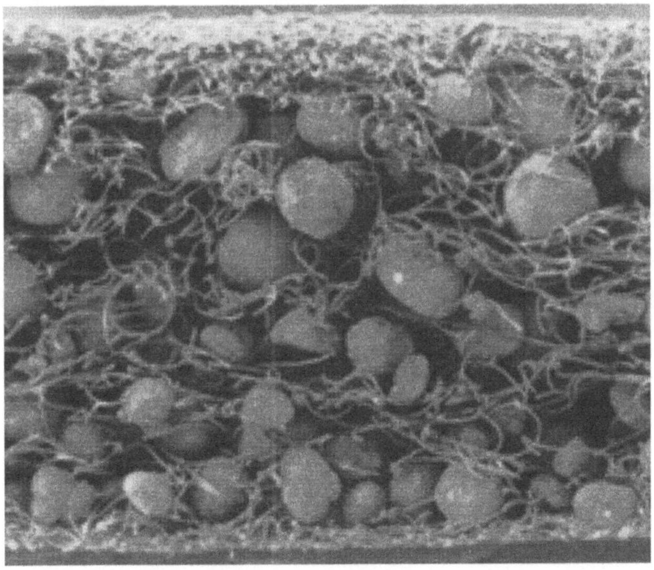

**Figure 3.18** MFM™ by H&V Patented Fiber Structure. *Reproduced by permission of Hollimgsworth and Vose Company, East Walpole, Massachusetts, USA.*

An interesting application of activated carbon is as a panel in men's (and woman's) underwear to reduce or eliminate flatulence odor. One such product is Shreddies© marketed by My Shreddies©, a UK-based intimate apparel company which has released a line of odor-control undies for both men and women. It neutralizes bad smells with a rear panel made of activated carbon cloth, the same sort of material as used in military gas masks to protect soldiers from chemical attacks. The name "Shreddies" comes from the British military servicemen's term for underwear, and is well known throughout England.

#### *3.10.1.2 Chemsorption*

Both activated carbon and activated alumina can be treated with chemicals that provide chemsorption properties. Examples of such chemicals include metal ions such as silver, copper, and palladium. In addition, commercial filters for air cleaning systems may contain activated carbon or alumina media impregnated with potassium permanganate, iodine, phosphoric acid, and potassium hydroxide. Choice would depend on the contaminant to be captured. For example, impregnation of the sorbent (activated carbon or activated alumina) with potassium permanganate would greatly enhance its ability to remove odorous sulfur compounds such as sulfur dioxide, hydrogen sulfide, and mercaptans. Iodine might be used where there is mercury vapor to be removed. There are a number of filter technologies where the different sorption technologies are combined into one medium. This includes filters combining activated carbon and activated alumina into one medium and with and/or without the chemsorption component. For example, Permatron Inc., Elk Grove Village, Illinois USA markets the CHARBXPL® brand of filters that contain activated carbon and activated alumina bonded to a polyester nonwoven substrate as the medium. Permatron also offers 5% potassium permanganate for odor, fume and gas removal (129).

In the market place there are pleated filters for residential air conditioning and furnace systems that combine fine particle removal with the odor removal properties of baking soda (sodium bicarbonate, $NaHCO_3$). Each fiber in the web is coated with a fuzzy covering of baking soda. The "fuzziness" increases odor removal and provides increased surface area for particulate removal.

### 3.10.2 Absorbtion media

Absorbent structures such as diaper fluff, hygienic pads, and wipes are not considered as filter media, because there is no separation involved. On the other hand, some materials are selective. Absorbent materials such as nylon, polypropylene and polyurethane are hydrophobic and reject water and yet have an affinity for oil and are often used to absorb oil from water. Booms containing polyolefin materials are often used to control oil spills at sea. Oil absorbing booms played an important role in controlling the spills from BP's Deepwater Horizon oil rig in the Gulf of Mexico in 2010. Nylon webs such as Cerex® are used as fences or barriers at oil spills to control the oil that spilled into the water. One form is marketed by Cerex as the "OIL SHARK" medium. It allows the passage of water while holding back or containing the oil or liquid hydrocarbon. See Section 3.14 for more discussion of fence and barrier materials.

Diatomaceous earth (DE) is an absorbent clay that contains a porous structure. It has application in some wet lay media used for water filtration, and for hot oil filtration. However it appears that the most important function of DE in a filter medium is not as a separation technology, but rather to increase the flux rate through the medium.

In many situations, absorbency is an undesired quality in a filter medium. The tendency of cellulose media to absorb water tends to soften and weaken the filter structure and shorten filter life. In addition, the absorption of moisture causes fiber swelling and interference with filtration performance. An undesired quality effect of a cellulose medium when it absorbs moisture is that it causes growth in the width direction resulting in slit width variations in finished rolls.

## 3.11 ANTIMICROBIAL MEDIA

Antimicrobial activity in a filter medium serves either or both of two basic purposes:

1. It protects the filter and filter medium from biological decay and resulting failure.
2. It prevents the passage of biological organisms to the downstream environment of the filter.

Price et al. (130) have shown that even in filters, such as HEPA filters designed to filter out microorganisms, arrested microorganisms can grow and in some cases "grow through" the filter medium. Some organic media such as cellulose media provide nutrition for microbiological growth. A function of activated carbon media are to adsorb organic material and this makes the medium more vulnerable to being a nutritional havens for microorganisms. In aqueous and high moisture air environments, biological activity can cause serious degradation of the filter media and be a source of contamination downstream of the medium.

Price et al. (130) demonstrated that an anti-microbial preservative can significantly inhibit fungal colonialization. In their studies, the preservative investigated was Intersept®[8], a phosphated amine antimicrobial complex.

A number of filter medium and filter manufacturers are now incorporating antimicrobial agents into their filter media. For example, some residential air filters are treated with disinfectants to inhibit biological growth.

[8] Intersept® is a trade name of Interface Research Corporation, Kennesaw, Georgia, USA.

In aqueous filtration, swimming pool and spa filters are logical candidates for antimicrobial filter media. PPG in Old Hickory, Tennessee offers two spunbond polyester lines (Reemay Advantage® and Reemay Freedom®) treated with Microban®[9] antimicrobial agent.

## 3.12 CATALYTIC MEDIA

Catalysis is a process whereby a chemical induces a chemical change in another chemical but does not change itself. In some cases catalysis is similar to chemsorption (see Section 3.10.1.2) and sometimes identified as such. In catalytic filtration, the chemical is built into the medium to initiate a chemical reaction to destroy the contaminant after it has contacted the medium. For example, The Pall Corporation is developing a Photo-Catalytically Regenerated Adsorber (PCRA) (131) device that uses a unique combination of adsorbent and catalyst technologies to remove particulate and gaseous contamination from cabin air. The operation comprises two phases:

- Phase one—the adsorbent adsorbs the gaseous contaminants.
- Phase two—the system oxidizes the contaminants on the adsorbent and they are purged overboard.

Photocatalytic Oxidation (PCO) is an oxidation process that uses ultra violet light. The process is diagrammed in Figure 3.19. PCO is used in some home air purifiers (see Chapter 8, Section 8.7). Chuaybamroong et al. (132) reported on the effect of PCO on HEPA media at two application levels of $TiO_2$. They indicated a 60−100% microbe reduction for a wide variety of organisms.

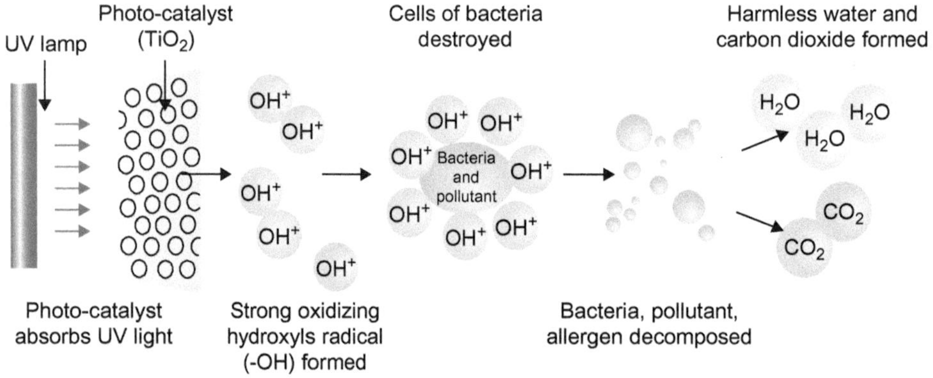


**Figure 3.19** Photocatalytic Oxidation Phenomena.

[9] Microban® is a trade name of Microban International, Ltd.

In water treatment, manganese greensand is the most common medium in use for removal of iron and manganese through pressure filtration. Greensand is a processed material consisting of nodular grains of the zeolite mineral glauconite. The material is coated with manganese oxide. The ion exchange properties of the glauconite facilitate the bonding of the coating. This treatment gives the medium a catalytic effect in the chemical oxidation-reduction reactions necessary for iron and manganese removal. This coating is maintained through either continuous or intermittent feed of potassium permanganate.

## 3.13 MEMBRANE FILTER MEDIA

### 3.13.1 Organic membranes

There are several constructions for organic membrane filters. This includes track edge membranes, solvent cast (phase inversion) film membranes, and stretched microporous membranes. Processes for producing these membrane types are discussed in Chapter 5, Section 5.9. In addition, inorganic membranes can be produced from sintered particles (processes discussed in Chapter 5, Section 5.9.4), the most common method for inorganic membranes. Nanofiber material such as from the electrospun and centrifugal spinning processes can be designed to have membrane like properties. DuPont™ is marketing a nanofiber media identified as DuPont™ Hybrid Membrane Technology (HMT). DuPont (133) describes this medium as "a "membrane-like" sheet product composed of continuous submicron fibers with resultant sub-micron to low micron pores and high surface area". The nanofiber media are believed to be a product of DuPont's electroblowing technology described in Chapter 5, Section 5.3.4. Currently, nanofiber membrane technology is limited to microfiltration and ultrafiltration separation phenomena.

Membranes that have the same chemical and physical structure throughout their thicknesses are referred to as microporous or isotropic membranes. If they have changing or different structures in the direction of their thicknesses they are called asymmetric or anisotropic membranes. Asymmetric membranes are like composite structures in that they have different layers of density and thickness. Sometimes the different layers are of different materials. The common forms of asymmetric membranes have very thin skins, of a highly selective material, supported on a thick porous substrate. Figure 3.20 diagrams an asymmetric membrane as it is used for separation purposes. The type of membrane depicted is typical of an ultrafiltration (UF) membrane (Crittenden et al. (134)).

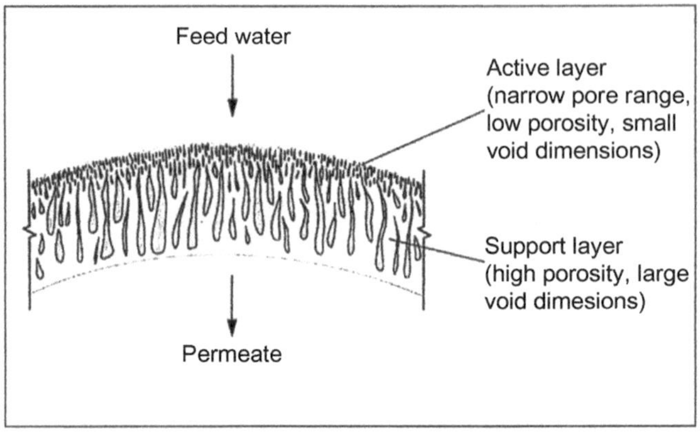


**Figure 3.20** Asymmetric Structure of a UF Membrane. *Reproduced with permission of John Wiley & Sons Inc., Hoboken, New Jersey USA.*

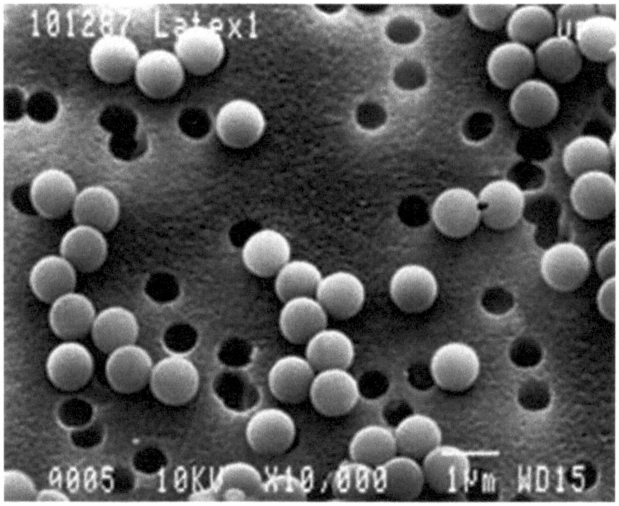

**Figure 3.21** Track Edge Membrane with Microscopic Latex Spheres. *Reproduced with permission of Simplex Corporation. Simferopol, Crimea.*

### *3.13.1.1 Track edge membranes*

Figures 3.21 and 3.22[10] provided by Simpex Corporation in Simferopol, Crimea, indicate the structure of track edge membranes. Figure 3.21 shows the surface structure. Pores are "drilled" by radiation beam and are uniform in size. The

[10] Figures 3.21 and 3.22 were kindly provided by Dr. Antipov Valery, General Director, Research and Production Enterprise, Simpex Corp.

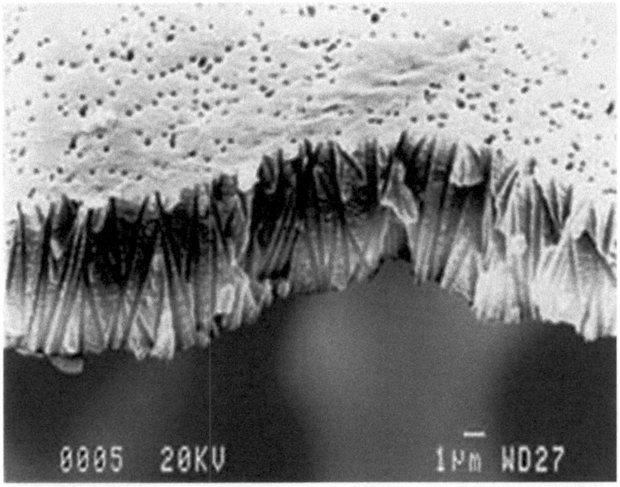


**Figure 3.22** Cross Section of a Track Edge Membrane. *Reproduced with permission of Simplex Corporation. Simferopol, Crimea.*

micro-photograph also shows retained latex spheres that were used to test the medium. The cross-sectional view of Figure 3.22 reveals that each pore is like a cylinder penetrating the complete thickness of the membrane. It can be seen from Figure 3.22, that the density through the thickness of the structure is uniform, the characteristic of an isotropic membrane.

#### *3.13.1.2 Solvent cast film membranes*

Solvent cast or phase inversion membranes, are generally asymmetric as diagrammed in Figure 3.20 above, are the main source of UF membranes. There are several types of polymers that are used in phase inversion membranes; polysulfone being among the most popular. UF membranes are typically hydrophilic (135).

#### *3.13.1.3 Stretched microporous membranes*

Stretched microporous membranes are made from uniaxially oriented film. Lamella and lamellae are surface characteristics of such film as shown in Figure 3.23. A stretching process is described by Wilkes (136) and discussed in Chapter 5, Section 5.9.3.

### 3.13.2 Inorganic membranes

Inorganic membranes are barriers composed of noncarbon containing compounds. They are used in applications where strong heat resistance and/or corrosion resistance

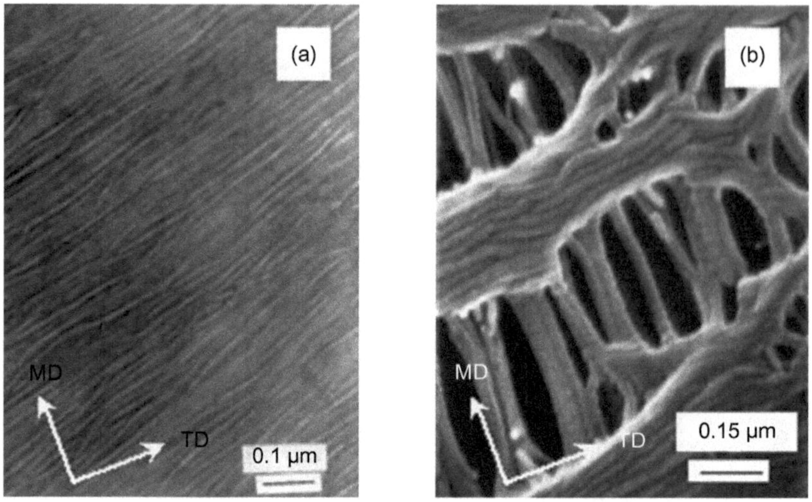


**Figure 3.23** (a) TEM Micrograph of a Uniaxially Oriented HDPE Film; (b) SEM Microphotograph of a HDPE Microporous Membrane. *Reproduced with Permission of Dr. Garth L. Wilkes, Blacksburg Virginia USA.*

is required. Inorganic materials used in inorganic membranes are listed in Chapter 4. Section 4.6. Processes for making inorganic membranes are discussed in Chapter 5, Section 5.9.4.

## 3.14 BARRIER MEDIA

Barrier fabrics or devices, such as oil booms, serve to control oil spills in fresh water or salt water by containing the spill, diverting to a collection point, or blocking the contaminating from a sensitive area.

Nylon is a polymer that has an affinity for oil, but unlike polypropylene and polyurethane is hydrophilic (loves water). Nylon webs contain and retain oil but allow water to pass through. Ortega et al. (137) have patented this method of separating oil from seawater by using strong nylon fabrics. They claim that a strong fabric such as a nylon spunbond (SPECTRAMAX®, CEREX®, PBN-II®, and ORION®) can absorb up to 10 times its weight in oil and still be strong enough to be recovered after saturation with oil and water. These types of fabrics can be used as barrier fabrics to contain oil spills in small bodies of water and prevent them from spreading with the water flow. Figure 3.24 depicts the "OIL SHARK" medium being used in such an application. In this case a train wreck created an oil spill into a nearby stream. The picture shows the barrier fence containing the oil and allowing the water to flow through, In addition, Cerex Advanced Fabrics, Pensacola, Florida has demonstrated that these fabrics can be made into

**Figure 3.24** Cerex OIL SHARK® Containing an Oil Spill in a Small Stream, the Oil Spill Resulted From a Nearby Train Wreck. *Reproduced with permission of Cerex Advanced Fabrics, Pensacola, Florida USA.*

**Figure 3.25** OIL SHARK® Skimmer Bag, Full of Oil, Being Removed From the Water and into the Boat. *Reproduced with permission of Cerex Advanced Fabrics, Pensacola, Florida USA.*

skimming bags that can be deployed from boats and other sea vessels to separate oil spills from oceans, seas and other bodies of water. Figure 3.25 is a picture of a bag that is full of skimmed oil being loaded into the skimming boat. It should be noted that the bag fabric (Cerex® nylon in this case) not only has to have the oil-water separation capabilities, but also be strong enough so as not to burst when lifted from the water.

# CHAPTER 4

# Raw Materials for Nonwoven Filter Media

## 4.1 INTRODUCTION

The raw materials for nonwoven filter media are divided into four categories:

1. Polymers
2. Fibers
3. Binders
4. Additives

The choice of materials for the design of nonwoven filter media depends on performance, processability, and economics. The web must have the right combination of physical properties such as thickness, strength, flexibility, and tear resistance to be processed into its final filter form. It must have a raw material structure that optimizes its performance as a filter medium. Finally, the medium must be composed of materials that give it durability, chemical resistance, and environmental stability in its intended application.

## 4.2 POLYMERS

Polymers are used for the fibers in nonwoven webs and for the resins and additives used to bond, coat, and provide other properties to the fiber formed web. Most membrane products are functioned from polymer sources, the exception being inorganic membranes.

In melt-blown, spunbonded, and nanoweb spinning processes the polymer is the fundamental raw material supplied to the process and the forming of filament fibers is an intermediate stage in the forming of the web. Web properties for these processes are determined by the appropriate choice of polymer based on its properties and selection of the process conditions for forming that polymer into a fibrous form.

On the other hand, formation of webs by the air laid, dry laid, and wet laid processes begins with a raw material that is already in fiber form. These are shorter nonfilament types of fibers provided by some form of outside supplier. The choice of fiber properties is the guiding criteria in the design of the web.

Organic fibers and resin binders are composed of polymers. Their properties and processability are very much dependent on the chemistry and properties of the polymers from which they were made. A listing of the structures and properties of many of the polymers used for nonwoven materials is tabulated in Table 4.1. Natural fibers



*Handbook of Nonwoven Filter Media*
ISBN: 978-0-08-098301-1

**Table 4.1** Structures of Polymers Used in Nonwovens

| Line | Polymer | Type(s) and trade names | Chemical name | Application | Structure |
|---|---|---|---|---|---|
| 1 | Cellulose | Wood pulp<br><br>Vegetable fibers<br>Abaca (*Manila hemp*), Hemp, Esparto Grass, Sisal, Jute, Kenaf, and Flax<br>Cotton | Repeat unit: anhydro-beta-cellobiose | Fiber | 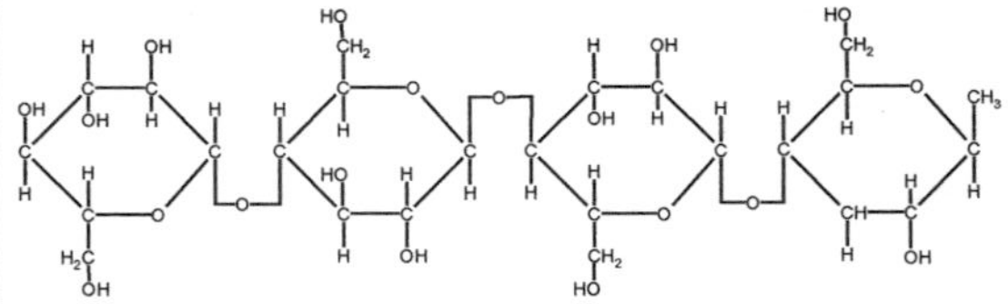 |
| 2 | Cellulose | Regenerated cellulose—Rayon | Repeat unit: anhydro-beta-cellobiose | Fiber | 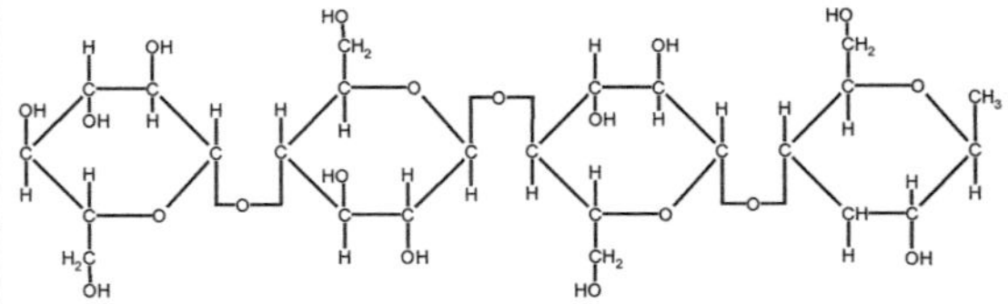 |
| 3 | Phenolic | Resole | Phenol formaldehyde | Resin—solvent based, some water based | Resole possible structures–includes novolak structure below<br>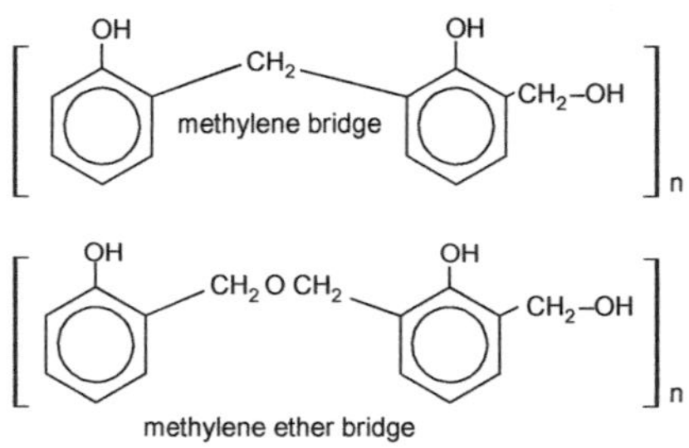 |

(*Continued*)

Table 4.1 (Continued)

| Line | Polymer | Type(s) and trade names | Chemical name | Application | Structure |
|---|---|---|---|---|---|
| 4 | Phenolic | Novolak | Phenol formaldehyde | Resin—solvent based, some water based | **Novolak structure** 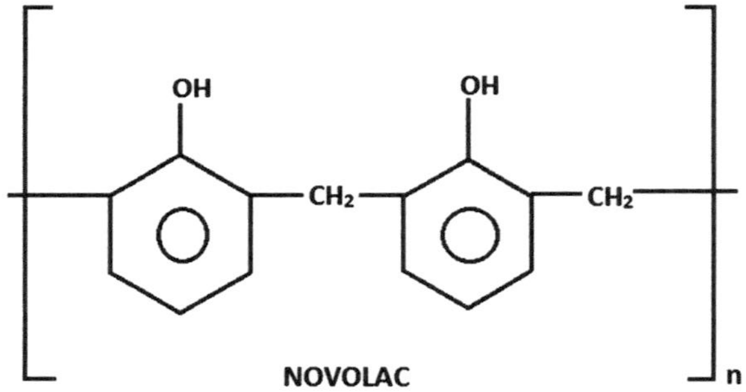  |
| 5 | Epoxy | Polyether resins | Polyether | Resin—water based and solvent based | **Epoxy prepolymer** 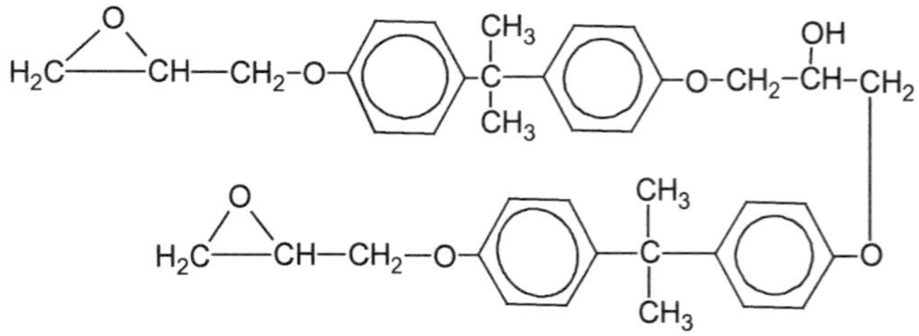 |
| 6 | Polyamide | Nylon 6,6 | Polycaprolactum | Fiber<br>Meltspun webs<br>Membranes | 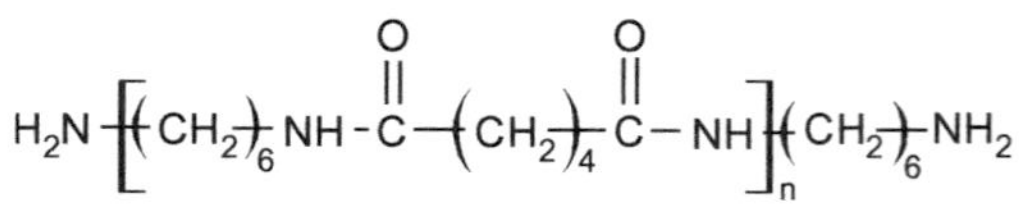 |

| | | | | | |
|---|---|---|---|---|---|
| 7 | Polyamide | Nylon 6 | Poly(hexa-methylene-adipamide) | Fiber<br>Meltspun webs<br>Membranes | 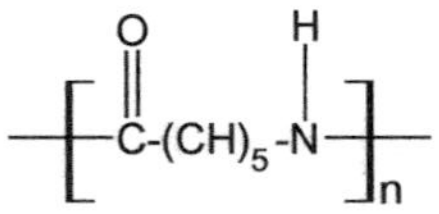 |
| 8 | Polyaramid | Kevlar®<br>Twaron® | Polyparaphenylene terephthalamide | Fiber | 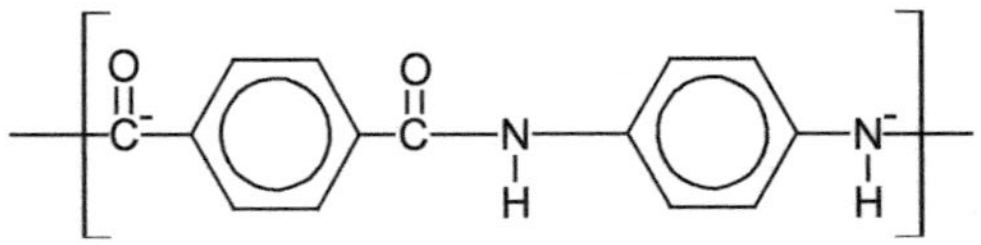 |
| 9 | Polyaramid | Nomex® | Polymetaphenylene isophthalamide | Fiber | 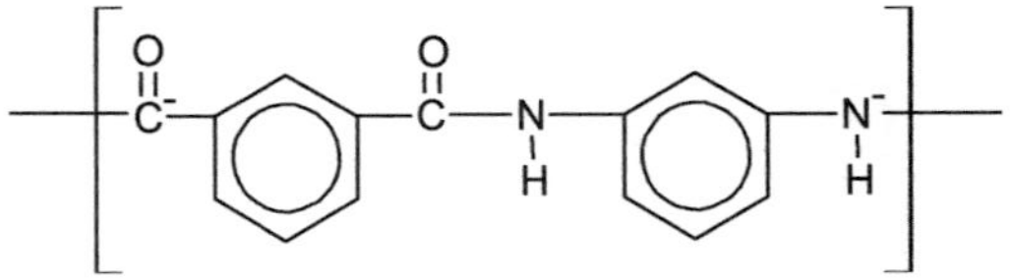 |
| 10 | Polyester | PET<br>Dacron<br>Trevira® | Polyethylene terephthalate | Fiber<br>Meltspun webs<br>Membrane | 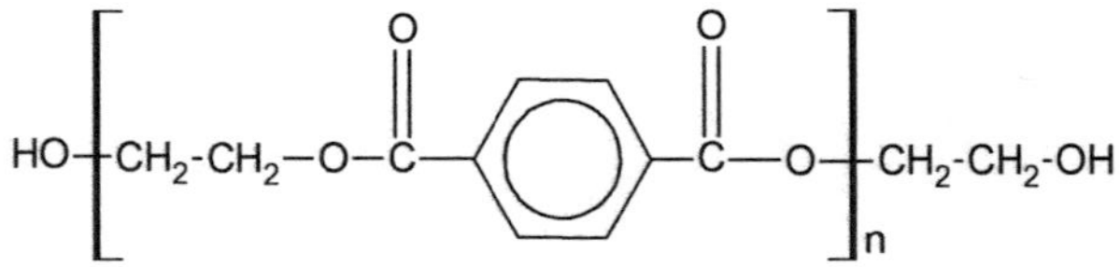 |
| 11 | Polyester | PBT | Polybutylene terephthalate | Fiber<br>Meltspun web | 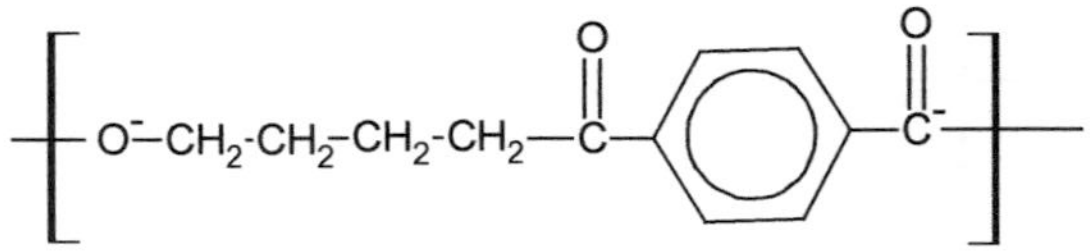 |

*(Continued)*

**Table 4.1** (Continued)

| Line | Polymer | Type(s) and trade names | Chemical name | Application | Structure |
|---|---|---|---|---|---|
| 12 | Polyimide | P84® | – | Fiber | 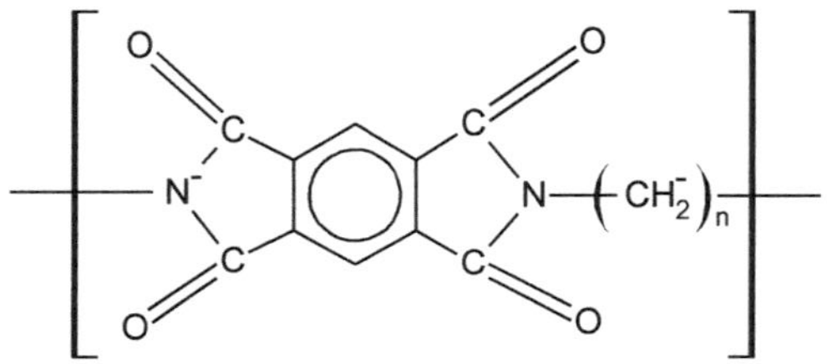 |
| 13 | Polylactic acid | PLA | Polylactide | Fiber<br>Meltspun webs | 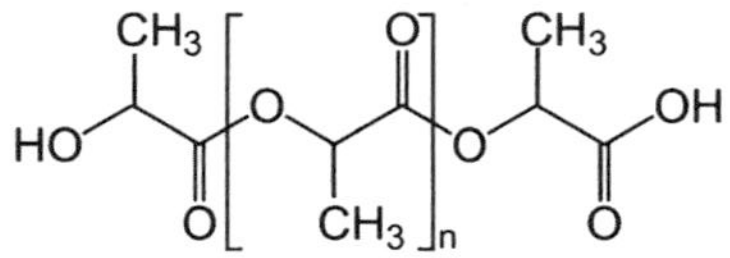 |
| 14 | Polyphenylene sulfide | Ryton® | Poly(phenylene sulfide) | Fiber<br>Membrane | 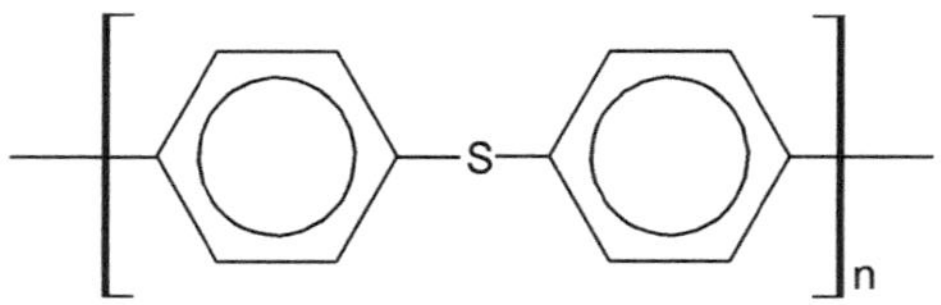 |
| 15 | Polyurethane | Spandex® | Polyurethane | Electrospun nanofibers | 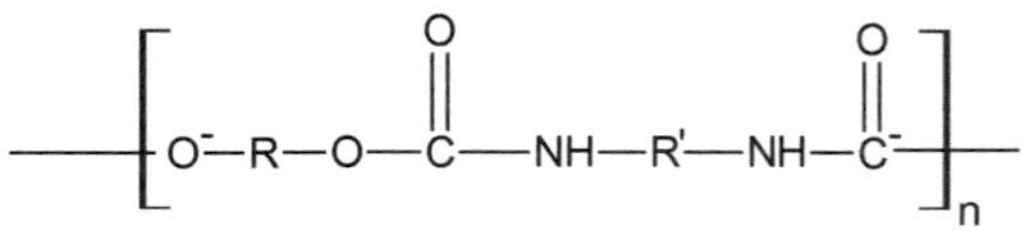 |
| 16 | Polycarbonate | Lexan® | Polycarbonate | Membrane | 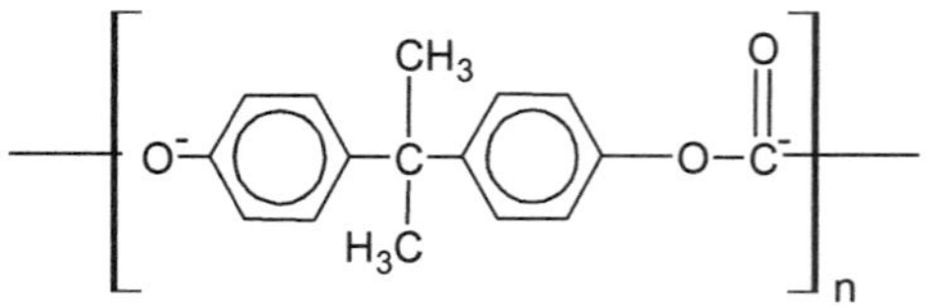 |

| | | | | | |
|---|---|---|---|---|---|
| 17 | Polytetra-fluoroethylene | PTFE<br>EPTFE<br>Teflon | Polydifluoro-methylene | Fiber additive<br>Membrane web coating<br>Filtercake release additive | 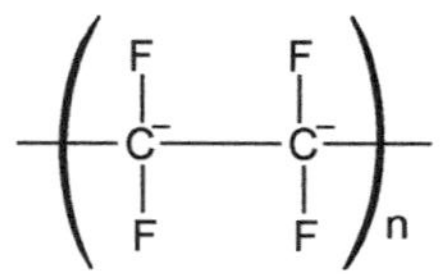 |
| 18 | Silicone | | Polysiloxane | Additive: water repellency and cake release | 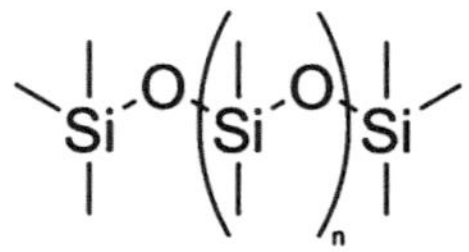 |
| 19 | Styrene Butadiene | SBR<br>Neolite | | Additive<br>Latex coating<br>Binder | 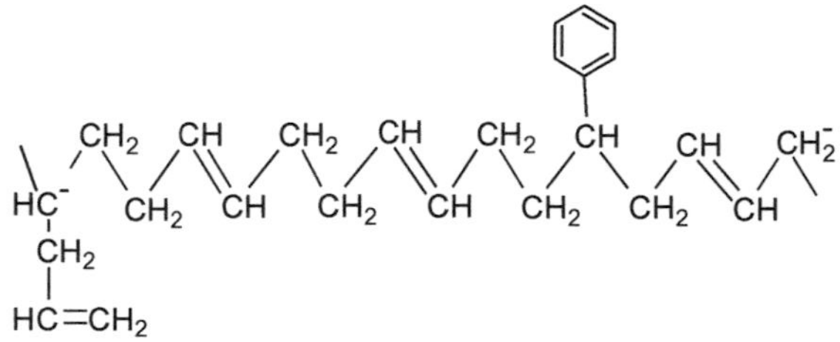 |
| 20 | Styrene-acrylic | Rhoplex®<br>Primal® | | Latex resin binder | 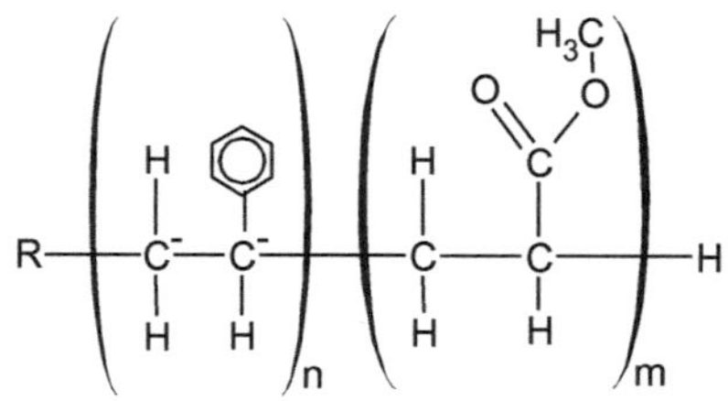 |
| 21 | Polyvinyl compounds | General | | Fibers<br>Meltspun webs<br>Flash spun webs<br>Resins<br>Membranes | 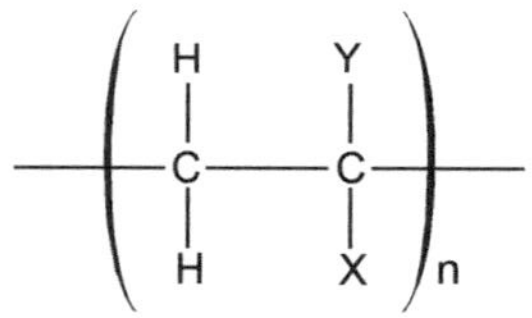 |

(Continued)

**Table 4.1** (Continued)

| Line | Polymer | Type(s) and trade names | Chemical name | Application | Structure |
|---|---|---|---|---|---|
| 22 | Polyethylene | HDPE<br>LDPE<br>UHMWPE<br>Tyvek® | Poly(methylene) | Fiber<br>Meltspun webs<br>Flash spun webs | **If Y = H and X = H**<br>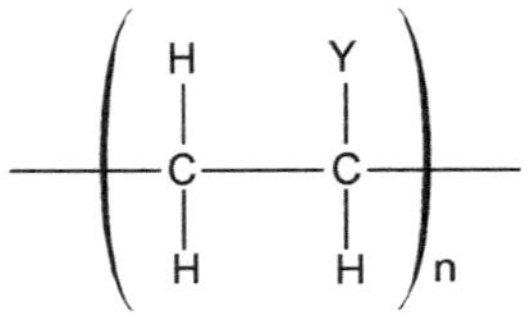 |
| 23 | Polypropylene | Typar®<br>Tekton® | Polypropylene | Fiber<br>Meltspun webs<br>Membranes | **If Y = $CH_3$ and X = H**<br>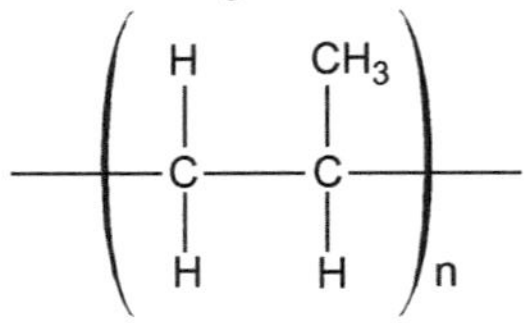 |
| 24 | Polyvinyl styrene | | Poly(1-hydroxy-ethylene) | Resin<br>Elelectrospun webs | **If Y =** 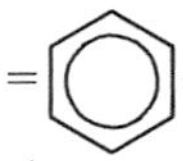 **and X = H**<br>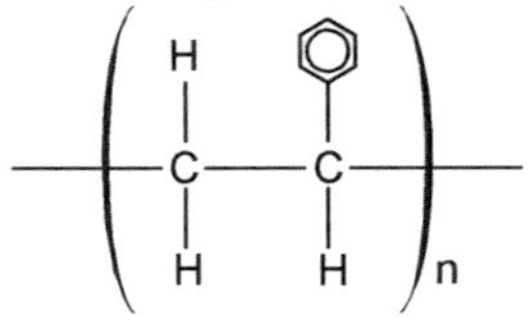 |
| 25 | Polyacrylonitrile | PAN<br>Modacrylic | Creslan 61 | Fiber<br>Resin<br>Source for carbon fiber | If Y = **C≡N** and X = H<br>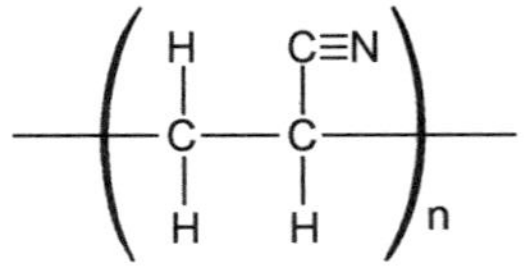 |

| | | | | | |
|---|---|---|---|---|---|
| 26 | Polyvinyl acetate | PVAC | | Latex Resin (sometimes solvent based) | If Y = $O^{-}-\overset{\overset{O}{\parallel}}{C}-CH_3$ and X = H 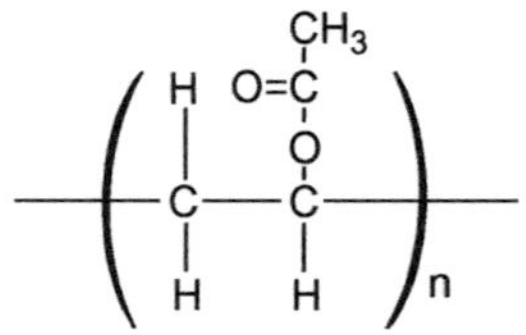 |
| 27 | Polyvinyl alcohol | Fully hydrolyzed<br>Intermediate hydrolyzed<br>Partially hydrolyzed<br>Specialty grades<br>Kuralon® | Poly(hydroxyl-ethylene) | Thermo-bonding<br>Fiber<br>Water-based resin solution<br>Thixotropic agent | If Y = OH and X = H 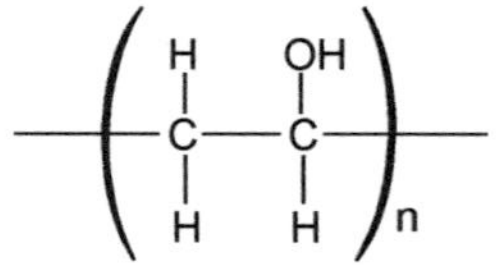 |
| 28 | Polyvinyl chloride | Latex resin | | Flame retardant latex resin | If Y = Cl and X = H 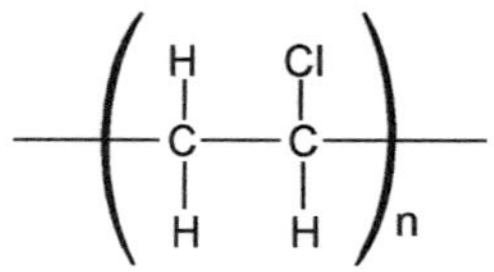 |

(Continued)

**Table 4.1** (Continued)

| Line | Polymer | Type(s) and trade names | Chemical name | Application | Structure |
|---|---|---|---|---|---|
| 29 | Polyvinylidene chloride | PVDC | Poly (1,1-dichloro-ethylene) | Flame retardant resin systems | **If Y = Cl and X = Cl** 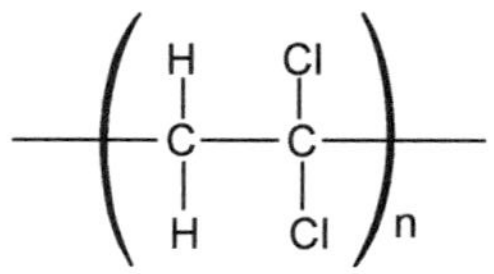 |
| 30 | Polyvinylidene difluoride | PVDF Kynar® | Polyvinylidene difluoride | Fiber<br>Meltspun and flashspun webs<br>Membranes<br>Coatings | **If Y = F and X = F** 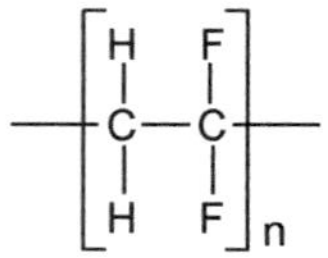 |

such as wood pulp, cotton, and other vegetable fibers are composed of cellulose, a natural polymer structure. Animal fibers such as wool and hogs hair are composed of keratin, also a natural polymer.

Important properties of polymers are melting point, glass transition temperature ($T_g$), and the degree of polymerization. According to Stille ((138) p. 31), the glass transition temperature is the temperature at which a polymer loses its hardness or brittleness, becomes more flexible, and takes on rubbery or leathery properties. Notable changes in specific volume, thermal conductivity, refractive index, stiffness, heat content, and dielectric loss are apparent. The degree of polymerization, dp, (Stille (138) p. 4) defines the molecular size of a polymer. It is the number of repeat units in the polymer chain.

Vinyl polymers are linear polymers containing ethylene as the repeat structure in its backbone. Important to the properties of vinyl polymers and the materials made from it are stereoregularity. This term refers to the regularity of the substituent groups on the main carbon chain. This regularity can greatly affect the properties of the polymer. Figure 4.1 illustrates this; A and B are hypothetical substituent groups on the carbon chain of the polymer. If all the substituent groups are in the same position on each repeating unit of the polymer chain, then the polymer is referred to as "isotactic." If they alternate on the polymer chain, then the polymer is "syndiotactic." If they are random along the chain, then they are "atactic."

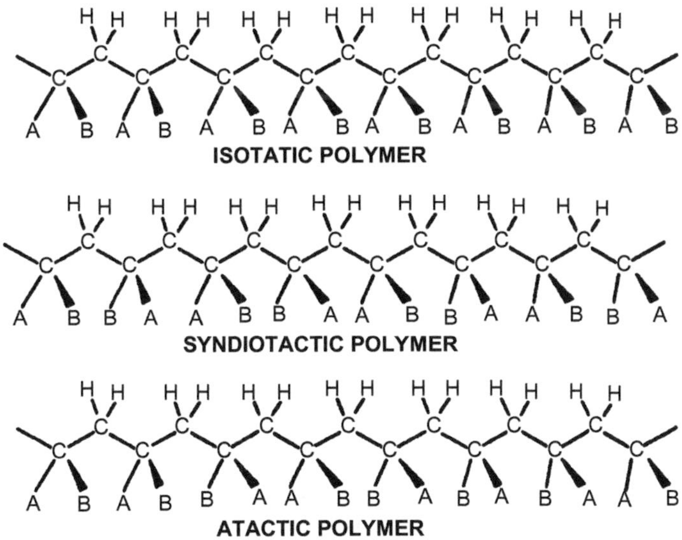


**Figure 4.1** Polymer Stereochemical Structures. *Reproduced with Permission of TAPPI, Atlanta, Georgia, USA.*

## 4.3 FIBERS

### 4.3.1 Classification

Fibers are classified by their source as follows:

1. Natural fibers
   a. Wood pulp
   b. Vegetable fibers
   c. Cotton
   d. Animal fibers (wool)
2. Biofibers
   a. Rayon (regenerated cellulose)
   b. Lyocell
   c. Polylactic acid (PLA)
3. Synthetic polymer fibers
   a. Polyester
   b. Nylon (polyamide)
   c. Polyaramid (e.g., Kevlar® and Nomex)
   d. Polyphenylene sulfide (PPS)
   e. Polyethylene
   f. Polyacrylic
   g. Polyimide (P84)
4. Inorganic and mineral fibers
   a. Glass
   b. Ceramic
   c. Carbon
   d. Asbestos
   e. Basalt
5. Binder fibers
   a. Bicomponent high-melt/low-melt polymer
   b. Polyester copolymer
   c. Polyvinyl chloride
   d. Polyvinyl acetate/polyvinyl chloride copolymer
   e. Polyamide copolymer
   f. Polyvinyl alcohol
6. Bicomponent fibers
7. Nanofibers.

### 4.3.2 Physical characteristics

The following physical characteristics of fiber are important for their use in filter media:

1. Diameter
2. Length
3. Aspect (length to diameter) ratio

4. Density
5. Linear density
6. Cross section shape
7. Length characteristics
   a. Crimp
   b. Curl
   c. Kinkiness
8. Internal structure (cellular or solid)
9. Strength properties
   a. Tensile strength and breaking length
   b. Stretch or elongation
   c. Young's modulus
   d. Stiffness
   e. Bending modulus.

### 4.3.3 Physical and chemical characteristics

Following physical and chemical properties of fibers are to be considered in their choice for use in filter media:

1. Glass transition temperature
2. Melting point
3. Additives, components, contaminants, and impurities
4. Bonding properties
5. Chemical composition
6. Chemical resistance
7. Corrosion resistance
8. Electrostatic charge
9. Water and moisture resistance
10. Moisture absorption
11. Moisture content
12. Water repellency
13. Polymer structure
14. Surface properties
15. Surface size
16. Surface tension and energy
17. Temperature and heat resistance
18. Volatile components.

### 4.3.4 Fiber properties for filtration media

The choice of fibers for nonwoven filter media depends on the properties of the medium, the process for producing the medium, and the application. Table 4.2 provided by MiniFIBERS, Inc. compares the properties of several fiber types used in filter media (139).

**Table 4.2** Comparison of Fiber Properties (MiniFIBERS, Inc.)

| Generic type | Types available | Deniers available (dpf) | Minimum cut length |
|---|---|---|---|
| Acrylic | Crimped | 1.5, 15.0 | 1/8″+ |
| | Uncrimped | 3.0, 15.0 | 1/8″+ |
| | Flock | 1.5, 15.0 | 0.3 mm+ |
| Aramid | Meta-aramid | 2.7 | 7 mm |
| | Para-aramid | 1.5 | 1/8″+ |
| Bicomponent | Sheath—core: S/C* °C | | 1/8″+ |
| | Polyester-polyester 110/250 | 2 | |
| | Ethylene Vinyl acetate-polypropylene 100/165 | 2 | 5 mm |
| | HDPE-polypropylene 130/165 | 2 | 5 mm |
| | *sheath melting point/core melting point | | |
| Nylon 6,6 | Mixed Luster - Regular Tenacity | 1.0, 3.0, 6.0, | 1/8″+ |
| | Bright - High Tenacity | 3.0 | 1/8″+ |
| Polyester | Blend | Blend | 1/8″, ¼″ |
| | Regular tenacity | 1.5 | 1/4″+ |
| | High tenacity | 3.0, 6.0, 12.0 | 1/8″+ |
| | High modulus | 3.0 | 1/8″+ |
| | Low shrink | 6.0 | 1/8″+ |
| | Undrawn (high shrink) | 25 | 1/8″+ |
| Polyethylene | Low Melt LLDPE | 6.0 | 1/8″+ |
| | High strength UHMW (special order) | 10.0 | 1/8″+ |
| Polylactic acid fibers | Fiber | 3.0 | 1/8″+ |
| | Flock | — | 0.3 mm+ |
| Polypropylene | Fiber | 1.5, 7.0, 17.0 | 1/8″+ |
| | Flock | — | 0.3 mm+ |
| Polyvinyl alcohol PVOH | | 1.8 | 1/8″+ |
| Rayon | Regular tenacity | 0.8, 4.5, 25.0 | 1/8″+ |
| | Regular tenacity trilobal | 3.0 | 1/8″+ |
| | High tenacity | 1.5 | 1/8″+ |
| | Flock | 0.8, 4.5 | 0.3 mm+ |
| Basofil® | | 2.4 | 5—10 mm |
| Melamine fibers | | | 10—25 mm |

Reproduced with permission of MiniFIBERS, Inc., Johnson City, Tennessee, USA.

The principal fiber properties for filtration performance are those designed to optimize bulk, air permeability, and pore size of the medium. Bulk and air permeability relate to the porosity (or solidity) of a web and affect the dirt holding capacity of the medium. Air permeability is important to filter design. It would determine the type, capacity, and size of the pump, blower, or fan that would be used to force fluid through the medium. Pore size relates to the size of the particle that is capable of penetrating the medium, in effect to its filtration efficiency. These properties, bulk, air permeability, and pore size, do not go hand in hand, but rather are in conflict with each other. Generally the goal is to maximize bulk and air permeability of the filter medium, and to minimize pore size. Unfortunately, the fiber properties one would choose to maximize bulk and permeability are not the properties that would reduce pore size. Fine fibers of low diameter will give high density, small pore size, and high filtration efficiency to filter media. On the other hand, coarse fibers will give bulky media and high permeability, but at the expense of pore size and filtration efficiency. Obviously, filter medium design requires compromise and imagination in the choice of fibers and their properties.

Other fiber properties have to be considered in the design of the medium. These include properties that effect medium strength, stiffness, chemical resistance, temperature resistance, and other properties that would relate to its processability and filtration. Following is a discussion of fiber properties and how they relate to filter media produced from these fibers.

#### 4.3.4.1 Diameter

In Chapter 2, a review of various filtration theories (e.g., Langmuir, Davies, Happel, Kuwabara, etc.) reveals that the fiber diameter along with porosity (or solidity) is a dominant factor affecting the filtration properties of a fibrous filter medium. In the models of Langmuir, Happel, and Kuwabara, porosity is assumed from the diameter of the fiber and the surrounding void space. Peart and Ludwig (140) reported that the number of fibers and number of pores in a medium can be related to fiber diameter as follows:

$$N_f = \frac{4M_s}{\pi \rho_f D_f L_f} \tag{4.1}$$

$$N_P = \frac{16M_s^2 L_f}{\pi^3 \rho_f D_f^4} \tag{4.2}$$

where:

$N_f$ is the number of fibers per unit area ($m^2$) of medium.
$N_P$ is the number of pores per unit area ($m^2$) of medium.
$D_f$ is the fiber diameter (m).
$M_s$ is the grammage of the medium ($g/m^2$).
$L_f$ is the fiber length (m).
$\rho_f$ is the density of the fiber ($g/m^3$).

Depending on porosity, fiber diameter is inversely related to the flow resistance of the web, i.e., the smaller the diameter, the greater is the resistance. The density of nonwoven webs tends to increase with decreasing fiber diameter.

On the other hand, reducing fiber diameter is the dominant way of increasing filtration efficiency of a nonwoven filter medium. High efficiency air filtration of nanoparticles is achieved by using nanofibers that are generally less than 0.5 μm (500 nm) in diameter. Nanofiber technology is discussed in Section 4.3.5.5.

As important as fiber diameter is to the filtration properties of the medium, it is also that difficult to measure, particularly if the cross section is not circular. Broughton and Brady (141) mention "the difficulty in measuring a fiber's physical dimensions (particularly diameter)." To further complicate the issue many natural fibers and some synthetic fibers are not round. For example, most wood pulp fibers are ribbon shaped. Often the measurement of a ribbon shaped fiber diameter is made by micrograph or image analysis techniques. In such cases the fiber diameter may be pseudo-determined from measuring the width of the ribbon. Another complication with natural fibers is that the diameter varies along the length of the fiber and from fiber to fiber. At best, some form of statistical average is necessary to express natural fiber diameters. To deal with this, weight-based systems such as linear density (see Section 4.3.4.5) have been devised for relating to a fiber's diameter.

#### 4.3.4.2 Length

The spun fiber types of webs (spunbond, melt-blown, flash-spun, centrifugal spinning and electrospun) are produced from filament fibers; therefore, fiber length is not a variable for these types of webs. Air laid, dry laid, and wet laid processes use short fibers or staple fibers of finite lengths, ranging from 25 μm, for glass microfiber, up to 12 cm, for staple fibers, used in the dry laid process. Following are length categories of fibers that are commonly used for nonwoven media.

Pulp fibers—Lengths vary based on plant source of the pulp. Hardwood pulps are generally 1−2 mm whereas softwoods are in the range of 3.5−5.0 mm. Pulped vegetable fibers such as abaca (*Manila hemp*) are as long as 6 mm. Natural fibers have a great deal of length variation, and at best the length of any pulp fiber species can only be a statistical average of what has been observed.

Staple fibers—Historically staple fibers are associated with short natural fibers such as wool or cotton that could be twisted into yarn. These types of fibers have a great deal of length variation depending on plant or animal species and growing location. In more modern times, staple fibers, including short cut synthetic fibers are recognized as those used in dry laid carding operations. Shorter lengths of staple fiber are also used in wet lay processing. In the carding process, they have to be long enough to satisfactorily entangle for web consolidation, and yet, short enough to be easily individualized by the carding action. In cases where needle punching or hydroentanglement is required, The fibers have to be long enough for desired entanglement bonding. Longer staple fibers will have more tendencies for machine direction orientation of the web. This is true in both carding and wet lay processes. Staple lengths can range from an estimated 25 to 200 mm.

Short cut fibers—Filament fibers that are cut to short lengths for wet laid or air laid purposes are also referred to as chopped fibers. Short cut lengths can vary from 3 to as long as 40 mm. Short cut fibers are differentiated not only from staple fibers by length, but also by the way they are cut. Generally, staple fibers will be cut from filament fibers as they are spun or produced in line. Short cut fibers are produced by off-line converting operations. The filament fiber is shipped to the converter in bale or spool form and precisely cut by the converter to the desired length. MiniFIBERS, Inc. is one such converter. Table 4.2 includes the type and minimum length of short cut fibers available from this supplier (142).

The length of fiber that can be used in these processes is not only process dependent, but also related to other characteristics of the fiber such as diameter and modulus. For example, the 40 mm limitation on the wet laid process is based on technology for dispersing rigid glass rovings. On the other hand, the same technology will have difficulty in dispersing much shorter 2 mm cotton linters, because of the extreme tendency of cotton to entangle in water.

The tendency to entangle, flocculate, and disrupt formation is a primary reason for process limitations of fiber length. Longer fibers tend to lay down in the machine direction (MD) of the forming process. This results in highly oriented webs, low in cross-direction (CD) strength properties.

The advantage of using longer fibers is that the entanglement effect, such as in needle punching and hydroentanglement, enhances better tensile and tear strength of the web. Webs made of longer fibers are more flexible and exhibit better resistance to flexing and tearing. Longer fibers also result in better elongation.

#### *4.3.4.3 Aspect ratio*

Aspect ratio is the ratio of fiber length to fiber diameter. It has been previously mentioned in Chapter 1 as being a condition for determining if a substrate should be called a nonwoven or a paper. It is also an important property that affects the quality and performance of the nonwoven web. Aspect ratio, even more than length is the variable that affects web formation and MD:CD orientation of the fibers. Table 4.3 reports the aspect ratios of several types of fibers used for wet lay media.

#### *4.3.4.4 Density*

The density of a fiber is the mass weight of the fiber divided by its volume ($g/cm^3$). Note in Table 4.4 the term specific gravity is used. Specific gravity is the relative density of a substance compared to the density of water at 20 °C which is 1.0 $g/cm^3$. As long as density is expressed as $g/cm^3$, the numbers will be the same or similar and there will be little confusion. However, if different units are used for density (example

**Table 4.3** Aspect Ratio and Modulus Properties of Fibers Used for Wet Lay Purposes

| Fiber | Description | Length (mm) | Equivalent fiber diameter (μm) | Aspect ratio (mm/mm) | Modulus (g/de) |
|---|---|---|---|---|---|
| Esparto | Grass | 1.3 | 9 | 144 | |
| Sisal | | 3.0 | 20 | 150 | 1300–2600 |
| Abaca | Manila Hemp | 6.0 | 24 | 250 | 1750 |
| Eucalyptus | Short fibered hardwood kraft | 1.0 | 13 | 77 | |
| Birch | Hardwood kraft | 1.9 | 28 | 68 | |
| Aspen | Hardwood kraft | 1.1 | 18 | 61 | |
| Beech | Hardwood kraft | 1.2 | 18 | 67 | |
| Redwood | Northern softwood kraft | 6.1 | 58 | 105 | 870 |
| Douglas Fir | Northern softwood kraft | 3.8 | 40 | 95 | 599 |
| West. Red Cedar | Northern softwood kraft | 3,5 | 35 | 100 | 900 |
| Slash Pine | Southern softwood kraft | 4.6 | 40 | 115 | |
| Loblolly Pine | Southern softwood kraft | 3.5 | 40 | 88 | |
| Cotton Linters | Cotton | 2.0 | 18 | 111 | 360–450 |
| 106 glass microfiber | Glass | 1.0 | 0.6 | 1667 | 1200–3000 |
| DE rovings | Glass | 6.4 | 6.5 | 985 | 1200–3000 |
| Rayon | 3d, ¼″ | 6.4 | 16 | 400 | 400–600 |
| Rayon | 3d, ½″ | 12.7 | 16 | 794 | 400–600 |
| Rayon | 1.5d, ½″ | 12.7 | 11.8 | 1076 | 400–600 |
| Rayon | 20d, ¾″ | 19.1 | 43 | 444 | 400–600 |
| PET (Polyester) | 3d, ¼″ | 6.4 | 18 | 356 | 300–600 |
| PET (Polyester) | 3d, ½″ | 12.7 | 18 | 706 | 300–600 |
| PET (Polyester) | 1.5d, ¼″ | 6.4 | 12.4 | 516 | 300–600 |
| PET (Polyester) | 1.5d, ½″ | 12.7 | 12.4 | 1024 | 300–600 |
| Nylon | 3d, ¼″ | 6.4 | 19.3 | 332 | 210–340 |
| Nylon | 3d, ½″ | 12.7 | 19.3 | 658 | 210–340 |

lb/gal), one must be careful not to use specific gravity when density was intended or vice versa.

Density of a fiber is used for the calculation of its linear density to be discussed in Section 4.3.4.5. It is also used for calculation of the porosity produced from that fiber. It will be noted that certain types of fibers such as cellulose fibers have densities different than the polymers from which they were composed. Cellulose fibers are cellular structures with lumen (openings) through the length of the fiber. Because of lumen and other parts of the cell structure, the cellular interfiber structure includes a

**Table 4.4** Specific Gravity and Moisture Content of Common Natural and Man-Made Fibers (10)

| Fiber | Specific gravity | Equilibrium moisture content (%) at 21 °C and 65% relative humidity |
|---|---|---|
| Acrylic[b] | 1.15 | 1–2 |
| Cellulose acetate | 1.32 | 6 |
| Cellulose triacetate[b] | 1.25 | 2.5–4.5 |
| Cotton | 1.54 | 7 (Commercial = 8.5) |
| Glass[b] | 2.54 | 0 |
| Polyamide (nylon 6 and nylon 66)[b] | 1.14 | 4.1–4.5 |
| Polyester[b] | 1.38 | 0.4–0.5 |
| Polyethylene[b] | 0.92 | 0 |
| Polylactic acid (PLA) | 1.25 | 0–1 |
| Polypropylene[b] | 0.90 | 0 |
| Polyurethane[b] | 1.21 | 1.0–1.5 |
| Polyvinyl chloride[b] | 1.38 | 0–1 |
| Polyvinylidene chloride[b] | 1.70 | 0 |
| Protein[b] | 1.25 | 10–18 |
| Silk | 1.37 | 9 |
| Viscose rayon | 1.51 | 13 |
| Wool | 1.32 | 13–15 |

[a]Man-Made Fiber and Textile Dictionary (10)
[b]Average of major commercial brands

considerable amount of air so that its apparent density may be considerably less than 1.55 g/cm$^3$ of pure cellulose polymer. Note that this is considered in the definition of fiber coarseness for cellulose pulp fibers. The coarseness is expressed in terms of "wall material per specified length of fiber" (Smook (143)).

### *4.3.4.5 Linear density*

The linear density of a fiber is an expression of its weight per unit length. In papermaking, the equivalent term is fiber coarseness. There are several terms for linear density and/or fiber coarseness:

1. Denier (de)—the weight in grams of 9,000 m of fiber. It is often expressed as denier per filament (dpf).
2. Tex—the weight in grams of 1,000 m of fiber.
3. Decitex—the weight in grams of 10,000 m of fiber.
4. Decigrex (dg)—papermaker's units for fiber coarseness (143): the weight in milligrams of 100 m of fiber or the weight in decigrams of 10 km of fiber.

**Figure 4.2** Southern Pine Wood Pulp Fiber. *Reproduced with Permission of Renewable Bioproducts Institute, Atlanta, Georgia, USA.*

5. Micronaire—term used by the American Cotton Industry (144): the weight in micrograms of one inch of fiber. Note: one micronaire = 2.82 de.
6. Fiber coarseness (pulp fiber)—the "weight of fiber wall material in a specified fiber length" (95), usually expressed as mg/100 m.

#### *4.3.4.6 Cross section shape*

It has been mentioned in Section 4.3.4.1 that fibers are not necessarily round in their cross section. Figure 4.2 is an electron microphotograph of a typical cellulose wood pulp fiber (Southern pine). It is quite apparent that the cross sections are anything but round. They tend to be flat or ribbon shaped. This is not a good shape for filtration purposes, because flat fibers tend to lie flat and close off pores. This increases the flow restriction of the medium. Other natural fibers such as abaca, sisal, and esparto (see Figure 4.17) tend to be of round cross sections and are more suitable for quality filter media. They also tend to be considerably more expensive than wood pulp fiber.

Synthetic fibers can be produced in ribbon shapes, cross shapes, trilobal shape, and multilobal shapes. Figure 4.23 in Section 4.3.5.3.8 shows the cross sections of P84™ polyimide fiber used in heat resistant needle felt filter bags. The advantage of trilobal and multilobal fiber in filtration applications is that it increases the filter surface area for filtration.

Trilobal fiber can be made from a number of polymers including polyester, nylon, carbon, etc. Dugan (145) discusses segmented bicomponent fibers (see Section 4.3.5.6)

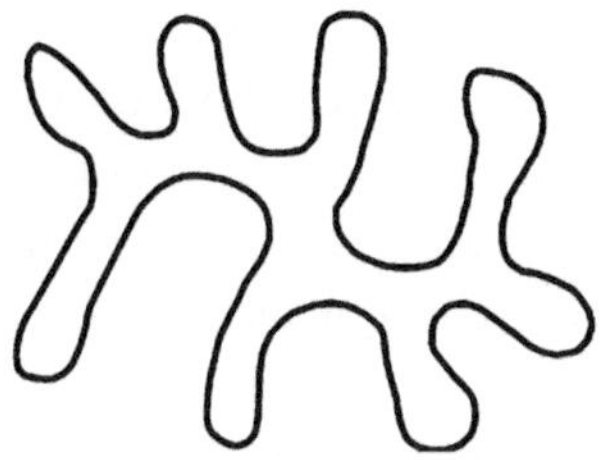

**Figure 4.3** Fiber Innovations Technologies 4DG™ Fiber Cross Section. *Reproduced with Permission of Fiber Innovations Technologies, Johnson City, Tennessee, USA.*

that can be split into smaller fibers. He mentions that these fibers when used in cross or trilobal shape provide for improved processability in carding operations.

Fiber Innovations Tecnologies, Johnson City, Tennessee, USA, markets a 4DG™ (deep groove) multilobal fiber that has cross section shape characteristics as shown in Figure 4.3. Fiber Innovation Technologies (146) indicates that, among other things, the deep groove properties provide for increased surface area and bulk—desirable properties for a filter medium.

#### *4.3.4.7 Moisture content and moisture absorption*

The moisture content of a fiber is related to its surface properties and physical properties. Hydrogen bonding, which will be described in the discussion of cellulose fibers below, is a major determinant of how much moisture a fiber can absorb. Hydrogen bonding sites include hydroxyls (−OH) and/or amides (−NH2). The greater the number of hydrogen bonding sites, the greater is the amount of moisture that the polymer might absorb.

Crystallinity also has an effect on moisture absorption. Usually moisture absorption will occur only in the amorphous regions of a fiber and will not invade the crystalline regions of a fiber. For highly amorphous fibers, water absorption can also cause swelling of the fibers. Broughton and Brady (141) suggest that an amorphous fiber having at least one hydrogen-bonding site for every two carbons in the polymer backbone can absorb up to about 12% of its weight in moisture. This may have undesirable effects in a filter medium. For example, fiber swelling may cause pore closure resulting in greater flow resistance and loss of filtration efficiency.

In a cellular fiber structure after wetting, moisture retained within the cellular void space of the fiber is difficult to remove by mechanical means such as pressing or vacuum. This part of the moisture content can best be removed by evaporation or drying.

The equilibrium moisture content of a fiber is that which is in equilibrium with its environment. This depends on the temperature and relative humidity of the environment. The equilibrium relative humidity is the relative humidity at a given temperature at which exposure of the fiber will not result in either moisture absorption or desorption. This will only happen if the fiber is already at the equilibrium moisture content for that humidity.

In measuring the equilibrium moisture content of a fiber, or for that matter any substance, one must be careful about hysteresis effects. This means that the equilibrium moisture content will be different depending on whether the substance is approaching equilibrium from a dryer state or from a wetter state. For example, fibers conditioned at 50% relative humidity will have greater moisture content when this environment is approached from a higher relative humidity than from a lower relative humidity. For this reason TAPPI T402 om-93 recommends that pulp products be preconditioned at a lower humidity prior to conditioning at conditions for equilibrium moisture content.

Table 4.4 lists the equilibrium moisture contents of several types of natural and man-made fibers.

#### *4.3.4.8 Thermal properties*

##### 4.3.4.8.1 Melting point

The melting point of polymer fibers has many ramifications to filter media. It is the temperature at which the crystallized segments of the polymer are transformed into noncrystalline forms. In other words, the point at which they soften and melt.

Fibers from high melting polymers will implicitly have better heat resistance and be used accordingly; for example, high-temperature baghouse operations. On the other hand there is a need for low melting polymers in thermobonding operations. Thermobonding fibers or the thermobonding components of bicomponent fibers have to be melted and liquefied enough to be able to flow into and conform to fiber intersections.

Some fibers such as cellulose have melting points, however, will undergo thermal and chemical decomposition at temperatures well below their melting point.

##### 4.3.4.8.2 Glass transition temperature ($T_g$)

Glass transition temperature ($T_g$) is a fiber property that can dramatically affect the performance of the fabric, because it relates to amorphous regions of the fiber. Bresee (147) points out that all polymeric objects contain at least some amorphous regions of noncrystalline material. These regions are stiff and brittle and become rubbery above their $T_g$. The cause is polymer segmental mobility. Some defects caused by physical handling and aging can be simply erased by heating the polymer above its $T_g$. An example of this is the ironing or textile fabrics to remove wrinkles. In Section 4.3.4.7, it was noted that amorphous regions have more tendency to absorb water. Water acts as a plasticizer and lowers the $T_g$ of the polymer. This is particularly true for natural fibers, such as cotton, that have high levels of moisture absorbency. For example, the ironing process, mentioned above, requires the use of steam and moisture.

Polymers do not crystallize below the $T_g$, but may crystallize above it. Some rayon fibers of low crystallinity will undergo distinct changes in properties, such as hardness and stiffness, when crystallized.

$T_g$ is almost always lower than the melting point. Bresee (147) reports the relationship to be

$$T_g = \frac{2}{3} T_m \qquad (4.3)$$

$T_g$ is the glass transition temperature, K.
$T_m$ is the melting point, K.

#### *4.3.4.9 Strength properties of fibers*

Two important properties of fibers are tenacity and modulus. Tenacity is the breaking strength of the fiber, usually expressed as g/de (denier). Modulus relates to the rigidity or flexibility of the fiber and is also expressed in units of g/de. High modulus fibers result in bulkier webs, however, they will be weaker because the bonding area at fiber intersections is reduced. High tenacity fibers do not necessarily produce stronger webs. This often depends on how well the fibers are bonded. If the fibers are well bonded, then the strength will be enhanced by the tenacity of the fibers. Table 4.4 lists modulus data for various fibers used in the wet laid process.

### 4.3.5 Fiber types

#### *4.3.5.1 Natural fibers*

##### 4.3.5.1.1 Wood pulp

Wood pulp fibers used for making paper are a dominant fiber material in paper filter media and in wet laid nonwoven filter media. All wood pulps consist of cellulose, although in varying degrees of purity and contamination. Contaminants such as lignins are important to the properties of the fiber, either because they are there or because they are not there. Filter media manufacturers are concerned with the purity of a pulp as measured by its α-cellulose content. High purity pulps do not bond very well and form weak, bulky structures. Yet, bulk is important to filter media, and to achieve it, wet laid filter media manufacturers will find ways to compensate for the lower strength properties. Caustic treated, mercerized pulps are of very high purity (99+% α-cellulose) and are known as high bulk pulps. They are important to wet laid filter media because they provide for bulk and permeability that cannot be achieved with less pure wood pulps. Several factors influence the properties of wood pulp fibers. These include:

1. Type of tree
2. Tree source
3. Growing season
4. Growing location
5. Pulping process.

Two basic types of trees from which wood pulp is made are hardwoods and softwoods. Hardwoods are broad leaf (deciduous or evergreen) trees. Hardwoods used for papermaking include Oaks, Gum trees, Birch, Beech, and Aspen. Eucalyptus is also a form of hardwood whose fiber is becoming increasingly popular in filter paper.

Softwoods or conifers are normally woods from what is known as pine trees. There is a wide variety of pine trees and each type offers specific properties to the wood pulp fiber. Softwoods used for papermaking include Spruce, Fir, Cedar, Hemlock, Redwood, Loblolly pine, and Slash pine.

Fibers produced from softwood trees are typically longer and of coarser diameter than fiber produced from hardwood trees. This is important because both types are incorporated into the design of filter media; softwood to provide bulk, strength, and permeability, whereas the finer fibered hardwoods provide for filtration efficiency. The following comparison illustrates the typical properties of the two types of fibers:

| Type | Length (mm) | Diameter (μm) | Aspect ratio |
|---|---|---|---|
| Softwood | 4.0 | 40 | 100 |
| Hardwood | 2.0 | 22 | 90 |

Northern softwoods have distinctly different properties than Southern softwoods. This is largely due to the difference in growing seasons. There are two parts to a growing season: spring when springwood or earlywood is grown and summer when summerwood or latewood is grown. The properties of summerwood are quite different from springwood and explain the visible ring structure in the trunks of trees. Tracheids are long tapering cells of wood that become the fibers. A tracheid consists of a wall structure and a long open canal called the lumen. Springwood tracheids are of lower density than summerwood tracheids and are characterized by thin wall structures and large lumen. On the other hand, summerwood may have two or three times the density of springwood and is characterized by thick wall structure and small lumen. Figure 4.4 is an electron micrograph of a wood specimen from a Southern yellow pine revealing the transition from earlywood to latewood. It reveals how the tracheid structure is changed in the transition. The transition from earlywood to latewood is the change at the bottom of the picture. Note how abrupt it is (Figure 4.4).

Southern softwoods have a longer summer growing season than do Northern softwoods; therefore, they have a higher percentage of thicker-walled latewood fiber. For this reason Southern softwoods provide for bulkier structures. This is illustrated in Figure 4.5. The thick-walled fiber structure being less conformable than the thin wall structure would form a less compressible and thereby bulkier paper medium. To further illustrate this effect, Figures 4.6 and 4.7 are microphotographs of handsheets made from earlywood and latewood fibers of Southern pine, respectively. The flat ribbon-like shapes of the earlywood fibers are in contrast to the more three-dimensional shape of the latewood fibers.

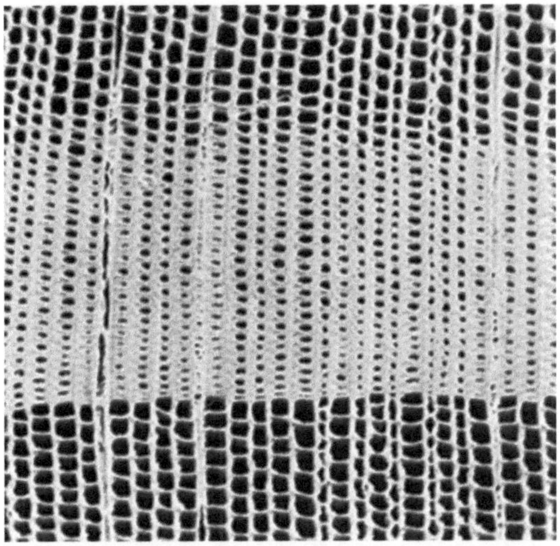

**Figure 4.4** Earlywood to Latewood Transition in a Southern Yellow Pine ((148), p. 9). *Reproduced by Permission of the Institute of Paper Science and Technology, Atlanta, Georgia, USA.*

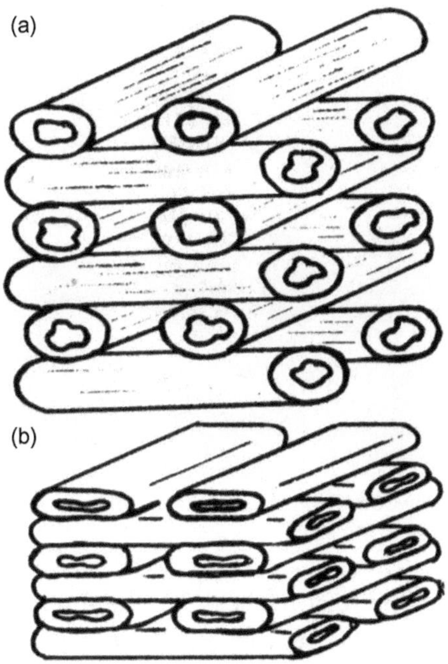


**Figure 4.5** Idealized Fiber Structures. The Thick-wall Fibers are Less Conformable than the Thin-walled Fibers. *Reproduced with Permission of Angus-Wilde Publications Inc. Vancouver, British Columbia, Canada (143).*

**Figure 4.6** Southern Pine Earlywood. *Reproduced with Permission of the Institute of Paper Science and Technology, Atlanta, Georgia, USA (148).*

**Figure 4.7** Southern Pine Latewood. *Reproduced with Permission of the Institute of Paper Science and Technology, Atlanta, Georgia, USA (148).*

It has been mentioned above that tracheids are the fundamental building blocks of a wood that is converted to a fiber form. Figure 4.8 illustrates the structure of a softwood tracheid. Note that the tracheid consists of a number of layers. The layers are as follows:

Middle lamella (ML)—bond between fibers, mostly lignin

Primary wall (P)—a thin, relatively impermeable covering about 0.05 μm thick

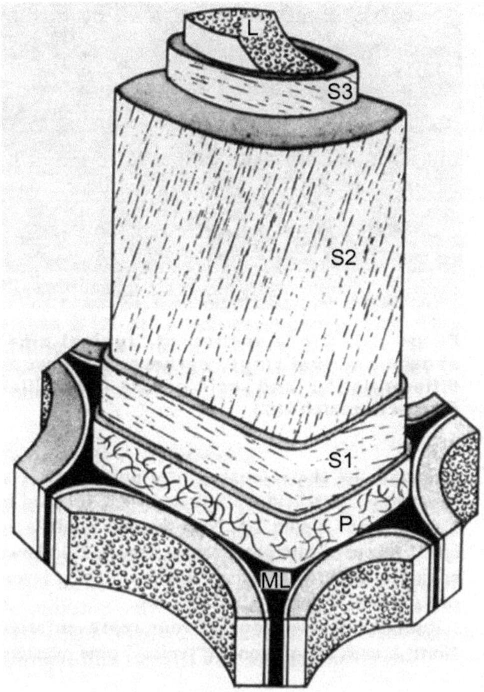


**Figure 4.8** Tracheid Cell Wall Organization. *Reproduced by Permission of Angus Wilde Publications, Inc., Vancouver, British Columbia, Canada (143).*

Secondary wall—makes up bulk of cell wall; forms three distinct layers characterized by different fibril alignments:

- S1 is the outer layer of secondary wall (0.1–0.2 μm thick)
- S2 forms the main body of the fiber and is from 2–10 μm thick
- S3 is the inner layer of the secondary wall (about 0.1 μm thick)

Tertiary wall (T)—same as S3

Lumen (L)—the central canal of the fiber.

The pulping process breaks down tracheids into individualized wood pulp fibers by destroying the lignin bonding that holds the fibers together in the ML layer. There are several types of pulping processes that includes mechanical pulping, chemical pulping, and semichemical pulping. Chemical pulping includes the alkaline kraft process and the acidic sulfite process. The kraft process uses sodium hydroxide (NaOH) and sodium sulfide ($Na_2S$) to break down the molecules of lignin into smaller segments. The sulfite process uses sulfurous acid ($H_2SO_3$) and bisulfite ($HSO_3^-$) ion to attack and solubilize the lignin.

The kraft process is by far the favorite process for producing filter pulps. The word "kraft" is German for strong. Accordingly, kraft pulps are characterized for being stronger. The wood raw material, depending on its source, is broken down into chips. The chips are then fed to a digester, which can be either a batch or continuous process. In the batch process the wood chips are fed to a large digester vessel. The cooking liquor containing the NaOH and $Na_2S$ is then added to completely cover the chips. The contents are then heated, usually by forced circulation of the cooking liquor through a heat exchanger. Maximum temperature is typically reached in 1.5 h. The cook is then maintained at its cooking temperature for the predetermined cooking time. Typical cooking conditions for a kraft cook are:

Temperature: 170–180 °C

pH: 13+

Cooking time: 2–4 h

When the cooking cycle is completed, the contents are discharged into a blow tank where the softened chips are disintegrated into fibers. The cooking liquor, now known as black liquor, is extracted from the pulp and sent to a chemical recovery system. The pulp is separated from the residual liquor in a process known as brown washing. The pulp is then screened and cleaned.

The pulping process may be followed by a bleaching process depending on whether bleached or unbleached pulp is desired. The washed pulp from the cooking process will have an inherently yellow or brownish color. The bleaching process makes it into the white color usually associated with high-grade paper. Most pulps used for filter media are bleached. Some filter media use unbleached pulp in applications where there are no quality demands for whiteness.

The secondary (S2) wall layers of Figure 4.8 are composed of microfibrils. Microfibrils are bundles of cellulose molecules. Their orientation relative to the axis of the fiber influences the characteristics of the fiber. This orientation is often referred to as the "fibril angle." The fibrils might be considered as a sheath around the fiber. If the fibril angle is low so that the fibrils circle around the fiber like a coil, then the fiber will be quite flexible, soft, and have a low modulus (e.g., cotton). On the other hand if the fibrils are straight and parallel to the fiber axis, then the fibers will tend to be rigid, stiff and have a high modulus (e.g., abaca).

Fibrillation is the shearing of wet wood pulp fibers to loosen the microfibrils from the fiber surface and fiber wall. In effect, the microfibrils are unraveled, and the fiber is "frayed" similar to the fraying of a cotton thread. Shearing the stock slurry of fibers between two steel surfaces causes fibrillation; one surface is part of a moving rotor, the other is static. Generally, there are two types of refiners used for the refining of cellulose filter media: disc refiners and conical refiners. Refining or beating is the process by which fibrillation occurs. These process steps are discussed in Chapter 5, Sections 5.6 and 5.6.1. Figure 4.9 is a micrograph of fibrillated fiber showing the unraveling of the microfibrils. The unraveled microfibrils are seen to entangle web across fibers and increase their bonding area. It will be noted in Section 4.3.5.5, that microfibrillated cellulose is a source of nanofibers. Microfibrils have diameters as low

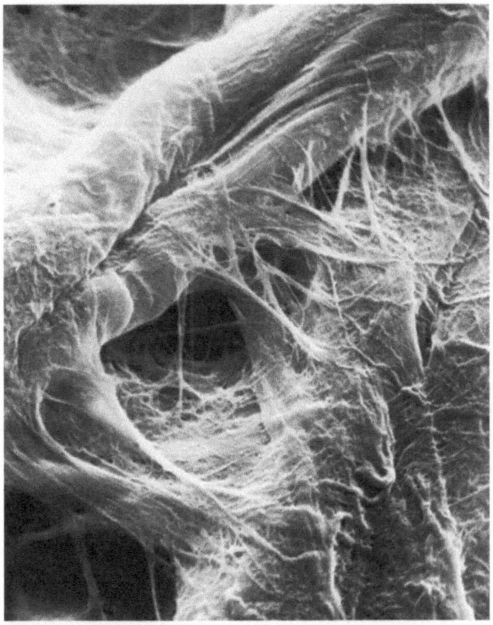

**Figure 4.9** Fibrillated Softwood. *Reproduced by Permission of the Institute of Paper Science and Technology, Atlanta, Georgia, USA ((148) p. 53).*

as 3.5 nm. Fibrillation by refining increases the strength of a wood pulp substrate. It also reduces the air permeability. Wet lay filter machines producing cellulose filter media will use refining as a fine tune control for controlling permeability.

#### 4.3.5.1.2 Handsheet testing of wood pulp fibers

Numerous tests can be conducted on wood pulp fibers. Among these are fiber length, diameter, coarseness, modulus and strength, freeness, and chemical purity. For the filter media manufacturer, laboratory handsheet testing is the most informative. Based on handsheet testing, the manufacturer can categorize and evaluate various pulps for their suitability in filter medium products. It is the basic experimental tool for furnish design and evaluation. Handsheets are tested for air permeability, pore size, strength properties, bulk, and other related properties. The handsheet procedure also provides a way of noting the dispersion qualities of the pulp. Handsheets are often used as base sheets for resin impregnation and subsequent resin studies. Handsheets, with and without resin treatment, can be subject to several types of flat sheet filter tests to estimate and compare furnish performance.

The TAPPI specified handsheet procedure is T 205 sp-12. Filter media laboratories will vary considerably from this procedure because the dense structure that results from this procedure is not consistent with the bulky structures desired for filter media. Filter medium companies have provincial procedures. There does not appear to be much standardization between companies. It is even likely that different mills in the

**Figure 4.10** Valley Beater—Shows the Beater Roll and its Protruding Bars. *Reproduced by Permission of Hollingsworth and Vose Co. Inc., East Walpole, Massachusetts, USA.*

same company will not have the same procedure. What is important is that the parallels be the same. Different mills may get different handsheet numbers for the same pulp, but the results have to correlate. Each mill designs its handsheet procedure based on the wet lay machine(s) at its' mill and the types of products they produce.

The procedure involves the following steps:

1. Dispersion of the fiber—usually done in a disintegrator of laboratory beater.
2. Forming the handsheet in a sheet mold.
3. Couching of the handsheet from the forming fabric.
4. Drying of the handsheet on a heated surface.

Disintegrators or beaters are used to disperse fibers into an aqueous slurry. Disintegrators are vessels with an agitator at the bottom and operate much like a Waring blender. Beaters are oval shaped vessels that circulate the slurry past a beater roll that has protruding bars to rub the fibers against a flat steel breastplate. A laboratory beater, commonly called a valley beater, is depicted in Figure 4.10.

A sheet mold is a vessel that has a wire or forming fabric at the bottom. It is either cylindrically shaped or square shaped. The vessel is filled with water, and a measured amount of stock from the disintegrator is added. The pulp is mixed, usually by a plunger with a perforated plate that fits closely to the dimensions of the sheet mold. The plunger is manually pumped up and down the inside of the sheet mold to adequately mix and disperse the fibers into the water. After the fibers are adequately dispersed a valve at the bottom of the mold is opened. The water drains through the forming fabric. The fibers are retained on the fabric to form the wet handsheet. A square 30.5 × 30.5 cm (12 × 12 in) sheet mold is depicted in Figure 4.11. The perforated plunger is shown resting on top of it.

**Figure 4.11** The 30.5 × 30.5 cm (12″ × 12″) Sheet Molds—the One on the Left is in the Closed Position. The One on the Right is in the Open Position and Reveals the Forming Wire Over the Water Drainage Area and the Up and Down Perforated Plate Plunger for Agitating the Fiber Dispersion. *Reproduced by Permission of Hollingsworth and Vose Co. Inc., East Walpole, Massachusetts, USA.*

The wet formed handsheet is removed from the wire by couching (pronounced kooching). The couching procedure involves mechanically removing enough water from the handsheet so that it can be peeled off the forming wire. The water is removed by sandwiching the wire and sheet between blotters and applying some form of pressure to help blot out the water. In some places it is done by vacuum suction from the wire side.

The couched sheet is dried on a surface heated dryer. Electric photo dryers work very well for this purpose. Once dried the handsheet is tested for basis weight, air permeability and a host of other properties. It is important that the handsheet be made to a specific basis weight target so that all handsheets are compared on the same basis.

Table 4.5 is a tabulation of handsheet data on a broad range of filter pulps. The tabulation was presented in a brochure by Buckeye Technologies, Inc.[1] of Memphis, Tennessee, USA. Note that the handsheet properties measured include air permeability (Frazier), tensile strength, density, bulk, and mean flow pore size. The table also lists fiber properties as determined by image analyzing techniques of the fiber slurries. Curl is a measure of the nonstraightness of the fibers. Weighted length is an average length of all the fibers tested in the sample. Coarseness was determined from the total weight of sample and the total measured length of all the fibers in the specimen.

[1] Buckeye Technologies, Inc. was purchased by Georgia Pacific, Atlanta, Georgia, USA, in 2013.

**Table 4.5** Wood Pulps Commonly Used in Filter Paper

| | Properties of unpressed TAPPI handsheets | | | | | | | |
|---|---|---|---|---|---|---|---|---|
| | Frazier (cfm/sq ft) | Tensile (g/in) | Density (g/cc) | Bulk (cc/g) | Mean pore size (μs) | CURL (%) | Weighted length (mm) | Coarseness (mg/100m) |
| HPZ III (Mercerized slash pine) | 455 | 52 | 0.141 | 7.09 | 58.6 | 88 | 1.80 | 49.6 |
| HPZ (Mercerized slash pine) | 308 | 80 | 0.157 | 6.37 | 46.6 | 50 | 2.37 | 42.5 |
| Buckeye cotton linters Grade 512 | 195 | 267 | 0.167 | 5.99 | 39.5 | 39 | 1.62 | 26.5 |
| PV-5 | 132 | 315 | 0.198 | 5.05 | 36.1 | 39 | 2.14 | 38.1 |
| HP-11 (Predominantly slash) | 64 | 1710 | 0.232 | 4.31 | 29.5 | 28 | 2.72 | 36.2 |
| Flash-dried Southern hardwood Kraft | 70 | 262 | 0.198 | 5.05 | 23.6 | 30 | 0.84 | 19.4 |
| Predominantly slash pine Kraft | 64 | 1635 | 0.243 | 4.12 | 28.8 | 25 | 2.62 | 35.4 |
| Slash/loblolly pine Kraft | 50 | 1954 | 0.254 | 3.94 | 29.1 | 23 | 2.73 | 33.3 |
| Esparto | 47 | 1570 | 0.225 | 4.44 | 19.1 | 18 | 1.02 | 9.1 |
| Eucalyptus | 12 | 1351 | 0.296 | 3.38 | 11.6 | 21 | 0.75 | 12.6 |
| Northern softwood Kraft | 10 | 2887 | 0.329 | 3.04 | 13.1 | 21 | 2.28 | 17.6 |

The fiber of different wood species varies considerably in such morphological features as fiber length, diameter and wall thickness. These differences affect paper properties such as porosity, tensile, bulk, and pore size. Special treatment in the pulping process can also affect papermaking properties. Shown above are the properties of fibers with a range of morphological and processing differences.

It should be noted that handsheet testing of pulp fibers is not limited to just wood pulps. It can also be used for vegetable fibers, synthetic fibers, inorganic fibers and any other fibers that can be dispersed into a fiber slurry and reasonably formed on a handsheet forming fabric. In the discussion of vegetable fibers to follow, some handsheet data is presented. Some fibers such as polyester may be difficult handsheets, because their wet web strength is too weak for couching and downstream handling.

#### 4.3.5.1.3 Vegetable fibers

A number of natural fibers used in nonwoven filter media, come from plants other than trees. They are referred to as vegetable or non-woody fibers. Typical length, diameter, and aspect ratios are listed in Table 4.6. Chemical compositions of some common vegetable fibers are listed in Table 4.7. It is important to note from which part of the plant the fiber came. This is indicated in Table 4.6. For example, cotton is a seed hull fiber. Fibers like flax, jute, hemp, and ramie are bast fibers; fibers obtained from the inner bark or phloem of a woody plant. Abaca and sisal are leaf fibers. Kenaf is listed as a core fiber, however, the bast fibers from this plant are of more interest to filter media. Esparto is a grass fiber that has important applications in filter media. Other vegetable fibers such as those from cane (bamboo and sugar) and stalks (rice, wheat, barley, oat, and rye) have utility as paper making pulps, however, they have not gained acceptance as fibers for filter media.

**Table 4.6** Average Length, Average Diameter and Length/Diameter Ratio of Various Vegetable Fibers

| Fiber source | Fiber type | Length (mm) | Diameter (μm) | Aspect Ratio |
|---|---|---|---|---|
| Woods | Coniferous (softwood) | 4.0 | 40 | 100 |
| | Deciduous (hardwood) | 2.0 | 22 | 90 |
| Straws and Grasses | Rice | 0.5 | 9 | 60 |
| | Esparto | 1.1 | 10 | 110 |
| | Misc. (wheat, rye, sabai) | 1.5 | 13 | 120 |
| Canes and Reeds | Bagasse (sugar cane) | 1.7 | 20 | 80 |
| | Miscellaneous | 1.2 | 12 | 100 |
| Bamboo | Several varieties | 2.8 | 15 | 180 |
| Woodystalks with bast fibers (jute, flax, kenaf, cannabis) | Woody stems | 0.25 | 10 | 25 |
| | Bast fibers[a] | 20 | 20 | 1000 |
| Bast fibers[a] | Linen (flax) | 55 | 20 | 2600 |
| | Ramie | 130 | 40 | 3500 |
| Leaf fibers | Abaca (Manila hemp) | 6 | 24 | 250 |
| | Sisal | 2.8 | 21 | 130 |
| Seed fibers | Cotton | 30 | 20 | 1500 |
| | Cotton linters | 20 | 20 | 1000 |

[a]Fibers obtained from inner bark.

Reproduced with permission of Angus Wilde Publlcations Inc., Vancouver, British Columbia, Canada.

**Table 4.7** Chemical Composition of Some Common Vegetable Fibers (% of Total)

| Type of fiber | Cellulose | Lignin | Pentosan | Ash | Silica |
|---|---|---|---|---|---|
| ***Stalk fiber*** | | | | | |
| Rice | 28–48 | 12–16 | 23–28 | 15–20 | 9–14 |
| Wheat | 29–51 | 16–21 | 26–32 | 4.5–9 | 3–7 |
| Barley | 31–45 | 14–15 | 24–29 | 5–7 | 3–6 |
| Oat | 31–48 | 16–19 | 27–38 | 6–8 | 4–6.5 |
| Rye | 33–50 | 16–19 | 27–30 | 2–5 | 0.5–4 |
| ***Cane fiber*** | | | | | |
| Sugar | 32–48 | 19–24 | 27–32 | 1.5–5 | 0.7–3.5 |
| Bamboo | 26–43 | 21–31 | 15–26 | 1.7–5 | 0.7 |
| ***Grass fiber*** | | | | | |
| Esparto | 33–38 | 17–19 | 27–32 | 6–8 | – |
| Sabai | – | 22.0 | 24 | 6 | – |
| ***Reed fiber*** | | | | | |
| Phragmites communis | 44–46 | 99–24 | 20 | 3 | 2 |
| ***Bast fiber*** | | | | | |
| Seed flax | 43–47 | 21–23 | 24–26 | 5 | – |
| Kenaf | 44–57 | 15–19 | 22–23 | 2–5 | – |
| Jute | 45–63 | 21–26 | 18–21 | 0.5–2 | – |
| Hemp | 57–77 | 9–13 | 14–17 | 0.8 | – |
| Ramie | 87–91 | – | 5–8 | – | – |
| ***Core fiber*** | | | | | |
| Kenaf | 37–49 | 15–21 | 18–24 | 2–4 | – |
| Jute | 41–48 | 21–24 | 18–22 | 0.8 | – |
| ***Leaf fiber*** | | | | | |
| Abaca (Manila) | 56–63 | 7–9 | 15–17 | 3 | – |
| Sisal (agave) | 47–62 | 7–9 | 21–24 | 0.6–1 | – |
| ***Seed hull fiber*** | | | | | |
| Cotton linter | 85–90 | 0.7–1.6 | 1–3 | 0.8–2 | – |
| ***Wood fiber*** | | | | | |
| Coniferous | 40–45 | 26–34 | 7–14 | <1 | – |
| Deciduous | 38–49 | 23–30 | 19–26 | <1 | – |

Reproduced by permission of TAPPI, Atlanta, Georgia, USA ((149), p. 78).

**Figure 4.12** Cotton Ball.

**4.3.5.1.3.1 Cotton** As a fiber for nonwoven purposes, cotton has the following properties: absorbency, breathability, high wet modulus (i.e., stronger when wet), biodegradability (under certain conditions), excellent wicking, wipe dry performance, sterilizable by all industrial methods, low static potential, printability, dyeability and chemical modifiability, and finally it is a renewable resource.

Cotton is the product of a woody shrub that is grown from an annual field crop. When it is harvested, the fiber and seed are removed from the plant as seed cotton. Ginning is the process by which the fibers are separated from the seed. The ginned cotton is compressed into bales and referred to as raw cotton.

The separated seeds are used for making cotton oil. They still contain short fibers known as linters. At the oil mill, linters are removed from the seed by delinting machines. The fibers removed are known as mill run linters. Usually the seeds are run through the delinting machines twice to produce first-cut and second-cut linters. Cotton staple and first-cut linters are a source of fiber for many nonwoven products including needle felt filter media. First- and second-cut linters are used in wet lay filter products. They also are an alternative to high alpha wood pulp for providing bulk and permeability to wet lay filter media. Actually, it is the other way around. The high alpha wood pulps were developed as a lower cost replacement for cotton to provide high air permeability to wet laid filter media.

Another by-product of the gin is motes. Motes are small, immature seeds with attached fiber that are removed at a different stage of the gin stand than the mature seeds. The fiber can be removed from the motes using a delinting machine. This fiber is called gin mote fiber and is used in nonwoven products.

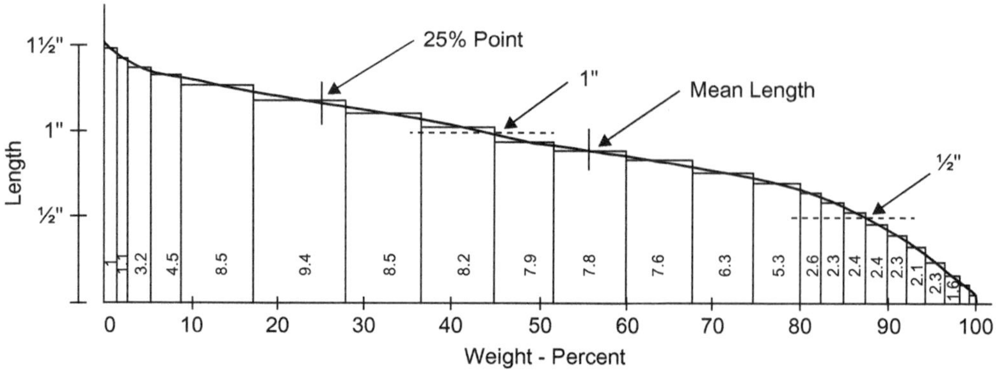


**Figure 4.13** Cotton Fiber Length Array. *Reproduced with Permission of Cotton Incorporated, Cary, North Carolina, USA.*

According to Cotton Incorporated (150) classification or classing of cotton is based on three prime parameters: micronaire, length, and strength. These and other properties are measured by a test instrument known as a High Volume Instrument (HVI). It is manufactured by Uster Technologies, Switzerland.

1. Micronaire—Micronaire already defined in Section 4.3.4.5 is determined by compacting a 2.34 g specimen to a specific volume in a porous chamber. Air is forced through the chamber and the resistance to airflow is proportional to the linear density of the fibers (expressed in μg/in). TAPPI 1205-42 (Boyer (144)) states "the micronaire of cotton is governed by the variety of cotton planted and varies within each variety due to fiber diameter and actual amount of cellulose present. In general, shorter staple length cottons tend to be coarser, and longer staple length cottons tend to be finer."
2. Length—Cotton fiber length varies genetically. It follows a fiber length array or fiber length distribution as demonstrated in the graph of Figure 4.13. American upland cottons are classified as short staple, medium staple, or long staple (see Section 4.3.4.2). Reportable length is the measure of the average length of the longer one-half of the fibers (upper half mean length), reported in both 100th and thirty-seconds (1/32) of an inch. Length uniformity is determined by dividing the mean length of the fibers by the upper mean length and reporting as a percentage. The higher the percentage, the greater the uniformity. If all the fibers in the sample were of the same length, the mean length and the upper half mean length would be the same, and the uniformity index would be 100. However, cotton fibers within a sample vary considerably, therefore, length uniformity will be less than 100.

   The length of most waste cotton, which includes linters, comber noils, strips, and motes, will be shorter than raw, unprocessed fiber. Comber noils are the short fibers removed by a combing operation. The machine doing this is called a comber. The waste fibers, usually 0.5 in (12.7 mm) or less in length are often used in nonwoven materials and filter media.

3. Strength—Fiber strength is one of the properties measured by the HVI tester. It is the force to break a bundle of fibers one tex unit (1 g per 1,000 m) in size. It is expressed in g/tex.
4. Nep count—Nep count is another property important to cotton. Neps are small tangled knots of fiber caused by mechanical processing. It is measured by AFIS-M tester also manufactured by Uster Technologies, Switzerland.

Raw cotton, as harvested and ginned, contains considerable contaminants and impurities, as well as oils and waxes that makes it hydrophobic. Raw cotton has some utility in industrial nonwovens where absorbency and aesthetics are unimportant, however, for applications where absorbency, whiteness, and purity are desired the cotton has to be scoured and bleached. Figure 4.14 is a block diagram of the bleaching process. Table 4.8 lists the properties of bleached cotton as published TAPPI TIP 1205-02 (144).

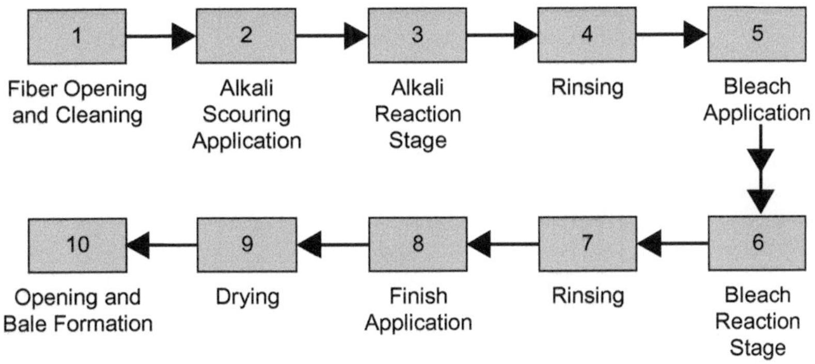


**Figure 4.14** Cotton Scouring and Bleaching. *Reproduced by Permission of Cotton Incorporated Cary, North Carolina, USA.*

Cotton is 99% pure cellulose. Cellulose is a polymer containing anhydro-beta-cellobiose as the repeat unit. The number of repeat units in a cellulose molecule is its degree of polymerization. The simplified chemical structure of cellulose is among those listed in Table 4.1. Wood pulp fibers and rayon are also polymers of cellulose. The difference is in their degrees of polymerization or polymer length and in their degree of crystallinity. Crystallinity indicates that the fiber molecules are closely packed and parallel to one another. Table 4.9 compares the degree of polymerization of cotton to rayon and wood pulp.

The shape of a cotton fiber is that of a convoluted ribbon. This twisted shape is well demonstrated in the micrograph of Figure 4.15. Because of the twists and convolutions, cotton is a very flexible fiber that can readily bend in any direction. The fiber has a high degree of curl and this contributes to bulking behavior in filter media.

**4.3.5.1.3.2 Abaca (*Manila hemp*)** Abaca is more commonly known as *Manila hemp*, the type used for making Manila rope. Abaca is not a true hemp, but comes from a plant related to the banana tree. Abaca's use in filter media are mostly from

**Table 4.8** Properties of Bleached Cotton

| | | |
|---|---|---|
| 1. Length in. | A. Commodity staple<br>B. Gin Motes<br>C. Comber<br>D. First-cut Linters | 0.75–1<br>0.5–0.75<br>Less than 0.5<br>0.25–0.5 |
| 2. Fiber diameter | A. Micronaire (μ g/inch)<br>B. Denier (g/9000 m) | 2.0–6.5<br>0.7–2.3 |
| 3. Elastic recovery | A. At 2% Extension<br>B. At 5% Extension | 74%<br>45% |
| 4. Breaking Elongation (dry) | | 3–9.5% |
| 5. Tensile Strength | A. Dry<br>B. Wet | 27–44 g/tex, 3.0–4.9 g/de<br>28–57 g/tex, 3.3–6.4 g/de |
| 6. Water absorbing capacity (USP method) | | Greater than 2.4 g water/g fiber |
| 7. Specific Gravity | A. Cellulose polymer<br>B. Cotton fiber | 1.54 g/cm$^3$<br>1.27 g/cm$^3$ |
| 8. Degree of polymerization | | 9,000–15,000 |
| 9. Crystallinity by X-ray diffraction (average) | | 73% |
| 10. Color (Hunter Collimator | A. Whiteness index<br>B. Blue reflectance | 90–100<br>75–85 |
| 11. Thermal resistance | A. long exposure to dry heat above 300 °F will cause gradual decomposition<br>B. Temperatures above 475 °F causes rapid deterioration | |
| 12. Acid resistance | A. Disintegration by hot dilute acids or cold concentrated acids.<br>B. Unaffected by cold weak acids | |
| 13. Alkali resistance | Swelling in NaOH above 18%, but no damage | |
| 14. Organic solvent resistance | Resistant to most common industrial and household solvents | |
| 15. Web formation | A. Dry lay (carding, garneting)<br>B. Air lay<br>C. Wet lay (linters only) | |
| 16. Bonding mechanisms | A. Needle punching<br>B. Hydroentanglement<br>C. Stitch bonding<br>D. Chemical bonding<br>E. Thermal bonding | |

Reproduced with permission of Cotton Incorporated, Cary, North Carolina, USA.

**Table 4.9** Comparison of Cellulose Polymer Fibers

| Fiber | Degree of polymerization |
|---|---|
| Cotton | 9000–15000 |
| Viscose Rayon | |
| Regular | 250–450 |
| High Tenacity | 500–650 |
| High Wet Modulus | 400–550 |
| Wood pulp | 600–1500 |

Reproduced by permission of TAPPI, Atlanta, Georgia, USA (144).

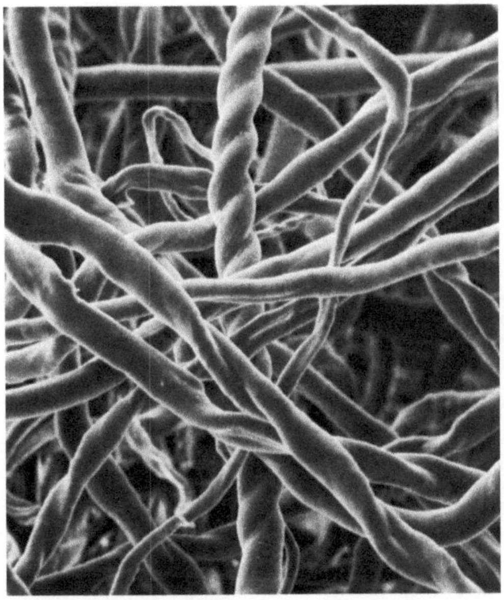

**Figure 4.15** Convoluted Cotton Fiber.[2] *Reproduced by Permission of the Institute of Paper Science and Technology, Atlanta, Georgia, USA (148).*

wet lay filter applications. Its mainstay use in filter products is teabag, however, it is also used in coffee filter, industrial filter, and vacuum bag. In addition, it finds application in medical gas masks and as a reinforcement for microglass air filters. 85% of the world output of abaca is grown in the Philippine Islands. It has been introduced into other locations including Borneo, Indonesia, and Central and South America (151).

The abaca fiber is derived from the leaf sheaths of Musa textilis. The plant consists of a cluster of sheath-like leaf stalks. The stalks often grow to heights as high as six meters or more. The fiber comes from the outer layer of the sheath-like leaves. At harvest time, the stalks are separated from the stem of the plant and grouped as follows: outside sheaths, adjacent to the outside sheaths, middle sheaths, and inner

[2] Parhwem and Kaustinen ((148), p. 44).

**Table 4.10** Main Hand Stripped Abaca Grades of the Phillipines and Its Principal End Uses

| Cleaning | Grade | Layer of leaf sheath | Uses |
|---|---|---|---|
| **Excellent cleaning** (non-serrated stripping knife) | AD | Middle area | Handicraft industry |
| | EF | Middle area | Handicraft industry |
| | S2 | Next to outside | Cordage, Teabags, Filters, Stencil, Casing Paper, Handicraft |
| | S3 | Outside | Cordage |
| **Good cleaning** (20 tooth/inch stripping knife) | I | Innermost and middle | Cordage, Teabags, Filters, Stencil, Casing Paper |
| | H | Outside | – |
| | G | Next to outside | Cordage, Teabags, filters, Stencil, Cigarette, Handicraft industry |
| **Fair cleaning** (17 tooth/inch stripping knife) | JK | All except outside | Cordage, Cigarette, Currency paper Cordage, Cigarette |
| | M1 | Outside | |
| **Coarse cleaning** (14 tooth/inch stripping knife) | L | Innermost and middle | – |
| **Damaged during cleaning** | Y1-Y2 | All | Cordage, Cigarette, Currency paper |

**Nomenclature**
**AD:** Superior Current; white to light yellow aspect; 5–8 mm fiber length.
**EF:** Mid-Current; light ocher brown or red with small quantities of ivory yellow; about 5 mm fiber length.
**S2:** Streaky Two.
**S3:** Streaky Three.
**I:** Fair Current; light brown, but darker than EF; about 5 mm fiber length.
**G:** Soft Seconds; light brown or red mixed with yellow ocher in variable proportions; 3–5 mm fiber length.
**H:** Soft Brown; Brown or dark red mixed with light purple or light burnt sienna; 3–5 mm fiber length.
**JK:** Superior-Medium Seconds; light brown, dingy white; 3–5 fiber length.
**M1:** Medium Brown; same as grade H but ranging to dark brown and black; 3–5 mm fiber length.
**L:** Coarse; same as grade I but ranging to darker brown; about 5 mm fiber length.
**Y1:** Damaged find from grades AD, EF, S2, I, S3 and G.
**Y2:** Damaged Medium Coarse from grades H, JK, MI and L.
**Note:** Spindle stripped abaca grades follow a similar designation but are preceded by an 'S'.
Reproduced by permission of TAPPI, Atlanta, Georgia, USA (152).

sheaths. The location and width of each sheath affects the yield and quality of fiber. The fiber layer is separated from the leaf sheath by a process known as stripping. In its simplest form, the fiber is peeled away from the rest of the plant with a serrated stripping knife. A more productive technique is to strip the fiber on a motor-driven tapered spindle called a "hagotan." In some places a "decorticator" is used which increases productivity, but produces a lower quality fiber.

Table 4.10 lists the various grades of Philippine hand stripped abaca. Note much of the grading depends on the type of stripping knife. If the fiber is spindle stripped the grade rating is similar, however, the designation is preceded by an "S."

**Table 4.11** Pulp Handsheet Properties of Some Abaca Grades

| Fiber grade | S2 | JK | Y2 | AB-EC[a] |
|---|---|---|---|---|
| Beating Time, min | 13 | 11 | 10 | 12 |
| Breaking Length, km | 5.3 | 5.1 | 5.0 | 4.5 |
| Permeability, $cm^3/min$ | 72,000 | 76,800 | 74,400 | 108,000 |
| Tear index, $mN*m^2/g$ | 43.4 | 42.3 | 40.8 | 45.9 |

All results tested at 27° SR (Shopper Riegler).
[a]Ecuadorian Grade.
Reproduced by permission of TAPPI, Atlanta, Georgia, USA (152).

Most abaca fiber used for filtration is processed into wet lay filter media. The fiber has to be pulped in a pulping process much like wood fiber. Table 4.11 lists handsheet properties of some abaca grades including on Ecuadorian grade (AB-EC). Note for permeability, the Ecuadorian grade is considerably more open and porous than the Philippine grades.

Abaca fibers are up to 6 mm long and 20–24 mm in diameter. The 6 mm length makes abaca just about the longest pulp fiber used in the wet lay process. The fiber has tapered ends. It also has large lumen and thin cell walls. Because of its length and good hydrogen-bonding qualities, abaca provides enhanced strength to wet laid pulp products.

**4.3.5.1.3.3 Sisal** Sisal (*Agave sisalana*) is a nonwood leaf plant of the *Amarillydaceae* family. It is a native of Mexico, however, the plant has thrived in semi-arid regions of Africa and South America. Other forms of sisal that are commercially available are *Agave cantala* and *Agave fourcroydes* (also known as henequen). The plant is characterized by rosettes of fleshy leaves, usually long and narrow, which grow out from a central bud. The fibers lie embedded longitudinally in the leaves, being most abundant near the leaf surfaces. Sisal grows all year and the first harvest can be made 2.5 years after planting. The productive period of sisal is 4–6 years.

The leaves contain about 90% moisture forming firm, fleshy pulp. Fiber removal is accomplished by scraping the pulpy material. This is done by a decorticating process. The process consists of pairs of metal drums on which scraping blades are mounted. The leaves are passed between the drums in such a way that the pulp is scraped away by the scraping blades. Water is sprayed on to the leaves to assist in the process. The wet decorticated fiber is washed and dried. The dried fibers are then brushed to remove clinging dust and bring out the luster.

Each leaf contains three main fiber zones: peripheral, median, and ground tissue (Martinez etc. (152)). They contain either ribbon fibers (median zone) or mechanical fibers (peripheral zone) or both (ground tissue zone). Ribbon fibers run the full length of the leaf and are coarser than mechanical fibers. They have a crescent-shaped cross section and are coarser than mechanical fibers. They are easily split in the longitudinal direction cross-mechanical ribbon fibers. Mechanical fibers are nearly round in cross section. They do not split, therefore, they determine the maximum diameter of the fiber.

A single sisal filament or fiber is constructed of numerous elongated cells with tapering ends. These are known as ultimates and are closely packed and bonded together so that there are no intercellular spaces. Ultimates range in length from 1.5 to 4.0 mm long and average approximately 3.0 mm. The cell diameters range from 20–30 μm.

Sisal fiber is graded according to country and district of growth and further subgraded according to color, cleanness, and length. Table 4.12 lists the classification of some grades from Brazil and East Africa. Sisal is known for its use in the cordage industry. Sisal strand is used in air laid and needle punch nonwovens. In pulp form, it has interesting properties for wet laid filter media. The fiber shape is very much like a shorter form of abaca discussed above. It offers 40–70% higher tear strength than softwood pulp. Because it is a stiff (high modulus) fiber and has a relatively narrow fiber diameter, it forms wet laid webs that are bulky, permeable, small in pore size, and strong. Some pulp handsheet properties for various grades of sisal are listed in Table 4.13.

**Table 4.12** Classification of Some Sisal Grades from Brazil and East Africa (Shaded Areas)

| **Length Quality** | EL Extra-long Type Superior | L-Long Type 1 | M-Medium Type 2 | C-Short Type 3 | Rejects or Refugo |
|---|---|---|---|---|---|
| | Grade 1 | Grades A and 2 | Grade 3L | Grade 3 | Grades UG and SCWF |
| **Residues from processing** | Bucha 1 | Bucha 2 | Bucha 3 | Aperas | |
| | Tow 1 | Tow 2 | – | – | – |
| **Residues from spinning** | Combing | Calander | Spinning | Brushing | Twining |

Surrounding sections are considered as equivalent.
– Grades UG, rejects, or residues from processing operations are normally used for papermaking.
– Superior grades are destined for cordage or handicraft works.
Reproduced by permission of TAPPI, Atlanta, Georgia, USA (152).

**Table 4.13** Handsheet Properties of Bleached Sisal Pulp

| Grade | Brazilian Type 3 long | Brazilian BUCHA | African grade 3L | African grade UG |
|---|---|---|---|---|
| **Beating Time, min** | 6 | 5 | 6 | 5 |
| **Breaking Length, km** | 3.5 | 3.0 | 3.9 | 3.7 |
| **Permeability ml/min** | 72,000 | 88,800 | 96,000 | 100,800 |
| **Tear Index mN*m$^2$g** | 26.5 | 23.5 | 22.9 | 22.4 |

All results tested at 25° SR (Shopper Riegler).
Reproduced by permission of TAPPI, Atlanta, Georgia, USA (152).

**4.3.5.1.3.4 Kenaf** Kenaf (*Hibiscus cannabinus*) is interesting because of its potential as a papermaking fiber and its applicability for wet lay filter media. Kenaf is an annual plant that requires regular amounts of water. It grows best in Southern climates. The plant reaches a height of approximately 5 meters and a diameter of approximately 10–25 cm. The plant provides two sources of fibers: the inner core which provides a short fiber much like hardwood, and the bast fiber which is considerably longer. These are depicted in Figure 4.16. Table 4.14 compares the two forms of kenaf fiber along with flax (see Section 4.3.5.1.3.5) to Northern softwood (NSW) and Southern hardwood (SHW). The advantage of the Kenaf fiber becomes apparent. It is as long as softwood fiber and has a narrower diameter. It is also a relatively high modulus fiber, therefore, it has the potential of providing bulk, air permeability, and strength to wet lay filter media. There is considerable development of Kenaf as an alternate fiber source to wood pulp in the United States, however, commercialization is still in the future. In Japan and Europe there has been some success in the development of Kenaf and flax as nonwoven reinforcements for automotive plastic components. Presently, there are no known filter applications.

**Figure 4.16** Kenaf Fiber: Bast Fiber on Left, Core Fiber on Right.

**Table 4.14** Kenaf and Flax Fiber Comparison

| Fiber | Length (mm) | Diameter (μm) | L/D |
|---|---|---|---|
| Flax, Bast | 30 | 20 | 1350:1 |
| Kenaf, Bast | 2.6 | 20 | 135:1 |
| Kenaf, Core | 0.6 | 30 | 20:1 |
| NSW | 2.7–4.6 | 32–43 | 100:1 |
| SHW | 0.7–1.6 | 20–40 | 40:1 |

Reproduced by permission of TAPPI, Atlanta, Georgia, USA (152).

**4.3.5.1.3.5 Flax** Flax is a group of the bast annual and perennial plants from the *Linacease* family that form the genus *Linum*. Long known for its use in producing the fabric known as linen, it is one of the oldest of cultivated plants. The primary fibers are from the bast layer. Ultimates are 10–65 mm long and 14–30 μm in diameter. The fibers are separated by a process known as retting. This is a natural, enzymatic process lasting approximately two weeks. In the process, the woody cores holding the fiber bundles together are broken down and the fibers are then easily separated. The longer high quality fibers are used for fabric applications including dry laid nonwovens. The shorter fibers, most of which comes from process residues are directed towards products of papermaking. Cigarette wrap is one of the larger markets for flax paper. There appears to be very little application for flax fibers in filter media markets. Table 4.15 lists some handsheet properties of flax pulp with various levels of shive content.

**Table 4.15** Evaluation of Some Bleached Pulps From Flax With a Different Quantity of Shive in the Original Raw Material

| Grade | Textile flax 1–2% shive | 15–20% shive | 30–35% shive | Seed flax 40–45% shive |
|---|---|---|---|---|
| **Beating Time (min)** | 31 | 26 | 21 | 20 |
| **Breaking Length (km)** | 6.8 | 6.7 | 6.6 | 7.3 |
| **Permeability (ml/min)** | 12.0 | 10.0 | 10.0 | 7.0 |
| **Tear Index (mN*m$^2$/g)** | 26.5 | 23.5 | 22.9 | 22.4**S** |

All results tested at 90° SR (Shopper Riegler).
Reproduced by permission of TAPPI, Atlanta, Georgia, USA (152).

**4.3.5.1.3.6 Hemp** Hemp is possibly the oldest cultivated fiber plant. However, it may be more famously known as the source of marijuana. Increased interest for its commercial usage in fiber products has developed because of the ongoing legalization of marijuana. True hemp (*Cannabis sativa*) is a completely different plant than the abaca plant. Abaca is not a hemp plant even though it is known as Manila hemp. The hemp fiber is produced by retting process similar to the flax operation. Hemp fibers are flat ribbons 1 to 5 m long and 0.5–5 mm wide. The ultimate cells are 5–55 mm long and 16–50 μm in diameter. The longer fibers are used for production of yarns and twine. Shorter fibers, which may contain an impurity called hurds, are used for specialty papers. There is no known specific use of hemp in nonwoven filter media. Table 4.16 lists handsheet properties of hemp pulp containing various degrees of hurd content.

**4.3.5.1.3.7 Esparto grass** Esparto is a coarse grass native to Southern Spain and Northern Africa. It is also known as "halfah grass" and "needle grass." There are two species: *Lygeum spartum* and *Stipa tenacissima*. The fiber length is less than 3 mm and averages 1.5 mm. The fiber diameter ranges from 5 to 15 μm. The average is

**Table 4.16** Evaluation of Some Bleached Pulps from Hemp with a Different Quantity of Shive in the Original Raw Material

| Grade | 25% hurd[a] | 35% hurd[a] | 75% hurd[b] |
|---|---|---|---|
| **Beating Time (min)** | 37 | 33 | 27 |
| **Breaking Length (km)** | 6.5 | 6.7 | 7.3 |
| **Permeability (ml/min)** | 10 | 8 | 3 |
| **Tear Index (mN*m$^2$/g)** | 8.5 | 8.1 | 5.8 |

[a]results tested at 90° SR (Shopper Riegler) (81, p. 127).
[b]results tested at 45° SR.
Reproduced by permission of TAPPI, Atlanta, Georgia, USA.

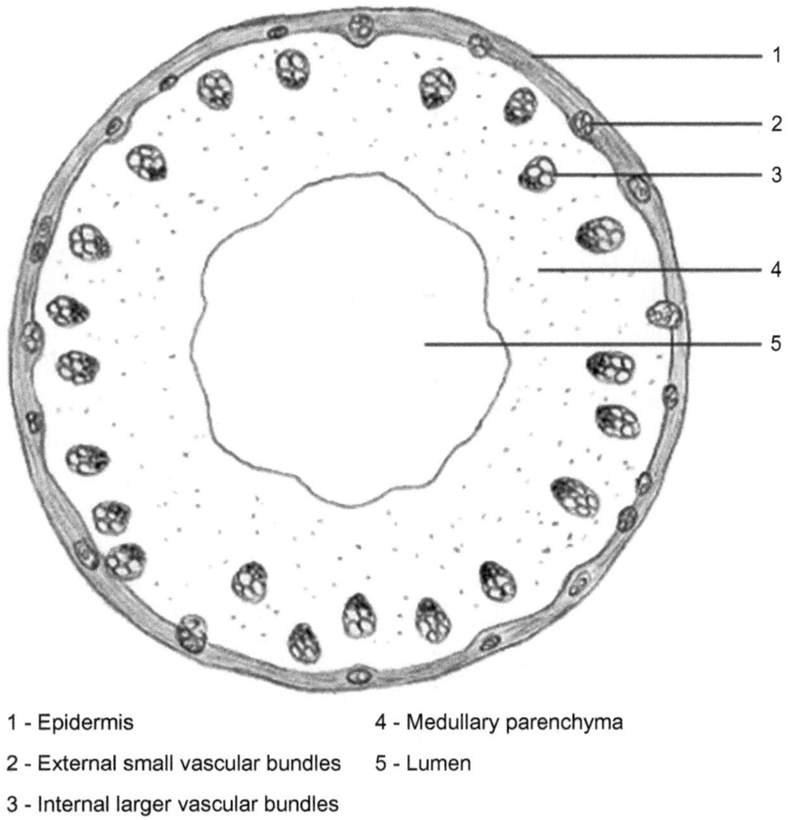


**Figure 4.17** Cross Section Through an Esparto Grass Stalk. *Reproduced with Permission © Gesamtverband der Deutschen Versicherungswirtschaft e.V. (GDV), Berlin, Germany (153).*

approximately 12 μm. The fiber has a fiber density of 15 million fibers per gram making it the highest of any paper pulp. The fiber cross section is round and the cell wall is thick. The cross section is pictured in Figure 4.17. The fiber is not straight, but has some curvature to it. This contributes to bulking qualities in a filter medium.

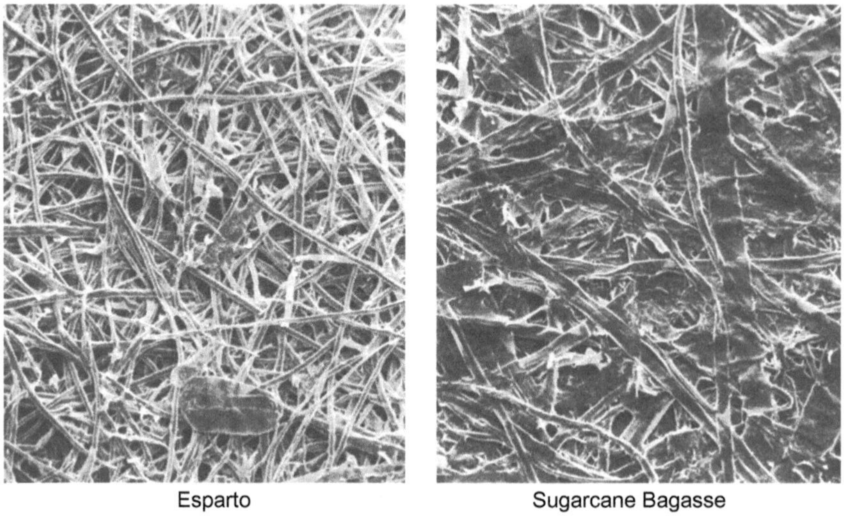


**Figure 4.18** Esparto–Bagasse Comparison. *Reproduced by Permission of the Institute of Paper Science and Technology, Atlanta, Georgia, USA ((148) p. 17).*

Esparto pulp, considerably more expensive than wood pulp fibers and yet, is broadly used as a fiber for wet laid filter media, particularly in those grades where high filtration efficiency is desired. Its fine fiber diameter combined with fiber curvature provides for high efficiency combined with relatively high bulk for the efficiency. The thick cell wall structure contributes to the bulk and permeability properties and adds dimensional stability to the sheet structure. The filtration properties of Esparto are better appreciated by the comparison of Figure 4.18. The figure shows side-by-side micrographs from handsheets of Esparto and Bagasse pulp (a sugar cane fiber). Note that the slender Esparto fiber structure allows for ample pore openings in the medium, whereas the broad flat ribbon structure of Bagasse fiber covers the openings and blinds the medium. Bagasse pulp has found some important niches in board and paper products; however, filter medium is not one of them.

**4.3.5.1.3.8 Coconut fiber** Coconut fiber, commonly known as coir, is obtained from the fibrous husk of the coconut. This is the thick fibrous middle layer (2) of the coconut sketched in Figure 4.19. To separate the fibers the coconut shells are cut in half and then retted. In the retting process, the coconut shells are buried in wet soil to allow microbial degradation of the softer tissues. The shells are then beaten and washed to readily separate the coir fibers. Incidentally, the hard inner layer, called the stone (3 in Figure 4.19), is the source of the coconut shell activated carbon that is very popular in adsorbent filter media.

Coir fiber has a high lignin content which makes it resilient, strong, and highly durable. It is graded into "bristle" fiber (combed, approximately 20–40 cm long) and "mattress" fiber (random fibers approximately 2–10 cm long). The fiber finds application in residential furnace and ventilation filter pads. It is often combined with hogs hair in "Natural Filters".

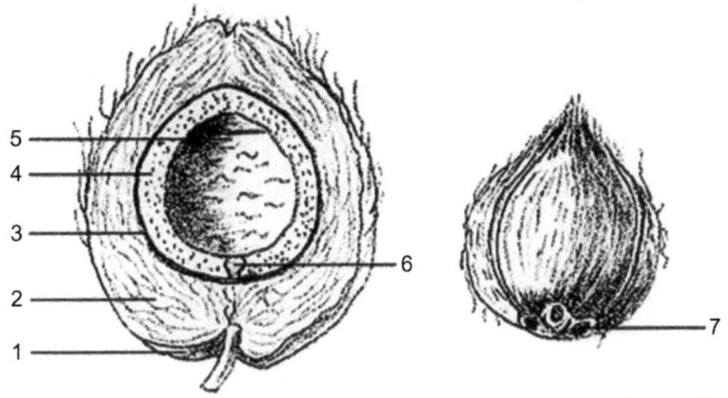


Longitudinal section through a cocount (left) and plan view of stone (right)

1 - thin, yellow-brown, watertight outer skin (exocarp)
2 - thick, fibrous middle layer (coconut fiber, coir, mesocarp)
3 - hard inner layer, the stone (endocarp)
4 - white, oily copra layer, 1 - 2 cm thick (solid endosperm)
5 - cavity filled with cocount milk (liquid endosperm)
6 - embryo
7 - "eyes" (3 germ pores set in pits)

**Figure 4.19** The Components of a Coconut. The Thick Fibrous Middle Layer 2 is the Source Layer or Coir (Coconut Fiber). The Hard Inner Layer 3 is a Source of Coconut Shell Activated Carbon. *Reproduced with Permission © Gesamtverband der Deutschen Versicherungswirtschaft e.V. (GDV), Berlin, Germany.*

**4.3.5.1.3.9 Others** There is a long list of nonwood natural fibers that find application in nonwoven and paper structures; however, like Bagasse they are not readily found in filter media. Included in this list are jute, various straw fibers, reed fibers, bamboo, grass fibers other than Esparto, various variations of hemp fibers, ramie, etc. Some of these fibers might be good filter medium fibers, however their use is limited because of availability or other economic considerations.

#### 4.3.5.1.4 Animal fibers

**4.3.5.1.4.1 Wool** Although sheep are generally perceived as the source of wool, the wool of other animals are also cultivated for their properties. This includes goat's wool (mohair) and alpaca wool from a camel related animal of South America. Although the usage of wool in many of its commercial applications is being displaced by synthetic fibers, it has unique characteristics for filter medium applications. It is most often used in needle felt form for bag filters. It is used in air filter applications, such as baghouse filtration, residential filters, vacuum cleaner bags, and industrial respirators. It is also used in conjunction with kitchen grease filters. In liquid filtration its most noted application is in aquarium filters.

Wool has some unique properties for filtration. Wool fibers are naturally crimped, offering bulk and loft. It can easily be electrostatically charged. It has good strength and dimensional stability, and it is durable. To illustrate, it can be flexed back and forth on itself more than 20,000 times without breaking as compared to 3,000 times for cotton and 2,000 times for silk and 75 times for rayon. It is flame resistant. Wool will char in the presence of a flame, however it will self-extinguish once the flame is pulled away. Wool will absorb up to 30% by weight of moisture. It absorbs over 40 times its weight in oil, more than polypropylene. Mats made from wool are very useful for the cleanup of oil and chemical spills. It is biodegradable and may help alleviate the problems of hazardous waste disposal.

**4.3.5.1.4.2 Silk** For centuries and centuries, silk has been the high-end fiber for luxurious fabrics and clothes. It is produced by the silkworm in preparing its cocoon. There has been and still is very little use of silk in filter media.

**4.3.5.1.4.3 Hogs hair** Hogs hair has been traditionally used as the bristles in hairbrushes and is often referred to as bristle. When carded into a filter medium and coated with a rubber coating it forms a bulky, high loft structure used as filter pads in residential and ventilation furnace filter systems. Sometimes the fiber composition includes coir or coconut fiber. Several companies manufacture hog hair filters for residential furnaces and air conditioning systems. Among these are Purolator of Henderson, North Carolina, American Air Filters International (AAF), Air Guard of Louisville, Kentucky, and Air Filters, Inc. They are often advertised as "natural" fiber filters.

### *4.3.5.2 Biofibers*

Biofibers include the natural fibers discussed in Section 4.3.5.1. They also include fibers synthesized from natural plant resources. This includes rayon, lyocell, and polylactic acid, all of which will be discussed in Sections 4.3.5.2.1–4.3.5.2.3. Concern about preservation of natural resources and recycling has led to the renewed interest concerning biomaterials with focus on renewable raw materials. Because of increasing environmental consciousness and demands of legislative authorities, interest in natural fibers and fibers derived from agricultural materials has been reinvigorated. Recent advances in natural fiber development, genetic engineering, and composite science offer significant opportunities for improved fibers from renewable resources with enhanced support for global sustainability. The important feature is that they can be designed and tailored to meet different requirements. A key requirement is biodegradability. Since natural fibers are inexpensive and biodegradable, the biodegradable composites from biofibers and biodegradable polymers will render a contribution in the twenty-first century due to serious environmental problems. Biodegradable polymers offer scientists a possible solution to waste-disposal problems associated with traditional petroleum-derived fibers.

### 4.3.5.2.1 Rayon

In the 1880s, a Frenchman named Hilaire de Chardonnet first spun fibers of "artificial silk" from a solution of cellulose. Two Englishmen, Charles Cross and Ernest Bevan, invented the "viscose" process in 1891. Courtalds, a British company started producing the viscose fiber in 1905. The first American producer of artificial silk was American Viscose Corporation (now Avtex Fibers) in 1910. In 1920, DuPont secured from the French, the American rights to produce artificial silk. In 1921, The DuPont Fibersilk Company produced the first yarn of what later became known as rayon. In the evolution, The DuPont Fibersilk Company became the DuPont Rayon Company, which was the predecessor for DuPont's Textile Fibers Department. In 2007, DuPont sold the Textile Fibers Department, which was reorganized into the company now known as Invista.

Four primary forms or modifications of rayon are:

**a.** Viscose rayon
**b.** High wet modulus (HWM) rayon
**c.** High tenacity rayon
**d.** Cuprammonium rayon.

HWM and high tenacity rayon are variations of the viscose process for producing rayon. HWM seeks to correct the wet strength deficiencies of regular viscose rayon. It is also known as "polynosic" rayon and as modal rayon. Modal rayon is used in high end apparel products. High tenacity rayon is a modification of regular viscose rayon to provide exceptional strength. It has twice the strength of HWM rayon. High tenacity rayon has been widely used in tire cord. Cuprammonium rayon is produced from a different chemistry.

The process for producing rayon involves preparing a viscous solution of cellulose and spinning it into fiber form. The cellulose source is either cotton linters or a highly purified form of wood pulp. In the viscose process, the cellulose is treated with caustic and carbon disulfide. The resulting material referred to as "yellow crumb" is a copolymer of cellulose and cellulose xanthate. The yellow crumb is dissolved in an aqueous caustic solution. The high viscosity of this solution lends itself to the term "viscose". The solution is aged, filtered, and further treated before being subject to a wet spinning process. The fiber precipitates as cellulose as it is spun into a solution consisting of sulfuric acid, sodium sulfate, and zinc ($Zn^{++}$) ions. The fiber filaments are drawn out of the solution, washed, and cut into staple form—if staple is the desired form.

Cuprammonium rayon fiber is produced from a solution of cellulose in cuprammonium hydroxide at low temperature and in a nitrogen atmosphere. The solution is wet spun into a precipitating solution of sulfuric acid. Although no longer produced in the United States, Rayon is still being produced in Europe. Accordis (formerly Courtalds) in England, Lenzing in Austria, and Mitsubishi Rayon in Japan are all major producers of rayon fiber.

Other forms of rayon are saponified rayon and lyocell. Saponified rayon is an acetate derivative of cellulose made by steeping cellulose in acetic acid, then treating it with acetic anhydride. Acetate rayon fiber has a higher degree of orientation than

Table 4.17 Properties of Viscose Rayon Fiber

| Property | Value or level |
|---|---|
| Melting point | >400 |
| Glass Transition | Indeterminate |
| Moisture Regain | 6–8% |
| Static Electricity | Does not accumulate except in very dry atmosphere |
| Dyeability | Excellent dyeability with a variety of different dye classes |
| Density | 1.5 g/cm$^3$ |
| Tenacity | 3–5 g/de |
| Elongation at Break | 8 to 25% |
| Modulus | Moderate |
| Abrasion Resistance | Good |
| Flammability | Burns quickly O.I. = 18 |
| Acid/Base Stability | Attacked by mineral acids, reasonably stable in bases |
| Oxidative/Light | Degraded by strong oxidants and UV light |
| Solvents | Strongly polar compounds, amine complexes with Cu and Cd |

Reproduced by permission of Dr. Roy M. Broughton, Jr. ((141), p. 97).

rayon fiber and is more durable. It is a popular fiber for cigarette filters because it is believed to have an affinity for certain contaminants in cigarette smoke.

Viscose rayon is the dominant form of rayon in the world. Rayon is a soft, highly absorbent fiber that drapes well and is easy to dye. It is heat resistant up to 149 °C. It will char and decompose at temperatures in the range of 175 to 200 °C. Unless modified the fiber has low wet strength. It has moderate chemical resistance. It does not build up static electricity. It is biodegradable which gives it an advantage in disposable filter products. It has poor dimensional stability when exposed to moisture. It will gradually degrade from sunlight and moisture in outdoor environments. The typical properties of viscose rayon are presented in Table 4.17.

Filter media are rarely composed of 100% rayon, but rather the rayon is blended with other fibers such as polyester in carded webs and wood pulp in wet lay substrates. It helps provide bulk to a filter medium. When used with polyester, it provides some moisture absorbency to alleviate the dryness and static electricity of 100% polyester. Carded webs containing rayon are most often needle punched and used in air filter applications. This not to say that rayon is not used in liquid filtration applications. National Nonwovens advertises a coolant filter medium, FIL003, which is chemically bonded and composed of 10% rayon and 90% polyester. There is still some use of rayon staple in wet laid filter media, however, most of this has been replaced by polyester.

#### 4.3.5.2.2 Lyocell

Lyocell is a solution spun fiber made from wood pulp or cotton cellulose. The cellulose is directly dissolved in an amine oxide solvent, N-Methylmorpholine N-oxide (NMMO). The NMMO solution is pumped through a spinnerette to produce the fibers, which are then washed, finished, and dried. The NMMO is recovered and recycled back into the feed stream.

Lyocell manufacture was licensed from AkzoEnka and developed by Lenzing Fibers, Austria and Courtaulds Research in Coventry, UK, who later became part of Acordis, the Netherlands. The first commercial-scale plant started manufacturing the product in 1992 under the trade name of Tencel®. The Tencel® business was subsequently acquired from Acordis by Lenzing A.G., Austria in 2004. Lenzing, now the only manufacturer of lyocell, continues to market under the trade name of Tencel®. Note that lyocell by itself is a not a trade name. It is a fiber designation as is nylon and rayon.

Generally lyocell finds its niche in apparel and home interior fabrics as well as in nonwovens wipes production. It also has utility in filter media. One important function is that it is a source of nanofiber. The fiber is composed of crystalline regions aligned parallel to the fiber axis. Schematic representation of the fiber cross section suggests an "island in the seas" macrostructure, although unlike splittable fibers, the composition is a single polymer (see Sections 4.3.5.5.5 and 4.3.5.6). Mechanical shear such as caused by refining or beating can fibrillate the fiber into individual macrofibrils as illustrated in Figure 4.20. Slater (154) reports that the fibers themselves have a filament diameter of 10–15 μm. The macrofibrils have a broad diameter range from 100 nm to 10–15 μm. The peak fibril diameter is in the range of 300–500 nm. The macro fibrils themselves are composed of nanofibrils that have nanofiber diameter dimensions. Although it is not a bicomponent fiber, the lyocell fiber microstructure can be modeled as comprising a number of macrofibrils within a single fiber cross section.

Lyocel fiber is characterized by its smooth, round cross section, excellent filament strength, high modulus, high crystallinity, and ordered microstructure. The wet abrasion of refining causes fibrils near the fiber surface to be released from the internal filament structure. Continuation of the refining action causes progressively more fibrils to become exposed from the main element. This is apparent in Figure 4.20.

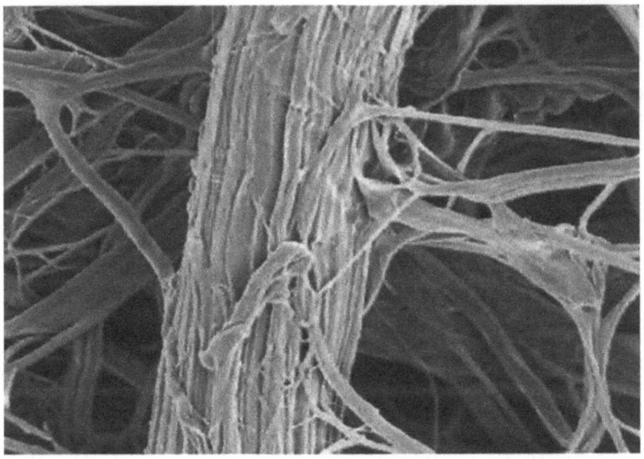

**Figure 4.20** Fibrillated TENCEL® (Lyocell) Fiber. *Reproduced by Permission of Lenzing A.G. Lenzing, Austria.*

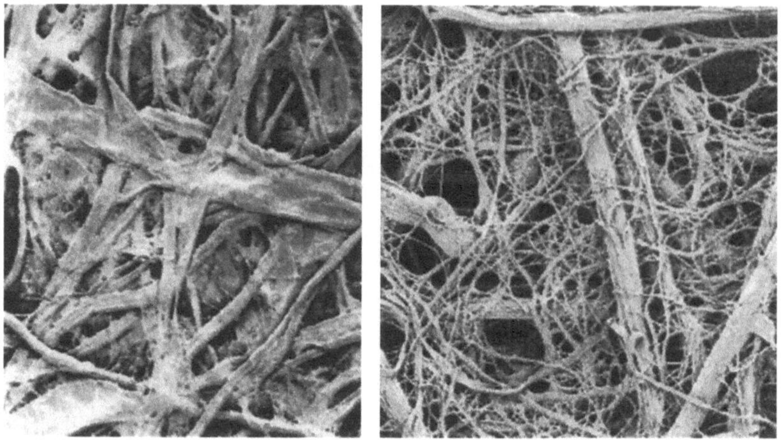

**Figure 4.21** Refined Wood Pulp (Left) and TENCEL® (Lyocell) (Right). *Reproduced by Permission of Lenzing A.G. Lenzing, Austria.*

Because of the fiber roundness, Lyocel papers are more uniform and have a finer pore structure than comparable wood pulp papers such as softwoods. This difference is evident in the comparison of Figure 4.21. The effect results in filter papers of higher porosity, higher bulk, and high filtration efficiency. Filter media applications include automotive oil and fuel filters, depth filter sheets for beverage filtration, and filter papers for low tar cigarettes.

#### 4.3.5.2.3 Polylactic acid (PLA)

Polylactic acid is a thermoplastic polymer sourced from renewable resources such as corn starch or sugar cane. Because of its thermoplastic nature it can be meltspun into fiber, used as a thermobond resin, and find other applications as a plastic material (e.g., disposable plates, cups, and packaging material). It is touted as a sustainable material because it is a renewable resource and is biodegradable. It is considered as an alternative to polyester and polypropylene because of these properties.

According to Wikipedia (155), Lactic acid, and the cyclic di-ester, lactide, are the most common monomers for the synthesis of polylactic acid. The most common route to PLA is the ring-opening polymerization of lactide with various metal catalysts. Lactic acid tends to be a chiral molecule (a type of molecule that has a non-superposable mirror image). Dugan (156) notes that unique polymer characteristics arise from the fact that the lactide dimer occurs in three forms: the L form, which rotates polarized light in a clockwise direction; the D form, which rotates polarized light in a counterclockwise direction; and the meso form, which is optically inactive. During polymerization, the relative proportions of these forms can be controlled, resulting in relatively broad control over important polymer properties. According to Martin and Averous (157) commercial grades of PLA are generally copolymers of poly(L-lactic acid) (PLLA) and poly(D,L-lactic acid) (PDLLA). PLLA and PDLLA are produced from L-lactides and D,L-lactides, respectively. The ratio of L- to D,L-lactides affects properties of PLA such as melting temperature and degree of crystallinity.

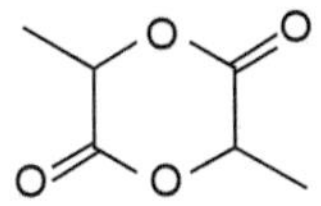

Lactide

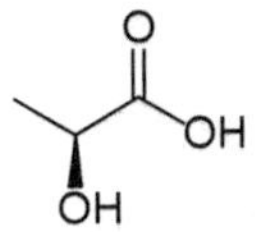

Lactic Acid

Typical properties of PLA are listed in Table 4.18.

**Table 4.18** Typical Properties of Polylactic Acid (PLA)

| Property | Value or level | Method or source |
|---|---|---|
| Molecular weight | >1.2 $(10)^5$ | Zhongyuan U. China |
| Melting point | 120–175 °C | – |
| Glass transition | 133–145 °F | DSC |
| Refractive index | 1.35–1.45 | – |
| Moisture regain | >1.0% | Minifibers |
| Specific gravity | 1.24–1.27 | ASTM D792 |
| Tensile str., break @ 73 °F | 2,860–10,400 psi | ISO-527-2 |
| Tensile modulus @ 73 °F | 96,500–516,000 psi | ISO 527-2 |
| Elongation, break @ 73 °F | 1.0–8.5% | ISO 527-2 |
| Abrasion resistance | Poor | – |
| Flammability | Good self-extinguishing characteristics | – |
| Acid/Base stability | Poor resistance to bases | – |
| Oxidative/Light | Good resistance to UV light | – |
| Solvents | Resistant to drycleaning solvents | – |

According to Wikipedia (155) the largest producer of PLA is NatureWorks LLC, headquartered in Minnetonka, Minnesota, USA. They market the product under the trade name of Ingeo™. They have a production facility in Blair, Nebraska that was slated to increase its Ingeo™ nameplate capacity to 150,000 metric tons in the first quarter of 2013. Another producer is Corbion Purac, Gorinchem, the Netherlands, a major producer of lactides. Corbion Purac partners with other companies to produce various forms of PLA.

Several companies market PLA as a short cut fiber. This includes MiniFIBERS, Inc. and Fiber Innovation Technologies.

Filtration areas of potential applicability for PLA fibers have appeared in recent years:

1. Triboelectric media to improve filtration efficiency.
2. A source of nanofibers.

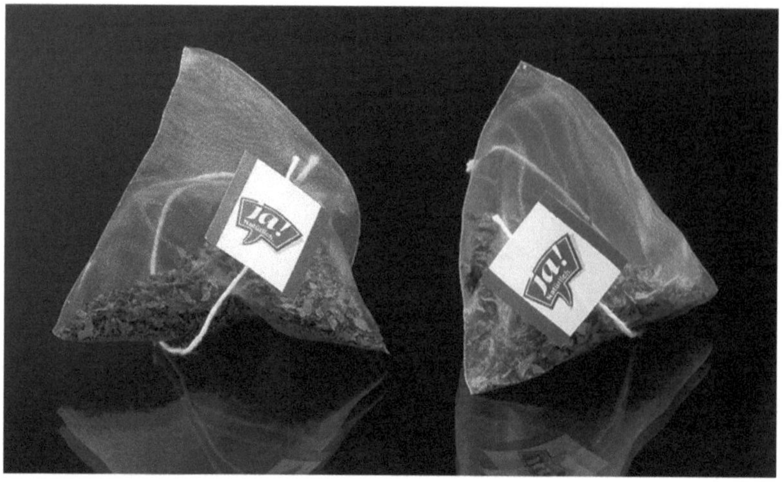

**Figure 4.22** Teabags of PLA. *Source: Wikipedia—Elke Wetzig.*

3. Filters in the automotive and chemical industries, as well as single use applications.
4. Teabag (see Figure 4.22). PLA is believed to be used as the sheath in a bicomponent sheath-core fiber where the low melting sheath provides the thermobonding component. PLA is already used in Ahlstrom BioWeb® teabags.

   PLA can be electrospun and as such is a source of a nanofiber web structures.

#### *4.3.5.3 Synthetic fibers—organic*

Almost all organic synthetic fibers are produced by a spinning process. One exception is fibrillated film such as that used in some of the Filtrete® filters produced by 3M Company (See Chapter 3, Section 3.7.1).

##### 4.3.5.3.1 Polyethylene

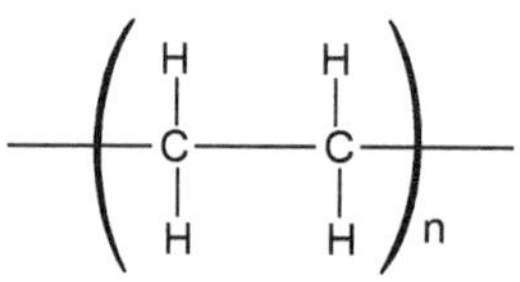

Polyethylene is produced through the polymerization of ethylene molecules. The reactions can be either free radical, coordination, or cationic. The free-radical reaction produces a low-density form of polyethylene (LDPE). It has low density because the reaction mechanism allows for the formation of branch chains on the polymer's backbone. The branch chains mechanically reduce the polymer's density. The cationic reaction produces a linear, denser, structure known as high-density polyethylene (HDPE). Catalyzed coordination polymerization is capable of producing an ultra-high molecular weight polyethylene (UHMWPE). UHMWPE is an extremely long linear polymer capable of inordinately high orientation and strength.

Polyethylene is composed of only carbon and hydrogen. It is the simplest of all polymer structures. It has a low polarity over a long chain length. The polymer is highly flexible. The repeating units are symmetrical with very few reactive functional groups. The HDPE and UHMWPE repeating units are highly regular, LDPE is less so. The degrees of polymerization may range from 500—10,000 for LDPE and HDPE, and may exceed 35,000 for UHMWPE.

LDPE has a low melting point. This makes it suitable for melt-blown webs. It is also suitable for the sheath in thermobonding sheath-core fiber. HDPE has a higher melting point than does LDPE. The glass transition temperature of polyethylene is extremely low (in the neighborhood of −115 °C). Polyethylene has an absence of hydrogen bonding sites, therefore, it is hydrophobic. Its equilibrium moisture content is very low. On the other hand, the polymer is oleophilic. It finds widespread use in operations to separate hydrocarbon liquids in a fluid stream. This would include coalescing separations. It is also used in oil spill clean-up situations ranging from oil spills at sea to the cleanup of oil spills in machine shops and metal working operations.

Because the fiber does not absorb water, it is very susceptible to static electricity. There are no chemically reactive functional groups on the polymer structure; therefore, polyethylene has relatively good resistance to chemicals. Although the polyethylene will burn, it resists flammability by melting away from a flame. Some of the key properties of polyethylene fiber are listed in Table 4.19.

**Table 4.19** Properties of Polyethylene Fiber (LDPE and HDPE)

| Property | Value or level |
|---|---|
| Melting point | 110 °C LDPE, 135 °C HDPE |
| Glass Transition | ca. −115 °C |
| Moisture Regain | <0.1% at standard conditions |
| Static Electricity | Accumulates |
| Dyeability | No dyeable product commercially available |
| Density | 0.90 g/cm$^3$ LDPE, 0.95 g/cm, HDPE |
| Tenacity | Low up to 6 g/den, depending on processing |
| Elongation at Break | High to low depending on previous treatment (600−50%) |
| Modulus | Low to high depending on orientation, tendency to creep at high stress |
| Abrasion Resistance | Excellent |
| Flammability | Melts and burns, Oxygen Index: ca. 15 |
| Acid/Base Stability | Excellent/good |
| Oxidative/Light | Can be etched by strong oxidants, relatively poor light stability |
| Solvents | Hydrocarbons at high temperature (>80 °C) |

Reproduced by permission of Dr. Roy M. Broughton, Jr. ((141), p. 76).

#### 4.3.5.3.2 Polypropylene

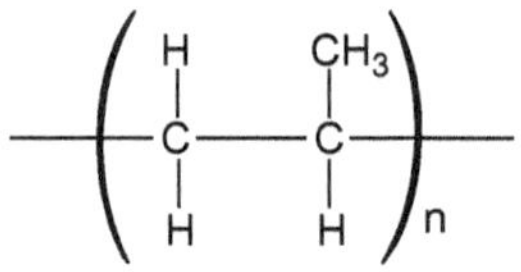

The polymerization of polypropylene from propylene is a coordination reaction. Depending on the parameters of the coordination process, the polymer can be produced in isotactic, syndiotactic, or atactic forms. The isotactic form is the one used in fibers. In the isotactic form polymer, repeating units do not vary in their symmetry. Average molecular weights are similar to polyethylene. Degrees of polymerization range from 500 to 10,000. Polypropylene is the fiber form most often used in melt-blown products. It is also used in spunbonded products such as Typar® and Tekton® produced by BBA Fiberweb.

Polypropylene differs from polyethylene in that it has a methyl group attached to its backbone. The steric hindrance of the methyl groups makes the polypropylene polymer less flexible than polyethylene. It also has a higher melting point and $T_g$. Nevertheless, these properties are still low enough that the fiber will avoid burning by melting away from a flame. In air filtration, the maximum operating temperature for polypropylene is 90–100 °C. The properties of polypropylene are tabulated in Table 4.20.

**Table 4.20** Properties of Polypropylene Fiber

| Property | Value or level |
|---|---|
| Melting point | 165 °C |
| Glass Transition | −15 °C |
| Moisture Regain | >0.1% moisture regain at standard conditions |
| Static Electricity | Accumulates |
| Dyeability | No dyeable polypropylene is commercially available |
| Density | 0.90 g/cm$^3$ |
| Tenacity | up to 7 g/den depending on processing |
| Elongation at Break | Drawn: <50%, undrawn: 500%+ |
| Modulus | Low to moderate depending on orientation, typically 20–50 g/den |
| Abrasion Resistance | Excellent |
| Flammability | Melts and burns |
| Acid/Base Stability | Excellent/good |
| Oxidative/Light | Can be etched by strong oxidants, relatively poor light stability, unless stabilized |
| Solvents | Soluble in hot hydrocarbons, chlorinated hydrocarbons |

Reproduced by Permission of Dr. Roy M. Broughton, Jr. ((141), p. 79).

### 4.3.5.3.3 Acrylic fiber

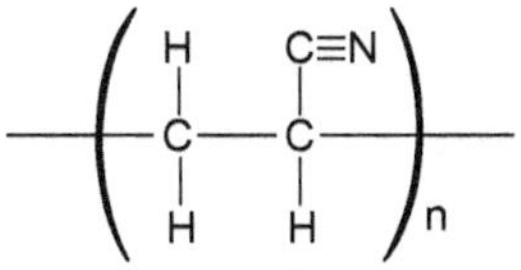

According to TAPPI TIP 1205-1 (158), the Federal Trade Commission defines acrylic fibers as containing access of 85% by weight of acrylonitrile monomers. At the higher acrylonitrile content the polymer is referred to as polyacrylonitrile (PAN). If the acrylonitrile content is 35–85%, the fiber is defined as modacrylic. Like polyethylene and polypropylene, the repeat unit is an ethyl group with one of its hydrogens substituted by a pendant nitrile (cyano) group. The cyano pendant is carbon triple-bonded to nitrogen. Unlike polyethylene and polypropylene, comonomers (up to the allowable 15%) are deliberately added to provide sites for the attachment of dye molecules. Comonomers can be methyl methacrylate, methyl acrylate, and vinyl acetate.

PAN can be produced by free radical, anionic, or coordination polymerization process chemistry. The free-radical mechanism is the one most commonly used for commercial production. This results in a mostly atactic structure. The melting point of PAN is usually higher than the temperature at which it begins to thermally decompose. To be spun into a fiber it has to be solution spun, or sometimes spun from a plasticized melt. The polymer is mixed with the solvent to form a very viscous solution called a "dope." There are a number of solvents that can be used including sodium thiocyanate, dimethylacetamide, dimethylfomamide, nitric acid, zinc chloride, dimethysulfoxide, and ethylene carbonate.

Acrylic fibers can be "wet spun" or "dry spun." In the wet-spinning process, the dope is spun through a spinnerette immersed in a spin bath containing a mixture of solvent and nonsolvent. The dope is coagulated (precipitated) into a fiber form. The fiber form is withdrawn from the solution and further processed to establish and reinforce fiber properties.

In the dry-spinning process, the viscous solution is spun with a stream of hot air into a vertical chamber. Filament formation occurs due to evaporation of the solvent. Dry-spun fibers normally have a dog bone cross section because of inherent factors of the process. Approximately 75% of world capacity of PAN is from the wet-spinning process and 25% from the dry-spinning process.

Acrylic fibers tend to be more brittle and less flexible than polyethylene and polypropylene fiber. On the other hand, they have better temperature resistance, good chemical resistance, and particularly good resistance to outdoor exposure—sunlight and microorganisms. For this reason they have a niche in outdoor applications such as tenting and awning fabrics. Acrylic fibers also have good wicking properties making them suitable for wipes and absorbent products.

The staple fiber form of the polymer is used in filter media in both dry laid and wet laid applications. The fiber has good bulking properties for filter media. Its temperature and chemical resistance makes it suitable in media for corrosive baghouse operations.

Special forms of the polymer exist. It can be fibrillated into a synthetic pulp for papermaking purposes. In this manner, it has nanofiber characteristics and serves as a binder fiber. Filter medium is one application of fibrillated acrylics. Bicomponent fibers can be produced by spinning two acrylic polymers with dissimilar copolymer levels. Unlike polyester, nylon, or polypropylene sheath-core fibers, the acrylic bicomponents tend to be side by side. Such a structure shrinks unevenly when exposed to high temperatures above 100 °C and form into a spiral or coil shape; very handy where high loft is desired. PAN fibers can be carbonized and is the basis for carbon fiber manufacture.

Table 4.21 is a summary of the properties of acrylic fibers.

**Table 4.21** Properties of Acrylic Fiber

| Property | Value or level |
|---|---|
| Melting point | Not very crystalline, decomposes before melting |
| Glass Transition | 105 °C |
| Moisture Regain | 1% to 2% |
| Static Electricity | Accumulates |
| Dyeability | Dyeable with disperse and cationic dyes |
| Density | 1.14 to 1.17 g/cm$^3$ |
| Tenacity | Low up to 5 g/den depending on processing |
| Elongation at Break | 10% to 50% |
| Modulus | Low to moderate, typically 5–10 g/den |
| Abrasion Resistance | Poor to fair |
| Flammability | Burns slowly, can evolve cyanide gas if no oxygen present |
| Acid/Base Stability | Excellent for most acids, moderate for bases |
| Oxidative/Light | Excellent |
| Solvents | Polar solvents such as DMF and DMSO and even water at elevated temperatures, are solvents for PAN |

Reproduced by Permission of Dr. Roy M. Broughton, Jr. ((141), p. 83).

#### 4.3.5.3.4 Polyester

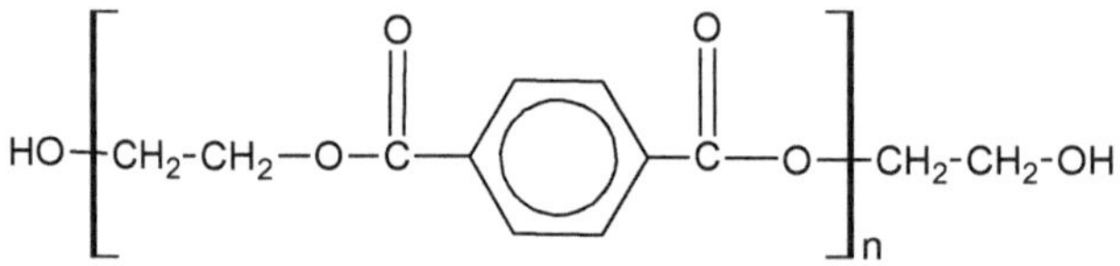

Polyethylene Terephthalate

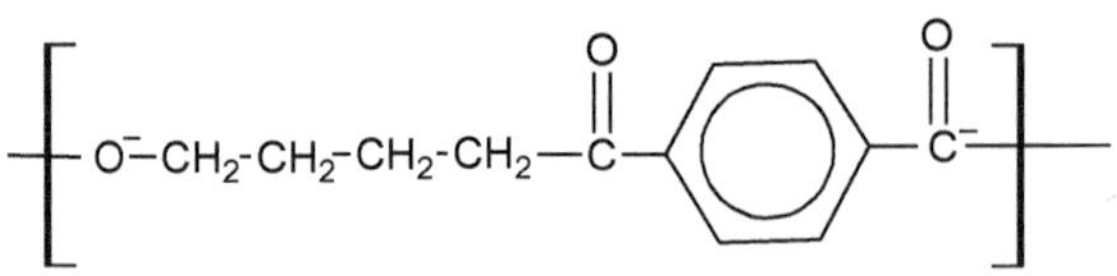

Polybutylene terephthalate

The development of polyester fiber began in the late 1930s, originally by W.H. Carothers of DuPont and later by a group of British scientists: J.R.Whinfield, J.T. Dickson, W.K. Birtwhistle, and C.G. Ritchie[3] (159). In 1941, the British produced the first polyester fiber called Terylene®. In 1946, Dupont purchased all legal rights to the technology. In 1953, they initiated the commercial production of the polyester fiber known as Dacron. In 1958, Eastman Chemicals Products, Inc. introduced another polyester fiber called Kodel®. Since then, the demand for polyester fiber has grown incessantly. It was reported (160) that in 2005, the global demand for polyester fiber exceeded that of cotton. According to Fiber Organon, a US-based consultancy, global production of polyester fibers rose to 37.3 million tons in 2010, making up 74% of the overall production of synthetic fibers (161).

Polyester is a condensation polymer composed of esters, 85% or more, that are attached to two aromatic rings (141).[4] There are several different chemistries for producing polyester, however, the one most commonly used is the reaction product of terephthalic acid and ethylene glycol. The polymer is polyethylene terephthalate (PET). PET is a relatively stiff polymer and has moderate polarity. It has a relatively high melting point of 260 °C, and therefore, it can be used in environments where heat resistance is desired. There are very few hydrogen bonding sites and these are at the end of the molecule—none in the repeating unit. The polymer has good resistance to moisture and very little moisture absorption. The low moisture allows the polymer to accumulate a static electricity charge.

The properties of PET fiber along with the continuous progress in its economics have made it a very popular fiber for nonwoven filter media. PET staple fiber is carded, either by itself or with other fibers (example cotton) to form needle-punched webs for air filtration applications. PET is the principal fiber in thermally bonded high loft webs for HVAC applications. It is also becoming a significant fiber for automotive cabin air filtration applications. Short cut PET fiber is widely used for wet laid filter applications. Six- and twelve-millimeter PET fiber are often mixed into cellulose slurries to offer improved wet laid, resin-treated filter media. The improvements being a web with a more open structure, higher tensile and tear strength, and greater flex resistance. In recent developments, blends of PET and nylon fiber have been shown to synergize the best properties of both fibers (see Section 4.3.5.3.5).

Spunbonded structures such as Reemay®, produced from PET, have found many applications in both air and liquid filtration.

A sister fiber to PET is polybutylene terephthalate (PBT). PBT has a lower melting point and a lower $T_g$ than does PET. It can be used in both spunbond and melt-blown processes. A melt blown medium for blood filtration is produced from PBT. Ticona, the technical polymers business for Celanese AG, produces a PBT polymer with trade name Celanex®. The polymer is also produced by DuPont and GE Plastics.

[3] Swicofil AG Textile Sevices.

[4] Broughton and Brady (141).

The properties of polyester fiber are summarized in Tables 4.22 (PET) and 4.23 (PBT).

**Table 4.22** Properties of PET Polyester Fiber

| Property | Value or level |
|---|---|
| Melting point | 260 °C |
| Glass Transition | 85 °C |
| Moisture Regain | <1% standard regain |
| Static Electricity | Accumulates |
| Dyeability | Moderate, requires disperse dyes, and assistants or pressure |
| Density | 1.43 |
| Tenacity | low—up to 9 g/den |
| Elongation at Break | 500%—15% |
| Modulus | Average modulus of 15 g/den typical, can be 100 g/den for industrial fibers |
| Abrasion Resistance | Excellent, not quite as good as olefin or nylon in flex abrasion |
| Flammability | Burns slowly with smoke, O.I. = 20 |
| Acid/Base Stability | Attacked readily by hot dilute bases, less affected by acids |
| Oxidative/Light | Reasonably good light stability |
| Solvents | Trifluro-acetic acid, Phenol/tetrachloro-ethylene mixed, and concentrated $H_2SO_4$ |

Reproduced by Permission of Dr. Roy M. Broughton, Jr. ((141), p. 90).

**Table 4.23** Properties of PBT Polyester Fiber

| Property | Value or level |
|---|---|
| Melting Point | 227 °C |
| Glass Transition | 66 °C |
| Moisture Regain | 0—0.5% |
| Static Electricity | Good anti-static properties |
| Density | 1.3 g/cm$^3$ |
| Tenacity | 2.8 cN/dtex convert to g/den |
| Elongation at Break | 5—300% |
| Abrasion Resistance | Good |
| Flammability | O.I. = 22 |
| Acid/Base stability | Excellent acid resistance, fair base resistance |
| Oxidative/Light | Exceptional resistance to thermal oxidative degradation |

### 4.3.5.3.5 Nylon

Nylon is the generic name for polyamide fiber. According to Fibersource (162)[5], the United States Federal Trade Commission defines nylon as "A manufactured fiber in which the fiber forming substance is a long-chain synthetic polyamide in which less than 85% of the amide-linkages are attached directly (—CO—NH—) to two aliphatic groups."

[5] Fibersource is an educational website maintained by the American Fiber Manufacturers Association, Inc.

There are two major commercial forms of nylon: nylon 6,6 and nylon 6. Nylon 6,6 was the first form invented by W.H. Carothers of DuPont in the 1930s and first commercialized by DuPont in 1939. By 1938, Paul Schlack of I.G. Farben in Germany had polymerized caprolactam and created Nylon 6. It was reported in Section 4.3.5.2.1 that the the former Textile Fibers Department of Dupont was sold in 2007 and reorganized into the company now known as Invista. Invista is also a major producer of nylon fiber.

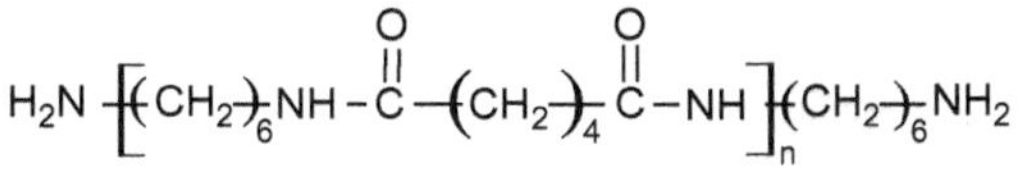

**Nylon 6, 6**

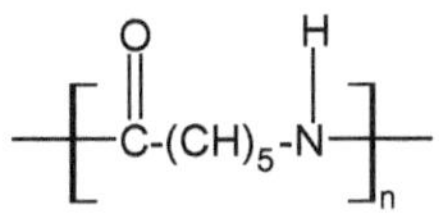

**Nylon 6**

Nylon 6,6 is the condensation product of hexamethylene diamine and adipoyle chloride. Nylon 6 is produced by the polymerization of caprolactam, a ring opening reaction. Nylon fibers are flexible and exceptionally strong. Hydrogen bonding sites every six atoms along the backbone allow for some moisture absorption. The presence of moisture affects the glass transition temperature, which is relatively low compared to polyester or acrylic fibers. The presence of moisture also makes the fiber less vulnerable to static accumulation, particularly in environments of high humidity. The low glass transition temperature combined with the flexibility in the chain structure allows for the easy orientation in the fiber. Orientation enhances the strength of the fiber, however the elongation will be low. On the other hand, unoriented nylon can be stretched to several times its original length. Nylon 6,6 has a higher melting point than nylon 6. Otherwise their properties are fairly similar.

Polyamides are not particularly stable to oxidants and ultraviolet light, however, Ortega (163) reports that the "the addition of copper and other antioxidants are known to provide improved heat and light stability performance to nylon". Seat belts made of Nylon yarn are a good example. Nylon is resistant to damage from oil and many chemicals. It is resistant to alkali solutions and weak acids. Nylon will dissolve in strong acids.

Nylon fiber is widely used in filter media. It is used in needle felt filter media for baghouse filtration applications. It is used in other carded web structures such as high loft and gradient density media for air and liquid filter applications. It is the fiber in spunbonded Cerex® and PBN-II webs produced by Cerex Advanced Fabrics, many of which have filter medium applications. Nylon fabrics also have good barrier properties in controlling oil spills in water (see Chapter 3, Section 3.14). Donaldson Company, Inc. uses nylon as the fiber material in their electrospun nanofiber filters.

Table 4.24 Properties of Nylon Fiber

| Property | Value or level |
|---|---|
| Melting point | Nylon 66: 260 °C, Nylon 6: 220 °C |
| Glass Transition | 50 °C for both |
| Moisture Regain | 4% standard regain |
| Static Electricity | Accumulates in dry atmosphere. Usually not a problem in moist environs. |
| Dyeability | Dyeable with direct, acid, disperse, and vat dyes. |
| Density | 1.12–1.15 g/cm$^3$ |
| Tenacity | low–10 g/den |
| Elongation at Break | 20% to 500% |
| Modulus | Low to moderate |
| Abrasion Resistance | Excellent |
| Flammability | Burns after melting |
| Acid/Base Stability | Degraded by bases, dissolves in strong acids |
| Oxidative/Light | Degrades in UV light and in oxidant solutions |
| Solvents | Strong mineral and organic acids, DMF and other polar organics |

Reproduced by Permission of Dr. Roy M. Broughton, Jr. ((141), p. 87).

Table 4.24 summarizes the typical properties of nylon fiber.

### 4.3.5.3.6 Polyaramid

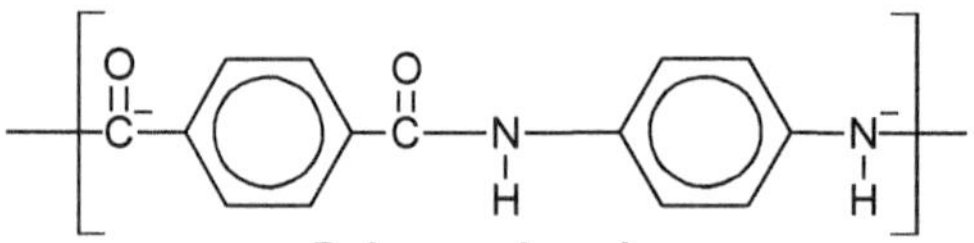

Polyparaphenylene terephthalamide (Kevlar®)

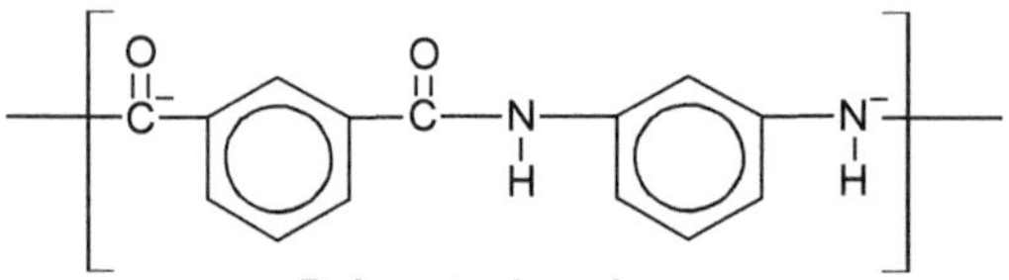

Polymetaphenylene isophthalamide (Nomex®)

There are two well-known forms of polyaramid fibers: para-aramid (polyparaphenylene terephthalamide) and meta-aramid (polymetaphenylene isophthalamide). The para-aramids are better known by the trade names of Kevlar® (trademark of DuPont) and Twaron® (trademark of Teijin Limited, Osaka, Japan). Likewise, the meta-aramids are commonly referred to by their trade names, Nomex® (Dupont) and

Teijinconex® (Teijin Limited—often referred to as Conex®). Both polymers are related to nylon in that they are forms of polyamide; the difference is that nylon has a carbon chain linkage with the amide structure whereas the polyaramids have a phenyl linkage. Referring to the structures above, the para-aramid connects at the para-position of the phenyl link whereas the meta-aramid connects at the meta-position of the phenyl linkage.

The fibers from both forms of the polymer are very heat resistant. They find application in heat resistant air filters. Generally they are suitable for applications up to 177 °C (350 °F).

The para-aramid fiber is tremendously strong. The tenacity can go up to 25 g/de when annealed under tension. It is a very rigid fiber of very high modulus. It's most notable application is in bullet-proof vests for police and military protection. When used in filter media it absorbs just enough moisture to prevent static charge buildup. Properties of para-aramid fiber are listed in Table 4.25.

**Table 4.25** Properties of Para-Aramid Fiber

| Property | Value or level |
|---|---|
| Melting Point | >500 C |
| Glass Transition | 375 C |
| Moisture Regain | 1 to 4% |
| Static Electricity | Low static build-up |
| Dyeability | Poor due to high $T_g$ and high crystallinity |
| Density | 1.44 g/cm$^3$ |
| Tenacity | 15 g/den as spun, 25 g/den when annealed under tension |
| Elongation at Break | 1 to 4% |
| Modulus | Very high |
| Abrasion Resistance | Poor compared to meltspun fibers, better than inorganics |
| Flammability | Does not burn under standard conditions, O.I. = 30 |
| Acid/Base Stability | Excellent |
| Oxidative/Light | Oxidized by bleaches. Degraded by Prolonged exposure to UV. |
| Solvents | Soluble in concentrated $H_2SO_4$ |

Reproduced by Permission of Dr. Roy M. Broughton, Jr. ((141), p. 92).

Table 4.26 reveals that the meta-aramid form of the polyaramid polymer is not as strong, but it is more flexible. Nomex® is widely used in needle felt filter media for high temperature baghouse operations. It resists attacks by mild acids, mild alkalis, and most hydrocarbons. It has flex resistance comparable to polyester fiber, and much higher temperature resistance. It has exceptional dimensional stability and will not stretch or shrink over 2% when exposed to temperatures up to 218 °C (425 °F).

**Table 4.26** Properties of Meta-Aramid Fiber

| Property | Value or level |
|---|---|
| Melting Point | 390 °C |
| Glass Transition | 280–290 °C |
| Moisture Regain | 4% |
| Static Accumulation | Static build-up possible in dry conditions |
| Density | 1.38 g/cm$^3$ |
| Tenacity | 2.6–2.9 g/den |
| Modulus | 8–14 N/tex |
| Elongation at Break | 19–22% |
| Abrasion Resistance | Excellent |
| Flammability | Does not burn in air, does not melt or drip |
| Acid/Base Stability | Fair in strong acids, good in alkalis and weak acids |
| Oxidative/Light | Yellows on prolonged light exposure |
| Solvents | Sodium hydroxide solution |

#### 4.3.5.3.7 Polyphenylene sulfide

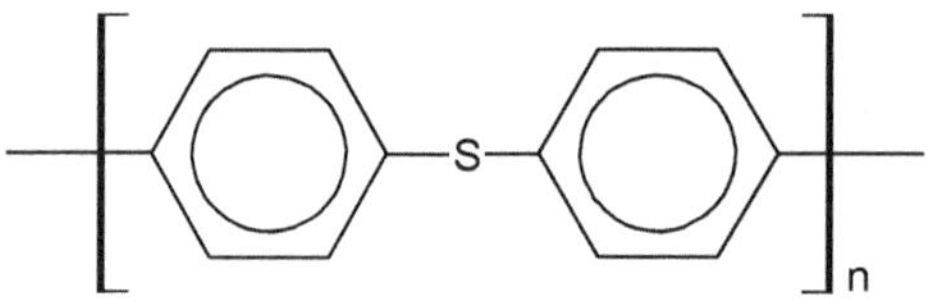


Polyphenylene sulfide (PPS) is a rigid, thermoplastic fiber that is used in felt filter media for high temperature air filter applications. The polymer is available in branched and linear forms. The fiber is produced from the linear form. It is a semi crystalline material (60–65% crystallinity). No solvent is known that will dissolve the fiber at temperatures less than 200 °C.

The fiber is ideal for filter media exposed to continuous operating temperatures up to 190 °C (375 °F). It has exceptional chemical resistance to most acids, alkalis, organic solvents, and oxidizers. It also has excellent resistance to gamma and neutron radiation. It is non-burning, thus increasing its desirability for high temperature applications.

The polymer is marketed under the trade names of Ryton® (Chevron Phillips Chemical Co., LP) and Fortron® (Ticona Technical Polymers). Toray, headquartered in Japan, is the world's leading producer of PPS fiber. Other fiber producers are Inspec Fibres, Amoco Fiber, and Fiber Innovation Technology, Inc. of Johnson City, Tennessee, USA. National Nonwovens Inc. of Easthampton, Massachusetts markets a needle punch felt media,TFIL17®, composed of 100% PPS fiber and reinforced with a PPS woven scrim.

The heat resistance, chemical resistance, and resistance to hydrolysis of PPS fiber makes it suitable for many liquid filtration applications, particularly where hot and/or

**Table 4.27** Properties of PPS Fiber

| Property | Value or level |
|---|---|
| Melting Point | 280 °C |
| Glass Transition | 90 °C |
| Moisture Regain | 4% |
| Density | 1.43 g/cm$^3$ |
| Tenacity | Very high |
| Modulus | Very high |
| Elongation at Break | 5% |
| Abrasion Resistance | Excellent |
| Flammability | Does not burn, O.I.: 47 |
| Acid/Base stability | Fair in strong acids, good in alkalis and weak acids |
| Oxidative/Light | Resistant to oxidation |
| Solvents | None |

corrosive liquids are filtered. The polymer can be spun into melt-blown and spunbond nonwoven fabrics.

Some of the properties of PPS fiber are listed in Table 4.27.

#### 4.3.5.3.8 Polyimide

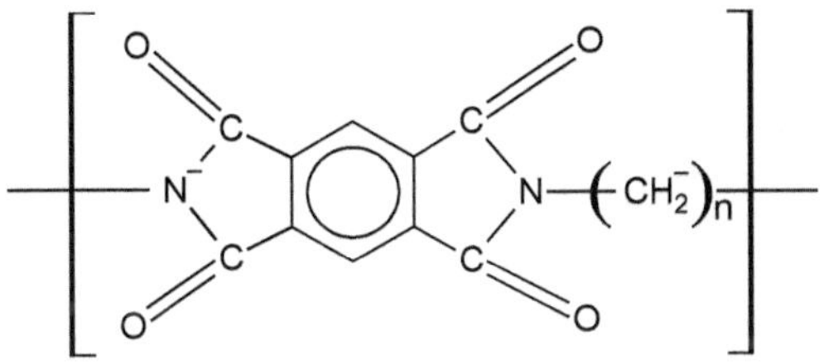


Polyimide fiber is often referred to by its trade name P-84®. The fiber was developed by Lenzing AG in Austria. It is marketed by Evonik Fibres GmbH in Austria. Evonik is the world's only producer of polyimide fiber. It has already been reported in Section 4.3.4.6 that the P-84 ® fiber has a trilobal cross section (see Figure 4.23), which enhances its surface area for filtration. The fiber provides unique heat resistance properties. It can be used continuously in temperature environments up to 260 °C. It also has very good flame retardant properties.

The fiber is manufactured as a staple fiber ranging from 0.6 to 8.0 dtex. It is best used in needle punch felts for high temperature air and fume applications. Evonik (164) mentions its application as a sealant in the manufacture of oil filter cartridges because of its stability at high temperatures and its resistance to hot oil. This is done by melting polyamide into a polyimide felt ring.

Some typical properties of P-84® polyimide fiber are summarized in Table 4.28.

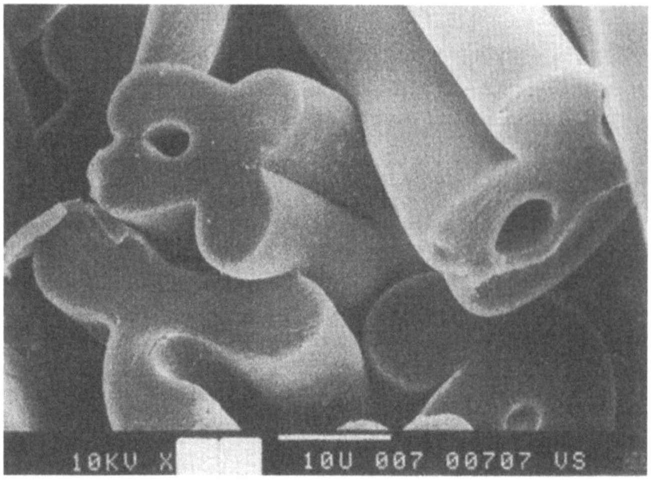


**Figure 4.23** The Trilobal Shape of P-84® Fiber. *Reproduced with Permission of Elsevier Science Ltd., Oxford, England, UK (18).*

**Table 4.28** Properties of P-84®. Polyimide Fiber

| Property | Value |
|---|---|
| Tenacity | 4.2 g/den |
| Elongation | 30% |
| Continuous operating temperature | 260 °C |
| Limiting oxygen index | 38% |
| Glass transition temperature | 315 °C |
| Density, | 1.41 g/cm$^3$ |
| Shrinkage | <1% at 250 °C, 10 min |
| Chemical resistance | Good |

#### 4.3.5.3.9 Polytetraflouroethylene (PTFE)

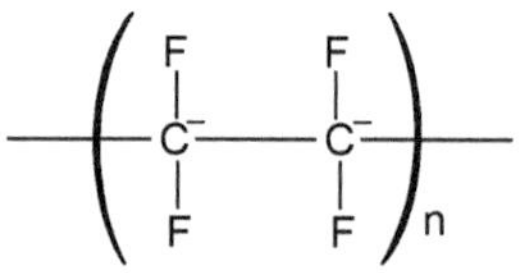

Polytetraflouroethylene (PTFE) is one of a group of fluoropolymers that follow the DuPont trade name of Teflon®. There are also fluoropolymer products produced by Asahi Glass (AGC) under the trade name of Fluon® and by Toray under the trade name of Toyoflon®. PTFE is the form that is most used in filtration applications. It is used as a membrane material, as a fiber coating, and as a fiber. The fiber is used in needle punch felt filter media for hot gas filtration, and in wet lay nonwovens for liquid filtration. Its monofilament form is also used for filtration products. Membrane filter materials are an important application of PTFE. Both W.L. Gore

**Table 4.29** Properties of PTFE Fiber

| Property | Value or level |
|---|---|
| Melting point | 327 °C |
| Moisture regain | < 0.1% |
| Density | 2.13–2.22 g/cm$^3$ |
| Tenacity | 2 g/den |
| Elongation at break | 25% |
| Abrasion resistance | Excellent |
| Flammability | Extinguishes when flame is removed. L.O.I.: 95 |
| Operating temperature | 260 °C |
| Acid/Base stability | Outstanding |
| Weather resistance | Rated at 20 years unaffected |
| Solvents | None |

(GORE-TEX® ePTFE) and Donaldson (TETRATEX®) market PTFE membranes laminated to a number of baghouse filter materials (see Section 8.1.1).

PTFE is unique in its combination properties of hydrophobicity, chemical resistance and thermal stability. "Nothing sticks to Teflon®" is also a very useful property for certain types of filter media. An example is baghouse pulse jet filtration. The presence of Teflon® fiber in needle punch and/or spunlace felts or the presence of PTFE film laminates enhance the dirt release characteristics of the medium. Table 4.29 lists some properties of PTFE fiber.

#### 4.3.5.3.10 Polybenzimidazole (PBI)

Polybenzimidazole is an organic fiber with excellent thermal resistant properties. It does not burn in air and it does not melt. Its high limiting oxygen index (LOI) coupled with good chemical resistance and good moisture regain makes this fiber a good candidate for flame resistant filter media such as needle felts. Its strength properties are relatively low, however, it blends well with other fibers such as carbon and polyaramid and is sometimes used this way in flame-resistant needle felts to give very high performance, sufficient to justify its extremely high cost. Property data on PBI is provided by Smith (165) in Table 4.30.

**Table 4.30** PBI Properties

| PBI property | Value |
|---|---|
| Tenacity g/de | 2.7 |
| Modulus g/de | 32 |
| Elongation (%) | 29 |
| Continuous operation temperature (°F) | 482 |
| Limiting Oxygen Index (%) | 41 |
| Chemical resistance | Good–Excellent |

Reproduced with Permission of Bill Smith (172) Industrial Textile Associates, Greer, South Carolina, USA.

### *4.3.5.4 Inorganic fibers*

Inorganic fibers, both metal and non metal, are more resistant, more rigid, higher melting, and more heat resistant than traditional organic fibers. NF New Fibres (166) identifies the following materials as being used for inorganic fibers: carbon, glass, metal, boron, silica, silica (silicone) carbide, and ceramic. They note that carbon fibers, although inorganic as such, are produced from organic fibers. Silicon carbide fibers are also sourced from organic fibers (see Section 4.3.5.4.3).

NF New Fibres (166) provides comparison of typical properties of some inorganic fibers in Table 4.31.

**Table 4.31** Comparison of Inorganic Fiber Properties

| Material | Density, g/cm$^3$ | Tenacity, MPa | Modulus GPa |
|---|---|---|---|
| Glass | 2.5–2.6 | 3400–4500 | 70 |
| Carbon | 1.8–2.1 | 2000–7000 | 240–700 |
| Silica Carbide | 2.6–3.5 | 2000–3700 | 200–420 |
| Oxides | 3.9 | 1200–1400 | 340–400 |

Source: Studio Tevnica, Florence, Italy.

#### 4.3.5.4.1 Glass fibers

Glass fibers are formed to a broad range of fiber diameters. They are classified by diameter and by chemical composition. They can be produced as microfibers and as chopped strand. The process is a melt-blown process related to the melt-blown process for producing polymer webs (see Chapter 5, Section 5.2.2). Two primary processes for producing glass microfibers are rotary attenuation and flame attenuation. Figure 4.24, provided by Lauscha Fiber International, Lauscha, Germany, is a picture of glass microfiber produced by both of these processes. In both processes, the fibers are flame attenuated to control their diameters. From these, Scheffel (167) describes four variations as follows:

**a.** Rotary Process—The rotary process is diagrammed in Figure 4.25. Molten glass is charged into a rotating basket-like device having a large number of holes as centrifugal openings. Because of the centrifugal forces arising due to the rotation, the glass melt flows to the circumferential wall and is driven out through the openings as primary monofils. A hot gas stream transverse to the direction of the monofilaments draws them to fine glass fibers.

**b.** CAT process—The controlled attenuation technology (CAT) process represents a modified rotary technology with less and thinner monofils and a significantly higher gas stream (Figure 4.26).

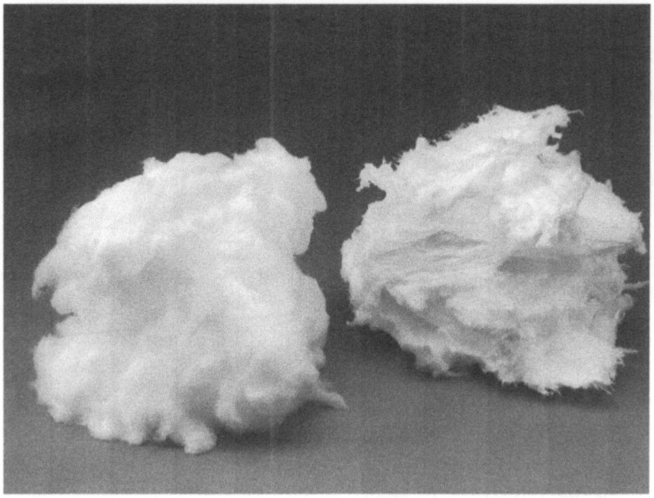

**Figure 4.24** Rotary and Flame-Attenuated Fibers. *Reproduced by Permission of Lauscha Fiber International, Lauscha, Germany.*

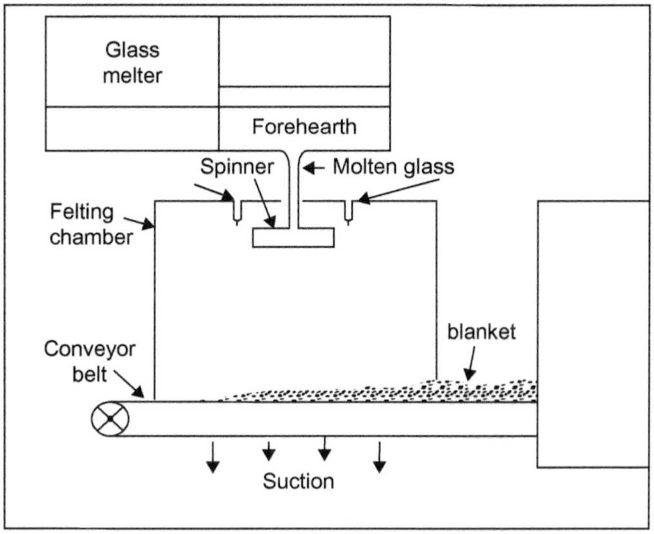


**Figure 4.25** Rotary Process. *Reproduced with Permission of the American Filtration and Separation Society, Nashville, Tennessee, USA (167), p. 188.*

**c.** Flame attenuation process—This process is characterized by primary monofils with diameters of 25 to 38 μm that are drawn from a platinum bushing and then fiberized by a highly accelerated hot gas stream (Figure 4.27).

**d.** Duplex process—The starting materials are glass rods from which primary monofils with diameters of 0.3 to 1.0 mm are continuously drawn mechanically and these are fiberized in a highly accelerated gas stream of high temperature (Figure 4.28).

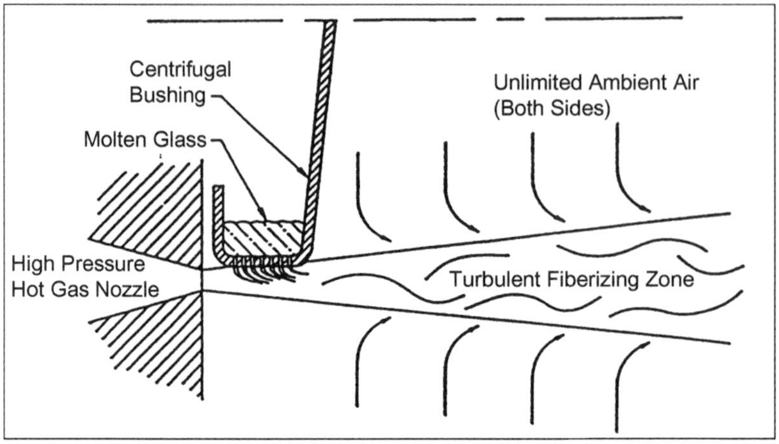


**Figure 4.26** CAT Process. *Reproduced with Permission of the American Filtration and Separation Society, Nashville, Tennessee, USA (167). p. 189.*

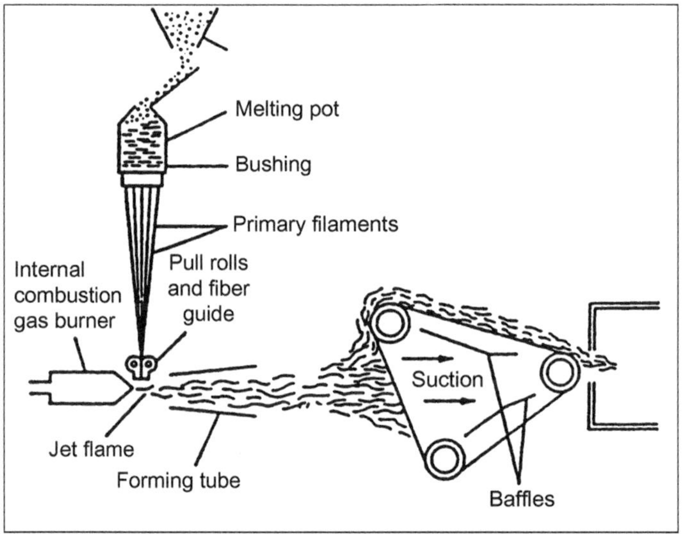


**Figure 4.27** Flame Attenuation Process. *Reproduced with Permission of the American Filtration and Separation Society, Nashville, Tennessee, USA Society (167), p. 190).*

Several glass compositions are available for the production of glass microfibers. Lauscha Fiber International (168) lists three compositions (A, B, and E) in Table 4.32. According to Lauscha, A-Glass is an alkali silicate glass composition developed to reduce possible contamination from boron outgassing in critical clean room filtration applications. This is particularly important in clean room filters in the electronic

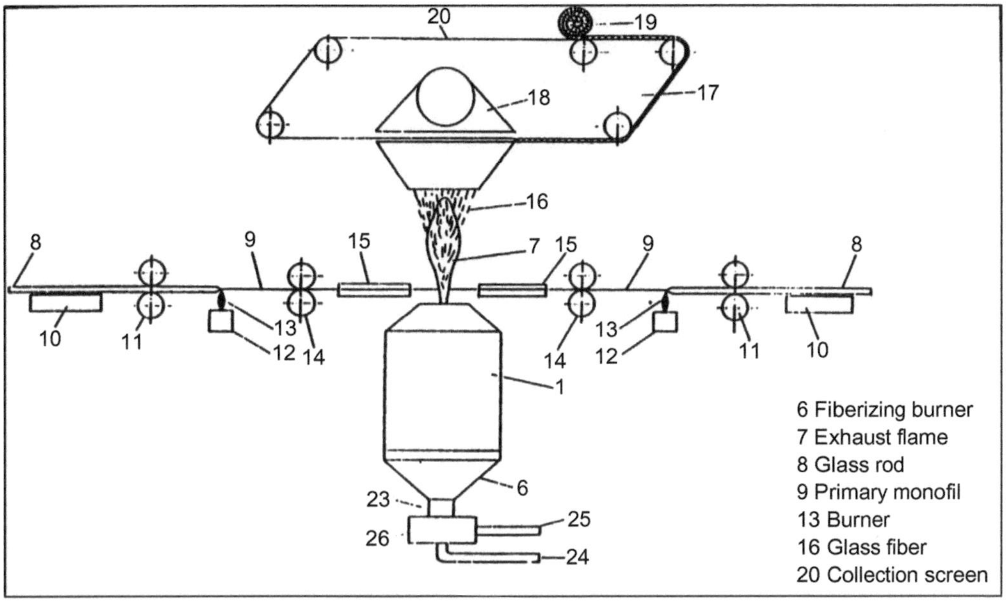


**Figure 4.28** Duplex Process. *Reproduced with Permission of the American Filtration and Separation Society, Nashville, Tennessee, USA (167), p. 191.*

**Table 4.32** Chemical Composition Lauscha Glass Microfibers

| Glass composition | A-Glass (%) | B-Glass (%) | E-Glass (%) |
|---|---|---|---|
| $SiO_2$ | 69.0–71.0 | 55.0–60.0 | 50.0–56.0 |
| $Al_2O_3$ | 2.5–4.0 | 4.0–7.0 | 13.0–16.0 |
| $B_2O_3$[a] | <0.09[a] | 8.0–11.5 | 5.8–10.0 |
| $Na_2O$ | 10.5–12.0 | 9.5–13.5 | <0.60 |
| $K_2O$ | 4.5–6.0 | 1.8–4.0 | <0.40 |
| CaO | 5.0–7.0 | 1.0–5.0 | 15.0–24.0 |
| MgO | 2.0–4.0 | <2.0 | <5.5 |
| $Fe_2O_3$ | <0.20 | <0.25 | <0.50 |
| ZnO | <2.0 | 2.0–5.0 | <0.02 |
| BaO | – | 3.0–6.0 | <0.03 |
| $F_2$ | <1.0 | <1.0 | <1.0 |
| $TiO_2$ | – | – | <1.0 |

[a]$B_2O_3$ contains 31.1% boron by weight. It follows that the maximum allowable boron content in A-Glass is 0.028%.
Reproduced by Permission of Lauscha Fiber International, Lauscha, Germany (168).

industry, where even a few molecules of boron contamination can destroy a large batch on high purity electronic components. B-Glass is a borosilicate glass chemistry designed for broad use in a full range of applications including air, gas, and liquid filtration. E-Glass microfiber is unique within the industry. Lauscha's technology

**Table 4.33** Chemical Composition Johns Manville Glass Microfibers

| Glass composition | 475 chemical composition for 100 series | 253 chemical composition for 200 series |
|---|---|---|
| $SiO_2$ | 58.3 | 65.5 |
| $Al_2O_3$ | 5.8 | 3.1 |
| $B_2O_3$ | 11.3 | 5.3 |
| $Na_2O$ | 10.1 | 16.0 |
| $K_2O$ | 2.9 | 0.7 |
| CaO | 1.8 | 5.9 |
| MgO | 0.3 | 3.0 |
| ZnO | 4.0 | – |
| BaO | 5.0 | 0.01 |

Reproduced by Permission of Johns Manville, Denver, Colorado, USA.

enables fiberization of electrical grade E-Glass to submicron levels with no shot and no organic contamination. E-Glass microfiber has specific applications in advanced analytical filtration and diagnosis. It is designated as an electrical grade glass because of its relatively high electric conductivity.

Johns Manville lists two: 475 chemical composition for their 100 series of glass microfiber and 253 low boron chemical composition for their 200 series. These compositions are listed in Table 4.33.

Fiber diameter determines the filtration efficiency of filter media made from glass microfibers; the finer the diameter, the more efficient will be the medium. It is common practice to grade glass microfibers with a numbering system: the lower the number, the smaller the fiber diameter. Lauscha uses a system that identifies the glass composition A, B, C, or E, and uses a two-digit number from 00 to 50 for the fiber diameter code. 00 is the smallest diameter available and 50 is the largest. Their grade numbering system (see Table 4.34) also includes a letter designation to identify the fiberization technology (F, flame attenuation; Fi, high speed F-technology; and R, rotary attenuation). For example, Grade B-06-F would be a glass microfiber composed of the B-Glass composition in Table 4.34, having a fiber diameter code of 06, and produced by flame attenuation (F) technology. Table 4.35 lists several of the Lauscha grades along with their fiber diameters and specific surface areas (SSA). Figure 4.29 relates Lauscha grades and SSA to handsheet pressure drop. It is apparent from the Lauscha data that the finer the fiber diameter, the larger is the SSA and the larger the SSA, the more resistant will be the medium to flow.

The Johns Manville system uses a digital number series: 90 and up for the 475 glass composition and 200, and up for the 253 glass composition. There is also a CX grade for the 5.5 μm diameter. Figure 4.30 plots the Johns Manville code versus fiber diameter.

**Table 4.34** Lauscha Product Designation System

The product grade designation system incorporates: four glass chemistries (A, B, C or E-Glass), various diameter codes, and three proprietary fiberizing processess (F, F(*i*), and R-technology):

| GLASS COMPOSITION | DIAMETER CODE | FIBERIZATION TECHNOLOGY |
|---|---|---|
| A-Glass (low boron alkali silicate)<br>B-Glass (borosilicate)<br>C-Glass (acid resistant borosilicate)<br>E-Glass (calcium aluminoborosilicate) | 00 - 50 | F-flame attenuation technology<br>F*i*-high speed F-technology<br>R-rotary attenuation technology |

Example:

**B — 06 — F**

Reproduced by Permission of Lauscha Fiber International, Lauscha, Germany ((169), p. 1).

**Table 4.35** Lauscha Glass Microfiber for Air and Liquid Filtration

| Microfiber chemistry | Grade | Nominal diameter μm | Nominal specific surface area m²/gm | Fiberization technology |
|---|---|---|---|---|
| B-Glass | B-X9 Fa | 0.26 | 6.20 | Flame Attenuated |
| | B-00-Fa | 0.33 | 4.80 | Flame Attenuated |
| | B-02-F | 0.46 | 3.50 | Flame Attenuated |
| | B-04-F | 0.53 | 3.00 | Flame Attenuated |
| | B-06-F | 0.65 | 2.47 | Flame Attenuated |
| | B-08-F | 0.80 | 2.00 | Flame Attenuated |
| | B-10-F | 1.00 | 1.60 | Flame Attenuated |
| | B-15-F | 1.48 | 1.08 | Flame Attenuated |
| | B-26-R | 2.44 | 0.66 | Rotary |
| | B-39-R | 3.20 | 0.50 | Rotary |
| | B-50-R | 4.10 | 0.42 | Rotary |
| | B-56-R | 5.00 | 0.35 | Rotary |
| A-Glass | A-04-F | 0.53 | 3.00 | Flame Attenuated |
| | A-06-F | 0.65 | 2.47 | Flame Attenuated |
| | A-26-F | 2.44 | 0.66 | Flame Attenuated |
| E-Glass | E-04-Fa | 0.56 | 2.88 | Flame Attenuated |
| | E-06-F | 0.65 | 2.47 | Flame Attenuated |
| | E-08-F | 0.80 | 2.00 | Flame Attenuated |

Reproduced by Permission of Lauscha Fiber International, Lauscha, Germany (167).

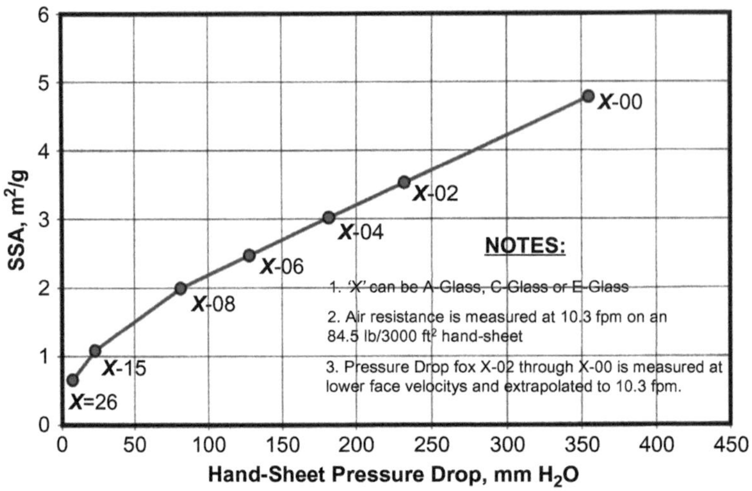


**Figure 4.29** Lauscha Relationship Between Fiber Grade, Pressure Drop, and Specific Surface Area (SSA).[6] *Reproduced by Permission of Lauscha Fiber International, Lauscha, Germany.*

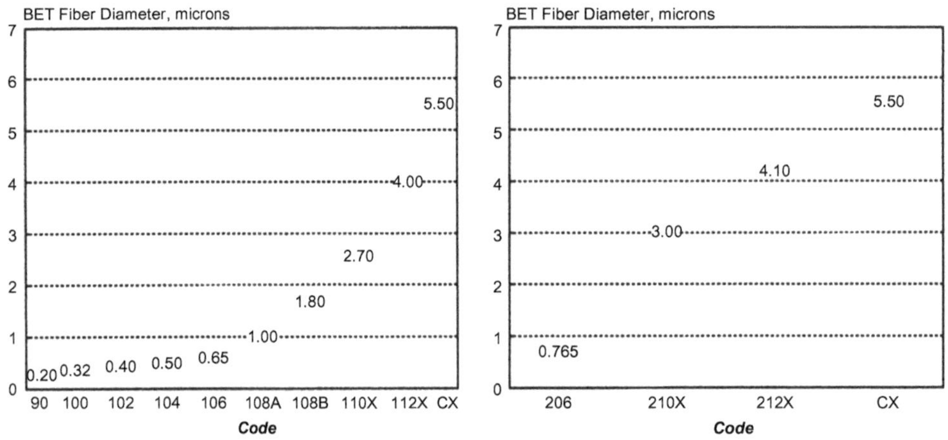


**Figure 4.30** Johns Manville Glass Microfiber Codes Versus Fiber Diameter. *Reproduced by Permission of Johns Manville, Denver, Colorado, USA.*

### 4.3.5.4.2 Ceramic fibers

Ceramic fibers, produced from metal oxides or refractory oxides, are resistant to high temperatures (1100–1700 °C). This class of fibers includes alumina, beryllia, magnesia, thoria, zirconia, silicon carbide, quartz, and high silica reinforcements. Technically, glass is also a ceramic, however, it is generally not included with the class of ceramic materials. Ceramic fibers are produced by chemical vapor deposition, melt drawing, spinning, and

[6] Data provided by Wolfgang Rodigas, Lauscha Fiber International, Lauscha, Germany.

extrusion. Refractory ceramic fibers, fibrous glass, and mineral wool belong to a group known as synthetic vitreous fibers. These are materials made from molten masses of raw material under highly controlled conditions. All of these fibers possess high strength and modulus. Ceramic fibers are used in felt type filter media where high temperature resistance is required. The fiber has also been developed for diesel fuel particulate filtration. 3M Company markets a line of ceramic fiber products under their Nextel™ trademark.

#### 4.3.5.4.3 Silicon carbide fiber

Nixdorf (170) of Industrial Ceramic Solutions LLC describes a patented "Microwave-Cleaned Particulate Filter System" based on silicon carbide fiber. This fiber has a unique ability to heat at a phenomenal rate in a microwave energy field. In applications such as diesel engine exhaust filters and restaurant grease filters, they can be made to be self-cleaning. When exposed to microwave energy, trapped particulate matter is oxidized to carbon dioxide and water.

The fiber is manufactured in two stages. First an organic fiber is carbonized in an inert atmosphere to yield a carbon fiber. The carbon fiber is then subjected to a high-temperature carbo-thermal reduction in the presence of silicon dioxide to form the silicon carbide fiber. A fiber has been produced, $16 \times 1000$ μm (1 mm), that is large enough to be nonrespirable and be safe for handling. It is also suitable for making ceramic paper for cartridge filter applications. Figure 4.31 is a scanning electron microphotograph of a silicon carbide fiber.

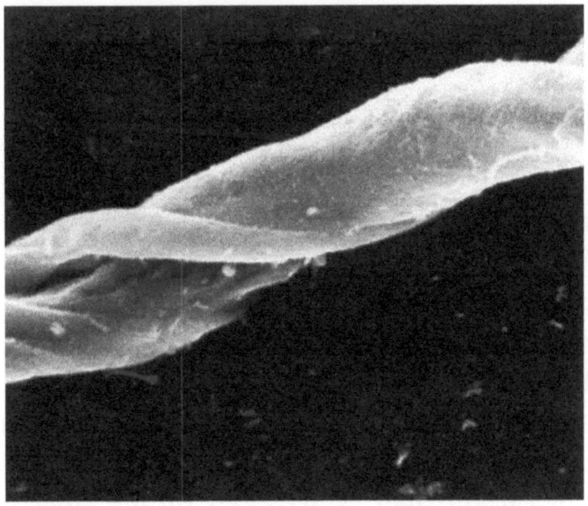

**Figure 4.31** Silicone Carbide Fiber (170). *Reproduced with Permission of the American Filtration and Separation Society, Nashville, Tennessee, USA.*

#### 4.3.5.4.4 Carbon fibers

According to NF New Fibres (166), carbon fibers were discovered in 1879 by Thomas Edison. They can be considered as the transition between inorganic and

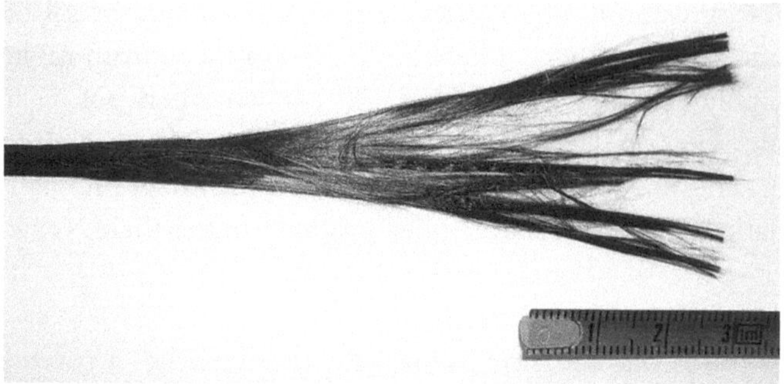


**Figure 4.32** Carbon Fibers. *Source Wikipedia.*

organic fibers. There are two sources of carbon fiber: the alteration of organic fibers such as rayon, acrylics etc., and the remains of petroleum or tar distillation. The former is referred to as PAN-based carbon and the latter is referred to as pitch-based. Table 4.36 compares the properties of these two fiber types.

**Table 4.36** Comparison of Properties of PAN-Based Carbon and Pitch-Based Carbon

| Characteristics | Fibers from PAN | Fibers from pitch |
|---|---|---|
| Tenacity (Gpa) | 1.8–7.0 | 1.4–3.0 |
| Modulus (Gpa) | 230–540 | 140–820 |
| Elongation at Break (%) | 0.4–2.4 | 0.2–1.3 |
| Density (g/cm$^3$) | 1.75–1.95 | 2.0–2.2 |

Reproduced by Permission of Bill Smith, Industrial Textiles Associate (ITA), Greer, South Carolina, USA.

PAN is most often used as the precursor of PAN-based carbon fibers. Carbon fibers are produced through heat treatment and pyrolysis of the precursor. High modulus carbon fibers are often referred to as graphite fibers. Graphite fibers may be a misnomer because they are not sourced from graphite, but rather are carbon fibers that have been heat treated to temperatures over 2,000 °C.

The PAN- and pitch-based carbon fibers are predominantly used as a reinforcing fiber in high strength structural composites. They find use in some filter applications where high chemical and temperature resistances are desired. Since carbon fibers are conductive fibers, they are sometimes used in felt media to dissipate electric charge.

A different form of carbon fiber is activated carbon fiber that is used in media filtration and separation applications because of its adsorption properties. It can be produced from cellulose (Rayon) or acrylic precursors. The fiber is produced by heat treatment at temperatures up to 300 °C. After the treated fiber is formed into a web (woven or nonwoven), it is oxidized at temperatures up to 1300 °C to activate the surface into a porous structure.

Manufacturers of carbon fibers include Cytec Industries, Formosa Plastics, Hexcel, Mitsubishi Rayon, SGL Carbon, Toho Tenax, Toray Industries, and Zoltek.

Another version of carbon fiber is vapor grown carbon fiber (VGCF) and its smaller size variant, vapor grown carbon nanofiber (VGCNF). They are among short carbon fibers that have drawn lots of attention for their potential thermal, electrical, frequency shielding, and mechanical property enhancements. According to Wikipedia (171) "they are being more and more utilized in different material systems like composites, thanks to their exceptional properties and low cost". Carbon nanofibers (CNFs, VGCFs) or VGCNFs are cylindrical nanostructures with graphene layers arranged as stacked cones, cups, or plates. Carbon nanofibers with graphene layers wrapped into perfect cylinders are called carbon nanotubes.

Carbon nanotubes are appearing in commercial applications of filter media. WAL SA, Geneva, Switzerland, is advertising a WMD 30 filter medium (172) that produces high efficiency water filtration based on carbon nanotube technology.

#### 4.3.5.4.5 Metal fibers

Metal fibers are produced as thin filaments with diameters ranging from 1 μm to 80 μm. Among the various forms are chopped fibers and needle felt. Chopped fibers can be used in wet laid operations. Needle felt forms composed of or containing metal fibers are often sintered to produce a stiff rigid medium. Metal fibers are produced from a variety of metal alloys. Included are stainless steel, nickel, nickel alloy, and high temperature resistant alloys. Table 4.37 provided by Mott Corporation (173) of Farmington, Connecticut, USA lists maximum operational temperatures of some alloys. Mott Corporation produces a variety of filters from fiber metal filter media.

Metal fiber medium are used in liquid and air filter applications where a high degree of heat and chemical resistance is required. They can be welded into filter shapes of high structural integrity. Metal fiber filters are cleanable and reusable. Global Material Technologies, Inc. of Palatine, Illinois, USA markets a metal fiber medium under the trade name of VersaFilter. Bekaert of Brussels, Belgium has developed a sintered metal fiber filter medium which they market under the

**Table 4.37** Maximum Operational Temperatures for Some Metal Fiber Alloys

| Available Alloys | AISI 316 SS | INCONEL®[a] 601 | ALLOY HR | FECRAALLOY®[b] |
|---|---|---|---|---|
| Maximum Operational Temperature | 350–360 °C<br>650–720 °F | 560 °C<br>1040 °F | 600 °C<br>1100 °F | 1000 °C<br>1800 °F |

[a]Inconel® is a trademark of International Nickel Co., Inc.
[b]Fecralloy® is a trademark of UKEA, United Kingdom
Reproduced with Permission of Mott Corporation, Farmington Connecticut, USA (173).

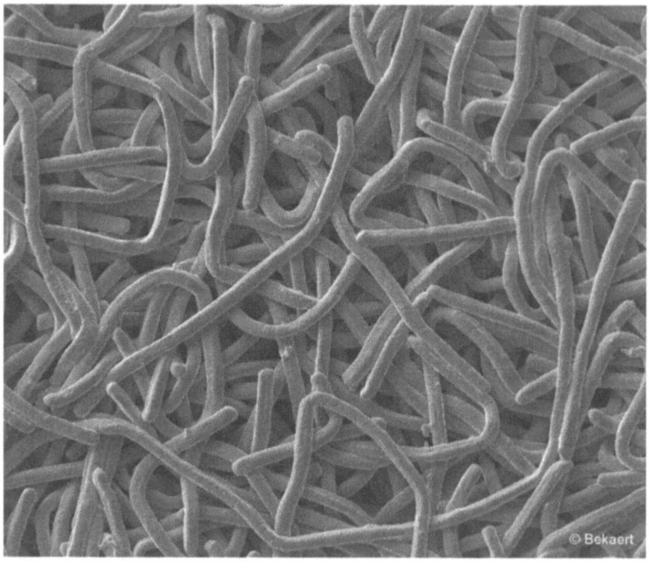


**Figure 4.33** Sintered Metal Fiber (174). *Reproduced with Permission of Bekaert, Brussels, Belgium.*

**Figure 4.34** Bekipore®ST Sintered Metal Fiber Filtration Medium (174). *Reproduced with Permission of Bekaert, Brussels, Belgium.*

trade name of Bekipor® ST.[7] Figure 4.33 is a photograph of sintered metal fibers in the medium. Figure 4.34 shows the Bekipore®ST medium in a pleated form.

### 4.3.5.4.6 Basalt fibers

Basalt is a material produced from volcanic rock. Lisakovsky et al. (175) reported that continuous basalt fibers were developed in the Ukraine in the mid-1970s. The fiber is

[7] Bekipore®ST is a trademark of Bekeart, Belgium.

highly heat and chemical resistant. Compared to glass microfiber, it is stiffer and has less elongation. It is mostly used in woven fabrics. In filter applications, AFFCO (American Felt and Filter Company) (176) in New Windsor, New York uses woven basalt fabrics as scrim reinforcement for needle punch Nomex® felts to provide improved heat resistance in hot air filtration applications. AFFCO and their "The Heat Shield" product is further discussed in Chapter 8, Section 8.1.1.

#### 4.3.5.4.7 Asbestos

Asbestos fibers are from natural minerals that are contained within natural rock. The minerals are in the form of masses of strong flexible fibers that can be separated into individualized fibers and thin threads. The threads can be woven into fabrics that are highly heat and chemical resistant. The fibers are very fine with nano-sized fiber diameters. They can easily be dispersed in the wet end of a wet lay forming system. For this reason, asbestos fibers were at one time very popular for high-efficiency filter applications. Some of the original developments for military gas mask media and HEPA media utilized asbestos fibers. In the 1970s the carcinogenic dangers of asbestos became apparent and they were replaced with other fibers—mostly glass microfibers. There is still some applications of asbestos in filter media, however, this is very specialized and involve very small quantities.

#### 4.3.5.4.8 Boehmite

Boehmite is an aluminum oxide hydroxide ($\gamma$–AlOOH) particle, containing extra hydroxyl groups on its surface. It is a component of the aluminum ore, bauxite. Ahlstrom has developed their Disruptor media based on the attachment of Boehmite fibrils to the surface of the glass fibers (see Figure 4.35). The Boehmite fibrils are 2 nm in diameter and 250 nm in length (Komlenic (177). Boehmite particles are

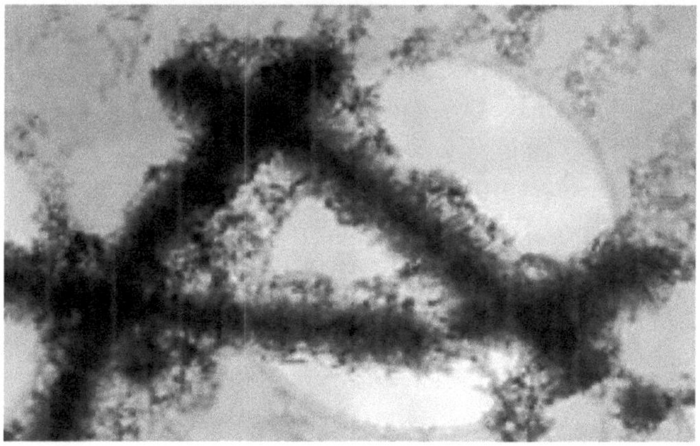

**Figure 4.35** Glass Fiber with Attached Boehmite Fibrils in Disruptor® Medium. *Photo courtesy of R. Ristau, IMS, University of Connecticut, USA.*

inherently electropositive and provide for an electrosorptive effect in aqueous filtration operations. The Disruptor® medium is discussed in Chapter 7, Section 7.2.8.

#### *4.3.5.5 Nanofibers*

Nanotechnology is one of the important emerging technologies of the twenty-first century. In filtration nanofibers are the important constituent of high-performance filter media. Nanofiber is a broad phrase generally referring to a fiber with a diameter less than 0.5 μm or 500 nm[8] (Graham et al. (29)). Although nanofiber is a relatively recent term for sub-micron fibers, some of the fibers now identified as nanofibers, such as inorganic fibers have been around for a long time.

For filter media purposes there are two basic sources of nanofiber: web formed and fiber formed.

1. Web formed—The web is formed as the fiber is formed. This type of formation includes meltspun and electrospun webs. Webs so formed contain fibers in the nanofiber range. Types of forming processes include melt-blown, centrifugal spinning, and electrospinning.
2. Fiber formed—The nanofibers are first prepared by the pertinent process and then used as a fiber raw material in the furnish or fiber blend used to prepare the web. Pertinent processes that might use nanofibers as the whole or part of their fiber mix are wet laid, air laid, and dry laid webs. The nanofiber slurry might be incorporated as a substrate coating as discussed below.

There are several ways by which nanofibers may be incorporated into a nonwoven filter medium:

1. Polymer sourced spun webs: The polymer is spun either as a melt or as a solution into a web containing filament fibers of nanofiber dimensions.
2. Fiber sourced webs: Raw material nanofibers are blended into the fiber mix or furnish and formed into a web-by-web forming process such as air laid, dry laid, or wet laid.
3. Layering and/or coating: Polymer sourced webs can be layered to a suitable substrate and form a composite structure filter medium. Another way to form the composite is to apply a nanofiber slurry as coating to a substrate. Coating is often the term used for applying a nanofiber structure to a substrate. This would also include the process whereby the polymer sourced web is formed directly on the substrate.

Table 4.38 lists the most common sources of nanofibers and typical diameters. Some of these that have already been discussed above includes inorganic fibers: asbestos (Section 4.3.5.4.7), boehmite (Section 4.3.5.4.8), carbon (VGCNF, Section 4.3.5.4.4), and glass microfiber (Section 4.3.5.4.1).

[8] A nanometer (nm) is one billionth of a meter.

**Table 4.38** Sources of Nanofibers

1. Fibers of spun webs
   a. Melt-blown: 300 nm and higher
   b. Centrifugal spinning
      i. Forcespinning™: 300 nm
      ii. DuPont melt spun nanofiber process: <1,000 nm
      iii. Electrospinning: 100–300 nm
2. Sources of raw material nanofiber
   a. Inorganic
      i. Asbestos: 30–1130 nm
      ii. Boehmite fiber: 2 nm
      iii. Glass microfiber: 20–10,000 nm
      iv. Carbon nanofiber
         - nanofiber 50–200 nm
         - nanotube 1–3 nm
   b. Splittable fibers (Islands in the sea): as low as 250 nm
   c. Cellulose sourced fibers
      i. Nanowhiskers: 10–30 nm
      ii. Microfibrillated cellulose: 3.5–20 nm
      iii. Nanofibrillated cellulose: 50–500 nm
   d. XanoShear™[a] (shear spinning technology): 200–2000 nm

[a]Trademark of Xanofi, Raleigh, North Carolina, USA.

Dugan and Homanoff (178) discuss the possibilities of splittable fibers as an alternative to glass microfibers for high-efficiency filtration. They note that splittable fibers can be split into nanofiber diameter dimensions (250 nm) and have an advantage over glass fibers in that they can be electrostatically charged.

The United States Environmental Agency is considering at least some nanoscale materials as "chemical substances" and pursuing a program to have these substances regulated by the Toxic Substances Control Act (TSCA). This would include premanufacture notification: significant new use rule, an information gathering rule, and a test rule. The regulatory activity of TSCA is further discussed in Chapter 10, Section 10.1.5.

#### 4.3.5.5.1 Spun webs

Recent advances in melt-blown fiber technology are producing melt-blown fiber webs approaching the nanofiber range. Even finer webs can be formed from electrospinning and centrifugal spinning processes.

#### 4.3.5.5.2 Electrospun webs

Electrospun webs are composed of a nanofiber web structure. There are a number of polymeric materials that used for electrospun nanofiber production. Donaldson

Corporation uses nylon in their Spider Web® and Ultra web® media. Schreuder-Gibson and Gibson (179) in their discussion of electrospun nanofibers for aerosol filtration conducted experiments with polyurethanes and Nylon 6. Formhals (26) conducted his studies with cellulose esters. Choi et al. (180) created electrospun nanofibers from silica using a sol-gel process. Huang et al. (181) identified 44 polymers that have been successfully electrospun from the solution process and 6 from the meltspun process.

Companies involved in selling machines or providing nanofiber-based solutions using electrospinning technology include Elmarco, Finetex, eSpin Technologies, Donaldson, Dienes Apparatebau GmbH, SNS Nanofiber Technology, Ahlstrom, and TopTec.

#### 4.3.5.5.3 Centrifugal spinning

Forcespinning™ by Fiberio and the Melt Spun Fiber Porcess (MSPN) by DuPont are two processes that produce nanosize fibers by centrifugal spinning. The processes are described in Chapter 5, Section 5.4.

#### 4.3.5.5.4 Splittable fibers

Certain melt-blown bicomponent fibers and multicomponent fibers can be configured to be split into fibers that have nanofiber dimensions. The splittable fiber structure is part of the discussion on bicomponent and multicomponent fibers in Section 4.3.5.6. Splittable fibers are a form of bicomponent and/or multicomponent fibers that can be split into sub fibers of microfiber and nanofiber dimensions. A good example are "islands-in-the-sea" cross section structures that can be split into microfibers and nanofibers by disintegrating the sea around the islands. The ways this may be done are by mechanical shearing, by melting the sea away from the higher melting island fiber components, or by dissolving the sea in a suitable solvent that does not affect the different polymer composition of the island polymer. A related fiber cross section, pie-shaped segment bicomponent fibers can also be subject to this type of action to produce microfibers and possibly nanofibers.

The island in the seas approach is most certainly not limited to producing round cross section nanofibers. Eastman Chemical Company in Kingsport, Tennessee, has developed a new technology platform for manufacturing microfibers for wet laid nonwovens applications including filtration, specialty papers, and battery separators. Eastman Cyphrex™ microfibers are made using islands-in-the-sea bicomponent fiber technology and a water-dispersible sea polymer. They are produced in a variety of shapes and sizes using a range of synthetic fiber polymers. One commercial product is a polyester fiber. It is a round, 2.5 μm (0.06 denier) × 1.5 mm length microfiber. The unique "islands in the sea' fibril structure of lyocel has already been discussed in Section 4.3.5.2.2.

#### 4.3.5.5.5 Cellulose sourced micro/nano fibers

**4.3.5.5.5.1 Microfibrillated cellulose** Cellulose fibers, such as wood pulp and vegetable fibers such as cotton are made up of fibrils on a macro, micro, and nanofiber scale that can be mechanically fibrillated into individualized fibrils (Homonoff et al. (182)). Biofibers such as rayon, lyocel, and PLA also have a fibril structure that is subject to fibrillation. It should be noted that there are synthetic polymer fiber sources such as acrylic fibers that are also fibrillatable.

**4.3.5.5.5.2 Nanowhiskers or nanocrystals** The website FrogHeart which offers commentary about nanotech, science policy and communication, society, and the arts, offers the opinion (183) that nanocrystaline cellulose (NCC), cellulose nanocrystals (CNC), cellulose whiskers (CW) and cellulose nanowhiskers (CNW) "are all the same stuff: cylindrical rods of crystalline cellulose (diameter: 5–10 nm, length: 20–1000 nm)". In contrast, cellulose nanofibers or nanofibrils (CNF) are less crystalline and are in the form of long fibers (diameter: 20–50 nm, length: up to several micrometers).

Cellulose nanowhiskers being a more crystalline form of nanocellulose are formed by the acid hydrolysis of native cellulose fibers commonly using sulfuric or hydrochloric acid. The amorphous sections of native cellulose are hydrolyzed and after careful timing, the crystalline sections can be retrieved from the acid solution by centrifugation and washing. Cellulose nanowhiskers are rod-like, highly crystalline particles (relative crystallinity index above 75%) with a rectangular cross section. Their dimensions depend on the native cellulose source material, hydrolysis time, and temperature.

#### 4.3.5.5.6 Shear-induced nanofibers

Xanofi, a company started in 2010 based on technology invented at North Carolina State University (NCSU), has recently developed XanoShear™ as a novel high-throughput method for the continuous production of staple polymeric nanofibers (Gangwal and Wright (184)). The technology is based on antisolvent-induced precipitation of polymer solvent droplets under shear. The polymer droplets extend in parallel under shear stress and precipitate due to the antisolvent in the bulk of the liquid, forming staple nanofibers of short lengths (10 mm–10 cm). These fibers can further be milled to shorter lengths allowing the fiber length to diameter (L/D) aspect ratios to be controlled.

The XanoShear™ process is similar to typical wet spinning where a polymer dissolved in a solvent is drawn through a solution containing an antisolvent. However, unlike wet spinning which can achieve minimum fiber diameters of ~10 μm, the ability to draw very thin fibers comes from the ultra-low interfacial tension between the polymer solution droplets and the medium, allowing a high degree of stretching of the droplets and generation of nanofibers.

### 4.3.5.6 Multicomponent and bicomponent fibers

Bicomponent and multicomponent fibers are spun fibers that have two or more different polymer phases in their cross section. Each phase has its own distinct properties and contributes something to the properties of the fiber that would otherwise would not be there. Figure 4.36 diagrams some of the more common fiber cross sections. Included are:

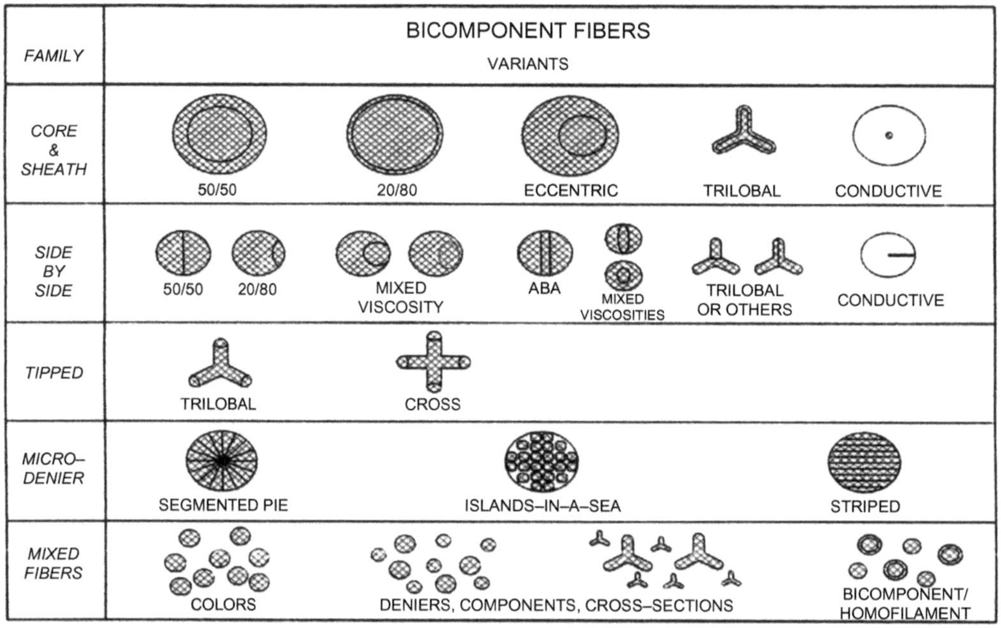


**Figure 4.36** Various Cross Sections of Bicomponent Fiber. *Reproduced by Permission of Hills, Inc., West Melbourne, Florida, USA.*

**a.** Sheath/core fibers most often used to make binder fibers.
**b.** Side by side fibers used mostly to produce bulky, self-crimping fibers.
**c.** Tipped products mostly used in special filtration products.
**d.** Segmented products that can be split into small individual fibers.
**e.** Islands-in-a-sea products wherein the sea is normally dissolved away to leave only the small islands, a source for nanofibers.
**f.** Mixtures of two or more fiber types to make specialized strands for yarns and fabrics having multiple cross sections.

The most common form of bicomponent fiber is the sheath/core fiber. In this cross section, one polymer forms an outer sheath around the inner core of the second polymer. Figure 4.37 is a photomicrograph example of a sheath/core fiber with a circular sheath. This particular fiber is melt-blown with a polyethylene sheath and used as thermobonding fiber. Thermobonding binder fiber is the most common use

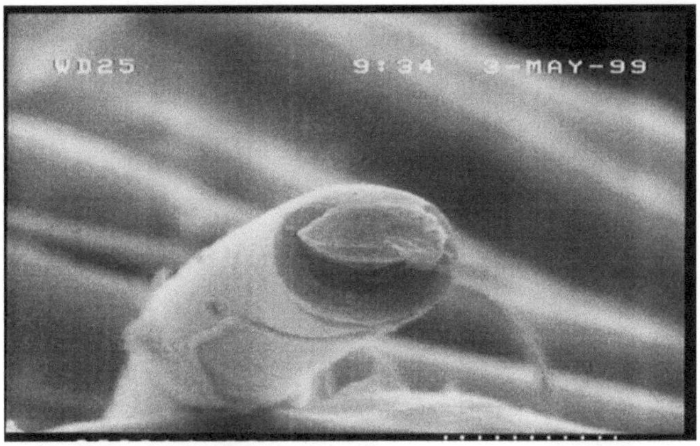


**Figure 4.37** Photomicrograph of a Sheath/Core Fiber. *Reproduced by Permission of Hills, Inc., West Melbourne, Florida, USA.*

**Figure 4.38** 16 Segment Pie Wedge Fibers. *Reproduced by Permission of Hills, Inc., West Melbourne, Florida, USA.*

of sheath/core fibers. A number of sheath/core variations are possible as demonstrated by the top row in the tabulation of Figure 4.36. Note that the cross sections are not necessarily circular as indicated by the inclusion of trilobal shapes.

It becomes apparent that segmented products and island-in-a-sea are a powerful source of microfibers for high-efficiency filtration. Figure 4.38 is an example of a

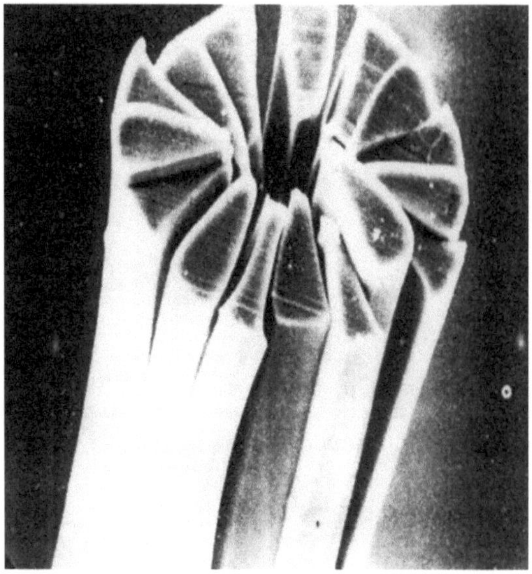

**Figure 4.39** Pie Wedge Segments Split into Wedge-Shaped Fibers. *Reproduced by Permisson of Woodhead Publishing Ltd., Cambridge, England, U.K.*

**Figure 4.40** 32 Pie Wedge Segment Fiber. *Reproduced by Permission of Hills, Inc., West Melbourne, Florida, USA.*

16-segment "pie wedge" fiber. Figure 4.39 shows how this fiber can be split into smaller wedge-shaped segments of much smaller fiber size. Figure 4.40 shows the splittable segments of the fiber. The splitting can be accomplished by thermal, chemical, or mechanical means. Dugan (145) suggests that for filtration media, a conventional purpose would be to form a carded web of 3 dpf (denier per filament) fibers

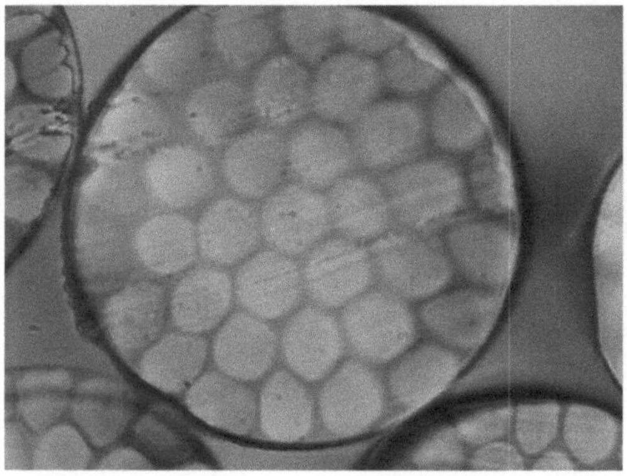

**Figure 4.41** 37 Islands-in-the-Sea. *Reproduced by Permission of Hills, Inc., West Melbourne, Florida, USA.*

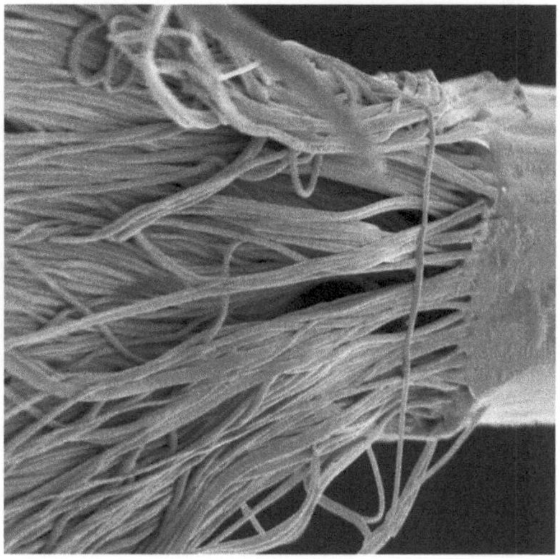

**Figure 4.42** 600 Islands-in-the-Sea. *Reproduced by Permission of Hills, Inc., West Melbourne, Florida, USA.*

and then pass the web under hydroentanglement jets (see Chapter 5, Section 5.1.2.3.2, which simultaneously split the fibers into individual wedges and entangle the fibers to give fabric strength and integrity. The fabric will contain fibers down to 0.2 dpf and yet still maintain much of the throughput and processing advantages of a 3-dpf fiber. Dugan also suggests that the alternating segments between polyester and nylon. These two polymers have little adhesion to each other and would more readily split in the hydroentanglement process.

Segmented fibers are certainly not limited to 16 segment pie wedges. Figure 4.40 is a microphotograph of a 32 segment pie wedge.

Figure 4.41 is a cross section of an "island-in-the-sea" type fiber. In this configuration, the micro-sized nanofibers are formed by dissolving away the sea. Usually, the island portions are polyester, and the sea is composed of a water-soluble polymer. Hagewood (185) reported that the island fiber can be produced in diameter ranges of 100 to 800 nm. Hagewood also indicates that the technology is capable of producing up to 1,000 islands and not even limited to that number. Figure 4.42 is a microphotograph of a 600 islands-in-the-sea fiber. A variation of the islands-in-the-sea technology are hollow islands having a wall thickness of approximately 40 nm and a diameter of 300 nm.

The islands-in-the-sea technology has been commercialized for staple and filament yarns. It has not yet been applied to melt-blown or spunbond nonwovens.

In Table 4.39, Hagewood (186) compared the microfiber capabilities of bicomponent and multicomponent fibers to the microfiber capabilities of some other

**Table 4.39** Microfiber Comparison

| Fiber I.D | Mfg. process | Fiber description | Fiber cross section | Fiber size (μ) | Fiber surf. area (Sq-mt/Gr) | Prod. rate (Gr. per min. per Spin hole) |
|---|---|---|---|---|---|---|
| ***CONVENTIONAL PROCESSES*** | | | | | | |
| 1 | Conventional Staple or Spunbond | One denier fiber, Homopolymer | Round | 10.1 | 0.3 | 0.67 |
| 2 | Conventional Meltblown | Two micron fiber, Homopolymer | Round | 2.0 | 1.4 | 0.5 |
| 3 | Conventional Electrospun | Size/shape as best reported | Round | 0.3 | 9.5 | 0.03 |
| ***SEGMENTED PIE PROCESSES*** | | | | | | |
| 4 | Segmented Pie Staple or Spunbond | One denier fiber, 32 Segment Pie | Pie Segments | Ea. Segments = 1.0 Arc X 2 × 5.1 Legs | Ea. Segment = 3.2 | 0.67 |
| 5 | Segmented Pie Meltblown | Two micron fiber, 16 Segment Pie | Pie Segments | Ea.Segments = 0.4 Arc X 2 × 1.0 Legs | Ea. Segments = 8.7 | 0.5 |
| ***ISLANDS-IN-A-SEA PROCESSES*** | | | | | | |
| 6 | Islands-in-a Sea Staple or Spunbond | One denier fiber, 50/50 Islands/Sea, 30 Islands | Round Islands | Ea. Island = 1.3 | Ea. Island = 2.2 | 0.3 |
| 7 | Islands-in-a Sea Staple or Spunbond | One denier fiber, 50/50 Islands/Sea, 600 Islands | Round Islands | Ea. Island = 0.3 | Ea. Island = 9.8 | 0.3 |
| ***NANOTUBE ISLANDS*** | | | | | | |
| 8 | Islands-in-a Sea Staple or Spunbond | One denier fiber, 50/50 Islands/Sea, 30 Islands | Microtube Islands, 50% Hole | Ea. Tube = 1.2 OD X 0.2 Wall | Ea. Tube = 7.5 | 0.15 |
| 9 | Islands-in-a Sea Staple or Spunbond | One denier fiber, 50/50 Islands/Sea, 600 Islands | Microtube Islands, 50% Hole | Ea. Tube = 0.2 OD X 0.04 Wall | Ea. Tube = 33.6 | 0.15 |

Reproduced by Permission of Hills, Inc., West Melbourne, Florida, USA.

technologies (staple, spunbond, melt-blown, and electrospun). Of the processes compared, only the electrospun process is capable of producing finer fibers, however at the expense of production rate per unit spin hole. It is presumed that centrifugal spun fibers also can be produced this fine.

## 4.4 RESINS AND BINDERS

Resins for filter media usually come in one of three forms: solvent-based solutions, water-based latex emulsions and dispersions, and water-based solution forms. Their main purpose is to strengthen the medium, however, they provide for other important properties. These include stiffness, tear strength, modulus, dimensional stability, flex resistance, water resistance, downstream processability, heat and temperature resistance, resistance to chemical and environmental exposure, and durability in the end use application. Resins can also affect filtration properties. For example, some film-forming resins are undesirable because the film may form across a pore and close it to the passage of fluid. Resins are available in thermoplastic form or thermosetting form. Very often, both properties are desired for processing. For example, a resin may be thermoplastic so that it can be thermally softened for the pleating operation and then afterwards heat cured and thermoset to provide for stiff rigid pleats in the end use application. The following is a discussion of some of the various types of resins used as binders for filter media.

### 4.4.1 Formaldehyde resin

Three major forms of formaldehyde resins are phenol formaldehyde and phenolics, melamine formaldehyde, and urea formaldehyde. In filter medium applications, all three are used as cross-linkers with other resin systems. Melamine formaldehyde, and urea formaldehyde play important roles as wet strength agents for wet laid media (see Sections 4.4.1.2 and 4.4.1.3). Solvent-based phenol formaldehyde resin is a very important saturant for wet laid filter paper particularly for engine lube oil filter applications.

#### *4.4.1.1 Phenolic resins*

Phenol formaldehyde resins and phenolic resins are not necessarily synonymous. Phenol formaldehyde resins are formed from the specific reactions of phenol with formaldehyde, whereas the definition of phenolic resins includes the reaction of phenol with other aldehydes as well. Most of the phenolic resins used for the treatment of filter paper are of the phenol formaldehyde form; however, there are some variations. Phenolic resins are thermosetting resins available in both solvent-and water-based forms.

Water-based phenolics are formulated with other water-based polymer resins to provide thermosetting and cross-linking properties. Usually they come in solution form and have a water tolerance or limited water solubility; meaning that if diluted beyond a certain point, the resin kicks out of solution. This may happen, even when a phenolic resin is formulated with other resins, and the whole resin mix is diluted. When this happens, the resin sometimes agglomerates into a very unusable sludge.

In saturating substrates, waterborne GP Georgia-Pacific Chemical Division (GP) resins are used for applications that include oil and air filters. The resin line features properties that include good wicking, abrasion resistance, moisture resistance and durability. Kolon, headquartered in Seoul, Korea produces the HiRENOL® line of phenolic resins, which includes solvent-based phenolic resins for filter media.

Solvent-based forms of phenolics can be built to higher molecular weights and impart far superior properties. The solvent-based phenolics are thermosetting resins and they are mainly available in two different chemical structures: resole and novolak. Both are used in filter medium applications, however, the resole form is the more widely used in the United States. Often the two are formulated together to optimize properties.

Resoles are formed from the base catalyzed reaction of phenol and formaldehyde and requires an excess of formaldehyde. For solution resins, they are reacted to a low molecular weight of 300–700 (Stille (138)), so that the polymer is still soluble. The polymer contains hydroxymethyl ($-CH_2OH$) side groups and possibly some ether linkages ($-CH_2-O-CH_2-$). Further heating will cause condensation reactions and result in a high molecular weight cross-linked polymer.

Novolaks are formed from acid catalyzed reactions requiring an excess of phenol and usually result in a linear polymer of molecular weight 1000–1500 (Stille (138)). Hydroxymethyl groups are not present because under the acid condition, they are unstable and decompose to a methylene group. The methylene groups react with phenol to give methylene bridges (Intelligensys Ltd (187)). Further cross-linking can only take place by the addition of more formaldehyde, usually in the form of hexamine. Novolaks can react with epichlorohydrin to give epoxy-novolak systems. The OH from the phenol forms an epoxide group, so there are generally between 3 and 6 epoxide groups per molecule.

In Figure 4.43, Bell and Cove (188) have compared the relative cure times of impregnated filter paper treated with resole and novolak resins. At 130 °C, the resole cures considerably faster than the novolak, however at 150 °C they both cure at roughly the same rate.

Phenolic resins have unique properties that make them ideal for filter paper used in hot oil filter applications such as the lube oil filters of automobiles. First,

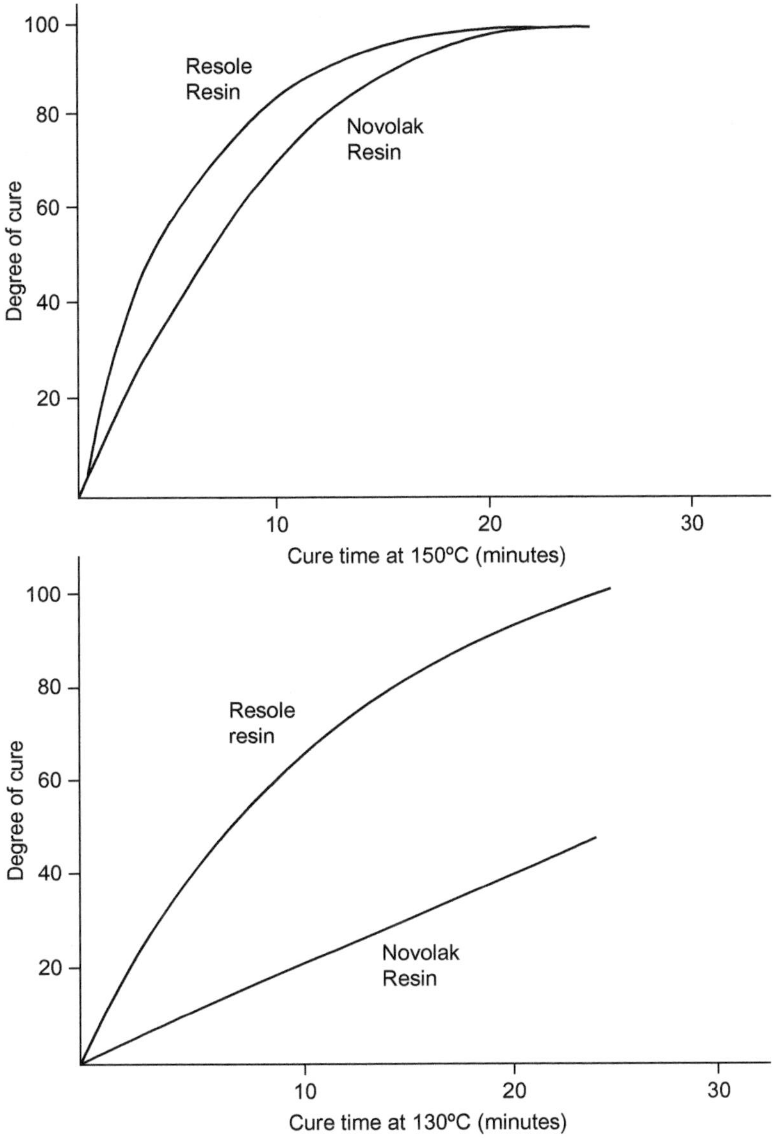


**Figure 4.43** Typical Cure Schedules of Impregnated Paper. *Reproduced with Permission of Elsevier Science Ltd. Oxford, England, UK.*

they are "B-stageable," meaning they can be applied to the base paper, then dried but not cured or thermoset. At this point the resin in the treated paper is considered as thermoplastic. It makes the paper amenable to such converting operations as corrugating and embossing. When shipped to the filter manufacturer, the paper is soft and pliable for the pleating operation. Once formed into pleated filter elements, the resin treated filter paper is heat cured to its final state. Full cure is essentially achieved in 10 min at 150 °C, lesser time at higher temperatures. The cured filter paper has good heat resistance up to 200 °C. The resin provides for good dimensional stability so that the medium will not seriously change dimensions with changes in humidity. It provides for hot oil stiffness, a property determined by measuring the stiffness of the paper after soaking in hot oil. The phenolic treated paper is well designed to stand up to the prolonged rigors of hot oil. Water resistance is excellent. Wet to dry ratio properties of tensile strength, stiffness, and burst strength are better than most other types of resins The solvent-based resin has little tendency to form films across the pores. Filter media made from phenolic resin generally has better filtration properties than compared to water-based resins.

Environmental concerns have stimulated new technologies in the manufacture and processing of phenolic resins. Manufacturers of filter media now use high tech incinerator technology to destroy methanol, formaldehyde, and phenol emissions (see Chapter 10, Section 10.1.2.1). Resin manufacturers have developed lower emission and faster curing resins. Dupré (189) reported that free phenol contents of solvent phenolics were reduced from approximately 10% in 1986 to 3% in 1996. Since then the ability of resin manufacturers to further reduce phenol and formaldehyde content has become understandably limited. In response, the medium manufacturers are devising ways to reduce phenol and formaldehyde in the resin treated filter media they are shipping their customers.

Hollingsworth and Vose (H&V) is not a resin supplier but offers a broad band of resin technologies for the resins they apply to their filter medium products. Included are a variety of phenolic resin technologies that can be chosen to best meet the needs and applications. The resin systems are designed for application, processability (pleating), environmental concerns, and cost. Table 4.40 published by H&V (190) is a tabulation of the various H&V resin systems. It includes solvent-based phenolic resins, water-based resins (acrylics and phenolics), and epoxies. Part A of Table 4.40 identifies whether a resin system is suitable for 5 filter applications and gives an asterisk rating *, **, and ***; three *** being best. Part B provides a best estimate of process conditions and gives asterisk ratings for processability. The table also lists resin systems for non-cure, standard cure and advance cure. One system A250 is offered at three levels of advance cure. A250 itself is not precured and requires a cure time of 5 minutes at 160 °C. A250/3 is advanced cured and requires a cure time of 3 minutes at 160 °C to complete the cure. A250/7 is sufficiently advanced cured to require no further curing.

**Table 4.40** Resin Applications and Process Settings

Part A

| H & V resin code | Description | Panel air | Heavy duty air gas turbine | Lube | Fuel CR diesel gasoline | EDM |
|---|---|---|---|---|---|---|
| ***Non-Cure Resin Systems*** | | | | | | |
| VH 134 | no-cure acrylic | ** | ** | not suitable | * | not suitable |
| VH 243 | water-resistant acrylic | ** | ** | not suitable | * | not suitable |
| VH 243 FR | flame-retardant, water-resistant acrylic | ** | *** | not suitable | not suitable | not suitable |
| VH 206 | epoxy resin | **** | ** | * | ** | * |
| ***Standard-Cure Resin Systems*** | | | | | | |
| VH 60 | resol phenolic | * | * | ** | ** | *** |
| VH 188 | flexible phenolic | ** | * | ** | ** | ** |
| VH 198 | flexible phenolic | ** | ** | ** | *** | ** |
| VH 255 | high-performance, flexible phenolic | *** | *** | *** | *** | ** |
| A 250 | high-performance, flexible phenolic | *** | *** | *** | *** | *** |
| F 260 | high-performance, flame-retardant phenolic | *** | *** | not suitable | not suitable | not suitable |
| ***Advanced Cure Phenotic Resins*** | | | | | | |
| A 250/3 | high-performance, flexible phenolic | **** | *** | *** | *** | ** |
| A 250/7 | high-performance, flexible phenolic | *** | ** | *** | *** | ** |
| ***Water-Based Resin Technology*** | | | | | | |
| | water-resistant acrylic | ** | **** | not suitable | * | * |
| | water-based phenolic (non-European production only) | not suitable | not suitable | ** | ** | ** |

Part B

| H & V resin code | Description | Pleatability | | Preheat surface temperatures | Element manufacture post-curing | Environmental emissions | |
|---|---|---|---|---|---|---|---|
| | | Curable Resin | Non-Curable Resin | | | Phenol | Formaldehyde To VDA 275 |
| ***Non-Cure Resin Systems*** | | | | | | | |
| VH 134 | no-cure acrylic | | ** | low preheat recommended | no cure required however, some post-pleating heat may be desirable to assist in setting pleats and reducing element springiness | | ** |
| VH 243 | water-resistant acrylic | | *** | 70–80 °C | | | * |
| VH 243 FR | flame-retardant, water-resistant acrylic | | *** | 70–80 °C | | | * |
| VH 206 | epoxy resin | | *** | 75–85 °C | no cure required | | **** |
| ***Standard-Cure Resin Systems*** | | | | | | | |
| VH 60 | resol phenolic | ** | | 70–85 °C | 10 min 160 °C | ** | ** |
| VH188 | flexible phenolic | ** | | 70–85 °C | 12 min 160 °C | ** | ** |
| VH 198 | flexible phenolic | ** | | 70–85 °C | 10 min 160 °C | ** | ** |
| VH 255 | high-performance, flexible phenolic | *** | | 70–85 °C | 5 min 160 °C | ** | ** |
| A 250 | high-performance, flexible phenolic | **** | | 70–85 °C | 5 min 160 °C | *** | *** |
| F 260 | high-performance, flexible-retardant phenolic | **** | | 70–85 °C | 5 min 165 °C | *** | *** |

Part B

| H & V resin code | Description | Pleatability | | Preheat surface temperatures | Element manufacture post-curing | Environmental emissions | |
|---|---|---|---|---|---|---|---|
| | | Curable Resin | Non-Curable Resin | | | Phenol | Formaldehyde To VDA 275 |
| ***Advanced Cure Phenolic Resins*** | | | | | | | |
| A 250/3 | high-performance, flexible-retardant phenolic | **** | | 75–90 °C | 3 min 160 °C | *** | *** |
| A 250/7 | high-performance, flexible-retardant phenolic | | **** | 75–90 °C | no cure required | **** | **** |
| ***Water-Based Resin Technology*** | | | | | | | |
| | water-resistant acrylic | | *** | 70–85 °C | No cure required | | * |
| | water-based phenolic (non-European production only) | *** | | 70–85 °C | 10 min 160 °C | ** | * |

Reproduced with Permission of Hollingsworth and Vose Co., East Walpole, Massachusetts, USA.

Advance cure resin systems greatly reduce phenol and formaldehyde emissions at customers pleating sites. In Figures 4.44 and 4.45, Murphey and Ravel (191) demonstrate how advance curing reduces free formaldehyde and free phenol levels.

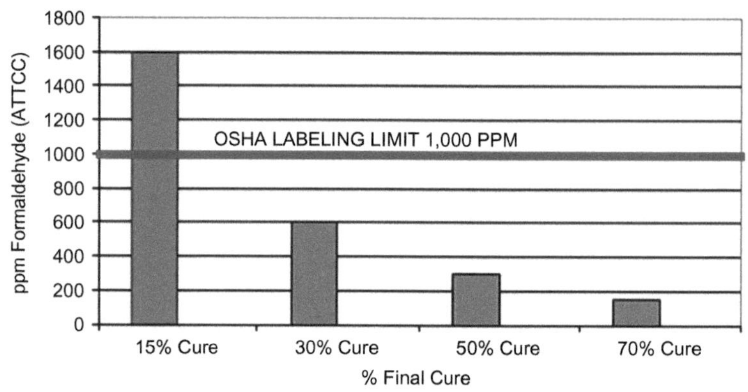


**Figure 4.44** Free Formaldehyde Levels. *Reproduced with Permission of INDA, Cary North Carolina, USA.*

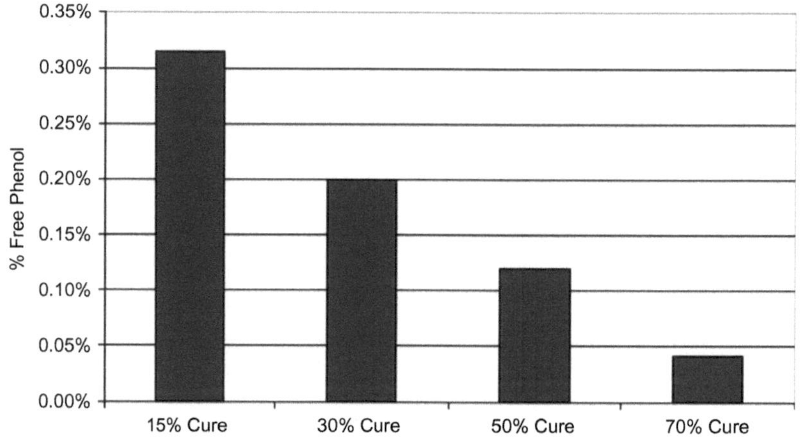


**Figure 4.45** Free Phenol Levels. *Reproduced with Permission of INDA, Cary North Carolina, USA.*

### *4.4.1.2 Melamine formaldehyde*

Melamine formaldehyde resins find use in filter media as a cross-linking agent for latex resins. Cationic forms of the resin are used in wet lay systems as wet strength agents. In this application, they are added to the fiber slurry.

### *4.4.1.3 Urea formaldehyde*

Like melamine formaldehyde, urea formaldehyde is used as a wet strength additive in wet lay systems and as a cross-linking agent in latex systems. Urea formaldehyde resin is often blended with acrylic resins in glass mat materials (Bell et al. (192)).

## 4.4.2 Latex resins

### *4.4.2.1 Introduction to latex resins*

Latex resins are emulsions of polymer particles in water. In addition to the base polymer, the emulsion will include a surfactant system to stabilize the emulsion. Other additives are added to enhance the polymer and provide for additional properties when applied to the nonwoven filter medium. Examples are pigments and dyes, water repellent agents, cross-linking mechanisms, flame retardants, viscosity control agents, and additives for pH stabilization. Often, more than one resin is blended together to achieve desired properties. When applied to filter media, latex resins are expected to impart fiber to fiber bonding strength and possibly stiffness. Stiffness is of particular importance to media that are pleated. The purpose is to provide pleat rigidity, high modulus, and resistance to pleat collapse when the pleated filter paper is used in its intended pleated filter operation. Generally resins with a high $T_g$ (25–60 °C) are preferred because these tend to be the higher modulus resins.

It should be noted that there are filter applications where low $T_g$ latices are desired. For example, industrial and medical face mask filters are bonded with softer resins to provide greater comfort for skin contact. Chemically bonded media in bag filters will utilize lower $T_g$ resins. This would include bag filters for industrial baghouse operations as well as bag filters for household vacuum cleaners.

Other properties that result from the latex resin system are water resistance, dimensional stability, long-term durability, chemical and environmental resistance, flame retardancy, resistance to oils and chemicals, heat resistance, and processability in downstream mechanical processing operations. In addition to pleating, downstream mechanical operations, which may be in-line or off-line, include corrugating, creping, and slitting.

Latex resins tend to be thermoplastic, however cross-linking agents are often added to provide cure related properties. Blending with a thermosetting resin such as a water-based phenolic or melamine formaldehyde greatly enhances the cured properties of the media. The improvements may show in medium strength, modulus, stiffness, and Mullen burst. The effects of curing are much more evident when measuring some form of water resistance. To illustrate, the uncured tensile strength of a latex treated medium may be little different than its cured tensile strength. On the other hand, the wet tensile strength of the cured media usually is much greater than the uncured wet tensile strength. The ratio of wet uncured property to wet cured property is sometimes used as an indicator of cure. If the ratio is one, the medium is presumed to be fully cured.

Sometimes the application of a resin to a nonwoven filter medium is referred to as a coating. This is exactly what a filter resin is not. The resin is applied as a saturant (sometimes referred to as an impregnate). Unlike a coating, which is intended to

**Table 4.41** OMNOVA Vinyl Acetate Copolymers for Filter Binders

| Product | $T_g$ °C | Viscosity (CPS) | Solids % | pH | Description |
|---|---|---|---|---|---|
| Sequabond® FVAC | +15 | 100 | 42 | 4.5 | Vacuum Bags<br>Improved tear resistance |
| Sequabond DSL-A50 | +30 | – | 50 | 4.0 | HVAC, Cabin Air<br>Firm hand, heat sealable, radio frequency bonding, pleatability |
| Suncryl RW-41 | +30 | – | 45 | 4.0 | HVAC, Cabin Air<br>Foamable, Firm hand, radio frequency bonding, pleatability |

Reproduced with Permission of OMNOVA Solutions Inc. Beachwood, Ohio, USA.

cover the surface of the substrate, a saturant is expected to penetrate the medium and form bond points in the interior as well as at the surfaces. Because of the need for penetration, low viscosity is a very desirable property in liquid resin systems for filter media. The latex producer has some control of this by his choice of surfactant systems, particle size, and polymer type. The user may reduce viscosity diluting the resin to a lower viscosity and/or by preheating the resin. Preheating the resin requires a strong temperature control system otherwise there is a potential for overheating. Overheated resin can agglomerate and solidify in the mix tank and cause a disastrous mess. Control of resin solids is probably the most common way to regulate viscosity. However, this approach has to be compromised with the need to maintain a certain resin content in the finished web. Diluting the resin with more water to maximize penetration will increase the drying burden for removing the water, resulting in increased energy costs and possible loss of productivity.

In most nonwoven applications, especially those for wipes and covers for absorbent materials, hand and softness are desired properties. For these products, a soft, low $T_g$ (glass transition temperature) resin is preferred. It has already been mentioned above that resins for many filter media are just the opposite, particularly if the medium is to be pleated. The resin has to impart stiffness and rigidity to the medium enabling the pleats to stand up to the "wind" force of the fluid flowing through it. Generally high $T_g$ resins, >25 °C, are the chosen binders for filter media. In bulkier media, such as high loft media, the resin would have to provide for compression resistance and compression recovery. To illustrate, Table 4.41 is a compilation of filter resin binders marketed by Omnova Solutions Inc., a major producer of latex binders for the filter medium industry. Note that one Sequabond® resins (Sequabond® FVAC ) are designed for vacuum bag applications and therefore, have a lower $T_g$ to provide for flexibility and resilience. The Sequabond® DSL-A50 and Suncryl® RW-41 resins, intended for more rigid media to be used in pleated applications, have higher glass transition temperatures.

The polymers listed in Table 4.41 are copolymers, that is, the polymer structure includes two monomer species. If there is only monomer species, the polymer is a homopolymer. Generally, the resins in filter media latices are copolymers, however there are exceptions. For example homopolymer PVAc latex (Polyvinyl acetate) has been widely used as a saturant for wet laid filter media (see Section 4.4.2.2.1). There are also terpolymers that are made up from three monomer species. Graft polymers are another way of combining different polymer species to achieve desired properties. A graft polymer is one polymer structure that is attached to the backbone of another polymer structure. A water-soluble polymer such as starch or polyvinyl alcohol (PVOH) is often used as the backbone polymer.

Table 4.41, in addition to $T_g$, also lists solids % and pH. These are important properties for latex binders. Solids content is of economic importance, because the more water in the neat resin, the more will be its shipping costs. Typical solids contents of latex resins shipped to the customer are 40–50%. Depending on the complexities of the resin system it may sometimes be higher and sometimes lower. Rarely is the neat latex resin used "as is" by the filter medium manufacturer. The media manufacturer will blend in other ingredients, such as other resins, cross-linking agents, water repellents, plasticizers, and flame-retardants. The resin blend will be diluted to a solids content as demanded by the process. The solids content determines the amount or percentage of resin content in the finished product. It affects the viscosity of the resin as applied, and the degree of penetration into the web. Lowering the solid content of the resin will enhance the ability of the resin penetration, however it will also increase the amount of water applied to the web and result in a higher thermal load for drying the web. This may be evidenced by increased energy costs and/or lower production rates. Dilution may also increase pot life, however, there are some resins such as water-based phenolics that have a water tolerance. They will kick out if diluted below a certain point.

pH helps to determine how other ingredients may be formulated with the latex. For example, all of the latex binders listed in Table 4.41 are acidic (low pH). If they are blended with basic (high pH) materials pot life may be shortened. At worst, they may kick out of emulsion and agglomerate into a gummy sludge.

Shelf life and pot life are two very important considerations in choosing a latex resin. Latex resins are limited in their stability. Shelf life is a measure of how long the resin will last in ambient storage conditions. This period can be a matter of weeks or it can be years. Generally if the latex has been around for over a year it is probably not too wise to use it. Storage instability will usually be indicated by a buildup of sludge in the bottom of the container or storage tank. The user has means to optimize shelf life.

1. Store in a cool place. Cold rooms are preferable, however, do not freeze.
2. Do not store outside in hot climates.
3. Use agitated storage tanks.
4. Take care that the order rate is consistent with usage.
5. Be aware of how long the resin has been stored at the supplier's site before it was shipped.
6. Be conscious of the weather conditions during shipment. For example, if the shipment is in winter during below zero weather, the possibility of the latex freezing has to be considered.

Pot life is how long the resin formulation will last after it has been prepared. The addition of reactive ingredients and possible heat may greatly increase resin instability. The criterion is that the resin has to remain stable till it is used in process. When considering this timing, allowance has to be made for unexpected process shutdowns.

Formaldehyde (HCHO) and alkyl phenol ethoxylate surfactants (APEs) are two environmental concerns related to latex emulsions. Formaldehyde is released by cross-linking components built into the latex system. Formaldehyde is a by-product of the cross-linking mechanism. It is reported to be a carcinogen and is of concern both as an environmental contaminant and as an as an occupational health and safety hazard. There is intense effort to produce low formaldehyde or formaldehyde free latices. Ways of doing this are:

1. Use nonformaldehyde crosslinking agents—see Section 4.4.4.
2. Incorporate a formaldehyde scavenger into the resin mix. These are specific chemical areas that bind to formaldehyde and prevent it from volatilizing.

APEs are nonionic surfactants that are primary components in latex surfactant systems. There is some evidence to suggest that they may "affect the reproductive or development health of wildlife or humans" (Mlynar (193)). There is a possibility that they may have an impact on life in aquatic systems.

#### *4.4.2.2 Polymer systems for latex resins*

Following is a discussion of various polymer systems used in latex emulsion resins.

##### 4.4.2.2.1 Polyvinyl acetate (PVAc)

PVAc latex emulsions are common and low cost. DeVry (164) advised that PVAc is not a popular binder for many nonwoven materials because of its stiffness. On the other hand, it is for this very reason of stiffness that PVAc is a desired latex resin for filter media. It is often used in wet laid media for pleated filters in engine air intake filtration applications. Often it is formulated with a cross-linking agent such as melamine formaldehyde. Since PVAc is alcohol soluble it is sometimes blended into phenolic resin solutions as a plasticizer to reduce the brittleness of the cured medium.

PVAc comes in many copolymer forms, however, a homopolymer version has been very popular for intake air filters for automotive and heavy duty vehicles. For many years Dur-O-Set® C310[9], a PVAc homopolymer latex, that has been and is still widely used as a saturant for wet laid filter media.

#### 4.4.2.2.2 Ethylene vinyl chloride (EVCL)

Ethylene vinyl chloride (EVCL) emulsions are terpolymers of vinyl chloride, ethylene, and a third monomer which imparts amide functionality.

EVCL emulsions are major players in the production of filter media, particularly those used in cartridges for high-efficiency air filter applications. They provide a high level of water and moisture resistance, alkali resistance, and dimensional stability. They can be readily formulated and compounded with plasticizers, coalescing agents, and the same types of solvents commonly used with acrylic and PVAc systems. They are film formers and provide good moisture barrier properties and abrasion resistance. Because they contain chlorine, they are an excellent polymer base for flame retardant formulations. They are environmentally friendly because they have low emissions of VOCs, phenols, and formaldehyde.

Wacker-Chemie GmbH produces the Vinnol® line of EVCL emulsions. Table 4.42 lists properties of five products; Vinnol® 4500, Vinnolx® 4514, Vinnol® 4530, Vinnol® CE 35, and Vinnol® CEN 2752. Table 4.43 illustrates how three of these latices (Vinnol® 4500, Vinnol®, and Vinnol® 4530) relate to each other.

#### 4.4.2.2.3 Polyvinyl chloride (PVC) and Polyvinylidine chloride (PVDC)

PVC and PVDC are of interest for filter media because of the inherent flame retardancy provided by their chlorine content (see Section 4.5.2). PVDC contains twice as many chlorine atoms per repeat unit of polymer than does PVC and therefore is even more flame retardant. Nevertheless, a filter medium containing PVC or PVDC will still not be flame retardant unless it contains excessive amounts of the resin or it contains additional amounts of supplementary flame retardant. In addition, increasing regulatory resistance to chlorinated flame retardants, because of the toxicity of their emissions when exposed to flame, has limited their marketability.

[9] The Dur-O-Set® resin line was originally developed by National Starch Corporation (now Ingredion) and sold to Celanese Corporation. The business unit responsible for the Dur-O-Set is now Celanese Emulsions, Bridgewater, New Jersey, USA.

Table 4.42 Typical Properties of VINNOL® EVCL Emulsions

| | VINNOL® 4500 | VINNOL® 4514 | VINNOL® 4530 | VINNOL® CE 35 | VINNOL® CEN 2752 |
|---|---|---|---|---|---|
| Copolymer type | Ethylene-vinyl chloride | Ethylene-vinyl chloride | Ethylene-vinyl chloride | Ethylene-vinyl chloride and vinyl acetate | Ethylene-vinyl chloride |
| ***Emulsion Properties*** | | | | | |
| Solids,[a] % | 50 | 50 | 50 | 50 | 50 |
| Viscosity,[b] cps | 25–150 | 25–150 | 25–500 | 20–80 | 50–350 |
| pH | 7–9 | 7–9 | 7–9 | 6–7.5 | 5–7.5 |
| Particle Size Range, μm | 0.11–0.19 | 0.11–0.19 | 0.11–0.19 | 0.15 | 0.2 |
| Surfactant type | Anionic | Anionic | Anionic | Anionic | Anionic |
| Density, lb/gal[c] | 9.2 | 9.2 | 9.4 | 9.4 | 9.3 |
| Mechanical Stability[d] (pH range of 4.0–11.0 | Excellent | Excellent | Excellent | Excellent | Excellent |
| MFFT, °C | 4 | 14 | 50 | 45 | 5 |
| pH Stability[e] | Excellent | Excellent | Excellent | Excellent | Excellent |
| Functional Group | Amide | Amide | Amide | Amide | Amide |
| ***Film Properties*** | | | | | |
| Flexibility | Excellent | Good | Fair | Fair | Good |
| Reactivity | Cross-linkable | Cross-linkable | Cross-linkable | Cross-linkable | Cross-linkable |
| Water Resistance | Good | Good | Excellent | Excellent | Good |
| Oil resistance | Excellent | Excellent | Excellent | Excellent | Excellent |
| Tg,[f] °C | 3 | 12 | 29 | 40 | 10 |

[a]Cenco Moisture Balance.
[b]Brookfield LVT Viscometer, 60 rpm, #2 spindle @ 25 °C.
[c]Foam free at 25 °C.
[d]Does not break after 30 minutes at high speed in a Hamilton mixer.
[e]If adjusted with dilute acids or alkalis.
[f]Differential thermal analysis.
Reproduced with Permission of Wacker Chemical Corporation, Adrian, Michigan, USA.

#### 4.4.2.2.4 Styrene butadiene (SBR)

Styrene butadiene (SBR) is one of several polymers with elastomeric or rubber-like properties. It is often referred to as synthetic rubber. Other "rubber" polymers are natural rubber from gum trees in Asia and Africa, polybutadiene rubber (BR), butyl rubber (IIR), hydrogenated nitrile butadiene rubber (NHBR), styrenic block copolymer (SBC), flouro elastomers, silicone rubber, nitrile rubber, and polyolefin elastomer. SBR has the largest volume. The latex form finds broad use in nonwovens and paper.

SBR is a copolymer of styrene and butadiene, the greater the proportion of styrene in the molecular chain, the stiffer is the resin. It finds application in both dry

**Table 4.43** Comparison of Vinnol EVCL Emulsions Resins

| VINNOL 4514 | VINNOL 4500 | VINNOL 4530 |
|---|---|---|
| ← | Increasing Ethylene Content | |
| | Increasing Tg | → |
| | Increasing Solvent Resistance | → |
| | Increasing Water Resistance | → |
| | Increasing Stiffness | → |
| | Increasing Tensile | → |
| | Increasing Flame Retardancy | → |
| ← | Increasing Gloss | |
| ← | Increasing Fold | |
| ← | Increasing Flexibility | |
| ← | Increasing Elongation | |

Reproduced with Permission of Wacker Chemical Corporation, Allentown, Pennsylvania, US.

laid and wet laid nonwovens. It is used as a binder for high loft filter media. It is used as the rubber coating on the fibers of natural filter media (hogs hair and coir). Tacky forms of SBR have been used as a sticky fiber surface to trap airborne particles.

#### 4.4.2.2.5 Acrylics

Acrylic polymers are loosely defined as being the polymers or copolymers formed from the esters of acrylic or methacrylic acid. Usually they are used in copolymer form with each other and with other monomers such as styrene which gives styrene (acrylic) and PVA which gives vinyl acrylics. Common acrylic monomers used in the copolymer systems of latex resins are ethyl acrylate, butyl acrylate, and methyl methacrylate. Acrylic and acrylate resins are considered the same thing. Polymers of acrylonitrile are often referred to as acrylics (see Section 4.3.5.3.3).

Styrene/Acrylate Copolymer is a polymer of styrene and a monomer consisting of acrylic acid, methacrylic acid or one of their simple esters. Styrene acrylate resins are one of the most popular resins for resin treated filter paper. They are of generally low cost and are usually offered at high $T_g$ (45–60 °C). They provide improved water resistance and water repellency to a wet laid medium.

Lubrizol Advanced Materials, Inc. is one of the largest suppliers of acrylic emulsions to the specialty paper and nonwovens industry. Table 4.44 is a listing of their resin emulsions recommended for filter media.

**Table 4.44** Lubrizol Acrylic Emulsions for Filter Media

| Identification | $T_g$ (°C) | Charge *A* or *A-N*[a] | Solids, (%) | pH | Viscosity, (cP) | Specific gravity | Heat reactive | Solvent resistance | Water resistance | Comment |
|---|---|---|---|---|---|---|---|---|---|---|
| Hystretch™ V-60 | −60 | A-N | 50.5 | 7.1 | 40 | 1.01 | Yes | G | G | Controlled tack; ultra-soft; dust holding performance in filtration |
| Hycar® 2679 | −3 | A | 49 | 3.7 | 100 | 1.06 | Yes | VG | G | Good general-purpose acrylic for saturation, filter paper |
| Hycar® 26120 | −11 | A | 50 | 3.8 | 115 | 1.07 | Yes | G | VG | Highly re-dispersible, excellent heat stability, saturation and wet end addition; medical |
| Hycar® 26084 | +8 | A | 48 | 6.1 | 110 | 1.07 | Yes | G | G | Heat seal adhesive with excellent solvent and plasticizer resistance; paper coating |
| Hycar® 26138 | +25 | A | 49 | 5.5 | 60 | 1.07 | Yes | VG | G | Durable, excellent paper coating binder, excellent solvent resistance |
| Hycar® FF26921 | +29 | A | 44 | 8.0 | 65 | 1.07 | Yes | VG | G | Formaldehyde-Free; excellent for micro-glass and synthetic media, fast curing and hot tensile strength. |
| Hycar® 26450 | +32 | A | 46 | 4.0 | 25 | 1.06 | Yes | VG | G | High crosslink density; oil resistant: FDA compliant for direct food contact. |
| Hycar® 26391 | +36 | A | 50 | 3.6 | 125 | 1.09 | Yes | G | VG | Highly water resistant; very good color |

| | | | | | | | | | | |
|---|---|---|---|---|---|---|---|---|---|---|
| Hycar® 26599 | +42 | A | 45 | 3.0 | 20 | 1.04 | Yes | G | VG | Firm, good wet strength, hydrophilic polymer; suitable for food grade filtration applications. |
| Hycar® 26315 | +55 | A | 49.5 | 2.1 | 36 | 1.07 | Yes | G | VG | Heat sealable; FDA compliant for direct food contact; excellent water resistance |
| Carbocure™ TSR-72 | +72 | A | 35 | 4.6 | 50 | 1.07 | Yes | VG | G | Stiff, high crosslink density polymer with excellent oil and solvent resistance; high temperature dimensional stability, moldability; saturation and spray bond |

[a]A = anionic, A-N = anionic/nonionic

Reproduced by Permission of Lubrizol, Inc. Cleveland, Ohio, USA.

**Table 4.45** OMNOVA Styrene Acrylics and Acrylics for Filter Binders

| Product | $T_g$ °C | Solids % | pH | Description |
|---|---|---|---|---|
| AcryGen F 116-50 | +53 | 50 | 5.0 | HVAC, Cabin Air, Engine Intake, Respiratory, Firm hand, binder with excellent balance of moisture resistance and pleatability |
| AcryGen 27 | +27 | 46 | 3.0 | Respiratory Firm hand, high tensile strength, white, moldable |

Reproduced with Permission of OMNOVA Solutions Inc. Beachwood, Ohio, USA.

Another important supplier of resins for filter media are OMNOVA Solutions Inc. OMNOVA has already been mentioned in Section 4.4.2.1 and Table 4.41 for their vinyl acetate copolymers. They also manufacture styrene acrylic and acrylics for their filter binder resin line. Two of these are tabulated in Table 4.45.

## 4.4.3 Solution polymers

Solution polymers are made by connecting monomer units together to make a long polymer chain that carries an ionic charge along the chain itself. During the manufacturing process, the chains curl around into one another. These products dissolve in water in fashion similar to salt and sugar. When water comes into contact with the molecule, the polymer chains uncurl and lengthen.

Solution polymers can be anionic, cationic, or nonionic depending upon the monomers used in the production process. They are available in a wide range of molecular weights and charge densities. They act as solid/liquid separating agents, otherwise known as flocculants. In wet lay operations they are used to enhance the strength of the medium. Certain types of charged polymers are added to the wet end of the wet lay process to provide for the retention of particulate matter such as diatomaceous earth or activated carbon.

Solution polymers usually form very viscous solutions even at low polymer contents in the water (<5%). They are most often used in the wet end of wet lay filter media. They are also used in saturation applications, usually as a component of the latex system.

Certain types of solution polymers can be compounded to become rheology modifiers and as such they can be used as dispersing agents. Such agents can be very useful in dispersing long fibers in wet lay systems. They do this by:

1. Increasing the viscosity of the dispersion water so that the fibers are less mobile and less likely to entangle with each other.
2. Forming hydrophyllic–hydrophobic surface effects on the fiber so that they reject contact with each other.

#### 4.4.3.1 Starch

Starch is a natural polymer with a very high molecular weight. Brouwer et al. (195) stated that it is the most widely used chemical in the paper industry. It is used in synthetic fiber manufacture as a finish for fibers. For wet lay systems it helps to make the fiber (synthetic and glass) more dispersible in water. It is also used as a finish for dry formed nonwovens. Starch is used as a backbone for grafted polymer systems.

Starch is sourced from agricultural plant products such as sugar cane, corn, or potato. Starch solution is prepared by dispersing starch powder in water and "cooking" the mixture at temperatures approaching 100 °C. The solution formed is very viscous and usually cannot be used at concentrations above 5–10%. Starch can be prepared in cationic, anionic, or nonionic forms. In wet lay systems it is mostly used in the cationic form, although the purpose of Brouwer et al. (195) was to justify some advantages of the anionic form. In wet lay systems, starch is mostly used as an additive to the fiber slurry. It can be used for adding dry strength to the wet lay product. The cationic form is also used in polymer retention systems to retain other additives such as activated carbon, fillers, wet strength agents, etc into the fiber structure.

#### 4.4.3.2 Polyvinyl alcohol (PVOH)

Polyvinyl alcohol (PVOH) is a water-soluble, film-forming resin that has very strong bonding properties for nonwoven filter media. Its biggest use is as oil and grease proof coatings for packaging materials such as paper and board. It is chemically resistant to oils, grease, hydrocarbon solvents and when fully hydrolyzed, highly resistant to water. It is available in white flake, powder, or pellet form. It is produced by first forming polyvinyl acetate and then hydrolyzing the resultant chemical structure. The properties of the polymer are very dependent on the degree of hydrolysis. The ranges for degree of hydrolysis[10] are as follows:

| | |
|---|---|
| Fully hydrolyzed | 98–99.5% |
| Intermediate hydrolyzed | 90–97% |
| Partially hydrolyzed | 86–89% |
| Specialty grades | 76–87.6% |

PVOH is also dependent on its degree of polymerization for its properties. Table 4.46 indicates how hydrolysis and degree of polymerization affect some of the properties of PVOH.

Preparation of PVOH solution is similar to the preparation of starch solution. The resin solids have to be added carefully to cold water and heated with agitation to >90 °C. Like starch, the solubilized resin is very viscous, even at solids contents are

[10] Hydrolysis is expressed as the mole percent of acetate groups hydrolyzed (dry basis).

**Table 4.46** The Effect of Degree of Hydrolysis and Degree of Polymerization on the Properties of PVOH Resin

| Property | Degree of hydrolysis | | Degree of polymerization | |
|---|---|---|---|---|
| Effect on | High | Low | High | Low |
| Solvent resistance | + | − | + | − |
| Water resistance | + | − | + | − |
| Solubility | − | + | − | + |
| Hygoscopicity | − | + | − | + |
| Tensile strength | + | − | + | − |
| Flexibility | − | + | − | + |
| Viscosity | + | − | + | − |

Note: + indicates the property gets higher or stronger. − indicates that the property gets lower or weaker.

less than 10%. Because of penetration difficulties, the high viscosity makes it difficult to use the resin as a saturant. Because of the high viscosity, wet pickup and corresponding dry pickup is high. This means the resin has to be applied at lower solids, and therefore more water to dry off.

When formulated with the right kind of cross-linking or thermosetting agent, PVOH can produce a web with outstanding heat resistance, water resistance and strength properties. The use of the thermoset curing mechanism allows the properties to be achieved even with a low hydrolysis PVOH. The polymer is very stable and shows little tendency to oxidize or degrade with time.

### 4.4.4 Cross-linking agents for water-based resins

Thermosetting resins such as phenolics described above have their own built in cross-linking mechanisms that create rigidity, strength, heat resistance, water resistance and chemical resistance to the medium when fully cured. Many water-based latexes and resins are formulated with cross-linking agents to impart these properties. Typical cross-linking agents are urea formaldehyde, melamine formaldehyde, water-based phenolics and various dialdehydes. Glyoxal, the simplest of the dialdehydes, is often used. There has been a movement to incorporate formaldehyde free cross-linking agents. Polyamide-epichlorohydrin resins and salts of zirconium ammonium carbonates are two that have been considered.

## 4.5 ADDITIVES AND FINISHES

### 4.5.1 Adsorbent materials

Following are some adsorbent materials that may be used in nonwoven filters:

**a.** Activated carbon.

**b.** Activated alumina—Lower surface area but good base for certain impregnations.

**c.** Zeolite–Man-made pore sizes can be designed for specialty separations.
**d.** Zeolite–Natural.
**e.** Ion exchange resins—Can be produced with controlled pore sizes and high chemical capacities.
**f.** Silica gel.
**g.** Sodium bicarbonate (NaHCO) for odor removal.
**h.** Potassium permanganate.

**Figure 4.46** Extended Surface—Activated Coconut Carbon Granule. *Reproduced by Permission of INDA, Cary, North Carolina, USA.*

Middlebrooks (128) suggests that activated carbon is by far the preferred adsorbent for nonwoven filter media applications. Generally it is the adsorbent with the largest total surface area depending on source. The most common sources are coal, wood, and coconut. Figure 4.46 is a microphotograph of the extended surface of an activated coconut carbon granule. Any organic material that can be charred and activated to a carbon can be used for producing some form of activated carbon. Synthetic polymer fibers such as PAN have been used for producing activated carbon fiber. The choice depends on pore size and distribution. Generally, coconut sourced activated carbons are used for air filtration applications while coal-based carbons are preferred for water filtration.

Many activated carbon media are impregnated with some form of chemical to develop and enhance chemsorption properties. One type is ASZM-TEDA carbon (196), which has been used in US military nuclear, biological, and chemical (NBC) filters since 1993. This material, manufactured by Calgon Carbon Corporation is a coal-based activated carbon that has been named for its impregnates: copper (A), silver (S), zinc (Z), molybdenum (M) compounds and triethylenediamine (TEDA).

**Table 4.47** Application of Activated Carbon Impregnates

| Impregnate | Chemical contaminant |
|---|---|
| Copper/silver salts | Phosgene, chlorine, arsine |
| Iron oxide | Hydrogen sulfide, mercaptans |
| Manganese IV oxide | Aldehydes |
| Phosphoric acid | Ammonia |
| Potassium carbonate | Acid gases, carbon disulfide |
| Potassium iodide | Hydrogen sulfide, phosphine, mercury, arsine, radioactive methyl iodide |
| Potassium permanganate | Hydrogen sulfide |
| Silver | Arsine, phosphine |
| Sulfur | Mercury |
| Sulfuric acid | Ammonia, amine, mercury |
| Triethylenediamine (TEDA) | Radioactive methyl iodide |
| Zinc oxide | Hydrogen cyanide |

Reproduced by Permission of INDA, Cary, North Carolina, USA.

ASZM-TEDA carbon provides a high level of protection against and a wide range of toxic chemicals.

Table 4.47 provides a list of chemical impregnates and the air contaminants against which they are effective. Potassium permanganate is becoming an important impregnate for activated carbon media used in residential air cleaning systems. Permanganate-treated carbon is particularly effective for removing odors in air. AAF International, Loisville, Kentucky, USA markets home ventilation filters (197) that contain activated alumina treated with potassium permanganate for effective odor removal.

## 4.5.2 Flame retardants

Flame retardants can be divided into four functional groups:

1. Halogenated
2. Phophorous-based
3. Nitrogen-based
4. Inorganic

A flame retardant is a chemical used to impart flame resistance to a nonwoven web. In many air filtration systems (including automotive) it is desired and/or required that the filter medium be flame resistant. The general requirement is that the medium not be able to sustain a flame after the heat or ignition source is withdrawn. Fiber materials such as polyester accomplish this by melting and shrinking away from the heat source so that combustion does not occur. Inorganic fibers such as glass are fire resistant and do not burn because they do not contain a component that will react with oxygen to produce a flame. Halogenated polymers such as PVC and PVDC are

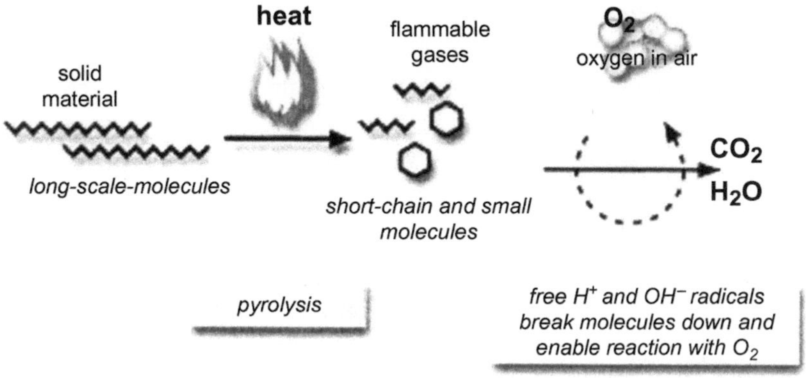


**Figure 4.47** The Burning Mechanism. *Source: European Flame Retardant Association (EFRA).*

inherently flame resistant because of their chlorine atom content. Most organic filter media, particularly cellulose will burn and require a flame retardant additive to inhibit the oxidation effect after the flame is withdrawn.

Flame retardants for filter media have to satisfy a number of challenging criteria. They must be environmentally friendly, non-hazardous to humans, animals and plants, and must not release any additional toxic fumes when they burn. These additives should not escape the finished product into the atmosphere, or when it comes into contact with water. And researchers must make sure the flame retardant does not react with the plastic or other components in unwelcome ways that might alter the material, influence its functionality, or affect its appearance. This includes the "stay-white" properties of a white air automotive air filter medium. Washability is an important factor for air filters for heavy duty vehicles such as large trucks, tractors, and off the road vehicles. The filters, often washed and reused, are expected to retain their flame retardancy, even after several washings.

The flame resistance of a filter material is usually determined by some form of ignition test (vertical, 45°, or horizontal). Generally a flame is applied to the material and observation is made as how far the flame spread before extinguishing. Another method of measuring flame resistance is the LOI, which determines the minimum oxygen content in air that will sustain combustion of a material.

Materials burn because they are decomposed by heat to release flammable gases. Visible flames appear when the gases combine with oxygen. The carbon-hydrogen chemical structure of the organic material reacts with oxygen in the atmosphere to produce carbon dioxide ($CO_2$) and water ($H_2O$). The oxidation is endothermic and releases energy. If sufficient energy is generated, the oxidation is maintained and burning will continue as long as the organic fuel supply lasts. The process is shown in Figure 4.47.

#### *4.5.2.1 Halogenated flame retardants*

Halogenated flame retardants act in such a way as to inhibit the oxidation reaction that causes flame. For example, halogenated compounds (containing chlorine or bromine) when exposed to high temperatures will release the halogen as a free radical ($Cl^-$ or $Br^-$). These will then react with hydrocarbon molecules to form HBr and/or HCl and inhibit the oxidation process. The effectiveness depends on the quantity of halogen atoms and how well their halogen release is controlled. For example the chlorine content of PVC resin may be enough to make the polymer flame retardant, however, if the resin is used as a binder in a filter medium it is insufficient to make the whole web flame retardant. Additional flame retardant will have to be added to the formulation. The advantage of a halogenated compound such as PVC is that it reduces the amount of additional flame retardant.

The market place has flame retardants based on bromine as well as chlorine. Brominated flame retardants are generally more expensive, but have the advantage of releasing over a narrower temperature range, thus resulting in more optimal concentrations in the flame zone.

Antimony trioxide ($SbO_3$) is not a flame retardant, however it has a synergistic effect for halogenated flame retardants. It serves as a catalyst, facilitating the breakdown of the flame retardants to active molecules. It also reacts with halogens to produce antimony-halogen compounds, which in themselves are effective flame retardants.

#### *4.5.2.2 Phosphorous-based flame retardants*

Phosphorous containing compounds usually act in the solid phase of burning materials. When heated these compounds break down to release polymeric forms of phosphoric acid that chars the medium. The charring effect inhibits the "pyrolysis" process that releases flammable gases to feed the flames.

Diammonium phosphate (DAP) is a common form of phosphorous containing flame retardant because it is inexpensive and it is effective. There are several caveats associated with its use:

1. It decomposes over time and accelerates the degradation and discoloration of the medium.
2. It has a relatively low decomposition temperature, therefore, media containing this additive will have very poor heat and temperature resistance.
3. It is very hygroscopic and will adversely affect the moisture resistance of the medium. The dimensional instability of the medium will be increased.
4. DAP is water soluble and cannot be used in washable filter media.

Other forms of phosphorous containing compounds include phosphate esters, phosphonates, and halogenated esters of phosphates, etc. Tetrakis (hydroxymethyl) phosphonium (THPC), the "Cadillac of fabric flame retardant finishes" (198), is a phosphine formaldehyde derivative precondensed with urea to make a water-soluble prepolymer. When applied as a finish to fabrics, it produces a flame retardant medium that is resistant to multiple launderings.

#### 4.5.2.3 Nitrogen-based flame retardants

Nitrogen flame retardants are believed to act in the following manners (198):

1. They form cross-linked polymer structures, stable at high temperatures, and thereby, physically inhibiting the decomposition of materials to flammable gases.
2. They release nitrogen which reduces the flammable gases and thus reduces flames.
3. They are synergistic with phosphorous containing flame retardants and reinforce their function.

Nitrogen-based flame retardants are used at high concentrations, or in conjunction with other flame retardants. Melamine-based products are the most widely used type of nitrogen flame retardants.

#### 4.5.2.4 Inorganic flame retardants

This category includes inorganic flame retardants and flame retardant synergists such as silicon dioxide, metal hydroxides (e.g., aluminum hydroxide and magnesium hydroxide), antimony compounds (e.g., antimony trioxide), boron compounds (e. g., zinc borate—which is often used as a synergist for both halogenated and non-halogenated flame retardants), and other metal compounds (molybdenum trioxide). As a group, these flame retardants require high loading levels to impact the desired fire safety. It has been mentioned above that antimony trioxide by itself is not a flame retardant, however, is invariably used as a synergist for halogenated flame retardants. There are a number of inorganic products that serve as flame retardants. These include aluminum trihydrate, magnesium hydroxide, boron compounds, and zinc borate. Other zinc and tin compounds act in a way to reduce smoke emissions.

#### 4.5.2.5 Flame retardant regulation

The use of flame retardants in fabrics, foams, and other applications (including filter media) are of great concern because of their toxic chemical implications. While it is difficult to find flame retardant standards and regulation that is specifically pertinent to filter media, one must consider the importance of flame retardant filter media as it relates to other fabrics or materials. It is the nature of flame retardants to release severely toxic chemicals to the environment in their flame retarding chemistry. This is particularly true of halogenated flame retardants. In many cases of fire, the death and injury from the toxic chemicals are more severe than from the fire itself. In addition, it is suspected that flame retardant fabrics, such as used in drapes and furniture will slowly release toxic gas and vapor emissions to the environment that may cause long term health effects. Of particular concern are foam products used in insulation and furniture and bedding. California recently passed a law (Technical Bulletin 117-2013 (199)) that upholstered furniture products be fire resistant without flame retardant. The law also applies to baby products. For example, furniture will be judged by the ability of an upholstery fabric to resist a smoldering cigarette. In many cases nonwovens can accomplish this by using fire resistant fibers such as glass or polyester.

Following are some regulatory activities, standards, and test methods concerning fire protection and flammability:

- AC 25.853-1—Flammability Requirements for Aircraft Seat Cushions
- ASTM E84—14: Standard Test Method for Surface Burning Characteristics of Building Materials
- ASTM E119—14: Standard Test Methods for Fire Tests of Building Construction and Materials
- ASTM E814—13a: Standard Test Method for Fire Tests of Penetration Firestop Systems
- 16 CFR 1633—Standard for the flammability (Open Flame) of Mattress Sets
- 29 CFR 1926.150—Safety and Health Regulations for Construction, Fire Protection and Prevention
- IEEE 634-2004—IEEE Standard Cable-Penetration Fire Stop Qualification Test
- ISO 6944—Fire Resistance Tests: Ventilation Ducts
- UL 72—Tests for Fire Resistance of Record Protection Equipment
- UL 94—Standard for Safety of Flammability of Plastic Materials for Parts in Devices and Appliances
- UL 1479—Fire Tests of Through-Penetration Firestops

Test methods for flammability are discussed in Chapter 6, Section 6.7.

### 4.5.3 Water repellents

Water repellency or hydrophobicity is important to certain types of filter media. For example, hydrophobic media are used in diesel and gasoline fuel filtration to reject moisture particles and prevent the moisture from penetrating the filter. High water repellency is also associated with low surface energies required in filter media used in surface filtration and dust collection systems where cake release, surface cleaning, and blow-back operations are required.

Water repellency helps prevent an air filtration medium, from being wetted out and damaged when exposed to water. For example, the Nuclear Air Cleaning

Handbook[11] (200) states that "water exposure is unquestionably an important factor leading to the deterioration of HEPA filters and their degradation to 0 percent efficiency when coupled with higher pressure drop. HEPA filters become weak and plug with water. One of the most common events is when people think no detrimental effects occur as a result of repeatedly wetting the filter and drying it. Tests have shown that repeat wetting and drying of a HEPA filter will cause the loss of half its strength". There also are very strong effects of operational time on the behavior of HEPA filters under wet conditions. Tests have indicated that the binder starts to get soft and dissolves at high differential pressures".

Water repellency is incorporated into a filter medium by the use of hydrophobic materials in the medium surface. This can be done by use of water repellent fibers or by water repellent coatings. Water repellent materials have low surface energies, meaning that the surface lacks the energy to hold on to a liquid droplet (see Section 3.9).

Polyester and polyolefin are examples of fibers that are hydrophobic and will help impart water repellency to a filter medium. Waxes, silicones and fluorocarbon resins are examples of water repellent coatings that can be applied to the fibers of a filter medium. Some binders such as styrene-acrylates also impart a high degree of water repellency. In addition to other properties, such as filtration efficiency and cake release, low surface energy membranes such as ePTFE membranes impart water repellency (see Chapter 8, Section 8.1).

### 4.5.4 Antimicrobial agents

Antimicrobial agents inhibit microbial growth (bacteria, fungus, mildew) on the surface and/or interior fiber surface of a filter medium. Their functions are as follows:

1. Protect the filter medium from biological degradation while in service.
2. Inhibit the biological growth of microbes downstream of the filter.
3. Serve as a means of odor control.

SpecialChem S. A (201), an online Internet service for specialty chemicals lists 162 biocide chemicals identified along with 13 polymer systems where they are used.

In general, antimicrobial agents either are organic or contain a metal component such as silver, zinc, or copper. Ciba Specialty Chemicals markets both, organic and inorganic, under their Irgaguard® trade name.

Antimicrobial agents can be incorporated into a medium as a surface treatment or they can be formulated into the fibers that make up the nonwoven. Foss

[11] A publication of the United States Department of Energy.

Manufacturing Co. Inc of Hampton, New Hampshire, USA has patented a fiber technology (Foss etc. (202)) incorporating a silver zeolite. The additive can be included in mono-component and bicomponent fibers. The technology is being traded as Fosshield®. Foss claims (203) that "Fosshield® protected HVAC filters buy helping them capture and neutralize a variety of potentially harmful and destructive microorganisms, while eliminating their offensive odors in indoor environments".

The EU Biocides Regulation 528/2012 (EU BPR) covers a very diverse group of products, including disinfectants, pest control products and preservatives.

## 4.5.5 Inks and colorants

Colored media are becoming a growing novelty in filter media. In addition it is becoming more and more necessary to add text, logos, barcodes etc for identification and instruction purposes. The coloring is for both aesthetic and operational purposes. Some reasons for this are as follows:

1. The colored medium makes the filter look more attractive.
2. The medium is color coded for operational use. For example, in a filter production operation the color coding will help the operator know to which line and to which product he sends the roll of medium.
3. Line marks on the medium help the operator identify the "wire" side and "felt" side, of the medium.
4. Color change of thermodegradable dyes or pigments during the curing process can help an operator determine when the medium is cured.

Both dyes and pigments are powerful colorants. Dyes are colorants available in liquid or solution form, whereas, pigments are insoluble solids or powders that are usually applied to a substrate as a colloidal suspension or as an emulsions. Pigments are thick and opaque. On the other hand, liquid dyes are thinner in consistency and more transparent.

### *4.5.5.1 Dyes*

Dyeing is the process of adding color to textile products like fibers, yarns, and fabrics. It is normally done in a special solution containing dyes and particular chemical materials. After dyeing, dye molecules have an uncut chemical bond with fiber molecules. The temperature and time controlling are two key factors in dyeing. There are mainly two classes of dye, natural and man-made.

Historically, the primary source of dye, has generally been natural, In most cases dyes were extracted from animals or plants. Since the mid-18th century, however, artificial dyes have resulted in a broader range of colors and dyes more resistant to washing and environmental exposure. Different classes of dyes are used for different types of fiber. For example, acrylic fibers are dyed with basic dyes, while nylon and protein fibers such as wool and silk are dyed with acid dyes. Polyester fiber is often dyed with disperse dyes. Cotton is dyed with a range of dye types.

#### *4.5.5.2 Pigments*

A pigment is a material that changes the color of reflected or transmitted light as the result of wavelength-selective absorption. Pigment particles tend to settle into the fibers of the nonwoven materials. When the ink dries the particles of the pigment get stuck in the on the fibers therefore the pigments tend to be more water resistant than the dies. In addition, there is much less chance for reflow with pigmented inks. It is highly possible to get more colorful results into pigment-based printouts than dye based. Pigment colors are tend to be more vibrant than dye-based colors. They are also more fade resistant. Pigments are more expensive because they more to produce.

#### *4.5.5.3 Dye application to nonwovens*

Dyes and pigments may be applied to nonwovens at different stages of the textile production process.

- Directly to the polymer raw material, molten polymer or polymer solution prior to being spun into fiber or web form.
- Directly to the fiber raw material or to the fiber slurry prior to web forming. Vat dying is a process for dying several forms of fiber (cotton, wool, rayon, etc.). In wet lay operations colorants may be added to and retained by the fiber slurry prior to forming operation.
- Directly to the formed web:
  - Saturating
  - Coating
  - Printing
- A popular way to add a colorant to a resin treated filter medium is by adding the dye to the resin mix and applying to the web along with the resin. It should be noted that dye or pigment compatability with the resin is an important concern. Note; It is very hard to find dyes and pigments that are compatible with solvent-based phenolic resin mixes.

#### 4.5.5.4 Factors in choosing a dye

Following are some of the concerns in choosing a dye or pigment for a nonwoven medium:

- Color match.
- Is the color reproducible from dye lot to lot or pigment lot to lot?
- If the dye is to be applied with a resin, is it compatible with the resin mix? Does the dye color change with pH? In some cases this is desirable e.g., litmus paper for pH determination.
- How well does the dyed finished medium stand up to environmental exposure; heat, temperature, moisture, water, etc.
- If the dye is a cure indicator does it degrade at the proper exposure to heat and temperature.
- How substantive is the dye or pigment to the fiber or nonwoven.

  Information in testing for color is presented in Section 6.13.

## 4.6 MATERIALS FOR MEMBRANE FILTERS

Filter membranes are made from both organic and inorganic materials. Following is a listing of some of the materials.

- Organic polymer
  - Cellulose acetate
  - Cellulose nitrate
  - Nitrocellulose
  - Nylon 6
  - Nylon 66
  - PFTE and ePTFE
  - Polyaniline
  - Polycarbonate
  - Polyester
  - Polyether sulfone
  - Polyethylene
  - Polylactic acid (PLA)
  - Polypropylene
  - Polystyrene
  - Polyurethane
  - Regenerated cellulose
- Inorganic
  - Aluminum oxide (Alumina)
  - Ceramic

- Glass
- Nickel
- Silver (SPI-Pore™)
- Stainless steel (type 316)
- Titanium dioxide
- Zirconium oxide

Generally, inorganic membranes are more durable, chemical resistant, temperature resistant, and heat resistant than organic membranes. They are also considerably more expensive. Processes for producing membrane filter media are described in Section 5.9.

# CHAPTER 5

# Processes for Nonwoven Filter Media

Chapter 1, Section 1.7 classifies nonwovens by process and presents brief descriptions of each process. More detailed discussions follow.

## 5.1 DRY FORMED

### 5.1.1 Air laid

The fundamental process for air laying a web is to hammer mill dry pulp fibers into an air stream and depositing it onto a moving collector fabric. The important products produced using this process are wipes and absorbent materials for hygienic products, incontinence pads, and diapers. The absorbent products may incorporate superabsorbent (SAP) materials to be blended with the fibers. Figure 1.12 of Chapter 1 illustrates the MTS pilot process for producing air laid produts. The web forming is done by a Dan-Web dual head former produced by Dan-Webformin in Denmark. The process also provides fiber metering systems to feed long synthetic fibers to the air laid forming device. SAP powder is added through "dosers" at the forming heads. A latex spray provides for web bonding. The wet resin-treated web is dried by a flat bed through-dryer.

Air laid webs can also be bonded by thermal bonding, hydroentanglement, and needle punching. Pittman and Irish (204) examined the effects of thermal bonding air laid webs with bicomponent sheath-core fibers.

At one time the air laid process was reported as producing an automotive carburetor air filter for Fram Filters ((205) and (206)). This product has since been discontinued. The author is unaware of any other filter medium produced by the air laid process.

A variation of the air laid process is the use of formers such as the Rando Webber® to form webs directly from bale opening equipment. Since this is more of a mechanical forming process rather than an aerodynamic process, the Rando Webber® is included as part of the discussion of dry laid processes in Section 5.1.2.1.

### 5.1.2 Dry laid webs and felts

#### *5.1.2.1 Forming of dry laid webs*[1]

Figure 1.3 (24) depicted in Chapter 1 is a layout of a typical dry laid line. The core of the dry laid process is the card or a related machine called the garnett. Cards and garnetts

[1] Much of the description of the dry laid process presented here is based on a publication by LeGrande Crook (24) and subsequent conversations with Mr. Crook.





*Handbook of Nonwoven Filter Media*
ISBN: 978-0-08-098301-1

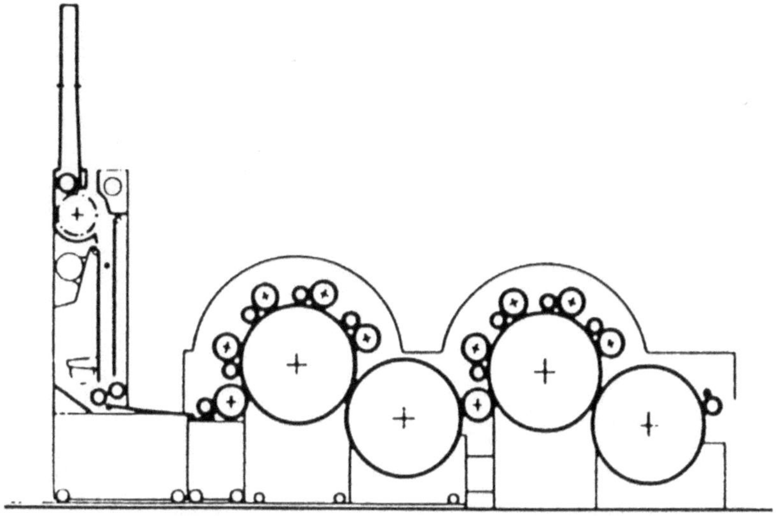

**Figure 5.1** Chute Feed and Roller Top Card. *Source: Reproduced by permission of TAPPI Press, Atlanta, Georgia USA ((24), pg 160).*

were originally designed to produce slivers for yarn manufacture. Modified versions were developed for nonwoven manufacture. Figure 5.1 is a diagram of a roller-top card. It consists of a series of rolls coated with spirally wound sawtooth wire often referred to as metallic clothing. The sawtooth points are engineered to have specified heights, pitches, and angles. Fiber is carded when it passes between two such rolls (or two metallic clothed surfaces) moving at different speeds. If the points of the metallic clothing are opposite, a combing action takes place that parallelizes the fiber. It is not necessary that both metallic coated surfaces be rolls. Generally one surface is the master cylinder roll, however, the other could be a stationary surface such as depicted in Figure 5.2.

Figure 5.3 is a diagram of a garnett. Note the similarity between the garnett and the roller-top card of Figure 5.1. In general, garnetts differ from cards in that they form more randomized webs. They also produce higher loft webs for filter applications. Garnetts are considered as an older technology. More and more of garnett-produced products are being replaced by carded webs and felts.

Following are the types of metallic-coated rolls found on a card or garnett:

1. Master cylinder—The master cylinder is the main roll in a card or garnett. Its function is to transfer the web from one point in the process to the next. It also provides the surface on which other rolls such as stripper rolls and worker rolls can do their work.
2. Likerin roll—The likerin roll accepts lap or batting that is condensed from fibers fed from a chute or hopper, opens them into small tufts, and feeds them to the master cylinder.

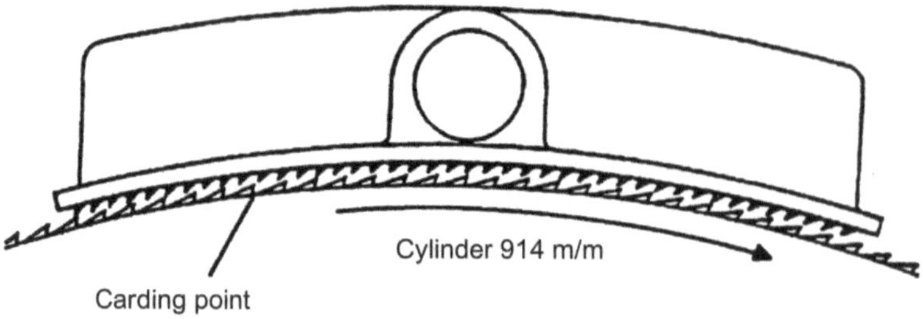


**Figure 5.2** Stationary Carding Action. *Source: Reproduced by permission of TAPPI Press Atlanta, Georgia USA ((24), pg 161).*

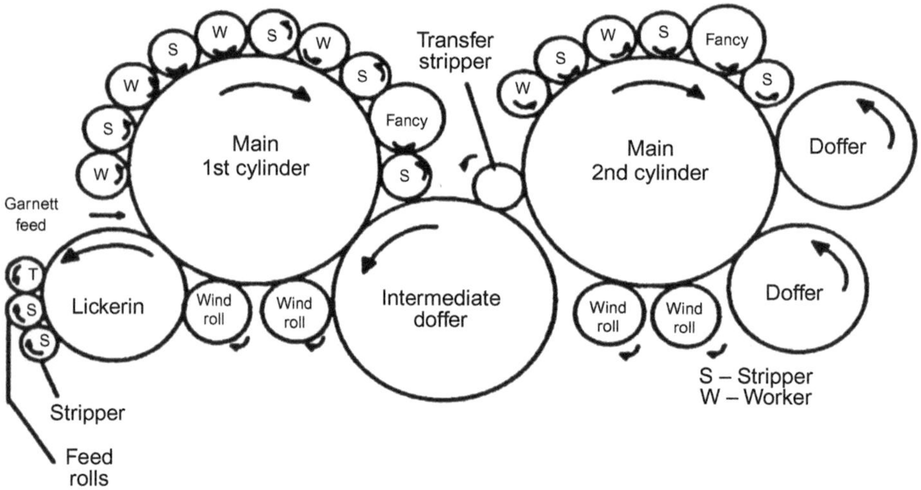


**Figure 5.3** Model 70 Garnett. *Source: Reproduced by permission of TAPPI Press Atlanta, Georgia USA ((24), p. 162).*

3. Stripper rolls—Stripper rolls have the wire points in the same direction as the master cylinder. They are used to transfer the web from one roll to another.
4. Worker rolls—Worker rolls act in conjunction with the the master cylinder to do the fundamental carding action. In Figure 5.4 carding takes place between the tangent point of the master cylinder and the worker roll. The stripper roll which has points in the same direction as the worker roll strips fiber tufts from the worker roll and puts the fiber back on the web ahead of the carding plane. This allows the fiber to be reworked again and again until the tufts are reduced to individualized fibers.
5. Doffer rolls—Doffer rolls are used to transfer the web from the master cylinder to a moving bed. They rotate counter to the direction of the master cylinder, although the points are in the same direction.

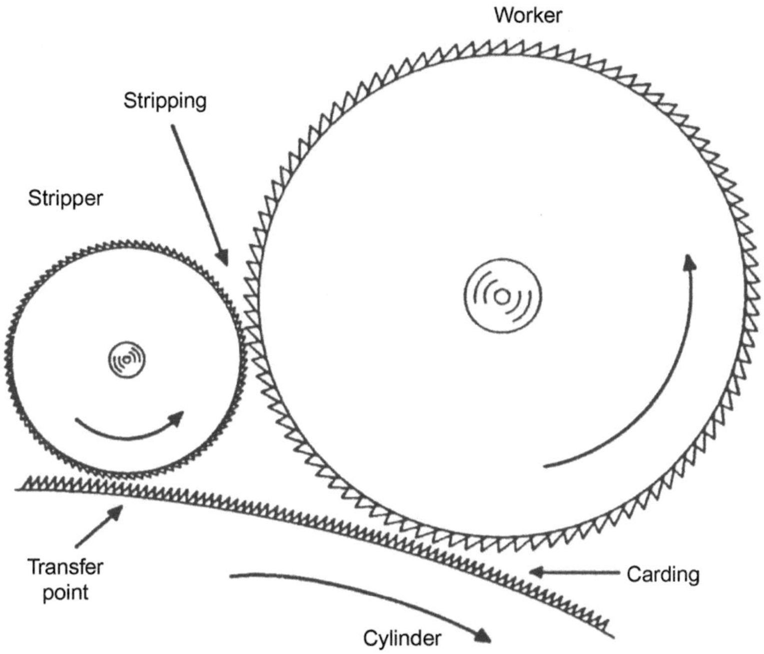


**Figure 5.4** Worker/Stripper Action. *Source: Reproduced by permission of TAPPI Press Atlanta, Georgia USA ((24), pg. 160).*

6. Condensing rolls—The carded webs off the master cylinder are very linear in the machine direction. The MD:CD ratio of fiber orientation is in the range of 10:1. Condensing rolls work with the doffer rolls to randomize the structure and put more fibers in the cross direction. This is done by using the condensing roll to doff fibers from the doffing roll. The condensing roll runs slower than the doffer roll and rotates in an opposite direction than the doffer roll and in the same direction as the master cylinder. Its wire points are in an opposite direction to both the wire points of the master cylinder and the doffer roll. The transfer action allows the doffing roll to push fiber into the teeth of the condensing roll. The result is a more isotropic web. Figure 5.5 demonstrates the activity between the master cylinder, the condensing roll and the doffer roll.
7. Randomizing rolls—Randomizing rolls are high-speed rolls inserted between the master cylinder and the doffing roll. The randomizing roll doffs the web from the master cylinder and by centrifugal force transfers the fibers to the doffing roll. Random webs with MD:CD ratios of 4:1 are obtainable with this arrangement.
8. Fancy rolls—Fancy rolls are rolls associated with a garnett to help produce high-loft mats.

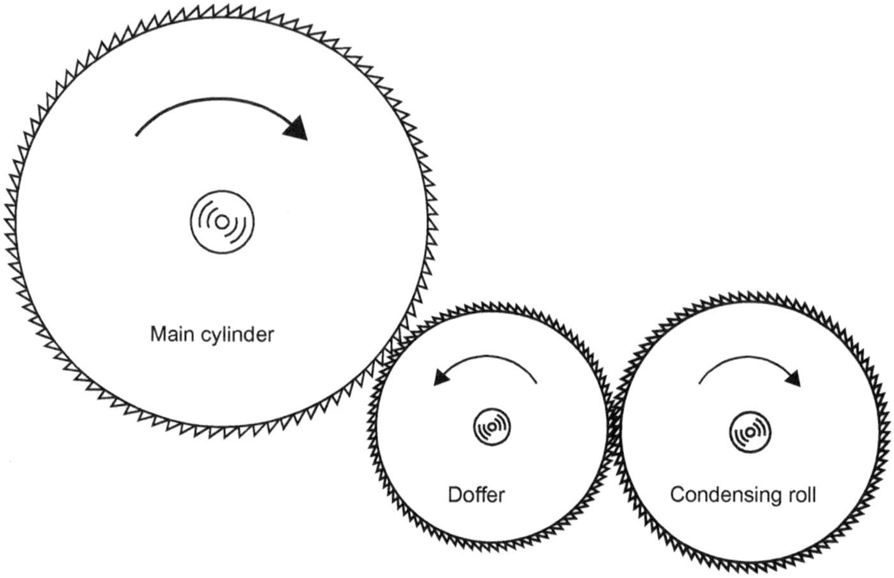


**Figure 5.5** Compensating (Condensing) Roll Action.

**Figure 5.6** Layering of Webs Coming From Cards. *Source: Reproduced by permission of TAPPI Press Atlanta, Georgia USA ((24), pg. 163).*

In addition to cards and garnets, a dry laid line will include bale opening machines, blending machines, and machines to feed the cards (or garnets). In a carding line there may be two or more cards as shown in Figure 5.6, each one forming a separate layer that is combined with the previously formed layers. The layering allows for thicker webs and higher grammages.

Cross layering is a process step for producing thicker and heavier webs by lapping them back and forth unto a conveyor moving at a right angle to the cross-layering motion. The cross-layering operation is diagrammed in Figure 5.7. Cross layering or cross lapping is the method of building a web for the needle punching operation described in Section 5.1.2.3.1.

The process requires a device for forming the final web. This could be the output of the cards or garnetts. There are also air laid formers fed by carded fibers from the cards. A very popular such former is the Rando Webber® produced by Rando Machine Corporation. This device is depicted in (Figure 5.8).

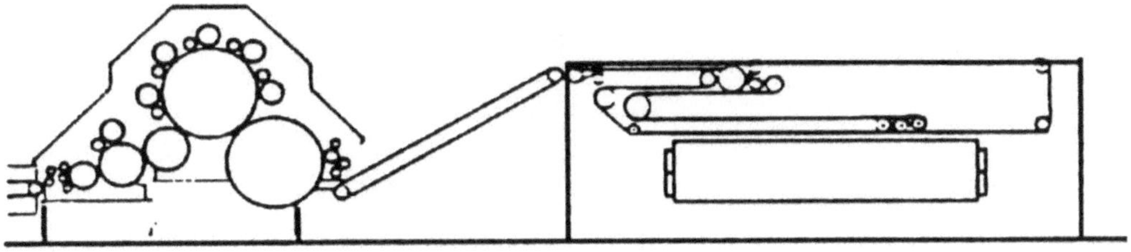

**Figure 5.7** Card and Cross Lapper. *Source: Reproduced by permission of TAPPI Press Atlanta, Georgia USA ((24) p. 163).*

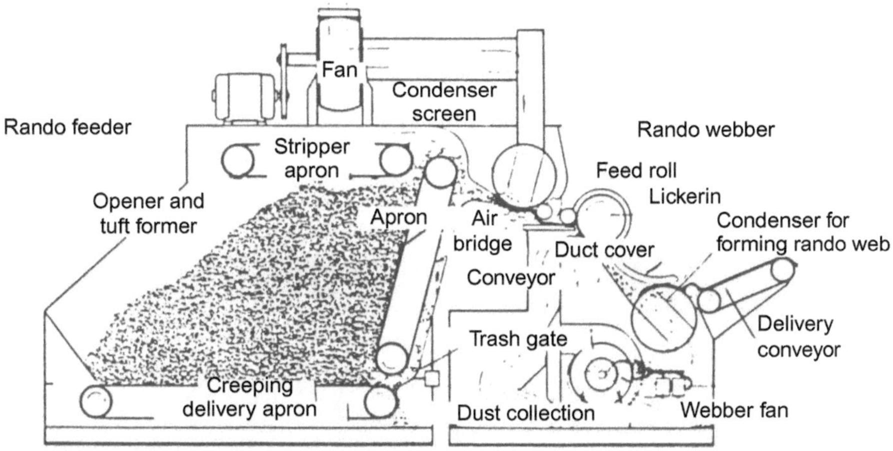


**Figure 5.8** Rando Webber. *Source: Reproduced by permission of TAPPI Press Atlanta, Georgia USA ((24), pg. 164).*

### *5.1.2.2 Bale opening, feeding, and blending*

A number of steps are necessary to process fibers on their way to the forming process. This includes bale opening, blending, and feeding. Figure 5.9 is a diagram of the head of a bale opening device called an OPTIMIX® bale opener. It traverses over the top of a letdown of bales of fiber and picks up the fibers with two opening rolls that create an air vacuum. The picked fibers are aerodynamically guided to a pneumatic plenum and channeled to a blending hopper.

The blending hopper feeds to a blend roll which opens the tufts of fiber and aerodynamically feeds them to a fine opener. The blend roll diagrammed in Figure 5.10 is a Kirschner beater type that has a three lag pin beater. There are also roll type beater rolls, which are fully metallic clothed.

The fine opener is depicted in Figure 5.11. This is a micro-tuft opener (MTO) that uses carding roll principles in further reducing the tuft size. The carding occurs

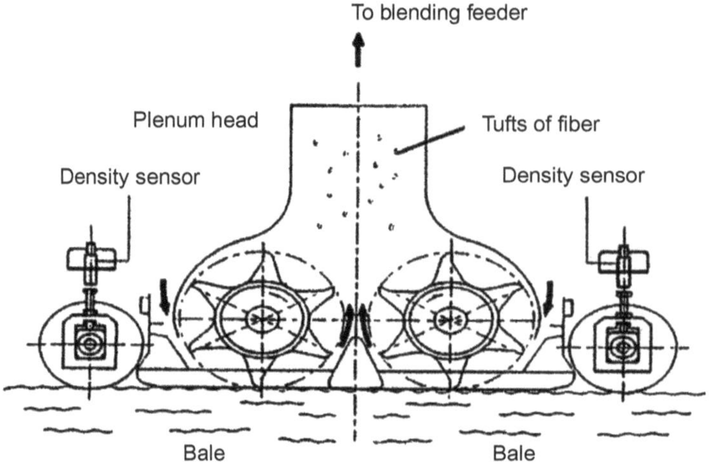


**Figure 5.9** OPTIMIX® Head. *Source: Reproduced by permission of TAPPI Press Atlanta, Georgia USA ((24), pg. 156).*

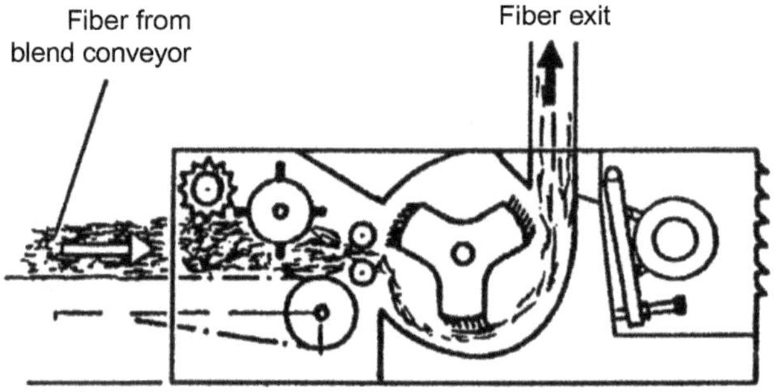


**Figure 5.10** Blend Roll. *Source: Reproduced by permission of TAPPI Press Atlanta, Georgia USA ((24), pg. 156).*

between rolls A and B and between rolls B and C, all of which are metallic clothed. Depending on the degree of tuft size reduction required, more openers may be added in line.

Once the fibers are satisfactorily opened, they are fed to the former. There are two types of feeding devices: a chute feeder and a hopper feeder. A chute feeder called the MASTERCHUTE® is depicted in Figure 5.12. Opened fibers blown into the transition, located at the top of the chute are fed into a feed roll and feed plate arrangement. This arrangement meters the fiber into the nip point of the pin beater roll and feed plate, which provides for gentle opening of the fiber. Air introduced at this point carries the fiber down into the forming chamber of the chute. A spanker plate at this point helps to distribute the fiber uniformly and evenly in both the MD and CD directions. A feed mat of uniform consistency is created to feed the web former.

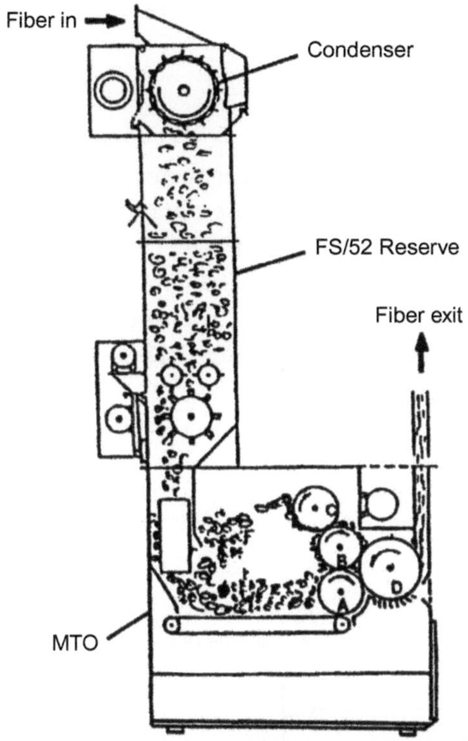


**Figure 5.11** Micro-Tuft Opener, FS/52, KD Condenser. *Source: Reproduced by permission of TAPPI Press Atlanta, Georgia USA ((24), p. 157).*

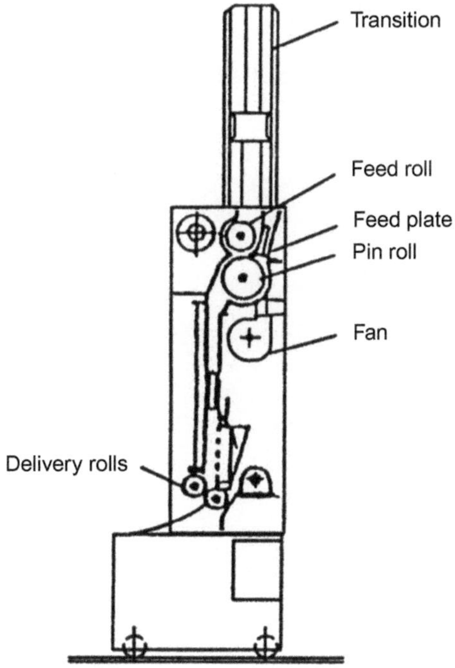


**Figure 5.12** MASTERCHUTE®. *Source: Reproduced by permission of TAPPI Press Atlanta, Georgia USA ((24), pg. 158).*

### *5.1.2.3 Bonding of dry laid webs*

#### 5.1.2.3.1 Needle punch

Needle punch webs are the most dominant form of nonwoven used as filter media. They are a prominent medium for baghouse filtration operations. Needle punching is a process by which fibers are entangled and mechanically interlocked by puncturing the web with a series of barbed needles. The density and strength of the web can be regulated by the strokes per minute of the web, the advance rate of the web, and the degree of penetration of the needles. Figure 5.13 is a diagram of the needle felting operation and of a typical needle.

Figure 5.14 is a side-view photograph of a needle felt web. The locations of the needle penetrations are quite visible.

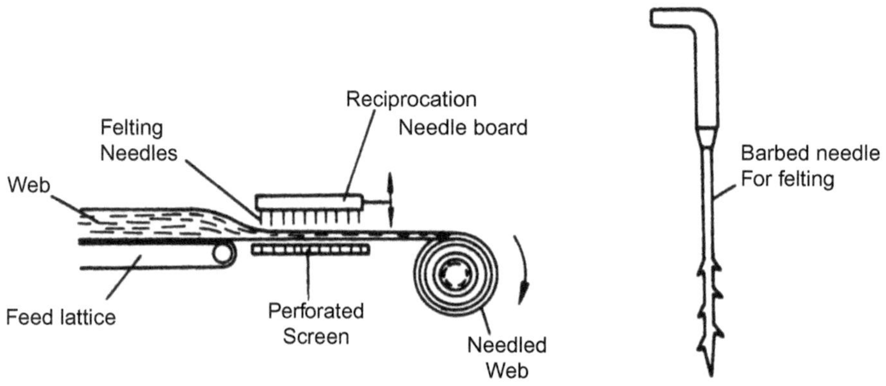


**Figure 5.13** Needle Felting and an Example of a Barbed Needle. *Source: Reproduced by permission of reifenhauser REICOFIL, GmbH & Co, KG Machine Fabric, Troisdorf, Germany.*

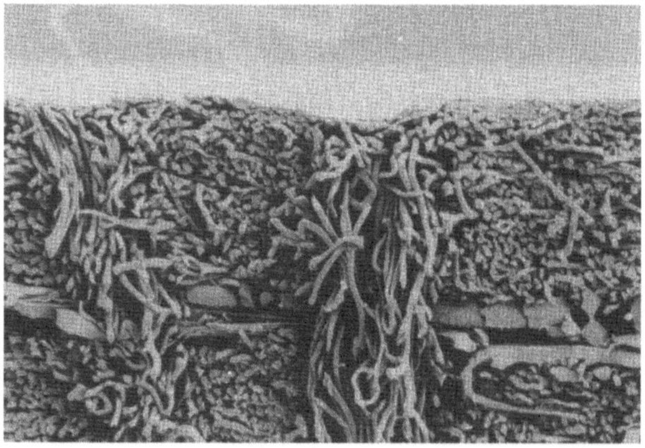

**Figure 5.14** Cross Section Through a Needle Felt, Showing the Scrim and also the Fiber Re-Orientation Caused by the Needling, at X62 Magnification. *Reproduced by permission of Elsevier Advanced Technology ((18), p. 87).*

Crook (24) describes 5 components to a needle loom.

1. Feed apron—carries the formed web to the needle punching operation.
2. Bed Plate—supports the web during the needling action. The bed plate has holes that match the needle pattern.
3. Stripper—strips the fiber off the needles if tagging occurs when the needle is withdrawn from the felted web. The stripper is a plate aligned with the bed plate. It also has holes that match the needle pattern.
4. Needle beam—houses the needle board and moves up and down to perform the needling action.
5. Draw rolls—control the advance rate of the needled fabric in relationship to the strokes of the needle board. These rolls are at the delivery point of the needled fabric.

#### 5.1.2.3.2 Spunlace (hydroentanglement)

Spunlace is the generic terminology for hydroentanglement. The process was first introduced by DuPont in 1973 along with the spunlaced product, tradenamed Sontara®. Figure 5.15 (Israel et al, (207)) is a general schematic of the process. The process uses very fine high velocity jets of water instead of barbed needles to entangle the fibers. The jets are applied from high pressure headers. The header pressures generally range from 0.5 to 25 MPa. The jet holes in the headers range from 100–120 μm in diameter. The jets are arranged in rows across the header. Spacing between rolls are 3–5 mm. Along each row there may be 30–80 holes per 25 cm.

The process for hydroentanglement involves the following steps:

1. The formation of a precursor nonwoven web.
2. Entanglement of the web with water jets.

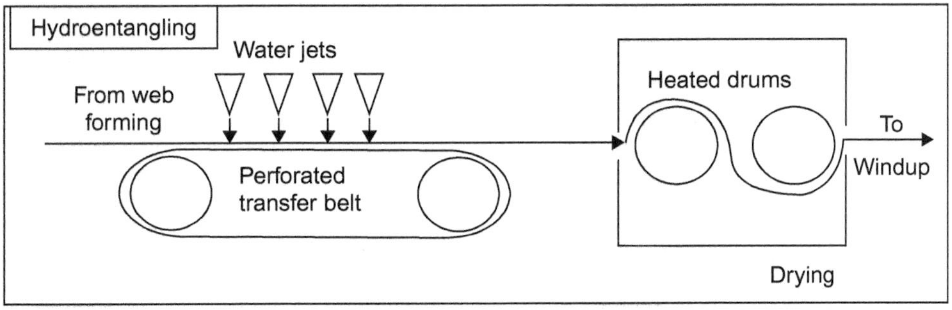


**Figure 5.15** The Spunlace (Hydroentangling) Process. *Source: Reproduced by permission of Donaldson, Inc. Minneapolis, Minnesota USA (207).*

3. A vacuum system to remove excess water.
4. Recirculation and filtration of the water.
5. Drying of the web.
6. Wind-up of the web into rolls for shipment or further processing.

The webs may be hydroentangled from one side or two sides. This usually is a stepwise, 2-stage process where the web is first entangled on one side and then entangled from the other side. Important to the product is the woven support fabric. The design of the fabric will affect where and how the web is hydroentangled. The mesh and structure of this fabric will affect the appearance and physical properties of the web. For example, a coarse mesh support fabric may result in a hydroentangled web patterned with a number of large holes whereas a fine mesh fabric may result in a larger number of smaller holes. It should be noted that the holes in the web are not formed at the holes in the support fabric, but rather at the knuckles created by the fiber intersections of the support fabric.

Other factors that may affect the properties of the hydroentangled web are jet velocity and pressure, web speed, the number of manifolds or headers, and the angle of the jets as they strike the web. The best web properties are not necessarily achieved from a vertical jet strike. Better properties may result if the jets are directed at some angle from the vertical when they strike the web.

Properties of the fiber in the web will affect how well they entangle. Generally soft longer fibers will entangle better than rigid shorter fibers. Thin fibers will entangle better than coarse fibers. The properties of the web will affect the entanglement performance. Basis weight, air permeability, and fiber orientation are factors that may affect how the web is entangled. Usually the best webs for hydroentanglement will have a grammage of 25–250 g/m$^2$ and composed of fibers in the range of 1.5–3.3 de.

The water recirculation and filtration system is critical to process performance. Depending on jet velocity and the types of fibers used, there will be some tendency for small fiber segments to break off as "fines." The fines are in the micron size category and could easily cause plugging of the water jet holes. A very efficient filtration system is required. It can be quite surprising as to how much fines may be generated. If the filtration system is not well designed, early plugging and short filter life will occur.

The sketch of Figure 5.15 indicates a flat hydroentanglement area. There are units in the market place based on a round configuration. Figure 5.16 illustrates a unit based on honeycomb hydroentanglement rolls. It is a diagram of a honeycomb hydroentanglement unit (207) produced by Valmet Inc. Honeycomb Division, Biddeford, Maine, USA.

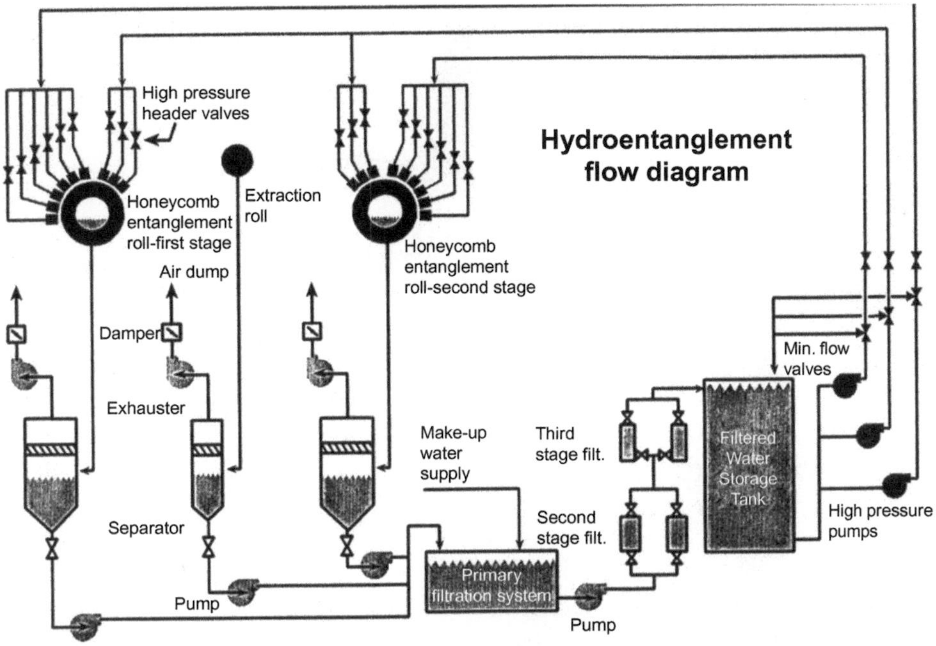


**Figure 5.16** Hydroentanglement with Honeycomb Rolls (208). *Source: Reproduced with permission of Valmet Inc. Honeycomb Division, Biddeford, Maine, USA.*

### 5.1.2.3.3 Stitch bonding

Stitch bonding is a mechanical method of consolidating fiber webs. It uses knitting elements with or without yarns or sewing threads to interlock the fibers. Usually a fleece is made by cross lapping the web from conventional cards. The fleece is then fed through the stitch area of the stitch bonding machine. Warp threads of either man-made or staple spun yarns are needle knitted through the web to consolidate it into a nonwoven fabric (see Figure 5.17). Stitch bonding can also be accomplished by meshing the fibers. Quilted structures are often produced by stitch bonding.

Sächsisches Textil Forschungs Institut (STFI) e. V., Chemnitz, Germany markets Hycoknit®, a stitch-bonded nonwoven composite, as a depth filter medium. Schmalz et al. ((209) and (210)) reports that the product is made in two steps. In the first step a pile stitch-bonded nonwoven with a one loop surface is manufactured. In the second step the one loop surface of pile stitch-bonded nonwoven is compressed by putting a hydroentangled nonwoven on it. The result is a gradient density medium as shown in Figure 5.18. The product is unique in that the fibers of the upper pile surface are oriented in the direction of flow. This is in contrast to conventional filter medium where the fibers are generally perpendicular to the direction of flow.

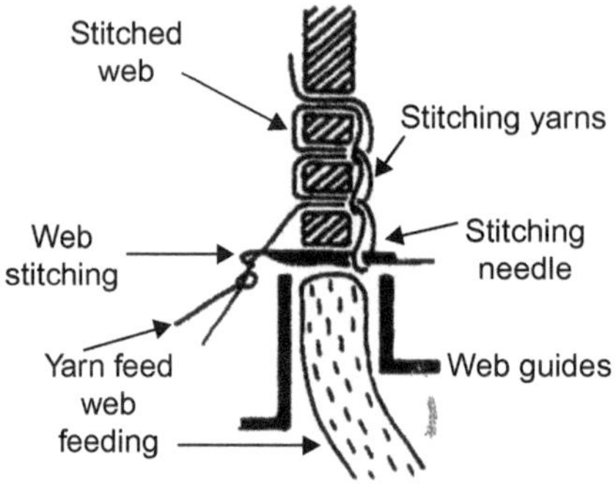


**Figure 5.17** Stitch Bonding. *Source: Reproduced by pernission of INDA, Cary, North Carolina USA ((1) (2), pg 62).*

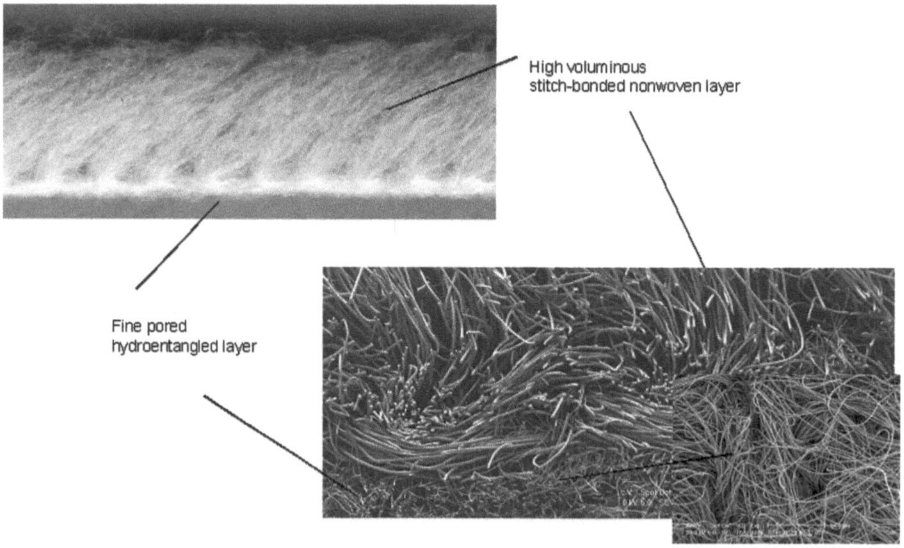


**Figure 5.18** Stitch-Bonded-Hydroentangled Nonwoven Composite Hycoknit®.[2] *Source: Reproduced by permission of Sächsisches Textil Forschungs Institut e.V., Chemnitz, Germany.*

### 5.1.2.3.4 Thermal bonding

Webs containing low melting thermoplastic material can be thermally bonded to consolidate the web. Usually the low melting material is fibers or bicomponent fibers, however, thermoplastic powders are used in some applications. Following are some forms of thermal bonding:

1. Calendering or embossing—The heated web is squeezed between two rolls (either smooth or patterned) which forces intimate contact of the molten fibers with

[2] Photographs and layout provided by Dr-Ing. Elke Schmalz of STFI.

other fibers in the web. If the rolls are smooth (calendering), the web is uniformly bonded. If the rolls are patterned, an embossed web results where bonding is most intense in the dense areas of the embossed structure. Embossing results in a softer web. It is also a way of providing spot bonding to the web.

2. Through-air bonding—The web is thermally bonded by passing the web through a through-air dryer or heater. This type of thermal bonding results in a bulkier and loftier web. High loft webs used in HVAC filters are often bonded in this manner.
3. Ultrasonic bonding—This relatively new technology utilizes high intensity ultrasonic energy to melt and form bonds from thermoplastic fiber. The ultrasonic bonding is like stitch bonding in that it takes place along narrow lines in the web. It is also a way of producing quilted structures.
4. Hot melt bonding—A hot melt thermoplastic is extruded onto the web and while still in the molten stage, penetrates and bonds the web.

#### 5.1.2.3.5 Chemical bonding

Chemical bonding involves the use of a chemical binder or resin to bond the web. It should be noted that the author will use the term "resin" as a generic for all chemical binders. There are two basic forms of saturating resins: water-based and solvent-based. Water based resins are the ones most frequently applied to dry laid nonwovens. They include latex binders, a form of resin emulsified in water, resin dispersions, and solution resins. Latex is the most commonly used resin binder for dry laid webs. Solvent-based resins are more likely to be used in wet laid products (see Section 5.6.3). It is implicit in a wet resin-bonding process, that a drying step follows the application of wet resin to the web. In addition, heat to cure or thermoset the resin may follow the drying step. Following are ways of applying the latex:

1. Saturation—To saturate a web with latex binder, it is usually applied from a padder, which uses an applicator roll to apply resin from a resin bath. One form of padder is illustrated in Figure 5.19 below. In this simple mode an applicator roll

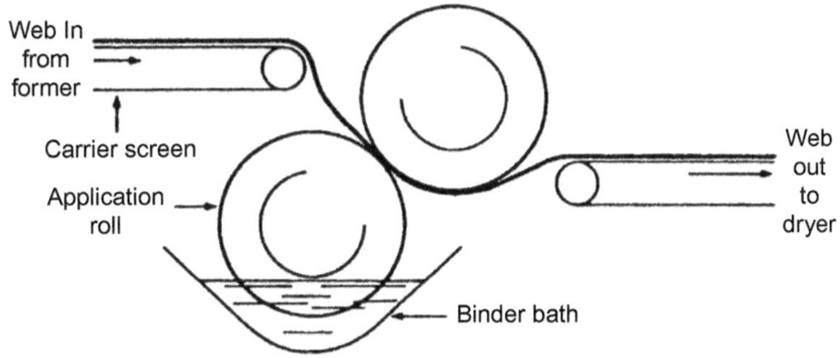


**Figure 5.19** Padder for Web Saturation with Latex Binder. *Source: Reproduced by permission of TAPPI Press ((219) p. 69), Atlalnta Georgia USA.*

passes through the bath and coats the resin unto the web as it passes through the squeeze rolls. The squeeze rolls force the resin to the interior of the web and helps achieve uniform depth bonding.

Another form of saturator is an impregnator, which immerses the web into the resin bath prior to the squeeze rolls. The squeeze rolls serve to remove excess resin from the web. This method of application assures a uniform resin distribution through the depth of the medium. However, the web pickup is far greater than other application methods such as padding. This forces a more dilute resin mix to achieve the desired web resin content. The more water, that has to be removed in the drying process, results in higher energy costs and possible lower production rates.

Saturators used for the impregnation of wet lay products are often referred to as size presses. They are discussed in more detail in Section 5.6.3.

2. Spray bonding—Spray nozzles are used to spray chemical binder unto the web. Spray bonding is an effective way of bonding very thin webs, or for low strength webs where bulk is desired, however, for thick webs the process becomes inefficient because of the difficulty of the sprayed resin droplets to penetrate into the interior of the web.
3. Gravure bonding—A gravure press is used to apply liquid binder to the web. The press roll is patterned to hold a fixed volume of liquid that is applied to the surface of the web. For saturated products, squeezing and resin migration into the web is required. In many applications two gravure rolls are necessary: one to apply resin to the bottom surface of the web and the other to apply it to the top surface.
4. Spot bonding—Gravure presses and rotary screen printers are used to apply resin to specific areas of the web and leave other areas unbonded. It is a way of providing strength to a nonwoven structure and still leave a bulky or lofty structure between bond points. The process is most often used for wipes, however it has application for medium used in filter bag applications.
5. Froth binder application (foam bonding)—Surfactants are applied to a latex so that it can be agitated to a stable froth or foam. The latex is applied as foam to the medium. One method of applying the foam is from a padder similar to the one depicted in Figure 5.19 above. Another method of application is to apply the foam from an applicator to the top side of the web and use a vacuum suction device from underneath the web to pull the foamed latex into and through the thickness of the web. The vacuum force also serves to collapse the foam bubbles.
6. Curtain coating—In this method of application the resin overflows from a reservoir and down an inclined flat surface. It then flows unto the web as a thin film or curtain. For web saturation, vacuum suction devices underneath the web may be necessary to pull the resin through the web. The process schematic is shown in Figure 5.20.

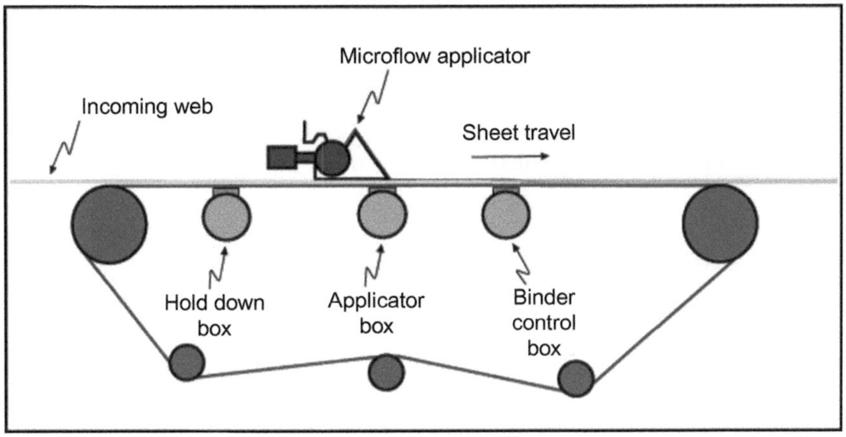


**Figure 5.20** Curtain Coater Schematic.

### 5.1.2.3.6 Wet mill process

Felts are probably the oldest form of nonwoven materials. There are numerous historical references to felts including the bible. Wet milling is one of several methods which can produce felt from wool and other animal fibres. In modern times synthetic fibers may be blended in. Often a carding machine is used to produce layers of felted material. Warm soapy water is applied to several layers placed at 90 degree angles to one another. Repeated agitation and compression causes the fibres to hook together into a single piece of fabric. After the wet felting process is complete, the felted material may be finished by fulling. Fulling is the process of cleansing, shrinking, and thickening felt with heat, pressure, and moisture.

Only certain types of fiber can be wet felted successfully. Most types of fleece, such as those taken from the alpaca or the Merino sheep, can be put through the wet felting process. These types of fibre are covered with tiny scales, similar to the scales found on a strand of human hair. Wetting and soaping the fleece causes the scales to open, while agitating them causes them to latch onto each other, creating felt. Plant fibres and synthetic fibres will not felt, however, they are often blended with materials that do felt. In modern times carded materials composed of synthetic fibers are often identified as felts.

## 5.2 MELTSPUN WEBS

Meltspun webs are produced from molten polymer forced through thousands of tiny holes in a spinnerette to emerge as a spray of continuous molten fiber. Convergence with one or more air streams cools the fiber streams, attenuates them, randomizes them, and forces them to become a random fiber web when they land on the collector. There are two predominant processes for producing meltspun webs: spunbond and melt-blown. Schematically, the processes for both product types look similar. However the webs produced have much different properties. Much of it due to the way the air streams converge around the fiber sprays out of the spinnerette. Following is a discussion of each of the processes.

### 5.2.1 Spunbonded webs[3]

Jirsák and Wadsworth (211) defines the spunbond process as consisting of the following integrated operations:

1. Polymer melting, transportation of polymer melt, and filtration of polymer melt
2. Filament extrusion
3. Filament drawing
4. Filament deposition
5. Bonding

The types of polymers used in the spunbond process include polyester, polypropylene, nylon, and any other polymer that can be considered as a fiber forming polymer. Flash spinning is a variation of the spunbond process in that the polymer is spun from a solution rather than a hot melt. High molecular weight polyethylene is used in this process to produce the spunbond known as Tyvek® that is manufactured by DuPont.

A schematic of the Reicofil® system developed by Reifenhauser GmbH of Germany is presented in Figure 5.21. Molten polymer is extruded and fed by a spin pump to a die block. One of the components of the die block or die assembly is the spinnerette. The spinnerette is a block of metal having several thousand drilled holes. Molten polymer forced through the spinnerette is formed into filament fibers. The spun fibers are rapidly cooled and drawn by a succession of air streams through a venturi. Downstream of the venturi the filaments pass through a distributing chamber which effects fanning and entangling of the filaments. The randomness of the web is greatly affected by turbulence of the air in the distributing chamber. The entangled fibers are then deposited by suction unto a moving foraminous belt. The movement of the belt and the suction below the belt enhances the random lay down of the web. The formed web then passes through a calender to achieve bonding.

Figures 5.22, 2A through 2O are a pictorial display illustrating its operation.

The process begins with polymer, in either powder, granular, or pellet form, that is pneumatically conveyed from its container through a long hose into the dosing

[3] A thorough discussion of both the spunbond and melt-blown processes by Malkan and Wadsorth (25) is found in "Nonwovens: Theory, Process, Performance, and Testing." Albin F. Turbak editor, TAPPI Press, 1993 Chapter 8: Polymer-laid Systems, Page 171. In addition, detailed discussions of both processes can be found in "Introduction to Nonwoven Technology" by Batra and Pourdeyhimi (16) DEStech Publications, Inc. 2012.

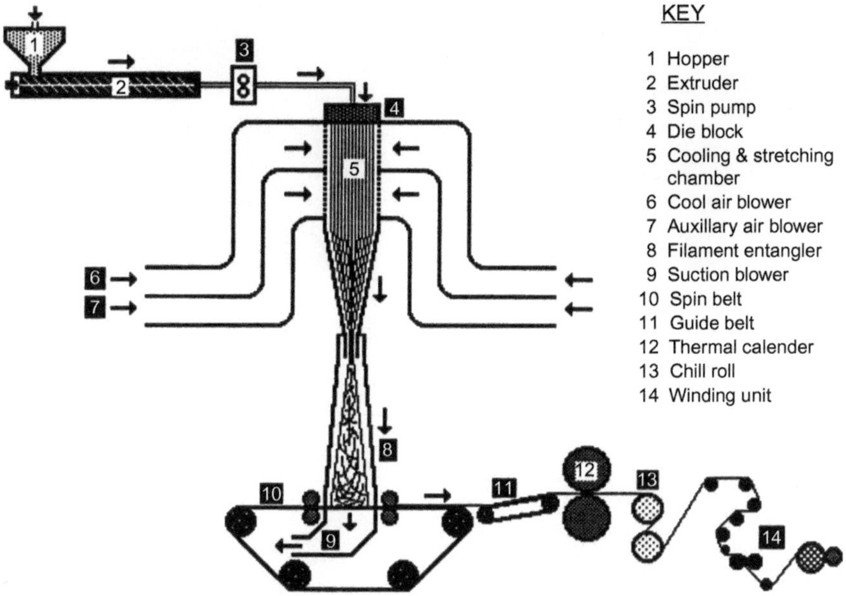


**Figure 5.21** Reicofil® Spunbond Process. *Source: Reproduced by permission of Reifenhäuser REICOFIL GmbH, Germany.*

system vessel of Figure 5.22A. The vessel is an air—solid separator. The pneumatic air carrying the polymer is exhausted to a vacuum pump through a tube at the top of the vessel. The heavier polymer gravity settles to the bottom of the vessel and into the hopper depicted in Figure 5.22B. The hopper is a water-cooled vessel that feeds the polymer into the throat of the extruder. After a run, the 45° tube that can be seen connecting to the right side of the hopper is used to feed hot air into the hopper and clean out any residual polymer. The extruder, depicted in Figure 5.22C is diagrammed in Figure 5.23. Inside is a rotating screw shaft that forces the polymer through four heating zones. The polymer is in a molten state by the time it reaches the filter screen inside the screen changer of Figure 5.22D. The filter screen removes unmelted contaminants from the molten liquid polymer. The molten polymer is then forced into the die block by the gear pump (often referred to as a spin pump) of Figure 5.22E. Figure 5.24 below is a schematic showing how the gear pump works.

The die block has two critical components: the polymer feed distributor and the spinnerette. The purpose of the distributor is to provide for a uniform feed of polymer across the entire block. The distributor balances both flow and residence time across the width of the die. Two types of distributors that have been employed in the spunbond process are T-type (tapered and untapered) and the coat hanger-type. The coat hanger-type is referred to as a coat hanger, because the parabolic shape of

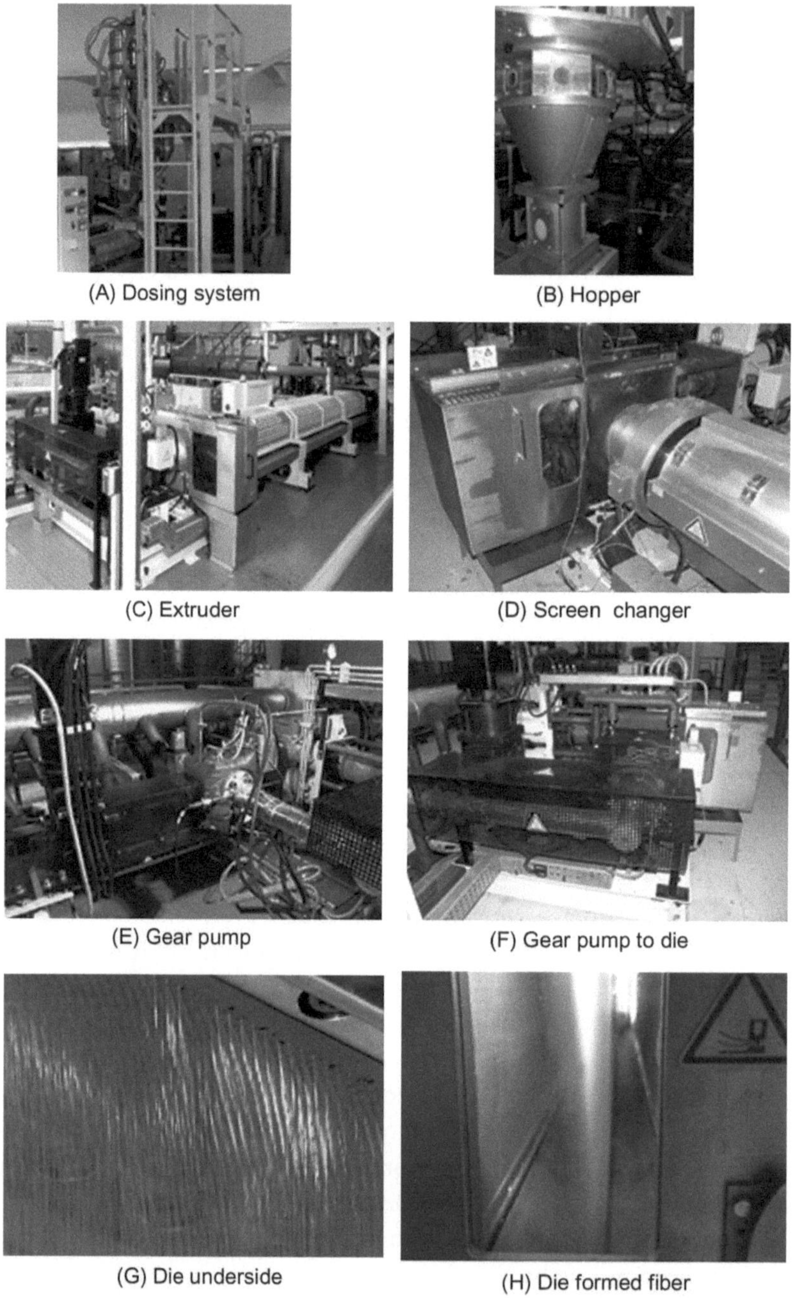


**Figure 5.22** Parts of a Reicofil® Spunbond Line. *Source: Reproduced by permission of Reifenhäuser REICOFIL GmbH, Germany.*

(I) Quench

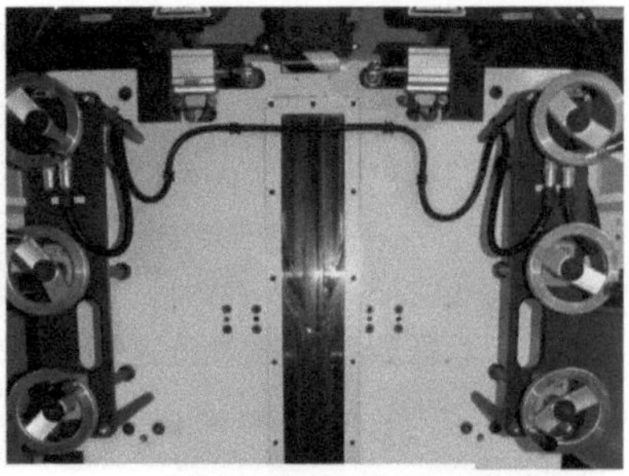

(J) Venturi entrance

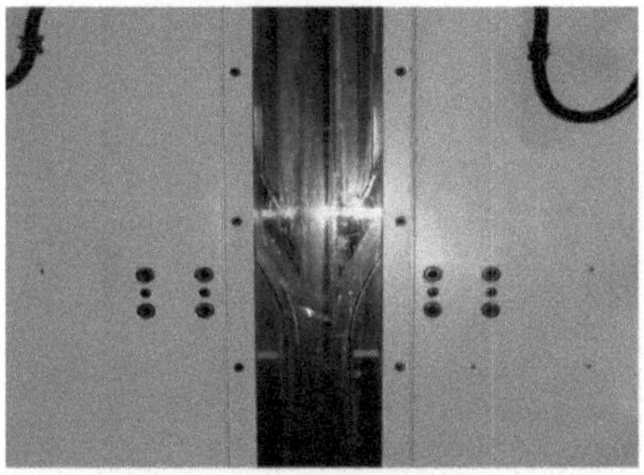

(K) Venturi close-up

(L) Contact roll

(M) Calender

(N) Chiller roll

(O) Winder

**Figure 5.22** Continued

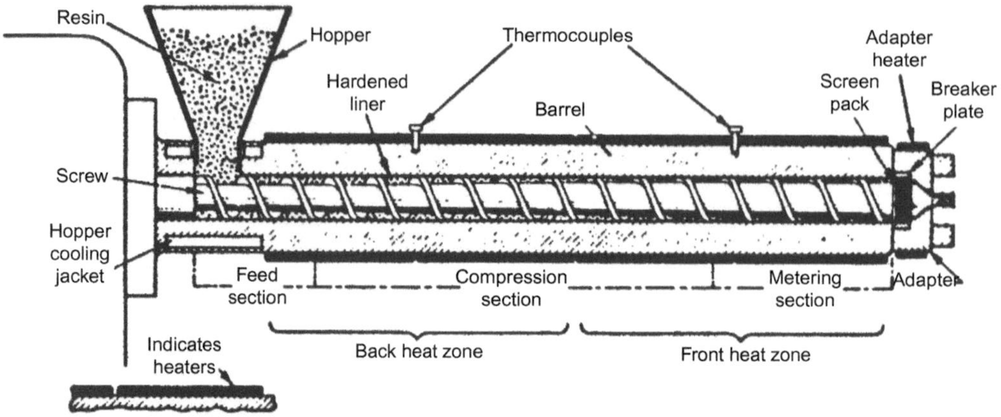


**Figure 5.23** Schematic of an Extruder. *Source: Reproduced by permission of Routledge/Taylor and Francis Group, LLC-Copyright© from "Principles of Polymer Systems" by E. Rodriguez (212).*

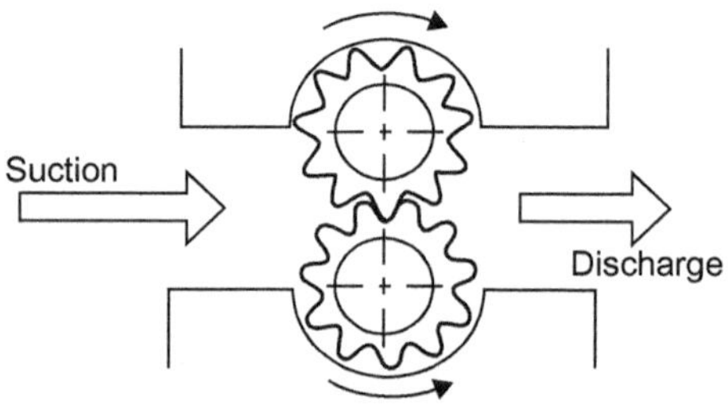


**Figure 5.24** Schematic of a Gear Pump. *Source: Reproduced by permission of TAPPI Press Atlanta, Georgia USA, ((25) p. 173).*

the wall around the distributor chamber resembles a coat hanger. Figure 5.29 below is a photograph of a coat hanger distributor used in a melt-blown line.

The spinnerette is the underside of the die block as depicted in Figure 5.22G. It is a block of metal having 4036 holes, 0.6 mm in diameter. The holes are usually circular, however they can also be rectangular. The molten polymer forced through the spinnerette is formed into filament fibers. The filaments then travel through a long, vertical 30-foot enclosed chamber. Airflow is forced through the chamber by a vacuum suction device underneath the web collection belt at the bottom or the chamber. A lot happens to the fibers as they travel through the chamber. First they are rapidly cooled by a cross flow of cold quenching air. The cross flow of air also creates a disorientation and randomization of the filaments. This causes the rather spectacular light effect shown in Figure 5.22H. This is visible through the lighted window of the chamber. There are two air-cooling zones. The first is the primary zone, just below the spinnerette, that quenches the filaments with conditioned air. Below the primary zone is an auxiliary zone that blows a stream or room temperature air (Figure 5.22I).

**Figure 5.25** Spunbonded Line Venturi.[4] *Source: Reproduced by permission of University of Tennessee Nonwovens Research Laboratory, University of Tennessee, Knoxville, Tennessee USA.*

Approximately 9 ft below the quench zones, the filaments are sucked through a pneumatic venturi that accelerates the filament flow and draws the fibers. The venturi chamber is depicted in Figures 5.22J and K. A clearer picture of the venturi structure is shown in Figure 5.25. The venturi consists of steel plates, 0.76 meters long, across the machine. Figures 5.22J and K only show the edges of these plates. The rotary wheel handles depicted in Figure 5.22J are used to make fine adjustments to the plates. In operation, the filament fiber coming down from the spinnerette is funneled through the converging plates of the venturi shown in Figure 5.25 and necked into the narrowed gap between the plates. On either side of the converging plates is a second set of converging plates forming an annulus through which air is introduced. The air merges with the filament flow and carries it into the distribution chamber beneath.

The air in the distribution chamber becomes turbulent and causes fanning and entangling of the filaments. The randomness of the web is greatly affected by the turbulence. The entangled fibers are then suction deposited unto a moving foraminous belt collector to form the web. The web-forming process is shown in Figure 5.22. What is pictured is the belt moving left to right between two support rollers. The entangled filaments lay down on the belt more or less vertically. They then leave the web-forming zone with the belt and pass underneath the contact roll on the right.

[4] Photograph provided by Chris Eash.

Figure 5.22L shows the exit side of the contact roll. The roll serves to consolidate the web and provide a seal to the negative-pressure forming zone.

The formed web is passed through heated calendar rolls to densify and bond it (Figure 5.22M). The heated web is transferred around a chiller roll (Figure 5.22N) and then to the winder (Figure 5.22O).

In addition to the Reicofil® system described above, there are other producers or licensors of spunbond lines. Lurgi Kohle & Mineralotechnik GmbH of Germany developed the Docan® system which like the Reicofol® system is based on melt spinning. Another melt spinning line is the Lutravil® system developed by Carl Freudenberg Co. of Germany.

## 5.2.2 Melt-blown webs

A schematic of the melt-blown process is diagrammed in Figure 5.26 below. The line as diagrammed is blowing the fibers in a horizontal path to the collector. Commercial lines with vertical flow are also in operation.

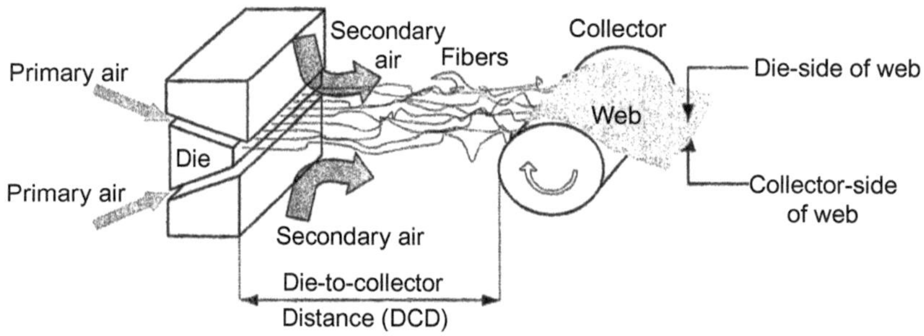


**Figure 5.26** The Melt-blowing Process[5]. *Source: Reproduced with permission of Dr. Randall R. Bresee, (213), University of Tennessee, Knoxville, Tennessee USA.*

The key difference between the spunbonded process and melt-blowing is in the die assembly. In the melt-blown process hot air converges with the fiber as it emerges from the die, whereas in the spunbond process the hot air flow is at a cross flow to the emerging fiber. The converging flow of the melt-blown process, diagrammed in Figure 5.27, serves to attenuate and draw the fibers so that the resulting web is composed of finer fibers than the fibers of spunbonded webs. The melt-blown web is softer, bulkier, and weaker. It has a smaller pore size and provides for better filtration efficiency. In most filter applications the medium has to be supported by another web, or used as part of a composite structure.

[5] First published by INDA in "Influence of Processing Conditions on Melt-blown Web Structure", Bresee, R.R., Qreshi, U.A., and Pelham, M.C., International Nonwovens Technical Conference (INTC), Conference Proceedings, Toronto, Canada, September 20–23, 2004.

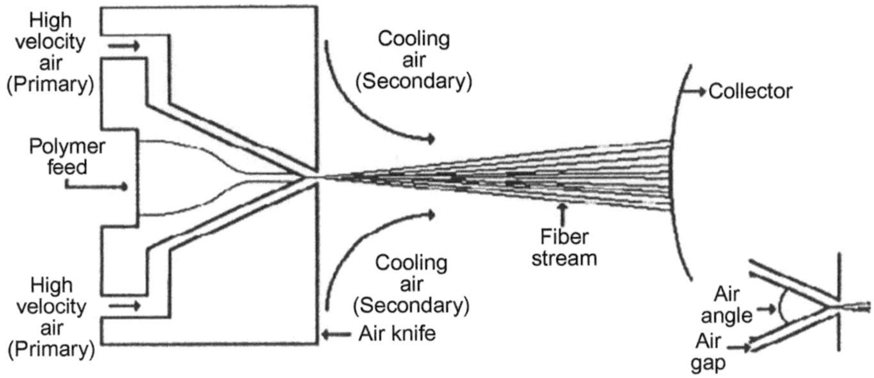


**Figure 5.27** Airflows in the Melt-Blown Process. *Source: Reproduced by permission of TAPPI Press Atlanta, Georgia USA ((25) p. 183).*

The collage of Figure 5.28 is a photo display of the 6-inch pilot melt blowing line at the University of Tennessee Nonwoven Research Laboratory (UTNRL), Knoxville, Tennessee, USA. The process starts with polymer in the form of pallets, granules, powder that is poured into the hopper of Figure 5.28A. From the hopper the polymer gravity drops into the feed throat of the extruder. Inside the extruder is a rotating screw shaft that forces the polymer through three heating zones that melts the polymer. The extruder is shown in Figure 5.28B. The operation of the extruder is similar to the one diagrammed in the schematic of Figure 5.22 above. In the case of the UTNRL pilot machine, the extruder provides sufficient pressure to force the molten polymer flow through the screen, through the die assembly and as fiber unto the collector. The screen changer is shown at the end of the extruder in Figure 5.28C. From the screen changer the polymer feeds into the back of the 6-inch die assembly, also shown in Figure 5.28C. The screens serve to separate solid contaminants and unmelted chunks from the molten polymer. There are two screens, so that when one becomes filled, it can hydraulically be removed from the polymer flow stream and the other one inserted in place.

The distributor to the die assembly is a parabolic coat hanger type similar to the one used in the spunbond line discussed above. The parabolic shape is evident in Figure 5.29, a photograph of the coat hanger taken during line disassembly for cleaning. Polymer enters from the feed pipe, at the top of the device, and exits through the melt-blown die at the bottom. The coat hanger configuration provides for even polymer distribution across the die.

Figure 5.28D shows the front of the die assembly where the fibers emerge. On both sides of the die, insulated piping can be seen. The insulated piping feeds hot air into two manifolds which are the pipe-like structures seen above and below the die opening. The manifolds regulate the high velocity streams of hot air that converge

**Figure 5.28** 6-Inch Meltblown Line at UTNRL. *Source: Reproduced by permission of University of Tennessee Nonwovens Research Laboratory, University of Tennessee, Knoxville, Tennessee USA.*

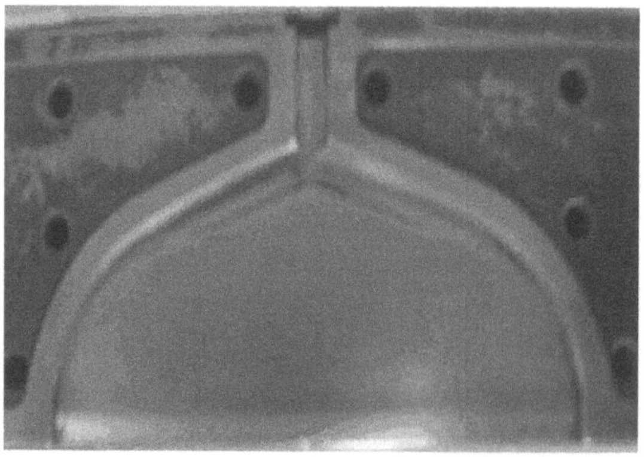

**Figure 5.29** Coat Hanger Distributor. *Source: Reproduced by permission of UTNRL University of Tennessee, Knoxville, Tennessee USA.*

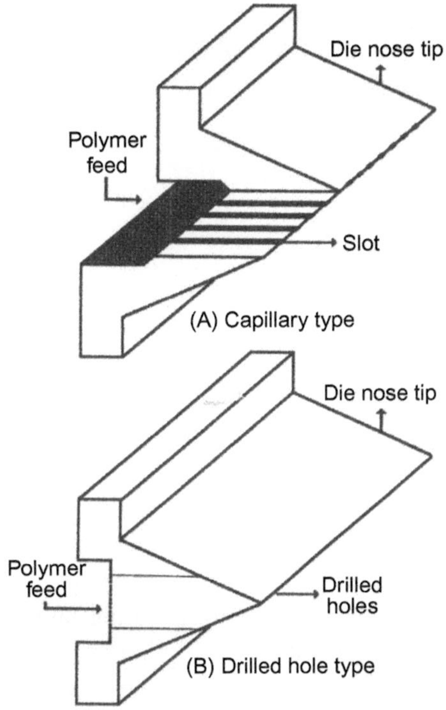


**Figure 5.30** Schematic of Die Nosepieces for Melt-Blown Lines. *Source: Reproduced by permission of TAPPI Press Atlanta, Georgia USA (25), p. 183).*

with the forming fibers as shown in Figure 5.24 above. Typical hot air temperatures are 215 °C to 340 °C. The air velocities through the air channels above and below the polymer feed die openings range from 0.5−0.8 times the speed of sound.

There are two types of die nosepieces: capillary type and drilled hole type. These are pictured in Figure 5.30 above. Capillary openings are actually slots that are milled into both sides of the flat mating surfaces that form the die nosepiece. This is shown in Figure 5.30A above. Figure 5.30B above illustrates the drilled hole types. The die holes are either mechanically drilled or drilled by electrical discharge machinery.

During processing, the entire die assembly is heated to 215 °C to 340 °C in order to produce uniform defect-free webs. Figure 5.28E shows the newly formed fiber emerging from the die on its way to the collector (Figure 5.28F). The fiber emerging from the die is traveling from left to tight whereas the collector is pictured from the other side of the machine so that the fiber approaching it in Figure 5.28F is coming from the right. The metal drum collector in Figure 5.28F, is rotating counterclockwise.

The melt-blown process is a major process for producing nonwoven filter media. The major development areas for the process are webs with finer fibers for greater filtration efficiency and webs containing bicomponent and multicomponent fibers. With respect to the latter, Hills Inc. of West Melbourne, Florida has developed proprietary melt-blowing technology to produce fibers with the following cross sections:

- Homopolymer
- Side by side
- Sheath/core
- Pie segments
- Islands in a sea

A detailed discussion of bicomponent and multicomponent fibers is presented in Chapter 4, Section 4.3.5.6.

## 5.3 ELECTROSPUN WEBS

In 1934, Anton Formhals (26) was the first inventor to patent an electrospinning phenomenon. Donaldson is one of the first companies to commercialize electospun webs for filtration purposes. Their Spider-Web® electrospun composites are directed toward gas turbine filtration, but are finding other applications in air filtration. Electrospinning is a process that produces nanofiber webs by applying a high voltage charge to a polymer solution or melt and using the charge to draw the polymer to a grounded collector. Voltages range from 5 to 30 kV, sufficient to overcome the surface tension forces of the polymer. The free surface of the charged polymer produces very fine jets of liquid that are rapidly drawn to a grounded collector. The highly attenuated fibers collect on a moving surface passing over the grounded collector and form an interconnected web of small filaments. The fiber diameters for filtration purposes are in the range of 0.25 μm. Since the webs are of very fine thickness (on the order of one micron or less) and since they have limited mechanical properties, the moving surface on which they are collected is usually a substrate that will provide strength and stability for subsequent processing. Very often the substrate will be a nonwoven chosen specifically for the filtration application. In Chapter 3, Section 3.5.1, Figures 3.4 and 3.5 show electrospun nanofiber structures on a polyester substrate and on a cellulose structure, respectively.

There are four known processes for forming electrospun fiber webs:

- Electrospinning with a needle
- Electrospinning with a free surface electrode
- Rotation spinning
- Electroblowing.

Most electrospun products are spun from a polymer solution, however, it is possible to produce meltspun products produced from a molten polymer version of the electrospun process.

### 5.3.1 Electrospinning with a needle or nozzle

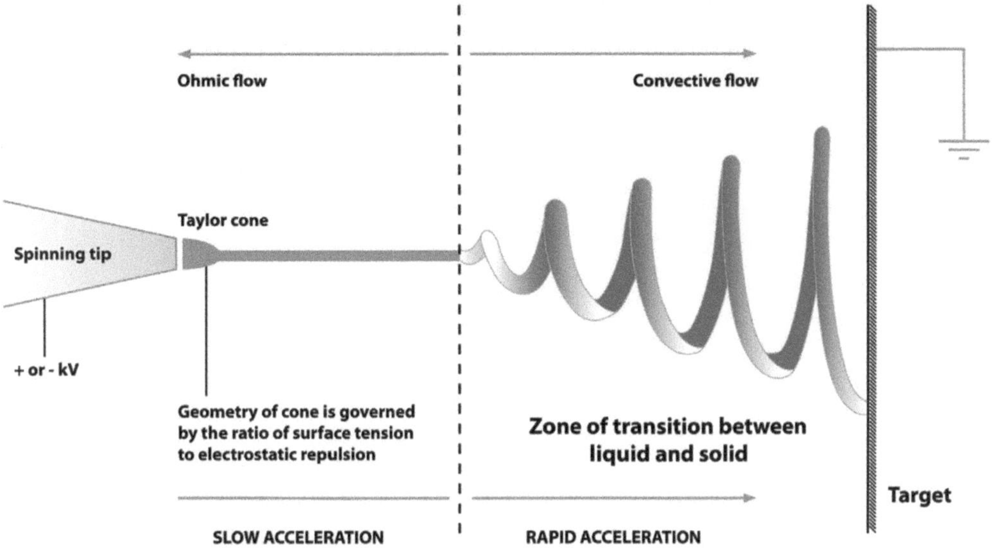


**Figure 5.31** Electrospun Nozzle Process Schematic.

Figure 5.31 above is a schematic of the "needle" (or nozzle) process. This is like the process schematic previously shown in Chapter 1, Section 1.7.1.2.3.1, Figure 1.18. The polymer solution is forced through a syringe-like needle or nozzle (also called a spinning tip) and emerges as a fiber stream. The effect is a result of the formation of a Taylor Cone. A Taylor Cone is like a small cone-shaped pimple that forms on the surface of the polymer solution at the tip of the nozzle under a high electric voltage charge. When the charge exceeds a certain voltage, the tip of the cone erupts and emits a fiber-like stream of polymer mass. The solvent is rapidly evaporated and the fiber stream becomes a very fine diameter filament. There is a point of bending instability in the path and the fiber path starts to rotate vigorously as it approaches the collector. This phenomenon benefits the formation of a random fiber web on the collector.

### 5.3.2 Rotation spinning

Elmarco, Inc. has developed a technology based on electrospun fibers from free surfaces. In contrast to needle or nozzle electrospinning, nanofibers can be spun from a high voltage charged surface coated with the polymer solution. Figure 5.32 shows this being done with a rotating cylinder. The cylinder rotates through a bath of the polymer solution to form a surface coating on the cylinder surface. The cylinder has a very high voltage charge and the voltage differential between the cylinder and the grounded collector forces the formation of many Taylor Cones, like pimples, on the

**Figure 5.32** Rotation Electrospinning of Nanofibers. *Source: Reproduced with permission of Elmarco s.r.o., Liberec, Czech Republic.*

surface of the polymer solution. When a certain voltage is reached the Taylor Cones erupt to emit a spray of very fine filament fibers. The filament fiber streams dry in transit and deposit on a suitable collector. Unlike the nozzle or needle approach discussed above there is no point of bending instability and the fibers spray streams stay relatively straight until striking the collector. The spray pattern off the rotating cylinder helps create the fiber randomness to form a web.

## 5.3.3 Electrospinning with a free surface electrode

Elmarco Inc. Nanospider™ technology is based on electrospun fibers from free surface electrodes. The process fundamental is based on the same technology as the rotation electrospinning discussed in Section 5.3.2, however, the surface is a long circular wire electrode that extends the length of the machine. The process is demonstrated in the photograph of Figure 5.33. The wire electrode itself is shown unwinding from the spool at the front of the machine and its direction is vertical to the plane of the photograph. The filament fibers erupting from the polymer-coated surface of the electrode are quite visible. The polymer coating is applied from polymer solution containing trolleys that traverse back and forth the length of the electrode.

Table 5.1 lists Nanospider™ technology parameters (214) based on the electrode process. It will be noted that the process is capable of producing webs with fiber diameters as low as 1 μm and thicknesses ranging from 500 μm down to 1 μm.

**Figure 5.33** Nanopspider™ Technology—Nanofibers Spun from a Free Surface Electrode. *Source: Reproduced with permission of Elmarco s.r.o., Liberec, Czech Republic.*

**Table 5.1** Nanospider™ Technology Parameters

| | **Measurement** | **Units** | **Values** |
|---|---|---|---|
| Fiber Metrics | Mean fiber diameter | Nanometers (nm) | 50, 100, 150, 200, 250 nm or higher |
| | Standard deviation of mean fiber diameter | standard deviation (given as % of mean) | Generally 30%, tolerances can be made wider depending on targets |
| | Linear mass density | denier (g/9,000 m) or dtex (g/10,000 m) | <0.0001 denier or dtex |
| Web parameters | Web area weight (grammage) | grams/square meter (gsm or g/m$^2$) | 0.03 g/m$^2$ is the lowest commonly used industrial application. 50 g/m$^2$ has been the highest used. |
| | Fiber packing density | meters of fiber/square meter (m/m$^2$) | Depends on mean fiber diameter, gsm, and polymer selection |
| | Web thickness | Micrometers (microns or μm) | 1–500 μm |
| Production Metrics | Production speed | meters/min (m/min) | Up to 60 m/min for production line |
| | Production width | meters (m) | Currently up to 1.6 m |
| | Annual volume | square meters/year (m$^2$/yr) | Up to 50 m$^2$/yr for PVA, with a mean fiber diameter of 200 nm, standard deviation of 30%, and grammage of 0.03 g/m$^2$ |

Source: Reproduced with permission of Elmarco s.r.o., Liberec, Czech Republic.

### 5.3.4 Electro-blowing process

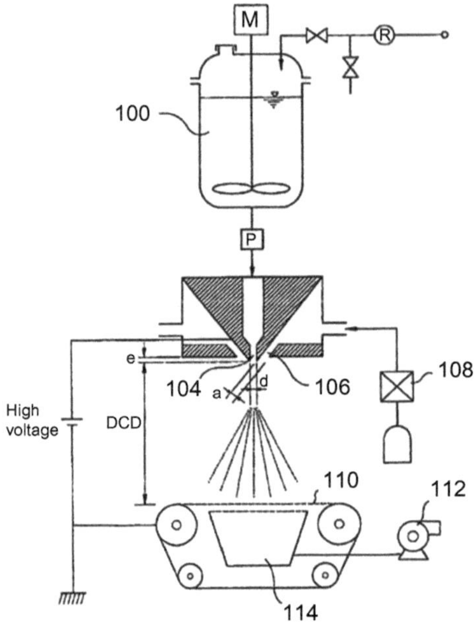


**Figure 5.34** Electro-blowing Process Patent. *Source: Reproduced with permission of DuPont, Wilmington, Delaware USA.*

The elecro-blowing process has already been referred to in Chapter 3, Section 3.5.1.2. It is a flash-spun process capable of producing nanofiber webs with membrane-like properties (DuPont™ Hybrid Membrane Technology). Patented by DuPont in 2009 (US 7618579 Kim etc. (27)), it is diagrammed in Figure 5.34 above. It produces webs that have fiber diameter dimensions ranging from 100 nm to 1 μm. DuPont (215) claims that the medium designed for liquid filtration offers improved flux/barrier properties when compared to other types of filter medium such as spunbond, melt-blown, wet laid glass, flash-spun olefin, and microporous membrane.

Referring to Figure 5.34 above, the process as described in the patent US 7618579 involves preparing a polymer solution in the storage tank 100 and discharging the solution through the spinnerette 104. Heated air from a heater 108 discharges from opening 106 and merges diagonally with the fibers formed from the polymer solution emerging from the spinnerette 104. The converging air stream dries solvent off the fibers and while doing so attenuates the solidifying fibers to microfiber dimensions. The nature or the airflow is to create a random turbulent motion to the filament fibers as they approach the collector 110. The result is the filaments form a random web as they lay on the collector. The driving force forcing the solution polymer through the spinnerette openings is a combination of mechanical pressure and a very high DC voltage. The voltage differential between the spinnerette and the collector can be as high as 300 kV. The voltage force also attenuates the filaments to nanofiber dimensions. The random web can be formed directly on the collector or it can be formed on a substrate coming through with the collector.

## 5.4 CENTRIFUGAL SPINNING

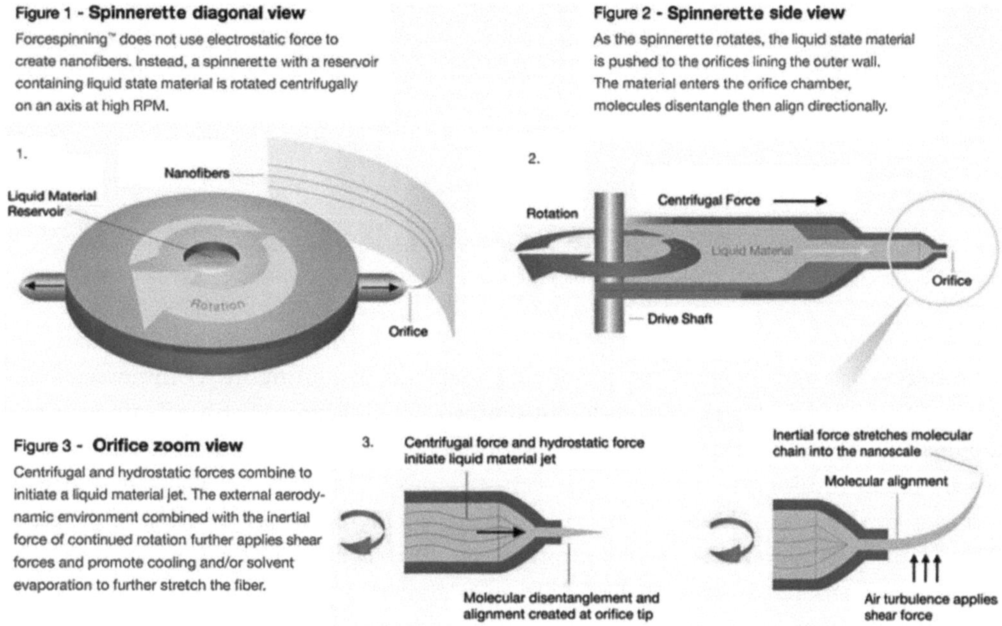


**Figure 5.35** The Forcespinning® Process. *Source: Reproduced with permission of FibeRio® Technology Corporation, Edinburg, Texas USA.*

Centrifugal Spinning has been briefly discussed in Chapter 1, Section 1.7.1.2.3.2. Two centrifugal processes are identified: Forcespinning® by FibeRio® and the Melt Spun Nanofiber Process (MSPN) by DuPont™.

### 5.4.1 Forcespinning®

Forcespinning® is a recent technology for producing nanofibers from a centrifugal spinnerette. The process is described in Figure 5.35 above. It has already been briefly discussed in Chapter 1, Section 1.7.1.2.3.2 and pictured in Figure 1.21. The Forcespinning® equipment, pictured in Figure 1.21, is a laboratory unit used for laboratory trials and experiments. It works well for making samples and process demonstrations, however, forming the web on a circular set of vertical bars is not a good way for commercial production.

**Figure 5.36** FibeRio Fire Engine® FX Forcspinning® Line. *Source: Reproduced with permission of FibeRio® Technology Corporation, Edinburg, Texas USA.*

FibeRio has two systems for producing webs on a continuous commercial basis: Fiber Engine® FS and the recently introduced Fiber Engine® FX. Both processes produce a web of continuous filament nanofiber, generally formed and coated on a substrate and resulting in a composite web structure. The systems are highly versatile and can produce nanofibers from both meltspun polymers and solution spun polymers. FibeRio reports that their equipment is capable of producing nanofibers with an average diameter of 250 nm for solution polymers and 500 nm for molten polymers. Both systems produce fiber in a centrifugal spinnerette above the substrate. The lay down on the moving substrate is like a cyclone of filament fiber.

The Fiber Engine® FS is 1.1 m wide. The Fiber Engine® FX depicted in Figure 5.36 above is offered in 1.1 m and 2.2 m widths. FibeRio reports that the FS version has a fiber forming capacity as high as 200 g/min and a line speed up to 200 m/min. Referring to Figure 5.36, the substrate is unwound from the left and conveyed into the centrifugal fiber spraying chamber. The Forcespinning® coated substrate emerges on the left and is wound into a jumbo roll.

## 5.4.2 Melt spun nanofiber process (MSPN) by DuPont®

The MSPN process only works with melt-based polymers. A simple diagram of the process has already been shown in Chapter 1, Figure 1.22. The centrifugal action of the spinnerette head lays down the filament fiber in a circle. The gap between circles caused by the advancing collector causes each circle to intersect with the previous circle and with the one after it. What is formed is an intersecting net of circular filaments. This action combined with the action of adjacent spinnerettes creates a nanofiber structure that has all the appearance of a random nonwoven web.

## 5.5 SOLUTION SPUN WEBS

Flash-spun webs or solution spun webs are webs where the fibers are formed from a polymer solution rather than from a hot melt. Note in Section 5.3 most electrospun webs are spun from polymer solutions.

Tyvek® by DuPont is a solution spun web where the fibers are spun from solutions and the solvent is evaporated as the fiber forms. This is in contrast to meltspun webs discussed in Section 5.2 that are spun from molten fibers. A variation of Tyvek® also from flash-spun technology is SoloFlo®, a medium used in some filtration applications (example: discharge water filtration).

## 5.6 WET LAY PROCESS

A wet lay process can be divided into two parts: the wet end and the dry end. The wet end is concerned with mixing fiber and water into an aqueous dispersion or slurry called stock and filtering the dispersion through a moving screen, called a Fourdrinier wire, and thereby forming a wet nonwoven sheet. The dry end uses various forms of mechanical and thermal devices to remove water from the formed sheet and to present a dry sheet at the end of the machine. In simplistic terms, the wet end puts water into the fiber and the dry end takes it out. Obviously, there is a lot more going.

In the wet end, additives other than fiber may be added to the stock as necessary for processability, product performance, and product quality. The wet end will include devices such as screens and cyclic cleaners to remove impurities and contaminants. Refiners will do work on fibers to help achieve the desired properties of the finished product. This includes control of air permeability: more refining gives a tighter medium (lower air permeability), less refining gives a more open medium (higher air permeability). The water hydraulic system before and into the former has to be very seriously managed to insure proper dispersion of fiber and formation of a uniform web.

In addition to removing water, the dry end provides for various forms of mechanical, chemical, and thermal treatments necessary for the performance of the web in its filtration application. Mechanical treatments may include wet pressing, corrugating, embossing, calendering, and hydroentanglement. Chemical treatments will include application of resin binders and other additives. Thermal treatment will include the control of cure on thermosetting and cross-linking resins.

### 5.6.1 Forming of wet lay webs

The wet lay process starts with a furnish, the papermakers term for the recipe of fibers and ingredients that is used to make the products of the wet lay process. The ingredients are added to water in a dispersion vessel called a pulper. The slurry so formed is called stock which is fed to wet lay machine. On the way, the stock may encounter several types of processing steps. These may include refining or fibrillation of the fibers to help them bond better and to control air permeability; cyclic cleaning, which uses devices that cycle the stock in a centrifugal manner to remove high density (or very low density) contaminants; and screens to remove large contaminants. The stock flow and its consistency (bone dry solids or fiber content) is precisely controlled to assure the targeted grammage. The stock is diluted with additional water just before it is pumped to the headbox of the paper machine.

Just as the card is the heart of the dry laid process, the headbox is the heart of the wet lay process. The headbox is a hydraulic device that maintains a uniform suspension or dispersion of fibers in the slurry and spreads the flow out evenly unto a moving screen called a Fourdrinier wire. The Fourdrinier headbox is the traditional papermaker's headbox. It is diagrammed in Figure 5.37. This type of headbox sends the slurry from a pond through a long, narrow, horizontal opening, called a slice. Among other things, the slice controls the height (usually 25 mm to 100 mm), and thereby quantity, of horizontal layer of aqueous slurry. The layer of slurry spreads out over the wire and drains through it. The fibers in the slurry are filtered by the wire and form a wet laid sheet on the wire surface. The velocity of the slurry relative to the speed of the Fourtdrinier wire is controlled by the slice opening and determines the fiber orientation and formation of the paper sheet. Admittedly, the headbox former as illustrated in Figure 5.37 is rather old fashioned, however it provides a convenient baseline for comparison to formers used for long fibered wet lay nonwovens.

The traditional Fourdrinier machine is used to produce some types of paper filter medium including high efficiency microglass fiber HEPA and ULPA medium. Some grades of coffee filter and vacuum bag are often produced on "slice-type" Fourdrinier machines. However, nonwoven medium produced from long fiber, synthetic fiber, and from difficult to disperse furnishes, require higher levels of dilution and headboxes specifically designed for this purpose. Two of the most common are the Rotoformer® headbox and the inclined wire (Deltaformer®) headbox. Both are produced by Glens Falls Interweb, Inc. (GFI) of Glens Falls, New York. They are depicted in Figures 5.38 and 5.39, respectively. Voith Paper in Germany also produces an inclined wire machine trade named the HydroFormer®. The HydroFormer®

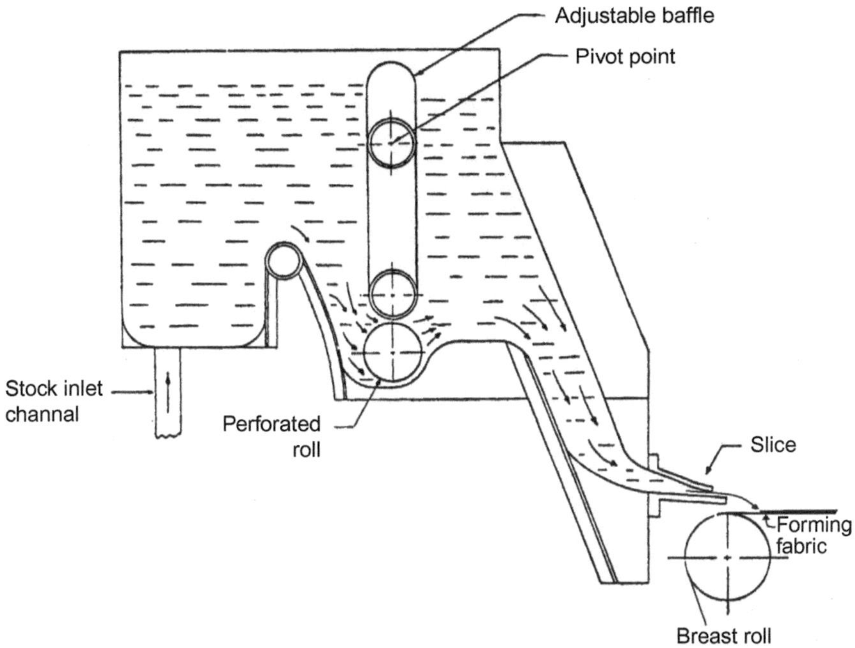


**Figure 5.37** Fourdrinier Headbox. *Source: Printed by permission of Angus Wilde Publications Inc. Vancover, British Columbia Canada (Smook (143), pg 231).*

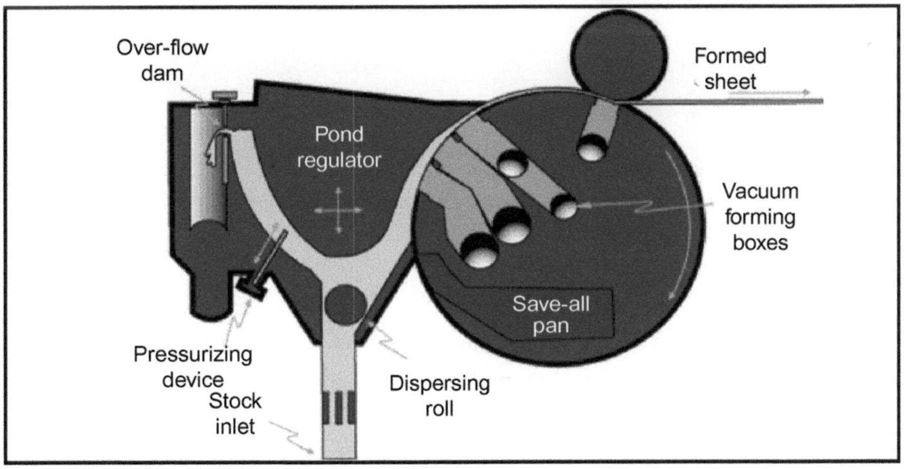


**Figure 5.38** Rotoformer ®. *Source: Reproduced by permission of Glens Falls Interweb, Inc. Glens Falls, New York, USA.*

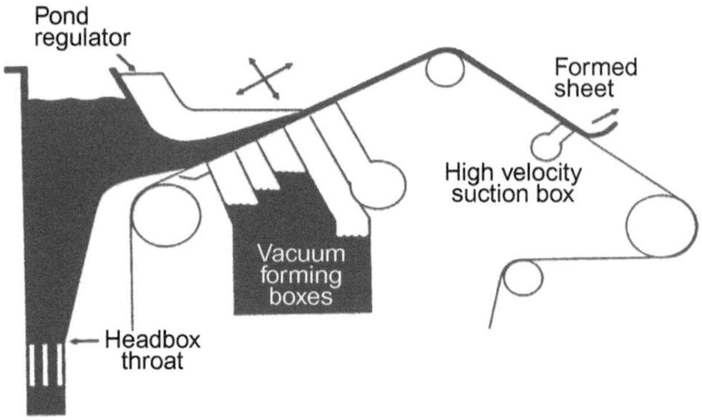


**Figure 5.39** Inclined Wire (Delta Former®) Headbox with Open Headbox. *Source: Reproduced by permission of Glens Falls Interweb, Inc. Glens Falls, New York, USA.*

**Figure 5.40** Voith Paper HydroFormer®. *Source: Reproduced by permission of Voith Paper Holding GmbH & Co. KG, Heidenheim, Germany.*

headbox is depicted in Figure 5.40 above. The Rotoformer® forms the sheet on a fabric covered rotating drum, whereas the inclined wire machines form the sheet on the incline of the forming fabric as it passes through the pond. Notice both differ from the traditional Fourdrinier headbox of Figure 5.37 above, in that the web is formed on the forming fabric within the *continuum of the pond*, whereas, the traditional Fourdrinier forms the web on the forming fabric after the slice and *outside the pond.* In the Rotoformer® and inclined wire formers the pressure head of water over the

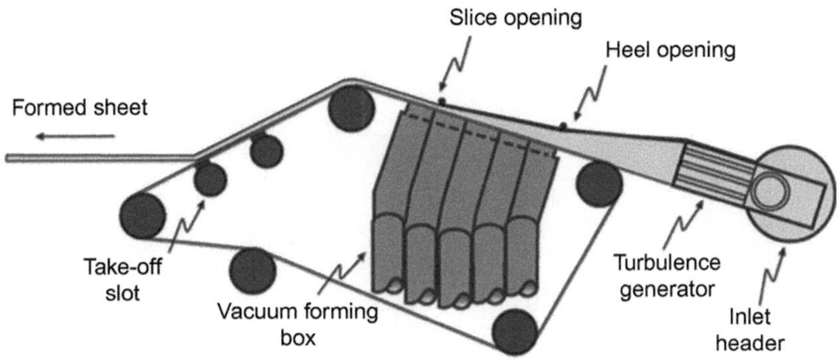


**Figure 5.41** Inclined Wire (Delta Former®) with Hydraulic Headbox. *Source: Reproduced by permission of Glens Falls Interweb, Inc. Glens Falls, New York.*

forming area can be as high as 100 cm, even higher if the headbox is enclosed and pressurized. In contrast, for traditional Fourdriniers, where formation takes place on the forming fabric after the slice. The gravity pressure head of water on the forming fabric can be no higher than the 25–100 mm (1–4″) height of the slice opening.

The inclined wire and Rotoformer® machines form the sheet on the incline of the forming fabric (straight incline for the inclined wire and circular incline for the Rotoformer®). Suction boxes underneath the forming fabric enhance the flow of water through the forming area. The inclined wire and Rotoformer® headboxes handle much more water than traditional paper machines. They are better suited for the low consistencies necessary for handling the longer fibers of a nonwoven web. The headboxes operate at much lower consistencies than do traditional paper machines. Traditional paper machines operate at consistencies in the range of 0.3 to 0.8%. The wet lay nonwoven machines can operate at consistencies below 0.1%, or at whatever is necessary to get good dispersion of the fibers.

Note from Figures 5.38 and 5.39, both the Rotoformer® and inclined wire machines have a pond regulator (sometimes called a boat slice) that acts like a slice. These usually come with adjustable controls for vertical, horizontal, and tilt positions. This device controls formation, fiber orientation, and other web properties. Figure 5.41 above diagrams a more up to date version of the Deltaformer® produced by GFI. The pond regulator is replaced with a hydraulic headbox so that the pond is closed and pressurized. Note there is no overflow dam. The overhead wall can be adjusted to regulate the pond in a fashion similar to the pond regulator.

Figure 5.42 is "Rotoformer® 2000" that has been introduced by GFI. The headboxes of most paper machines as well as nonwoven wet laid machines are fed by a tapered header designed to provide a uniform flow into the headbox. The "Rotoformer® 2000" replaces the tapered header with a radial distributor (sometimes

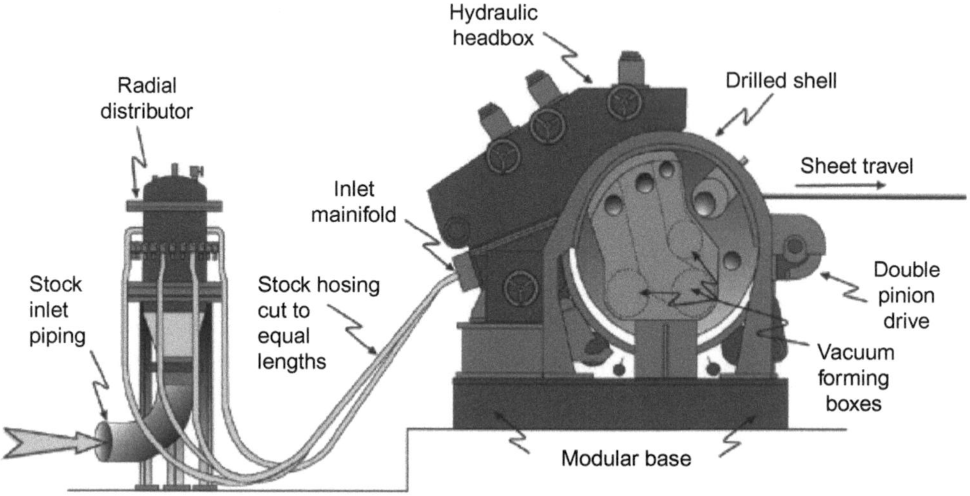


**Figure 5.42** Rotoformer®2000 with Radial Distributor. *Source: Reproduced by permission of Glens Falls Interweb, Inc. Glens Falls, New York.*

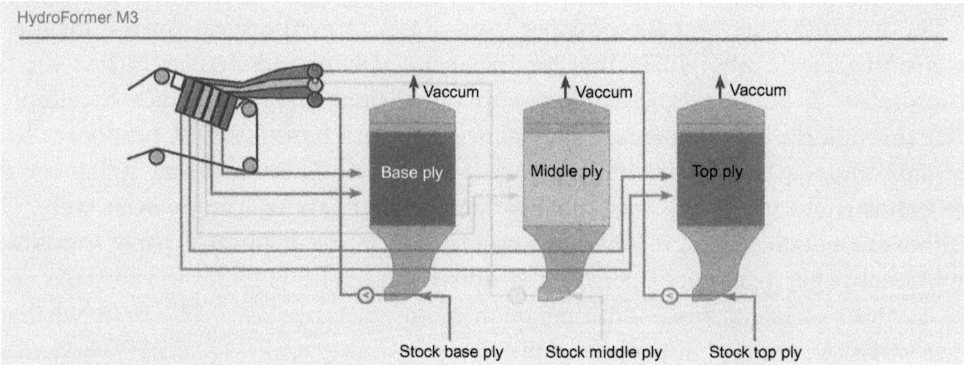


**Figure 5.43** Voith HydroFormer, Multiply Wet Lay Formerdation. *Source: Reproduced by permission of Voith Paper Holding GmbH & Co. KG, Heidenheim, Germany.*

referred to as an octopus header). This device provides a way to obtain more uniform basis weight profiles in the cross-direction of the web.

The Voith HydroFormer® has been described above in Figure 5.40. A three ply version is diagrammed in the simplified diagram of Figure 5.43. Three stock systems are prepared: one for the top ply, one for the middle ply, and one for the bottom or base ply. All three stock systems are fed to the headbox. They are separated by lamellae as they enter the headbox. The lamellae keep the three flows separate and the resulting stratified flow patterns keep the three flows from mixing as they approach the wire or forming fabric. The base ply flow is the bottom flow in the headbox and the first to

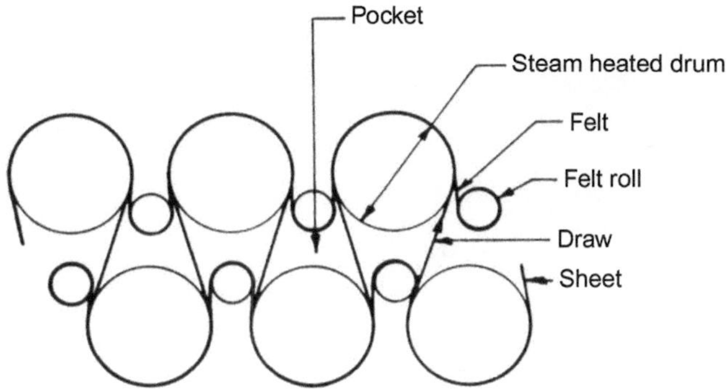


**Figure 5.44** Two-Tier Cylinder Drying Configuration. *Source: Printed by permission of Angus Wilde Publications Inc. Vancover, British Columbia, Canada (Smook, (149) pg 266).*

reach the wire, thereby it forms the first layer on the wire. The middle ply flow is the middle flow to the headbox and forms on top of the first ply. The top ply flow to the headbox and is the last to reach the wire and forms a layer on the first two plies.

There are other ways of forming three ply medium at former of a wet end. Another way is two use a two-ply former to form the first two plies and a secondary headbox that more or less pours a third ply on the first two plies after they come out of the primary headbox.

## 5.6.2 Drying of wet lay webs

Once the wet lay sheet is formed a lot happens to it on its way to the windup reel. Most of this are concerned with removing water and drying the sheet. Mechanical devices such as wet presses and suction slots are often used. However, at best these can only get the moisture content down to 50–60%. Most wet laid filter media products are not wet pressed, because the wet pressing action densifies the sheet and makes it less bulky. Without wet pressing, moisture contents of the sheet of 70% are generally the lowest achievable before the dryers. This is important. Consider that if a wet pressed sheet can be wet pressed to 50% moisture content, then the dryers only have to remove one kg of water per kg of bone dry fiber. On the other hand, at 70% moisture content the dryers have to remove 2.3 kg of water for every kg of bone dry fiber. The thermal requirements of the dryers more than double if the medium is not wet pressed.

After mechanical dehydration, the rest of the water is removed by thermal drying. This can be in the form of steam heated can dryers, hot air dryers, infrared devices, microwave heating or very often some combination thereof. Steam heated can dryers (illustrated in Figure 5.44 above) and hot air dryers are the most common means of drying a wet laid nonwoven. Dryer felts are woven fabrics used with heated can dryers that assist in the thermal drying process by pressing the wet web against the surface of the dryer can. They are porous enough to allow for the escape of moisture. They also

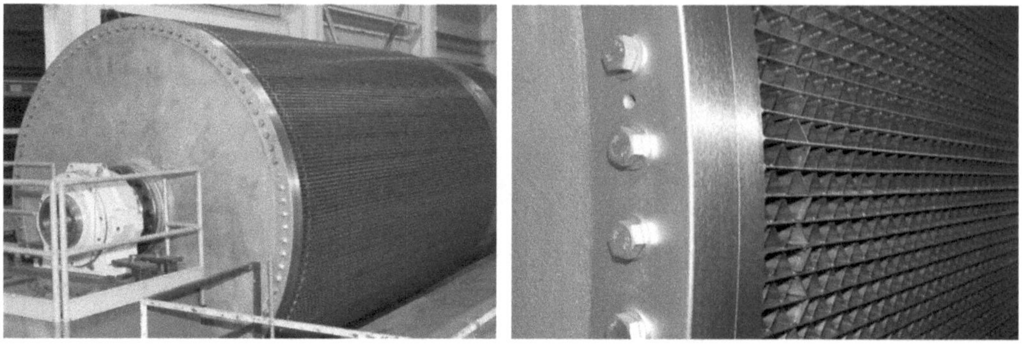

**Figure 5.45** Valmet Honeycomb Roll. *Source: Reproduced by permission of Valmet Inc., Biddeford, ME, USA.*

make it easier for web handling over the dryer cans and allow for easier threading of the machine at startups. The pressure on the web between the felt and dryer surface does compress it, resulting in less bulk. For this reason some wet lay filter mills will operate without dryer felts, even though this results in far less efficiency and production speed in the thermal drying operation.

For filter medium hot air is preferred, because it allows for bulkier structures. Hot air dryers can operate either by impingement air or by through-air drying. Through-air drying can be very effective for highly porous webs, however, limited in the drying of very tight low porosity webs. Most often the air dryer is a long, straight, flatbed dryer. The dryer may be sectioned for different levels of drying and for possible curing of the resin in the web. Honeycomb rolls are often used as through-air dryers. Figure 5.45 above is an example of a honeycomb through-air drier produced by Valmet Inc. Honeycomb Division, Biddeford, Maine.

## 5.6.3 Bonding of wet laid nonwovens

Wet laid nonwoven webs can be bonded in both the wet and dry ends of the wet lay machine. Wet end bonding can take place by hydrogen bonding of cellulose pulp fiber webs, by refining the pulp, and/or by adding a bonding agent to the fiber slurry before the web is formed (often referred to as beater addition or beater-ad). The availability of low melting polymer fibers and/or bicomponent fibers is making the thermal bonding option more attractive. The thermal bonding fibers are part of the fiber furnish. Wet end bonding mechanisms whether they are wet strength resins, beater-ad latexes, or thermo-bonding fibers will require dry end heat to become effective. To illustrate, wet strength resins are thermosetting and have to be heat cured to become effective. Thermal bonding fibers should have sufficient heat to melt the thermoplastic component.

Dry end bonding of wet lay webs is mostly accomplished by some form of chemical treatment. Most wet lay filter medium, even when resin treated in the dry end will have some form of applied wet end bonding mechanism. For example, medium to be dried and then treated with a latex treatment will usually contain a wet strength agent to hold the web together as it is wetted out with the aqueous resin.

#### *5.6.3.1 Wet end bonding*

In Section 5.6.3, various chemistries for wet end bonding are mentioned. They are discussed in more detail as follows:

**a.** Hydrogen bonding—Cellulose pulp fibers in water bind to each other in a chemical mechanism known as hydrogen bonding. Hydrogen bonding is the fundamental bonding mechanism in the making of paper. It is one of the primary mechanisms by which wet lay cellulose fibers adhere to each other in the dry state. This is possible because cellulose and hemicellulose, two of the main components of papermaking fibers, are covered with hydroxyl groups. The oxygen atoms in these groups are able to bond to hydrogen atoms on adjacent fibers or water molecules. Drying causes some fiber-to-fiber hydrogen bonds to take the place of fiber-to-water hydrogen bonds. The mechanism is diagrammed in Figure 5.46 below. If no other bonding mechanism is present recycling is relatively easy because addition of water reverses this process. Hydrogen bonds have only about 5% of the energy content of a covalent bond. Most wet laid nonwovens contain some percentage of hydrogen bonding cellulose pulp to provide strength and integrity to the web for downstream processing.

**b.** Wet end binders—Several types of resins or chemical agents can be added to the fiber slurry or stock to enhance the strength properties of the web. Most often, these are cationic agents attracted to the negatively charged anionic fibers. If the additive is a latex binder, a complex system of retention aids is used to retain the latex particles on the fiber surfaces. Similar chemistry can also be used to retain solid fillers such as clays and activated carbon powder. Cationic wet strength agents are the most frequently used wet end additives. They serve two functions: one is to give the medium enough strength to be transported through the wet lay

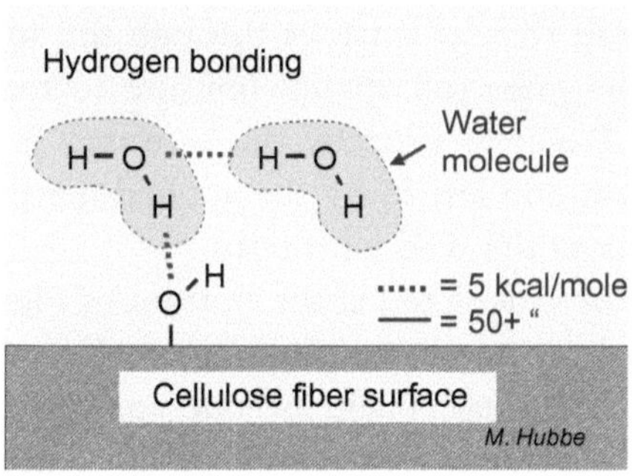


**Figure 5.46** Hydrogen Bonding. *Source: Reproduced with permission of Dr. Martin Hubbe, North Carolina State University, Raleigh, North Carolina, USA.*

machine and through subsequent aqueous treatments. The other is to make the medium strong enough to stand up to an aqueous filtration application. Sheeted filter medium used in plate and frame industrial filtration applications are often referred to as high wet strength filter paper. Coffee filter paper is another example of filter medium bonded with a wet strength agent. Common types if cationic wet strength agents are urea formaldehyde and melamine formaldehyde. Polyamide-epichlorohydrin, PAE, is a very popular wet strength agent because it does not emit formaldehyde fumes when the web is dried. Ashland Inc. Covington, Kentucky USA markets a PAE wet strength agents under the trade name of Kymene™ 557h.

**c.** Binder fibers—These are thermoplastic fibers that can be blended and mixed in with the fiber slurry. Their melting point is low enough to soften in the drying section and then meltbond the fiber structure on cooling. Binder fibers include polyvinyl alcohol type fibers such as Kuralon®[6] manufactured by Kuraray Co., Ltd., Japan and low melting thermobonding fibers. Bicomponent sheath-core fibers with a low melting outer core are very often employed for this purpose.

### *5.6.3.2 Resin application to the web*

Most wet laid nonwovens used for filter medium applications are resin treated after the web is formed. The resin treatment is intended to provide the desired structural properties so that the web can be converted to the end use filter application and have the integrity to stand up to the hostile challenges of the filtration environment.

Resin can be applied to the web either in-line or off-line. Either way, the most common form of resin applicator is the size press. There are many versions of size presses. Two of the simplest types are depicted in Figure 5.47. The top diagram is a vertical size press, the bottom is a horizontal size press.

In the vertical size press operation, the web passes underneath a bead roll and then through a set or press rolls where the resin is applied. The bottom roll is the applicator roll whose bottom is below the liquid resin level in the resin pan. The applicator roll applies the resin to the base web as it passes through the nip between the applicator roll and the press roll. The press roll applies pressure to force the resin to penetrate and impregnate the web. Resin pick-up by the base web is controlled mostly by the resin solids in the resin bath and less so by the applied nip pressure. It is the intent of this operation to fully saturate the web with resin.

The horizontal size press is depicted at the bottom of Figure 5.47. The rolls of this configuration are along a horizontal line. Sometimes the two outside rolls (metering rolls) are at 45° with the horizontal. There are two applicator rolls, two metering rolls and two resin baths. The metering rolls act to fix the film thickness of the resin on

[6] Kuralon® is a trade name of Kuraray Co. Ltd., Tokyo, Japan.

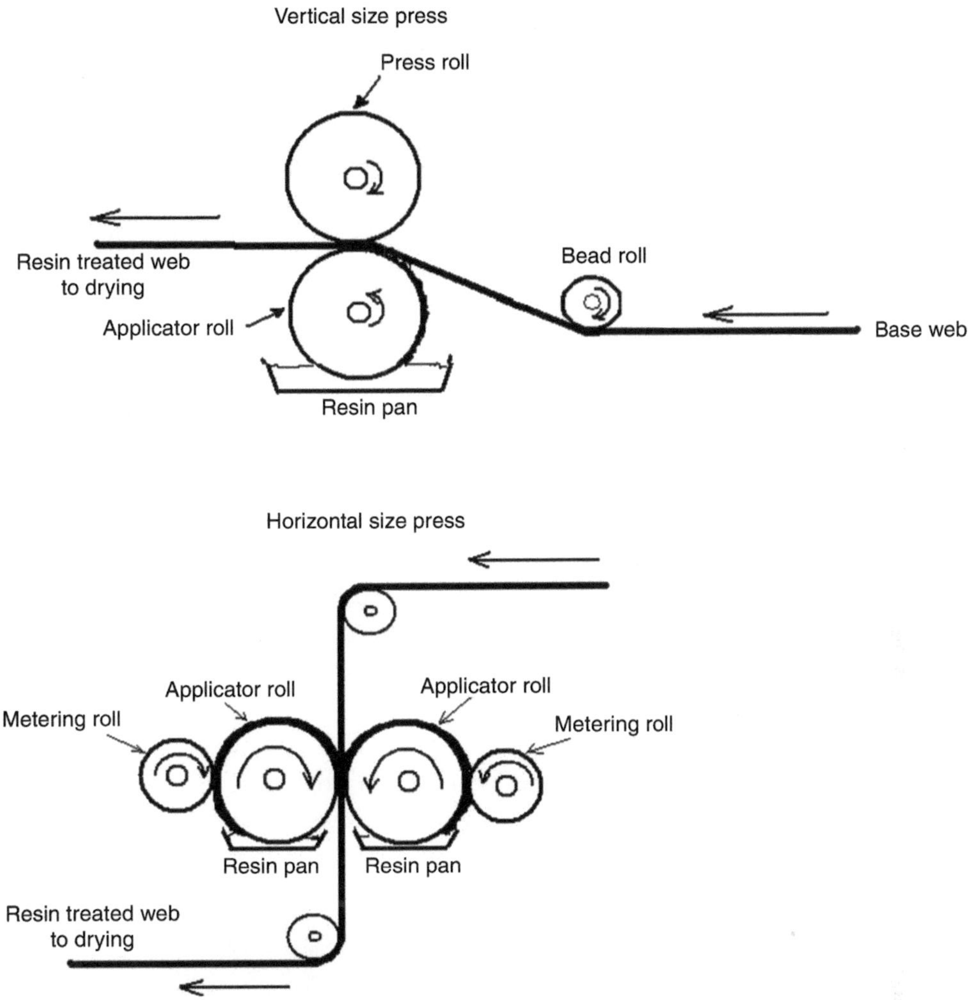


**Figure 5.47** Vertical and Horizontal Size Presses for Applying Resin to a Wet Laid Web.

the applicator rolls. In this configuration shown in the base web is traveling downward through the nip between the two applicator rolls. The applicator rolls pick up resin by rotating through the resin bath. The surface of the metering rolls move contrary to the surface of the applicator rolls, thereby shear off the excess resin on the applicator rolls. The nip between the two applicator rolls is not a pressure nip. It is gapped at some distance that is close enough for both applicator rolls to touch the web and apply resin. The resin is coated on the base web. The ability of the web to become fully impregnated depends on how well the resin can migrate to the web interior. This depends on the capillary properties of the webs fiber structure and the viscosity of the liquid resin mix.

The gap between the metering roll and the applicator roll is adjustable so that the film thickness can be varied. This provides for two levels of control on resin pickup: solids content of the liquid resin bath and film thickness. The advantage of controlling resin pickup by film thickness is that it is much quicker than trying to make changes to the solids content of the liquid resin bath. Controlling by film thickness and applying resin to both sides of the base web allows for liquid resin to be applied at higher solids content than the vertical size press. This reduces energy costs in drying and may allow for higher production rates. It also reduces solvent costs if the resin is being applied from a solvent base.

There are other types of applicators for saturating a wet laid web. These include gravure presses, spray systems, foam bonding, and curtain coaters.

## 5.6.4 Converting processes for wet lay webs

Wet lay medium are subject to a number of converting processes as necessary for its end use filtration performance. These include corrugating, slitting and rewinding, creping, sheet cutting, die cutting, pleating, bag making, and other operations that alter the mechanical structure and geometrical surface of the web. In addition, chemical treatments other than resin are applied to the web. This includes flame retardants, water repellents, surfactants, electrostatic treatments, adsorbents, antimicrobial agents, etc.

### *5.6.4.1 Corrugating*

Corrugation is accomplished by embossing the web between the two corrugating rolls of a corrugating press. The press embosses a sine wave configuration across the web. Figure 5.48 is a photograph of a sample of a pleated medium with the corrugations

**Figure 5.48** Corrugated Filter Medium.

**Figure 5.49** Corrugated Filter Medium Emerging from the nip of the Corrugating Press. *Source: Reproduced with permission of Hollingsworth and Vose Company, Inc., East Walpole, Massachusetts USA.*

very evident. The corrugations are in the machine direction of the sheet. The corrugating rolls are ridged so that there is a sine wave configuration to their surfaces. The rolls are set up so that when they are brought close together the ridges mesh; hill into valley. When a sheet is transported through the nip, the pressure forces the sheet to take on the corrugated cross section. The photograph of Figure 5.49 shows the corrugated medium emerging from the nip of the corrugating rolls.

The corrugating effect is to densify the sheet in its thickness and stretch it in the cross section. The degree of corrugation is limited by how much the medium can be stretched in the cross direction before cracking and tearing occur. For cellulose media used in engine filtration applications, the stretch is approximately 4% depending on furnish and resin treatment. Densification of the sheet by corrugation causes a loss in permeability. This is somewhat compensated for by the gain in surface area.

For the medium to be corrugatable, it has to be very plastic. Wet lay media are corrugated after the resin has been applied, but before the sheet is fully dried. The volatiles content of the sheet at the corrugating press is 5–20%, whether it be water or organic solvent. The sheet is usually preheated prior to corrugation in order to utilize the thermoplasticity of the resin system. There are cases where the fully dried sheet is corrugated. This can only be done with a low melting thermoplastic resin. The degree of corrugation varies, depending on the resin, the nature of the fibers making up the media, and process conditions. Basis weight and thickness are important properties that affect the medium's ability to be corrugated. The corrugation generally ranges from 200 to 400 μm. The maximum achievable on a consistent production basis is 500 μm.

Dense structures such as those used for heavy duty air filter are easier to corrugate than bulky media. Media composed of glass microfibers are very difficult to corrugate because the fibers are so short and the medium has very little stretch. A medium containing a high percentage of polyester does not corrugate as well as a pure cellulose medium. The long-fibered polyester allows the medium to corrugate without cracking, however, the fiber does not readily mold to the sign wave mesh of the corrugating rolls.

### *5.6.4.2 Creping*

Some industrial filter medium grades and some vacuum bag grades are creped to increase the surface area for filtration. Usually the creping is done in some form of press rolls (wet press or size press) by pushing the medium against a creping doctor so that it bunches up against the blunt edge of the doctor as shown in Figure 5.50. Machine speed downstream of the creping doctor and the creping press roll is reduced so as not to pull out the crepe.

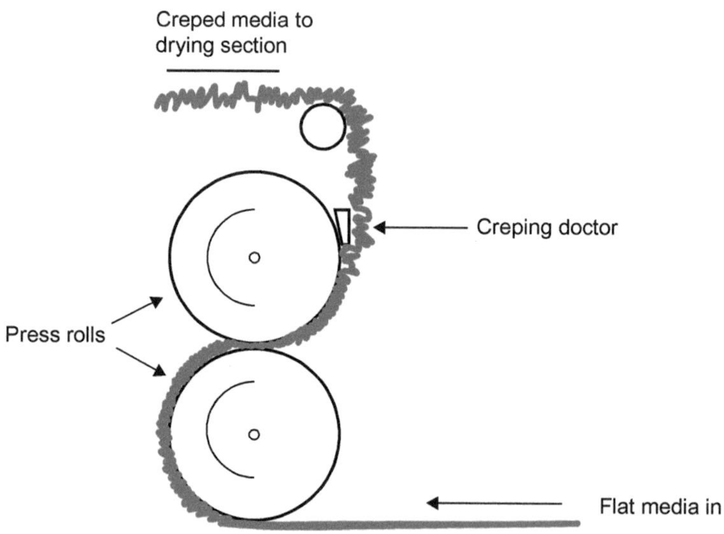


**Figure 5.50** Creping Press.

### 5.6.4.3 Rewinding and slitting

At the end of the wet lay process is a reel. The reel winds up jumbo rolls of the finished paper unto cores. The jumbo rolls are then sent to a rewinder where they are rewound into uniform rolls of a specified diameter. The rewound rolls may be trimmed jumbos and/or split rolls for later processing on a slitter, or finished rolls for the customer. The rewinder can also be used for other purposes such as combining different medium into a composite laminate.

Very often, rolls off the rewinder are sent to a slitter to be slit into finished rolls for the customer. A slitter is a form of rewinder that contains rotary cutting blades to slit the jumbo roll into specific roll widths as specified by the customer. A slitter contains a number of rotary cutting blades so that multiple rolls can be cut with each cutting. The blades are set at very carefully measured distances to insure that each roll is cut to the customers tolerances. Modern slitters have mechanized computer control to set the cutting blades for accurate slit widths. Slitters can custom cut a roll to widths ranging from 2.5 cm to approximately 250 cm (if the machine is wide enough). The maximum practical width is usually based on the trailer size of the truck that will be transporting the rolls to the customer. Wider rolls are usually ordered by customers who do their own slitting to allow more flexibility in their production schedule.

Rewinding and slitting are critical parts of the wet lay process because it is where most of the waste occurs. The largest cause of waste is the trim, the extra width of paper in the jumbo roll that contains the deckle edge from the paper machine. Generally a mill will allow for approximately 1.2 to 2.5 cm of trim on either side of the jumbo. Some wet lay machines are 200 cm or less in width. For such machines the trim waste can be as high as 3%. This is further complicated by the customer's order. Although wet lay machines have flexibility in adjusting the deckle width at the former, they are committed to the width for which they are designed. If the finished rolls do not trim the machine deckle then there may be considerable excess trim. To illustrate, the machine deckle width is 200 cm and the customer orders 60 cm wide finished rolls. The machine can produce medium wide enough for three roll cuts totaling 180 cm. This leaves 20 cm of the 200 cm deckle that has to be trimmed, either at the former, at the slitter, or at some other suitable place in the process. Usually the trim can be recycled, however, there is a loss in production rate.

There are a number of issues that increases waste because of quality rejections. The most serious of these is slit width variations. The roll as shipped to the customer is too narrow or too wide for the customers operation. Mills can do a good job of measuring and controlling slit width at the slitter; however, they have little control of the environment and its effect on dimensional change when the rolls leave the mill.

#### *5.6.4.4 Stamping and dye cutting*

Filter sheets and industrial filter paper produced for industrial filters such as filter presses have to be carefully cut to sizes and shapes that fit the filter. The shapes are die cut or stamped in converting mills dedicated to this purpose. Die-cutting is like using a cookie cutter. A die-cutting machine uses steel rule die shapes to cut through a wide range of filter materials quickly and easily. A typical converter may have a library of over 200 hundred dyes to handle a large variety of customer orders.

#### *5.6.4.5 Bag making*

There is a huge variety of filter bag and pocket filter operations, many of which are very proprietary. Among the steps in a bag or pocket filter operation are folding, sewing, glueng, trimming, shaping, and packaging. If the bag is to be used for a medical or food processing filtration and separation, then there are also hygienic and regulatory issues in process control and packaging.

A variation of bag filters is pocket filters. Pocket filters are pleated media with pleat edges sealed so that each pleat fold becomes a pocket.

#### *5.6.4.6 Pleating*

It is estimated that over 80% of resin treated filter paper production will go to a customer's pleating line. Pleating lines have been adapted to pleat dry formed filter media as well as wet laid media, including composite structures and media supported by a support structure such as a metal wire mesh screen. A detailed discussion of the pleating process is presented in Section 5.8.

## 5.7 COMPOSITE STRUCTURES

Composite structures were defined in Chapter 1, Section 1.4.11 as " Two or more different web technologies that are combined to optimize the characteristics of a filter medium".[7] There are several ways of combining web structures. Unique ideas for new ways are constantly emerging. For example, Graham, Schreuder-Gibson, and Gogins ((216) describe a number of composite structures that combine various substrates with electrospun webs. This includes combination with woven webs, spunbonded nylon, and spunbonded polyester webs. Laminates increase the number of possible combinations. For example, an electrospun-spunbonded web was laminated to itself so that the two electrospun layers were on the inside and the two spunbonded layers were on the outside. Another combination was to laminate the electrospun-spunbonded to another spunbonded web so that the electrospun layer was sandwiched between the two spunbonded layers. The advantage of having the electrospun layer

[7] R.W. Murphey (23)—personal conversation.

between the spunbonded layers is that the outer layers protect the nanofiber layer from abrasion and other mechanical damage. It is interesting to note that the electrospun-spunbond combinations were explored for protective apparel that would allow for the high air permeability and yet prevent the penetration of hazardous and toxic aerosols. In such cases, the apparel is the filter and the fabric is the filter medium.

Following are brief descriptions of some of the processes for producing composite structures.

### 5.7.1 SMS process

U.S. Patent 4041203 (217), assigned to Kimberly-Clark Corporation in 1977, describes a product obtained by combining "an integrated mat of generally discontinuous, thermoplastic polymeric microfibers and a web of substantially continuous and randomly deposited, molecularly oriented filaments of a thermoplastic polymer." It is recognized that the microfiber mat was a melt-blown and the continuous filament web was a spunbond. Bonding of the layers was accomplished by passing them through a set of heated rollers: one of which was smooth and the other containing a plurality of raised points so that the bonding effect was patterned and in discrete points. In one embodiment of this invention the melt-blown web is sandwiched between the two spunbond webs so as to form what is now known as the SMS composite. In the example, the inventors (Brock and Meitner (126)) note that the composite has "especially desirable strength characteristics."

### 5.7.2 Multi-ply forming

Some forming processes are capable of forming multilayered medium. Carding lines with successive carding machines are certainly capable of doing this. Some wet lay machines are capable of forming dual-ply and triple-ply media (see Section 5.6.1 and Figure 5.43). Ahlstrom uses a three-ply wet lay machine to produce a three-layered filter medium composite identified as their Trinitex® technology.

### 5.7.3 Laminating

The simplest way to form a composite structure is to combine two or more layers of medium in a laminating line. The layers are glued to each other by latex adhesives or hot melt glues. In some cases Van der Waals forces or electrostatic attraction is sufficient to hold the layers together for the intended downstream processing. Usually the different layers come off separate unwind stands and are then wound together on a wind-up stand.

### 5.7.4 Entanglement

Hydroentanglement, needle punching, and stitch bonding are other ways of combining the individual layers. Lang and Schmalz (217) described a nonwoven composite structure consisting of two nonwoven webs that were hydroentangled with a scrim material. The scrim material is sandwiched between the two nonwoven webs. The process concept is illustrated in Figure 5.51 below. The scrim material, usually a woven textile material, provides for improved strength. The product has been developed by Jacob Holm industries GmbH, Germany. It is marketed under the trade name of Norafin®.

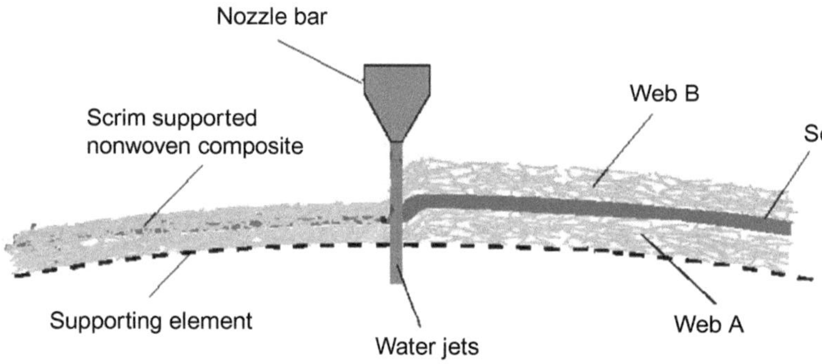


**Figure 5.51** Composite Structure by Hydroentangling Two Nonwoven Webs to a Scrim. *Source: Reproduced by permission of INDA, Cary, North Carolina USA.*

### 5.7.5 Collectors

In some cases, a substrate or nonwoven layer serves as the collector for forming another type of nonwoven layer. The layer being formed on the substrate is often referred to as a coating. Nanofiber webs from the electrospun process and melt-blown processes are usually formed this way. Cellulose wet lay webs, spunbonded webs, and needle-punched and spunlaced webs are most often used as substrate collector material. A more recent product, nanofiber membranes, is the result of nanofiber webs being formed on a membrane material.

## 5.8 THE PLEATING PROCESS[8]

It has been mentioned in Section 5.6.4.6 that most resin treated paper produced for filter applications will be pleated into one form or another. In addition, many dry formed nonwovens are pleated into some form of cartridge or panel shape. In almost

[8] Much of the material for this section was provides by Charley Wood (deceased) (218) who at the time was an employee of Hollingsworth and Vose Co., Inc.

all cases, the medium is in roll form and fed to the pleating process from an unwind stand. Often there is a slitter in line with the unwind stand that will slit the medium into to two or more widths. This allows more than one filter to be made at the same time and optimizes the width capacity of the filter line. There is often a dye marking mechanism to indicate where to cut the pleated medium for proper length in the pleat pack. The cutting of pleat packs into specified lengths can be done either by a person with a knife or by a sophisticated, automated, cutting apparatus.

There are many types of pleating machines including "home-made" types created by some filter manufacturers. Most are variations on four fundamental pleating mechanisms; gear pleater, blade pleater, knife pleater, and rotary pleater.

## 5.8.1 Gear pleaters

The gear pleater diagrammed in Figure 5.52 is perhaps the simplest pleating method available short of hand pleating. A flat sheet of filter medium passes between two oppositely rotating matched gear rolls, which fold the sheet into a pleat form and unto a conveyor belt for further processing. The height of the pleat pack is fixed and is determined by the length of gears. Typically, some sort of heat is applied to the flat medium ahead of the pleater head. Change in pleat height can only be accomplished with a new set of matched gears at a relatively high cost and downtime.

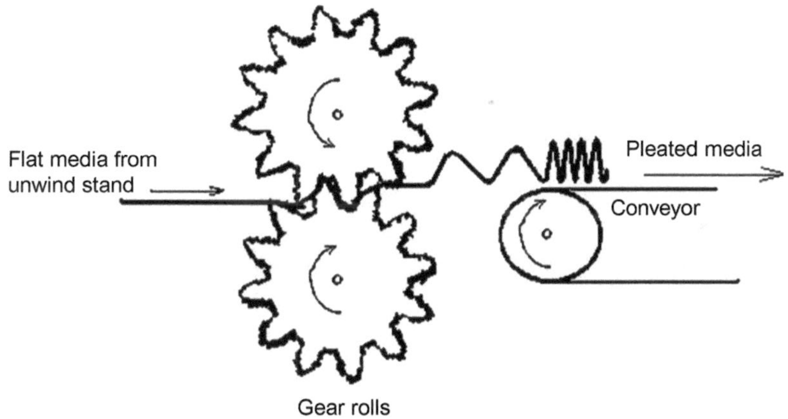


**Figure 5.52** Gear Pleater. *Source: Drawing provided by Charley Wood (deceased) formerly of Hollingsworth and Vose Company, West Groton, Massachusetts USA.*

## 5.8.2 Blade pleaters

Blade pleaters are often referred to as Chandler pleaters, since they were originally produced by the Chandler Pleating Co. of Ayers, Massachusetts. JCEM GmbH of Fulenbach, Switzerland has since bought out Chandler. JCEM has greatly modernized the original Chandler pleater with extensive automatic and digital controls and higher

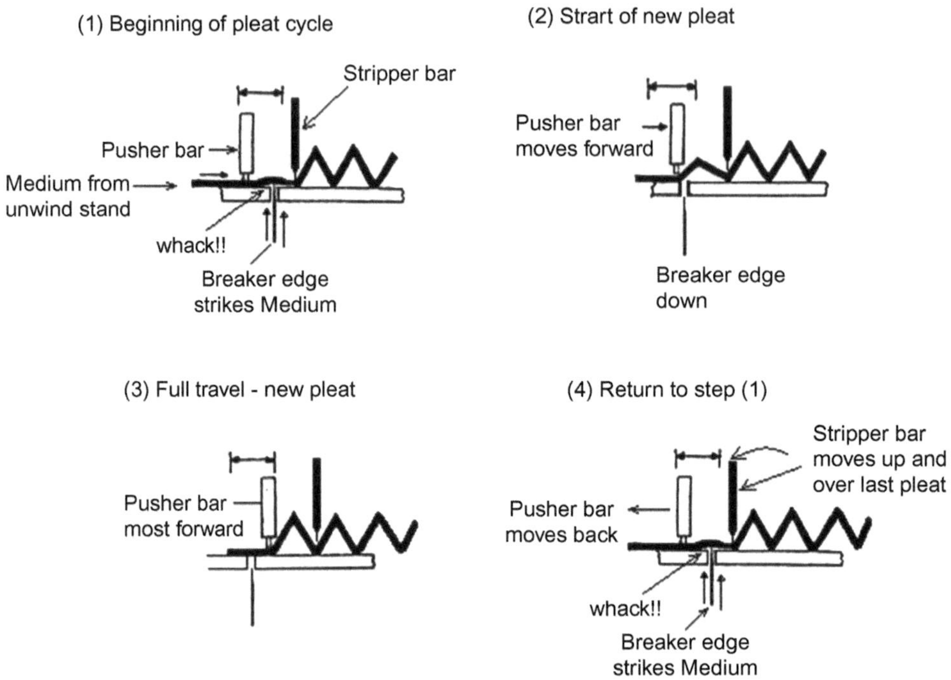


**Figure 5.53** Blade Pleater Operation. *Source: Drawing provided by Charley Wood (deceased) formerly of Hollingsworth and Vose Company, West Groton, Massachusetts USA.*

speed performance. JCEM claims that their digital pleater (DP) can handle "virtually all filtration related materials" (219). Their APCA system (A = auto, P = pleat height, C = control, A = adjust) allows for automatic pleat height control and adjustment and operator free operation when APCA is activated.

The Chandler type of blade pleater operates in a repeating stroke cycle by use of a pusher bar that compresses the medium, a breaker edge that forms the pleat and a stripper bar that holds back the formed pleat. The operation is illustrated in Figure 5.53. The pusher bar clamps down on the flat sheet and pushes the paper forward toward the stripper bar. The breaker edge, in a raised position to form the pleat, drops down to let the pusher bar passes. The stripper bar starts up and down in front of the newly created pleat and holds the formed pleat pack back. The pusher bar then lifts up and back to start the next stroke cycle. The stripper bar starts back up to assist in the formation of the next pleat. Changes in pleat height are accomplished by changing the cam which determines the length of each stroke. Pleat heights up to 8 cm are possible on the blade pleater.

Blade pleaters are also produced by a2z Filtration Specialties, New Delhi, India. Figure 5.54 is a diagram of the a2z pleater. Figures 5.55 and 5.56 are examples of pleat

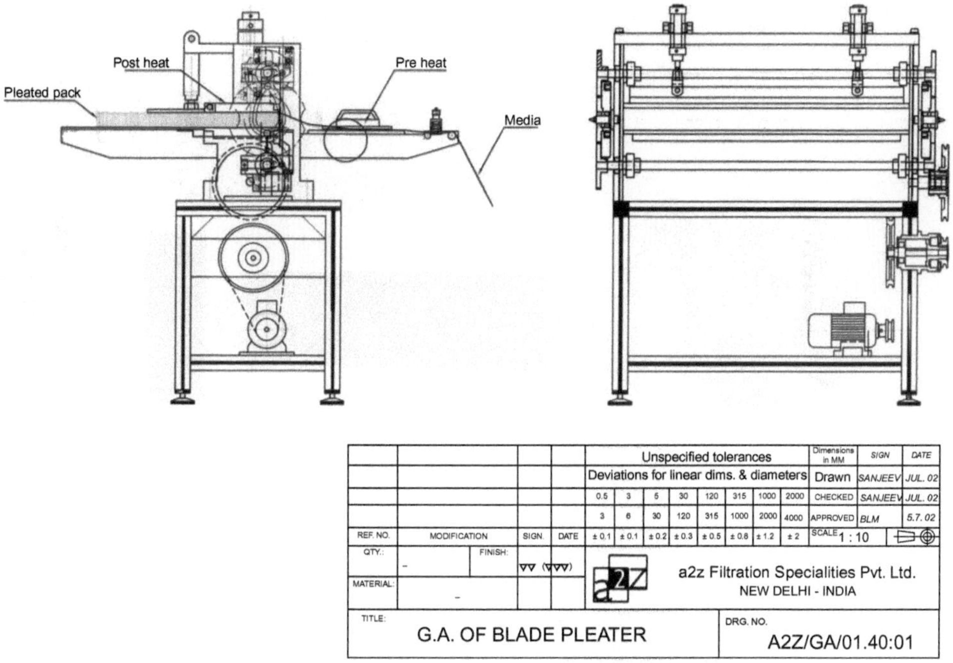


**Figure 5.54** Blade Pleater by a2z Filtration Specialties. *Source: Reproduced by permission of a2z Filtration Specialties Pvt, Ltd., New Delhi, India.*

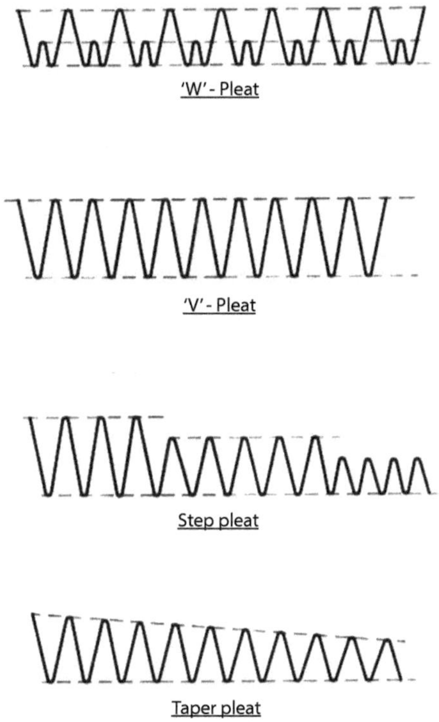


**Figure 5.55** Pleat Styles. *Source: Reproduced by permission of a2z Filtration Specialties Pvt, Ltd., New Delhi, India.*

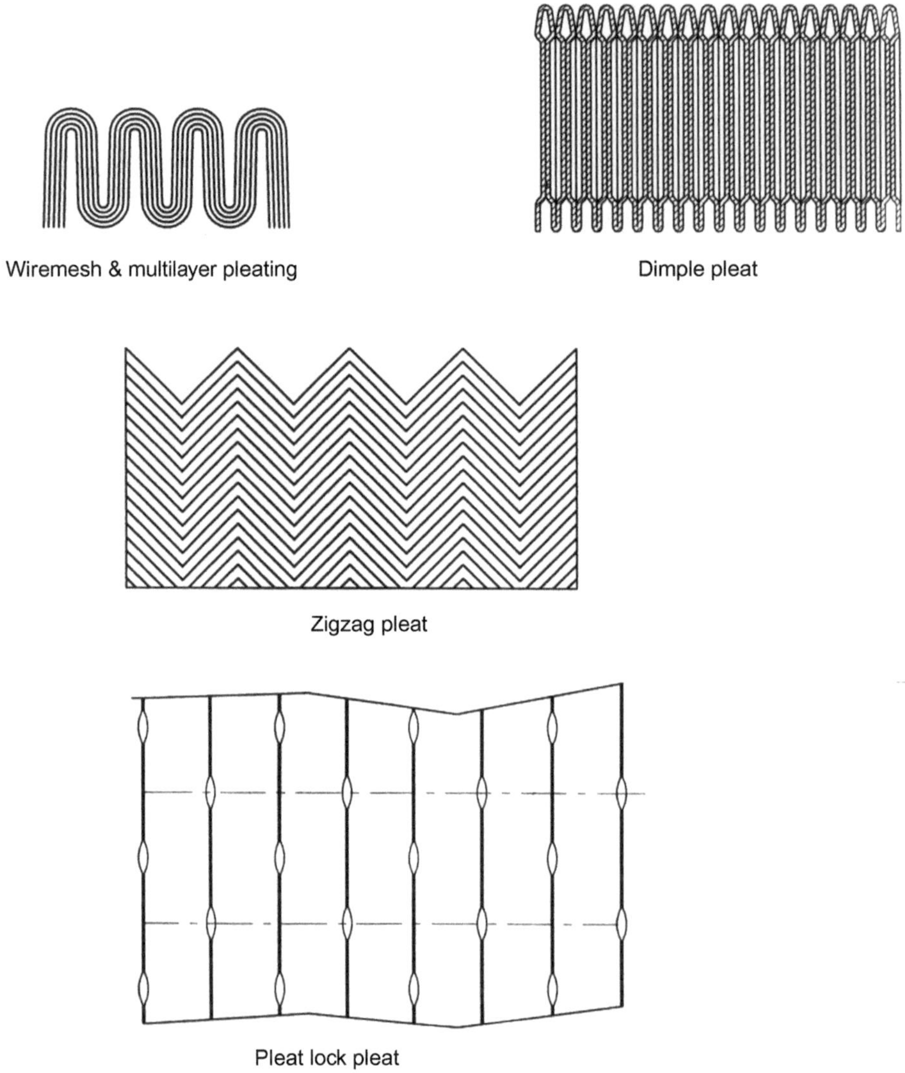


**Figure 5.56** Pleat Styles. *Source: Reproduced by permission of a2z Filtration Specialties Pvt, Ltd., New Delhi, India.*

styles that can be achieved with a2z pleaters. One of the pleat styles in Figure 5.56 is a pleat lock pleat. The pleat lock is an indentation in the pleat fold that locks it in place after it is folded.

### 5.8.3 Knife pleaters

Knife pleaters differ from blade pleaters in that they operate with a set of top and bottom knifes. The knives work in sequence to first compress the sheet to form a pleat

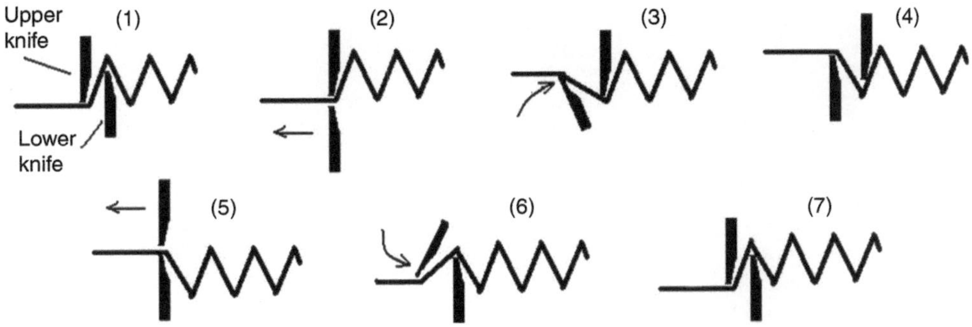


**Figure 5.57** Knife Pleater Operation. *Source: Drawing provided by Charley Wood (deceased) formerly of Hollingsworth and Vose Company, West Groton, Massachusetts USA.*

**Figure 5.58** Inlet Side of a Knife Pleater Showing Wire Mesh and a Glass Microfiber Filter Media Feeding into the Pleating Head. *Source: Reproduced by permission of Kaydon Custom Filtration Corporation, LaGrange, Georgia USA.*

and then hold the pleat pack down as the companion knife repeats the sequence. The sequence illustrated in Figure 5.57 shows the top knife (1) pushing the medium against the backstop of the bottom knife to form a top pleat fold. The bottom knife then moves under (2) and in back of the top knife (3) to start a bottom pleat fold (4). The top knife then moves back and over (5) the bottom knife to start a top pleat fold (6) and the sequence repeats itself (7).

Figures 5.58 and 5.59 are photographs of a knife pleater in action. The medium being pleated is a glass microfiber HEPA medium supported by a wire mesh.

**Figure 5.59** Exit Side of a Knife Pleater Showing Wire Mesh and a Glass Microfiber Filter Media Combined and Pleated, Emerging from the Pleating Head. *Source: Reproduced by permission of Kaydon Custom Filtration Corporation, LaGrange, Georgia USA.*

Figure 5.58 shows the HEPA and the wire mesh feeding to the inlet side of the pleating head. Figure 5.59 pictures the pleated medium emerging from the exit side of the pleating head. Both the wire mesh and the HEPA medium have been consolidated into a single pleated configuration. The knife pleater is most versatile for adjusting pleat height but at the expense of sacrificing production speed.

Traditional mechanical knife pleaters adjust pleat height by the use of cams. Karl Rabofsky GmbH, Berlin, Germany has introduced electronic knife pleaters, whereby electronic controllable servomotors replace the cams. Table 5.2 is a summary of knife pleaters offered by Karl Rabofsky.

### 5.8.4 Rotary pleaters

The rotary pleater, or more accurately the rotary score pleater is the most frequently used type of pleater, because it allows for high production rates. It typically uses two sets of rolls through which the medium travels. The first set of rolls contains score bars on the top and/or bottom roll to score the sheet using the other roll as an anvil. The illustration of Figure 5.60 shows all the score bars in the top roll. Figure 5.61 is a photograph of a scoring roll in its pleating line assembly. In many applications the score bars are placed in both rolls to alternatively score the sheet top then bottom.

**Table 5.2** Rabofsky Knife Pleating Machines Comparison Chart

| Characteristics | M2000 | ME2000 | SE2000 "Pleat Assistant" | E2000 "Pleat Master" | HE2002 "High Speed Pleater" |
|---|---|---|---|---|---|
| **Maximum physical speed** | – | – | Approx. 130 pl/min | Approx. 185 pl/min | Approx. 350 pl/min |
| **Set-up time (min)** | **15** | **8** | **3** | **1** | **1** |
| **Pleat height range (mm)** | **5–50 (5–100 optional)** | **5–50 (5–100 optional)** | **5–95** | **5–104** | **3–150** |
| **Intermediate pleats** | No | No | Yes | Yes | Yes |
| **Pleat programs** | No | No | M or W pleats | Free programmable | Free programmable |
| **Files to store** | No | No | 50 configurations | 1.000.000 pleat programs | 1.000.000 pleat programs |
| **Automatic recall of set values** | No | No | Almost completely | Completely | Completely |
| **Software-upgrade** | No | No | Yes | Yes | Yes |
| **Operation** | User elements | User elements | User terminal | Standard PC | Touch panel PC |
| **Network capability** | No | No | No | LAN capable | LAN capable |
| **Entry table positioning** | No | No | Manual | Automatic | Automatic |
| **Pressure plate positioning** | Manual, per scale | Manual, per scale | Automatic as option | Automatic | Automatic |
| **Heating system** | Quick or standard heating | Quick or standard heating | Quick or standard heating | Quick or standard heating | Quick or standard heating |
| **Number of drives** | 1 central drive | 1 central drive | 3 servo drives | 6 servo drives | 8 servo drives |
| **Slide guides** | Self-lubrication | Self-lubrication | Self-lubrication | Self-lubrication | Self-lubrication |
| **Material guide** | Option | Standard | Standard | Standard | Standard |
| **Standard counter pressure** | Pneumatic cylinders | Pneumatic cylinders | Pneumatic cylinders | Pneumatic cylinders | Pneumatic cylinders |
| **Pre-heating** | Option | Option | Option | Option | Option |
| **Pleat height adjustment** | Manual | Quick adjustment | Automatic | Automatic | Automatic |
| **Soft-touch** | No | No | Option | Yes | Yes |
| **Integrated pleat/pack counter** | Option | Option | Yes | Yes | Yes |
| **Flexibility/efficiency** | Normal | Normal | High | Very high | Extremely high |
| **Special packages** | HDI-Il | HD I-il | HDI-II | HD I-III, HS I | HS I as standard |
| **Remote service** | No | No | No | Option | Standard |

Source: Reproduced by permission of Karl Rabofsky Gmbh, Berlin, Germany.

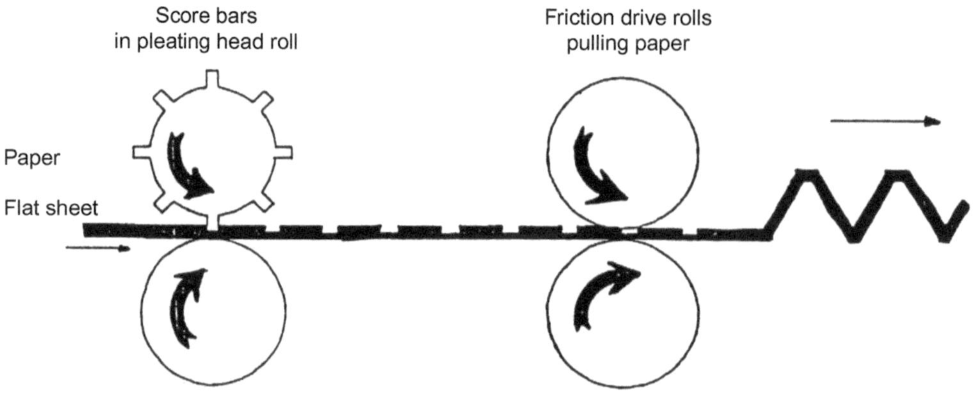


**Figure 5.60** Rotary Pleating Head Operation. *Source: Drawing provided by Charley Wood (deceased) formerly of Hollingsworth and Vose Company, West Groton, Massachusetts USA.*

**Figure 5.61** Rotary Pleating Head Roll. *Source: Reproduced by permission of Kaydon Custom Filtration Corporation, LaGrange, Georgia USA.*

Also depicted in Figure 5.60 is a second set of rolls. These are friction drive rolls that pull the sheet through the pleating operation. Back pressure after the drive rolls causes folding of the paper at the score marks. To a limited extent, pleat height can be adjusted by changing drive gears, which in turn change the relative rotational speed

**Figure 5.62** Rotary Pleater With Embossing Roll. *Source: Reproduced by permission of Manufacturing Machine Corp., Pawtucket, Rhode Island USA.*

between the score and drive rolls. Larger changes in pleat height are accomplished by changing the number of score bars active in scoring the flat sheet. This is done either by timing changes on the same set of score rolls or by replacing the score roll assembly with another assembly that has a different number of score bars. Modern rotary pleaters have rubber inserts on the score rolls that act as backing anvils for the score blades. This improves medium flexibility and reduces medium damage from over-scoring.

Figure 5.62 is a rotary pleating line produced by Manufacturing Machine Corp. (MMC) of Pawtucket, Rhode Island USA. Visible on the inlet side of the pleating head is the embossing roll that pressures dimples into the medium. The dimples serve as pleat separators in the pleat package. The MMC pleater also has the capability to apply hot melt along the pleat edges that will seal the folded edges and form pleat pockets. Figure 5.63 is an example of pleated medium produced by the MMC machine. Note that the pleat package contains embossed dimples to serve as pleat separators. Also note that the pleat edges are sealed and each pleat fold thus becomes a pocket pleat (that is, with radial flow from inside to out).

Figure 5.64 is rotary mini-pleat pleating line by MMC that is designed to pleat ASHRAE and HEPA glass microfiber filter media. It is capable of providing hot melt beads to both sides of the medium. The beads serve as pleat separators for the mini-pleat configuration.

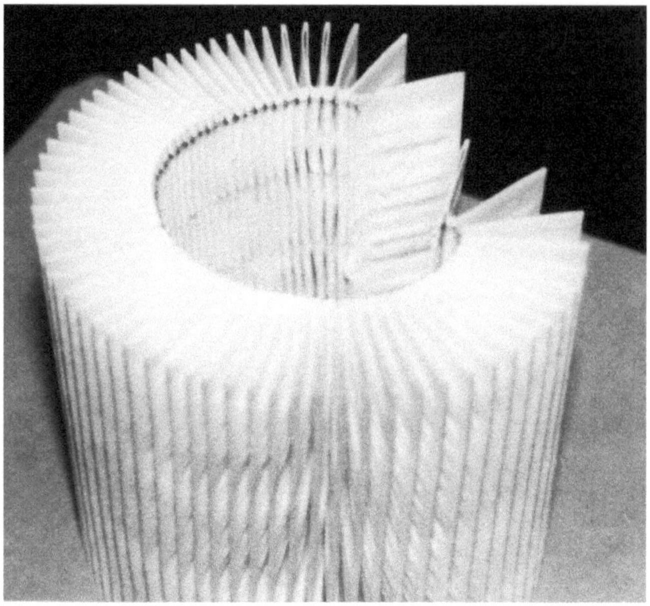

**Figure 5.63** Pleated Filter Medium with Dimples and Pocket Filter Configuration. *Source: Reproduced by permission of Manufacturing Machine Corp., Pawtucket, Rhode Island USA.*

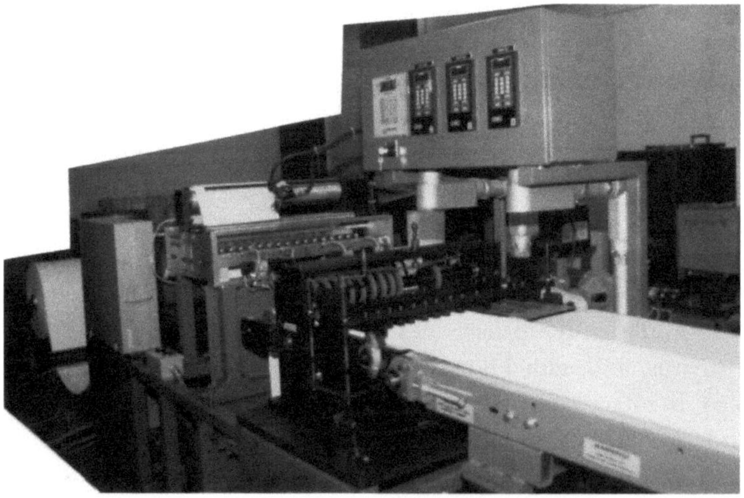

**Figure 5.64** Rotary Mini-Pleat Line for ASHRAE and HEPA Glass Microfiber Media. The Line is Capable of Applying Hot Melt Beads to Both Sides of the Medium—the Beads Serving as Pleat Separators. *Source: Reproduced by permission of Manufacturing Machine Corp., Pawtucket, Rhode Island USA.*

## 5.9 MEMBRANES

### 5.9.1 Track edge membranes

Photographs of track edge membranes are shown in Figures 3.21 and 3.22 of Section 3.13.1.1.

Sterlitech Corp. manufactures polyester track edge (PETE) membranes and describes their process as follows (220):

> *In the first step of the PETE membrane production process, thin polyester film is exposed to collimated, charged particles from a nuclear reactor. As these particles pass through the polyester material, they leave sensitized tracks. In the second step, the polymer tracks are dissolved with an etching solution to form cylindrical pores. Varying the temperature and strength of the etching solution and the exposure time to it ultimately produces precisely controlled pore sizes in the PETE membrane film.*

Sterlitech also reports that "The resulting membrane is a thin, translucent, microporous polyester film with a smooth, flat surface containing pores of controlled diameter and number." They also indicate that the diameter of the holes may vary .05–3.0 microns depending on the conditions of etching.

### 5.9.2 Cast membranes

EMD Millipore[9] is a major producer of filter membrane materials. On their website (221) they illustrate three types of film casting: Air casting, solvent or emersion casting and thermal casting. The three types are described below in Sections 5.9.2.1–5.9.2.3, respectively. They are also described Figures 5.65, 5.66, and 5.67 respectively. The captions, of each figure, briefly describe the pertinent process.

In addition there is a process for making stretched membranes described in Section 5.9.3. The production of inorganic membranes is discussed in Section 5.9.4.

#### *5.9.2.1 Air casting*

A polymer solution is extruded unto a chilling roll where it or partially solidifies into a cast film. The film is transported to an extraction and drying chamber where the solvent is evaporated in a controlled manner (Figure 5.65).

#### *5.9.2.2 Solvent or emission casting*

Solvent or emission casting takes advantage of the phenomena that if a polymer solution is introduced into a non-solvent liquid, the polymer will precipitate into a gum-like mass that becomes micro or nano porous as the solvent tries to escape through the polymer mass. By laying down the polymer solution unto a smooth surface before contact with the non-solvent a porous film or membrane can be formed (Figure 5.66).

[9] EMD Millipore is a division of Merck KGaA, Darmstadt, Germany.

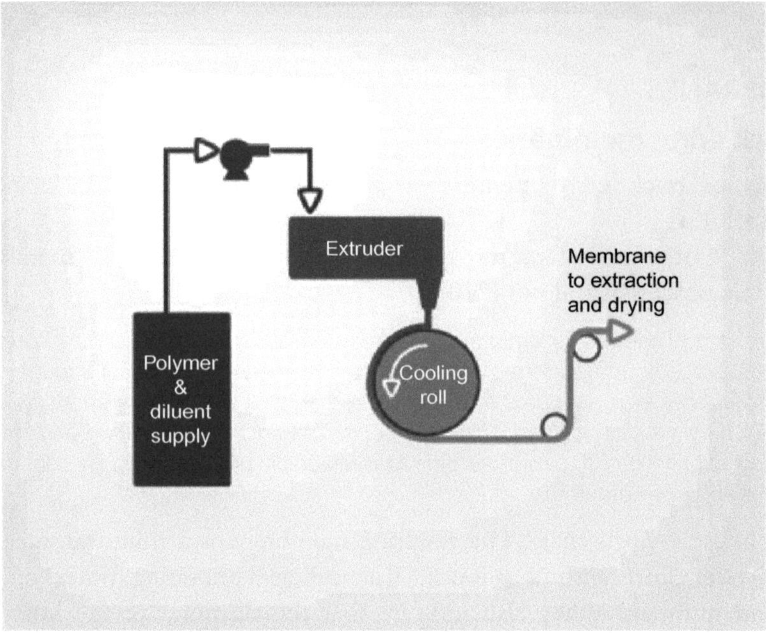


**Figure 5.65** Air Casting: The Dissolved Polymer Solution Passes Under a Series of Air Flow Ducts that Control the Evaporation of the Solvents in a Very Slow Manner. *Source: Reproduced with permission of EMD Millipore Corporation, Billerica, MA USA.*

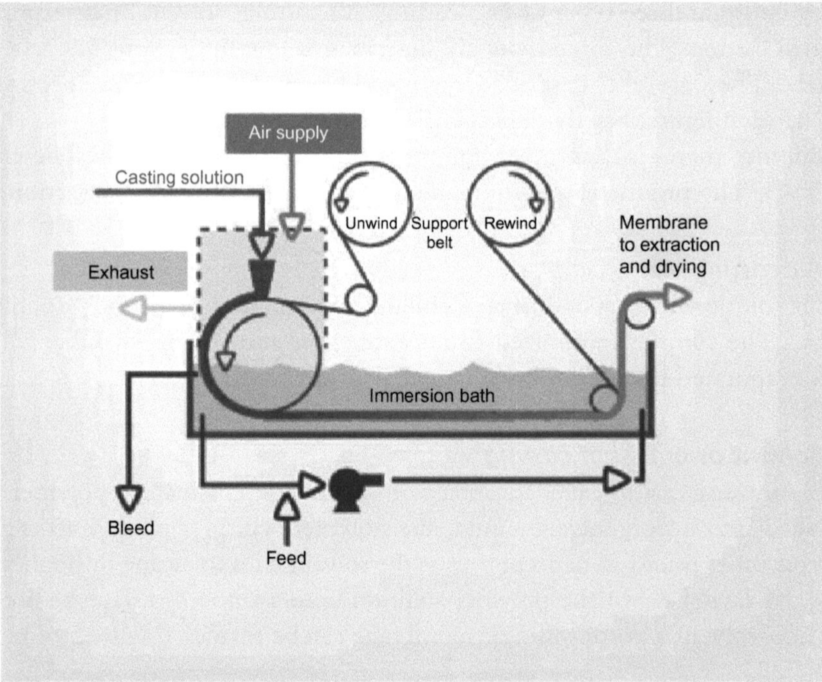


**Figure 5.66** Solvent or Emersion Casting: The Dissolved Polymer is Spread onto a Moving Belt and Run Through a Bath of Liquid; the Liquid in the Bath Exchanges with the Solvent in the Lacquer and Causes the Formation of the Pores. *Source: Reproduced with permission of EMD Millipore Corporation, Billerica, MA USA.*

### 5.9.2.3 Thermal casting

In thermal casting, heat and subsequent cooling is used to precipitate the polymer from the polymer solution and form it into a porous membrane form (see Figure 5.67 below). The plasticized membrane is then goes through extraction and drying steps.

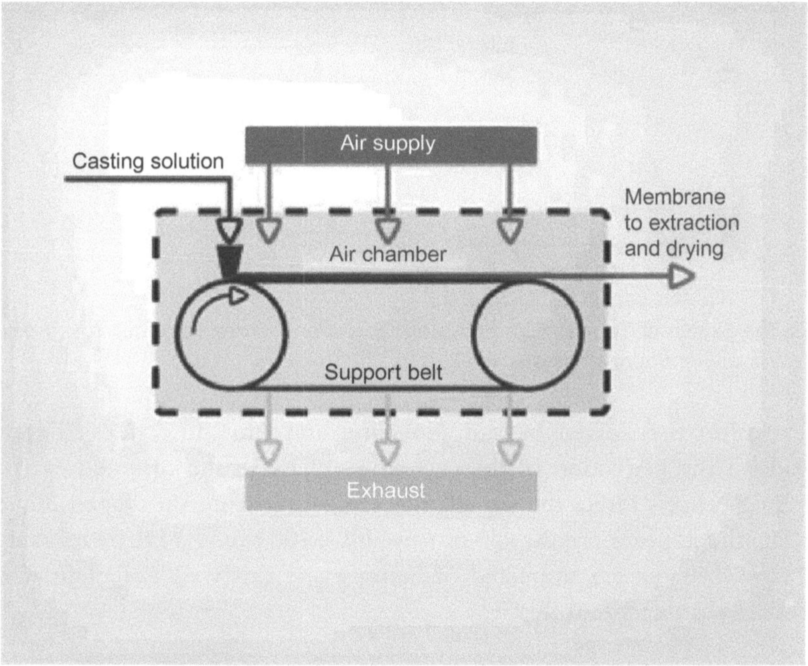


**Figure 5.67** Thermal Casting: Heat is Used to Drive the Solubility of the Polymer in a Given Solvent System. The Lacquer is then Cast Out onto a Moving Belt that is Being Cooled. Quenching the Heat in the Lacquer Causes Precipitation to Start and the Pores to Form. *Source: Reproduced with permission of EMD Millipore Corporation, Billerica, MA USA.*

## 5.9.3 Stretched membranes

Wilkes (136) describes a procedure of making microporous membranes by the uniaxial stretching method. It include three different stages: 1. extrusion, 2. annealing, and 3. stretching. In the first stage, a tubular film extrusion operation, as shown in Figure 5.68, is utilized to melt extrude raw resins into uniaxially oriented films. The film so formed is collapsed between collapsing guides and then flattened between nip rolls. The resulting morphology and orientation depends on the processing conditions and the characteristics of the raw resins. It is crucial to achieve the stacked lamellar morphology in this process because only stacked lamellae are able to open during the process of deformation.

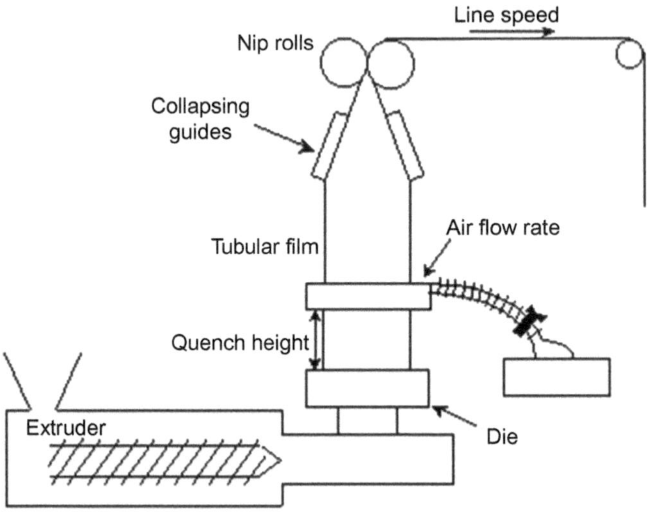


**Figure 5.68** The Sketch of Tubular Film Extrusion Operation. *Source: Reproduced with Permission of Dr. Garth L. Wilkes, Blacksburg Virginia USA.*

The extrusion is followed by the annealing and stretching steps. The annealing process leads to the perfection of the crystalline phase in the oriented extruded film. The stretching process forces the lamellae to separate leaving slit shaped micropores in between. Slit shape pores are 20–40 nm by 200–400 nm. PTFE membranes such as Goretex® and Tetratex are suspected of being made this way. Polyolefin Celgard®[10] membranes are manufactured this way.

## 5.9.4 Inorganic membranes

A good description of the process techniques for manufacturing inorganic membranes is provided by Kang Li (222). Following is a brief summary of the basic membrane processes.

### *5.9.4.1 Slip casting*

The most common technique for preparing inorganic membranes is the slip casting of suitable clay particles followed by a sintering process to form a cast medium. Slip is the slurry of clay that is poured into a designed mold. It is specially formulated with particles of the preferred clays and inorganic ingredients. Other ingredients will include dispersing and deflocculating agents, and possibly, a carbonaceous combustible material. The slip is then poured into suitable molds porous enough to absorb and extract the liquid solvent from the slurry. The particulate matter forms a layer or gel along the surface of the mold. Once dried in the mold the cast is removed as a weak

[10] Celgard® is a tradename of Celgard LLC, Charlotte, North Carolina, USA.

integrated solid piece. It is then furnace baked at very high temperatures, high enough to bond and sinter the inorganic particles and burn off any carbonaceous material that might have been added to the slip. The burning off of the carbonaceous material helps contribute to the porous structure. Usually a slip concentration is in the range of 40–60%. Furnace temperatures are in the range of 1400 °C. Membrane properties such as permeability and pore size are determined by the size of clay particles used.

### 5.9.4.2 Tape casting

A variation of slip casting for membrane manufacture is tape casting. The process is diagrammed in Figure 5.69. A slip slurry is coated unto a moving tape or carrier and passes under a knife edge to form a layer on the tape. Thickness of the layer is controlled by distance between the knife edge and the moving tape. The thickness is usually around 250—the layer then goes through a heating stage (Figure 5.69).

### 5.9.4.3 Disc membranes

Disc membranes may often be composite structures of more than one layer, with each layer possibly being of a different material. The bottom layer or downstream layer is often used as a support layer. Khallilia et al. (223) describe pressing as a commonly used method for preparation of disk inorganic membrane supports. A special press machine is used to apply more than 100 MPa pressure to press powders into a compacted disk. After the powder compaction, membranes supports are sintered at high temperatures in order to make a rigid and resistant structure (mechanically). For support purposes, additives which are called binders should be added to the initial powder. The role of these binders is to keep together the particles of this initial powder and to get defect-free membranes.

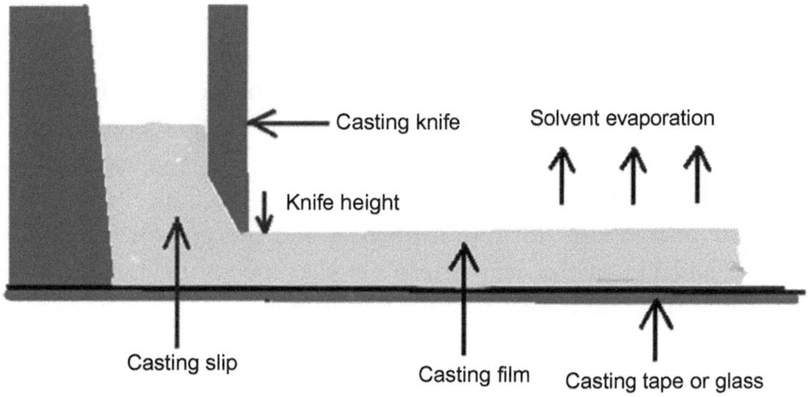


**Figure 5.69** Tape Casting. *Source: Reproduced with permission of John Wiley & Sons, Inc.*

#### 5.9.4.4 Extrusion

The extrusion process is a way of making membrane support tube media. It utilizes a stiff paste of the particulate matter and forces it through an extrusion nozzle. The tubular extrusion process is similar in principle as the extrusion process for stretched membranes described in Section 5.9.3. A die in the center of the nozzle forces the paste through the annular opening between the die and the nozzle opening. This forces the paste to extrude into a tubular geometry. Placing the tubular support material, so formed, on rotating rollers causes the tubular paste to rotate and to dry. The ceramic tube is then fired to remove all organic binders and chemicals utilized in the ceramic manufacturing process, such as the starch, amijel and methocel. Firing also remove any residual water.

#### 5.9.4.5 Sol gel process

Sols are colloidal dispersions of inorganic particles or polymers prepared in alcohol water solutions by the hydrolysis and condensation of metal alkoxide precursors. Inorganic membranes are prepared by deposition of dilute sols on porous inorganic supports. One way of doing this is by dip-coating. In the dip-coating process, the support is immersed in the sol and then withdrawn vertically at a constant rate. During immersion, capillary suction causes flow of the sol into the pores of the support. During withdrawal, additional sol is metered onto the support surface. The depositing sol experiences evaporation leading to aggregation and gelation of the entrained inorganic constituents. Beyond the gel point further evaporation causes the development of a tensile capillary stress in the gel pore fluid. The capillary stress causes film shrinkage resulting in a reduction in pore volume and pore size. Shrinkage of the external membrane layer causes a reduction in film thickness. Following withdrawal the membrane is subject to a series of finishing procedures which includes aging, drying, dehydration, and densification.

#### 5.9.4.6 Anodization

Anodic alumina nanotemplates have been investigated extensively since the 1960s and they have been manufactured by different methods. The general fabrication method starts with a film such as aluminum being exposed to an acidic medium, generally sulfuric, oxalic, phosphoric acid, or some combination thereof. A voltage is then applied. Because the medium is acidic, an oxide layer forms over the film. During the anodization, the applied voltage leads to pore formation and the growth of the oxide layer. While the type and pH of the acid determine the inter-pore distances, the applied voltage determines the pore diameter of the templates.

CHAPTER 6

# Testing of Nonwoven Filter Media

## 6.1 GRAMMAGE OR BASIS WEIGHT

Grammage usually follows the metric (SI) system. Its unit is g/m$^2$. Basis weight usually refers to United States customary units such as reams of material, lb/1,000 ft$^2$, or g/yd$^2$. Ream units are used by the American paper industry. The most common form for filter media is lb/3,000 ft$^2$. This is developed from the weight of a ream containing 500 sheets, 24 in × 36 in (24 × 36−500). Filter medium basis weights have also been based on ream sizes of 2,880 ft$^2$ (24 × 36−480). Sheeted filter media used for industrial filtration such as plate and flame filters are often listed at ream sizes of 1,389 ft$^2$ (20 × 20−500). In referring to basis weights based on ream sizes, one must make sure that the ream size area is understood. The basis weight unit of lb/1,000 ft$^2$ is used by the American paperboard industry. The units of oz/yd$^2$ and g/yd$^2$ are popular with the American textile industry. These are units often applied to nonwovens produced by dry formed processes. Because of the confusing nature in which basis weights and grammages are reported in the United States, the following (Table 6.1) is offered. As much as possible, grammage and the units of g/m$^2$ will be used in this handbook.

Grammages of nonwoven filter media can range from as high as 900 g/m$^2$ for certain types of needlefelt media to as low as 0.1 g/m$^2$ for supported melt-blown webs. The electrospun layers of high efficiency composites are in the nano-grammage range. Petrik et al. (224) report grammages as low as 0.01−0.1 g/m$^2$.

## 6.2 VOLATILES AND MOISTURE CONTENT

Moisture or volatiles content of the web is part of the "as is" basis weight. They affect the economics as well as the properties of the medium. Moisture content of

**Table 6.1** Conversion of Basis Weight (U.S. Custom Units) to Grammage (gm/m$^2$)

| U.S. custom unit | Ream size | Conversion to g/m$^2$ |
|---|---|---|
| lb/3,000 ft$^2$ | 24 × 36−500 | 1.628 |
| lb/2,880 ft$^2$ | 24 × 36−480 | 1.695 |
| lb/1,389 ft$^2$ | 20 × 20−500 | 3.515 |
| gm/yd$^2$ | – | 1.196 |
| oz/yd$^2$ | – | 33.93 |

*Handbook of Nonwoven Filter Media*
ISBN: 978-0-08-098301-1

**Table 6.2** Polymer Specific Gravities and Equilibrium Moisture Contents of Various Fibers Used for Nonwoven Media

| Fiber | Specific gravity[a] | Moisture content (%)[b] |
|---|---|---|
| Acrylic[a] | 1.15 | 1–2 |
| Cellulose acetate | 1.32 | 6 |
| Cellulose triacetate[a] | 1.25 | 2.5–4.5 |
| Cotton | 1.54 | 7 (commercial = 8.5) |
| Glass[a] | 2.54 | 0 |
| Polyamide[a] (nylon 6 and nylon 66) | 1.14 | 4.1–4.5 |
| Polyester[a] | 1.38 | 0.4–0.5 |
| Polyethylene[a] | 0.92 | 0 |
| Polypropylene[a] | 0.90 | 0 |
| Polyurethane[a] | 1.21 | 1.0–1.5 |
| Polyvinyl chloride[a] | 1.38 | 0–1 |
| Polyvinylidene chloride[a] | 1.70 | 0 |
| Protein[a] | 1.25 | 10–18 |
| Silk | 1.37 | 9 |
| Viscose rayon | 1.51 | 13 |
| Wool | 1.32 | 13–15 |

[a]Specific gravity is the weight of one cubic centimeter of polymer at 20°C and at atmospheric conditons. It is not necessarily the specific gravity of the fiber, since cellular fibers would have to account for cellular voids in the cellular structure.
[b]Equilibrium moisture contents at 21°C and 65% RH.

the web will relate to the equilibrium moisture content of its raw materials and the temperature and humidity to which it has been exposed. Equilibrium moisture content is the moisture content in equilibrium with the temperature and relative humidity of the exposure atmosphere. Table 6.2 includes equilibrium moisture contents of several fiber types used in nonwovens and textile materials. The equilibrium conditions are 21°C and 65% RH (relative humidity). Test methods such as TAPPI T550 om-13 requires a preconditioning period (usually 24 h) at a specified temperature and relative humidity followed by a conditioning period at 23°C and 50% RH (TAPPI T402). Media containing cellulosic fibers will have a hysteresis effect when conditioned. The equilibrium moisture content will be different for a specimen conditioned from a wetter stage than the same sample conditioned from a drier stage.

Volatiles refer to the total volatiles from the web including moisture and organic volatiles. This is not to be confused with VOC (volatile organic content), the environmental term that refers to the volatile organic only content in the web. Resin-bonded webs often contain both moisture and organic volatiles, such as formaldehyde, that are released as a result of the curing process. When reporting grammage or basis weight it is important to note whether it is reported on as is basis (moisture and/or volatiles included) or on a bone dry basis.

For resin-treated webs, the volatiles content is often a specified and measured property of the web. It affects other properties including stiffness and dimensional stability. For thermosetting webs it is sometimes used as measure and control for degree of cure. Volatiles such as VOC and formaldehyde are strongly regulated by government regulatory agencies (see Section 6.3 below for formaldehyde content).

The simplest method to determine volatiles is to weigh the medium specimen, then heat it in an oven at a specified temperature and time period. The difference in weight before and after heating is the volatiles. 150°C for 10 min is a frequently used heating period for determining volatiles content. The author has not found any standard method for volatiles content other than TAPPI T1012 om-10, Moisture Content of Fiberglass Mats.

## 6.3 FORMALDEHYDE CONTENT

Major environmental and workplace safety concerns deal with formaldehyde emissions from the filter medium. This is usually due to a formaldehyde-type resin or cross-linker in the medium's resin formulation. While the author has been unable to find a formaldehyde standard or test specific for filter media, one must assume that those that apply to fabrics or textiles also apply to filter media.

Fretz (225) identified the following test methods to test for formaldehyde in both materials and polymer dispersions.

- ISO 15373-2001 (Method A): "Plastics—Polymer dispersions—Determination of free formaldehyde"
- ISO 15373-2001 (Method B): "Plastics—Polymer dispersions—Determination of free formaldehyde," equivalent method: ASTM D5910-05
- ASTM D5910-05: "Determination of free formaldehyde in emulsion polymers by liquid chromatography," equivalent method: ISO 15373-2001 (Method B)
- ISO 14184-1 (Method A): "Textiles—Determination of formaldehyde—Part 1: Free and hydrolyzed formaldehyde, water extraction method," equivalent methods: EDANA ERT 210.1.99 and Japanese Law 112
- ISO 14184-1 (Method B): "Textiles—Determination of formaldehyde—Part 1: Free and hydrolyzed formaldehyde, water extraction method," equivalent method: EDANA ERT 212.0.96
- ISO 14184-2: "Textiles—Determination of Formaldehyde—Part 2: Released formaldehyde (Vapor absorption method)," equivalent methods: EDANA ERT 211.1.99, AATCC 112, and KS (K)
- AATCC Test Method 112-2008: Formaldehyde release from fabric, determination of: sealed jar Method

AATCC Test Method 112-2008 and Japanese Ministerial Ordinance #112 (also known as Japanese Industrial Standard JIS L 1041) are probably the most common methods for determining formaldehyde release from a fabric. In the AATCC method a weighed fabric specimen is suspended over water in sealed jar. The jar is placed in an oven at a controlled temperature for a specified length of time. The amount of formaldehyde absorbed by the water is then determined colorimetrically. The 29 CFR 1910.1048(m)(1)(iv) (US) specifies that if the free formaldehyde content of a medium is greater than 1,000 ppm (0.1%) then the package must be labeled for formaldehyde (see Chapter 10, Section 10.1.4.1.3).

The Japanese method is similar to the AATCC method, however the United States Government Accountability Office (GAO) (226) suggests that the AATCC test results in higher formaldehyde levels than the Japanese test except at low formaldehyde levels, such as 30 parts per million or less—although the difference between the formaldehyde levels measured by the two tests depends, in part, on the type of fabric and resin used.

## 6.4 CALIPER OR THICKNESS

Caliper and thickness are synonymous terms referring to the vertical surface-to-surface distance of the web. The thickness measurement is simple enough. Usually the web is placed under the presser foot of a thickness gauge, either manually operated or automatic, and the thickness is measured when the foot comes to rest on the surface of the medium. There are problems. Filter media by their design are bulky structures that are compressible. If the presser foot is exerted with too much force it can sink into the compressible web and give a false low reading. On the other hand, if the presser foot is loaded with too little force, a few loose fibers at the surface, can cause a misleading high caliper measurement. Any test method for caliper or thickness must specify the size and shape of the presser foot and the loading applied. Almost any organization that specifies test methods for nonwovens or paper will have a test method for caliper. TAPPI's method is T411 om-10. INDA–EDANA-harmonized methods are WSP 120.1R4(12) for irregular nonwovens and WSP 120.2R4(12)for highloft nonwovens. The ISO Method is 9073:1995, Part 2, has provisions for testing bulky nonwovens up to 20 mm, and bulky nonwovens greater than 20 mm.

Since thickness is measured over very small areas of the web, variations in the web can effect the thickness measurement. If the presser foot happens to be placed over a thick spot in the web it will give a different measurement than if it is placed over a thin spot in the web. Likewise, surface roughness will cause variations in the measurement. Bumpy surfaces will give more variation than smooth surfaces.

Corrugated filter media are widely used in pleated filter cartridges. They require special measurements as depicted in Figure 6.1. Overall caliper is the distance from bottom to top of the sine wave configuration. Thickness is the vertical distance

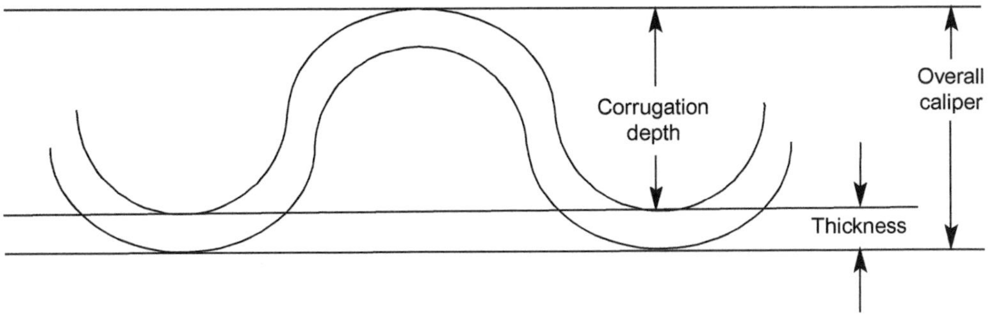


**Figure 6.1** The Measurement of Corrugated Filter Media.

between parallel surfaces of the medium. Corrugation height is the distance from the tops of the hills to the bottoms of the valleys of the corrugations.

Note from Figure 6.1 that measurement of two of the corrugation properties fixes the third. An ordinary caliper with a flat presser foot can be used to measure overall caliper, however a wedge-shaped presser foot that can fit between corrugations is necessary for the measurement of thickness and corrugation depth. Often an image-projecting device such as an optical comparator or a shadowgraph is used to measure corrugation.

## 6.5 AIR PERMEABILITY

### 6.5.1 The testing and measuring of air permeability

Air permeability[1] of a nonwoven filter medium is the measured airflow through a specified area of filter medium at a specified pressure drop. There are other methods that fix the flow and measure the pressure drop. However, the most popular way for a filter medium is to fix the pressure drop and measure the airflow. The most commonly used standard test method is ASTM D737-04. The instrument is most frequently identified as The Frazier Air Permeameter, manufactured by the Frazier Precision Instrument Company, Inc, Hagerstown, Maryland, USA. It is diagrammed in Figure 6.2. It has been used for years for this type of measurement. The air is sucked through the clamped medium specimen into an enclosed chamber and out through an orifice that measures the flow. A manometer across the medium gives the pressure drop and a manometer across the orifice gives the pressure drop across

[1] A related term is air permeance, usually applied to the measurement of air flow through paper and paper board. Testing is in accordance with ISO 5636-3 and SCAN-P 85:02 (Bendtsen Method), ISO 5636-4 and TAPPI T547 om-2 (Sheffield Method), ISO 5636-2 (Schopper Method), or ISO 5636-5 (Gurley Method). Units are measured in terms of air flow (per unit area) per unit of pressure drop per unit of time, that is, micrometres per Pascal second (μm/Pa · s).

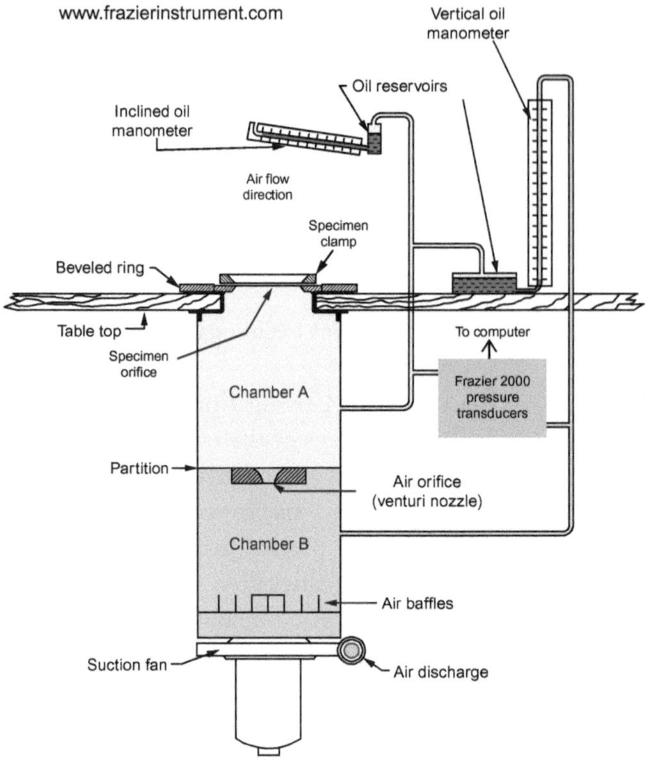


**Figure 6.2** Frazier Differential Pressure Air Permeability Tester. *Reproduced by Permission of the Frazier Precision Instrument Company, Inc., Hagerstown, Maryland, USA.*

the orifice for calculation of air flow. The manufacturer provides data tables for converting orifice pressure drop to air flows. Figure 6.3 is a photograph of the Frazier 2000®—Air Permeability instrument. This modern version implements electronic pressure sensing as well as computerized logging and manipulation of the electronic measurements.

The air flow of the Frazier type permeameter measured is based on Darcy's Law discussed in Chapter 2.

$$v_0 = -\frac{k}{\mu}\frac{\Delta p}{L} \tag{6.1}$$

In any measurement the pressure drop, $\Delta p$, and the viscosity of air, $\mu$, is fixed. The measured air flow, $v_0$, then reflects on the permeability constant, $k$, and the thickness, $L$, of the media. The permeability constant relates to the structure of the media.

Traditionally the air permeability is expressed in units of $ft^3/ft^2$ @ $\Delta p = 0.5$ in $H_2O$ and simply reported as cfm. The equivalent in S.I. units is $m^3/m^2$ @

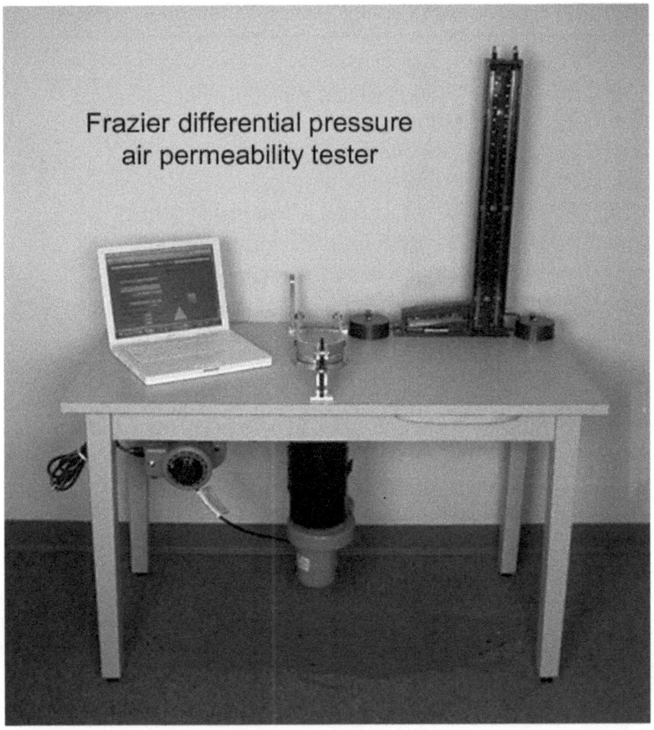


**Figure 6.3** Frazier 2000TM-Air Permeameter. *Reproduced by Permission of the Frazier Precision Instrument Company, Inc., Hagerstown, Maryland, USA.*

$\Delta p = 12.7$ cm $H_2O$. Often it is reported as $m^3/m^2$ @ $\Delta p = 20$ cm $H_2O$. Table 6.3 below, from the website of Frazier Instrument Company (227) lists many of the conversions for the various units of measurement.

## 6.5.2 The cost of permeability

Permeability costs money. This is illustrated in Figure 6.4, which shows generic cost of a presumably 130 $g/m^2$ wet lay medium. Note the minimum cost of this type of medium is at roughly 20 cfm, because it can be produced from essentially 100% hardwood pulp, the lowest costing of filter fibers. Moving in either direction from this point to the other results in a requirement for more expensive fibers. For lower permeabilities, one must resort to more expensive fine fibers such as nanofiber structures and glass microfiber. On the other hand, higher pemeability will require more expensive softwood and high bulk pulp fibers. At very high permeabilities very expensive high denier synthetics such as 12 and 20 denier rayon may have to

**Table 6.3** Conversion Factors for Permeability Testing Differential Pressure – (Inches of Water)

| | |
|---|---|
| 1 | (Inches of water) × 0.0833 = (feet of water) |
| 2 | (Inches of water) × 25.4 = (millimeters of water) |
| 3 | (Inches of water) × 2.54 = (centimeters of water) |
| 4 | (Inches of water) × 0.254 = (decimeters of water) |
| 5 | (Inches of water) × 0.0254 = (meters of water) |
| 6 | (Inches of water) × 0.0024582 = (physical atmosphere (atm)) |
| 7 | (Inches of water) × 0.0025399 = (technical atmosphere) |
| 8 | (Inches of water) × 0.073554 = (inches of mercury) |
| 9 | (Inches of water) × 1.8683 = (millimeters of mercury (torr)) |
| 10 | (Inches of water) × 0.036126 = (pounds per square inch (psi)) |
| 11 | (Inches of water) × 5.2022 = (pounds per square foot (psf)) |
| 12 | (Inches of water) × 25.399 = (kilograms per square meter) |
| 13 | (Inches of water) × 0.0025399 = (kilograms per square centimeter) |
| 14 | (Inches of water) × 249.08 = (pascals (Pa) or (newtons per square meter)) |
| 15 | (Inches of water) × 0.24908 = (kilopascals (kPa)) |
| 16 | (Inches of water) × 2.4908 = (milibars) |
| 17 | (Inches of water) × 0.0024908 = (bars) |
| Air Flow – (cubic feet per square feet of sample per minute) or (feet per minute) or (CFM) | |
| 1 | (CFM) × 0.3048 = (cubic meters per square meter of sample per minute) or (meters per minute) |
| 2 | (CFM) × 0.00508 = (cubic meters per square meter of sample per second) or (meters per second) |
| 3 | (CFM) × 30.48 = (cubic centimeters per square centimeter of sample per minute) or (centimeters per minute) |
| 4 | (CFM) × 0.508 = (cubic centimeters per square centimeter of sample per second) or (centimeters per second) |
| 5 | (CFM) × 3.048 = (cubic decimeters per square decimeter of sample per minute) or (liters per square decimeter of sample per minute) or (decimeters per minute) |
| 6 | (CFM) × 0.0508 = (cubic decimeters per square decimeter of sample per second) or (liters per square decimeter of sample per second) or (decimeters per second) |
| 7 | (CFM) × 18.288 = (cubic meters per square meter of sample per hour) or (meters per hour) |

Reproduced by permission of the Frazier Precision Instrument Company, Inc., Hagerstown, Maryland, USA.

be worked into the fiber composition. Of course, there are other ways to adjust permeability in a filter medium. For wet lay media, refining can be used. In other forms of media, adjustments to permeability can be accomplished by increasing or decreasing basis weight and/or density, provided these degrees of freedom are allowed (Table 6.3).

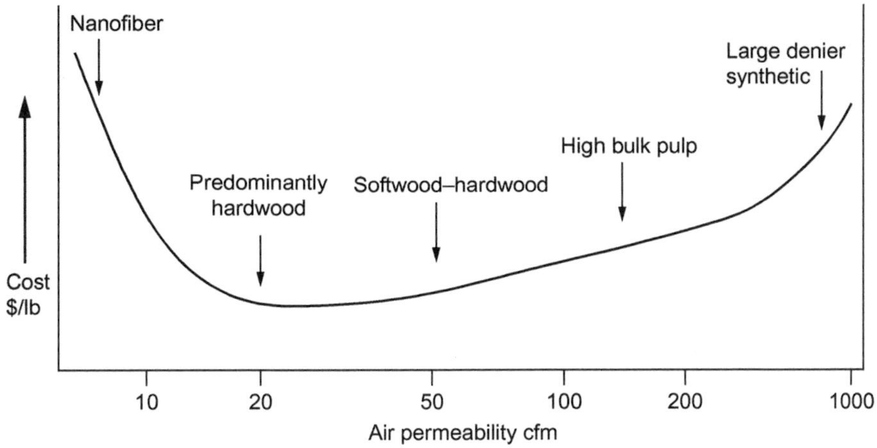


**Figure 6.4** The Cost of Permeability.

## 6.6 DENSITY AND BULK

Density and bulk are properties of nonwoven filter media that are related to its filtration properties particularly dirt holding capacity (DHC). Density, $\rho_w$, is the mass per unit volume of the filter medium in g/cm$^3$ and is calculated from the grammage and thickness as follows:

$$\rho_w = \frac{G}{t} \tag{6.2}$$

$\rho_w$ is the density of the web in g/cm$^3$.
$G$ is the grammage of the web in g/m$^2$.
$t$ is the thickness of the web in μm.
Bulk, $\beta_w$, is the reciprocal of the density and is expressed as cm$^3$/g of media.

$$\beta_w = \frac{1}{\rho_w} = \frac{t}{G} \tag{6.3}$$

Table 6.4 lists some conversion factors if the grammage or basis weight is expressed in units other than g/m$^2$ and/or the thickness is expressed in units other than cm.

## 6.7 SOLIDITY AND POROSITY

In simplifying terms solidity can be identified as that part of the medium that is something, whereas, porosity is identified as that part of the medium that is not nothing. Solidity, $\chi$ (often referred to as packing density), is the volume of solid material in a unit volume of media. It can also be expressed as the volume fraction of the solids in the medium. In a nonwoven, the solids will include fiber, resin, additives, and any other ingredient that contributes to the solid volume of the structure. Porosity,

**Table 6.4** Conversion Factors for Calculating Density from Various Units of Grammage and/or Basis Weight and Caliper

| If the units for grammage, G, are: | And the units for Caliper, t, are: | Then to get density in g/cm$^3$ multiply G/t by the value below: |
|---|---|---|
| gm/m$^2$ | μm | 1 |
| lb/3,000 ft$^2$ | mil | 0.06414 |
| lb/3,000 ft$^2$ | in | 0.00006414 |
| lb/3,000 ft$^2$ | μm | 1.628 |
| lb/2,880 ft$^2$ | mil | 0.0668 |
| lb/1,389 ft$^2$ | mil | 0.1385 |
| gm/yd$^2$ | mil | 0.0471 |
| gm/yd$^2$ | in | 0.0000471 |
| gm/yd$^2$ | μm | 1.196 |
| oz/yd$^2$ | mil | 1.333 |
| oz/yd$^2$ | in | 0.001333 |

Printed with permission of the Frazier Precision Instrument Company, Inc., Hagerstown, Maryland, USA.

$\varepsilon = (1 - \chi)$, is the three-dimensional volume void fraction of the medium. Porosity is not to be confused with air permeability. Permeability as described above is the flow rate of fluid through the medium at a given pressure drop. Both the solidity, $\chi$, and the porosity, $\varepsilon$, are important properties of filter media that are very evident in the theoretical discussions of Chapter 2. They can be "apparently" calculated from basis weight and caliper of the medium and the density of its materials as follows:

$$\chi = \frac{G}{t\rho_m} \tag{6.4}$$

$$\varepsilon = 1 - \chi = 1 - \left(\frac{G}{t\rho_m}\right) \tag{6.5}$$

$\chi$ and $\varepsilon$ are the solidities and porosities of the web, respectively, in cm$^3$/cm$^3$.

$\rho_m$ is the weight averaged density of the fibers, resins, and other additives making up the web in g/cm$^3$.

Note from Eq. (6.4) the solidity or the web is calculated from the density of the web, $\rho_w$, as determined from Eq. (6.2) divided by the density, $\rho_m$, of its fibers and materials. For a filter medium, it is important to note the difference between solidity and density and between porosity and bulk. The terms are sometimes confused. The following illustrates the difference:

- Density and solidity
  - Density is the weight or mass of fiber per unit volume of medium (g/cc).
  - Solidity is the volume of fiber per unit volume of medium (cc/cc).
- Bulk and porosity
  - Bulk = 1/density: is the volume of medium per unit weight or mass of fiber (cc/g).
  - Porosity = 1 − solidity: is the volume of void space per unit volume of medium (cc/cc).

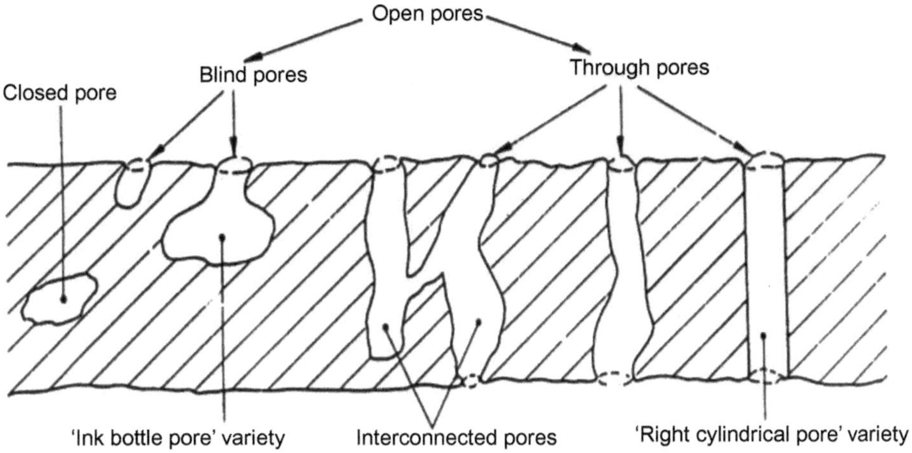


**Figure 6.5** Schematic of Various Pore Types. *Printed with Permission of Elsevier Advanced Technology, London, England.*

**Table 6.5** Porosity Definitions

| **Term** | **Definition** |
|---|---|
| Total porosity, $\varepsilon = 1 - \chi = 1 - \left(\frac{G}{t\rho_m}\right)$ | Volume fraction of pore space, or 1 − volume fraction of solid matrix |
| Effective flow porosity, $\varepsilon_{eff}$ | Volume fraction of pore space open to flow. |
| Relative flow porosity, $\varepsilon_{rel}$ | Fraction of total pore space open to flow $\varepsilon_{rel} = \varepsilon_{eff}/\varepsilon$ |
| Extrafiber porosity, $\varepsilon_0$ | Volume fraction of pore space between swollen fibers, or one-volume fraction of swollen fibers |

Reproduced with permission of TAPPI, Atlanta, Georgia, USA.

The reader should exercise caution in using Eqns (6.4) and (6.5). Chapter 2 (Section 2.2.2.3) mentions how the complexities of cellular fiber structure, closed pores, and blind pores effect permeable void volume. Sutherland and Purchas (18) provided Figure 6.5 above. to illustrate the various pore complexities.

Chapter 2, Section 2.2.3.3 discussed Lindsay's (47) approach to this problem for water in pulp mats. He refers to the fiber cell structure that is not available to fluid flow as being swollen fiber. He tabulates the definitions of the various porosities in Table 6.5 above.

In Chapter 2, Section 2.2.3.3, Lindsay defined $\varepsilon_0$ in terms of $\alpha$ and $c$ (see Chapter 2, Eq. (2.25)).

$$\varepsilon_0 = 1 - \alpha\chi \tag{6.6}$$

$\alpha$ is the volume of swollen fiber per unit mass of fiber in the medium, cm$^3$/g, and $c$ is the mass concentration of fiber per unit volume of medium, g/cm$^3$.

Rearrangement with the Kozeny–Carman equation yields (see Chapter 2, Eqns (2.23) and (2.26)):

$$(kc^2)^{1/3} = \left(\frac{1}{5.55S^2}\right)^{1/3}(1 - \alpha\chi) \tag{6.7}$$

Equation (6.7) is the most general form of the Kozeny–Carman equation used for wood fiber mats (Singh (228)).

$S$ is the flow-exposed surface of the fibers per unit mass, commonly called the specific surface area.

$$S = \alpha S_0 \tag{6.8}$$

Lindsey measured $\chi$ and $k$ values on fibrous webs under various degrees of compression. Using this data to plot $(k\chi^2)^{1/3}$ vs $c$ in Eq. (6.7) he was able to determine $\alpha$ from the slope and $S$ from the intercept. The relative flow porosity was then calculated:

$$\varepsilon_{\text{rel}} = \frac{\varepsilon_0}{\varepsilon} = \frac{1-\alpha}{\varepsilon} \tag{6.9}$$

Lindsey's experiments indicated that values of $\varepsilon_{\text{rel}}$ range from as low as 45% up to 70% and higher.

## 6.8 PORE SIZE AND PORE STRUCTURE

Pore size, pore structure, along with porosity are related properties that help define a nonwoven filter medium. Porosity relates to the void volume within a filter medium and its capacity to hold dirt. Pore size and pore structure relate to the particle size that can be arrested. Almost all pore size measurements assume cylindrical capillaries through the medium and express the pore size in terms of the diameters of those capillaries. Two primary pore size measurements of a filter medium are as follows: maximum pore size (often referred to as bubble point) and mean flow pore size. The test methods are described in ASTM F316-03(2011) (229). Although this particular standard is specified for membrane filter media, the technique has been used by nonwoven and paper filter manufacturers for many years.

### 6.8.1 Bubble point (maximum pore size)

Figure 6.6 is a schematic of a very simple filter media holder for determining maximum pore size or bubble point. The medium is clamped in place over a pressurizing chamber. Above the medium is a reservoir, which is filled with a test fluid of known density and surface tension. The pressurizing chamber is connected to a source of pressurized air and to a manometer. The pressure increases when airflow is introduced into the pressurizing chamber. Eventually the air forces its way through the medium

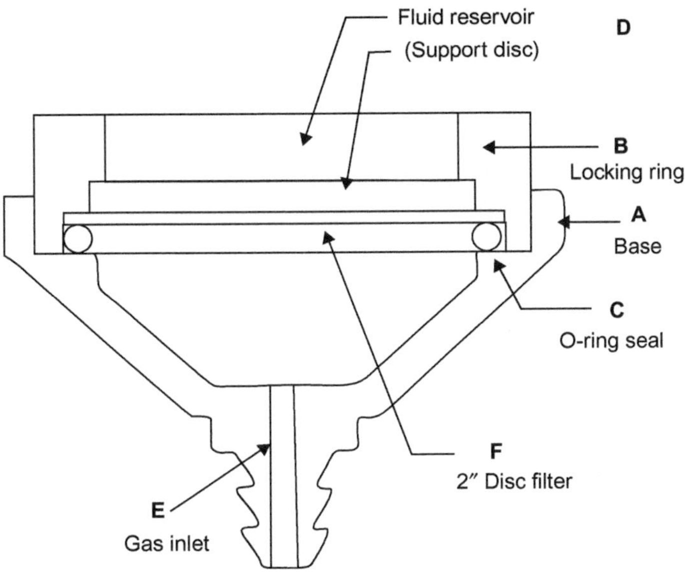


**Figure 6.6** Bubble Point Filter Holder. *Repreoduced with Permission of ASTM International, West Conshocken, Pennsylvania, USA.*

and bubbles through the liquid reservoir. The manometer pressure is recorded at the time the first bubble is observed rising through the reservoir. The bubble point or maximum pore size is calculated as follows[2]:

$$d = (4\sigma\cos\theta/p) \times 10^6 \tag{6.10}$$

$d$ is the maximum pore diameter in μm.
$\sigma$ is the surface tension, at 20°C, of the reservoir fluid in N/m.
$p$ is the pressure when the first bubble was observed in Pa.
$\theta$ is the contact angle between the reservoir liquid and the pore wall in degrees (°).

In the filter holder diagrammed in Figure 6.6 above, the pressure, $p$, in Eq. (6.10) has to be corrected for the height of the reservoir.

$$p = p_0 - \rho_L g h \tag{6.11}$$

$\rho_L$ is the density of the reservoir fluid, at 20°C, in g/cm$^3$.
$h$ is the height of the reservoir in cm.
$g$ is the gravitational constant (981 cm/s$^2$).

[2] ASTM F316-03(2011) applies a capillary constant $B$ to Eq. (6.10). It has the value of 0.715, so that the ASTM version of the equation is $d = [4B\sigma\cos\theta/p] \times 10^6 = [2.86\sigma\cos\theta/p] \times 10^6$. The capillary constant is a tortuosity factor developed in a publication by H. Bechold *The Permeability of Ultrafilters* written in 1908. Many companies and organizations who use the bubble point technique do not make this correction.

**Table 6.6** Reservoir Fluids for Pore Size Testing

| Test liquid | Density (g/cm$^3$) at 20°C | Surface tension (N/M) at 20°C |
|---|---|---|
| Methanol | 0.79 | 0.0225 |
| Ethanol (95%) | 0.805 | 0.0230 |
| Isopropanol | 0.79? | 0.0215 |
| Carbon tetrachloride | 1.59 | 0.0270 |
| A.C. oil, petroleum distilate | 0.78 | 0.0300 |
| Porewick®[a] | 1.90 | 0.0160 |
| Porofil®[b] | 1.88 | 0.0160 |
| Water | 1.00 | 0.0720 |

[a]Test fluid recommended for PMI capillary flow porometers.
[b]Test fluid recommended for Coulter-I Porometer®.

The height of the reservoir depicted in Figure 6.6 is 0.5 cm (0.2 in). Donaldson Corp. in Minneapolis, Minnesota, USA specifies a device that has a reservoir of 1.5 cm (0.6 in).

Table 6.6 above is a tabulation of various liquids that might be used as bubble point reservoir fluids.

In addition to ASTM F316 other standard test methods for pore size are ASTM E128-99(2011) and ISO 4003:1977. Both of these test methods are intended for stiff rigid porous materials such as sintered metal filter media.

## 6.8.2 Mean flow pore size

The mean flow pore size is the pore diameter at which the flow through a wetted medium is 50% of the flow through the dry medium at the same presspure drop. It is not the mean pore size because the flow through large diameter pores can be disproportionately larger than the flow through small diameter pores. Figure 6.7 diagrams the apparatus used for measuring mean flow pore size. Two setups are shown: a one holder apparatus and a two holder appatatus. In the one holder setup the dry filter specimen is placed in the holder and air pressure is applied in increasing increments. The flow is measured at each pressure and the flow—pressure relationship is plotted.

The filter specimen is then removed from the holder and completely wetted with a liquid of known surface tension and density or specific gravity. The liquids listed in Table 6.6 are possibilities. The wetted specimen is then placed in the holder and the flow—pressure plot is repeated on the wetted specimen. The bubble point is determined at the pressure at which air is first seen bubbling through the bubbler (bubble point detector) displayed in Figure 6.7. The pressure at which the flow rate through

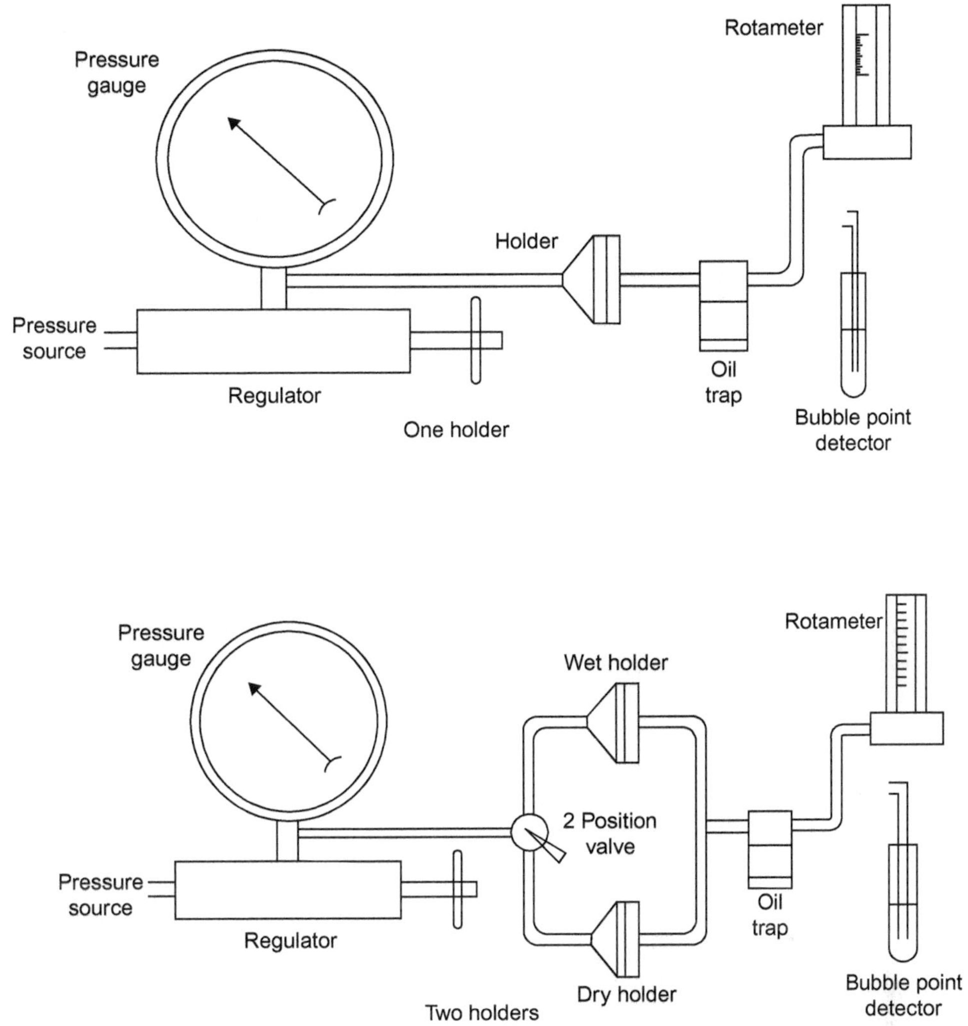


**Figure 6.7** ASTM F316 Apparatus for Mean Flow Pore Size and Bubble Point. *Reproduced with Permission of ASTM International, West Conshocken, Pennsylvania, USA.*

the wetted specimen is exactly one half of the flow through the dry specimen is the pressure at which the mean flow pore size is determined. Figure 6.8 illustrates the graphic procedure for determining mean flow pore size. Equation (6.10) above is applied to calculate the mean flow pore size[3].

[3] ASTM F316-03(2011), which applied the capillary constant, *B*, to the bubble point test, Eq. (6.10) above also applies this constant to the calculation for mean flow pore size.

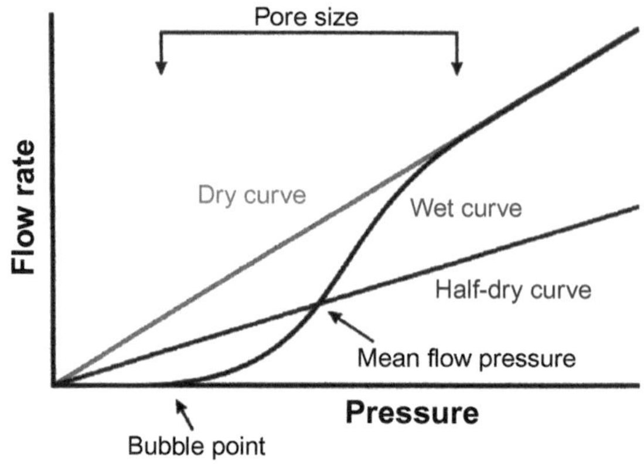


**Figure 6.8** Determination of Mean Flow Pore Size. *Reproduced by Permission of Porous Materials Inc., Ithaca, New York, USA (Jena and Gupta (230) Figure 2).*

The measurement as described in ASTM F316-03(2011) and illustrated by Figure 6.7 above is a form of extrusion flow porometry. The extrusion flow porometry procedure utilizes air pressure to force a fluid of known surface tension from the pores of the medium. The bubble point (or maximum pore size) mean flow pore size and minimum pore size can be determined in accordance with Figure 6.8 above and by using Eq. (6.10) above. The technique measures the most restricted part of the pore (see Figure 6.9 below). In addition to the ASTM apparatus described here, there are very sophisticated instruments for measuring the pore structure of filter media. At one time Beckman Coulter, Inc. marketed a Coulter-I Porometer® to perform the test. Updated versions of this instrument are marketed by Quantachrome Instruments in Boynton Beach, Florida and are referred to as the POROMETER 3G series.

Porous Materials, Inc, Ithaca, New York, USA currently markets several forms of porometers including the Capillary Flow Porometer®. In this instrument a wetting liquid is allowed to spontaneously fill the pores in the sample. Then, a nonreacting gas displaces the liquid from the pores. The gas pressure and flow rates through wet and dry samples are accurately measured. The instrument software automatically plots the flows in accordance with Figure 6.8 above. The pressure at which flow is first detected is used to determine the bubble point. The point at which the wet curve crosses the half-dry curve is the pressure at which the mean flow pore size is detected. The minimum pore size is determined from the pressure at which the wet curve merges with the dry curve. In addition to providing the maximum, mean flow, and minimum pore sizes, the instrument is also capable of providing the pore diameter range, the pore size distribution, the gas permeability, and the distribution function that relates the wet flow to the dry flow as the pore diameter increases.

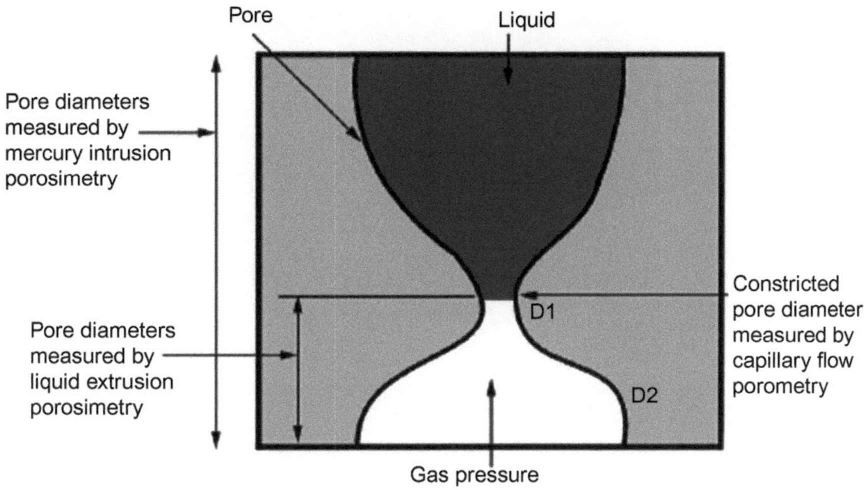


**Figure 6.9** Constricted Pore Diameter. *Reproduced by Permission of Porous Materials Inc., Ithaca, New York, USA (Jena and Gupta (230)).*

## 6.9 OTHER TECHNIQUES FOR MEASURING POROSITY, PORE SIZE, AND STRUCTURE

In addition to the porometry procedures there are other techniques for measuring porosity, pore size, and pore structure in filter media. Following is a discussion of those that may apply.

### 6.9.1 Image analysis

Computer techniques allow for scanning a web and conducting image analysis to determine pore size distribution and properties, however they are limited to relatively thin webs such as melt-blown structures and low basis weight spunbond webs. Nevertheless, a discussion of this approach is worthwhile, because it provides valuable insights to the nature of pore structure. Dr Randall Bresee (231), formerly of the University of Tennessee in Knoxville, Tennessee, provided Figure 6.10, images of spunbond webs, to illustrate this approach. The image shows the actual pore structure of the web (the diamond-shaped areas in the web are thermal spot bonds). For theoretical treatment, pores are assumed to be cylinders that have perfect circles as cross sections, even though a view of Figure 6.10 indicates that very few pores even approach the shape of a circle. Bresee's technique (232) includes an analysis of pore aspect ratios[4] presented in Figure 6.11. In this analysis, the mean ratio for spunbond web 2 is 2.39 and for spunbond web 4 is 2.34.

[4] Pore aspect ratio, as measured by Dr Bresee's technique, is based on the pore cross section. It is determined from the best fit ellipse that can be fitted into the pore cross section. The pore aspect ratio is determined from the long axis of the ellipse divided by the short axis. If the ellipse happens to be a perfect circle, then the aspect ratio will be 1.0. Interestingly, a perfectly square pore would also have an aspect ratio of 1.0.

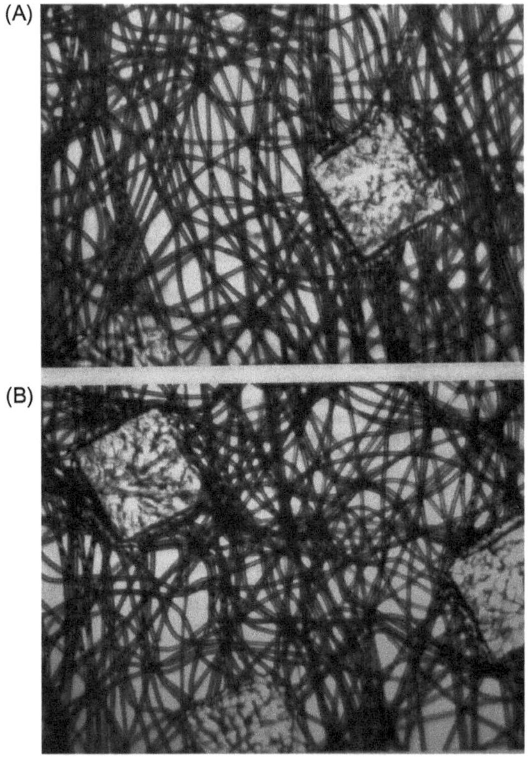


**Figure 6.10** Images of the Pore Structure of a Thermally Spot Bonded, Spunbonded Web (A) Spunbond Web 2, (B) Spunbond Web 4. *Courtesy Dr Randall R. Bresee (231).*

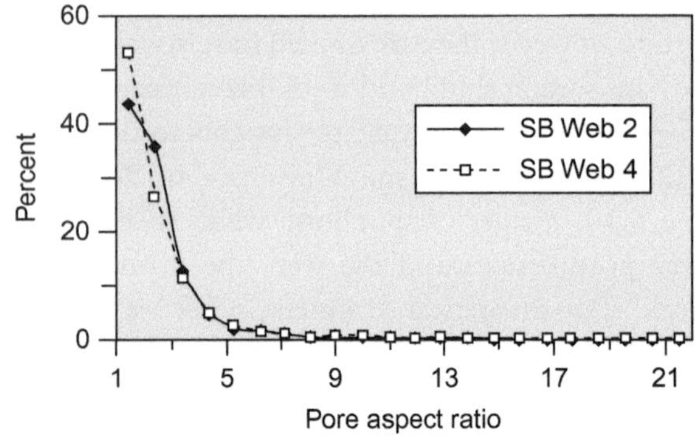


| Property | SB Web 2 | SB Web 4 |
|---|---|---|
| No. pores | 501 | 507 |
| Mean aspect ratio | 2.39 | 2.34 |
| Max aspect ratio | 13.82 | 19.32 |
| Cv % | 56.3 | 67.6 |
| Mean pore orientation | 25.4 | 34.4 |

**Figure 6.11** Pore Aspect Ratio Distribution for Spunbond Filter Media. *Courtesy Dr Randall R. Bresee (231).*

The property table alongside Figure 6.11 includes a row for mean pore orientation. The orientation of a pore is measured relative to the machine direction (MD) of the web. It is determined from the angle that the long axis of the best fit ellipse makes with the MD direction. If the long axis is parallel to the MD direction the pore

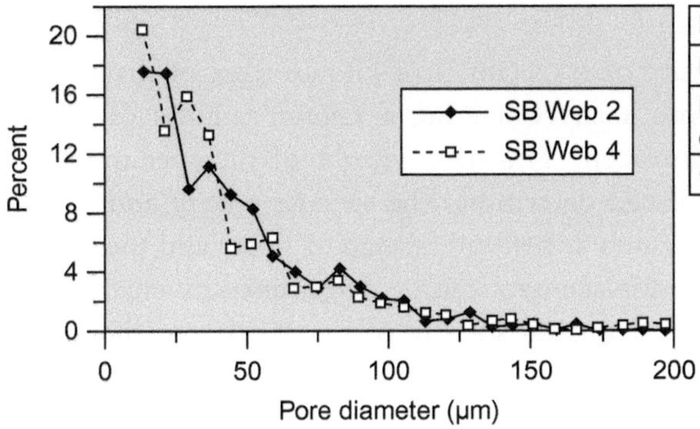


| Property | SB Web 2 | SB Web 4 |
|---|---|---|
| No. pores | 501 | 507 |
| Mean pore diameter | 44.4 μm | 43.0 μm |
| Cv % | 67.8 | 75.2 |

**Figure 6.12** Pore Diameter Size Distribution for Spunbond Filter Media. *Courtesy Dr Randall R. Bresee (231).*

orientation is 0°. On the other hand, if the axis is perpendicular to the MD direction (or parallel to the cross direction, CD) the pore orientation is 90°. If the oreiantation were purely random in all directions, then the mean orientation is 45°. Spunbond web 2 of Figure 6.11 had a mean pore orientation of 25.4°, and spunbond web 4 has an orientation of 34.4°. This suggests that the pores of spunbond web 2 are more MD orientated. It should also be noted that in both cases the pore orientation is less than 45° and therefore, MD orientated. This is an expected characteristic of most random formed nonwovens that are formed as a continuous sheet. It reflects on the fact that in most forming processes, fibers tend to form more in the MD direction than the CD direction.

Figure 6.12 above is an analysis that calculates the pore size distribution (based on circular pores). It indicates that the mean pore diameters are 44.4 μm for spunbond web 2 and 43.0 μm[5] for spunbond web 4.

## 6.9.2 Microscopy

According to Ptak and Cady (233), "Microscopic measurement is the most direct method of measuring the pore structure." However it can only be applied to media with straight pores perpendicular to the filter surface. Ptak and Cady determined that perforated sheet metal and woven fabrics were the only filter media for which reliable results could be obtained.

[5] Caution! The mean pore size as determined by Bresee's technique does not have the same statistical meaning as the mean flow pore size that is determined by the test of ASTM F316 and by Figure 6.8. Dr Bresee's mean pore size is a numerical mean of all pores, whereas, the mean flow pore size is based on the pore size for 50% flow through the medium. The mean flow pore size of ASTM F316 is based on a power distribution that gives more weight to larger pores. This makes the mean flow pore size a larger number than the mean pore size.

### 6.9.3 Pycnometry

A pycnometer determines the density of a specimen of known mass of matter. This is done by inserting the specimen into a jar containing a known volume of water and measuring the volume of water displaced. The mass weight of the specimen divided by the volume (mass) of water displaced determines the specific gracity and/or density of the specimen. If the fluid is a gas (such as helium) instead of water and the volume of the specimen is known then the gas displacement can be determined by change in pressure drop. The gas displaced is only from the solidity of the specimen, therefore, the solidity is determined by the volume of gas displacement divided by the volume of the specimen. Of course, the porosity can then be determined from one minus the solidity.

Sreedhara and Tata (234) used a pycnometer to measure the porosity of electrospun nanofiber mats. It is interesting to note that the nanofiber mats were so thin that they had to resort to FESEM (field emission scanning electron microscopy) to determine mat thickness for the volume calculation.

Sreedhara and Tata used an AccuPyc® 1330 pycnometer marketed by Micromeritics Instrument Company, Norcross, Georgia, USA. The AccuPyc® works by measuring the amount of displaced gas (helium). The pressures observed upon filling the sample chamber and then discharging it into a second empty chamber allow computation of the sample solid phase volume. Gas molecules rapidly fill the tiniest pores of the sample; only the truly solid phase of the sample displaces the gas.

### 6.9.4 X-ray microtomography

X-ray microtomography is a radiation technique that can reproduce graphic three-dimensional images of small sections of nonwoven and paper structures. Research on this technique is being conducted at the European Synchrotron Radiation Facility in Grenoble, France. Papers on the technique include Bloch (235) and Thibault et al. (236). Lin and Miller (237) used high-resolution X-ray microtomography to analyze the pore structure network of a filter cake.

The technique involves directing a very high intensity X-ray beam at various angles through successive layers of a rotating specimen and reconstructing the layers into a three-dimensional image. The layers can also be investigated independently. Bloch indicates that from the image analysis, the following properties can be calculated: porosity, mean chord, mean intercept numbers, specific surface, mean hydraulic diameter, and anisotropy factors. Figures 6.13 (blotter paper) and 6.14 (nonwoven structure) are examples of the structures that can be reproduced.

### 6.9.5 Liquid extrusion porosimetry

Liquid extrusion porosimetry is similar to liquid extrusion porometry in that gas pressure is used to force liquid from the pores of a medium, except, in this test the

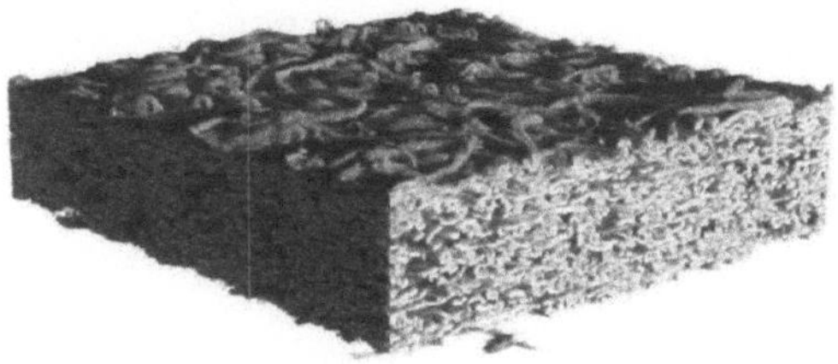

**Figure 6.13** Blotter Paper (2 μm). *Reproduce with Permission of the American Filtration Society, Nashville, Tennessee, USA.*

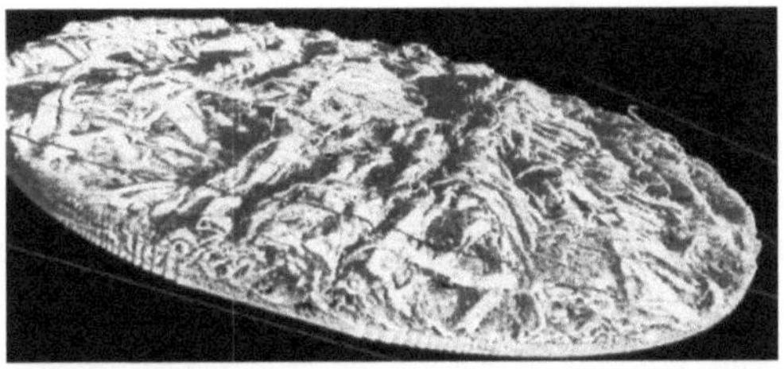

**Figure 6.14** Nonwoven Structure (2 μm). *Reproduced with Permission of the American Filtration Society, Nashville, Tennessee, USA.*

medium is supported by a membrane whose pores are much smaller than the pores of the medium. The pores of both sample and membrane are filled with a wetting liquid. Sufficient pressure is applied on the sample side so that liquid is forced from the pores of the sample but not from the pores of the membrane. The liquid forced from the sample is extruded through the membrane and its cumulative flow volume is measured. The gas pressure at each point of cumulative volume measured is converted to an equivalent pore diameter by Eq. (6.10). These data are used to calculate pore volume, pore volume distribution, and surface area through pores as demonstrated by Figure 6.15.

## 6.9.6 Liquid intrusion porosimetry

Liquid intrusion porosimetry uses nonwetting liquids to force a gas from the pore structure of a porous medium. This is in contrast with extrusion porometry and extrusion porosimetry techniques that use gas to force liquid from the pore structure. The intrusion technique can provide pore diameter distribution, pore volume, and pore surface area. It determines the pore structure from the volume of liquid used to displace the gas and from the pressure employed. The pressure to force the liquid into the medium combined with the surface tension of the penetrating liquid is used to determine pore size in accordance with Eq. (6.10). Mercury intrusion porosimetry and nonmercury intrusion porosimetry are two forms liquid intrusion porosimetry.

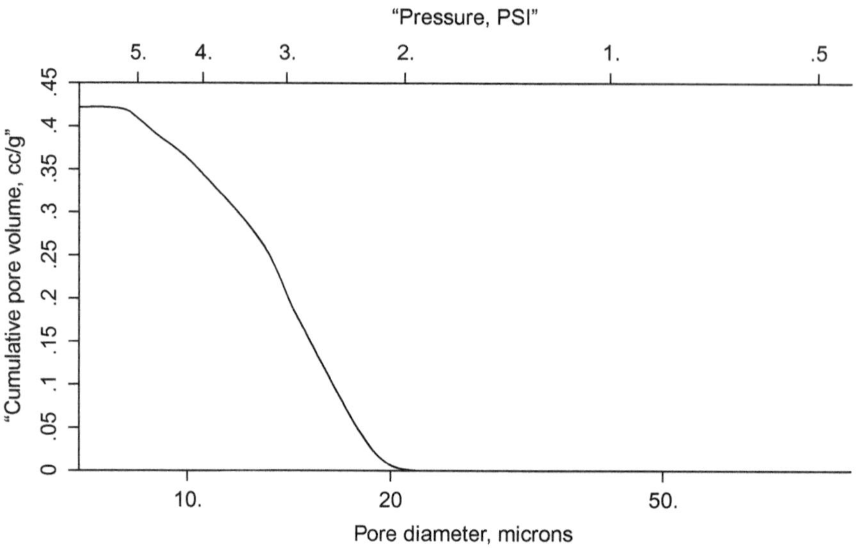


**Figure 6.15** Cumulative Pore Volume of Nanofiber Mat Measured in the PMI Liquid Extrusion Porosimeter. *Reproduced with Permissionn of Porous Materials, Inc., Ithaca, New York, USA (Jena and Gupta (230) Figure 2).*

**a.** Mercury intrusion porosimetry—Mercury intrusion is a technique for measuring pore volume and diameter by forcing mercury into the pore structure of a vacuum evacuated fabric. The intrusion technique can provide pore diameter distribution, pore volume, and pore surface area. Mercury is a nonwetting liquid and its contact angle with the medium surface is greater than 90°. As a result, mercury will not readily flow into the pores unless forced to do so by an outside pressure. This pressure is related to pore size. The work to force mercury into the pores is equated to the increase in the free energy of the surface. Pore diameter is determined from the negative form of Eq. (6.10):

$$d = -[4\sigma\cos\theta/p] \times 10^6 \tag{6.12}$$

The surface area is determined from the change in volume as pressure is increased:

$$S = [1/(-\sigma\cos\theta \times 10^6)] \int p\,dV \tag{6.13}$$

Pore volume is determined from the volume of mercury forced into the structure.

**b.** Nonmercury intrusion porosimetry—Nonmercury intrusion works just like mercury intrusion, except a nonwetting liquid, other than mercury, is used as the intrusion liquid. A nonwetting oil is an example. This technique has the advantage of avoiding a toxic material like mercury. It also greatly reduces the pressure necessary to force the fluid into the pores of the web.

### 6.9.7 Gas/vapor adsorption (BET)

BET (Brunauer, Emmet, Teller) (238) theory is based on the attraction of an inert gas to the surface of the substrate being tested. Usually the gas is nitrogen ($N_2$). The testing is conducted at or near the temperature of liquid nitrogen. The amount adsorbed on a pore surface is a function of its vapor pressure as related to its equilibrium vapor pressure. The BET isotherm is as follows:

$$\frac{p_v}{W(p_0 - p_v)} = \frac{1}{W_m C} + \frac{(C-1)}{W_m C}\left(\frac{p_v}{p_0}\right) \tag{6.14}$$

$W$ is the amount of adsorbed gas in moles.
$W_m$ is the amount of gas to form a monolayer in moles.
$p_v$ is the gas vapor pressure in Pa.
$p_0$ is the equilibrium vapor pressure in Pa.
$C$ is a constant related to the adsorption energy of the gas to the solid substrate compared to the liquefaction energy of the gas. If the adsorption energy is high compared to the liquefaction energy, then the gas has a high affinity for the solid substrate, and vice versa if the adsorption energy is low compared to the liquefaction energy, then the gas has a low affinity for the solid substrate.

The specific surface area of the substrate, $S$, is determined by first plotting $\frac{p_v}{W(p_0 - p_v)}$ against $\frac{p_v}{p_0}$. At low vapor pressures ($0.05 \leq \frac{p_v}{p_0} \leq 0.35$) the plot is usually linear. This allows the calculation of $W_m$ from the slope and intercept of the line. Figure 6.16 is an example of a BET plot.

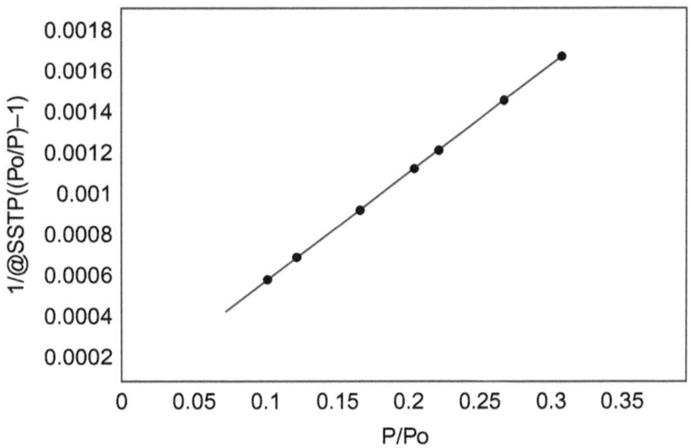


**Figure 6.16** BET Isotherm. *Reproduced wirh Permission of Porous Materials, Inc., Ithaca, New York, USA (Jena and Gupta (46)).*

The specific surface area, $S$, in $cm^2/g$, is then computed from $W_m$ as follows:

$$S = W_m N \alpha / m \tag{6.15}$$

$N$ is the Avagadro number: $6.02 \times 10^{23}$.

$\alpha$ is the cross-sectional area of the adsorbed gas molecule ($cm^2$).

$m$ is the mass of the specimen being tested (g).

The BET technique is generally recognized as being the most accurate technique for measuring surface area of samples with large specific surface areas.

The BET is also used for pore size analysis based on vapor condensation. The principal is that it is easier for vapor to condense in small pores than large pores. The smaller the pore the lower is the vapor pressure at which condensation will occur. The pore size is then related to the ratio of the condensation pressure to the equilibrium vapor pressure, $\frac{p_v}{p_0}$. The phenomenon is described by Kelvin's Law.

$$\ln\left(\frac{p_v}{p_0}\right) = \frac{2\overline{V}\sigma\cos\theta}{dRT} \tag{6.16}$$

$\overline{V}$ is the molar volume of the condensed vapor, $m^3$.

$d$ is the pore diameter in m.

$\sigma$ is the superficial tension, N/m.

$p_v$ is the vapor pressure at which condensation occurs, Pa.

$p_0$ is the equilibrium vapor pressure in Pa.

$\theta$ is the wetting angle, deg. If the condensed liquid perfectly wets the surface, $\cos\theta = 1$.

$R$ is the universal gas law constant, 8.31 J/mole-K.

$T$ is the absolute temperature, K.

### 6.9.8 Summation

Porous Materials Inc. of Ithaca, New York, USA (239) has published Table 6.7. It lists the various porosity measurement technologies. Included are listings of properties that can be measured, applications, capabilities (for example, pore size range), and operating conditions. It is a useful guide for picking the best technology for any particular application.

## 6.10 STRENGTH PROPERTIES

There are many components to the strength properties of nonwoven filter media. Tensile properties include tensile strength, stretch or % elongation, tensile energy absorption (TEA), and tensile stiffness. These are in-plane properties of the web. Out of plane properties include bending stiffness, burst strength, internal bond strength,

## Table 6.7 Porosity Characterization Chart

**Porosity Characterization Chart**

Porous Materials, Inc.
20 Dutch Mill Rd
Ithaca, NY 14850
USA

Fax: 607-257-5639
Phone: 607-257-5544
US & CA: 1-800-TALK-PMI
info@pmiapp.com
www.pmiapp.com

| | | Porometry (extrusion) | | | | | | Porosimetry (Extrusion) | Porosimetry (Intrusion) | | Gas Adsorption | | Permeametry | | | | | Pycnometry | |
|---|---|---|---|---|---|---|---|---|---|---|---|---|---|---|---|---|---|---|---|
| | | Capillary Flow Porometer | Cyclic Compression/ Compression Porometer | Integrity Analyzer | Bubble Point Tester | Filtration Media Analyzer | Complete Filter Cartridge Analyzer | Liquid Extrusion Porosimeter | Mercury/Nonmercury Porosimeter | Aquapore | BET Sorptometer | BET Liquisorb Sorptometer | Liquid Permeameter | Gas Permeameter | Envelope Surface Area Analyzer | Diffusion Permeameter | Water Vapor Transmission Analyzer | Gas Pycnometer | Mercury Pycnometer |
| Pore Structure Characteristics | Mean Pore Size | X | X | X | | X | X | X | X | X | X | X | | | | | | | |
| | Pore Size Distribution | X | X | X | | X | X | X | X | X | X | X | | | | | | | |
| | Total Pore Volume | | | | | | | X | X | X | X | X | | | | | | | |
| | Bubble Point (Largest Pore) | X | X | X | X | X | X | | | | | | | | | | | | |
| | Integrity | X | | X | | X | X | | | | | | | | | | | | |
| | Gas Permeability | X | X | X | | X | X | | | | | | | X | | | X | | |
| | Liquid Permeability | X | X | | | X | X | X | | | | | X | | | | | | |
| | Frazier Permeability | X | X | | | X | X | | | | | | | | | | | | |
| | Diffusion Permeability | | | | | | | | | | | | | | | X | X | | |
| | Hydro-Head Test | X | | | | X | X | | | | | | | | | | | | |
| | Single Point Surface Area | | | | | | | | | | X | X | | | | | | | |
| | Multi-Point Surface Area | | | | | | | | | | X | X | | | | | | | |
| | Envelope Surface Area | X | X | | | | | | | | | | | | X | | | | |
| | Porosimetry Surface Area | | | | | | | X | X | X | | | | | | | | | |
| | Adsorption & Desorption Isotherms | | | | | | | | | | X | X | | | | | | | |
| | Chemisorption | | | | | | | | | | X | X | | | | | | | |
| | Liquid Vapor Adsorption | | | | | | | | | | | X | | | | | | | |
| | Bulk Density | | | | | | | | X | | | | | | | | | | X |
| | Absolute Density | | | | | | | | X | | | | | | | | | X | |
| | Particle Size Distribution | | | | | | | | X | | | | | | | | | | |
| Operating Conditions | Compression/Cyclic Compression | X | X | X | X | X | | X | | | | | X | X | | | | | |
| | Tension | X | | | X | X | | | | | | | X | X | | X | | | |
| | Elevated Temperature | X | | X | X | X | X | X | X | X | X | | X | X | X | X | X | | |
| Sample Characteristics | Pore Size Range | 0.013 - 500μ | | | | | | 2000 - 0.05μ | 0.0035 - 500μ | 0.001 - 20μ | 20x10³-3Å | | N/A | N/A | N/A | N/A | N/A | Down to 2 Å | N/A |
| | Surface Area Range (in m²/g) | | | | | | | | 1-100 | 1-100 | > 0.01 | | $10^{-3}$ - 50 | | 0.1 - 10 | $10^{-12}$ - $10^{-6}$ | | N/A | N/A |
| | Permeability Range (in darcies) | $10^{-3}$ - 50 | | $10^{-3}$ - 50 | N/A | $10^{-3}$ - 50 | | | N/A | N/A | N/A | N/A | N/A | N/A | N/A | | | N/A | N/A |
| | Permeability Range (Microflow in cc/sec/m/torr) | $10^{-10}$ - $10^{-6}$ | | N/A | N/A | $10^{-10}$ - $10^{-6}$ | N/A | | N/A | N/A | N/A | N/A | | | | | | N/A | N/A |
| | Dead-End & Through-Pores | | | | | | | | X | X | X | X | | | | | | | |
| | Through-Pores Only | X | X | X | X | X | X | X | | | | | X | X | X | X | X | | |
| | Special Sample Requirements | | | | | | Complete Filters | | N/A | Hydro-phobic | N/A | N/A | No Powders | | N/A | | | N/A | N/A |
| Features | Nondestructive | X | X | X | X | X | X | X | | X | | | X | X | X | X | X | X | |
| | Test Substance | Air or Noncorrosive Gas | | | | | | | Mercury | Water | Any Pure Gas | Water, Hydro-Carbon, Solvents | Air or Noncorrosive Gas | | | | | Helium or Nonreactive Gas | Mercury |
| | Automated Control | X | X | X | X | X | X | X | X | X | X | X | X | X | X | X | X | X | |
| Applications | Automotive Industry | X | X | X | X | | X | X | X | X | X | | X | X | | | | X | |
| | Battery/Fuel Cell Industry | X | X | X | X | | | | X | X | X | | X | X | X | X | X | X | X |
| | Ceramic Industry | X | | X | X | | X | X | X | X | X | X | X | X | X | X | | X | X |
| | Chemical Industry | X | | | | | | | X | X | X | X | | | X | X | | X | |
| | Filtration Industry | X | X | X | X | X | X | X | X | | X | | X | X | | X | | | |
| | Geotextiles/Textiles Industry | X | X | X | X | | | X | | | | | X | X | | X | X | | |
| | Nonwovens Industry | X | X | X | X | | X | X | | | | X | X | X | | X | X | X | |
| | Paper Industry | X | X | X | X | | | | X | X | X | | X | X | X | X | X | | |
| | Pharmaceutical Industry | X | | X | X | X | | X | X | X | X | X | | | X | X | X | X | |
| | Powder Metallurgy Industry | X | | | | | | X | X | X | X | X | | | X | X | | X | |

Reproduced with permission of Porous Materials, Inc., Ithaca, New York, USA (Jena and Gupta (46)).

and Z-direction compression. In almost all cases the properties are anisotropic, different in the MD of the web than in the CD. Following are the strength related properties of nonwoven filer media.

## 6.10.1 Dry Tensile Properties

Tensile properties are measured by some form of tensile tester. The most common form of tester is a constant-rate-of-extension (CRE) tensile testing machine as defined by TAPPI T494 om-13 (240) or INDA IST No. 110.1-.4. There are a number of test methods worldwide. Among them are ASTM D 828 and ISO1924-2:1995. The measurement is usually defined in terms of the specified width of the specimen. In the United States, this is most often a 1 in wide specimen, although sometimes it is 2 in. European standards most often specify a 15 mm wide specimen.

The test involves clamping both ends of the test specimen to mechanically operated clamps separated by a specified distance (see Figure 6.17 below). One clamp moves and stretches the specimen till it breaks or ruptures. The moving clamp is connected to a load cell that measures the force exerted on the specimen. The tester is also capable of measuring the stretching distance traveled by the moving clamp.

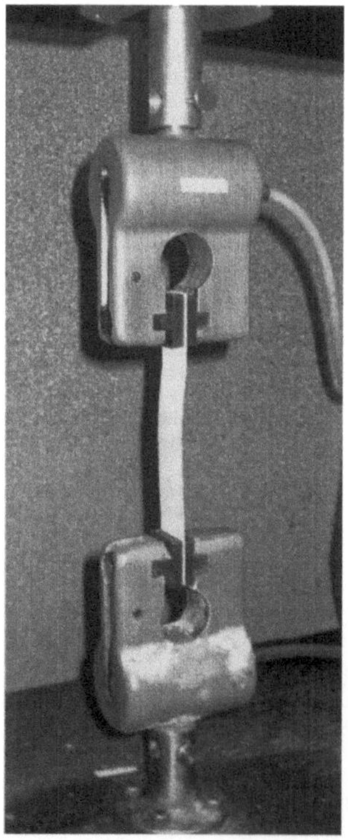

**Figure 6.17** Clamps of a CRE Tensile Strength Tester—the Strip is 1 in (2.54 cm) Wide.

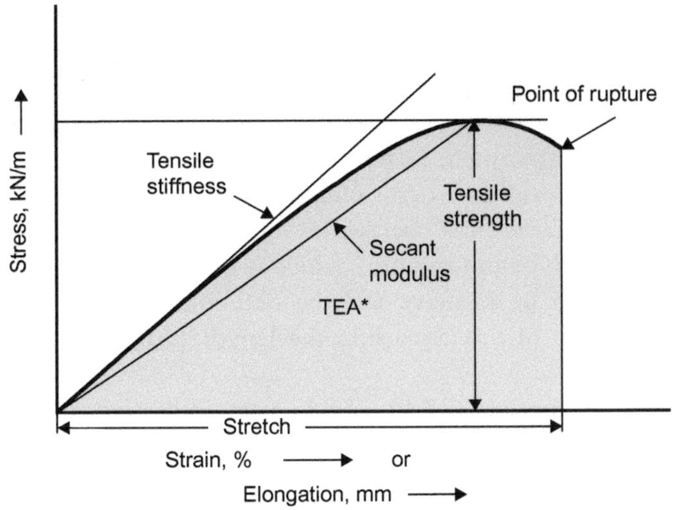


**Figure 6.18** The Stress–Strain Curve.

Specified conditions that have to be programmed into the tester are rate of strain and specimen length. Modern day testers contain processors that will calculate tensile strength, stretch and/or elongation, TEA, tensile stiffness, and secant modulus.

The tensile properties are based on the stress–strain or load–elongation curve illustrated in Figure 6.18 above. Note the curve has two possible horizontal axis: strain and elongation. Strain is the % change in length of the specimen as it is being stretched. Elongation is the change in the length dimension over its original length. The stress–strain is the curve when strain is being used. It is the load–elongation curve if elongation is the X-direction parameter. The various properties are defined as follows:

1. Stress—The load applied to a nonwoven specimen, ever increasing, as the tension increases. It is measured in force per unit width of specimen. In SI units this is often kN/m and sometimes kN/15 mm. In English units, the most common form is lb/in.
2. Tensile strength—The maximum stress or load before rupture of the specimen. Frequent units are kN/m, kN/15 mm, or lb/in. It should be noted from Figure 6.18 that the tensile strength or point of maximum stress and the point of rupture are not necessarily the same points on the stress–strain curve. Many elastic materials, including most nonwovens, can be stretched beyond their maximum stress before rupture occurs.
3. Strain—The increase in length of the specimen as stress is applied. Strain is measured as a percent of the original length before the load was applied.

$$\text{Strain} = (\Delta L/L) * 100 \tag{6.17}$$

4. Stretch—The strain at the point of rupture. Like strain, it is measured as percent of original specimen length. It is often referred to as % elongation.
5. Elongation—Elongation is a confusing term because there appears to be several definitions or understandings of the term. It can be used to express the change in length ($\Delta L$) of the specimen, in mm or inches. It is also used as a synonym for strain or percent of original length. Elongation is also expressed as the fraction of original length. TAPPI T404[6] defines elongation as the tensile strain "at maximum tensile strength before rupture" which is more or less synonymous with the definition of stretch in 4. above and the definition of stretch given in TAPPI T494. For purposes of this discussion the length ($\Delta L$) definition of elongation will be used.
6. Tensile energy absorption (TEA)—TEA is an indication of the toughness of the medium. It is the integrated area under the load–elongation curve up to the point of rupture. Referring to Figure 6.18 above, it is the gray area under the load–elongation curve when elongation is used as the horizontal axis. The units are in $J/m^2$. The equation for TEA in accordance with TAPPI T-494 is:

$$\text{TEA} = 1 \times 10^6 A/LW \tag{6.18}$$

where
TEA is the tensile energy absorption in $J/m^2$.
$A$ is the area under the load–elongation curve in J (Joules).
$L$ is the initial test span in mm.
$W$ is the specimen width in mm.

7. Tensile stiffness—Tensile stiffness is determined from the stress–strain curve. It is an indication of the stiffness or rigidity of the medium. It is the tangent to the elastic portion of the curve up to the elastic limit, i.e., the point where the curve starts to become nonlinear. It is determined from two points on the curve; the first point being at least 5% of the elastic limit and the second being no more than 75% of the elastic limit. The calculation is

$$S_i = (\Delta f/w)/(\Delta L/L) \tag{6.19}$$

where
$S_i$ is the tensile stiffness.
$\Delta f$ is the difference between the two force levels.
$w$ is the initial width of the initial test specimen.
$L$ is the initial test length.
$\Delta L$ is the change in length corresponding to $\Delta f$.

[6] TAPPI T404 "Tensile Breaking Strength and elongation of Paper and Paperboard (Using a Pendulum Type Tester)" is a tensile strength test using a constant rate of loading. This is an older test method that is now seldom used for nonwoven materials.

Note that $\Delta L$ is the elongation at rupture and $\Delta L/L$ is the stretch divided by 100. The tensile stiffness can be converted to Young's modulus, $E$, the modulus of elasticity, by dividing by the thickness, $t$.

$$E = S_i/t \tag{6.20}$$

8. Secant modulus—Secant modulus is the slope of the line drawn from the origin of the stress–strain curve to a point of interest on the curve. In Figure 6.18 above, the point of interest is the point of maximum stress where the tensile strength is determined. The secant modulus is then defined as the maximum tensile strength divided by the stretch. It is a useful number because it relates to the rigidity of the medium. A high secant modulus indicates a stiff, rigid medium, whereas a low secant modulus indicates a soft, stretchy medium. At one time, a major filter manufacturer specified secant modulus as an indicator of pleatability.
9. Breaking length—Breaking length is the length of a tensile strip whose weight, if hung from one end, would be sufficient to break the strip. It is related to tensile strength as follows:

$$\mathrm{BL} = 102,000\,(T/R) = 10,974\,(T'/R') \tag{6.21}$$

where

BL is the breaking length in m.
$T$ is the tensile strength in kN/m.
$R$ is the grammage[7] in g/m$^2$.
$T'$ is the tensile strength in lbf/in.
$R'$ is the basis weight[7] in lb/3,000 ft$^2$.

10. Tensile index—Tensile index (TI) is the tensile strength in N/m divided by grammage in g/m$^2$.

$$\mathrm{TI} = 1,000\,(T/R) \tag{6.22}$$

11. Tensile ratio (MD:CD)—Almost all nonwoven media are anisotropic, having more fiber orientation in the MD than the CD. The ratio of MD tensile strength to CD tensile strength is often used as a quantitative measure of the anisotropy. Terms often used are MD:CD ratio and fiber orientation ratio. A related term is grain ratio and this is often expressed in terms of the CD:MD ratio.

### 6.10.2 Wet tensile strength

Wet tensile strength is measured in accordance with TAPPI om-10 on an the same apparatus similar to the dry tensile strength apparatus of TAPPI T494 (240). Prior to testing, the sample is soaked in water to measure its strength on a wet basis.

[7] In accordance with TAPPI T-494 grammage or basis weight is calculated on an air dry basis, that is, the moisture content is in equilibrium with the atmospheric conditions to which the specimen has been exposed. For cellulose media the equilibrium moisture content is in the range of 3–9%.

Sometimes a surfactant is added to the water for materials that are difficult to wet out. Wet strength is an important property to nonwoven filter media because:

1. It is an indicator of water resistance.
2. It insures processability in wet operations such as saturation in a chemical binder application.
3. It is a necessary property in aqueous filtration operations.
4. It can be used as an indication of cure for media bonded with thermosetting resins.

Tensile strength is not the only test for determining wet strength-related properties of nonwoven filter media. Others are wet burst and wet stiffness.

For a medium bonded with a themosetting resin, the wet strength property is very much affected by the degree of cure. For example, a specimen treated with a thermosetting phenolic resin will have very little wet strength if the specimen is not cured and a much higher strength after curing. The ratio of the uncured wet strength to the cured wet strength is an indicator of how much the specimen has been cured. Of course, if the ratio is 1:1, the specimen is presumed to be fully cured.

## 6.10.3 Other strength related properties

### *6.10.3.1 Grab tensile strength*

Grab tensile strength (ASTM D4632) differs from tensile strength test by the way it is held in the CRE apparatus. It can be seen in Figure 6.19 that the clamps are circular and only grip a portion of the width of the specimen, whereas the clamps of the tensile strength test grip the full width of the specimen as shown in Figure 6.17 above. The grab tensile is the force to rupture the specimen. It is reported in kN or lb (force). The test also measures the elongation to the point of break as a percent of the original length.

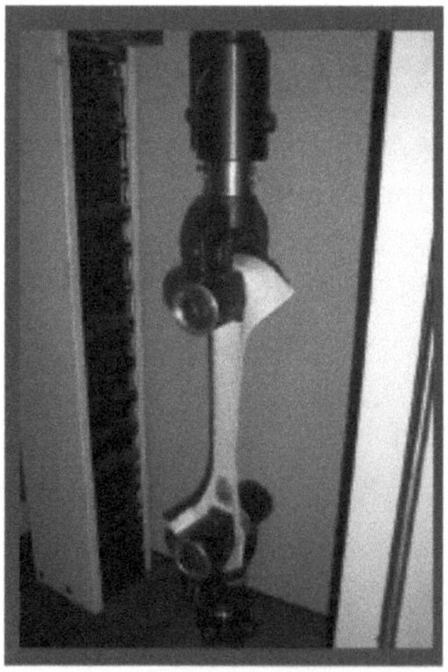

**Figure 6.19** The Grab Tensile Test in a CRE Tester.

### *6.10.3.2 Burst strength*

Because of its simplicity and quickness, burst strength is a commonly used tests for measuring the strength of nonwovens. It is an inexpensive test because it is quick, simple, and inexpensive. The Mullen burst tester is one of the most popular instruments for measuring this property. Among the numerous standards for this test are TAPPI T403, ASTM D774, British BS 3137, WSP 030.1 R3, and ISO 2758. Figure 6.20 below is a picture of a Mullen burst tester produced by Thwing Albert Instrument Company, West Berlin, New Jersey, USA. The test specimen is held between annular clamps and subject to an increasing pressure from a rubber diaphragm pressed against it by a hydraulic pressure. The pressure is increased at a controlled rate till the specimen bursts. The burst strength is the maximum pressure up to the point of burst. Note that, unlike the tensile tests above, the force for the burst test is an out of plane pressure applied perpendicular to the plane of the specimen. Nevertheless the failure at burst is mostly an in-plane failure. This because the pressurized diaphragm, as it pushes against the specimen, stretches and bulges it into a bubble-shaped configuration. When the specimen cannot be stretched any further, it bursts by failure of its in-plane properties.

The burst strength is expressed as a pressure force. Units are usually in kPa or in $lb/in^2$ (psi). Burst strength may also be reported as:

Burst index = burst strength, kPa/grammage, $g/m^2$

Burst ration = burst strength, psi/basis weight, lb/ream

Burst factor = burst strength, $g/cm^2$/grammage, $g/m^2$ (usually oven dry).

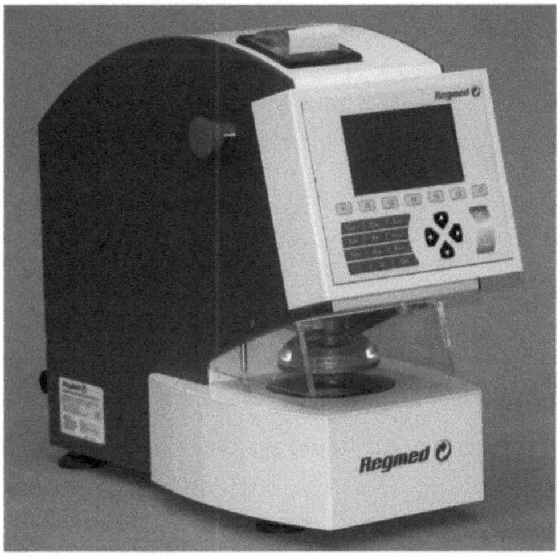


**Figure 6.20** Thwing Albert B-21 Burst Tester. *Reproduced with Permission of Thwing Albert Instrument Company, West Berlin, New Jersey, USA.*

#### *6.10.3.3 Tear strength*

Tear strength is the force to begin or continue a tear in a fabric under specified conditions. It can be related to filter media, in process and in the field, in its ability to resist rupture and mechanical failure. There are several types of tear mechanisms. In most of these, a slit is cut into the specimen and a force applied to the specimen in such a way that a continuing tear from the slit is propagated. The force may be applied so that it is in plane or out of plane depending on the test method. Tear strength is orientation dependent, so both MD and CD specimens are tested. Some examples are:

1. Tongue tear (ASTM D2261)—The specimen is cut as shown in Figure 6.21 below. A 3 in (7.6 cm) slit is cut down the middle of a 3 × 8 in (7.6 × 20.3 cm) specimen. The specimen is tested on a CRE tensile test machine. One side of the slit is fastened into an upper clamp; the other side turned down and fastened into a lower clamp. The tear strength is the force necessary to propagate the tear from the slit (Figure 6.22).

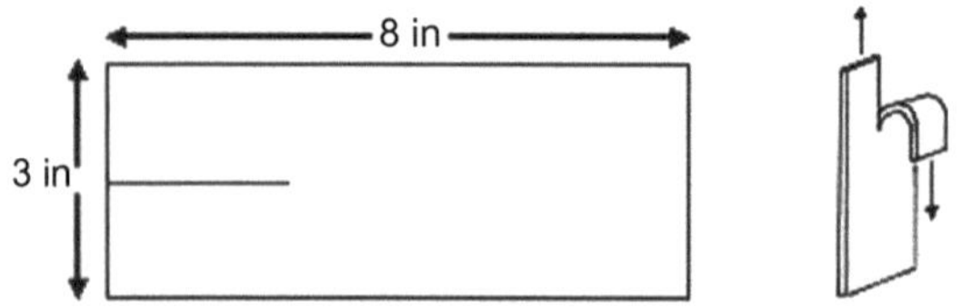


**Figure 6.21** Tongue Tear Test.

2. Trapezoid tear (ASTM D5587)—The trapezoid tear test uses a trapezoid-shaped test specimen as shown in Figure 6.22. The specimen is clamped in the CRE tester so that the diagonal sides of the specimen are in the clamp fixture. The pull is in plane and normal to the direction of the slit cut at the top of the specimen. The force to tear is reported as kN or lb (force).

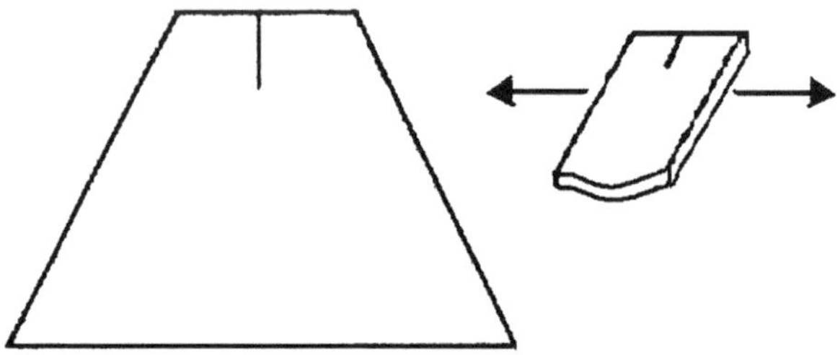

**Figure 6.22** Trapezoid Tear Specimens.

3. Elmendorf tear—The Elmendorf tear test is described by TAPPI T414. Although the test is generally designed for testing several plies of a paper medium it is often used for single ply tests on nonwoven media including filter media. The test

**Figure 6.23** Thwing Albert Elmendorf Tear Tester. *Reproduced with Permission of Thwing Albert Instrument Company, West Berlin, New Jersey, USA.*

apparatus involves a swinging pendulum. The specimen is slit so that one side of the slit is attached to a stationary clamp. The other side of the slit is attached to the moving pendulum. When the pendulum is released it continues the tear along the slit. The loss in potential energy of the pendulum is the work done to continue the tear. The force is measured in mN (millinewtons) or gf (gram force). If more than one ply is used then the force measured is divided by the number of plies. Some other related methods for Elmendorf tear are BS EN 21974 and ISO 1974. ASTM D5734 and INDA-EDANA WSP100 are pertinent to the testing of nonwoven fabrics by the Elmendorf apparatus. Figure 6.23 above is a photograph of Thwing Albert's Electronic Elmendorf ProTear tester.

#### *6.10.3.4 Stiffness*

Stiffness is an important property for resin-bonded filter media used in pleated filter configurations. It measures the resistance to bending of the medium. It is a direction-orientated property so that both MD and CD direction measurements are usually desired. There are many types of stiffness testers. Two of the most notable are the Gurley stiffness tester and the Taber stiffness tester.

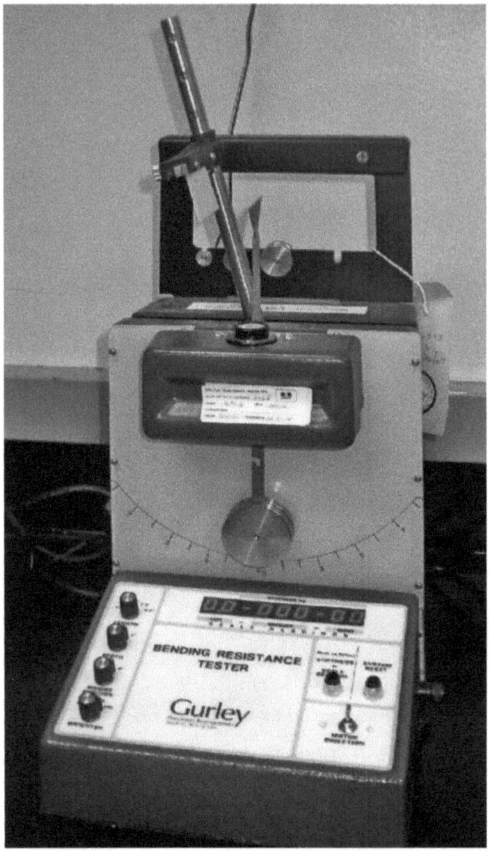


**Figure 6.24** Gurley Stiffness Tester. *Reproduced by Permission of Hollingsworth & Vose Company, Inc., East Walpole, Massachusetts, USA.*

The Gurley stiffness tester is depicted in Figure 6.24 above. TAPPI T543 is an applicable standard. The tester consists of a balanced pendulum or pointer pivoted at its center of gravity. The lower end is pointed and moves parallel to a graduated scale mounted on the base of the instrument. The upper end of the pointer is a triangular vane that makes contact with the moving specimen. The specimen clamp is attached to an arm that rotates about the same geometrical center as the pointer. When the specimen arm rotates, the specimen comes in contact with the triangular vane and deflects the pointer until the specimen bends sufficiently to release the pointer. The rotational distance traveled by the pointer is indicated by the graduated scale at base of the instrument. This distance is an indication of the bending stiffness of the specimen. The clamped specimen rotates back and forth to 20° from the vertical center. Two measurements are averaged. The first measurement is as the specimen arm rotates from left to right and pushes the specimen against the vane in this direction. The arm is then rotated

**Figure 6.25** Taber Stiffness Tester. *Reproduced by Permission of Applied Paper Technology, Atlanta, Georgia, USA.*

right to left to push the specimen against the vane from the opposite side for the second measurement. The two-test average is the average of the bending stiffness from left to right and the bending stiffness from right to left. Weights of 5, 25, 50, and 200 g can be attached to the pointer to allow measurements of specimens with higher stiffness. The weight is chosen to be consistent with the anticipated stiffness of the specimen. The stiffness is reported as Gurley units: one Gurley unit is equivalent to one milligram of force (mgf). In terms of force units, one millinewton (mN) is equivalent to $9.807 \times 10^{-3}$ Gurley units. Modern electronic testers provide digital readouts of the bending stiffness.

The Taber stiffness tester is depicted in Figure 6.25 above. The tester is manufactured by Taber Industries in North Tonawanda, New York, USA. Applicable standards include TAPPI T489, ISO 2493, BS ISO 2493, SCAN P-29:95, and APPITA P431. The test has been around for over 50 years. The test instrument involves a pendulum containing a clamp for attaching the top of the specimen. The lower edge of the clamp coincides with the center of the pendulum bearing. Behind the pendulum is a vertical disc that rotates on the same axis as the pendulum. The disc contains an attachment of two parallel rollers so that the lower edge of the specimen is fitted between these rollers. When the motor-driven disc rotates, the rollers push against the specimen and try to bend it in the direction of rotation. The pendulum to which it is clamped rotates with the disc depending on the bending stiffness of the specimen. Since the specimen does bend, the pendulum does not rotate as much as the disc. How much it rotates is a measure of the bending stiffness of the specimen. The reading is taken when the specimen has bent 15° with respect to the rotating disc (7.5° for stiff specimens). This is indicated when the line mark at the top of the pendulum matches up to the 15° or 7.5° line marks on the rotating disc. At this point the line mark at the top of the pendulum is matched up to a scale on a fixed annular disc surrounding the rotating disc. The scale reading is the measurement of the bending stiffness. The stiffness reported is the average of two readings (left and right deflection) as gram centimeters (g-cm).[8]

#### 6.10.3.5 Internal bond strength

Internal bond strength is like an out of plane tensile strength determined by pulling on the surfaces of the medium so as to try to pull them apart. It is sometimes referred to as the ply bond strength and as the Z-direction strength. Although the tests were developed for paperboard and multiply corrugated boars they have utility for nonwoven filter media. For resin-treated media the internal bond strength may be an indicator of resin penetration into the internal structure. If the resin has not penetrated well, then the internal bonding at the center of the medium will be weak and the surfaces will readily delaminate or pull apart. The internal bond test is also used to measure the laminating strength between layers of a composite structure and determine how well they adhere to each other. There are two types of internal bond tests. In both tests the substrates are sandwiched between two metal platens. Two-sided tape is used to stick each surface to its respective platen. The medium is first compressed by the platens to secure a better adhesive bond. The platens are then separated in a direction normal to the surfaces of the substrate so as to force the surfaces to separate.

TAPPI T541 and SCAN P80 utilize a compression tester to first vertically compress the taped substrate between the platens and then reverse direction to delaminate the specimen. The force or delaminating pressure required to pull apart the surfaces is reported. The units are lb/in$^2$ or kPa.

TAPPI T833 utilizes a pendulum-type tester often referred to as a Scott Internal Bond tester. Figure 6.26 is a picture of the instrument marketed by Thwing Albert. The results of this test are often referred to as the Scott bond. The specimens are taped to a stationary bottom platen on one side and a right-angled metal plate on the top side. The right-angled plate is forced against the specimens for 1–2 s and then the pendulum is released. The right-angled plate is struck by the moving pendulum with a force sufficient to delaminate the specimens.

[8] The following equation can be used to convert Gurley stiffness units to Taber stiffness units:

$$S_t = 0.01419 S_g - 0.935$$

where $S_t$ is the Taber stiffness and $S_g$ is the Gurley stiffness.

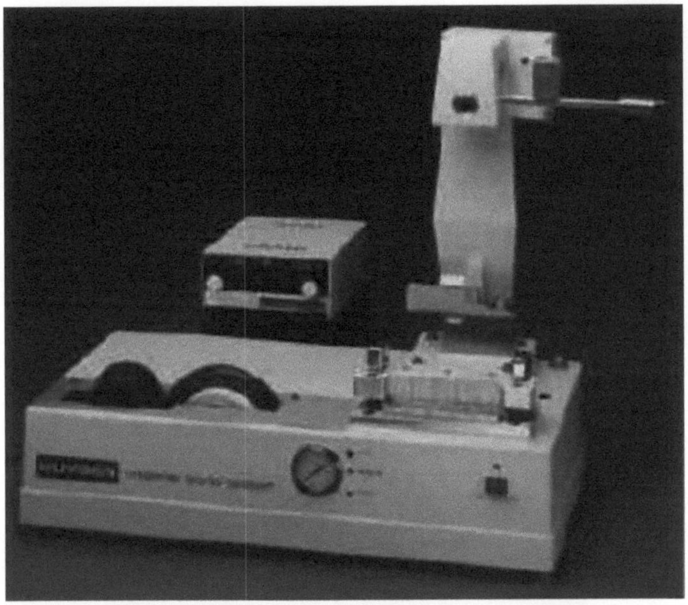

**Figure 6.26** Thwing Albert Model 1314 Internal Bond Tester. *Reproduced with Permission of Thwing Albert Instrument Company, West Berlin, New Jersey, USA.*

A pointer attached to the pendulum and a corresponding scale indicates the loss in kinetic energy of the pendulum. This is the energy necessary to delaminate the specimens. It is an indicator of the internal bond strength. The delaminating energy is reported in $kJ/m^2$ or $ft\text{-}lb/in^2$.

### *6.10.3.6 Fold endurance*

Filter media in use are subject to mechanical variations that may affect their endurance. These include pulsations in pressure drop of the fluid flowing through the medium and mechanical vibrations in the equipment where the filter and its medium are housed. Fold endurance is one way to evaluate a medium's ability to endure these pulsations and vibrations. The general technique to attach a strip of medium to two clamps, one which rotates back and forth at prescribed frequency. This is illustrated in the fold endurance tester of Figure 6.27.

## 6.11 WATER REPELLENCY AND WATER/MOISTURE RESISTANCE

Water repellency relates to the ability of the medium to resist wetting and penetration by water. The property is determined by the hydrophobicity of the medium. An important property if the medium serves as a moisture barrier in its filtration application. Water repellency and water and/or moisture resistance are different properties.

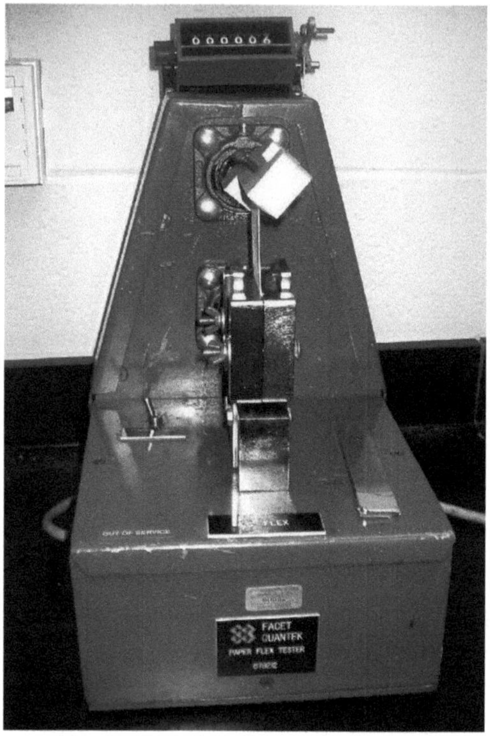

**Figure 6.27** Fold Endurance Tester. *Reproduced with Permission of Hollingswortb and Vose Company Inc., East Walpole, Massachusetts, USA.*

Water resistance relates to the ability of the medium to maintain its properties and structural integrity after exposure to water or moisture. Moisture resistance is generally perceived as the ability of the medium to stand up to environmental variations of humidity in the atmosphere.

### 6.11.1 Water repellency

There are several ways to measure water repellency. Some are locally improvised. TAPPI T558 describes a method whereas a drop of liquid is carefully placed on the substrate and the contact angle is measured by an automated contact angle tester. Another method is to carefully place a drop of liquid on the substrate and measuring the time to soak into the substrate. A related procedure is to place drops of mixtures of alcohol and water. The higher the alcohol content the more readily the drop will soak into the substrate. Water repellency is then rated by the percent of alcohol in and how well it soaks into the medium over a short period of time such as 10 s. Absorbency is also an indicator of water repellency. In soak tests, a water repellent material will absorb less water than a nonwater repellent material.

Other test methods for water repellency include hydrostatic head methods, spray methods, and shower methods.

1. Hydrostatic head methods are sometimes referred to as water resistance tests, however they are in reality a water repellency test. The measurement is based on the head of water the medium will withstand before penetration occurs. It is a water barrier property. AATCC 127 known as the Suter test, is one of several methods that measure this property. ASME AG-1, Section FC, Paragraph FC-I-3240, (241) requires hydrostatic head testing of HEPA media to be used in nuclear installations.
2. Spray methods—AATCC 22 is an example of a spray test. A small body of water placed about 4 in (10 cm) over the fabric is sprayed on the fabric through spreader nozzle. The fabric is at 45° to the direction of spray. The spray duration is 30 s. The fabric is visually rated, on a scale of 0–100, on how well the water beads up; 0 being the worst and 100 being the best (no wetting). The method provides a set of photographs or spray test ratings as standards by which to judge the repellency.
3. Shower methods—Shower methods, also referred to as rain tests, are more severe and demanding than spray methods. The Bundesmann test (ISO 9865) showers water into the fabric specimen from a height of 4 ft (1.2 m). The duration of the shower is 10 min. The fabrics are clamped to cups and angled 10° to the direction of spray. During the test an X-shaped squeegee exercises the fabric from behind to simulate fabric movement during a rainstorm. When the test is completed the beading of water on the fabric is compared to visual standards to obtain a rating of 1–5; 5 being the best. The method can also be used to measure penetration by measuring the amount of water that has penetrated the fabric into the support cup. In addition, water absorption of the fabric can be determined by weighing the fabric before and after the test.

## 6.11.2 Water resistance

Water resistance is a measure of the medium's ability to maintain properties after being immersed in water or wetted out with water. The primary concerns are media disintegration, structural failure, and fiber swelling. Filter media used in filtering aqueous fluids need to have sufficient wet strength so as not to disintegrate when exposed to the aqueous environment. A good example is teabag paper, which has to survive in a hot water environment and be burst resistant, even when pressed with a teaspoon. Pleated air filters such as engine intake air filters are vulnerable to softening and pleat collapses if they are contaminated with water, say from a rainstorm. Ratio tests of a medium's wet and dry strength properties are an indicator of its ability to maintain strength properties. The ratio of wet to dry stiffness is a good indicator of pleated media structural integrity.

Cellulose fibers in particular will swell when exposed to water. This will cause dimensional change in the medium. Dimensional change may contribute to loss of structural integrity in a filter. ISO 5635:1978, designed for paper and paperboard, is a test that measures this change by immersing a specimen in water and measuring its dimensional change.

In a filter medium, fiber swelling may cause pore closure, increased flow resistance, and poorer filtration performance. The author is unaware of any standard test methods that relate fiber swelling to filtration performance. There have been attempts to run flow through tests on wetted filter media in an effort to compare the pressure drop loss to that of dry media. However, the results from such experimentation have been inconsistent.

There are medium tests where the the medium is tested as an element in a filter housing. Water is injected as a contaminant into the test fluid. Generally, filter failure occurs as a result of media dimensional change and structure collapse.

### 6.11.3 Moisture resistance

Moisture resistance is mainly concerned with a medium's ability to maintain dimensional stability when exposed to high humidity and changes in humidity. Dimensional stability is related to the ability of a medium to maintain dimensions when exposed to some adverse condition such as moisture content, heat, the application of load, and/or passage of time. Hygroexpansivity is the change in dimensions due to changes in humidity. Creep, also known as "cold flow," is the change in dimensions with time of a medium that is under constant load. In a pleated filter, the combination of hygroexpansivity and creep sometimes results in pleat deformation and even pleat collapse. Figure 6.28 illustrates a pleated filter element that has "S-shaped" or curly pleat folds, undoubtedly a result of hygroexpansivity or creep after the element was manufactured.

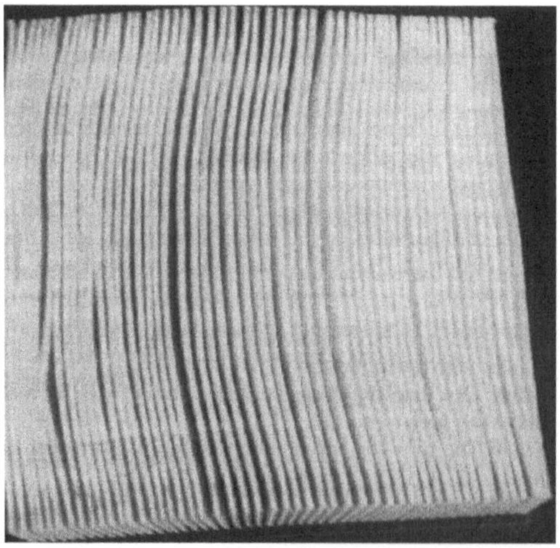

**Figure 6.28** Filter Element with Distorted Pleats Due to Hygroexpansivity.

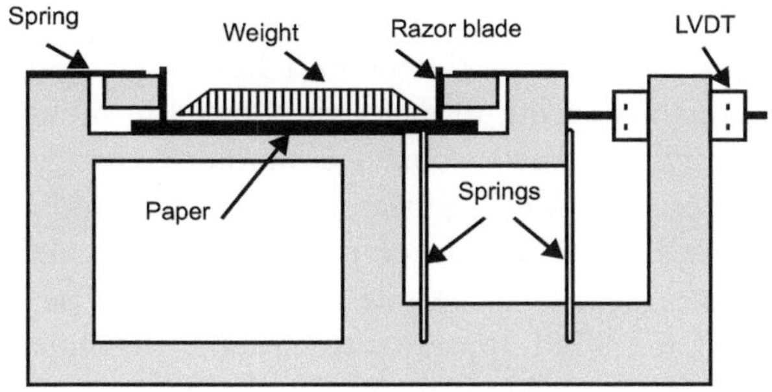


**Figure 6.29** Schematic of Hygroexpansivity Meter. *Reproduced with Permission of STFI-Packforsk, Stockholm, Sweden.*

The quality problems associated with dimensional change due to hygroexpansivity are probably the most serious in the nonwovens industry, particularly if the medium contains a moisture-absorbing material such as cellulose fiber. Media treated with hygroscopic flame retardants are even more vulnerable to the undesirable effects of hygroexpansivity. For example, slit width variations of finished rolls are most often caused by dimensional change in their width due to exposure to weather and climate variations after the rolls have been manufactured and slit.

The problems of hygroexpansivity and creep are certainly not limited to filter media. It has been long established that paper creeps more when subject to cyclic humidity conditions than at constant humidity. There is reason to believe that this is also true for filter media. Habeger and Coffin (242) claim that this "accelerated creep" is due to sorption-induced stress concentrations and a nonlinear creep response. They also argue that it is a phenomenon that is still not well understood.

Despite all the concerns of hygroexpansivity there are very few published standards and test methods related to this property and what few there are appear to be related to paper and paperboard. Among these are ISO 8226-1:1994 (up to 68% RH) and ISO 8226-2:1990 (up to 86% RH).

Most test methods are variations on the Neenah Multiple Specimen Paper Expansimeter. The development of this apparatus traces back to the 1940s. It is still advertised by Adirondack Machine Company in Queensbury, New York, USA. The specimens are enclosed in a controlled humidity chamber. They are 1 in strips that are clamped on either end and in a vertical position. One end is fixed and the other end is attached to a movable clamp that moves with the dimensional change of the specimen as it is exposed to higher and higher levels of humidity. The dimensional changes are measured by built-in micrometers. The instrument allows up to 20 specimens for each test.

A variation of the method is an STFI apparatus that was originally marketed by Packforsk[9] in Stockholm, Sweden. The tester is depicted in Figure 6.29 above.

[9] Packforsk has since been purchased by Innventia AB, Stockholm, Sweden.

Whereas the Neenah technique had the clamped specimens in a vertical position, the STFI method positions the specimens in a horizontal position. Razor blades are used for the clamps. A spring load design allows the movable razor clamp to move with the expansivity of the specimen as humidity changes. The dimensional change is measured by an LVDT (linear variable differential transformer). An LVDT is an electromechanical transducer that produces an electrical output proportional to the displacement of a separate movable core. The core is attached to the movable clamp of the STFI apparatus so that the movement of the core and related electrical outputs corresponds to the hygroexpansion of the test specimen. Pulkinnen et al. (243) conducted experiments with a modified STFI unit that utilized a laser beam reading device to measure the movement of the movable clamp. It should be noted that a weight is placed on the specimens when measurements are recorded. This is to flatten out any out of plane deformities caused by the hygroexpansion.

Apparently Innventia has improved on the test equipment. They report (244) on new test equipment to measure the hygroexpansion of paper materials. This equipment is both labor-effective and time-effective, since it is fully automatic and allows 30 independent hygroexpansion tests to be run at the same time. The equipment is connected to a moisture generator, which can be programmed to make measurements possible under optional humidity histories.

In Grenoble, France, Techpap SAS produces the Varidim®[10] Expansimeter. It is depicted in Figure 6.30. Visually, the humidity chamber of the Varidim® apparatus is similar in appearance to the Neenah machine. The differences are that the Varidim® has magnetic clamps and utilizes LVDT devices to measure hygroexpansive change. Techpap reports (245) that the Varidim® unit can perform a variety of tests to measure expansive properties of paper. The system makes a continuous measurement of up to ten paper strips in a humidity-controlled chamber with relative moisture between 15% and 90%. The system is PC driven and saves the data output so that it can be easily worked with. The set point number and duration can be saved along with the humidity for each set point. It is easy to achieve a complete range of results as tables or curves and all of the raw data can be transferred to a standard spreadsheet. The Young modulus can be calculated for sample elasticity under each humidity level (Figure 6.31).

Pukkinen et al. (243) report on the OPTIDIM method of measuring hygroexpansivity. In this test a specimen lies flat inside a humidity chamber and covered with glass weights when dimensional measurements are made. The measurements are made by utililizing a CCD camera.

[10] Varidim® is copyrighted by Techpap SAS, Grenoble, France.

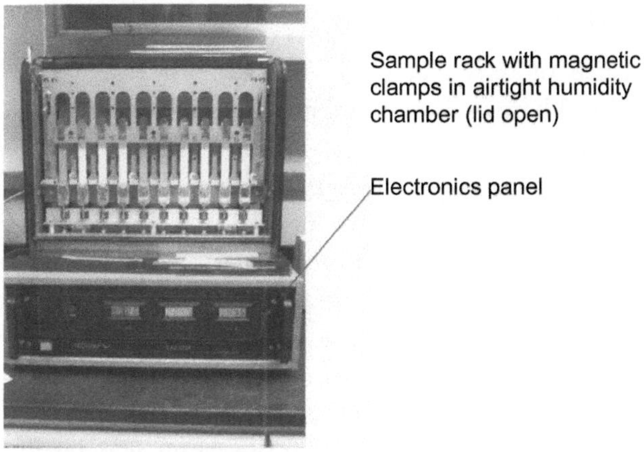


**Figure 6.30** Varidim® Expansimeter. *Printed with Permission of Georgia Tech Renewable Bioproducts Institute (formerly IPST), Atlanta, Georgia, USA.*

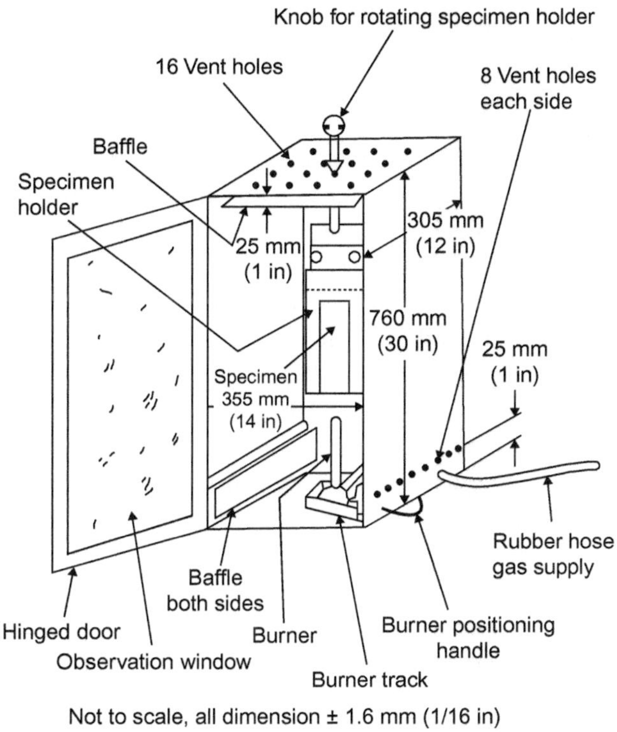


**Figure 6.31** TAPPI T-461 Vertical Flame Test Configuration. *Printed with Permission of TAPPI, Atlanta, Georgia, USA.*

## 6.12 FLAMMABILITY

There is an abundance of tests that might be used for testing nonwovens and filter media for flammability. There is also a number of tests that are applied to filters containing nonwoven filter media. For filter media, TAPPI T-461 and National Fire Protection Agency NFPA-701 (National Fire Protection Association) are vertical flame tests most commonly used. The TAPPI T-461 test configuration is depicted in Figure 6.31 above. The method includes a water leach test procedure for possible nondurable flame retardants. Papers not subjected to the leach test are identified as Type I papers. Papers subjected to the leach test are identified as Type II papers. The flame is applied, for 12 s, to the bottom of a specimen, 70 mm × 210 mm (2.75 in × 8.25 in), and then withdrawn. The following properties are measured and recorded:

1. Char length—the length of char caused by the flame as determined by poking the char away with a 6 mm rod.
2. Flaming time—the time the flame continues to burn or be visible after the igniter flame has been removed.
3. Glowing time—the time the specimen continues to glow after the flame has gone out.

NFPA-701 has two test methods: Test Method 1 and Test Method 2. Both are vertical tests, however they differ in the design and dimensions of the flame chamber. In Test Method 1, a vertical flame is applied directly to the bottom of the vertical test specimen so that the flame is propagated upward. In Test Method 2 the flame is applied at a 25° angle to the vertical specimen. Test Method 1 is designed for fabrics used in window treatments, such as curtains, table linens, wall hangings, etc. It does not apply to fabrics having a grammage greater than 700 g/m$^2$. Test method 2 is intended for heavier fabrics having a grammage greater than 700 g/m$^2$. This includes materials such as multilayered fabrics, awnings, tent materials, tarps, and vinyl-coated fabrics. For most filter media, Test Method 1 would be the more appropriate method. The criteria, for passing Test Method 1, are:

1. Mass loss due to burning is less than 40%.
2. Fragments or residues that fall to the floor of the chamber shall not burn for more than 2 s (average of 10 specimens).

NFPA 701 is recognized as a national standard and a fabric is certified as flame retardant if it has been certified to have been tested and passed the NFPA 701 test.

ASTM D-2863-13 is also a vertical flame test that defines an "Oxygen Index" (O.I.) to measure flammability. A small test specimen is supported vertically in a mixture of oxygen and nitrogen flowing upward through a transparent chimney. The upper end of the specimen is ignited and the subsequent burning behavior of the specimen is observed to compare the period for which burning continues, or the length of specimen burnt, with specified limits for each burning. By testing a series of specimens in different oxygen concentrations, the minimum oxygen concentration is

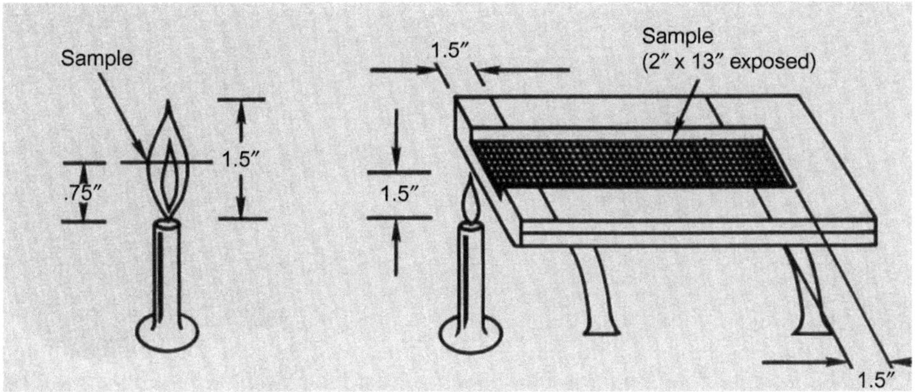


**Figure 6.32** MVVS-302 Horizontal Flame Test.

determined. The O.I. is that minimum concentration. The higher the O.I. of a material, the lower is its flammability. Mango (246) noted that previous versions of ASTM D-2863 were used to measure flammability of polymers and this may differ significantly from the flammability of that polymer in a nonwoven substrate. It is noted that ASTM 2863-13 and ISO 4589-2 are equivalent tests.

There are 45° versions of the vertical flame test. Among these are 16 CFR Part 1610, and ASTM 1230 (these are not identical tests). MVVS-302 (Motor Vehicle Safety Standard, U.S. Department of Transportation) is a horizontal test for interior materials of transportation vehicles. The configuration is illustrated in Figure 6.32.

UL 94 is a standard for the flammability of plastic materials for parts and devices. It is the Standard for Flammability of Plastic Materials for Parts in Devices and Appliances.

## 6.13 COLOR

Most filter medium manufacturing operations are very fundamental in their ability to test for color. In most cases some internal "color expert" places the test sample next to a color standard or a color match sample and makes an eyeball judgement as to whether this is "good enough." Fortunately, this type of testing is usually "good enough." Filter manufacturers may be very stringent in specifying a color for their filter media, however, they often are not very disciplined in enforcing limits on color match or color reproducibility. "If it looks green that's good enough." This attitude will probably change as the industry becomes more sophisticated in color technology.

There are more sophisticated techniques to help the "color expert." One is to use a light booth such as pictured in Figure 6.33. This particular light booth (247) is produced by GTI, Graphic Technology, Inc., Newburgh, New York, USA, is their Model CMB 2540. The operator does his or her visual color judgements by placing

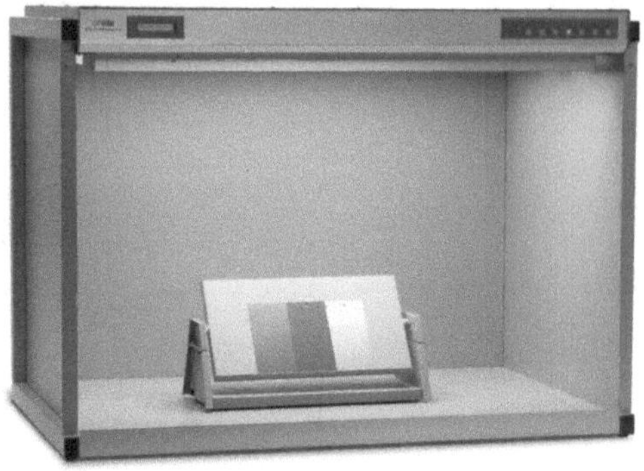

**Figure 6.33** GTI Graphic Technology Light Booth. *Reproduced with Permission of GTI Graphic Technology Inc., Newburgh, New York, USA.*

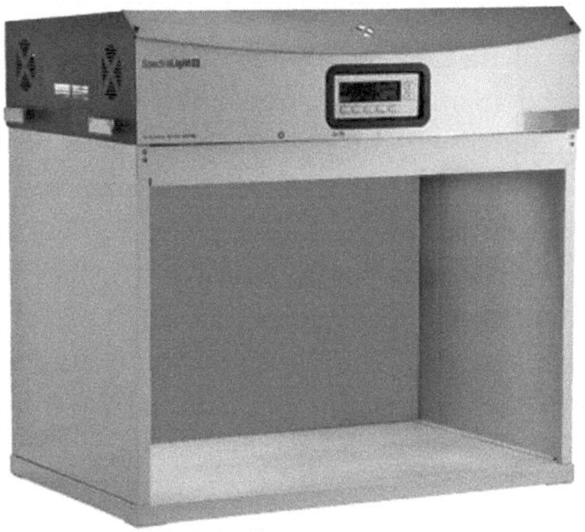

**Figure 6.34** SpectraLight QC Light Booth. *Reproduced with Permission of X-Rrite, Grand Rapids, Michigan USA.*

the samples in the light booth. The advantage of the booth is that it provides a constant and reproducible light source for comparing color specimens. Modern light booths also provide a choice of light sources. The light booth in Figure 6.34 is the SpectraLight QC produced by X-Rite. This particular unit has seven available light sources including daylight, cool white fluorescent, horizon,[11] incandescent, and ultraviolet.[12] ASTM D1729 provides guidances and best practices for doing color measurements in a light booth. Light booth color matching and color development of filter media are usually done with a daylight light source.

[11] Horizon simulates early morning sunrise and/or late afternoon sunset.

[12] Ultraviolet light is useful for identifying certain types of defects and contaminants in filter media that are not readily visible in regular light.

There are instruments to measure color and color differences. Two types available for these purposes are colorimeters and spectrophotometers. The colorimeter provides tristimulus values for red, green, and blue directly calculated from reflectance measurements on the sample. The spectrophotometer provides percent reflectance data at distinct intervals throughout the visible spectrum. The data are then converted to tristimulus values. Either way, the color is expressed as coordinates in various color space systems. Following is a listing of the most commonly used color space systems:

1. Hunter L, a, b
2. CIE (International Commission on Illumination) L*, a*, b*
3. L*, C*, h
4. Dominant wavelength, purity, luminosity
5. Color difference [ΔE, ΔE*, ΔE*94, ΔE(CMC)].

TAPPI T1215 (248) provides a thorough explanation of the measuring system, the equations, and how the calculations are done. The calculations are complicated and tedious. Fortunately, most modern measuring instruments are equipped with a microprocessor or can readily be connected to appropriate software for rapid calculation. TAPPI has two test methods (TAPPI T524 and TAPPI T527) for the measurement color of paper and paperboard. They supposedly give the same results, but differ on the geometry of how the source illumination reflects of the specimen and is measured by the receptor. T524 uses indirect lighting and the rays are reflected diffusely into the receptor. In T527 the light source is illuminated diffusely.

## 6.14 FILTER MEDIUM FILTRATION TESTING

A filtration test on a nonwoven filter medium is generally designed to evaluate and rate the the medium for some form of filtration efficiency, dirt or dust holding capacity, and/or pressure drop. Tests may be run on flat sheet media or on standard filters or filter elements made from the medium. Filtration results on a flat sheet medium do not necessarily parallel the field performance of a filter made from this medium. In addition, standardized filter element testing may not accurately predict field performance. Many variables in filter design, construction, and quality can easily offset the flat sheet performance of a filter medium. In addition, the challenge conditions by which a filter is tested may not effectively represent the environmental conditions a filter will encounter in real life applications. The test dirt or challenge aerosol of a test may not simulate the contaminants in real filter environments. Climate, weather, temperature, humidity, aging, chemical degradation, etc., are all factors that can affect a filter contrary to test performance expectations. Nevertheless, tests and standards are necessary for filter media. It would be practically impossible to test and choose a medium for any given filter application unless there were test data to support it. Powder Technology

Incorporated (PTI) of Burnsville, Minnesota, USA, provides test dust for many of the test procedures and methods. These are listed in Table 6.10 below (Section 6.14.5).

Following is a discussion of the various factors that are considered in the filtration testing of a medium.

## 6.14.1 Efficiency

There are a number of test methods and methodologies to evaluate filtration efficiency. Following are some of them.

### 6.14.1.1 Filtration efficiency

Filtration efficiency simply rates a medium by the percentage of contaminant removed by the filter medium as expressed in Eq. (6.23). The quantity can be by weight or mass, by number of particles, by volume, and interestingly enough by cross-section area of the particles.[13]

$$e = 100\left(\frac{M_u - M_d}{M_u}\right) \tag{6.23}$$

$e$ is the filtration effiiciency, %.

$M_u$ is the upstream quantity of the contaminant.

$M_d$ is the downstream quantity of the contaminant.

Penetration or leakage is the percentage of contaminant that penetrates the medium:

$$e = 100\left(\frac{M_d}{M_u}\right) \tag{6.24}$$

In some situations penetration might be the more pertinent perspective. Consider a case, where the efficiency has to be improved from 98% to 99%. This appears to be a mere 1% increase in efficiency and would appear to require only a minor modification of the filter medium. On the other hand, the improvement has to reduce penetration from 2% to 1%, a 50% reduction. This would suggest a major redesign of the medium. The author's experience suggests the latter is the more realistic perspective.

### 6.14.1.2 Single pass efficiency

In a single pass efficiency (SPE) test the particle contaminant in the fluid passes through the filter once. Any particles that penetrate are trapped by an "absolute" filter downstream of the test filter. After the test is completed, the absolute filter is weight analyzed to determine the penetration from Eq. (6.23). Usually the test contaminant is controlled to a specified particle size or particle size range. An example of a SPE test is SAE HS806 which includes a single pass particle retention capability test for

[13] Several types of image analyzer particle counters do not measure the three-dimensional size of the particle but rather the area of its shadow.

engine full flow lube oil filters. The HS806 test also allows for cumulative efficiency testing (see Section 6.14.1.3 below), multipass efficiency (Section 6.14.1.5), and DHC (Section 6.14.4).

#### 6.14.1.3 Cumulative efficiency

Cumulative efficiency is determined by making efficiency measurements at various intervals during a filter life test. In SAE HS806, the specified test dust contaminant is continuously added to the oil, which is continuously recirculated through the test filter. The oil is sampled at selected intervals during the test and analyzed for dust contaminant content. This is compared to the cumulative quantity of dust fed to system at the time of sampling. The efficiency so determined is the cumulative efficiency up to that point in the test. Note that unlike the SPE test, any contaminant that penetrates or leaks through the filter comes around for another chance to be filtered. The test is a multipass test, however the term "multipass testing" usually refers to testing where particle analyzers and counters are used to determine efficiency by particle size (see Section 6.14.1.5 below).

#### 6.14.1.4 Micron rating

Micron rating is intended to be an indicator of the size of the particles that might be filtered by the medium. It is generally applied to liquid filters and their media. Supposedly a 5-μm filter would be capable of filtering particles 5 μm and larger in diameter. This type of filter rating is weak because it does not indicate the degree of efficiency for the rating, and allows for wide variance in filter performance of different filters having the same micron rating. Nominal and absolute micron ratings were adapted by some manufacturers. According to the Filter Manufacturers Council (249), nominal ratings are based on 50% filtration efficiency for the particle size characterized and absolute ratings are based on 98.7%. Even this rating system is nebulous because test conditions are not well defined. The Filter Manufacturers Council encourages the use of the Beta rating as discussed in Section 6.14.1.6 below.

#### 6.14.1.5 Multipass efficiency

The multipass procedure uses particle counteranalyzers to determine filtration by particle size. It is used to test actual filter elements and flat sheet media. Figure 6.35, provided by Southwest Research Institute in San Texas, San Antonio, Texas, USA, is a picture of the test stand. Efficiencies are measured at selected intervals during the test and for a range of selected particle sizes. The efficiencies are cumulative and reported as "β (beta) ratios" as defined in Section 6.14.1.6. ISO 16889 developed for hydraulic fluid filters is one method based on the multipass procedure. It replaces previous method ISO 4572. It is discussed in more detail in Chapter 7, Section 7.4.2. The test dust requirements and complications are included in the test dust discussion

**Figure 6.35** Multi-Pass Test Stand at Southwest Research Institute (SwRI). *Reproduced with Permission of SwRI (Southwest Research Institute), San Antonio, Texas, USA.*

of Section 6.14.5 below. Other standards based on multipass testing include ISO 19438 for diesel fuel and ISO 4548-12 for engine lube oil filters. The ISO 16889 procedure, itself, may be used for several types of liquid fluids in addition to hydraulic fluids, for example, engine fuels.

The $\beta$-ratio is the ratio of upstream particles to downstream particles. The more efficient is the filter, the higher is the $\beta$-ratio.

$$\beta = \frac{N_{d,U}}{N_{d,D}} \tag{6.25}$$

$N_{d,U}$ is the upstream particle count for particles of diameter d or greater.

$N_{d,D}$ is the downstream particle count for particles of diameter d or greater.

The $\beta$-ratio is related to filtration efficiency, $e$, as follows:

$$\beta = \frac{1}{1-(e/100)} \tag{6.26}$$

In multipass testing, for example, in accordance with ISO 16889, subscripts are attached to $\beta$ to indicate the particle size for which the rating is being reported and to indicate whether the test dust is in accordance with ISO 16889 standards. For example, a $\beta$-ratio of $\beta_{20(c)} = 200$ indicates that the $\beta$ value for 20 μm particles was

200 and the test dust was in accordance with ISO 16889 standards and National Institute of Technology and Science (NIST) traceability. A more detailed explanation of the $\beta$-ratio is provided in Chapter 7, Section 7.4.2.

### 6.14.1.6 Fractional efficiency

Fractional efficiency is a term generally applied to filtration efficiencies based on particle size or particle size groups. The $\beta$-ratio discussed above is a form of fractional efficiency, however the term, fractional efficiency is more often applied to air filtration standards. For example, ASHRAE 52.2 assigns MERV ratings (minimum efficiency reporting values) to fractional efficiency groups of different particle size ranges (see Section 6.14.1.7 below).

### 6.14.1.7 MERV rating

MERV ratings are minimum efficiency reporting values that are particular to ASHRAE 52.2-2012 (250). They are determined by subjecting a filter to fractional efficiency tests in accordance with test procedures defined by ASHRAE 52.2. The procedures and methodology are discussed in Chapter 8, Section 8.2.2.2. A MERV rating is assigned based on its performance in the tests. MERV ratings range from MERV 1 for very coarse filters to as high as MERV 16 for high efficiency filters. ASHRAE 52.2 also includes MERV ratings (MERV 17–MERV 20) for HEPA and ULPA filters, although the ratings are based on the testing of the Recommended Practices (RPs) of the Institute of Environmental Science and Technology (see Chapter 8, Section 8.3.5.3). It should be noted that the lowest MERV ratings (MERV 1–MERV 4) are not determined from fractional efficiency testing, but from ASHRAE arrestance testing as prescribed in ASHRAE 52.2[14] (Section 6.14.1.8 below).

### 6.14.1.8 Arrestance

Arrestance is a mass or weight efficiency as determined by Eq. (6.23) above. The ASHRAE procedure determines "average ASHRAE arrestance" of a filter based on measurements made at prescribed intervals during the life of the test.

CEN (Comité de Europêen Normalisation) Standard EN779 (251) utilizes an arrestance procedure for its G-rated filter classes, as described by Section 6.14.1.11 below. EN 779 is also further discussed in Chapter 8, Section 8.2.2.1.

### 6.14.1.9 ASHRAE dust-spot efficiency

The ASHRAE dust-spot efficiency test, originally part of the ASHRAE 52.1-1992, was eliminated when ASHRAE 52.1 was retired.

[14] The ASHRAE arrestance test and ASHRAE Dust-Spot Efficiency test procedures were specified by ASHRAE 52.1 until 2007. The ASHRAE arrestance test has since been incorporated into ASHRAE 52.2. The dust-spot test has been eliminated. In addition, the dirt holding capacity part of ASHRAE 52.1 was also incorporated into ASHRAE 52.2. Accordingly, ASHRAE 52.1 has since been retired.

#### 6.14.1.10 CEN ratings and eurovent 4/11

CEN Standard EN779:2012 (251) is the European counterpart to ASHRAE 52.2. The standard rates filters according to their efficiencies by class, the higher the class number the more efficient is the filter. The coarse filters are rated G1–G4 and are tested by an arrestance test similar to the ASHRAE arrestance discussed in Section 6.14.1.9 above. Classes M5–M6 (medium) and classes F7–F9 (fine) are tested by 0.4 μm liquid aerosol particles. The efficiency is based on the percent of particles removed. Note that this differs from the arrestance, which is based on the weight percent of particles removed. The standard was revised and updated in April 2012. Changes included the introduction of minimum efficiencies for classes F7–F9 and the renaming of classes F5 and F6 as M5 and M6. A more detailed discussion of EN779 will be found in Chapter 8, Section 8.2.2.1. The method was also connected to Eurovent 4/11, an energy classification system also discussed in Section 8.2.2.1. Eurovent 4/11 is a way of rating and certifying filters based on both: EN779 filtration performance and energy consumption performance.

#### 6.14.1.11 HEPA and ULPA ratings

European standards for rating the efficiencies of HEPA and ULPA filters and media are prescribed by EN 1822:2009 (261). They are based on the filtration efficiency of the most penetrating particle size (MPPS) as discussed in 6.14.1.12. For filter testing, the MPPS is determined by flat sheet testing in accordance with EN 1822-3. The filter is then tested for that particle size and at the flow rate at which the MPPS was determined. The challenge contaminant is either a monodisperse aerosol or a polydisperse aerosol. The ratings are an extension of EN 779. Ratings are H10–H14 for HEPA filters and U15–U17 for ULPA filters. IEST RP Classification.

The United States counterpart to EN 1822 are the RPs of the Institute of Environmental Science and Technology (IEST). IEST-RP-CC001.4 is the RP that classifies 11 different filter types from Type A to Type K. The classifications are not necessarily in the order of efficiency. The efficency includes a leak test and a penetration test. The penetration test is based on the efficiency of the medium against an 0.3 μm liquid aerosol, usually dioctylphthalate (DOP). Polyalpha olefin (PAO) has been approved as an alternative to DOP.

#### 6.14.1.12 Most penetrating particle size

The concept of MPPS has been discussed in Chapter 2, Section 2.1.2. It is the particle size that has the highest penetration through the medium. Usually it is in the range of 0.1–0.4 μm. In the EN 1822 procedure, the efficiency is determined by number count. If the aerosol is quasi-monodispersed, the aerosol generated has to have a particle size near MPPS. If the aerosol is polydispersed its fractional efficiency counting is done to determine MPPS.

#### 6.14.1.13 Alpha value

The alpha (or gamma) value is an air filtration quality factor that has been previously mentioned above in Chapter 3, Section 3.7.2 and defined as $q_f$ in Eq. (3.3) is a way of assigning a relative performance to HEPA media. It is defined as:

$$\text{Alpha } (\alpha) = (-\text{log10 (Penetration}/100) \times 100)/\text{Resistance} \tag{6.27}$$

where

*Penetration* (*P*) is in %.

*Resistance* is in mm $H_2O$.

The alpha (or gamma) value is the air filtration counterpart to the beta rating (Section 6.14.1.5). It is often referred to as the figure of merit.

According to Lydall Filtration/Separation Inc (253): "Alpha is an index used to characterize and compare filtration media and finished filters. The index displays the relationship of penetration (%) to resistance (mm $H_2O$). Improving alpha can be accomplished by lowering the penetration (how many particles pass through the filter) while maintaining the resistance to air flow or by reducing the resistance to air flow while maintaining the penetration. In general, a higher alpha value indicates a superior filter."

"One must be careful to compare similar media or filters when using alpha. Specifically, electrostatically charged filter media provide a high initial alpha but may not offer the lifetime performance required by a given application."

#### 6.14.1.14 Aerosol efficiency penetrometers

There are several types of filter testers that are used to test HEPA and ULPA filter media. The original Q127 penetrometer tested in accordance with MIL-STD-282 utilized 0.3 μm DOP smoke.[15,16] British Standard BS 3928 based its efficiency on 0.6 μm sodium chloride (NaCl) particles which, unlike DOP and PAO, is a solid

[15] Vijayakumar, points out that the DOP particle size of 0.3 μm, as determined by a photometer, is a mass median diameter. The count mean diameter (CMD) based on the average particle size of the number distribution is quite a bit smaller and is closer to the MPPS for HEPA filters. CMD is defined as the average calculated diameter in a population of particles in a gas or liquid phase above which there are as many particles with larger diameters as there are particles below it with smaller diameters (Encyclo).

[16] A test method will not only specify the particle size of the test contaminant, but also its concentration. The concentration for aerosols is expressed either in ppmv (parts per million by volume) or in $mg/m^3$ (milligrams per cubic meter). The conversion is

$$mg/m^3 = (ppmv)(12.187)(\text{MW})/273.15 + T$$

where MW is the molecular weight of the aerosol and $T$ is the ambient temperature in °C.

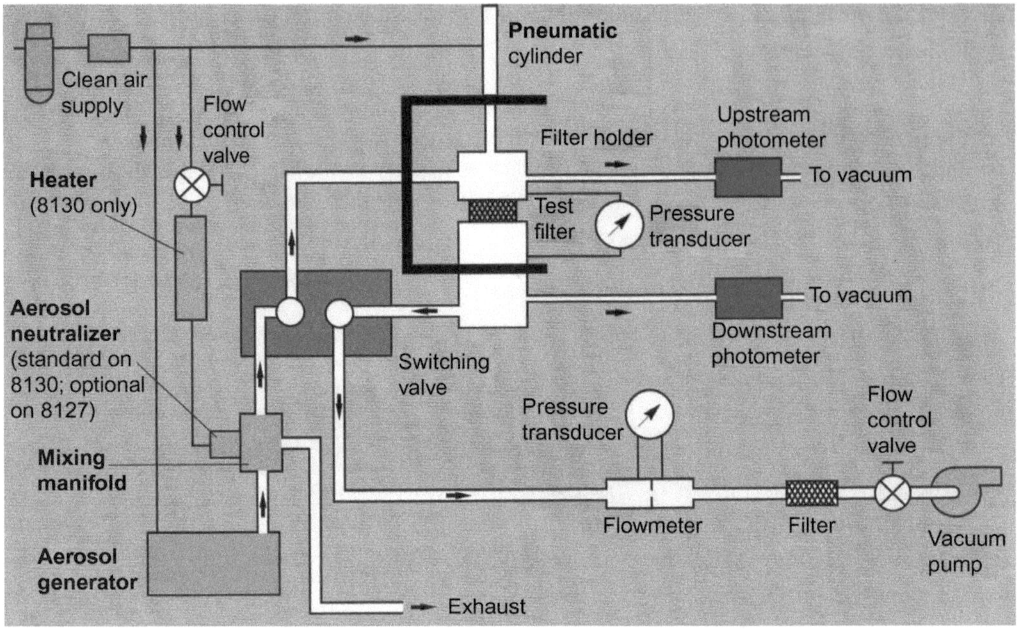


**Figure 6.36** Schematic of TSI Models 8127 and 8130 Filter Testers. *Reproduced by Permission of TSI Incorporated, St. Paul, Minnesota, USA.*

aerosol. Although the Q127 penetrometer is still in use, more modern versions have been developed. TSI Incorporated in St. Paul Minnesota, USA has developed several automated filter testers and associated equipment capable of testing flat sheets, face mask filters, and other forms of air filters. The Model 8127 tests filters by oil aerosol only whereas the Model 8130 has the capability to test by either oil aerosol or salt aerosol. There is also the Model 3160 tester that also generates salt and oil aerosols, but over a range of particle sizes, so as to allow fractional efficiency and MPPS determination. Figure 6.36 is a flow diagram schematic of the Models 8127 and 8130 filter testers. Note that they both use photometers to measure particle counts by light scattering. The Certitest® Model 3160, depicted in Figure 6.37 uses condensation particle counters.

Air Techniques International (ATI) also measured filter testers for high-efficiency air filtration testing. Their Model TDA-100P automatic filter tester is pictured in Figure 6.38. Shown is the unit in manual operation configuration and full automatic configuration. The unit can use either DOP or PAO as challenge aerosols.

### *6.14.1.15 Particle filtration efficiency, bacterial filtration efficiency and viral filtration efficiency*

Particle filtration efficiency (PFE), bacteria filtration efficiency (BFE), and viral filtration efficiency (VFE) are tests that are set up to evaluate surgical face mask and other medical filtration materials (see Chapter 8, Section 8.4.3).

1 Filter Holder

2 Access Doors

3 Flow Meter

4 Power Switches

5 Flow Control Valves

6 CPCS

7 System Computer

8 Printer

9 Drawer

10 Filter Holder Controls

11 Electrostatic Classifier

12 Diluter

**Figure 6.37** TSI Certitest® Model 3160 Automatic Filter Tester. *Reproduced by Permission of TSI Incorporated, St. Paul, Minnesota, USA.*

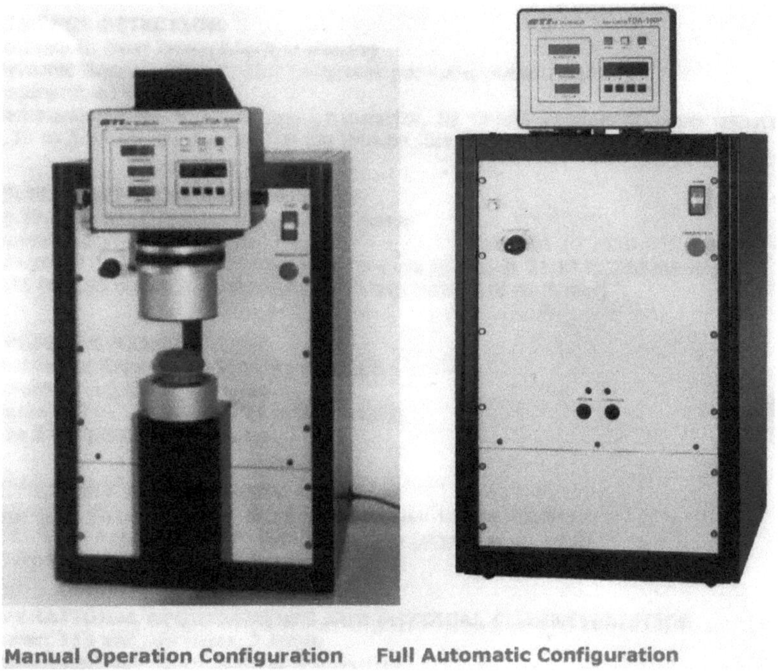


**Figure 6.38** ATI Model TDA-100P Automatic Filter Tester. *Reproduced by Permission of Air Techniques International, Owings Mills, Maryland, USA.*

The PFE test in accordance with ASTM F 2299-03 uses a aerosolized latex particles, usually 0.1 μm, as the challenge aerosol. Control counts are performed without a test sample in the system, then the sample is tested. Efficiency is determined by comparing the counts through the sample to the control counts.

ASTM F2101-01 is a standard using an aerosol of *Staphylococcus aureus (S. aureus)*. to evaluate the BFE of face mask materials. Other types on bacterial organisms may be used, however they may give different results because of differences in shape and size. *Staphylococcus aureus* has an approximate diameter of 0.8 μm. The organisms are generated into an aerosol with a nebulizer. As in the PFE test, the control is determined from counts without the test specimen in the system.

The VFE test procedure is adapted from the Military standard MIL-M-36954C and ASTM F2101. The challenge aerosol is bacteriophage OX174. The bacteriphage is one of the smallest viral organisms and has a diameter of 0.025–0.027 μm. The testing is very similar to the BFE test, except the smaller organism is used. The VFE procedure has become of interest because of the developing concerns about viral deseases such as avian bird flu and Ebola.

#### 6.14.1.16 Log reduction value

The log reduction value (LRV) is used to evaluate biological membrane filtration and the capability of the separation system to remove microorganisms. It can be calculated from Eq. (6.28):

$$\begin{aligned}\text{LRV} &= \text{Log}_{10}(\text{concentration of microorganisms in feed})/\text{Log}_{10}(\text{concentration} \\ &\quad \text{of microorganisms in filtrate}) \\ &= \log_{10}A_f/\log_{10}B_f = \log_{10}A_f - \log_{10}B_f\end{aligned} \tag{6.28}$$

where

$A_f$ is the concentration of microorganisms in the feed.

$B_f$ is the concentration of microorganisms in the filtrate.

LRV can be related to efficiency ($e$) by Eq. (6.29):

$$\text{LRV} = \log_{10}\left[1/(1-e)\right] \tag{6.29}$$

### 6.14.2 Pressure drop

Most filtration test methods require a pressure drop at a given flow rate. The pressure drop relates to the size of the pump or blower necessary to force fluid through the filter or filter medium. It also relates to the breathability of gas mask and face mask materials.

The pressure drop is measured by some form of differential pressure device with probes upstream and downstream of the medium. In some cases this is no more than a simple u-tube manometer. Modern test stands have more sophisticated devices with electronic processors and digital readout capability. Usually the filter is pressure drop rated before it has been loaded with test contaminant. In some cases, the rating is established after a specified load has been applied to the filter.

### 6.14.3 Filter life testing

Filter life is determined by running a test to a specified pressure drop and measuring the time to reach that pressure drop. The time to reach the specified pressure drop is often demonstrated by a plot. When using this criteria, one must be careful that the flow rate and rate of contaminant feed is controlled very carefully and reproducible from test to test. It should be noted that filter life test results on finished filters may not necessarily reflect on the medium, but may possibly reflect on the filter construction. For example, the pleat count and pleat spacing may have more of an effect than the medium effect. Pleat quality, such as uniformity of pleat spacing can also be important.

### 6.14.4 Dust holding capacity

The DHC usually is defined by the amount of test dust that is retained by the filter. It does not include the amount of test dust that passes or leaks through the filter. In the typical test procedure such as ASHRAE 52.2 the DHC is determined by the amount of dusts trapped in a final filter downstream of the test filter plus the amount of dust retained in the test duct downstream of the test filter and before the final filter. This sum amount is substraced from the total dust fed to determine dust holding capacity. Often the DHC is prorated to area of filter medium in the filter and reported as $g/m^2$ of medium or in some other form of mass:area ratio. This is important when comparing different media.

The DHC is probably a more informative test than the time to pressure drop life test. The two do not necessarily correlate with each other. For example, a low efficiency may have a longer life than a high efficiency filter because so much of the dust passes through it and pressure drop buildup is slow. On the other hand, the high efficiency filter better retains dust and as a result, pressure drop builds-up more rapidly. At the end of the test the high efficiency filter may have retained more dust, even though its life was shorter.

Like filter life testing, DHC may not necessarily reflect on the medium, but may be a result of filter construction.

### 6.14.5 Test dusts and contaminants

Practically every filter medium or filter test or standard requires some form of specified test dust and/or contaminant to determine filtration efficiency and filter life. These test dusts and contaminants have to be carefully controlled for particle size, chemistry, and other properties that might affect the performance of the medium. Variations in the test dust and/or contaminant can cause significant variations in the test performance of the filters or

**Table 6.8** Test Dusts Specified by ISO 12103-1

| ISO designation | Common name | Particle size range (μm) | Other names |
|---|---|---|---|
| ISO 12103–A1 | ISO ultrafine test dust (ISO UFTD) | 0–10 | PTI* 0–10 μm test dust |
| ISO 12103–A2 | ISO fine test dust (ISO FTD) | 0–80 | PTI* fine test dust<br>SAE fine test dust |
| ISO 12103–A3 | ISO medium test dust (ISO MTD) | 0–80* ISO MTD has lower 0–5 μm content than ISO FTD | PTI* 5–80 μm test dust<br>SAE 5–80 μm test dust |
| ISO 12103–A4 | ISO coarse test dust (ISO CTD) | 0–180 | PTI* coarse test dust<br>SAE coarse test dust |

**Table 6.9** Graphical Comparison of ISO Test Dust Grades

**Volume fraction, %**

| Size μm | A1 ultrafine | A2 fine | A3 medium | A4 coarse |
|---|---|---|---|---|
| 1 | 1.0–3.0 | 2.5–3.5 | 1.0–2.0 | 0.6–1.0 |
| 2 | 9.0–13.0 | 10.5–12.5 | 4.0–5.5 | 2.2–3.7 |
| 3 | 21.0–27.0 | 18.5–22.0 | 7.5–9.5 | 4.2–6.0 |
| 4 | 36.0–44.0 | 25.5–29.5 | 10.5–13.0 | 6.2–8.2 |
| 5 | 56.0–64.0 | 31.0–36.0 | 15.0–19.0 | 8.0–10.5 |
| 7 | 83.0–88.0 | 41.0–46.0 | 28.0–33.0 | 12.0–14.5 |
| 10 | 97.0–100 | 50.0–54.0 | 40.0–45.0 | 17.0–22.0 |
| 20 | 100 | 70.0–74.0 | 65.0–69.0 | 32.0–36.0 |
| 40 | – | 88.0–91.0 | 84.0–88.0 | 57.0–61.0 |
| 80 | – | 99.5–100 | 99.0–100 | 87.5–89.5 |
| 120 | – | 100 | 100 | 97.0–98.0 |
| 180 | – | – | – | 99.5–100 |
| 200 | – | – | – | 100 |
| Bulk Density kg/m$^3$ | 500 | 900 | 1,025 | 1,200 |

Based on ISO 12103-1 specified limits median (cumulative volume).
Reproduced with permission of Powder Technology Inc., Burnsville, Minnesota, USA.

media being measured and cause difficulties in comparing filters and validating performance. For this reason most filter standards require test dusts based on the specifications of ISO 12103-1. Table 6.8 above lists four test dusts specified by ISO 12103-1 including particle size ranges and other designations. Table 6.9, provided by Powder Technology Inc. (PTI) of Burnsville, Minnesota, USA, lists particle size distributions for each of the ISO 12103-1 test dusts. In addition, PTI markets a number of other test dusts and contaminants that are listed in Table 6.10. Particle Technology Limited, Derbyshire, England is also a provider of test contaminants for filter and filter medium testing. Mention Duke.

**Table 6.10** PTI Test Dusts and Contaminants

| | |
|---|---|
| **ISO TEST DUST**<br>**ISO 12103-1 Test Dust Grades**<br>**ISO 12103-1, A1 Ultrafine Test Dust**<br>**ISO 12103-1, A2 Fine Test Dust**<br>**ISO 12103-1, A3 Medium Test Dust**<br>**ISO 12103-1, A4 Coarse Test Dust** | **QUARTZ TEST CONTAMINANTS**<br>**Graded Crushed Quartz Fractions**<br>**Quartz, Gypsum, Calcite, Salt Blend**<br>**ECE R16 Quartz Test Dust**<br>**Custom Quartz Distributions Upon Request** |
| **ARIZONA TEST DUST FRACTIONS**<br>**0-Specified Test Dust Grades**<br>**Intermediate Test Dust Grades**<br>**Typical Chemical Analysis for Arizona Test Dust Products**<br>**Miscellaneous Test Dust Grades** | **MILITARY STANDARDS TEST DUST CONTAMINANTS**<br>**MIL E-5007 and MIL-AV-E-8593F**<br>**MIL-STD-810E Silica Dust and Silica Sand**<br>**MIL-STD-810G Blowing Dust and Blowing Sand** |
| **ARAMCO TEST DUST**<br>**90% Arizona Test Dust, 10% Salt** | **IRON OXIDE TEST CONTAMINANTS**<br>**0-Specified Iron Oxide Grades**<br>**Intermediate Iron Oxide Grades** |
| **JIS TEST DUST**<br>**JIS Z8901 Classes 7, 8 & 11 Kanto Loam Test Dusts**<br>**Product List JIS II—Classes 1, 2 & 3**<br>**Product List JIS 5—Class 5** | **METAL CHIPS & POWDERS**<br>**Metal Chips**<br>**Metal Powders** |
| **ASHRAE TEST DUST**<br>**ASHRAE Test Dust #2 Per ANSI/ ASHRAE 52.2P**<br>**ASHRAE Test Dust #1 Per ANSI/ ASHRAE 52.1 and 52.2** | **MISCELLANEOUS TEST CONTAMINANTS**<br>**CHINA DUST**<br>**EMS 452 TEST DUST MIX**<br>**SOFTC-2A Oil Filter Test Contaminant**<br>**Second Cut Cotton Linters**<br>**Fly Ash**<br>**Talc Powder**<br>**Portland Cement**<br>**Carbon Black**<br>**Pulverized Vinsol Resin (Dark Pine Resin)**<br>**Olivine Sand** |

Reproduced with permission of Powder Technology Inc., Burnsville, Minnesota, USA.

It should be noted that test dusts and contaminants may serve one or both of two purposes in a filter test:

1. They serve as challenge particles for filtration efficiency determination.
2. They are the dirt source for pressure drop buildup to determine filter life and DHC.

In some cases the test dust serves both purposes. For example, ISO 16889 specifies the ISO 12103-1 test dust for both purposes: efficiency and dirt holding. On the other hand, ASHRAE 52-2 specifies a synthetic test dust for pressure buildup and dust holding capacity and specifiies a KCl aerosol for efficiency challenge tests[17] (see Chapter 8, Section 8.2.2.3).

In multipass testing of hydraulic filters and media one of the complications in upgrading the filtration standard from ISO 4572-81 to what is now ISO 16889:2008 was that the original test dust, AC Fine Test Dust (ACFTD) prescribed for ISO 4572-81 was no longer being produced by what was then the AC Rochester Division of General Motors. This was one of the reasons for upgrading the test method. The new test dust is designated as ISO Medium Test Dust (ISO MTD). ISO MTD (also identified as ISO 12103-A3 in Table 6.8 above) is similar to ACFTD in that it is silica based and from the same Arizona source as ACFTD, however it is processed differently and has a different particle size distribution. ISO MTD-A3 is supplied by Powder Technology Incorporated (PTI). ACFTD was ball milled and then classified into a convenient particle size distribution. There were two forms: ACFTD and AC Coarse Test Dust (ACCCT). The particle size distributions were done with microscopic particle counting. The new ISO dust instead of being ball milled is jet milled. The National Institute of Science and Technology (NIST) has taken on the responsibility of certifying ISO MTD as a traceable test dust.

Figure 6.38 provided by Lifshutz compared the old with the new. The curves criss-cross at 10 mm. Below 10 mm, the new size is larger than the old. Above 10 mm, the new size is smaller than the old. One result of this is that the definition of particle size was changed under the new NIST calibration procedure. For example, 2 mm under the old procedure now became 4.6 mm(c), the (c) referring to the NIST calibration. On the high side 15 mm becomes 13.6 mm(c). The ISO 16889 procedure also changed the definition of particle diameter and this may be at least part of the reason for the change in particle sizes from old to new. Although particles are not perfect spheres, their size is based on the assumption that they are. In the old definition, the particle diameter was based on the longest chord through the particle. Under the new definition the diameter is based on the sphere that has the same surface area as the particle. This is illustrated in Figure 6.39 provided by Lifshutz (254).

SAE AIR 5455 (255), SAE Aerospace Information Report (AIR), discusses the impact of the ISO Test Dusts, chosen as replacement contaminants for the Arizona Test Dusts (AC Test Dusts), and the new ISO calibration procedure; ISO 11171,[18] for automatic particle counters. ISO 11171 replaces the previous calibration procedure ISO 4402 (1991), used for laboratory performance of filter elements utilized in

[17] ASHRAE 52.2 does use the synthetic test dust for determining arrestance on low-efficiency air filters that do not test well with the KCl aerosol.

[18] ISO standard for calibration of automatic particle counters for liquids.

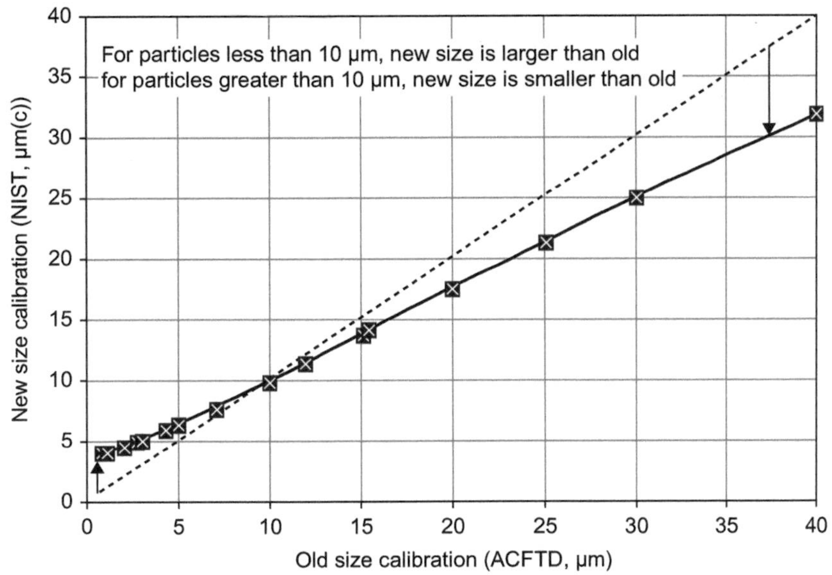


**Figure 6.39** New NIST Versus Old ACFTD Particle Size Distributions. *Reproduced with Permission of INDA, Cary, North Carolina, USA.*

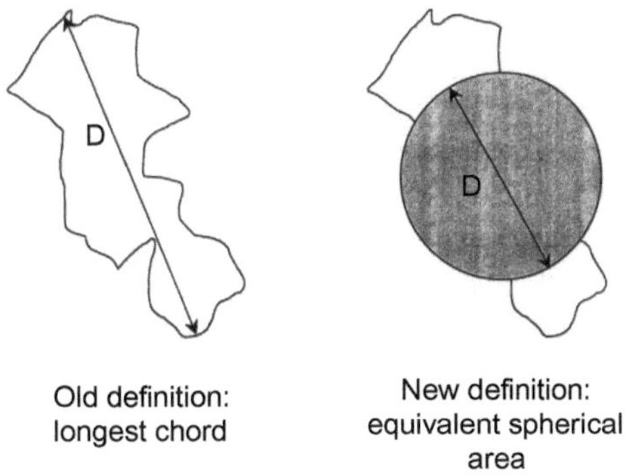


**Figure 6.40** Definition of Particle Size. *Reproduced with Permission of INDA, Cary, North Carolina, USA.*

aerospace lubrication, hydraulic and fuel systems, and fluid cleanliness levels determined with automatic particle counters.

PTI has published Table 6.11 (256), a list of filter test standards and indicates the type of contaminant used for each test.

## 6.14.6 Particle size counting

If one is to measure efficiency by such factors as $\beta$ ratio or MERV rating one must be able to measure the particles in the upstream and downstream fluids by number and

**Table 6.11** Filtration Standards & Specifications

| Filtration standards, procedures, & specifications | Contaminants required | Comments & information |
|---|---|---|
| **ANSI B93.31:** Multipass Method for Evaluating the Filtration Performance of a Fine Hydraulic Fluid Power Element | ISO 12103-1, A2 Fine Test Dust (AC Fine Test Dust per standard is obsolete) | B93.31: Hydraulic Filter Multipass Method is cancelled. ISO 16889 has replaced the ANSI B93.31 & NFPA T3.10.8.8 documents |
| **ISO 2941:** Hydraulic fluid power-filter elements-verification of collapse/burst pressure rating | ISO 12103-1, A2 Fine Test Dust and A3 Medium Test Dust | Test method for element strength & maximum anticipated differential pressure |
| **ISO 4020:** Road Vehicles – Fuel Filters for Diesel Engines | ISO 12103-2, M1 and M2 Aluminum Oxide Test Dusts | An Organic Contaminant is also required, see 4020 standard for details |
| **ISO 4548-4:** Initial Particle Retention Efficiency, Life & Cumulative Efficiency (gravimetric method) | ISO 12103-1, A2 Fine Test Dust ISO 12103-2, M2, M3, M4 and M5 Aluminum Oxide Test Dust | Life & Cumulative Efficiency uses A2 Fine Test Dust Initial Particle Retention uses M series of Aluminum Oxide Test Dust |
| **ISO 4548-3:** Resistance to High Differential Pressure and Elevated Temperature Section 10.1: Test for ability to withstand High Differential Pressure. | ISO 12103-1, A2 Fine Test Dust | Test Method for Element Strength & Maximum Anticipated Differential Pressure (Collapse) |
| **ISO 4548-12:** Filtration Efficiency using Particle Counting, & Contaminant Retention Capacity | ISO 12103-1, A3 Medium Test Dust | SAE J 1858 Oil Filter Multipass Method is cancelled. ISO 4548-12 has been adopted by SAE |
| **ISO 4572:** Hydraulic Fluid Power-Filters – Multipass Method | ISO 12103-1, A2 Fine Test Dust (AC Fine Test Dust per std. is obsolete) | ISO 4572 Hydraulic Filter Multipass Method is cancelled. ISO 16889 has replaced 4572 |
| **ISO 5011:** Inlet Air Cleaning Equipment for Internal Combustion Engines & Compressors | ISO 12103-1, A2 Fine Test Dust and A4 Coarse Dust | SAE J 726 Air Cleaner Test Code is cancelled. ISO 5011 has been adopted by SAE |
| **ISO/TS 11155-1:** Road Vehicles – Air Filters for Passenger Compartments Part 1: Test for Particulate Filtration | ISO 12103-1, A2 Fine Test Dust | SAE J 1669-1: Cabin Particulate cancelled. ISO/TS 11155-1 has been adopted by SAE |

(*Continued*)

**Table 6.11** (Continued)

| Filtration standards, procedures, & specifications | Contaminants required | Comments & information |
|---|---|---|
| **ISO/TS 13353:** Diesel Fuel & Petrol Filters for Internal Combustion Engines – Initial Efficiency by Particle Counting | ISO 12103-1, A3 Medium Test Dust | ISO/TS 13353 is cancelled & withdrawn. ISO 19438 has replaced 13353 |
| **ISO 16889:** Hydraulic Fluid Power Filters – Multipass Method for Evaluating Filtration Performance of a Filter Element | ISO 12103-1, A3 Medium Test Dust | ISO 16889 has replaced 4572 Hydraulic Filter Multipass Test Method |
| **ISO/PWI 23369:** Hydraulic Fluid Power – Multipass Method of Evaluating Filtration Performance of a Filter Element under Cyclic Flow Conditions | ISO 12103-1, A2 Fine Test Dust | New Test Method Under development by ISO/TC 131/SC6 (BSI) |
| **ISO 19438:** Diesel Fuel & Petrol Filters for Internal Combustion Engines – Filtration Efficiency using Particle Counting & Contaminant Retention Capacity | ISO 12103-1, A3 Medium Test Dust | ISO 19438 has replaced 13353 Fuel Filter Initial Efficiency |
| **ISO/TS 19713-1:** Inlet Air Cleaning Equipment for Internal Combustion Engines & Compressors – Fractional Efficiency Testing with Fine Particles | ISO 12103-1, A2 Fine Test Dust and A4 Coarse Test Dust | This Standard has been recently published as a TS document by ISO/TC22/SC7/WG3 |
| **ISO/TS 19713-2:** Inlet Air Cleaning Equipment for Internal Combustion Engines & Compressors – Fractional Efficiency Testing with Coarse Particles | ISO 12103-1, A2 Fine Test Dust and A4 Coarse Test Dust | This Standard has been recently published as a TS document by ISO/TC22/SC7/WG3 |
| **JIS D1608:** Fuel Filter Test Method for Automotive Gasoline Engines | JIS Z 8901, Class 8 (fine) Test Dust | ISO 12103-1, A2 Fine Test Dust may be substituted if listed in test report and/or with customer agreement |

(*Continued*)

Table 6.11 (Continued)

| Filtration standards, procedures, & specifications | Contaminants required | Comments & information |
|---|---|---|
| **JIS D1611:** Automotive Parts – Test Methods of Lubricating Oil Filters | SOFTC-2A | Gravimetric Test Method |
| **JIS D1612:** Test Methods of Air Cleaners for Automobiles | JIS Z 8901 Class 7 (coarse) Test Dust and/or Class 8 (fine) Test Dust | ISO 12103-1, A4 Coarse Test Dust and/or A2 Fine Test Dust may be substituted if listed in test report and/or with customer agreement |
| **JIS D1617:** Automotive Parts – Fuel Filters for Diesel Engines Test Methods | An Organic Contaminant is also required, see D1617 standard for detail. ISO 12103-2 Aluminum Oxide Test Dusts JIS Z 8901 Class 8 (fine) Test Dust | ISO 12103-1, A2 Fine Test Dust may be substituted for JIS Z 8901 Class 8 Test Dust if listed in test report and/or with customer agreement |
| **NFPA T3.10.8.8:** Hydraulic Fluid Power-Filters – Multipass Method for Evaluating Filtration Performance | ISO 12103-1, A2 Fine Test Dust (AC Fine Test Dust per standard is obsolete) | T3.10.8.8 Hydraulic Filter Multipass Method is cancelled. ISO 16889 has replaced the NFPA & ANSI B93.31 documents |
| **SAE J 726:** Air Cleaner Test Code | ISO 12103-1, A2 Fine Test Dust and A4 Coarse Test Dust | SAE J 726 Air Cleaner Test Code is cancelled. ISO 5011 has been adopted by SAE |
| **SAE HS- 806:** Oil Filter Test Procedure Section 6: Filter Capacity & Contaminant Removal Characteristics of Full-Flow Oil Filters | ISO 12103-1, A2 Fine Test Dust SOFTC 2A | Gravimetric Test Method for Overall Efficiency & Capacity |
| **SAE HS- 806:** Oil Filter Test Procedure Section 7: Single Pass Particle Retention Capability Test | Broad Range within 29–470 μm of glass beads<br>Narrow Range in 10 μm fractions of glass beads<br>Natural Contaminant: ISO 12103-1, A4 Coarse Test Dust in full range or fractions thereof | Gravimetric Test Method for Single Pass Efficiency |
| **SAE HS- 806:** Oil Filter Test Procedure Section 9: Collapse Test for Lube Oil Elements | ISO 12103-1, A2 Fine Test Dust | Test method for element strength & maximum anticipated differential pressure |

(*Continued*)

**Table 6.11** (Continued)

| Filtration standards, procedures, & specifications | Contaminants required | Comments & information |
|---|---|---|
| **SAE J 905:** Fuel Filter Test Methods Section 4: Filter capacity & contaminant removal characteristics | ISO 12103-1, A2 Fine Test Dust | Gravimetric Test Method for Efficiency & Capacity |
| **SAE J 905:** Fuel Filter Test Methods Section 6: Collapse test for fuel filter elements | ISO 12103-1, A1, A2, A3, A4 Test Dusts | Test method for element strength & maximum anticipated differential pressure |
| **SAE J 1669-1:** Passenger Compartment Air Filter Test Code – Particulate Filtration | ISO 12103-1, A2 Fine Test dust | SAE J 1669-1 is cancelled. ISO 11155-1 has been adopted by SAE |
| **SAE J 1858:** Full-flow Lubricating Oil Filters Multipass Method for Evaluating Filtration Performance | ISO 12103-1, A3 Medium Test Dust | SAE J 1858 is cancelled. ISO 4548-12 has been adopted by SAE |
| **SAE J 1985:** Fuel Filters – Initial Single Pass Efficiency Test Method | ISO 12103-1, A2 Fine Test Dust | Particle Counting Fuel Filter Test Method – Fractional Efficiency |
| **SAE J 2709:** Coolant Filter Test Procedure – Section 7: Filter Collapse Test | ISO 12103-1, A1, A2, A3, A4 Test Dust | Test Method for Element Strength & Maximum Anticipated Differential Pressure |
| **SAE J 2793:** Fuel dispensing filter test methods | ISO 12103-1, Arizona Test Dust A3 Medium Test Dust | New test method recently published for gasoline and diesel fuel dispensing filters by the SAE Filter Test Methods committee |

Reproduced with permission of Powder Technology Inc., Burnsville, Minnesota, USA.

by particle size. Particle size analyzers and particle count analyzers determine the size and count of particles in a sample. Following are some principal techniques for particle size analyzers.

1. Light extinction—This system is common for automated particle counting. The fluid is transported through a small transparent tube located between a light source on one side of the tube and an optical sensor on the other side. The sensor emits a specified electrical signal. The tube is narrow enough so that any particle in the fluid passing through the light source/sensor line will block some of the light to the sensor, the amount of light blockage depending on the size of the particle. The "shadow" so created on the sensor causes a change in the emitted electrical

signal, which can be related to the size of the particle. The number of changes will be a count of the particles.

2. Dynamic light scattering—Also known as photon correlation spectroscopy (PCS), this technique determines hydrodynamic particle sizes by measuring the speed at which particles diffuse in the base liquid. The suspended particles are subjected to laser beams. The sum of the dispersed beams is measured. The method gives the the mean particle diameter and the range of the particle size distribution.
3. Condensation nuclei counter (CNC)—This type of particle counting subjects an air stream to alcohol (or water) vapor and then cooling with a cool air stream to cause condensation of particles. An optical particle counter is then used to count the particles.
4. Electrosensing-zone technique—Originally developed for sizing blood cells, this technique provides a number-based distribution. Particles are suspended in a weak electrolyte and drawn through a small aperture. The width of the aperture, sized for the particle sample in question, can range from 0.4 to 1,200 μm. An "electrosensing" zone is created when current flows between two electrodes located on either side of the aperture. Particles passing through the aperture produce a current that is proportional to their particle size. This allows the electrosensing-zone to count each particle.
5. Laser diffraction—Also called low angle light scattering (LLAS), laser diffraction uses a light-scattering phenomenon coupled with mathematical algorithms to calculate the volume-based particle size distribution. They are reported as volume-based distribution of spherical particles. The particle size range is from 0.05 to 2,000 μm.
6. Microscopy—Allows for direct measurement of the particles in question. Manual particle size counting is tedious and subject to large variation because of small sample size. Modern computerized techniques allow for image analyzers to greatly speed-up and simplify the count procedure. An advancement of the microscope technology is the use of scanning electron microscopes (SEM). A problem with microscopic techniques in filter test stand studies is that a bottle is used to take the sample. Unless the bottle is ultracleaned, residual dust inside the bottle can greatly contaminate the analysis.
7. Sedimentation—Applies Stoke's law to determine particle size based on the particle settling rates. Centrifugation may be used to assist the settling rate. This method requires that the contaminant be denser than the fluid in which it is suspended. It also requires close temperature control for viscosity control of the fluid. The method is popular as a quality tool in some industries such as paint and ceramics, however it is not very well suited for filtration analysis.
8. Sieving—Useful for analyzing suspensions or slurries with large particle sizes such as pulp fiber length analysis. It does not lend itself to filtration analysis.

CHAPTER 7

# Liquid Filter Applications

## 7.1 INTRODUCTION TO LIQUID FILTER APPLICATIONS

Liquid filtration applications are concerned with the filtration of either aqueous or hydrocarbon fluids. One of the largest uses of liquid filter media made from nonwovens and/or filter paper is in engine filtration applications. Engine filtration including both air and liquid filtration is treated as a separate subject in Chapter 9, "Engine Filtration (Transportation)."

The types of nonwovens used in non-automotive liquid filter applications include needle felts, hydroentangled webs, paper sheets, filter papers, resin bonded paper, dry laid webs, spunbond products, and membrane media. There are also important applications for nonwoven supported membrane filters. Nanofiber webs or composites containing a nanofiber layer are finding increased use in liquid filter applications.

Large-scale liquid filtration is accomplished mostly by cake filtration. Many forms of filter equipment use a filter fabric to support the cake. Often, the fabric is a woven filter cloth[1], however there are applications where the fabric is a needle felt. In recent years, there has been an emergence of spunlace (hydroentangled) felts and fabrics for industrial filter applications. Nonwoven and paper filter media are popular in filter presses.

Membrane filtration has developed very important roles in liquid filtration. This includes water desalinization, beverage and juice filtration, pharmaceutical filtration, and medical fitration such as blood filtration.

Replaceable filter elements are used in many filter applications. Two of the most common types are liquid bag filters and cartridge filters. Liquid bag filters are often made of woven fabric filter media, however there are an increasing number of applications where a nonwoven or a nonwoven composite material is utilized. A well-known form of a bag filter is the teabag composed of wet lay Manila hemp (abaca) fibers.

One of the biggest uses of nonwoven filter media in liquid filtration is in cartridge filters. Figure 7.1 is an example that shows the components of a spin-on engine oil filter cartridge. This particular element contains a pleated medium consisting of resin treated filter paper. A perforated center tube (internal support frame) provides mechanical support for the pleated media. Oher components include a cover plate, a by-pass valve, a spring to operate the by-pass valve, anti-drain valves and a sealing

[1] Many forms of media used for cake filtration are often referred to as "filter cloths," even though they are not necessarily woven fabrics. In addition to woven fabrics, nonwoven and wet-laid paper media are often referred to as filter cloths when used in cake filtration applications.

*Handbook of Nonwoven Filter Media*
ISBN: 978-0-08-098301-1

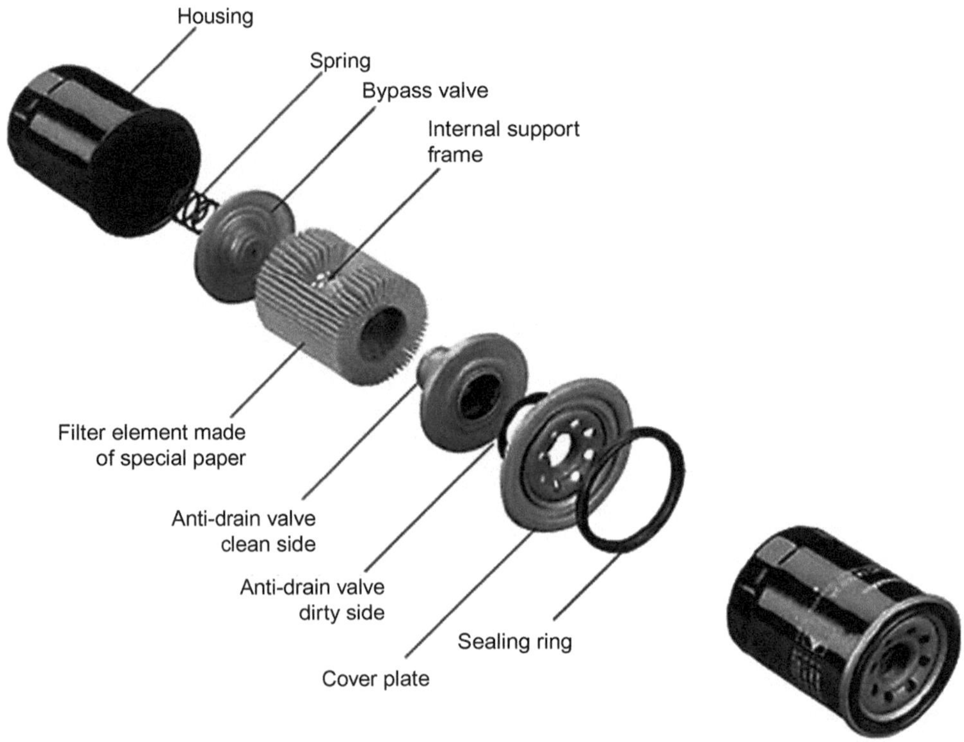


**Figure 7.1** Structure of a Spin-On Oil Filter. *Reproduced by Permission of Mahle Industrie,Inc. Troy, Michigan USA.*

ring. Inlet flow is through the circle of small holes in the cover plate and then through the pleated filter element from outside-in. Exit flow of the filtered liquor is through the center tube and out through the large center hole of the cover plate. The by-pass valve is for cold start-ups where the oil may be too thick to flow through the filter media. The pressure build-up allows the spring loaded valve to open and allow flow into the center core from the top of the pleated filter element. The by-pass valve also protects the engine in case of a filter pluggage, by allowing the flow to by-pass the filter element. The thinking is that it is better to allow a continuous flow of unfiltered oil to the engine rather than the possible catastrophic cosequences if the engine is starved of oil.

Cartridge filters are used as polishing filters in industrial applications, for applications where high purity is demanded or where ultra-high efficiency is expected. Examples are the health care and industries. Cartridge filters are also used in consumer filter applications such as home water filters and swimming pool filters. Pleated resin-treated filter paper is the filter medium for many cartridge filter applications. Filtration of hydrocarbon fuels and lubricating oils are the biggest use of resin-treated paper in liquid filter applications.

Following are applications of nonwoven filters in liquid filtration:

Filtration of aqueous fluids

1. Water treatment plants
2. Water discharge treatment plants
3. Industrial process streams
4. Drinking water filtration
5. Beverage filtration
   a. Coffee filter
   b. Non-alcoholic beverages
   c. Wine filtration
   d. Beer filtration
   e. Teabag paper
6. Swimming pool filters
7. Filtration of aqueous foods
   a. Sugars
   b. Food processing plants
   c. Fruit juice processing
8. Medical filtration applications
   a. Blood filters
   b. Water purification
9. Pharmaceutical applications
10. Industrial chemicals
    a. Paints
    b. Latices
    c. Electroplating solutions
    d. Printing inks

Filtration of hydrocarbon fluids

1. Industrial hydrocarbon processing
   a. Petrochemicals
   b. Resins (solvent base)
   c. Lacquers
2. Aviation and diesel fuel filtration
3. Hydraulic fluids
4. Machine tool coolant and cutting oil filtration
5. Edible oil filtration
   a. Fast food restaurants
   b. Food processing
6. Engine Filtration—to be discussed in Chapter 9.

## 7.2 NONWOVEN FILTER MEDIA FOR LIQUID FILTER APPLICATIONS

### 7.2.1 Entangled fiber

#### *7.2.1.1 Needle Felt*

Although needle felt filter media is are more often used in air filter applications, there are a number of liquid applications. They are a popular choice for bag filters and pocket filters. They are also used in vacuum belt filters. Because of the broad range of fibers that can be chosen for needle felt production, the medium can be designed for filtration efficiency, temperature, and desired chemical exposure. The following indicates the choice of needle felt fibers and what temperature and chemical exposures they can be used for:

1. Cotton—for alkalis and organic solvents up to 95°C
2. Rayon—for aliphatic and aromatic solvents up to 150°C in a closed system and up to 120°C in an open system
3. Polyamide (nylon)—for alkalis and solvents (except for alcohols) up to 150°C
4. Polypropylene—for acids and alkaline solutions, and aliphatic solvents up to 110°C
5. Polyester—for acids, oxidizing agents, and most organic solvents up to 150°C
6. Polyaramid (Kevlar®)—for acids and alkalis up to 220°C
7. Polyphenylene sulfide (PPS)—for acids, alkalis and organic solvents up to 200°C
8. Polytetrafluoroethylene (PTFE)—for acids, alkalis and organic solvents up to 260°C
9. Polyimide (P84)—for acids up to 260°C.

For example, Table 7.1 is a typical property list of some needle felt products identified for liquid filtration.

**Table 7.1** Needle Felt Properties for Liquid Filtration

| Fiber | Weight ($g/m^2$) | Air permeability ($m^3/m^2/min$@ $\Delta P = 20$ mm $H_2O$) | Thickness (mm) | Tensile strength (N/m) | |
|---|---|---|---|---|---|
| | | | | MD | CD |
| Polyester | 640 | 1.6 | 1.35 | 30 | 55 |
| | 640 | 1.0 | 1.1 | 30 | 55 |
| | 800 | 9.5 | 3.2 | 90 | 55 |
| | 800 | 3.0 | 3.0 | 90 | 55 |
| Polyamide | 830 | 4.4 | 2.6 | 55 | 75 |
| | 830 | 2.0 | 2.7 | 55 | 75 |
| | 830 | 1.3 | 2.2 | 60 | 80 |
| | 830 | 1.5 | 2.4 | 60 | 80 |
| | 1,050 | 3.5 | 3.3 | 65 | 90 |
| | 1,075 | 2.2 | 3.0 | 65 | 90 |
| | 1,050 | 2.5 | 3.4 | 65 | 90 |
| | 1,050 | 1.5 | 2.8 | 60 | 85 |
| Polypropylene | 400 | 8.9 | 1.5 | 60 | 50 |
| | 500 | 9.8 | 2.4 | 60 | 35 |
| | 500 | 6.0 | 1.75 | 60 | 35 |
| | 610 | 0.8 | 1.6 | 30 | 50 |
| | 640 | 20.0 | 3.8 | 50 | 75 |

#### *7.2.1.2 Hydroentangled webs*

Hydroentangled or spunlace webs are finding increasing use in liquid filter applications. Spunlace, pioneered by DuPont™ is used for coolant filtration and other liquid filter applications. Table 7.2 is a tabulation of some of the spunlace grades DuPont™ uses for filtration purposes. Note that there are two spunlace media types: a 100% polyester nonwoven and a blended nonwoven consisting of polyester and a wood pulp (Table 7.2).

### 7.2.2 Melt-blown media

Melt-blown nonwoven media are used in a number of liquid filter food and drink applications. This includes coffee bags, milk filter, and edible oil filtration. They are used in high efficiency reverse osmosis (RO) and ultrafiltration (UF) filters. Melt-blown media is most often used in composite form, often combined with spunbond media in SMS type structures.

### 7.2.3 Spunbond media

Spunbonded media are used in almost every possibility of liquid filtration applications. An important one is in roll filters used in machine shops for removing metal shavings from coolant fluids and cutting oils.

### 7.2.4 Flash spun media

SoloFlo® is a flash spun spunbonded developed by DuPont™ and is composed of HDPE (high-density polyethylene). SoloFlo® is designed for microfiltration (MF) of aqueous fluids. This includes drinking water as well as industrial streams such as waste water treatment and plating solutions. The development and properties of the SoloFlo type media are well discussed by Warren et al. (257). Mayer and Warren (258) reported on AC Fine Test Dust (ACFTD) challenge trails using an Oberlin pressure filter. The results were comparable to competitive media such as melt-blown polypropylene, PTFE membrane laminate, microporous membrane, and other flash spun media. In addition to SoloFlo®, DuPont™ markets five other grades of flash spun filtration media for liquid filtration applications (259).

### 7.2.5 Filter papers

Wakeman and Tarleton (39) sub-classify filter papers as follows:

1. Industrial papers
2. Laboratory papers
3. Filter sheets

**Table 7.2** Dupont™ Spunlace Filtration Media

**Typical material properties**

| Material properties | Units | Test method | Polyester (PET) style | | | Woodpulp/polyester style | | | |
|---|---|---|---|---|---|---|---|---|---|
| | | | 8004 | 8005 | 8017 | 8100 | 8801 | 8818 | 8838 |
| Basis weight | osy | ASTM D3776 | 1.8 | 2.0 | 0.7 | 4.0 | 2.0 | 1.9 | 1.5 |
| | gsm | EN ISO 9864 | 61 | 68 | 24 | 136 | 68 | 63 | 51 |
| Thickness | Mils at 0.6 psi | ASTM D5729 | 11 | 11 | 5 | 26 | 9 | 9 | 8 |
| Air permeability | $Cfm/ft^2$ at 125 Pa | ASTM D737 | 223 | 166 | 709 | 148 | 85 | 49 | 121 |
| Grab tensile strength, MD | $lb_f$ | ASTM D5034 | 45 | 64 | 15 | 103 | 36 | 39 | 32 |
| Grab tensile strength, CD | $lb_f$ | ASTM D5034 | 21 | 27 | 7 | 57 | 18 | 20 | 16 |
| Trap tear strength, MD | $lb_f$ | ASTM D5733 | 8 | 11 | 3 | 21 | 4 | 5 | 5 |
| Trap tear strength, CD | $lb_f$ | ASTM D5733 | 18 | 28 | 6 | 43 | 8 | 10 | 7 |
| Mean pore size | microns | ASTM E1294 | 51 | 39 | 61 | 30 | 35 | 14 | 24 |
| Bubble point | microns | ASTM E1294 | 99 | 83 | 191 | 76 | 73 | 67 | 108 |
| Mullenburst | psi | ASTM D3786 | 61 | 92 | 16 | >120 | 37 | 40 | 30 |

® E.I. du Pont de Nemours and Company or its affiliates. Reprinted with permission.

**Table 7.3** Ahlstrom Filter Media for Edible Oil

| Grade | Composition | Surface | Basis weight[a] lb/1,389 $ft^2$ ($gm/m^2$) | Thickness mils | Rapidity[b] ml/min | Micro retention[c] μm | Wet strength[d] in $H_2O$ |
|---|---|---|---|---|---|---|---|
| 933 | Cellulose | Creped | 70(246) | 40.0 | 435 | 27 | 120 |
| 963 | Cellulose | Creped | 54(190) | 28.5 | 440 | 30 | 100 |
| 967 | Cellulose | Creped | 70(246) | 40.0 | 475 | 33 | 100 |
| 968 | Cellulose | Creped | 50(176) | 27.5 | 600 | 40 | 75 |
| 1278 | Cellulose | Unfinished | 35(123) | 19.5 | – | 45 | 250 |
| 1384 | Cellulose | Creped | 44(155) | 23.0 | 450 | 31 | – |
| 8301 | Cellulose/ synthetic blend | Smooth | 14(49) | 12.4 | – | – | – |

[a]Basis weight based on a ream size of 500 sheets, 20 in × 20 in = 1389 $ft^2$.
[b]Flow of water through the media at a continuous column height of 100 cm.
[c]Particle size at 98% retention per Coulter Counter®.
[d]Burst strength: pressure in inches of water to rupture a two-inch diameter sample.
Reproduced by permission of Ahlstrom Filtration LLC, Mt. Holly Springs, Pennsylvania, USA.

#### *7.2.5.1 Industrial filter papers*

Industrial filter papers are filter media intended for general liquid filter applications. They are often die cut or stamp punched to the desired shape for the intended application. Very often this is a filter press. They are also sold in roll form for roll filters and downstream converting. Typical applications are beverage filtration, pharmaceuticals, light oils, syrups, and vegetable oil filtration in food processing and restaurants. Rolls of media are used in roll filters for removing metal filings from machine shop coolant and cutting oils.

Some grades are identified as "wet strength" papers because they contain a wet strength such as melamine formaldehyde or polyamide epichlorohydrin to provide durability and structural support in aqueous filtration operations. They are sometimes creped to provide increased surface area for filtration. Creping also imparts more stretch and flexibility when such properties are desired. Most industrial filter papers are made from wood pulp fibers, however for fine filtration operations, they are made with glass microfibers. In such cases a resin binder, such as polyvinyl alcohol, polyvinyl acetate, and acrylic polymers may be employed to add strength to the sheet. Industrial filter papers can be impregnated with adsorbent or absorbent powders such as activated carbon, diatomaceous earth, and perlite clay.

Ahlstrom is a major producer of industrial filter papers at their mill in Mount Holly, Pennsylvania. Table 7.3 above is a summary of their industrial filter papers used in the fast food industry for filtering edible oils. Note that this American segment of the filtration industry still uses basis weight based on a 500 sheet ream, 20 in × 20 in (1,389 $ft^2$).

**Figure 7.2** Pre-pleated Laboratory Filter. *Reproduced by Permission of GE Healthcare Life Sciences, Piscataway, New Jersey, USA.*

### 7.2.5.1.1 Laboratory papers

Laboratory papers are well recognized as the circular filter sheets used in laboratories around the world. They are also marketed in fluted (folded and prepleated), cone, and bag form, according to the application. Figure 7.2 is an example of a prepleated form of Whatman[2] filter paper produced by GE Healthcare Life Sciences, Piscataway, New Jersey, USA. There are two categories of laboratory papers: qualitative and quantitative. According to Sutherland and Purchas (18) "Qualitative filter papers are for use in qualitative analytical techniques aimed at identifying materials; they are accordingly also suitable for general use. Quantitative filter papers are for use in analytical techniques intended to quantify the composition of materials; where purity and composition of the filter paper are of crucial importance." Qualitative filter papers are characterized by their flow rates and relative efficiencies. Quantitative filter papers are also characterized by relative flow and efficiency and also by chemical composition. It is of critical importance that quantitative papers be of low ash content. For very high efficiency filtration, laboratory papers composed of glass microfibers are used.

It is mentioned above that GE Healthcare Life Sciences is the producer of Whatman filter papers. They produce both cellulose and microglass fiber filter paper. Typical properties of these filter papers are tabulated in Table 7.4. Note that the properties include "nominal particle retention in liquid" and "filtration speed." The particle retention number is based on the particle size at which 98% is retained. The Herzberg method of determining filtration speed is based on the initial flow rate at which distilled water is filtered through a sample of filter paper. The flow rate is measured at a constant hydrostatic head of 5 cm. Results are reported as milliliters per minute through an area of 100 $cm^2$.

[2] Whatman PLC, the original producer of Whatman filter papers was purchased by GE Healthcare Life Sciences in 2008. The filter papers are still marketed under the Whatman tradename.

**Table 7.4** Typical Properties of Whatman Filter Papers

| Grade | Nominal particle retention in liquid (μm) | Filtration speed (approx) Herzberg (S) | Typical thickness (μm) | Basis weight ($g/m^2$) | Grade for pre-pleated version | Flow-aspect |
|---|---|---|---|---|---|---|
| ***Standard qualitative cellulose filter papers*** | | | | | | |
| 1 | 11 | 150 | 180 | 88 | | Medium |
| 2 | 8 | 240 | 190 | 103 | 2 V | Medium |
| 3 | 6 | 325 | 390 | 187 | | Medium-thick |
| 4 | 20–25 | 37 | 205 | 96 | | Very fast |
| 5 | 2.5 | 1,420 | 200 | 98 | 5 V | Slow |
| 6 | 3 | 715 | 180 | 105 | | Medium to slow |
| 595 | 4–7 | 80 | 150 | 68 | $595^{1/2}$ | Medium to fast-thin |
| 597 | 4–7 | 70 | 180 | 85 | $597^{1/2}$ | Medium to fast |
| 602 h | <2 | 375 | 160 | 84 | 602 $h^{1/2}$ | Slow |
| ***Qualitative wet strengthened cellulose filter papers*** | | | | | | |
| 113 | 30 | 28 | 420 | 125 | 113 V | Fast—creped |
| 114 | 25 | 38 | 190 | 77 | 114 V | Fast—smooth |
| 91 | 10 | 70 | 205 | 71 | | Creped |
| 1573 | <2 | 700 | 140 | 92 | $1,573^{1/2}$ | Slow |

**Quantitative Filter Papers**

| Grade | Nominal particle retention in liquid (μm) | Filtration speed (approx) | Typical thickness (μm) | Basis weight ($g/m^2$) | Ash content | Flow-aspect |
|---|---|---|---|---|---|---|
| ***Ashless quantitative cellulose filter papers*** | | | | | | |
| 40 | 8 | 340 | 210 | 95 | 0.007% | Medium |
| 41 | 20 | 54 | 220 | 85 | | Fast |
| 42 | 2.5 | 1870 | 200 | 100 | | Slow |
| 43 | 16 | 155 | 220 | 95 | | Medium to fast |

*(Continued)*

**Table 7.4** (Continued)

**Quantitative Filter Papers**

| Grade | Nominal particle retention in liquid (μm) | Filtration speed (approx) | Typical thickness (μm) | Basis weight ($g/m^2$) | Ash content | Flow-aspect |
|---|---|---|---|---|---|---|
| 44 | 3 | 995 | 180 | 80 | | Slow to medium |
| 589/1[a] | 12–25 | 25 | 190 | 80 | 0.01% | Fast |
| 589/2[a] | 4–12 | 70 | 190 | 85 | | Medium to fast |
| 589/3 | <2 | 750 | 150 | 84 | | Slow |
| ***Hardened low ash quantitative cellulose filter papers*** | | | | | | |
| 50 | 2.7 | 2,685 | 115 | 97 | 0.015% | Slow |
| 52 | 7 | 235 | 175 | 101 | | Medium |
| 54 | 22 | 39 | 185 | 92 | | Fast |
| ***Hardened ashless quantitative cellulose filter papers*** | | | | | | |
| 540 | 8 | 200 | 115 | 88 | 0.006% | Medium |
| 541 | 22 | 34 | 175 | 82 | | Fast |
| 542 | 2.7 | 2,510 | 185 | 93 | | Slow |

**Glass fiber filters**

| Product | Filtration speed | Particle retention in liquid (μm) | Typical thickness (μm) | Basic weight ($g/m^2$) |
|---|---|---|---|---|
| Grade GF/A | Fast | 1.6[b] | 260 | 53 |
| Grade GF/B | Medium to fast | 1.0[b] | 675 | 143 |
| Grade GF/C™ | Medium to fast | 1.2[b] | 260 | 53 |
| Grade GF/D | Fast | 2.7[b] | 675 | 121 |
| Grade GF/F | Medium | 0.7[b] | 420 | 75 |
| GMF-150 1 μm—multilayer | Medium to fast | 1.2[b] | 730 | 139 |

[a]Pre-pleated versions available
[b]Particle retention rating at 98% efficiency
Reproduced by permission of GE Healthcare Life Sciences, Piscataway, New Jersey, USA.

### 7.2.5.2 Filter sheets

Filter sheets, such as the depth filter sheets depicted in Figure 7.3 produced by ErtelAlsop, Kingston, New York, USA, are made on a special type of wet lay machine. They are thick enough to be used for depth filtration purposes. Depth filter sheets range from 3 to 4 mm thick and are composed of cellulose fibers and a variety of filter aids such as diatomaceous earth or perlite clay. They can also be made to include particles of activated carbon to remove taste, odor, and other undesirable organic contaminants from water. They can also be formulated to possess a cationic charge, so as to provide for electrokinetic filtration in aqueous systems. At one time they were composed of asbestos. However, the use of asbestos has been discontinued because of its hazardous nature. Depth filter sheets are mostly used in plate and frame-type filter presses. Circular disc forms of filter sheets are used in lenticular filters.

Figure 7.4 is a diagram of a typical plate and frame filter press. A filter fabric is inserted between each plate and frame. Inlet flow is into the frames, through the fabric, and through the foraminous fabric support structure of the plate. The filtrate is discharged from the plate. The cake builds up on the filter fabric until the filter cycle is terminated. When the press is opened, the cake is then discharged to some sort of collection device such as a conveyor or collection bin. Filter fabrics can be woven cloth, nonwoven fabrics such as needle felt or spunbond, and industrial filter papers. Filter sheets are used when the effect of depth filtration is desired.

Lenticular filters are composed of parallel discs, each disc shaped like a double convex lens. Filter media covering both sides of the convex disc allows for inward filtration through both faces. They can be individually stacked inside the filter housing allowing for a choice of filters for the specific application Figure 7.5 is a photograph of lenticular filters produced by ErtelAlsop in Kingston, New York, USA.

**Figure 7.3** Depth Filter Sheets by ErtelAlsop. *Reproduced with Permission of ErtelAlsop, Kingston, New York, USA.*

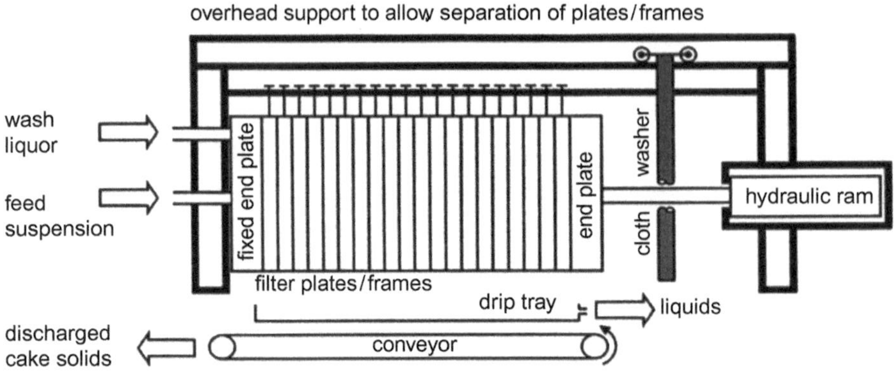


**Figure 7.4** Plate and Frame Filter Press. *Reproduced by Permission of Elsevier Science Ltd.*

**Figure 7.5** Lenticular Filters by ErtelAlsop. *Reproduced with Permission of ErtelAlsop, Kingston, New York, USA.*

3M Corporation markets a broad range of lenticular filters and other types of stackable filter elements. 3M utilizes Zeta Plus® filter medium, a rigid depth filter sheet that has been chemically bonded with a positive electrolytic electrokinetic charge to enhance its ability to attract particulate matter from a liquid. Zeta Plus® depth filters are composed of cellulose and diatomaceous earth. The positively-charged medium traps contaminants smaller than the rated port size. Zeta Plus® filters are capable of capturing submicron contaminants yet still achieve high flow rates with low differential pressure. The A series filters are designed for general industrial filtration. 3M also produces ZetaCarbon® cartridges containing activated carbon that remove trace organics without the cleanup problems associated with bulk activated carbon. Other Zeta Plus® filter products are conveniently formulated for small volume filtration, food and beverage filtration, pharmaceutical filtration, cosmetics, dielectric oil, and resin filtration. In 2009, 3M (Wang et al. (260)) launched the Zeta Plus™ Encapsulated System, a single-use depth-filtration system that is capable of processing up to 2,000 L cell harvest streams per system assembly. In bioprocessing applications, depth filters are often employed at the primary recovery step to remove cells and cell debris from cell culture media.

### 7.2.6 Resin treated filter paper

Resin treated filter paper, well described in Chapter 3, Section 3.6.2, is the medium of choice in most pleated filters. Liquid filter applications include lube oil filters, fuel filters, fuel filter separators, swimming pool filters, home water filters, and hydraulic filters.

### 7.2.7 Composite structures

In liquid filtration, composite structures are often used to provide gradient density filtration. They are also used as prefilters to membrane filtration and RO applications and also provide support structure to membrane and nanofiber filter media layers. Needle felts for liquid filter applications are usually scrim reinforced to provide adequate strength.

Another application for composite structures is taste and odor adsorbing media in drinking water filters. These contain an activated carbon layer (see Chapter 3, Section 3.10.1). It should be noted that single-layer filter media continuing particles of activated carbon are also used in home water filters. This type of medium is not considered a composite structure according to the definition of this book; nevertheless, it combines two separation technologies (filtration and adsorption) into one medium.

Composite media find broad applications in engine filtration applications. This includes the filtration of lube oil, diesel fuels, and aviation fuels.

### 7.2.8 Multi-separation technology media

Composite structures are a way to combine different separation technologies into one medium. Another way is to combine several technologies into a web of one layer. This is done by the combining fibers, additives, and impregnants into one structure. Zeta Plus® media discussed above has these qualities by combining particle filtration properties from fiber structure, adsorption from activated carbon particles, and electrokinetic properties from cationic impregnants.

Another product that has multiple separation capabilities is Disruptor® media produced by Ahlstrom. This medium has been previously mentioned in Chapter 4,

Section 4.3.5.4.8. Targeted strictly for water filtration applications, the uniqueness of this medium structure is that it contains cationic boehmite fibrils attached to glass microfibers as depicted in Figure 4.35 of Chapter 4. This combination gives the medium the power to separate particles with its fine-fiber microglass fiber structure and to separate other particles by electrokinetic attraction. Table 7.5 is a summary of Disruptor® grades. In addition to glass fiber and boehmite, the product can be made to contain activated carbon and silver as an microbial additive. Note from Table 7.5 that Grade 5289 contains four separation technologies in one piece of filter medium; particle filtration, adorption, antimicrobial, and electrokinetic.

## 7.3 FILTERS THAT USE NONWOVEN FILTER MEDIA

### 7.3.1 Roll filters

#### *7.3.1.1 Description of roll filters*

Roll filters are used in applications where the objective is to remove a waste material from a liquid stream. Examples are filters for sludge removal from waste water discharges and removal of metal particles in the coolant oils of machine shop operations. Roll filters are filters that unwind a roll of media and pass it through a filter chamber where it does the filtering. They generally operate on an indexing system that advances the filtering section of the medium when it starts to build up with filter cake and the flow rate becomes slow. Roll filters come in a variety of shapes and mechanisms. Two types are deep bed filters and pressure filtration systems filters.

Figure 7.6 is a picture of the deep bed filters marketed by Industrial Filters Company (IFC) in Fairfield, New Jersey, USA. The medium unwinds unto a conveyor belt which carries the medium through the deep bed filtration area. The movement of the medium is automatic indexing based on slowdown of the fluid flow rate. Solids removal is by gravity filtration. The medium and cake are discharged together into the sludge box. This type of filter is for filtration of cooling oil from grinding operations and machine shop operations, although it is used in other places such as food processing operations. Figure 7.7 is a schematic of its use in waste sludge dewatering operations.

Oberlin Filter Company in Waukesha, Wisconsin, USA is a manufacturer of pressure filtration systems.

Figure 7.8 below is a schematic of the Oberlin system used for coolant filtration. The roll filter medium unwinds into a filter compartment which closes around the medium and becomes pressurized by air and dirty coolant pumped into the compartment. The coolant is filtered through the medium until the cake build-up is such that the flow rate becomes too low or the pressure drop becomes too high. At this time, the medium is blasted with air to remove excess fluid. The chamber then opens and fresh medium is advanced into the chamber for the next filtration cycle. The cake on the dirty medium, advancing out of the chamber, is doctored off the medium and into a tote box.

**Table 7.5** Ahlstrom Disruptor® Grades

**Typical properties of Disruptor® grades**
**Updated July 23, 2014**

| Grade | | 5283 | 5284 | 5287 | 5288 | 5289 | 5292 |
|---|---|---|---|---|---|---|---|
| Typical data revision | | Rev 2 | Rev 2 | Rev 1 | Rev 2 | Rev 2 | Rev 1 |
| Revision date | | 11/14/2013 | 11/14/2013 | 11/14/2013 | 11/14/2013 | 11/14/2013 | 11/14/2013 |
| Description | | HS Disruptor® | Carbon HS Disruptor® | Carbon Disruptor® with silver | HS Disruptor® with silver | Carbon HS Disruptor® with silver | Disruptor® high flow with silver |
| Weight | lbs/1,389 $ft^2$ | 90.70 | 90.70 | 90.70 | 90.70 | 90.70 | 90.70 |
| Weight | oz/yd2 | 9.40 | 9.40 | 9.40 | 9.40 | 9.40 | 9.40 |
| Weight | g/m2 | 318.00 | 318.00 | 318.00 | 318.00 | 318.00 | 318.00 |
| Thickness | mils | 39.00 | 34.50 | 37.50 | 39.00 | 34.50 | 47.70 |
| Thickness | mm | 0.99 | 0.88 | 0.95 | 0.99 | 0.98 | 1.21 |
| Rapidity | mls/min | 8.00 | 9.00 | 5.20 | 8.00 | 9.00 | 49.40 |
| Frazier permeability @ 0.5″ | cfm/$ft^2$ | 0.60 | 0.60 | 0.40 | 0.60 | 0.50 | 5.50 |
| Tensile strength | lbs/in MD | 18.00 | 12.00 | 12.60 | 18.00 | 12.00 | 15.40 |
| Mean flow pore | Micron | 1.2 | 1.4 | 1.4 | 1.2 | 1.4 | 7.31 |

The information supplied in this document is for guidance only and should not be construed as a warranty.
All implied warranties are expressly disclaimed, including without limitation any warranty of merchantability and fitness for use.
All users of the material are responsible for assuring that it is suitable for their needs, environment, and end use.
All data subject to change as Ahlstrom deems appropriate.
*While antimicrobial product protection can prevent the growth of odor-causing microbes, protection is limited to the treated article.
It does not further enhance the Disruptor® filter's ability to protect against disease.
Reproduced by permission of Ahlstrom Filtration LLC, Mt. Holly Springs, Pennsylvania, USA.

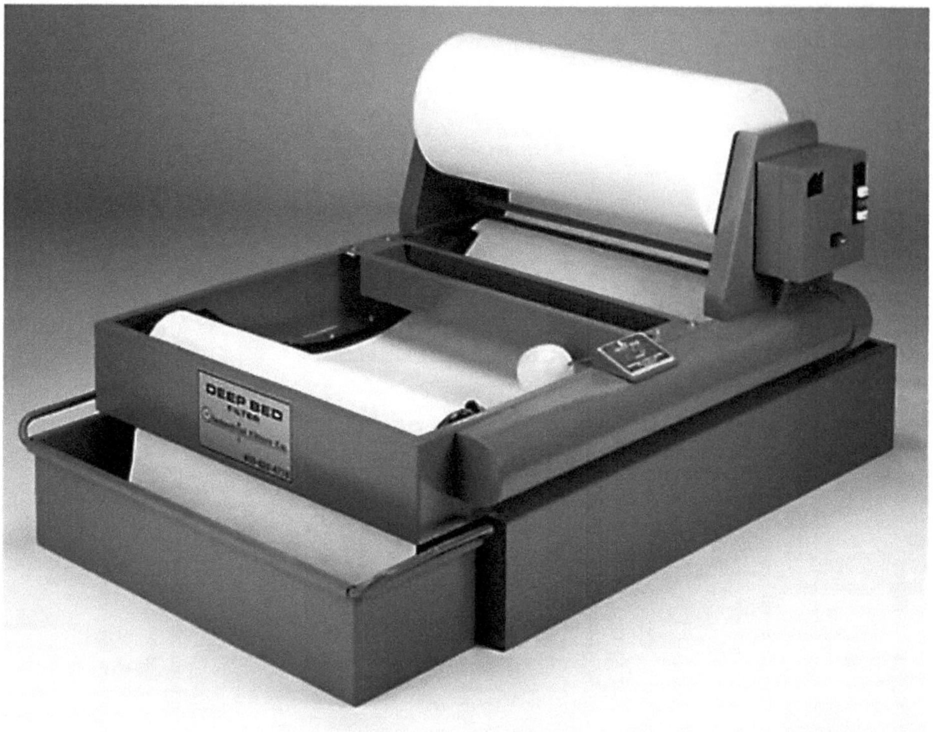


**Figure 7.6** Industrial Filters Corp. Deep Bed Filter. *Reproduced with Permission of Industrial Filters Company, Fairfield, New Jersey, USA.*

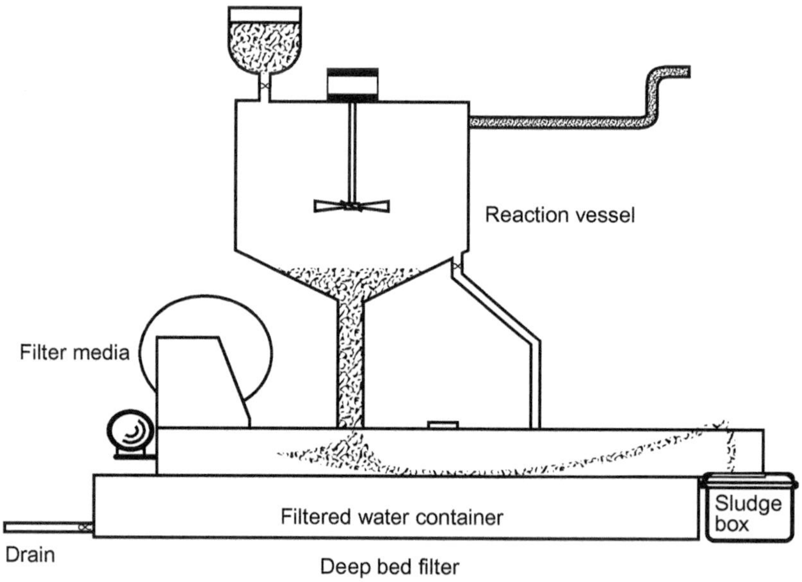


**Figure 7.7** IFC Deep Bed Filter for Sludge Dewatering. *Reproduced with Permission of Industrial Filters Company, Fairfield, New Jersey, USA.*

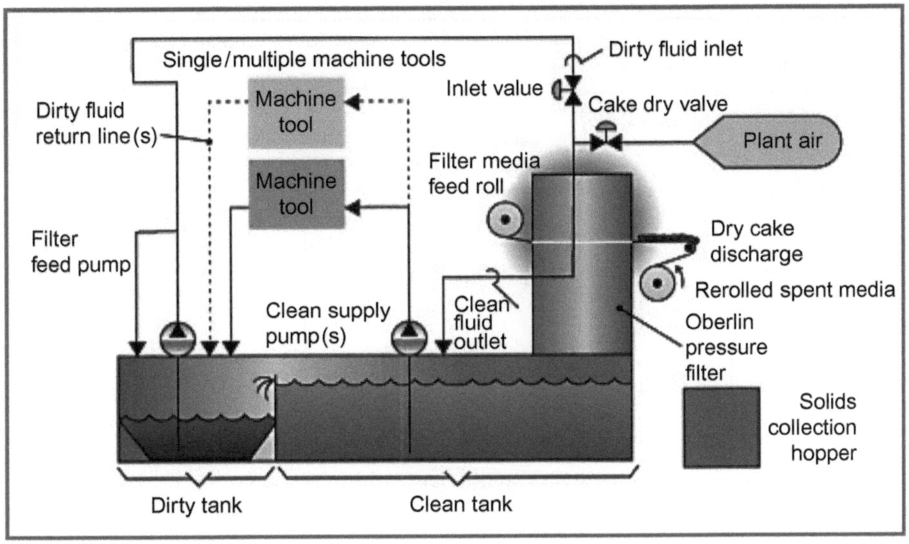


**Figure 7.8** Coolant Filtration Operation (Pressure Filtration). *Reproduced with Permission of Oberlin Filter Company, Waukesha, Wisconsin, USA.*

### *7.3.1.2 Media for roll filters*

Spunbonded polypropylene is probably the most widely used media for coolant filtration. This is one of the important applications for the Typar® and Tektron® spunbonded polypropylenes produced by PGI (Polymer Group International). Table 3.3 in Chapter 3 summarizes their typical properties.

Roman (261) identifies polyester blends and pattern bonded nylon as two frequently used media for roll filters used in cutting oil filtration. Polyester has the advantage of not extracting oil. Media made from polypropylene and nylon fibers do have this tendency and are more prone to blinding and premature indexing. On the other hand, the ability to absorb and separate oil from water makes polypropylene and particularly nylon spunbond webs useful barrier materials in off-shore oil spill clean-ups.

Tyvek®, a solution spun polyethylene is generally too tight and too dense to be used as a filter medium, however a variation to this is a series of filter grades, marketed by DuPont, known as SoloFlo®. Meyer (262) has reported on this medium in combination with Oberlin automated pressure filters (see Figure 7.8 above) to remove heavy metals from wastewater streams. In addition to SoloFlo®, DuPont also markets several other grade variations of Tyvek® for liquid filtration applications.

Kimberly Clark markets Powerloft®, a bicomponent polyolefin (polypropylene/polyethylene) media primarily used for roll filter applications. It is a high loft or medium loft, gradient density, spunbonded media produced by a proprietary Kimberly–Clark process. Typical properties are listed in Table 7.6 below.

Table 7.6 Powerloft® Filter Media by Kimberly Clark

| Grade | Grammage (basis weight) | | Permeability | | Color | Caliper | | CD tensile strength | |
|---|---|---|---|---|---|---|---|---|---|
| | oz/yd$^2$ | g/m$^2$ | cfm/ft$^2$ @ 0.5″ $H_2O$ | m$^3$/m$^2$-min @12.7 mm $H_2O$ | | in | mm | lb/in | kN/m |
| 2.5 osy medium loft | 2.5 | 85 | 420 | 2.1 | White/blue | 0.080 | 2.03 | 28.9 | 5.0 |
| 3.0 osy medium/high loft | 3.0 | 102 | 385 | 2.0 | White/blue | 0.110 | 2.79 | 32.0 | 5.6 |
| 4.0 osy medium loft | 4.0 | 136 | 310 | 1.6 | White/blue | 0.122 | 3.1 | 42.0 | 7.3 |



## 7.3.2 Nonwovens for replaceable filter elements

### *7.3.2.1 Explanation of replaceable filter elements*

The largest use of nonwoven filter media is in replaceable filter elements. Purchas and Sutherland (18) identify three types of replaceable filter elements:

1. panel or cassette filters mostly used for air filtration
2. bag or pocket filters
3. cylindrically shaped cartridge filters.

It should be noted that cartridge filters are not always cylindrically shaped. Generally, they consist of a housing and a filter element. The author also chooses to redefine "filter element" as that part of the filter that fits inside the housing and contains the medium that does the actual filtering. To illustrate, Figure 7.9 is a photograph of a diesel fuel filter. The pleated medium component on the left is the element and the entire assembly on the right, which includes the outside housing and the inside element, is the cartridge.

It is noted above that panel or cassette filters are used mostly for air filtration. This form of cartridge filter will be discussed in Chapter 8. The discussion in this chapter will be concerned with bag filters, and cartridge filters used for liquid filtration.

Cartridge filters are that part of a filtration system that are removable from the main body of the system for disposal and replacement, or for cleansing and reuse. Filters that are removed and cleansed for reuse are referred to as "washable filters." In some applications, they can be cleansed in place by procedures such as reverse flow and reused without being removed. There are filter cartridges that are reclaimed for reuse. The cartridge or element is shipped to the manufacturer who strips it down and installs a fresh filter medium.

**Figure 7.9** Filter Element and Housing.

Cartridge or elements filters that are removed for cleansing or reclaim are often specialized as "service-cleanable" cartridges or elements. These types of cartridges or elements are used in the manufacture and processing of polymers. Examples of "reclaimable" elements are filters used in high-pressure hydraulic systems that require very expensive housings. In this situation, it would be justifiable to reclaim the element and save the housing.

#### 7.3.2.1.1 Cartridge filters construction and assembly

Just about every type of nonwoven filter medium is used in filter cartridges. Most cartridge filters are cylindrically shaped. Some liquid filter cartridges have noncylindrical shapes. Figure 7.9 above is an example of a noncylindrically shaped cartridge. Note that the housing is not a cylinder and the pleated filter element is oval shaped. Engine filtration (lube oil and fuel), one of the biggest users of liquid filter cartridges, will be discussed in Chapter 9. However, it is difficult to have a discussion of the various types of filter cartridges without including engine filters. Therefore, the general construction, shape, and nature of liquid engine cartridge filters will be included in this discussion of cartridge filters.

There is a wide variety of cartridge filters and a wide variety of filter media used in these cartridges. Included are woven fabric media, sintered metal media, membranes, wound yarn and/or filament, mesh screen materials, and other novel media. This discussion will focus on cartridges containing nonwoven media.

#### 7.3.2.1.2 Pleated filter elements

The most common form of cartridges are those containing elements of pleated filter media. The pleating process is discussed in Chapter 5, Section 5.8. Figure 7.10 below

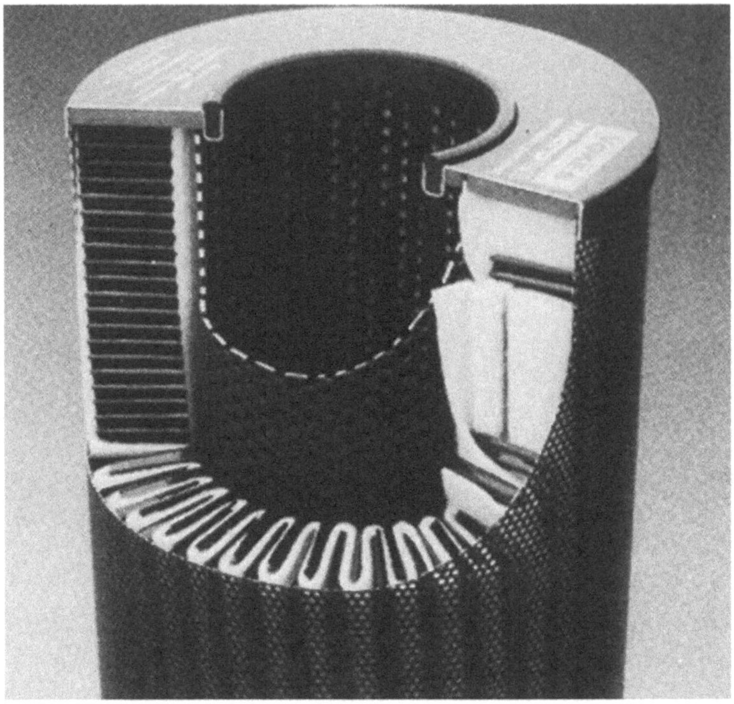

**Figure 7.10** Pleated Filter Element. *Reproduced by Permission of Elsevier Advanced Technology, Oxford England (18).*

is a picture of a pleated element construction using a felt medium. Note from the cut-away that the pleated felt medium is supported by a rigid pleated support mesh to prevent pleat collapse and pleat pinching. The outside and inside of the pleated filter cylinder is supported by concentric perforated cylinders. The inside cylinder is the core of the filter element. At the top of the element is an end cap that embeds and seals the pleated medium edges with the edges of the inner and outer cylinders. This is necessary to prevent leakage and by-passing of the filter medium. Note that the construction of Figure 7.10 is referred to as a filter element because it will be installed in some form of housing to become a cartridge.

Other forms of filter elements are illustrated in Figures 7.11 and 7.12. Figure 7.11 is a general purpose filter element containing multilayers of pleated filter media. Versions of this cartridge incorporate various types of filter media including membranes of polypropylene, nylon, and PTFE sandwiched between protective layers of spunbonded media.

Figure 7.12 is a filter element produced by Schroeder Industries, Leetsdale, Pennsylvania, USA. It is used for the filtration of hydraulic fluids and lube oils. The filter features Excellement® Z-Media®, which Schroeder (263) identifies as a

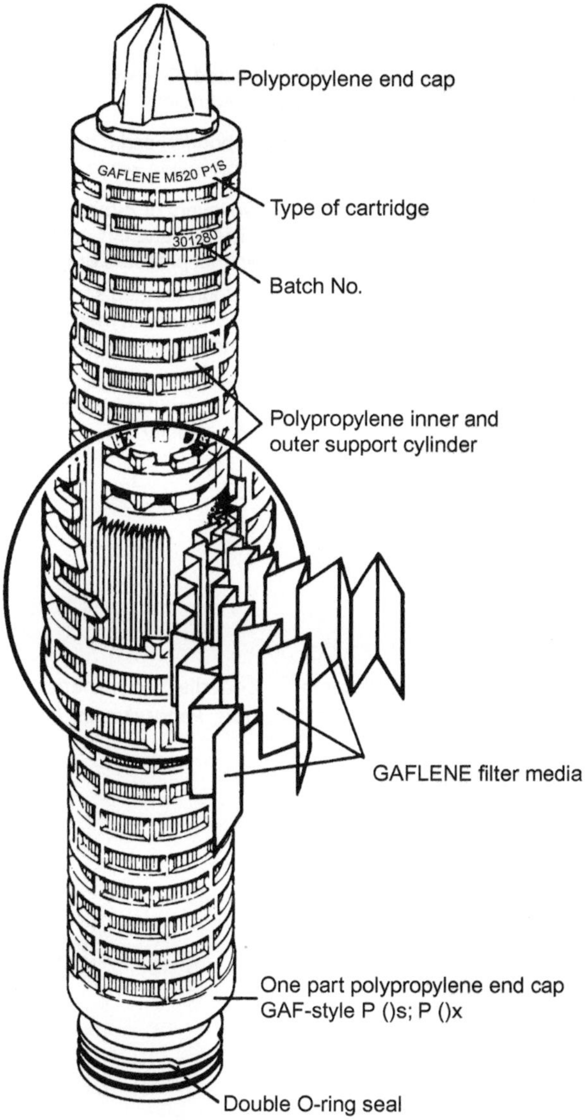


**Figure 7.11** All-Polypropylene Multilayer General Purpose Filter Element. *Reproduced by Permission of Elsevier Advanced Technology, Oxford England (18).*

"Synthetic microglass filtration media." The medium package also includes spun-bonded scrim and epoxy-coated steel wire fabric.

Figure 7.13 shows the cutaway components of a typical spin-on automotive engine oil filter. Figure 7.14 shows the filter in assembled form. The assembly is a cartridge filter containing both the filter element and the housing.

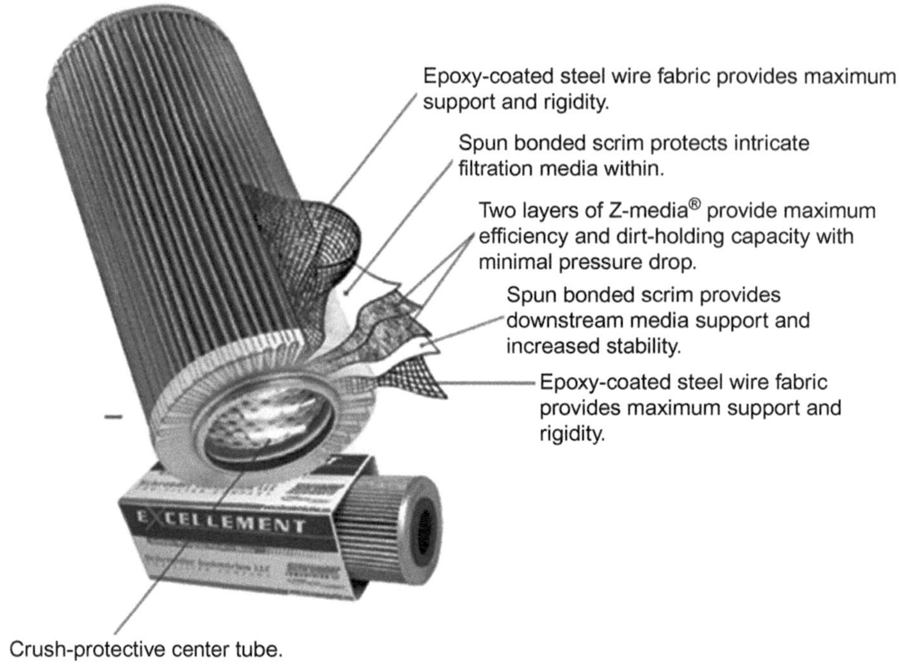


**Figure 7.12** Schroeder Industries Excellement® Z-Media™ Synthetic Replacement Element. *Reproduced by Permission of Schroeder Industries, Leetsdale, Pennsylvania, USA.*

**Figure 7.13** Engine Oil Filter Components. *Reproduced by Permission of Hollingsworth and Vose Co. Inc. East Walpole, Massachusetts, USA.*

**Figure 7.14** Cutaway of Engine Oil Filter Assembly. *Reproduced by Permission of Hollingsworth and Vose Co., Inc. East Walpole, Massachusetts, USA.*

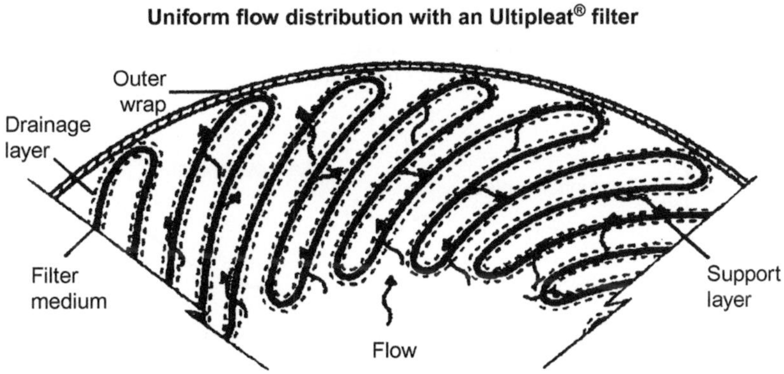


**Figure 7.15** Pall Ultipleat® Configuration and Flow Patterns. *Reproduced by Permission of Pall Corporation, East Hills, New York, USA.*

Although most pleat configurations are in a zigzag shape, there are other technologies that are particularly interesting. For example, Pall Corporation, East Hills, New York, USA has patented the Utipleat® filter technology (US Patents 5,543,047 and 5,690,765), a filter construction using crescent shaped pleats. The curved pleat configuration results in a more stress resistant filter medium. Figure 7.15 illustrates the pleat configuration and the fluid flow patterns. Figure 7.16 is an illustration of the pleat shape in the Pall Ultipeat® filter construction. Figure 7.17 demonstrates the improvement in filter life when compared to conventional pleats.

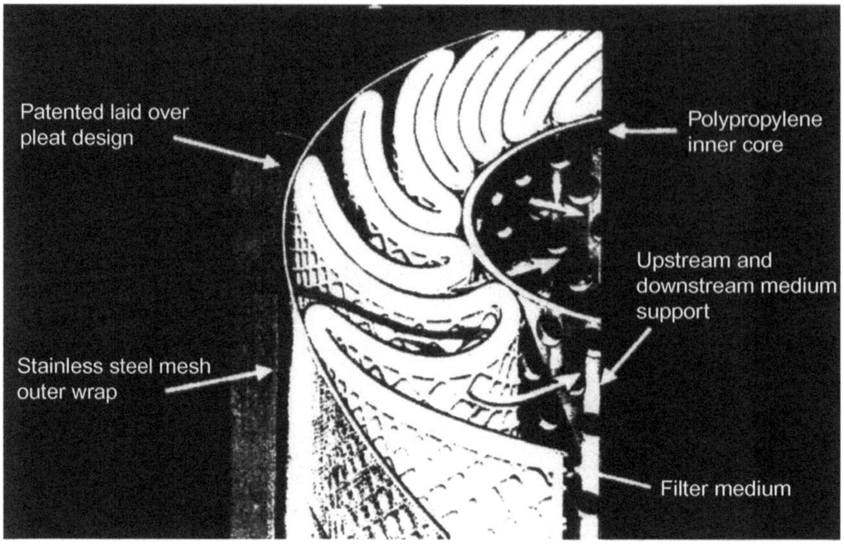


**Figure 7.16** Ultipleat® Pleat Configuration in Pall Septra® Filters. *Reproduced by Permission of Pall Corporation, East Hills, New York, USA.*

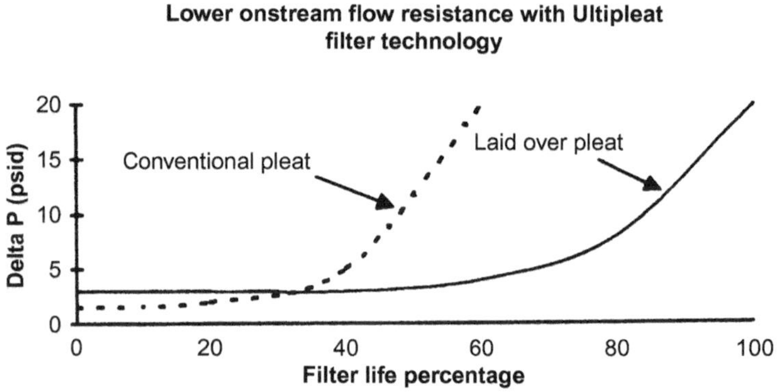


**Figure 7.17** Pall Ultipleat® Versus Conventional Pleat—Effect on Filter Life. *Reproduced by Permission of Pall Corporation, East Hills, New York, USA.*

### 7.3.2.1.3 Dirt holding capacity in cartridge filters

Disposable filter cartridges or elements are sufficiently inexpensive, so that any attempt to clean and reuse them would not be economically justified. Depth filter cartridges usually cannot be readily cleansed because the dirt is so well entrapped in the medium that cleansing would provide little benefit in restoring the cartridge for reuse. Eventually, even cleanable cartridges will have to be replaced. It is evident that dirt-holding capacity is an important requirement on filter cartridges and the filter medium they contain.

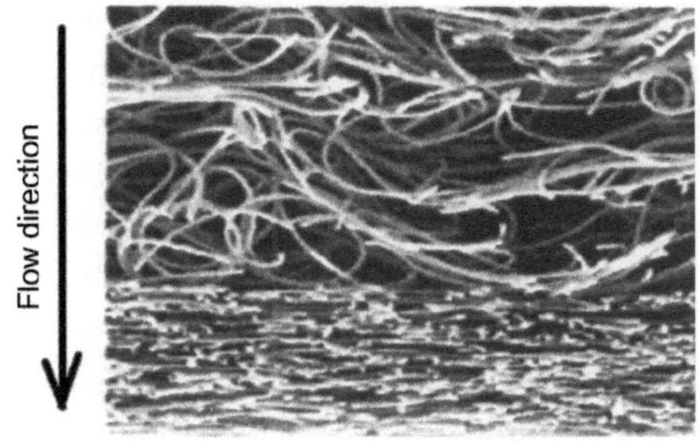


**Figure 7.18** Cross Section of a Gradient Density Filter Media. *Copyright Kimberly Clark Worldwide, Inc. Reprinted with permission.*

Ways to increase dirt holding capacity are:

1. Increase the surface area of the medium—pleating is the most popular way of increasing the surface area of a medium. By folding the medium back and forth, more of it can be fitted within the confines of a filter housing. Pleat shape is also an important factor. The wave pleat design of the Pall Ultipleat® mentioned above increases surface area and correspondingly dirt-holding capacity. Increasing the number of pleats is often an effective way of increasing the dirt-holding capacity of a filter element, however caution should be exercised. If there are too many pleats then the pleats may bunch together and result in pinching of the fluid flow between pleats. This can have a negative effect on dirt-holding capacity. Corrugation and creping are other ways to enhance the surface area of a filter medium.
2. Use a prefilter—the concept here is that the prefilter would filter out the rocks and bolts and the final filter would take out the fine stuff that passes through the prefilter. Several arrangements are used for doing this. One is to use two filters or two banks of filters depending on the system. The first filter or filter bank would be the prefilter(s) and the second filter or bank of filters would be the final filter(s). There are filter designs where the prefilter medium and the final filter medium are contained within the same filter housing. There are ways of designing prefiltration and final filtration effects into a single medium such as a gradient density filter medium discussed below.
3. Gradient density filter media—a gradient density medium has a change in density as the flow passes from the upstream surface to the downstream surface. Figure 7.18 is a cross section photograph of a gradient density filter medium which is used in liquid filter bag applications. The change in density allows for a form of depth filtration so that larger particles are removed at the low density upstream surface and finer particles are removed in higher density interior of the medium.

There are several ways of achieving gradient density in a filter medium. Wet lay filter paper is inherently a gradient density structure because as it forms the side against the wire or forming fabric (the wire side) is denser than the side away from the forming fabric (the felt side). Multilayer forming of a fabric, where each layer has a successively lower density or successively larger pore structure, is another way of forming gradient density media. This can be done with multi-card machines in the dry laid process and with multi-ply headboxes in the wet lay process. Kaukopaasi and Shah (264) reported on a study of resin treated multilayer wet lay media made out of wood pulp. Their report suggested that multilayering could as much as double dirt-holding capacity while at the same time increase filtration efficiency.

Dual layer structures can be designed to be gradient density media by combining a high porosity layer on the upstream side with a high filtration efficiency layer on the downstream side; in effect a prefilter and final filter within the same medium.

A variation of gradient density filter media is a composite medium containing a "capacity layer" upstream followed by an "efficiency layer" downstream. The capacity layer is intended to be a porous bulky substrate designed to trap large quantities of large particle dirt. It may be a spunbond or cellulose medium. The efficiency layer is usually a high efficiency filter medium such as a membrane or nanofiber web and is designed to trap the fine particles that penetrate the capacity layer. The layout is reverse of a typical substrate-supported membrane or nanofiber web where the substrate is downstream of the flow and is there to provide structural support to the efficiency layer.

4. Use bulkier and thicker media—the bulkiness of a filter medium is related to its dirt-holding capacity. Bulkiness can be increased by increasing the grammage of the filter medium or by increasing its pore structure or pore volume. Pore volume can be increased by using coarser fibers, crimped fibers, and higher modulus (stiffer) fibers.

   There are caveats in these approaches. A bulkier medium may have a snugger fit in the filter cartridge. If it is a pleated medium cartridge, the additional bulk may force the pleats to pinch and restrict flow resulting in a loss in dirt-holding capacity rather than the desired increase. One may correct for this by reducing the number of pleats in the cartridge, however this reduces the surface area available for filtration and then what is gained by increased bulk is lost by reduced surface area. Increasing bulk by increasing pore volume of the filter medium may increase dirt-holding capacity. However, filtration efficiency may be compromised.

5. Use electrokinetic positively charged media—an example of this are the Zeta Plus® filters and cartridges marketed by 3 M and discussed in Section 7.2.5.3. The filter sheets used in these filters and cartridges have been specially treated to give a positive ionic (cationic) charge.

#### *7.3.2.2 Wound filter elements*

Wound filter elements are sometimes used in liquid coalescing separations such as separating water from fuel streams. They are analogous to paper rolls, in that the medium is wound around a mandrel. Kaydon Custom Filtration Corporation, LaGrange, Georgia, USA, produces two such filters illustrated in Figure 7.19. The filter on the

**Figure 7.19** Wound Filter Elements—the Filter on the Left Contains a Wound Filter Element Produced from Heavily Impregnated Resin Treated Filter Paper, the Filter on the Right Contains a Medium from Wound Layers of Glass Fiber Mat. *Reproduced by Permission of Kaydon Custom Filtration Corporation, LaGrange, Georgia, USA.*

left contains a reddish filter cylinder made from filter paper heavily impregnated with a phenolic resin. The paper was wound around a mandrel and then heated to fuse, bond, and cure the layers to each other. The result is a filter cylinder that appears to be a solid mass of resin impregnated fibers.

The filter, on the right, also a liquid coalescing filter contains layers of wound glass fiber mat. This is the yellow layer contained between the inner core and outer cylinder housing.

#### *7.3.2.3 Bag filters*

Bag filters are used for a number of liquid filter applications including bulk chemicals, food industry applications such as vegetable oil, coolants, cleaning fluids, paints, varnishes, waxes, and plastisols. Bag filters may be as simple as a filter medium bag attached to the end of a pipe. It will probably be more secure if it is attached by means of an adapter as shown in Figure 7.20 below. For higher-pressure and larger volume applications, the bag is inserted into a housing and supported by a mesh basket as displayed in Figure 7.21 below.

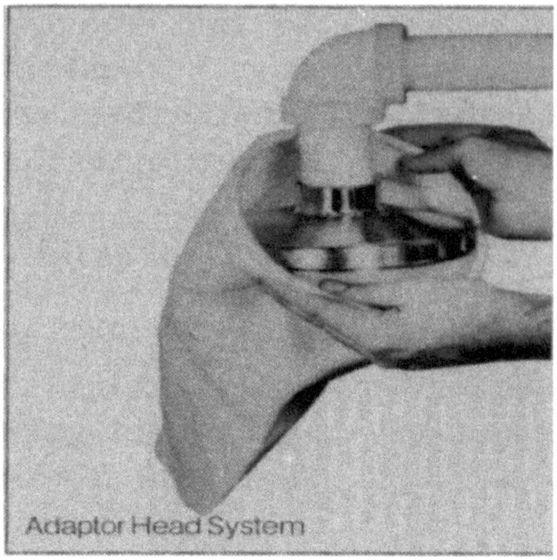


**Figure 7.20** Filter Bag Adapter. *Reproduced by Permission of Elsevier Advanced Technology, Oxford England (18).*

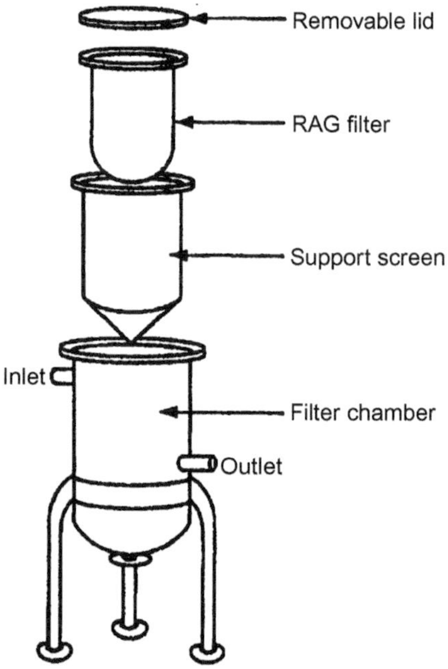


**Figure 7.21** Filter Bag Housing.

Traditionally, filter bags are composed of woven materials, however nonwoven materials have made a significant entry into this market. Nonwoven material types used in bag filters include needle felts, various forms of carded nonwovens, meltblowns, spunbonded webs and composite structures. Gradient density filter media, such as depicted in Figure 7.18 above, are often designed for liquid bag filtration.

## 7.4 TESTING OF LIQUID FILTERS

The testing of liquid filters has been previously discussed in Chapter 6. The rating of filters and filter media for filtration performance is at best a difficult and treacherous procedure. The two important factors in a filter test are efficiency assessment and dirt-holding capacity. Both of these factors can be very problematic. In multiple pass testing important considerations of a test are the quality and type of test dust (discussed in Chapter 6, Section 6.14.5) and the ability to measure particle size and count (discussed in Chapter 6, Section 6.14.6).

### 7.4.1 Efficiency assessment

Testing for efficiency has already been discussed above. For liquid filtration simplest measure of efficiency is by gravimetric means. One determines the mass of contaminant upstream of the filter and downstream of the filter. The efficiency is

$$e = 100\left(\frac{M_u - M_d}{M_u}\right) \tag{7.1}$$

$e$ is the filtration effiiciency, %.
$M_u$ is the upstream mass of the contaminant.
$M_d$ is the downstream mass of the contaminant.

Equation (7.1) has been previously used in Chapter 6, Section 6.14.1.1. The problem with the gravimetric approach is that it gives no concept of particle size. The sensitivity of our times demands that the filter user be able to assess not only the overall efficiency of the filter, but also the particle size on which this efficiency is based on. More modern filtration testing techniques deal with this requirement.

### 7.4.2 Multi-pass test stand

The mainstay test method for hydrocarbon filters (hydraulic fluids, diesel fuels, lubricating oils, etc.) is the multi-pass test stand (reviewed in Chapter 6, Section 6.14.1.5). Originally standardized by ISO 4572-81, the standard has since been replaced by ISO16889:2008. This test is primarily specified for the filtration of hydraulic fluids, but can also be used for diesel and aviation fuels, lubricating oils, and other types of hydrocarbon liquids. The test measures the particle count of a carefully calibrated liquid contaminant upstream and downstream of the filter. It is a fractional efficiency test because it measures the filtration efficiency of the contaminant at various particle sizes. For filtration efficiency, it uses the Beta ($\beta$) ratio (see Chapter 6, Section 6.14.1.6). The $\beta$-ratio is the ratio of upstream particles to downstream particles. The more efficient is the filter, the higher is the $\beta$-ratio.

$$\beta = \frac{N_{d,U}}{N_{d,D}} \tag{7.2}$$

$N_{d,U}$ is the upstream particle count for particles of diameter $d$ or greater.

$N_{d,D}$ is the downstream particle count for particles of diameter $d$ or greater.

The $\beta$-ratio is related to filtration efficiency $e$ by the formula:

$$\beta = \frac{1}{1-(e/100)} \tag{7.3}$$

$e$ is the filtration efficiency, %.

The $\beta$-ratio is reported with a subscript indicating the particle size for which the ratio is being reported. The test is cumulative. The $\beta$-ratio being reported is for all particles equal or greater than the particle size indicated by the subscript. The subscript will also include (c) to indicate if the testing was done by the ISO16889 procedure with a National Institute of Standards and Technology (NIST) certified and traceable test suspension. This is important, because there are still organizations that still use the old ISO 4572 method. To illustrate, $\beta_{10(c)} = 25$ indicates that the ratio for 10 μm particles or greater was 25 (equivalent to 96% filtration efficiency) and that the NIST certified test suspension was used.

Table 7.7 relates $\beta$-ratio to filtration efficiency based on an upstream particle count of 10,000 particles.

Figure 7.22 below is a diagram of a multi-pass test stand. Figure 7.23 is a photograph of a test stand used by Pall Corporation. The specified concentration of contaminant is injected into the injection reservoir upstream of the test filter at a known rate. Automated particle counters measure the particles by size and by count per volume of fluid both

**Table 7.7** Comparison of $\beta$-Ratio and Filtration Efficiencies

| Upstream particle count | Downstream particle count | Filtration efficiency | $\beta$-ratio |
|---|---|---|---|
| 10,000 | 5,000 | 50% | 2 |
| 10,000 | 1,000 | 90% | 10 |
| 10,000 | 500 | 95% | 20 |
| 10,000 | 100 | 99% | 100 |
| 10,000 | 10 | 99.9% | 1,000 |
| 10,000 | 1 | 99.99% | 10,000 |

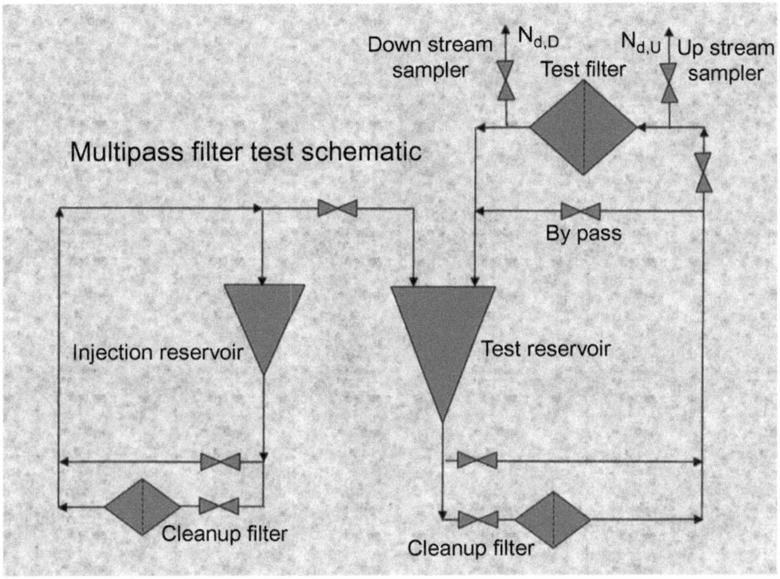


**Figure 7.22** Multi-Pass Filter Test Schematic. *Reproduced by Permission of INDA, Cary North Carolina, USA.*

**Figure 7.23** Multi-Pass Test Stand. *Reproduced by Permission of Pall Corporation, East Hills, New York, USA.*

upstream and downstream of the filter being tested. The fluid downstream of the test filter and is recyled and continuously injected with contaminant throughout the life of the test. In accordance with ISO16889 the following three contaminant levels are allowable: condition 1—3 mg/l, condition 2—10 mg/l, and condition 3—15 mg/l (Bensch (265)). The test is continued until a prescribed pressure drop is reached, at which point the test is terminated. The particle counts are recorded at 10 equal intervals during the test. For example, if the test is to be run for 100 min, particle size analysis will be conducted every 10 min. If the test is to last 1 h, then the analysis will be for every 6 min. Since the particle counting is continuous and the test stand is computerized, the computer is programmed to calculate the particle counts at 10 equal time-intervals.

One of the complications and reasons for upgrading from ISO 4572 to ISO16889 was that the original test dust, ACFTD prescribed for ISO 4572-81 was no longer being produced by what was then AC Rochester Division of General Motors. The complications that resulted, already been reported in Chapter 6, Section 6.14.5. was one of the reasons for upgrading to ISO 16889. The new test dust is designated as ISO medium test dust (ISO MTD). Chapter 6, Section 6.9.5 is a detailed discussion of test dusts for filtration purposes.

### 7.4.3 Cleanliness levels

The goal of a filter is to raise the purity of a fluid so that it can be used for its intended application. In fluid power systems and lubricating oil systems, particulate contamination causes abrasive wear and deterioration on the parts they are in contact with. Cleanliness levels for hydraulic fluids have been established to identify the quality of an oil or hydraulic fluid. Schroeder Industries LLC, Leetsdale, Pennsylvania, USA, in their Product Catalogue "filter systems" provide a very thorough and detailed report on contamination management (266). In it they identify the following standards for cleanliness levels: ISO 4405:1991, ISO 4406:1999, NAS 1638, and SAE AS 4059. They are summarized in Table 7.8.

ISO 4405 is a gravimetric standard for relatively heavily contaminated fluids. The others rate cleanliness classes by particle size and count. NAS 1638 (National Aerospace Standard—US) has been largely superseded by SAE AS 4059F (2013). In fact, the most recent versions of NAS 1638 states "Inactive for new designs after May 30, 2001". One of the biggest weaknesses of NAS 1638 is that it specifies particle counting by the tedious microscope procedure. Nevertheless it is still not necessarily dead. According to Day (267), NAS 1638 can be used for existing systems where correlation with earlier data is necessary. Table 7.9 tabulates the cleanliness classes of NAS 1638.

This SAE Aerospace Standard (AS4059F) defines contamination classes and levels for particulate contamination of hydraulic fluids and includes methods of reporting related data in Appendix A of this standard. The AS4059F cleanliness codes are tabulated in

**Table 7.8** Cleanliness Standards

| Standard | ISO 4405 | ISO 4406:1999 | NAS 1638 | SAE AS 4059 |
|---|---|---|---|---|
| Application | Highly contaminated media, e.g., washing media, machining fluids | Hydraulic fluids Lubrication oils | Hydraulic fluids Lubrication oils | Hydraulic fluids Lubrication oils |
| Parameters | (mg/liters of fluid) | Number of particles >4 μm (C) >6 μm (C) >14 μm (C) | Number of particles 5–15 μm 5–25 μm 25–50 μm 50–100 μm >100 μm | Number of particles >4 μm (c) >6 μm (c) >14 μm (c) >21 μm (c) >38 μm (c) >70 μm (c) |
| Analysis methods | In this lab method, 1 L of the fluid undergoing analysis is filtered through a prepared membrane, which is then weighted | 1. Manual evaluation: The fluid undergoing analysis is filtered through a prepared membrane and the cleanliness class (contamination rating) estimated or counted by hand using a microscope. 2. Automated particle counting: The fluid undergoing analysis is conducted through a particle counter, which tallies the particle fractions. | | |
| Remarks | Very time-consuming method | 1. Manual evaluation: Very time-consuming, not very exact. 2. Automated particle counting: Result available almost immediately | | |

Reproduced by permission of Schroeder Industries LLC, Leetsdale, Pennsylvania, USA.

Table 7.10. The contamination levels selected are based on the widely accepted NAS 1638 cleanliness classes. AS4059F defines the conversion from NAS 1638 cleanliness class specifications to AS4059 class specifications. The comparison of the NAS 1638 classes to AS4059 classes and levels is provided. Differences are defined and explained in Appendix B of AS4059. NAS 1638 classes based on weight of particles are not applicable to these classes and are not included. NAS 1638 has added a contamination code describe the contamination levels of the fluid at the specified particle size ranges.

Note particle sizes determined by two different standards for calibration: ISO 4402 (ACFTD) or ISO 11171 (MTD), do not give the same numbers. The reasons for this are due to the difference processing of ACFTD versus MTD, different methods of measurement, and by the revised definition of particle size. This has all been previously discussed in Chapter 6, Section 6.9.5.

**Table 7.9** NAS 1638 Cleanliness Code Classes

| Class | Maximum particles/100 ml in specified size range (μm) | | | | |
|---|---|---|---|---|---|
| | 5–15 | 15–25 | 25–50 | 50–100 | >100 |
| 00 | 125 | 22 | 4 | 1 | 0 |
| 0 | 250 | 44 | 8 | 2 | 0 |
| 1 | 500 | 89 | 16 | 3 | 1 |
| 2 | 1,000 | 178 | 32 | 6 | 1 |
| 3 | 2,000 | 356 | 63 | 11 | 2 |
| 4 | 4,000 | 712 | 126 | 22 | 4 |
| 5 | 8,000 | 1,425 | 253 | 45 | 8 |
| 6 | 16,000 | 2,850 | 506 | 90 | 16 |
| 7 | 32,000 | 5,700 | 1,012 | 180 | 32 |
| 8 | 64,000 | 11,400 | 2,025 | 360 | 64 |
| 9 | 128,000 | 22,800 | 4,050 | 720 | 128 |
| 10 | 256,000 | 45,600 | 8,100 | 1,440 | 256 |
| 11 | 512,000 | 91,200 | 16,200 | 2,880 | 512 |
| 12 | 1,024,000 | 182,400 | 32,400 | 5,760 | 1,024 |

Table shows the cleanliness classes in relation to the particle concentration determined.

**Table 7.10** SAE AS4059 Cleanliness Codes
**Maximum contamination limits (particles/100 ml)**

| ACFTD<br>MTD | | >2 μm<br>>4 μm(c) | >5 μm<br>>6 μm(c) | >15 μm<br>>14 μm(c) | >25 μm<br>>21 μm(c) | >50 μm<br>>38 μm(c) | >100 μm<br>>70 μm(c) |
|---|---|---|---|---|---|---|---|
| Classes | 000 | 195 | 76 | 14 | 3 | 1 | 0 |
| | 00 | 390 | 152 | 27 | 5 | 1 | 0 |
| | 0 | 780 | 304 | 54 | 10 | 2 | 0 |
| | 1 | 1,560 | 609 | 109 | 20 | 3 | 1 |
| | 2 | 3,120 | 1,220 | 217 | 39 | 6 | 1 |
| | 3 | 6,250 | 2,430 | 432 | 76 | 11 | 2 |
| | 4 | 12,600 | 4,860 | 864 | 152 | 22 | 4 |
| | 5 | 25,000 | 9,730 | 1,730 | 306 | 45 | 8 |
| | 6 | 50,000 | 19,500 | 3,460 | 612 | 90 | 16 |
| | 7 | 100,000 | 38,900 | 6,920 | 1,220 | 180 | 32 |
| | 8 | 200,000 | 77,900 | 13,900 | 2,450 | 360 | 64 |
| | 9 | 400,000 | 156,000 | 27,700 | 4,900 | 720 | 128 |
| | 10 | 800,000 | 311,000 | 55,400 | 9,800 | 1,440 | 258 |
| | 11 | 1,600,000 | 623,000 | 111,000 | 19,600 | 2,880 | 512 |
| | 12 | 3,200,000 | 1,250,000 | 222,000 | 39,200 | 5,760 | 1,024 |

Table shows the cleanliness classes in relation to the particle concentration determined.
ACFTD is determined in accordance with ISO 4402—particle size based on longest dimension.
MTD is determined in accordance with ISO 11171—particle size based on diameter of projected area-equivalant circle.

Regardless of which method is used, the particles are divided into six size ranges: A, B, C, D, E, and F. There are 15 coding classes ranging from 000, the cleanest, to 12, the dirtiest.

There are three reporting options:

1. Specifying at single sizes: for example, if the particle size of interest is 0.15 m or 0.14 m(c), and no more than 3,460 particles per 100/ml larger than this are specified, then the class code would be *AS4059 6C*.
2. Specifying for specific sizes: to illustrate, if the particle sizes of interest are 0.5 μm, 0.15 μm, and 0.25 μm, (0.6 μm, (c), 0.14 μm, (c), and 0.21 μm, (c)) and the maximum number of particles greater than these sizes are 2430, 217, and 20, respectively, then the rating is *AS4059 3B/2C/1D*. Note that not all particle sizes have to be specified.
3. Specifying the highest cleanliness class measured: for example, if all six particle size ranged are of interest and the rating is *AS4059 6A through F*, then the maximum number of particles exceeding any size cannot be greater than the limits for that size as listed in the line of size code 6.

ISO 4406-1999 is similar in principle to SAE AS 4059, however, it has different numbers. Unlike AS 4059, which allows particle counting by either the longest chord or the diameter of the area-equivalent circle, ISO 4406 specifies the equivalent circle technique. To see how ISO 4406 works, note in Table 7.11 below, that 18 cleanliness classes are shown (6–24). Three particle size ranges are 4 μm(c), 6 μm(c), and 14 μm(c). Note that the subscript (c) is there to show that the MTD test dust is specified. Let's suppose a test was done and the counts were 2,820 ppm for 4 μm(c), 582 ppm for 6 μm(c), and 98 ppm for 14 (c). Since the 2,820 for 4 μm is higher than 2,500 ppm but less than 5,000 ppm it gets a rating of 19. Likewise the 582 count at 6 μm is higher than 320 ppm, but less than 640 ppm, it gets a rating of 16. In a similar manner, the 98 count at 14 μm gets a rating of 14. The result is reported as 19/16/14 where 19 is the rating for 4 μm(c), 16 is the rating for 6 μm(c) and 14 is the rating for 14 μm(c).

## 7.5 MEMBRANE FILTRATION

In Chapter 2, Section 2.1.4, the four basic technologies of membrane separation are described as: MF, UF, nanofiltration, and RO. These are summarized in Figure 7.24. The figure identifies the micron range of each process, the types of contaminants removed at each micron level, and some of the aqueous separation processes at each micron level. Also note that from MF to RO, the pressure driving force increases from 1 to 4 bars for MF, to as high as 60 bars for RO.

Other liquid separation technologies, discussed in Chapter 2, Section 2.1.4.5 are electrodialysis and pervaporation.

**Table 7.11** ISO 4406 Code Chart

| | | Particles per milliliter | | |
|---|---|---|---|---|
| | | Range code | More than | Up to/including |
| | | 24 | 80,000 | 160,000 |
| | | 23 | 40,000 | 80,000 |
| | | 22 | 20,000 | 40,000 |
| | | 21 | 10,000 | 20,000 |
| | | 20 | 5,000 | 10,000 |
| 2,820 count at 4 μm | → | 19 | 2,500 | 5,000 |
| | | 18 | 1,300 | 2,500 |
| | | 17 | 640 | 1,300 |
| 582 count at 6 μm | → | 16 | 320 | 640 |
| | | 15 | 160 | 320 |
| 98 count at 14 μm | → | 14 | 80 | 160 |
| | | 13 | 40 | 80 |
| | | 12 | 20 | 40 |
| | | 11 | 10 | 20 |
| | | 10 | 5 | 10 |
| | | 9 | 2.5 | 5 |
| | | 8 | 1.3 | 2.5 |
| | | 7 | 0.64 | 1.3 |
| | | 6 | 0.32 | 0.64 |

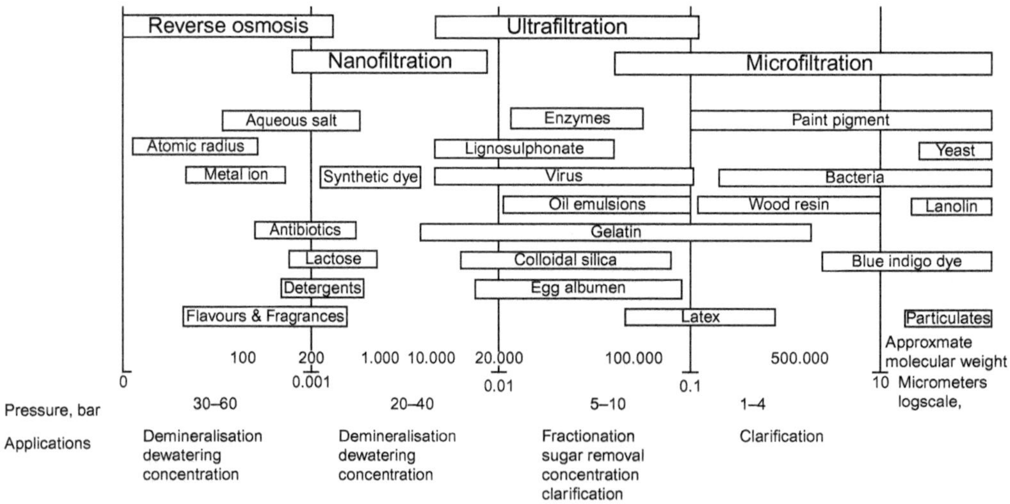


**Figure 7.24** Membranes for Water Treatment Processing. *Reproduced with per Permission of Elsevier Science Ltd. Oxford, England UK.*

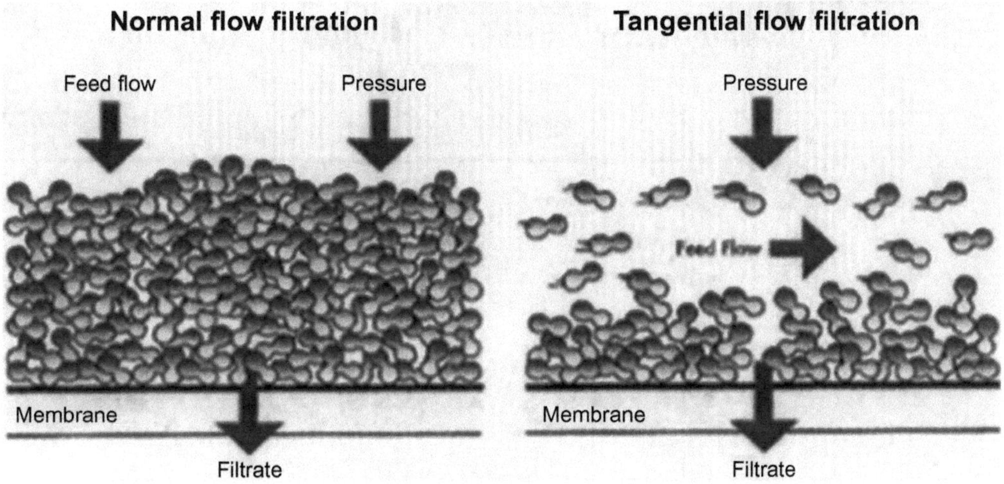


**Figure 7.25** Membrane Filtration: Normal Flow Filtration (NFF) and Tangential Flow Filtration (TFF).

There are two modes of membrane filtration: normal flow filtration (NFF) and tangential flow filtration (TFF). These are illustrated in Figure 7.25 above NFF is the case where the fluid flow is more or less perpendicular (allowing for pleated media varaiation) to the medium and it passes directly through. The contaminants and particles are trapped on the surface of the medium and build up a cake until pressure drop limitations are exceeded. NFF is used in small scale situations such as laboratory disc filters. Generally, NFF filters are used when clarification or bioburden reduction is desired in relatively low solid streams, such as during final polishing of a product.

TFF is the condition where flow is tangential to the medium and passes through in a tangential manner. The contaminants and particles are separated at the medium surface but are continually washed away by the tangential flow condition. The advantage of TFF is that there is little or no cake build-up on the membrane surface because the particles are washed away by the tangential flow. It is also fits better with the design of membrane systems to process aqueous fluids.

TFF is used in larger scale operations. TFF is the mechanism of flat sheet membrane modules, hollow fiber modules, tubular modules, and of spiral wound membrane modules. The various module configurations are discussed below.

## 7.5.1 Membrane filter modules

### *7.5.1.1 Flat sheet modules*

Flat sheet tangential flow modules are similar in design to that of plate and frame filter presses. The basic flat sheet flow unit, depicted in Figure 7.26 (268) consists of a shallow channel with rectangular flat sheet membranes on one or both sides. The membranes are supported on grooved plates which allow easy collection of permeate.

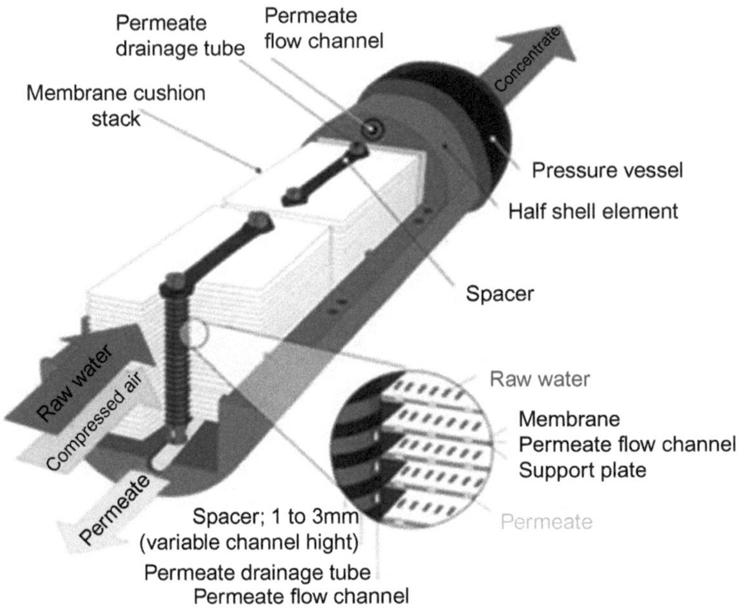


**Figure 7.26** Flat Sheet Tangential Flow Module (268). *Source: Docstoc.*

The feed is pumped into the channel and the permeate, passing through the membrane, is collected by the grooved plates. Most flat sheet membrane modules are based on membrane cassettes which contain basic flat sheet membrane units connected in series. A production scale membrane module typically contains several of these cassettes within it connected on parallel. This, like the lenticular filters, is demonstrated in Figure 7.5 above. One of the main advantages of the flat sheet tangential flow membrane module is the ease of cleaning and replacement of defective membranes. Other advantages include ability to handle viscous feeds and feed with high levels of suspended solids. The biggest disadvantage is the relatively low packing density as measured by the membrane area to module volume ratio.

### *7.5.1.2 Spiral wound modules*

The design of spiral-wound modules is similar in principal to that of flat sheet modules. A schematic of the spirally wound module is depicted in Figure 7.27 (Sutherland (269) p. 222). Two membrane sheets are cast on a porous polyester support sheet and separated by a mesh-like spacer, the active membrane sides facing away. The membrane can be made from a cellulosic polymer or from a thin film polymertype applied to a microporous polymer film, which in turn is bonded to the support sheet. Three edges of the two membrane sheets are glued together with the fourth edge open to a perforated center tube for the permeate removal. On the other two sides of "the

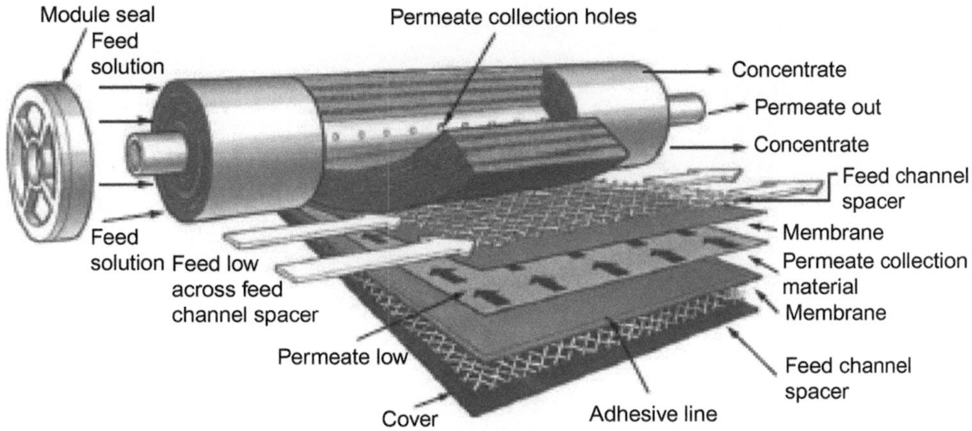


**Figure 7.27** Spirally Wound Membrane Module Ref. Sutherland p. 222. *Reproduced with Permission of Elsevier Inc. Oxford, England.*

envelope," another two mesh-like spacers with thicknesses in the range of 0.5–0.6 mm are placed as the feed channel spacers. The whole assembly is spirally wound around the perforated center tube in a spiral configuration. The characteristics of spiral-wound modules are as follows:

1. Spiral-wound modules are operated in the turbulent flow region because of the presence of feed spacers.
2. Due to the additional drag generated by feed spacers, the pressure drop in spiral-wound modules is relatively high.
3. Spiral-wound modules have fairly high surface area volume ratio and among all the four kinds of membrane modules are lowest in terms of capital cost.
4. Suspended particles can easily block the mesh-like spacers and partially block the feed channel. Therefore, spiral-wound modules require relatively clean feed with a minimum content of suspended particles. Often, pretreatment is needed.

#### *7.5.1.3 Tubular modules*

Tubular modules contain membrane tubes ranging from 5 to 25 mm in diameter and 0.6–6 m in length. The unit demonstrated in Figure 7.28 is cross section cut to reveal the tubular bundle structure. Cui et al. (270) report that tubular modules are composed of a number of membrane tubes assembled in a shell-and-tube arrangement. The membrane tubes are usually made up of porous fabric or plastic support with selective membranes on the inside. Tubular modules have some important characteristics:

1. Due to their large internal diameters, tubular modules are capable of dealing with the feed stream containing fairly large particles. In addition they can be easily cleaned by using either mechanical or chemical cleaning methods.

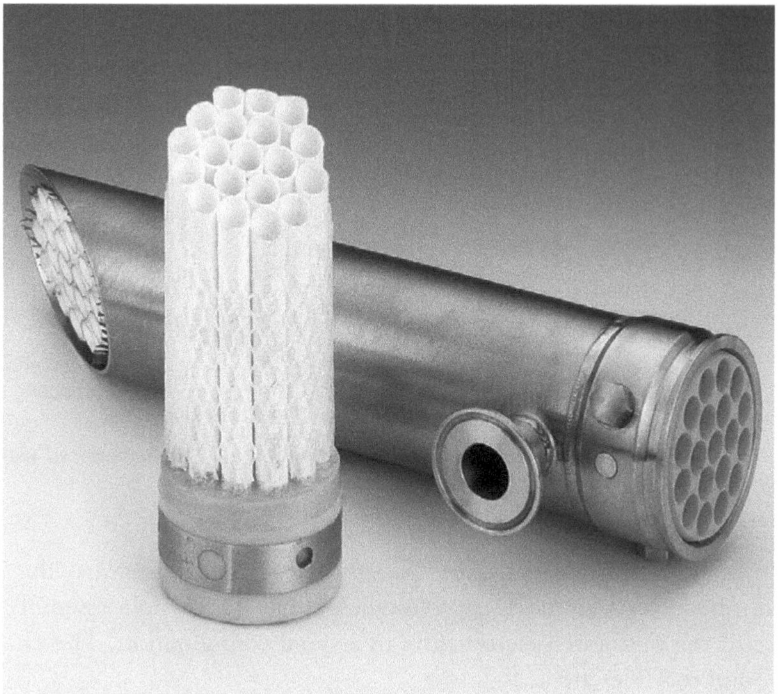

**Figure 7.28** Tubular Membrane Module (Cui and Muralidhara (271) Figure 3.2). *Reproduced with Permission of Elsevier Inc. Oxford, England.*

2. They need large pumping capacity, because they are usually operated under the turbulent flow conditions with the Reynolds numbers greater than 10,000.
3. They have the lowest surface area-to-volume ratio among all the four membrane configurations. The holdup volumes of tubular modules are also high requiring large floor space to operate.

#### *7.5.1.4 Hollow fiber modules*

Hollow fibers membranes are actually "thin" tubular membranes, the tubes being of fiber like dimensions. The hollow fibers are pictured in the SEM (scanning electron microscope) of Figure 7.29. Cui et al. (270) state that the fiber diameters typically range from 0.2 to 3 mm (except those used in RO, which may be as thin as 0.04 mm). They are used in compact modules that provide the form of self-support that enables them to withstand high backpressure. Normally, hollow fiber modules are composed of 50–3,000 individual hollow fibers, bundled and sealed together at each end with epoxy. The fiber lengths range from 18 to 120 cm. In MF and UF, hollow fiber modules are operated in the inside-out mode with selective skin layers on the inner sides of the fibers, while in RO, they are operated in the outside-in mode with

**Figure 7.29** Hollow Fiber Membranes. *Printed with Permission of Matrix Membranes, Inc. Oceanside, California, USA.*

selective skin layers on both sides of the fibers. Hollow fiber modules have some very different characteristics from tubular modules:

1. They are recommended to operate with the Reynolds numbers in the range of 500–3,000, therefore, most of them run in the laminar flow region. Additionally, the pressure rating of hollow fiber modules is low, normally with a maximum of 2.5 bar.
2. Due to the combination of low cross-flow rate and low pressure drop, hollow fiber modules are one of the more economical modules in terms of energy consumption.
3. Hollow fiber modules have the highest surface area-to-volume ratio among all the four membrane configurations, and their holdup volumes are low.
4. Hollow fibers are self-supported. The modules have good backwash capacity and are easy to clean.
5. One distinct disadvantage of hollow fiber modules is that, their thin fibers are susceptible to being blocked by feed with large particles, especially when they are operated in the inside-out mode. Pretreatment to reduce particle size to 100 μm is often required for hollow fiber modules.

## 7.5.2 Membrane filter applications

Most membrane filter applications are directed toward aqueous fluids. This includes water desalinization, water treatment and purification, tertiary sewage treatment, aqueous food and beverage processing, medical blood and body fluid filtration, pharmaceutical applications, biofuel production, precious metal recovery, clays and

**Table 7.12** Typical Examples of Membrane Applications in Food and Beverage Processing

| Application | Product | Membrane process | Industry |
|---|---|---|---|
| Cold sterilization | Beer, wine, milk | MF | Diary, beverage |
| Clarification | Wine, beer | MF/UF | Beverage, corn milling |
| Drying/thicken | Whey | UF/RO | Potato, diary |
| Desalting | Water, cheese | RO/ED | Diary, beverage |
| Concentration | Fruit juice | RO | Beverage, sugar refining |
| Dealcoholization | Wine | PV | Beverage |
| Fractionation | Egg | UF | Poultry |
| Product recovery | Lactic acid, citric acid | UF/ED | Biotech |

MF—microfiltration;
UF—ultrafiltration;
RO—reverse osmosis;
ED—elecrtrodialysis;
PV—pervaporation.
Reproduced with Permission of Elsevier Ltd. Oxford, England, UK.

**Figure 7.30** Membrane Ultrafiltration Plant (Ref. Cui et al. (271) Fig. 3.3). *Reproduced with Permission of Elsevier Inc. Oxford, England.*

minerals processing, cosmetics processing, and paints and coatings processes. Table 7.12 lists some membrane applications in aqueous food and beverage processing.

Figure 7.30 above depicts the ultrafiltration membrane system for apple juice filtration. There are several filter lines in parallel and each line consists of a series of tubular membrane modules. In the line series each module discharges the concentrate as the feed for the next module.

# CHAPTER 8

# Air Filter Applications

Important purposes of air filtration are to clean-up workplace and living environments and to control contaminants flowing to the atmosphere. People are protected by filters in heat ventilation and air conditioning (HVAC) and air cleaner systems to keep living and working space free of annoying air contaminants such as irritating dust, microorganisms, allergens, and other annoying contaminant chemicals. Industrial discharges to the environment are limited by baghouses, fume demisters, activated carbon adsorption, and other separating technologies. In Chapter One (Section 1.1.3) Kalil estimated that by 2018 air filtration will be greater than 70% of the North American market for total filtration sales.

A most useful document for exploring air filtration technology is the "NAFA Guide to Air Filtration". The 4th Edition (272), published in 2007[1], was the most recent edition as this book was being written.

## 8.1 INDUSTRIAL AIR FILTRATION

### 8.1.1 Dust collection systems

Dust collection systems are a big user of nonwoven filter media. The filter housings, often referred to as baghouses, are very large and usually consist of a large number of filter elements. The filter elements are usually fabric bags or fabric envelopes. Cartridge filter elements are also used in these systems. The fabrics for the bag or envelope filter elements are either woven fabrics or nonwoven felts. It is estimated that the market split between these two filter medium types is 50:50. Dickenson (273) argues that the filters operate by sieving the particles from dust-laden air and building a filter cake on the medium. The filter cake actually does most of the filtering. The fabric serves as a support structure. Figures 8.1 and 8.2 are diagrams of two types of modular baghouses; modular baghouse top (MBT) and modular baghouse walk-in (MBW). The MBT module is one where the bags are replaced from the top. The MBW module provides a walk-in space for replacing the bags.

Table 8.1 below lists the characteristics of fabric-type collectors. Note that nonwovens are predominantly used in the "continuous operation any loading" type of systems. The advantage of using felts rather than woven fabrics, in bag and envelope filters, is that felts are the more efficient filter medium as determined by efficiency ratings. Because felts are more difficult to clean, they are more commonly used in pulse-

[1] Published by National AIR Filtration Association (NAFA), Virginia Beach, Virginia, USA.

*Handbook of Nonwoven Filter Media*
ISBN: 978-0-08-098301-1

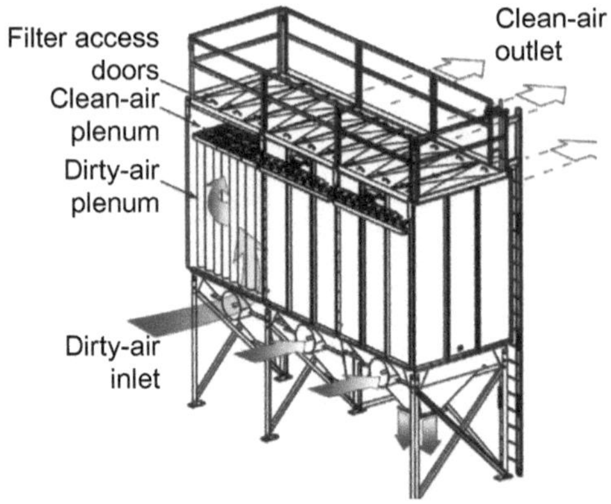


**Figure 8.1** MBT Normal Operation. *Reproduced with Permission of Donaldson Company, Inc. © Donaldson Company, Inc. All rights reserved.*

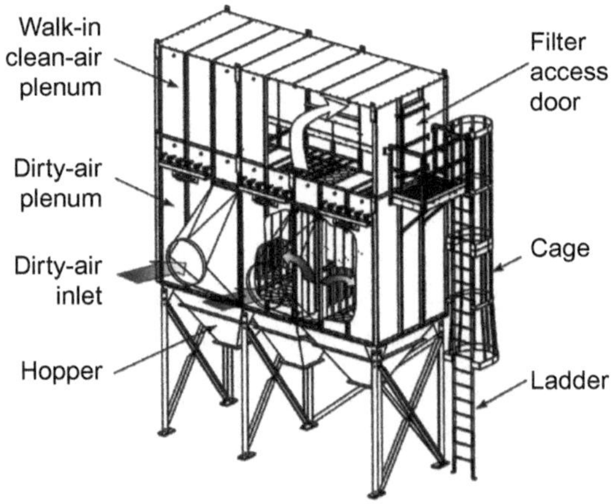


**Figure 8.2** MBW High Inlet Normal Operation. *Reproduced with Permission of Donaldson Company, Inc. © Donaldson Company, Inc. All rights reserved.*

jet type of filters. Often a surface treatment is applied to give it better cake release properties. This includes calendering, gazing, and singeing, previously discussed in Chapter 3, Section 3.2.1.

Economics requires that the filter elements be able to be reconditioned or cleaned. The most common methods for doing this are mechanical shaking, low-pressure reverse air, and high pressure-reverse jet (pulse jet). Figure 8.3 depicts a pulse-jet filter. Figure 8.4 diagrams the normal filtering mode and the reverse jet cleaning operation of a reverse jet cleaner.

The efficiency and cake release properties of a filter fabric can be improved by applying a membrane to the surface of the medium. Often these are applied to needle

**Table 8.1** Characteristics of Fabric-Type Collectors

| | Interruptible operation—light to moderate loading | Interruptible operation—heavy loading | | Continuous operation—any loading | |
|---|---|---|---|---|---|
| Fabric reconditioning requirement | Intermittent | Continuous | | | |
| Type of reconditioning | Shaker | Shaker | Reverse air (low pressure) | Reverse pulse (high pressure) pulse jet or fan pulse | |
| Collector configuration | Single compartment | Multiple compartments with inlet or outlet dampers for each. | | Single compartment | |
| Fabric configuration | Tube or envelope | Tube or envelope | Tube | Tube or envelope | Pleated cartridge |
| Type fabric | Woven | Woven | | Nonwoven felt | Nonwoven paper mat |
| Air flow | Highly variable | Slightly variable | | Virtually constant | Virtually constant |
| Normal rating (filtration velocity, ft/min) | 1–6 ft/min | 1–3 ft/min | | 5–12 ft/min | <1–2.5 ft/min |



felt fabrics. Figure 8.5 is a depiction (W.L. Gore (274)) of cake release with and without a membrane. The fibrous medium can be treated in such a way as to have good cake release properties, however, as shown in Figure 8.5A considerable dirt penetrates the medium and leaves it in a dirtier condition after each cleaning cycle. This is evident from dirt still on the medium surface after cleaning. This results in higher pressure drop and reduced filter life. On the other hand Figure 8.5B shows that if a ePTFE membrane is attached, the dirt does not penetrate the surface membrane. Cleaning is easier and the filter lasts longer.

W.L. Gore and Associates has pioneered this approach with their GORE-TEX® ePTFE membranes laminated to a number of baghouse filter fabrics. Table 8.2 lists properties of some of the grades GORE-TEX® needle felted laminates.

Donaldson also produces and markets membrane-felt composites under the trade name of Tetratex®. Summaries or their products and properties are tabulated in Tables 8.3 (low temperature), 8.4 (high temperature), and 8.5 (industrial wet).

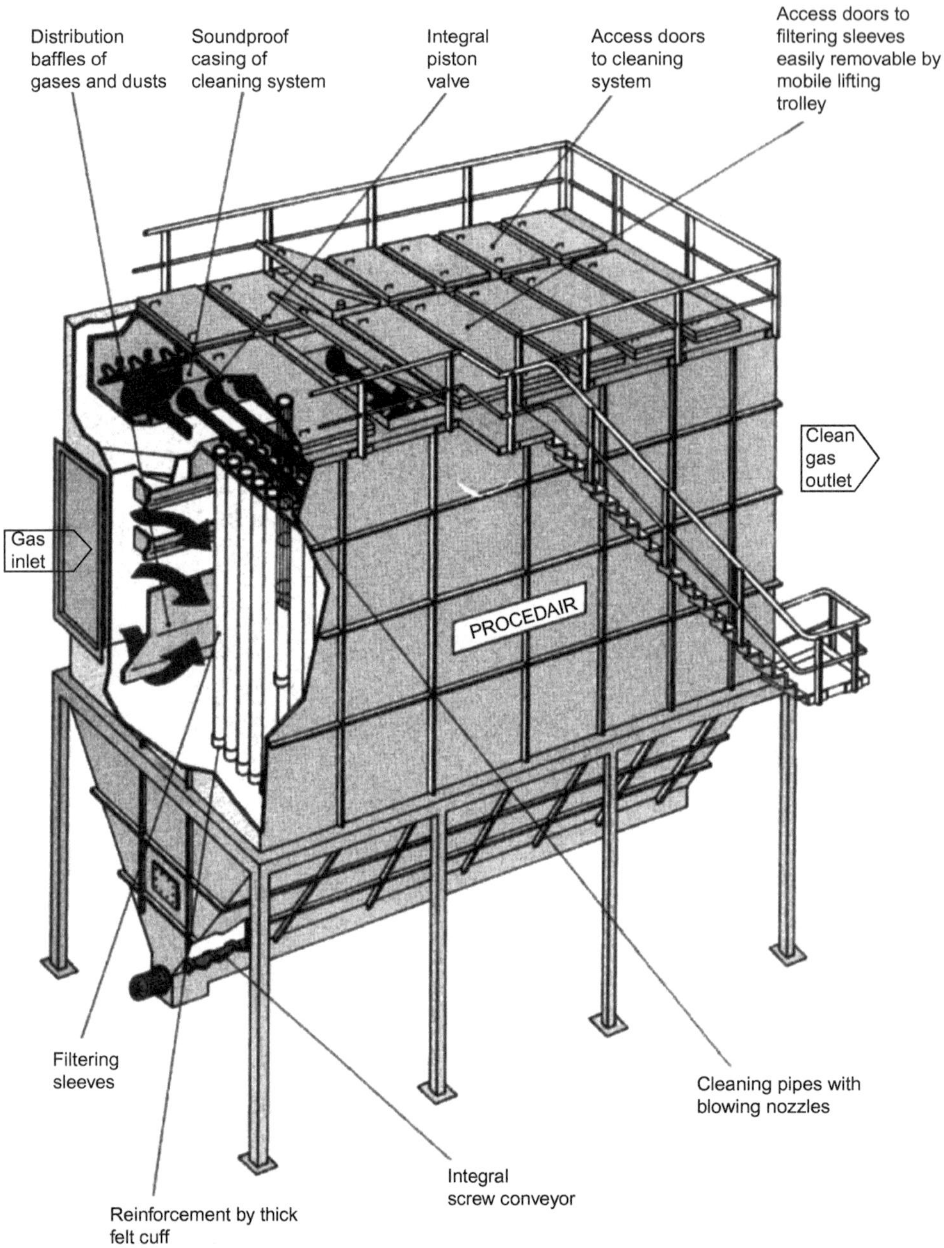


**Figure 8.3** Pulse Jet Filter. *Reproduced with Permission of Elsevier Advanced Technology, Oxford, England, UK.*

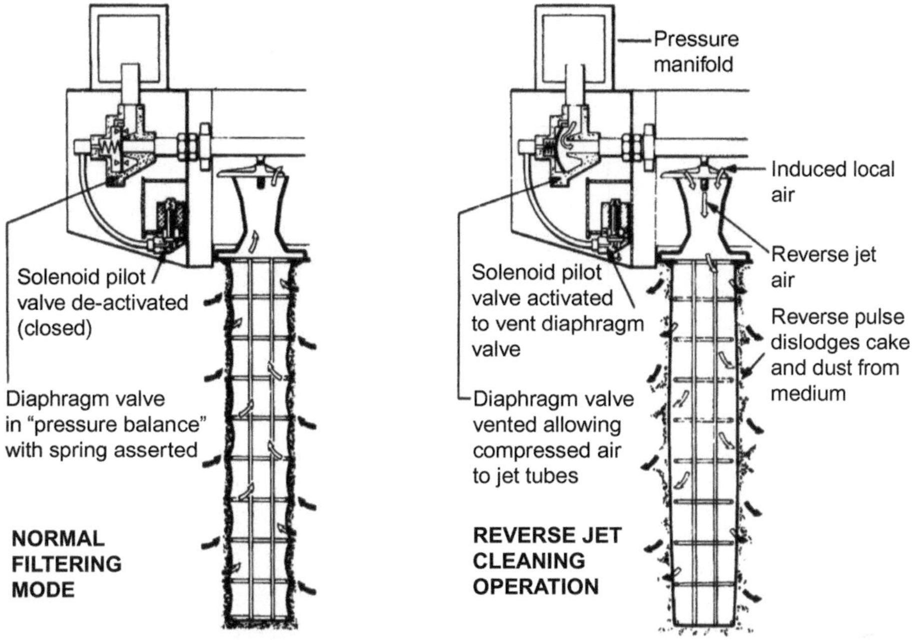


**Figure 8.4** Normal and Reverse Jet Cleaning Operation. *Reproduced with Permission of Elsevier Advanced Technology, Oxford, England, UK.*

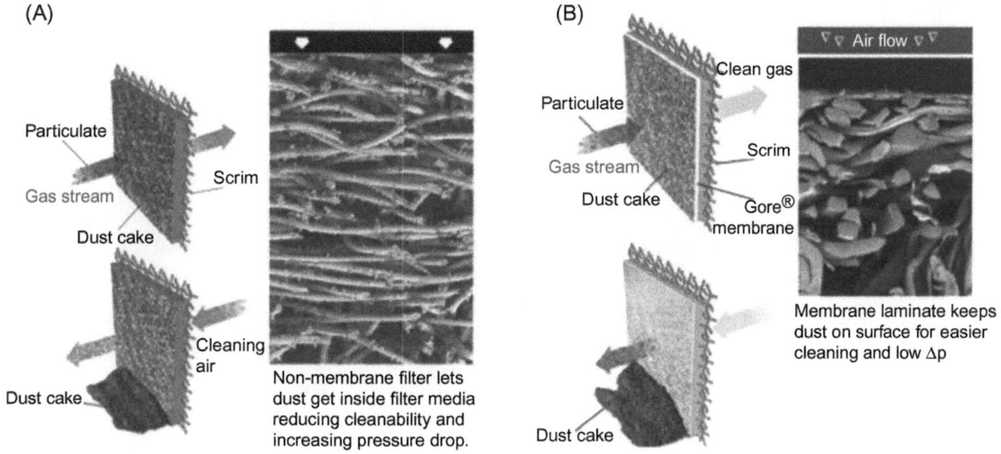


**Figure 8.5** Cake Release from Fibrous Felts—with and without Membrane Layer. *Reproduced with Permission of W.L. Gore & Associates, Newark, Delaware, USA.*

Note that both the Gore and Donaldson tables list temperature and chemical resistance as criteria in choosing the desirable medium. A variety of fibers are available for optimizing the choice of heat and chemical resistance. Table 8.6 lists temperature, physical resistance, and chemical resistance of a number of fibrous materials used in baghouse filter

**Table 8.2** GORE-TEX® Membrane—Felt Filter Bags

| GORE® membrane-needlefelt supported filter bags | Thickness mm (in) | Continuous operating temperature | Acid resistance | Alkali resistance |
|---|---|---|---|---|
| Polypropylene felt, 407 g/m$^2$ (12 oz/yd$^2$) | 2.1 (0.083) | 71 °C (160 °F) | Excellent | Excellent |
| Acrylic felt, 475 g/m$^2$ (14 oz/yd$^2$) | 3.0 (0.118) | 127 °C (260 °F) | Excellent | Good |
| Polyphenylene sulfide felt, 543 g/m$^2$ (16 oz/yd$^2$) | 2.2 (0.087) | 191 °C (375 °F) | Excellent | Excellent |
| Acid-resistant aramid felt, 475 g/m$^2$ (14 oz/yd$^2$) | 1.78 (0.070) | 204 °C (400 °F) | Very good | Good |
| Polyester felt, 407 g/m$^2$ (12 oz/yd$^2$) | 1.55 (0.061) | 135 °C (275 °F) | Fair | Fair |
| Polyester Felt, 543 g/m$^2$ (16 oz/yd$^2$) | 2.20 (0.087) | 135 °C (275 °F) | Fair | Fair |
| Antistatic polyester felt, 475 g/m$^2$ (14 oz/yd$^2$)—polyester ring spun scrim | 1.52 (0.060) | 135 °C (275 °F) | Fair | Fair |
| Polytetrafluoroethylene felt, 830 g/m$^2$ (24.5 oz/yd$^2$)—woven-expanded PTFE scrim | 1.07 (0.042) | 260 °C (500 °F) | Excellent | Excellent |
| Antistatic polytetrafluoroethylene felt 847 g/m$^2$ (24.5 oz/y oz/yd$^2$)—woven-expanded PTFE scrim | 1.24 (0.049) | 260 °C (500 °F) | Excellent | Excellent |

Reproduced with permission of W.L. Gore & Associates, Newark, Delaware, USA.

fabrics. Godfrey (275) published Table 8.7, a useful data guide on the various environmental resistance properties of polymers used in the fibers for dust collection filter media and systems for removal of obnoxious fume and vapors. Note that P84® is the most temperature- and heat-resistant polymer, whereas PTFE is the most chemically resistant.

The environmental and performance requirements for dust collection systems require continual and unique technical developments. For example the introduction of membrane technology is greatly affecting the performance of baghouse filter media. Another filter medium technology now being developed for dust collection systems is nanofiber webs. These are composite structures where the nanofiber web is coated on or combined with a substrate. One problem with these structures is that they can be designed for good cake release properties, however, they are often poorly adhered to the substrate and have a tendency to blow or peel off when reverse air cleaning is applied. Hollingsworth and Vose Company (H&V), East Walpole, Massachusetts,

**Table 8.3** Tetratex® Membrane Products—Felt Low Temperature

**Donaldson Tetratex® Membrane Products**
**Industrial filter applications**

**Reference guide**

**Felt low temperature**

| Product | Description | Width | Industry basis weight | | Thickness nominal | | Air permeability | | Continuous service temperature | |
|---|---|---|---|---|---|---|---|---|---|---|
| | | In (cm) | $g/m^2$ | oz/yd | mm | in | $l/min/dm^2$ @ 20 mm wg | $CFM/ft^2$ @ 0.5″ wg | °C | °F |
| 5212 | Tetratex® bonded to 16 oz. Self-supported polyester | 83 (211) | 540 | 16.0 | 2.00 | 0.080 | 33–58 | 7–12 | 135 | 275 |
| 6212 | Tetratex® bonded to 16 oz. Polyester | 83 (211) | 540 | 16.0 | 2.00 | 0.080 | 33–58 | 7–12 | 135 | 275 |
| 8212 | Tetratex® extreme bonded to16 oz. Polyester | 83 (211) | 540 | 16.0 | 2.00 | 0.080 | 48–86 | 10–18 | 135 | 275 |
| 8005 | Tetratex® high efficiency bonded to 16 oz. Polyester | 83 (211) | 540 | 16.0 | 2.00 | 0.080 | 16–34 | 3.5–7.0 | 135 | 275 |
| 6272 | Tetratex® bonded to 14 oz. Polyester epitropic | 83 (211) | 475 | 14.0 | 1.78 | 0.070 | 33–58 | 7–12 | 135 | 275 |
| 6222 | Tetratex® bonded to 15 oz. Polypropylene | 83 (211) | 510 | 15.0 | 2.29 | 0.090 | 33–58 | 7–12 | 90 | 190 |
| 6235 | Tetratex® bonded to 15 oz. Homopolymer acrylic | 83 (211) | 510 | 15.0 | 2.16 | 0.085 | 33–58 | 7–12 | 125 | 260 |
| 8135 | Tetratex® high efficiency bonded to 15 oz. Homopolymer acrylic | 83 (211) | 510 | 15.0 | 2.16 | 0.085 | 14–38 | 3–8 | 125 | 260 |

**Table 8.4** Tetratex® Membrane Products—Felt High Temperature

**Donaldson tetratex® membrane products**
**Industrial filter applications**

**Reference guide**

**Felt high temperature**

| Product | Description | Width | Industry basis weight | | Thickness nominal | | Air permeability (l/min/dm² @ 20 mm wg) | | Continuous service temperature | |
|---|---|---|---|---|---|---|---|---|---|---|
| | | In (cm) | g/m² | oz/yd | mm | in | (l/min/dm² @ 20 mm wg) | CFM/ft² @ 0.5″ wg | °C | °F |
| 5244 | Tetratex® bonded to 14 oz. Self-supported aramid | 83 (211) | 475 | 14 | 2.16 | 0.085 | 33–58 | 7–12 | 190 | 375 |
| 6244 | Tetratex® bonded to 14 oz. Aramid | 83 (211) | 475 | 14 | 2.16 | 0.085 | 33–58 | 7–12 | 190 | 375 |
| 8244 | Tetratex® extreme bonded to 14 oz. Aramid | 83 (211) | 475 | 14 | 2.16 | 0.085 | 38–98 | 12–20 | 190 | 375 |
| 8144 | Tetratex® high efficiency bonded to 14 oz. Aramid | 83 (211) | 475 | 14 | 2.16 | 0.085 | 19–48 | 4–10 | 190 | 375 |
| 6262 | Tetratex® bonded to 16 oz. PPS | 83 (211) | 540 | 16 | 1.91 | 0.075 | 33–58 | 7–12 | 190 | 375 |
| 8162 | Tetratex® high efficiency bonded to 16 oz. PPS | 83 (211) | 540 | 16 | 1.91 | 0.075 | 17–46 | 3.5–9.5 | 190 | 375 |
| 6232 | Tetratex® bonded to 14 oz. P84 | 83 (211) | 475 | 14 | 2.54 | 0.100 | 33–58 | 7–12 | 240 | 460 |

**Table 8.5** Tetratex® Membrane Products—Industrial Wet

**Donaldson tetratex® membrane products**
**Industrial filter applications**

**Reference guide**

**Industrial wet**

| Product | Description | Width | Industry basis weight | | Thickness nominal | | Air permeability | | Continuous service temperature | |
|---|---|---|---|---|---|---|---|---|---|---|
| | | In (cm) | ($g/m^2$) | lb/3000 $ft^2$ | mm | in | (l/min/$dm^2$ @ 20 mm wg) | CFM/$ft^2$ @ 0.5″ wg | °C | |
| 8771 | Tetratex® membrane (0.7 μm) bonded to 12.5 oz. Polypropylene felt | 42 (107) | 425 | 12.5 | 2.16 | 0.085 | Nominal 7 | Nominal 1.5 | 90 | |
| 8911 | Tetratex® high flow membrane (1 μm) bonded to 12.5 oz. Polypropylene felt | 83 (211) | 425 | 12.5 | 2.16 | 0.085 | Nominal 24 | Nominal 5 | 90 | 190 |
| 8016 | Tetratex® membrane (2 μm) bonded to 15 oz. Polypropylene felt | 83 (211) | 510 | 15.0 | 2.29 | 0.090 | Nominal 43 | Nominal 9 | 90 | °F<br>190 |

**Table 8.6** Properties of Fiber Materials Used in Filter Bags

| Generic name | Maximum temperature °C (°F) | | Physical resistance | | | | | Chemical resistance | | | | |
|---|---|---|---|---|---|---|---|---|---|---|---|---|
| | Continuous | Intermittent | Dry heat | Moist heat | Abrasion | Shaking | Flexing | Mineral acid | Organic acid | Alkalies | Oxidizing | Solvents |
| Cotton | 80 (180) | – | G | G | F | G | G | P | G | F | F | E |
| Polyester | 135 (275) | – | G | F | G | E | E | G | G | F | G | E |
| Acrylic | 235 (275) | 140 (285) | G | G | G | G | E | G | G | F | G | E |
| Modacrylic | 70 (160) | – | F | F | F | P – F | G | G | G | G | G | G |
| Nylon (polyamide) | 115 (240) | – | G | G | E | E | E | P | F | G | F | E |
| Nomex[a] | 205 (400) | 230 (450) | E | E | E | E | E | P – F | E | G | G | E |
| Polypropylene | 95 (200) | 120 (250) | G | F | E | E | G | E | E | E | G | G |
| PTFE (fluorocarbon) | 260 (500) | 290 (550) | E | E | P – F | G | G | E | E | E | E | E |
| Fluorocarbon | 230 (150) | – | E | E | P – F | G | G | E | E | E | E | E |
| Vinyon | 175 (350) | – | F | F | F | G | G | E | E | G | G | P |
| Glass | 290 (550) | 315 (600) | E | E | P | P | F | E | E | F | E | E |
| Wool | 100 (215) | 120 (250) | F | F | G | F | G | F | F | P | P | F |

E, excellent; F, fair; G, good; P, poor.
[a]Du Pont de Nemours International S.A. trademark.
Reproduced with permission of Elsevier Advanced Technology, Oxford, United Kingdom (282).

**Table 8.7** Environmental Resistance of Various Polymers

| Fiber type | | | Polypropylene | Polyester | Homopolymer acrylic | Aramid | Polyphenylene sulfide | Polyimide | PTFE | Spunbond polyester |
|---|---|---|---|---|---|---|---|---|---|---|
| **Trade name** | | | | | Dralon[a] | Nomex®[b] | Ryton® | P84® | | |
| Temperature °F (°C) | Continuous operating temperatures (dry)[c] | | 194 (90) | 275 (135) | 266 (130) | 374 (190) | 356 (180) | 356 (180) | 464 (240) | 140 (60) |
| | Media | Continuous dry | 212 (100) | 302 (150) | 284 (140) | 392 (200) | 374 (190) | 500 (260) | 482 (250) | 140 (60) |
| | | Continuous moist | 212 (100) | 212 (100) | 257 (125) | 356 (180) | 374 (190) | 356 (180) | 482 (250) | 140 (60) |
| | | Peak | 230 (110) | 356 (180) | 302 (150) | 482 (250) | 446 (230) | 572 (300) | 536 (280) | 176 (80) |
| | | Melting point or decomposition | 320 (160) | 482 (250) | 482 (250) | 752 (400) | 545 (285) | 842 (450)[c] | 554 (290) | 464 (240) |
| Supports combustion | | | YES | YES | YES | NO | NO | NO | NO | YES |
| Abrasion resistance | | | Very good | Excellent | Good | Excellent | Good | Very good | Moderate | Moderate |
| Resistance to hydrolysis | | | Excellent | Poor | Good | Good | Excellent | Very good | Excellent | Poor |
| Resistance to acids | | | Excellent | Good | Good | Very good | Excellent | Very good | Excellent | Very good |
| Resistance to alkalies | | | Excellent | Good | Very good | Very good | Very good | Very good | Excellent | Good |
| Resistance to oxidizing agents | | | Moderate | Very good | Very good | Very good | Poor | Very good | Excellent | Very good |
| Resistance to organic solvents | | | Good | Very good | Very good | Excellent | Excellent | Very good | Excellent | Very good |
| Relative cost vs polyester | | | ×2.0–2.5 | ×1 | ×2.5–3.0 | ×5 | ×7 | ×10 | ×25 | ×0.85 |

Note: Temperatures listed are in °F and (°C) in parenthesis.
[a]Dralon is the only true 100% homopolymer acrylics.
[b]Nomex is a registered trademark of E.I. DuPont de Nemours & Co., Inc.
[c]Begins to carbonize above this temperature.
Reproduced with permission of International Filtration News, Charlotte, North Carolina, USA.

USA has introduced Nanoweb® (276), a nanofiber coating technology that provides better adhesion to a substrate and therefore, more amenable to reverse air cleaning. H&V recommends that the Nanoweb® better adheres to a cellulose substrate.

Another example is a unique medium marketed by American Felt and Filter Company (AFFCO) in New Windsor, New York, USA called the Heat Shield". The "Heat Shield™" is a Nomex® needle punch felt reinforced with a woven basalt scrim. In addition to Nomex® the basalt reinforcing scrim can be used with other fibers such as PPS, P-84®, and Teflon®. It is being promoted for high temperature baghouse operations such as might be found in a hot mix asphalt plant. AFFCO (277) claims when a Nomex® felt was placed in a 1000 °F (523 °C) for 60 minutes it was "destroyed beyond recognition." On the other hand when the felt was reinforced with the basalt scrim there was "zero shrinkage."

At one time woven media were the principal fabric for baghouse filtration systems. Over the years needle felts have grabbed a share of the market position of woven medium. In more recent years spunlaced (hydroentangled) felts have encroached into the market position of needle punched felts. To illustrate, Donaldson Co., Inc., of Minneapolis has introduced Dura-Life™, a line of hydroentangled felts for baghouse operations. Donaldson reports twice the filter life (Fulko (278)). Israel et al. (207) provide comparisons of the hydroentangled medium to needle felts. Their data shows that the hydroentangled material with finer fibers gave a more uniform material with smaller pore size. This resulted in better surface loading and less depth loading, which in turn gives "better pulse cleaning = lower pressure drop = longer life."

### 8.1.2 Fume and vapor emissions

Federal regulation, good safety practice, worker protection, and protection of the environment demand that fumes and vapors from industrial operations be removed and separated from the exhaust air stream. Fumes and vapors can emerge from welding operations, paint spraying, machine tools (oil mists), chemical laboratories, etc. Dickenson (273) identifies two types of industrial fumes, "dry smoke" and "wet smoke." Dry smoke usually consists of solid particles in air such as that emitted by welding operations or certain types of machining operations. Wet smoke are aerosols such as oil mists. Aerosols are liquid droplets dispersed in air having a particle size range of 0.5–10 μm. The fumes of both dry and wet are removed from the workplace by air extraction devices such as hoods, canopied hoods, ventilated booths, ventilated enclosures, and slotted air extraction devices (also known as fish tails).

Once removed from the workplace, the fume particles have to be separated from the exhaust air stream. Separation devices include electrostatic and/or electrodynamic means, centrifugal impaction units, and paper- and bag-type filters. Nonwoven media

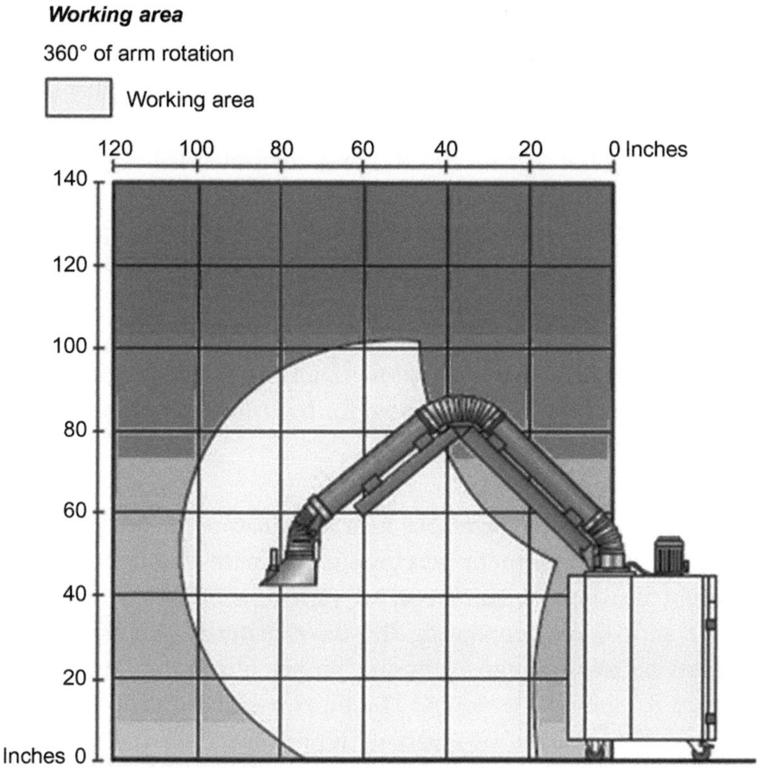


**Figure 8.6** Optiflo® WFE 150P Portable Weld Fume Collector. *Reproduced with Permission of AAF International, Louisville, Kentucky, USA.*

is certainly used in the paper and bag type filters. It is also sometimes used as the medium for prefilters to the other types of separation devices.

An example of fume filters are the Optiflo® weld fume filters (279) produced by AAF International, Louisville, Kentucky, USA. These are self-contained units that capture welding smoke and fumes at the welding site source. They consist of long duct arms that can be positioned over the localized work site and extract smoke fumes. This is illustrated in the schematic of Figure 8.6. The unit diagrammed is the Optiflo® WFE 150P. There is also an Optiflo® WFE 300P that contains two arms. The arms can be rotated 360 ° and extended to approximately 100 in. Two filters located in the main cabinet are high efficiency cartridge filters rated at 99.9% filtration efficiency. Activated carbon filters are also available.

In addition to workplace, protection gas cleaning devices are necessary to reduce or eliminate vapors and dust emissions from incinerators, power stations, cogeneration plants, and production facilities of the paper, cement, metallurgical, and other industries. This is to be in compliance with the environmental regulations of many countries. The systems include scrubbing systems, carbon adsorption, catalytic oxidizers, ceramic filters, and oxidizer systems. Some facilities use biofiltration, a technique to remove odorous and other obnoxious components from a gas stream by passing the stream through a medium that has been inoculated with microorganisms that break down the offending components.

The fumes are often hot (up to 900 °C) and corrosive. Fabric bag filters are often included as a prefilter to the cleaning system usually. According to Dickensen (273) "the development of heat and corrosion resistant fabric media, particularly modern treated needle felts, has been one of the most significant advances in the dust extraction and treatment of hot gases." Needle felts composed of polyamides, polypropylene, PTFE, aramids, glass, and mineral fibers have successfully been used in a wide range of hot gas filtration applications. Table 8.8 below (Dickenson (273) p. 676) lists the properties of various filter fabrics that might be used for hot fume filtration.

### 8.1.3 Air demisters

One form of air demisters are circular mist eliminators as depicted in Figure 8.7 below. The coalescing filter element uses glass media with fine fiber diameter distribution. The medium is a bed or thick mat of glass fiber, thick enough for a reduced amount of water and/or oil penetrating the filter material. Air velocity through the medium also must be low enough to ensure proper filtration. The circular demisters depicted in Figure 8.7 are upside down. The pipes sticking out of top of the filters are actually drain pipes for the flow of separated liquid. Installation would be at the bottom of the filters as shown in the diagram of a FLEXIFIBER® Mist Eliminator in Figure 8.8 below supplied by Koch-Glirtsch, LP, Wichita, Kansas, USA. The filters are mounted vertically so that coalesced mist droplets drain to the bottom of the filter housing where they are discharged. Ideally this drainage should be as a film falling along the discharge side (inside) of the bed; however, there is often gravity transport through the fiber structure itself.

### 8.1.4 Candle filters

Candle filters are a form of cylindrical filter cartridge that are mounted vertically. Very often there will be a multitude of these candles in an enclosure. They are generally used in air filtration applications. The two most common applications are ceramic, glass, or metal filters for hot gas filtration (up to 1,000 °C) and for air defogging operations (oil and/or moisture).

## 8.2 HEAT VENTILATION AND AIR CONDITIONING SYSTEMS

HVAC systems control air quality in residences and the work place. They also provide for clean air in sensitive work areas where dust or microbial contamination may be a problem. Examples are manufacturing of sensitive electronic components, pharmaceutical

## Table 8.8 Properties of Various Filter Fabrics

| Fabric | Nylon (polyamide) | PP (olefins) | Dacron (polyester) | Orlon® (acrylic copolymer) | Dralon® (acrylic homopolymer) | Nomex® (m-aramid) | Ryton® (poly-phenylene sulfide) | Glass | P84® | PTFE (polytetra-fluoroethylene) |
|---|---|---|---|---|---|---|---|---|---|---|
| Recomm. Usage | 195 °F | 195 °F | 265 °F | 230 °F | 260 °F | 390 °F | 355 °F | 500 °F | 500 °F | 500 °F |
| Temp.] °F]/] °F] | 90 °C | 290 °C | 130 °C | 110 °C | 125 °C | 200 °C | 180 °C | 260 °C | 260 ° | 260 °C |
| Short period | 250 °F | 220 °F | 295 °F | 250 °F | 285 °F | 430 °F | 375 °F | 555 °F | 500 °F | – |
| Excursions] °F]/] °F] | 120 °C | 105 °C | 145 °C | 120 °C | 140 °C | 220 °C | 190 °C | 290 °C | 260 °C | |
| | | | | | | | | | Amorphous | |
| Resistance against alkali | 2 | 1 | 2 | 2 | 1–2 | 1–2 | 1 | 1 | 1–2 | 1 |
| Mineral acids | 4 | 1 | 2 | 2 | 2 | 3 | 1 | 1–2 | 3 | 1 |
| Organic acids | 3 | 1 | 2 | 2 | 1–2 | 1–2 | 1 | 1 | 1–2 | 1 |
| Oxidizing agents | 3 | 2 | 2 | 2 | 2 | 4 | 4 | 1–2 | 2 | 1 |
| Organic solvents | 2 | 1 | 1–2 | 1–2 | 1–2 | 1–2 | 1–2 | 1–2 | 1–2 | 1 |
| Dry heat | 2 | 2 | 2 | 2 | 2 | 1–2 | 1–2 | 1–2 | 1 | 1 |
| Moisture | 2 | 3 | 3 | 2 | 1–2 | 1–2 | 1 | 1–2 | 2 | 1 |

1 excellent.
1–2 good.
2 very good.
3 fair.
4 poor.
Reproduced with permission of Elsevier Advanced Technology, Oxford, United Kingdom.

**Figure 8.7** Examples of Large Industrial Mist Eliminators.

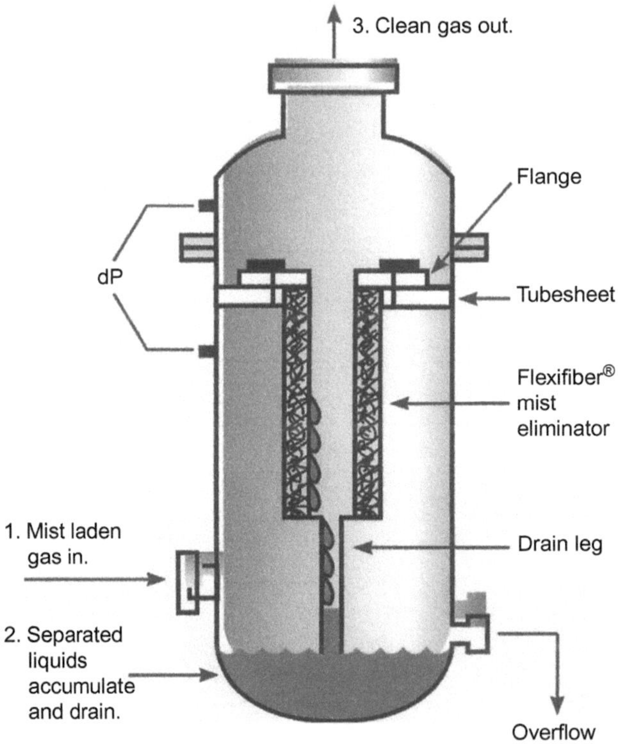


**Figure 8.8** FLEXIFIBER® Mist Eliminator. *Reproduced with Permission of Koch-Glirtsch, LP, Wichita, Kansas, USA.*

manufacture, and hospital operating rooms. Clean rooms using high efficient particulate air (HEPA) and ultra low penetration air (ULPA) filters are often designated for dust sensitive work areas. Residential ventilation filters usually do not require such high degrees of filtration efficiency. Likewise, routine workplace areas such as office locations usually do not justify such levels of filtration efficiency. However, there is consideration and concern for people with serious allergies and health problems that may need protection from the finer and smaller contaminants in the air environment. In recent times, an increasing concern about protection against CBR (chemical, biological, and radiation) agents has arisen.

## 8.2.1 Filters for air cleaning systems

### *8.2.1.1 Panel filters*

Square or rectangular panel filters are the most frequently used ones for air cleaning systems in the residences and in the work place. This does not include air freshener type filters that are becoming increasingly popular in homes and residences.

Panel filters can be as simple as mats of high loft media that a residence owner can cut from a roll of the medium and fit into his filter dimensions. Often referred to as cut-n-fit mats, these are also available in precut filter sizes. Usually the mats are a high loft medium composed of fiberglass or polyester. Some mat filters are composed of natural fibers. Hogs hair filters are often advertised as "natural fiber filters."

There are also more complicated minipleat panel filters that are most often used for high efficiency air filter applications (see Section 8.3.2.2).

Kinzer and Moreno (280) identified six main categories of in-duct air cleaning devices for residential air handling systems:

**a.** Fiberglass filters—These are glass mat filters composed of 15–60 μm glass fibers with a high porosity. These are the most common filters used in residential furnace and air conditioning systems. They are low cost. Their purpose is to protect the system from debris that might damage components of the air systems such as fans, motors, cooling coils, and heat exchangers. These filters are ¾ to 1 in (1.9–2.5 cm) thick and available in standard and special sizes up to 20 × 30 in (50 × 75 cm). Manufacturers recommend that these filters be changed once a month during seasons of normal use.

**b.** Pleated filters—Pleated filters are composed of nonwoven mats made from synthetic and/or natural fibers. Like the fiberglass filters they are highly permeable. The fibers most frequently used are cotton and polyester. The fibrous mat is the filter medium. It is usually supported by a wire frame in the form of V-shaped pleats. These filters are also ¾ to 1 in thick and come in a variety of sizes. The pleating increases the exposed surface area of the medium and extends filter life. The manufacturers recommend that these filters be replaced every two to three months during seasons of normal use.

**c.** Washable/reusable filters—These are panel filters that can be cleaned and reused. They are also highly permeable. The filter medium is usually a combination synthetic reticulated foam between layers of woven polyester or polypropylene mats. Another version consists of a layer of a nonwoven polyester sandwiched between layers of woven polyester or polypropylene mats. The layers are usually contained between metal or plastic grids. They are also very permeable. The thicknesses are ¾–1 in. Some filters have adjustable frames so that they can be cut to fit a variety of heating and cooling systems. Understandably, these filters are considerably more expensive than the fiberglass and pleated filters described above.

**d.** Electret filters—These filters use a filtering medium consisting of a mat of electrostatically charged fibers. In addition to simple mechanical particle capture, these filters can also trap particles by electrostatic attraction. The filtration efficiency is substantially improved compared to comparable uncharged filters and at no increase in resistance to air flow. The filters are ¾–1 in thick and come in a variety of sizes. They are disposable and priced higher than fiberglass and pleated filters but less than washable/reusable filters. Their recommended replacement life is three months during seasons of normal use.

**e.** Deep pleated filters—These filters are up to 5 in (13 cm) thick and usually require special support structure in the duct work. The medium is similar to that used in pleated filters discussed above. These filters are relatively expensive and pricing is in the range of washable/reusable filters. In some designs the frames are reusable and only the medium needs be replaced. Other designs have the filters made of rigid cardboard frames and the whole unit is disposable. The large surface area due to deep pleating greatly extends the filter life. Manufacturer's recommended replacement is once or twice a year.

**f.** Electronic air cleaners—Electronic air cleaners do not use a filter medium to collect particles. Rather, they use a high voltage ionization field to charge the particles with a negative or positive charge. The charged particles are then deposited on collector plates. These filters can be quite efficient as indicated below, however, they are quite expensive. The collector plates need to be cleaned periodically. The cleaners require regular maintenance every one to two months.

Kinzer and Moreno (280) tested 27 filters in a fractional efficiency procedure based on ASHRAE 52.2P, a drafted standard that later became ASHRAE 52.2-1999[2]. The 27 filters were divided into the six categories as discussed above. One of the results they presented was a comparison of the filtration efficiencies for 0.3–1.0 μm particles presented in Figure 8.9 below. Their data indicated that best performance comes from the electrostatic (electric) and electronic filters.

Following are some of the types of media used in indoor air filtration applications:

**a.** Fiberglass
**b.** High loft polyester (PET)
**c.** Cotton/polyester
**d.** Needle felt
**e.** Fibrillated film

[2] Since updated to ANSI/ASHRAE Standard 52.2-2012.

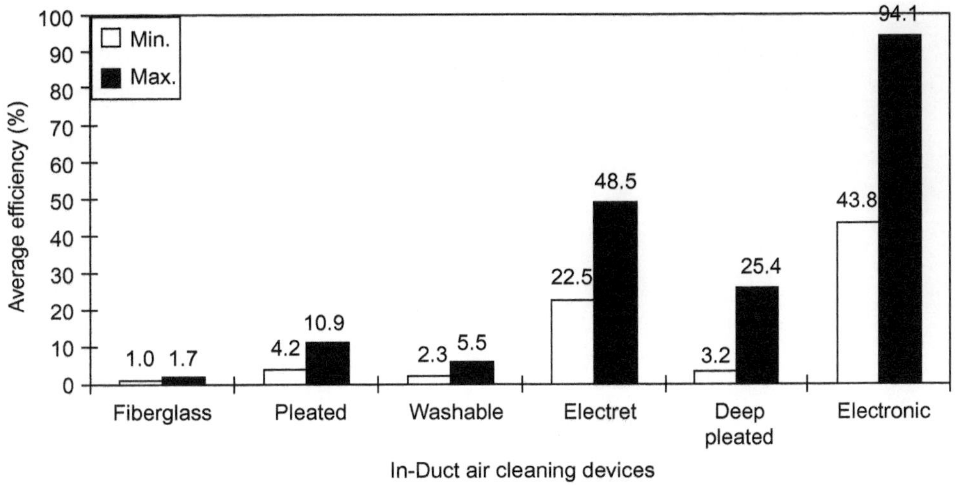


**Figure 8.9** Minimum and Maximum Efficiency Over the 0.3 to 1.0 Micron Particle Size Range for Each of the Six Residential In-Duct Air Cleaning Device Categories. *Reproduced with Permission of the American Filtration Society, Nashville, Tennessee, USA.*

**f.** Meltblown
**g.** Bicomponent spunbond
**h.** Mat rolls for cut-n-fit
**i.** Nanofiber media.

Composite structures, fillers, impregnants, and additives allow for filter medium design with more than one separation technology. It is quite common to combine mechanical filtration with adsorption, electrostatic effect, antimicrobial properties, and gradient density filtration. Other properties such as flame retardancy, water repellency, and color can be included and often are.

Gradient density media as related to liquid filtration have already been discussed in Chapter 7, Section 7.3.2.1.3. Martin (281) advances this discussion for mechano-electret gradient density structures used for indoor air quality (IAQ). He states that:

> *Not all mechano-electret filters are created equally. That is why it is important to look for an electret-treated medium filter that has depth-loading medium with a gradient density structure in which the medium's fibers are more loosely packed on the upstream side and more densely packed on the downstream side. This structure helps to reduce airflow resistance, enhance dust loading, and prevent face loading of the filter.*

### 8.2.1.2 Pocket filters

Pocket filters are one form of bag filters. They offer extended surface areas and increased dust holding capacity. Figure 8.10 below provided by Filtration Group, Joliet, Illinois, USA, shows how their pocket filter is constructed. The construction includes ultrasonic welding, separator strips, dual fasteners, and soft edge headers to

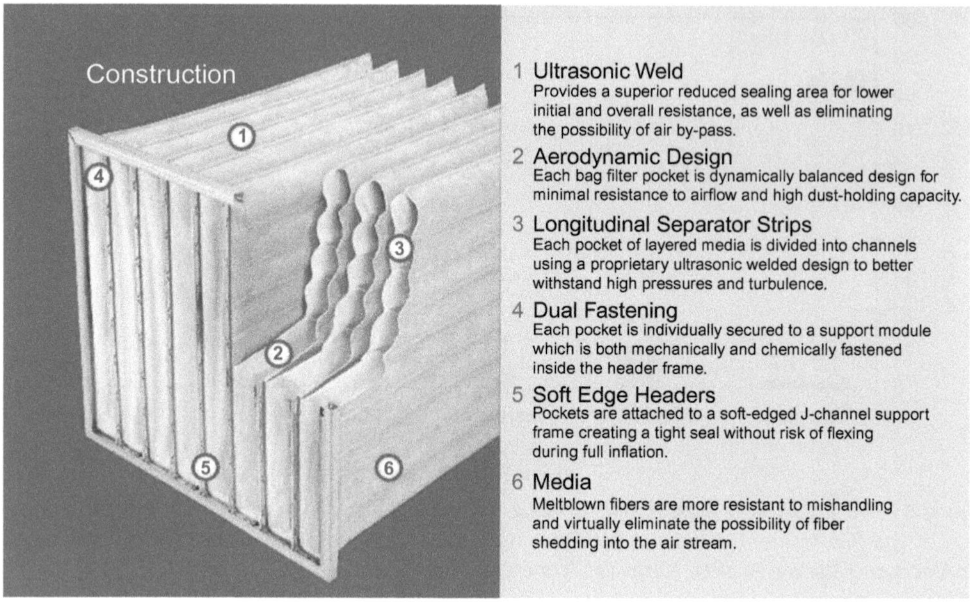


**Figure 8.10** Pocket Filter for A Ventilation System. *Reproduced with Permission of Filtration Group, Joliet, Illinois, USA.*

maintain pocket integrity and shape throughout the filtration cycle. The filter medium is a high bulk composite structure with a melt-blown fiber layer.

There are a number of manufacturers of pocket filters. Included are AAF International (Dri Pak), Air Filters Inc. (Flo-Pak™), Filtrair, Camfil, and Clarcor.

## 8.2.2 HVAC standards

In Europe, the primary standard for air filters is EN 779. This standard is based on air duct testing of filters and is controlled by CEN (Comité de Européen Normalisation), or in English, European Committee for Standardization. In the United States, the development of standards for ventilation filtration are being jointly sponsored by the United States Environmental Protection Agency and American Society of Heating, Refrigeration, and Air Conditioning Engineers (ASHRAE). The key standard is ANSI ASHRAE 52-2.

### *8.2.2.1 CEN EN779:2012 and Eurovent 4/11*

#### 8.2.2.1.1 EN779

EN779 (251), the European standard for HVAC filters, was revised in 2012. Changes in the standard were considerable in that it added the medium M class group and it added minimum efficiency standards for the F class ratings. In addition, it is connected to an energy efficiency certification system under the guidelines of Eurovent 4/11 (discussed in Section 8.2.2.1.2). This guideline includes an energy level classification label to verify Eurovent certification.

The EN779 ranking system ties to EN 1822:2009 for HEPA and ULPA filters discussed in Section 8.3. Referring to Table 8.9, EN779 rates nine classes of filters divided into three groups: G1–G4 for "coarse" filters, M5-M6 for "medium" filters, and F7–F9 for "fine" filters. Higher ratings for HEPA and ULPA filters are provided for in EN 1822. The G classes of filters are tested to a pressure drop of 250 Pa and are challenged with ASHRAE synthetic arrestance dust composed of Arizona road dust–72%, carbon black–23%, and cotton linters–5% (see Chapter 6, Section 6.14.4). The rating is based on the average arrestance, $A_m$, of the test dust. Arrestance is the weight fraction percentage of the test dust retained by the filter. The M and F classes of filters are tested to a pressure drop of 450 Pa. The rating is based on the average efficiency, $E_m$, of the test aerosol at 0.4 μm. The test aerosol is diethylhexylsebacate liquid (DEHS) or equivalent. DEHS has a particle size range of 0.2–10 μm.

Note that efficiency is not the same as arrestance. Efficiency is based on the number fraction of 0.4 μm particles retained on the filter. Arrestance is based on the weight percent of synthetic test dust retained by the filter. A tested filter is rated and classified by its arrestance or efficiency in accordance to Table 8.9. A new feature introduced in EN779:2012 is minimum efficiency (ME) values for the F Class ratings. ME is the lowest value of three different tests for 0.4 μm particles: initial efficiency, efficiency throughout the test's loading procedure, and efficiency after discharging. If a filter satisfies a rating based on average efficiency ($E_m$) it still might receive a lower rating based on ME. The purpose of ME is to prevent an unsatisfactory performance period of a filter during its filter test cycle, usually due to degradation of the filter as the test proceeds.

**Table 8.9** Classification of Air Filters According to EN779

| Group | Class | Final pressure drop test (Pa) | Average arrestance ($A_m$)[a] of synthetic dust % | Average efficiency ($E_m$) for 0.4 μm particles % | Minimum efficiency[b] for 0.4 μm particles % |
|---|---|---|---|---|---|
| Coarse | G1 | 250 | $50 \le A_m < 65$ | – | – |
| | G2 | 250 | $65 \le A_m < 80$ | – | – |
| | G3 | 250 | $80 \le A_m < 90$ | – | – |
| | G4 | 250 | $90 \le A_m$ | – | – |
| Medium | M5 | 450 | – | $40 \le E_m < 60$ | – |
| | M6 | 450 | – | $60 \le E_m < 80$ | – |
| Fine | F7 | 450 | – | $80 \le E_m < 90$ | 35 |
| | F8 | 450 | – | $90 \le E_m < 95$ | 55 |
| | F9 | 450 | – | $95 \le E_m$ | 70 |

[a]The characteristics of atmospheric dust vary widely in comparison with those of the synthetic loading dust used in the tests. Because of this, the test results do not provide a basis for predicting either operational performance or service life. Loss of media charge or shedding of particles or fibers can also adversely affect efficiency.
[b]Minimum efficiency is the lowest of any of the following three values: initial efficiency, discharged efficiency, or efficiency throughout the test's loading procedure.

#### 8.2.2.1.2 Eurovent 4/11

Associated with EN779:2012 is the Eurovent Energy Efficiency Classification system identified as Eurovent 4/11 (282). Introduced in 2011, these guidelines attempt to motivate filter manufacturers and correspondingly filter medium manufacturers to reduce energy costs associated with the filters. This is done by assigning an energy rating to the filter. AAF International (283) reports that "fans in air handling systems utilize 40% of the total energy requirements of a building. The air filters used in those systems consume 25% of their energy demand." It is evident that energy saving benefits can be accrued if the filters were less resistant to air flow. The rating requires EN779 testing and is limited to M and F class filters. The energy consumption W is determined from the dust loading portion of the test and is calculated from Eq. 8.1 below. The calculated value of W is compared to Table 8.10. For example, if the EN779 rating of a filter is F7 and W is calculated to be 1350 kWh then the energy rating in accordance with Table 8.10 is B. The energy ratings rank from A is best to F is worst[3].

$$W = \frac{q_v \cdot \overline{\Delta p} \cdot t}{\eta \cdot 1000} \tag{8.1}$$

where

$W$ is the energy consumption, kWh.

$q_v$ is the air flow, $m^3/s$.

$\overline{\Delta p}$ is the average pressure drop during the test life, Pa.

$t$ is the operation time, h.

$\eta$ is the fan efficiency, %.

In the test procedure $q_v$ is controlled at 0.944 $m^3/s$, $t$ is based on a 6000-h filter life, and $\eta$ is the fan efficiency arbitrarily set at 0.5 (50%). $\overline{\Delta p}$ is determined from a complex polynomial formula. It requires at least five measurements of pressure drop during the dust loading portion of the test. The curve developed is best fitted with a polynomial equation. The equation is then integrated and the result divided by minimum dust holding factor in Table 8.10 (250 g for M class filters ($M_m$), 100 g for F class filters ($M_f$))[4]. $\overline{\Delta p}$ is thus calculated.

If the filter testing is conducted by a qualified independent laboratory, the rating can be certified by Eurovent and a certification label can be placed on each filter package. The label used by AAF International is pictured in Figure 8.11. Note that in this case the label gives the AAF DriPak® product an F7 rating for EN779 performance and an A rating for energy performance.

[3] Originally there were seven energy levels (A–H) in Eurovent 4/11. In 2014, this was reduced to the five levels of Table 8.10.

[4] If the dust holding capacity in the EN779 test is less than the $M_m$ or $M_f$ value listed in Table 8.10, then the filter is regulated to a G4 classification.

**Table 8.10** Eurovent Energy Classification Ratings

| Filter class EN779:2012 | M5 | M6 | F7 | F8 | F9 |
|---|---|---|---|---|---|
| ME | – | – | ME ≥ 35% | ME ≥ 55% | ME ≥ 70% |
| Minimum dust holding | 250 g ASHRAE ($M_m$) | | | 100 g ASHRAE ($M_f$) | |
| A | 0–650 kWh | 0–800 kWh | 0–1200 kWh | 0–1600 kWh | 0–2000 kwh |
| B | >650–950 kWh | >800–1400 kWh | >1200–1700 kWh | >1600–2300 kWh | >2000–3000 kWh |
| C | >950–1250 kWh | >100–1400 kWh | >1700–2200 kwh | >2300–3000 kWh | >3000–3700 kWh |
| D | >1250–1550 k kWh | >1400–1700 kWh | >2300–3700 kWh | >3000–3700 kWh | >4000–5000 kWh |
| F | >1550 kWh | >1700 kWh | >2300 kWh | >3700 kWh | >5000 kWh |

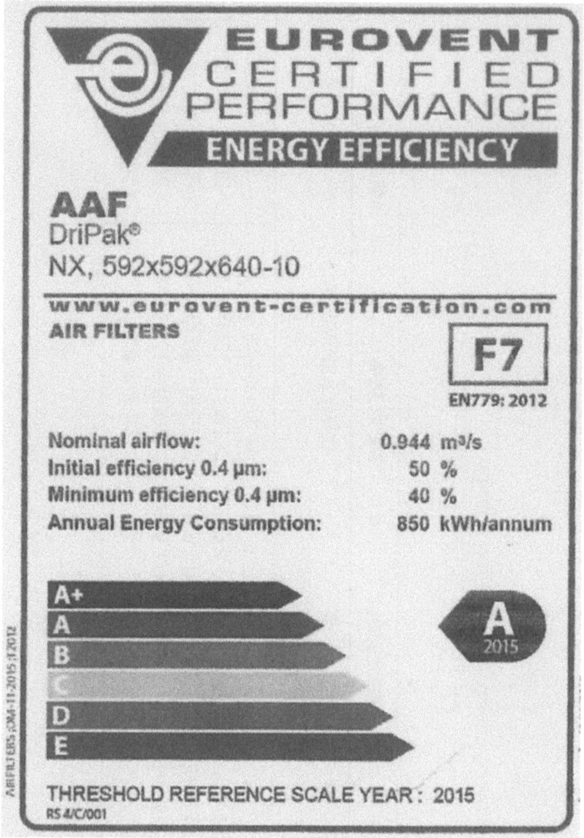


**Figure 8.11** Eurovent Energy Efficiency Label. *Reproduced with Permission of AAF International, Louisville, Kentucky, USA.*

### *8.2.2.2 ANSI/ASHRAE 52.2-2012*

#### 8.2.2.2.1 MERV ratings

ASHRAE takes a different approach than Eurovent. ANSI/ASHRAE 52.2-2012 (250) defines a minimum efficiency reporting value (MERV). This is determined by subjecting a filter to testing in accordance with test procedures defined by ASHRAE 52.2-2012. A filter is assigned a rating based on its performance in the tests. There are 20 possible MERV ratings listed in ASHRAE 52.2: MERV 1 being the least efficient and MERV 20 being the most efficient. MERV ratings 1–4, for the least efficient filters, are determined by the ASHRAE arrestance test specified in ASHRAE 52.2. MERV ratings 16–20 for high efficiency HEPA and ULPA are not really part of the test protocol of ASHRAE, but rather are rated based on the recommended practices (RPs) of IEST (Institute of Environmental Science and Technology) (see Section 8.3.5.3). MERV ratings 5–16 are for medium efficiency filters and are determined by a fractional efficiency test protocol in accordance with ASHRAE 52.2. Table 8.11 is an application guide for

**Table 8.11** Application Guidelines for ASHRAE 52.2-2012 MERV Ratings

| Std. 52.2 minimum efficiency reporting value (MERV) | Approx. Std. 52.1 results | | Application guidelines | | |
|---|---|---|---|---|---|
| | Duct spot efficiency | Arrestance | Typical controlled contaminant | Typical applications and limitations | Typical air filter/cleaner type |
| 20 | n/a | n/a | **≤ 0.30 μm Particle size** | Clean rooms | **HEPA/ULPA filters** |
| 19 | n/a | n/a | Virus (unattached) | Radioactive materials | ≥99.999% efficiency on 0.10–0.20 μm particles, IEST Type F |
| 18 | n/a | n/a | Carbon dust | Pharmaceutical manufacturing | ≥99.999% efficiency on 0.30 μm particles, IEST Type D |
| 17 | n/a | n/a | Sea salt | Carcinogenic materials | ≥99.99% efficiency on 0.30 μm particles, IEST Type C |
| | | | All combustion smoke | Orthopedic surgery | ≥99.97% efficiency on 0.30 μm particles, IEST Type A |
| | | | Radon progeny | | |
| 16 | n/a | n/a | **0.30–1.0 μm particle size** | Hospital inpatient care | **Bag filters** nonsupported (flexible) microfine fiberglass or synthetic media. 300–900 mm (12–36 in) deep, 6–12 pockets |
| 15 | > 95% | n/a | All bacteria | General surgery | **Box filters** rigid style cartridge filters 150–300 mm (6–12 in) deep may use lofted (air laid) or paper (wet laid) media |
| 14 | 90–95% | > 98% | Most tobacco smoke | Smoking lounges | |
| 13 | 80–90% | > 98% | Droplet nuclei (sneeze) | Superior commercial buildings | |
| | | | Cooking oil | | |
| | | | Most smoke | | |
| | | | Insecticide dust | | |
| | | | Copier toner | | |
| | | | Most face powder | | |
| | | | Most paint pigments | | |

*(Continued)*

Table 8.11 (Continued)

| Std. 52.2 minimum efficiency reporting value (MERV) | Approx. Std. 52.1 results | | Application guidelines | | |
|---|---|---|---|---|---|
| | Duct spot efficiency | Arrestance | Typical controlled contaminant | Typical applications and limitations | Typical air filter/cleaner type |
| 12<br>11<br>10<br>9 | 70–75%<br>60–65%<br>50–55%<br>40–45% | >95%<br>>95%<br>>95%<br>>90% | **1.0–3.0 μm particle size**<br>Legionella<br>Humidifier dust<br>Lead dust<br>Milled flour<br>Coal dust<br>Auto emissions<br>Nebulizer drops<br>Welding fumes | Superior residential<br>Better commercial buildings<br>Hospital laboratories | **Bag filters** Nonsupported (flexible) microfine fiberglass or synthetic media. 300–900 mm (12–36 in) deep, 6–12 pockets<br>**Box filters** rigid style cartridge filters 150–300 mm (6–12 in) deep may use lofted (air laid) or paper (wet laid) media |
| 8<br>7<br>6<br>5 | 30–35%<br>25–30%<br><20%<br><20% | >90%<br>>90%<br>85–90%<br>80–85% | **3.0–10.0 μm particle size**<br>Mold<br>Spores<br>Hair spray<br>Fabric protector<br>Dusting aids<br>Cement dust<br>Pudding mix<br>Snuff<br>Powdered milk | Commercial buildings<br>Better residential<br>Industrial workplaces<br>Paint booth inlet air | **Pleated filters** disposable, extended surface, 25–125 mm (1–5 in) thick with cotton–polyester blend media, cardboard frame<br>**Cartridge filters** graded density viscous-coated cube or pocket filters, synthetic media<br>**Throwaway** disposable synthetic media panel filters |
| 4<br>3<br>2<br>1 | <20%<br><20%<br><20%<br><20% | 75–80%<br>70–75%<br>65–70%<br><65% | **>10.0 μm particle size**<br>Pollen<br>Spanish moss<br>Dust mites<br>Sanding dust<br>Spray paint dust<br>Textile fibers<br>Carpet fibers | Minimum filtration<br>Residential<br>Window air conditioners | **Throwaway** disposable fiberglass or synthetic panel filters<br>**Washable** Aluminum mesh, latex-coated animal hair, or foam rubber panel filters<br>**Electrostatic** self-charging (passive) woven polycarbonate panel filter |

Note: A MERV for other than HEPA/ULPA filters also includes a test airflow rate, but it is not shown here because it has no significance for the purposes of this table.
Reproduced with permission of ASHRAE, Atlanta, Georgia, USA.

**Table 8.12** Particle Size Ranges and Groupings of ASHRAE 52.2

| Range | Lower limit, μm | Upper limit, μm | Average minimum PSE[a] designator group |
|---|---|---|---|
| 1 | 0.30 | 0.40 | E1 |
| 2 | 0.40 | 0.55 | |
| 3 | 0.55 | 0.70 | |
| 4 | 0.70 | 1.00 | |
| 5 | 1.00 | 1.30 | E2 |
| 6 | 1.30 | 1.60 | |
| 7 | 1.60 | 2.20 | |
| 8 | 2.20 | 3.00 | |
| 9 | 3.00 | 4.00 | E3 |
| 10 | 4.00 | 5.50 | |
| 11 | 5.50 | 7.00 | |
| 12 | 7.00 | 10.00 | |

[a]Particle size removal efficiency.

the various MERV ratings. Note that Table 8.11 published in ASHRAE 52.2-2012 lists dust spot efficiency and arrestance in accordance with ASHRAE 52.1. ASHRAE 52.1 has been retired and the dust spot efficiency test thus made obsolete. The arrestance test is now incorporated into ASHRAE 52.2-2012.

The fractional efficiency test for the MERV is a filtration test where by the filter medium is evaluated over a range of particle sizes. Twelve ASHRAE 52.2 particle size ranges are listed in Table 8.12. They are organized into the three particle size removal efficiency (PSE) groups shown in the last column of Table 8.12: E1 (0.3–1.0 μm), E2 (1.0–3.0 μm), and E3 (3.0–10.0 μm). The test duct for the ASHRAE 52.2-2012 procedure is depicted in Figure 8.12. It contains an aerosol generator, a dust feeder, and manometers to measure pressure drop across the filters. In addition to the test filter, the test duct also has a location for installing an absolute filter downstream of the test filter. The absolute filter is used for the dust holding capacity and arrestance part of the test.

In the test procedure, a filter is challenged with a laboratory-generated aerosol of potassium chloride (KCl). Particle size counters measure upstream and downstream counts for each of the 12 particle size ranges. Fractional efficiency of each particle size range is thereby determined.

The test begins when air flow through the test duct is initiated and reaches the prescribed flow rate. The initial KCl efficiency test is then conducted. The filter is then loaded with an increment of synthetic test dust and the KCl efficiency test is repeated. Each additional incremental loading of the synthetic dust is followed by another KCl efficiency test and the determination of KCl efficiency for each particle size range. The three KCl efficiencies of each PSE group (E1, E2, and E3) are averaged. During

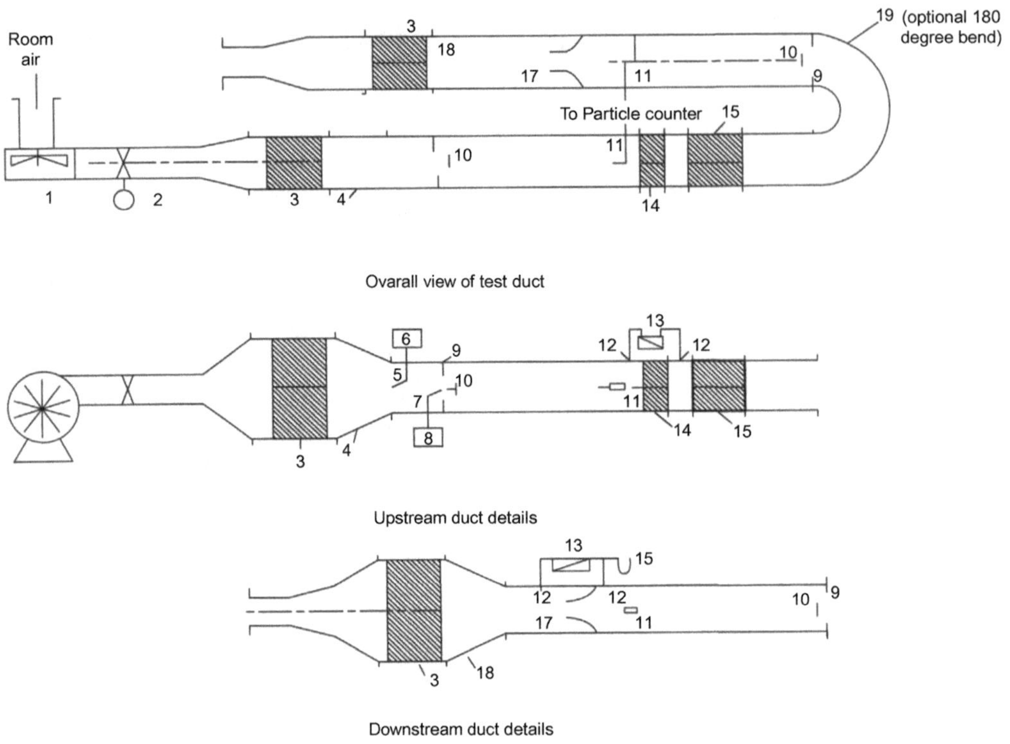


1. Blower
2. Flow control valve
3. HEPA filter bank
4. Transition, if required
5. Aerosol injection tube
6. Aerosol generator
7. Dust feed pipe
8. Dust feeder
9. Mixing orifice
10. Perforated diffusion plate
11. Location of sample probe
12. Static tap
13. Manometer
14. Air cleaner device and transition, if any
15. Final filter (installed only during dust loading)
16. 16 Vertical manometer
17. Main flow measurement nozzle
18. Transition, if needed
19. Bend, optional

**Figure 8.12** Schematic Diagram of ASHRAE 52.2 Test Duct. *Reproduced with Permission of ASHRAE, Atlanta Georgia, USA.*

the course of the test there will be a measurement for each particle size range that gives the lowest or minimum efficiency. The three minimum efficiencies for each PSE group are averaged and reported as E1, E2, and E3 accordingly. These are the numbers that determine the MERV rating from the criteria of Table 8.13.

To illustrate, presume that test determined minimum efficiency values are E1 = 33.5%, E2 = 66.3%, and E3 = 86.8%. Referring to the E1 column of Table 8.13, E1 at 33.5% fits the criteria of E1 < 75, therefore its MERV rating is 13 or lower pending the results of the other columns. In the E2, 1–3 μm column, E2 at 66.3% satisfies the criteria (65 ≤ E2 < 80) for a MERV rating of 11. In the E3,

**Table 8.13** Minimum Efficiency Reporting Values Parameters

| Minimum efficiency reporting value (MERV) | Particle size efficiency (%) | | | Average arrestance efficiency (%) |
|---|---|---|---|---|
| | E1 0.30–1.0 μm | E2 1.0–to 3.0 μm | E3 3.0–10.0 μm | |
| 1 | n/a | n/a | E < 20 | <65 |
| 2 | n/a | n/a | E < 20 | 65 ≤ E < 70 |
| 3 | n/a | n/a | E < 20 | 70 ≤ E < 75 |
| 4 | n/a | n/a | E < 20 | 75 ≤ E < 80 |
| 5 | n/a | n/a | 20 ≤ E < 35 | n/a |
| 6 | n/a | n/a | 35 ≤ E < 50 | n/a |
| 7 | n/a | n/a | 50 ≤ E < 70 | n/a |
| 8 | n/a | n/a | 70 ≤ E | n/a |
| 9 | n/a | E < 50 | 85 ≤ E | n/a |
| 10 | n/a | 50 ≤ E <65 | 85 ≤ E | n/a |
| 11 | n/a | 65 ≤ E <80 | 85 ≤ E | n/a |
| 12 | n/a | 80 ≤ E | 90 ≤ E | n/a |
| 13 | E < 75 | 90 ≤ E | 90 ≤ E | n/a |
| 14 | 75 ≤ E < 85 | 90 ≤ E | 90 ≤ E | n/a |
| 15 | 85 ≤ E < 95 | 90 ≤ E | 90 ≤ E | n/a |
| 16 | 95 ≤ E | 90 ≤ E | 95 ≤ E | n/a |
| 17 | 99.97 < E IEST Type A | n/a | n/a | n/a |
| 18 | 99.99 < E IEST Type C | n/a | n/a | n/a |
| 19 | 99.999 < E IEST Type D | n/a | n/a | n/a |
| 20 | 99.999 < E IEST Type F | n/a | n/a | n/a |

3–10 μm, column, E3 at 86.8% is also consistent with the MERV rating of 11 (85 ≤ E3). The highest rating that satisfies the criteria in all three columns is the one assigned to the filter. For the filter of this example, the rating is eleven and the filter is identified as a MERV 11 filter.

It should be noted that when a MERV rating is reported, the flow rate is also reported. In the hypothetical example above, let us presume that the flow rate was 492 ft/min (2.50 m/s), one of the standard flow rates. The filter is then reported as a MERV 11 @492 ft/min or MERV 11 @ 2.50 m/s. It should be also noted that the test requires a minimum resistance after the final incremental loading of the synthetic test dust.

### 8.2.2.2.2 ASHRAE arrestance test and dirt holding capacity

**8.2.2.2.2.1 Arrestance** The ASHRAE arrestance test previously specified by ASHRAE 52.1:1992 is now incorporated into ASHRAE 52.2:2012. It is intended to provide a way to assign a MERV rating to low efficiency filters that do not test well under fractional efficiency type testing. Arrestance and dust holding capacity are determined as part of the ASHRAE 52.2 test protocol. The test challenges the test filter with a prescribed "ASHRAE synthetic test dust" and utilizes an absolute filter to catch all the test dust that penetrates the test filter, other than that which deposited in the test duct between the test filter and the final filter. After each increment of dust fed, the dust deposited in the test duct is collected and weighed. This amount is added to the weight increase of the absolute filter to determine the amount of dust penetrating the filter. This amount, $W_d$, is divided by the increment of dust fed, $W_u$, and subtracted from one to determine the incremental arrestance, $A_i$.

$$A_i = 100\left(1 - \frac{W_d}{W_u}\right) \tag{8.2}$$

where

$W_d$ is the incremental weight of dust passing through or penetrating the filter (g).

$W_u$ is the incremental weight of dust fed (g).

$A_i$ is the incremental arrestance (%).

Note that $A_i$ is the incremental arrestance. The average arrestance, $A_{avg}$, is determined by averaging out the incremental arrestance measurements in accordance with Eq. 8.2.

$$A_{avg} = \left[1/W\sum_{i=1}^{f}[W_iA_i]\right] \tag{8.3}$$

where

$A_{avg}$ is the average arrestance for the life of the test (%).

$W$ is the total dust fed during the life of the test (g).

$W_i$ is the amount of dust fed during the increment $i$ (g).

$A_i$ is the arrestance during the increment $i$ (g).

$f$ is the final increment.

**8.2.2.2.2.2 Dirt holding capacity** For each dust loading increment, $i$, the weight of dust retained by the filter, $DHC_i$ is determined:

$$DHC_i = W_u - W_d \tag{8.4}$$

The total dust holding capacity for the filter life is then determined by summing up all the incremental $DHC_i$ values.

$$DHC = \sum_{i=1}^{f} DHC_i \tag{8.5}$$

where $DHC$ is the total dust holding capacity for the life of the test (g).

Note that the dust holding capacity is based on the weight of test dust retained by the filter. It does not include the dust that penetrated or passed through the filter.

#### 8.2.2.3 Fire and flame resistant standards

Test methods for media are discussed in Chapter 6, Section 6.12. Standards for fire or flame resistance of air filters are established by UL 900. This is a standard of Underwriters Laboratories, Northbrook, Illinois, USA. UL 900 The standard classifies two types of air filters as follows:

1. Class 1 air filters—Those that, when clean, do not contribute fuel when attacked by flame and emit only negligible amounts of smoke.
2. Class 2 Air Filters—Those that, when clean, burn moderately when attacked by flame or emit moderate amounts of smoke or both.

The test involves a wind tunnel in which the filter is installed. A flame is applied to the filter medium and the area as measured below the smoke density time curve is determined. If the area is less than 16.1 $cm^2$ (2.5 $in^2$), the filter may qualify for a Class 1 air filter rating. If it is less than 58 $cm^2$ (9 $in^2$), it may qualify for a Class 2 filter rating.

In addition, the filters are judged by how well they sustain flame. A Class 1 air filter rating requires that when subjected to the UL Spot Flame Test, the upstream face shall not continue to flame after removal of the test flame. The Class 2 air filter rating requires that the filter shall not produce flame or extended sparks (25 or more) which are sustained beyond the end of the test duct when subjected to the flame-exposure test.

ASME AG-1, Section FC to be discussed in Section 8.3.5.2 below, has provisions for HEPA filter resistance to heated air (paragraph FC-5150) and resistance to the spot flame test (paragraph FC-5160). The exposure to heated air is conducted in a test chamber at a rated flow of heated air at 3700 °C $\pm$ 250 °C (7000 °F $\pm$ 500 °F) for no less than five minutes. After exposure the filter is tested for aerosol penetrations (not to exceed 3%). In the flame test a Bunsen burner flame is directed at several points on the upstream face of the filter and applied for five minutes. There is to be no sustained flaming on the downstream face of the unit after removal of the flame. The code states that an Underwriters' Laboratories label (UL 586) is acceptable evidence of compliance with paragraphs FC5150 and FC 5160.

Another standard pertinent to air filters is NFPA 90A, "Standard for the Installation of Air-Conditioning and Ventilation Systems." This standard which is applied to HVAC systems in general has provisions governing the filters in these systems.

## 8.3 HIGH EFFICIENCY AIR FILTRATION

### 8.3.1 A brief history of high efficiency air filtration

The early developments of high efficiency air filters were designed to remove aerosol particles in the submicron range began with military gas mask canisters designed to protect military personnel from the toxic smokes of chemical and biological warfare. The earliest forms of the medium were developed by Germany and contained asbestos fiber mixed with esparto pulp. During World War II, the United States Army Chemical Warfare Service (CWS) in conjunction with the United States Naval Research Laboratory reproduced the German medium and had it manufactured by Hollingsworth and Vose Company in East Walpole, Massachusetts. The gas mask media produced for the US Army was identified as CWS type 6.

The National Defense Research Council (NDRC) solicited the assistance of university and industrial scientists to develop better smoke filters. The theoretical work of Irving Langmuir has already been discussed in Chapter 2. LaMer and Sinclair (284) developed a filter test based on dioctyl phthalate (DOP) smoke particles. The fundamentals of this test are still in practice today, however, the test stands for this type of test have been considerably modernized.

After World War II the development of filters to protect against chemical warfare expanded to large volume filters to protect operational headquarters where wearing of a gas mask was impractical. The developments of nuclear processes and materials at this time created a natural tie-in because of the necessity of controlling radioactive particles in air. The acronym, HEPA, was coined by Humphrey Gilbert from a 1961 Atomic Energy Commission report "High-Efficiency Particulate Air Filter Units, Inspection, Handling, Installation" (285).

In the 1950s, The US government sponsored development of new materials to replace the CWS medium. The concern was that the esparto and asbestos then used in the medium was imported from foreign countries and left the government vulnerable to shortages in these materials. This concern was fortunate, not because of the availability of the materials, but because afterwards the health effects of asbestos demanded that this fiber not be used. The replacement developed by Arthur D. Little, Inc. was a filter containing a filter medium composed of glass microfibers. Media composed of glass microfibers are still mainstream for the high efficiency air filters of today.

In the mid-1950s, media composed of Fiberfax® fibers were produced by Hollingsworth and Vose Company, Hurlbut Paper Company, and Flanders Filters. Fiberfax® was a ceramic fiber produced by Carborundum Corporation. The medium had excellent heat resistance in excess of 1000 °C. However, the fibers could not be produced with the fine diameters required for the high efficiency filtration requirements.

Arthur D. Little started the first high efficiency filter manufacturing company, the Cambridge Filter Company[5], in the early 1950s. By 1957, three firms were in the business. Early quality difficulties led to a quality assumption inspection program and the establishment of a Government-Industry Filter Committee. Military standards were established. These included MIL-F- 51068, "Filter, Particulates, High-Efficiency, Fire Resistant" (286) and MIL-F-51079, "Filter Medium, Fire Resistant, High Efficiency" (287). MIL-F-51068 was concerned with filters and MIL-F-51069 was concerned with the medium. The main purpose of these standards was to define filter media and filters to be used as protection against radiation leakage from nuclear facilities. These standards were active until 1994 when they were incorporated into ASME AG-1, Section FC (241) which is administered by the Committee on Nuclear Air and Gas Treatment (CONAGT). For many years, the military standards were the world-wide guide for HEPA filters and HEPA media. This was despite the fact that HEPA filters were finding more and more use in non-nuclear and non-military applications.

In the 1960s HEPA filters were finding clean room applications in the electronics industry, in the pharmaceutical industries, and in hospital operating rooms. More recently HEPA filters have become popular for protecting people against allergens, disease and harmful microbes in the workplace, in residences, in office buildings, and even in automobiles. Modern day vacuum cleaners room air purifiers often contain HEPA filters to collect the fine particles of annoying dust that leak through the filter bag. Mechanical room air purifiers often use HEPA filters; sometimes by themselves and sometimes in combination with an electronic separation.

In some applications the demand for filtration efficiency was even more restrictive than the 99.97% efficiency (0.3 μm DOP smoke) required of HEPA filters. Higher efficiency filters were first produced in 1961 and had an efficiency of 99.999%. These filters latter became known as ULPA (Ultra Low Penetration Air) filters.

In many cases, the standards were becoming too restrictive. That the heat and radiation resistance requirements of the MIL standards were limiting HEPA and ULPA media to glass microfiber construction. On the other hand, in some applications the specifications were unnecessary. For example, it would be indeed an unusual situation where a vacuum filter or air purifier would have to meet the radiation resistance requirements of MIL-F-51068 and MIL-F-51079. The same would be true of clean room filters for hospital operating rooms and pharmaceutical manufacture.

[5] The successor to this company is Camfil USA, which is now part of the Camfil Group headquartered in Stockholm, Sweden.

New standards needed to be developed. One result, EN1822-1, first introduced in 1998 (now updated to EN1822:2109 (32)) was developed by the Comité de Européen Normalisation and is discussed in Section 8.3.5.1 below. In addition IEST established RPs for the testing of HEPA and ULPA filters (see Section 8.3.5.3 below). ASHRA 52.2 standards for HVAC filters include MERV ratings for HEPA and ULPA filters.

The development of standards for applications other than military and nuclear applications allowed for the development of non-glass or partial glass high efficiency filter media. Composite structures containing nanofiber or membrane layers began finding a niche in high efficiency air filtration. This trend is continuing as this book is being written.

In addition, the global commitment to cleaner energy and less energy usage is motivating filter and filter media manufactures to develop media and filters with less resistance to air flow and thereby reduced energy usage. The certification program of Eurovent 4/11, discussed in Section 8.2.2.1.2 above is contributing to this motivation.

## 8.3.2 HEPA filters

### *8.3.2.1 Traditional panel filters*

The original HEPA filter design was a square or rectangular panel filter utilizing the medium in a deep pleated form. The traditional design of a HEPA filter is as depicted in Figure 8.13. The medium is a pleated microglass fiber fabric using corrugated metal

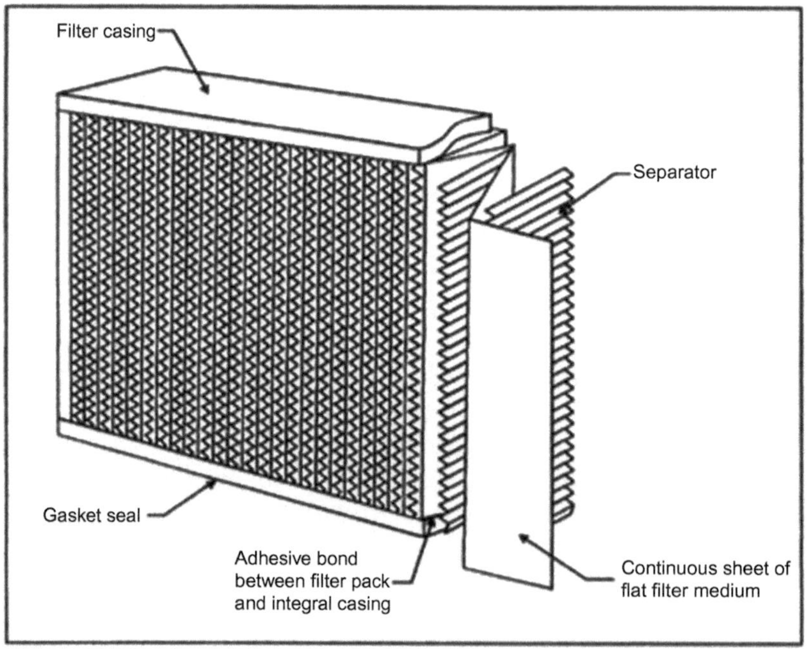


**Figure 8.13** Traditional HEPA Panel Filter.

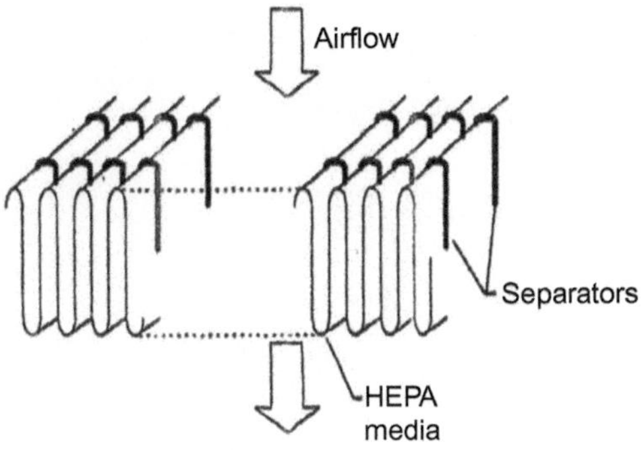


**Figure 8.14** String Separators for HEPA Filters. *Reproduced with Permission of NAFA, Virginia Beach, Virginia, USA.*

separators to keep the pleats apart and not pinching. Leakage by-pass around the medium is prevented by gasket seals and by adhesive bond between the filter pack and the integral casing.

There have been significant improvements in the design of panel HEPA filters. One of them is the replacement of the corrugated separator with other separator techniques. The purpose is to reduce the spacing between pleat folds and allow more pleats, more surface area, and more air flow. The separator techniques include string or ribbon separators (see Figure 8.14), hot melt beads, hot glue beads, and medium embossments (see Chapter 3, Figure 3.10).

New filter designs have been developed. They include mini-pleat, V-pack design, and round HEPA filters.

#### 8.3.2.2 Mini-pleat V-pack

Deep pleated panel filters, with or without separators, are based on a 6–12 in pleat design. Mini-pleat filters utilize pleat folds that are 7/8–1¼ in (22–32 mm). Abutting folds are separated by threads, ribbons, tapes, strips of medium, or continuous beads of glass, foam, or plastic spaced across the width of the medium. By organizing the panels into a "V"-array of sub-components (V-pack filter), as depicted in Figure 8.15 below, greater surface area of medium can be incorporated into the panel volume. This allows for higher flow volume, lower pressure drop, and greater filter life (dust holding capacity). The mini-pleat filters pictured in Figure 8.15 were produced by TDC Filters, Bolingbrook, Illinois, USA.

#### 8.3.2.3 Round HEPA filters

Round HEPA filters, often referred to as radial filters, are finding utility in air vacuum cleaning systems and in room air purifiers. In addition they are becoming prevalent in nuclear facilities. ASME recently added AG-1:2012 to Section FK of the "Code on Nuclear Air and Gas Treatment." ASME AG-1:2012 provides requirements for round

**Figure 8.15** HEPA Filters V-Pack Mini-Pleat. *Reproduced with Permission of Midwesco Filter Resources/TDC Filter, Bolingbrook, Illinois, USA.*

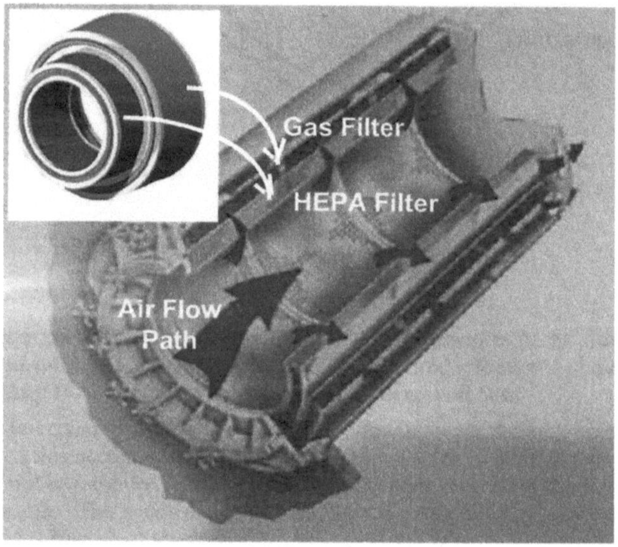


**Figure 8.16** Round HEPA Filter for CPS Systems. *Source: U.S. Army Natick Soldier Systems Center (288), Natick Massachusetts, USA. http://jocotas.natick.army.mil/ColPro_Papers/Larzelere.pdf.*

filters. Figure 8.16 above is a diagram of a round 200 CFM HEPA filter used in nuclear collective protection systems (CPS) (288). The air flow is from the center outward radically. These filters can be stacked up to provide more air as required. The medium is a glass and polymer fiber blend, and is pleated to provide more filter material in a smaller space. Examples of circular HEPA filter cartridges used for vacuum systems are discussed and depicted in Section 8.6. Kuile and Doig (289) indicate the following advantages to circular filters over square or rectangular HEPA filters.

1. Higher airflow and lower pressure drop
2. Easier installation
3. Less maintenance
4. Improved operation
5. Easier disposal—the spent filters can be crushed to disc like dimensions.

### 8.3.3 ULPA filters

ULPA filters are defined by the U (U15, U16, and U17) designations of Table 8.16 below and in accordance with EN 1822. They are also assigned MERV 19 and MERV 20 ratings by ASHRAE 52.2. ULPA filters are defined as being >99.999% efficient against 0.2 μm particles. ULPA filter media are usually wet laid and formed from glass microfibers of 0.2 μm diameter or less. In EN 1822, ULPA filters are tested with a different challenge aerosol than HEPA media.

### 8.3.4 Media for high efficiency air filters

High efficiency air (HEA) is a recent term that includes high efficiency air filters, not only in the HEPA and ULPA range but also in the sub-HEPA range. This would include all the H and U class filters of EN1822 and filters with MERV ratings of MERV 15 and higher. Generally filters with efficiencies of 85% or better are considered as HEA filters.

Traditionally wet laid media produced from glass microfibers have been the dominant medium choice for HEA filters. The fibers are available in a broad range of fiber diameters (see Chapter 4, Section 4.3.5.4.1). The expert filter medium technologist can formulate these fibers to give the desired filtration properties. Because the fibers are of glass, the medium are inherently fire resistant and heat resistant. They also provide a great deal of resistance to corrosive chemicals. In Chapter 3, Table 3.10 lists some of the HEPA grades produced by Lydal Inc., Manchester, Connecticut, USA. Note that most of the grades are listed as high alpha grades (see Chapter 6, Section 6.14.1.13) to designate their ability for achieving efficiency targets at lower pressure drop.

On the down side, microglass fiber medium tends to be brittle-weak. Glass fiber media are more difficult to process, particularly in pleating machines. In field operation, microglass fiber is vulnerable to mechanical disturbances that cause punctures, tears, and cracks. Solutions to these problems include improvements to micro glass structures as well as synthetic fiber alternatives to glass microfiber.

Hollingsworth and Vose has introduced its new PerForm™ line with significantly improved processability. The improvements reduce pleater machine setup time, increase running speeds, give better pleat formation, and reduce pleater scrap (290). These

Table 8.14 Overview of H&V PerForm™ Grades

| PerForm™ grade | HC3393 | HC4393 | HB5793 | HB5493 | HB5593 | HB5693 | HB5893 | HA8393 |
|---|---|---|---|---|---|---|---|---|
| Grammage, $g/m^2$ | 70 | 70 | 70 | 70 | 70 | 70 | 70 | 75 |
| Air resistance, mm $H_2O$ | 210 | 235 | 260 | 280 | 295 | 315 | 360 | 385 |
| DOP penetration, % | 0.100 | 0.30 | 0.015 | 0.008 | 0.005 | 0.002 | – | – |
| CNC[a], $cm^3/s$ | – | – | – | – | – | – | 1.3 | 1.7 |
| CNC[a], penetration, % | – | – | – | – | – | – | 0.0002 | 0.00011 |

[a]Condensation Nuclei Counter

Reproduced with permission of Hollingsworth and Vose Co. Inc. East Walpole, Massachusetts, USA.

improvements in pleatability provide for better filter performance with lower pressure drops for energy cost reduction. Table 8.14 is an overview of the PerForm™ grades.

Developments in microfiber, nanofiber, and membrane technologies are providing alternate filter medium possibilities for high efficiency filtration. The efficiencies can be further enhanced by electrostatic charging. In addition, new media structure ideas are providing ways to greater dirt holding capacity and filter life and at lower pressure drop. These developments are providing near HEPA quality filtration where previously such capability did not exist. Table 8.15 lists some of the emerging media concepts that either is being used in high efficiency air and in HEPA/ULPA applications or have the potential for doing so.

## 8.3.5 HEPA and ULPA test methods and standards

HEPA media were originally defined by MIL STD 282 as having 99.97% efficiency against 0.3 μm DOP smoke particles. The definition has been expanded by more recent standards that include ULPA media. Among these are EN 1822, ASME AG-1, and the RPs of the IEST. The ASME (American Society of Mechanical Engineers) standard has a section (FC-I-4000) for flat sheet testing of the medium. In addition, the United States Department of Energy (DOE) has a document DOE-STD-3020-2005 that provides guidance to DOE contractors for procurement and required testing of High Efficiency Particulate Air (HEPA) filters used in DOE nuclear facilities.

### *8.3.5.1 EN 1822*

The latest version of EN 1822 is EN 1822:2009. It has five parts:

1. EN 1822-1 Classification, Performance Testing, and Labeling
2. EN 1822-2 Aerosol Production, Measuring Equipment, and Particle Count Statistics
3. EN 1822-3 Testing Flat Sheet Filter Media. Note: EN 1822-3 is used for flat sheet testing to determine most penetrating particle size (MPPS).
4. EN 1822-4 Determining Leakage of Filter Elements
5. EN 1822-5 Determining the Efficiency of Filter Elements.

**Table 8.15** Emerging Media Technology Concepts for High Efficiency Air

| Manufacturer | Product | Description | Capability | High efficiency applications |
|---|---|---|---|---|
| Donaldson | Ultra-web® | Electrospun nanofiber on a substrate | High efficiency <HEPA ASHRAE MERV Ratings: 14 and 15 | Turbine air filters<br>Dust collection systems |
| Donaldson | Donaldson ePTFE HEPA | Membrane film on a support substrate. Reusable and washable | EN 1822 Ratings: E10-H13 | Dust collection systems<br>Vacuum systems |
| H&V | Nanowave | Wave shaped nanofiber/ microfiber layer embedded in a coarse fiber layer. Combined with a support structure. Electrostatic charge optional | ASHRAE 52.2 MERV Ratings: 11–15 | HVAC bag applications<br>Commercial HVAC systems<br>Aircraft paint spray facilities |
| H&V | Technostat® | Electrostatic charged felt material | HEPA rated grades available | Pleated filters for clean rooms and vacuum cleaning systems |
| Ahstrom | Flow 2 Save | Gradient density media | EN 1822 rated grades available | HVAC systems |

EN 1822-1 is based on an MPPS. MPPS has been previously discussed in Chapter 2, Section 2.1.2, and in Chapter 6, Section 6.14.1.12. It applies to ULPA filters as well as HEPA filters. The efficiency is determined by particle number counting. If the aerosol is quasi-monodispersed, the aerosol generated has to have a particle size near MPPS. If the aerosol is polydispersed its fractional efficiency counting is done to determine MPPS. MPPS is determined from flat sheet testing of the filter medium in accordance with EN 1822-3. The filter is tested at this predetermined particle size.

**Table 8.16** EN 1822 Efficiency Classes

| Fiber class | Overall value (%) | | Local value (%)[a] | |
|---|---|---|---|---|
| | Efficiency | Penetration | Efficiency | Penetration |
| E10 | 85 | 15 | – | – |
| E11 | 95 | 5 | – | – |
| E12 | 99.5 | 0.5 | – | – |
| H13 | 99.95 | 0.05 | 99.75 | 0.25 |
| H14 | 99.995 | 0.005 | 99.975 | 0.025 |
| U15 | 99.9995 | 0.0005 | 99.9975 | 0.0025 |
| U16 | 99.99995 | 0.00005 | 99.99975 | 0.00025 |
| U17 | 99.999995 | 0.000005 | 99.999975 | 0.000025 |

[a]Local values are results of scan tests to determine filter leakage.

The efficiency classes are presented in Table 8.16. Note the filter classes are an extension of EN 779 above (refer Section 8.2.2.1.1, Table 8.9). When EN 1822:2009 was introduced, the E classes (E10, E11, and E12) were designated as EPA (Efficient Particulate Air) grades. These are media that have high filtration efficiencies (>85%), but are in the sub-HEPA in their ratings (<99.95%). EN 1822:2009 continued to rate H13 and H14 as HEPA class filters and U15, U16, and U17 as ULPA class filters.

### *8.3.5.2 ASME AG-1, section FC*

Section FC of the "Code on Nuclear Air and Gas Treatment," ASME AG-1-2012 (291) provides for standards on HEPA filters directed toward Nuclear Facilities. Section FK sets the standards for round HEPA filters (the "Code", previously identified by Military Standards MIL-STD-51068 for HEPA filters and MIL-STD-51079D for HEPA filter media, was administered by the United States Department of Energy (DOE)). The administration of this code has since been turned over to the ASME (American Society for Mechanical Engineers). The ASME AG-1 Code "provides requirements for the performance, design, construction, acceptance testing, and quality assurance for high-efficiency particulate air (HEPA) used in nuclear safety-related air or gas treatment systems in nuclear facilities." Although the AG-1 Code is written for nuclear power and nuclear fuel cycle facilities, the design requirements may be referenced for other types of facilities where high reliability and performance are required. Appendix FC-I of this code identifies the requirements and testing of filter media.

Section FK (292) mentioned in Section 8.3.2.3 is a recent addition to ASTM AG-1. It is concerned with radial HEPA filters.

#### 8.3.5.3 IEST recommended practice

The IEST issues RPs for testing HEPA and ULPA filters.

Four pertinent practices are:

1. IEST RP-CC-001.5 "HEPA and ULPA Filters": provides glossary, definitions, and classifications of HEPA and ULPA filters, test instruments, and requirements for other RPs. This RP describes 11 levels of filter performance and six grades of filter construction. The 11 levels of filter performance are tabulated in Table 8.17.
2. IEST RP-CC-007.2 "Testing ULPA Filters": provides the test methodology for the overall efficiency of ULPA filters using particle counter(s). The penetration range of this procedure is 0.001−0.0001%.
3. IEST RP-CC-021.3: discusses test methods for physical and filtration properties of high-efficiency particulate air (HEPA) and ultra-low penetration air ULPA) filtration media.
4. IEST RP-CC-034.3 "HEPA and ULPA Leak Tests": covers definitions, equipment, and procedures for leak-testing HEPA filters and ultralow-penetration air (ULPA) filters in the factory as they are produced, at the job site before they are installed, and after they are installed in clean rooms and in unidirectional-flow, clean-air devices.

The testing of HEPA filters is different than ULPA filters. The RPs recommend that the less efficient HEPA filters be tested with a photometer and the efficiency based on 0.3 μm particles. The assumption is that the particles are more of less the same diameter.

**Table 8.17** IEST Classification of HEPA and ULPA Filters

| Filter type | Test method | Aerosol | Minimum efficiency rating |
|---|---|---|---|
| HEPA (Type A) | MIL-STD_282 | Thermal DOP | 99.97% |
| HEPA (Type B) | MIL-STD_282 | Thermal DOP | 99.97% |
| HEPA (Type C) | MIL-STD_282 | Thermal DOP | 99.99% |
| HEPA (Type D) | MIL-STD_282 | Thermal DOP | 99.999% at 0.3 μm |
| HEPA (Type E) | MIL-STD_282 | Thermal DOP | 99.97% |
| ULPA (Type F) | IEST RP-CC007 | Open | 99.999% at 0.1−0.2 or 0.2−0.3 μm |
| Super ULPA (Type G) | IEST RP-CC007 | Open | 99.9999% at 0.1−0.2 or 0.2−0.3 μm |
| HEPA (Type H) | IEST RP-CC021 | Open | 99.97% at 0.1−0.2 or 0.2−0.3 μm |
| HEPA (Type I) | IEST RP-CC007 | Open | 99.97% at 0.1−0.2 or 0.2−0.3 μm |
| HEPA (Type J) | IEST RP-CC007 | Open | 99.99% at 0.1−0.2 or 0.2−0.3 μm |
| HEPA (Type K) | IEST RP-CC007 | Open | 99.995% at 0.1−0.2 or 0.2−0.3 μm |

The RPs recommendation for ULPA filters is that they be tested for fractional efficiency using particle counters. The size range for the particles is 0.1–0.2 μm. There is also a filter medium identified as a Super ULPA whose efficiency is rated at 99.9999%. MPPS is recommended as the rating method for this type of filter. Table 8.17 above lists the HEPA and ULPA filter levels defined by IEST RP-CC-001. Wang and Winters (293) provide a detailed discussion comparing the RPs to EN 1822.

### 8.3.6 Clean rooms

Clean rooms are defined by International Standard ISO 14644-1 (294) as follows:

"A room in which the concentration of air-borne particles is controlled and which is constructed and used in a manner to minimize the introduction, generation, and retention of particles inside the room, and in which the relevant parameters, e.g., temperature, humidity, and pressure are controlled as necessary."

Following are a listing of some of the applications where clean rooms may be required:

**a.** Hospitals
**b.** Biomedical
**c.** Pharmaceutical
**d.** Genetic research
**e.** Universities
**f.** Laboratories
**g.** Food processing
**h.** Photo Processing
**i.** Semiconductor
**j.** Industrial processing systems.

There are two types of clean rooms: non-unidirectional airflow type (also known as non-laminar flow types or conventional clean rooms) and unidirectional airflow type (also known as laminar flow clean rooms). Figure 8.17 is a schematic of a non-unidirectional airflow type of clean room. Figures 8.18 and 8.19 are schematics of typical unidirectional airflow types of clean rooms. Figure 8.18 illustrates a vertical laminar flow type of clean room and Figure 8.19 illustrates a horizontal laminar flow type of clean room.

In the non-unidirectional airflow type of clean room HEPA filters are downstream of the blowers so that only HEPA-filtered air enters the ductwork connecting the mechanical equipment to the clean room. In the unidirectional vertical airflow type of clean room the HEPA filters consist of the ceiling while prefilters filters the air as it passes through the grated floor and is recycled. In the horizontal type of room, the prefilters and HEPA filters form one wall of the room. The horizontal air flow enters the room through these banks of prefilters and HEPA filters. Of course, there are variations in these designs.

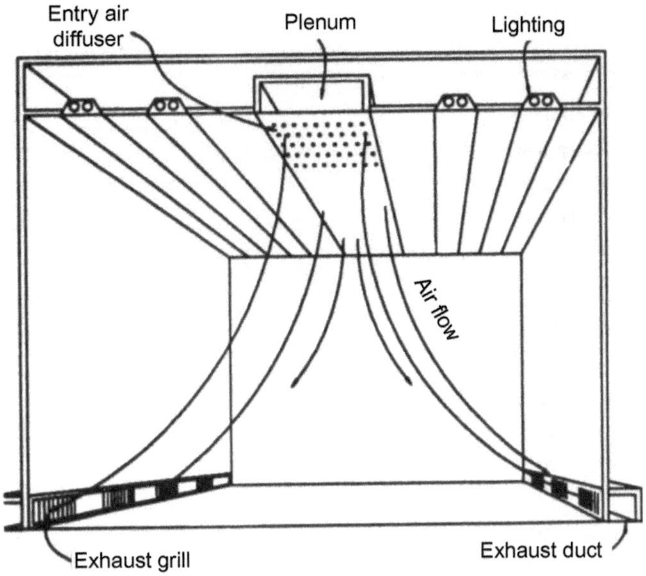


**Figure 8.17** Conventional Clean Room Air Diffuser. *Reproduced with Permission of NAFA, Virginia Beach, VA, USA.*

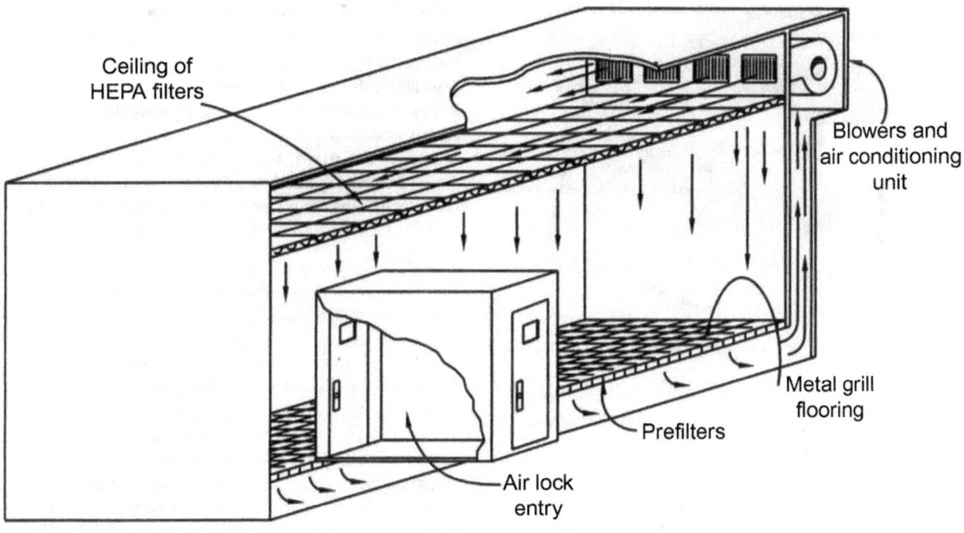


**Figure 8.18** Laminar Down Flow Room with Grated Floor. *Reproduced with Permission of NAFA, Virginia Beach, VA, USA.*

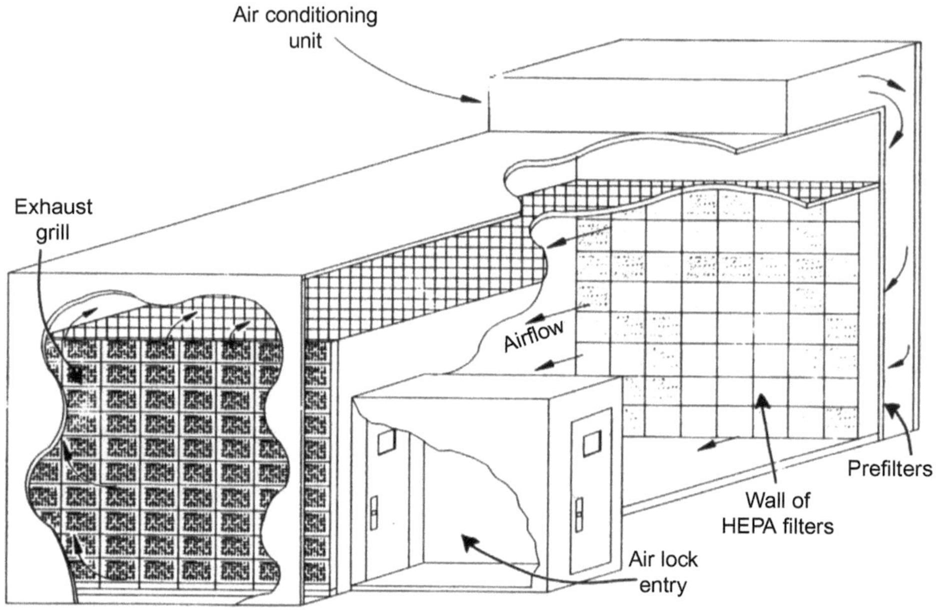


**Figure 8.19** Horizontal Laminar Flow. *Reproduced with Permission of NAFA, Virginia Beach, VA, USA.*

Originally, clean rooms were classified in accordance with Federal Standard 209E since replaced by ISO 14644-1 (295). ISO14664-1 identifies nine classes of cleanliness identified by an ISO number from ISO 1 through ISO 9; The lower the class, the cleaner is the clean room. The particle criteria are S.I. (metric). The classes are based level of cleanliness and a maximum number of allowable particles in accordance with their particle size. Table 8.18 below lists the ISO 146441-1classifications. The classifications are based on the maximum number of particles per cubic meter for that particulate size. The particle sizes are 0.1 μm, 0.2 μm, 0.3 μm, 0.5 μm, 1 μm, and 5 μm.

## 8.3.7 Paint spray booth filters

Paint spray booths are built with either downdraft or crossdraft ventilation flow. This is analogous to the down flow and laminar horizontal flow clean rooms depicted in Figures 8.18 and 8.19 above. Spray booths require two types of filter assemblies; intake filters and exhaust filters. The intake filter filters the air coming into the spray booth. The exhaust filter captures excess paint particles from air emerging from the spray booth.

Intake filters are the first line of defense against large particles that might possibly mar the painted surface. They are either panels or pads. Crossdraft spray booths usually require panel mounted along one wall of the spray booth. Downdraft booths usually use pads mounted into the ceiling of the spray booth. Filter pads are often referred to as diffusion media for intake filtration. The pads are usually a high loft

**Table 8.18** Airborne Particle Cleanliness Level, ISO 14644-1

**ISO/TC209 14644-1**

| Class | Number of particulates per cubic meter by micrometer (micron) size | | | | | |
|---|---|---|---|---|---|---|
| | 0.1 μm | 0.2 μm | 0.3 μm | 0.5 μm | 1 μm | 5 μm |
| ISO 1 | 10 | 2 | | | | |
| ISO 2 | 100 | 24 | 10 | 4 | | |
| ISO 3 | 1000 | 237 | 102 | 35 | 8 | |
| ISO 4 | 10,000 | 2,370 | 1,020 | 35 | 83 | |
| ISO 5 | 100,000 | 23,700 | 10,200 | 35,20 | 8,32 | 29 |
| ISO 6 | 1,000,000 | 237,000 | 102,000 | 35,200 | 8,320 | 293 |
| ISO 7 | | | | 352,000 | 83,200 | 2,930 |
| ISO 8 | | | | 3,520,00 | 832,000 | 29,300 |
| ISO 9 | | | | 35,200,000 | 8,320,000 | 293,000 |

Reproduced with permission of NAFA, Virginia Beach, VA, USA.

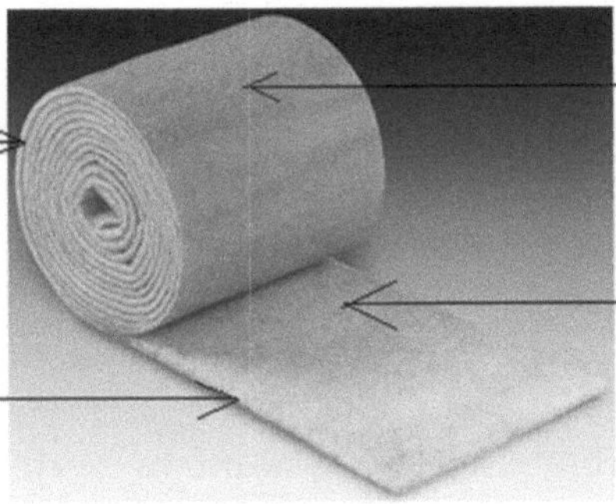


**Figure 8.20** Filtration Group EC-XL® Overspray Medium. *Reproduced with Permission of Filtration Group, Joliet, Illinois, USA.*

medium, composed of synthetic fibers. They are available in either roll form or pads custom sized to the users application. An example of EC-XL® roll form media by Filtration Group, Joliet, Illinois, USA, is depicted in Figure 8.20 above. In roll form, the user will cut the roll into pad sizes suitable for the application. Panel filters are typically made from media pads, separated by a metal support. In some panel filters the upstream pad is more porous and acts to catch the larger particles. The downstream pad is tighter and designed to trap smaller particles.

Exhaust filters are often referred to as paint arrestors. The filters are usually in the form of pads. Crossdraft and semi-downdraft booths use cut pads, 20″ × 20″ (51 × 51 cm) or 20″ × 25″ (52 × 63.5 cm) that are mounted into the wall of the booth. Downdraft booths usually use a long pad that is placed in a pit at the floor of the booth. Some downdraft booths use filter bags, cubes, and other products. Some examples of materials used for paint arrestor filter pads are fiberglass with polyester backing, various forms of fiberglass pads with a skin backing for additional efficiency, and three layer polyester providing three layers of gradient density filtration. Usually the pads are 2.5 in (6.35 cm) thick. Polyester medium is often the material used for cube filters. Sometimes an added layer provuides for increased efficiency. Polyester fiber with a scrim backing is a popular choice for bag filters. Hollingsworth and Vose Company NanoWave® filter media discussed in Chapter 3, Section 3.5.3, has been successfully applied to aircraft paint spray booth facilities.

Heated downdraft paint booths often have prefilters. These are generally installed in the ductwork bringing air into the booth. Filters are usually panel filters, however, dual ply polyester pads are also used. Media for panel filters can be a pleated high loft synthetic, high loft thermo-bonded polyester pad enclosed within a support frame, and 2-ply panel filters using an embedded wire frame for support.

## 8.4 GAS TURBINE AIR INTAKE FILTERS

### 8.4.1 Turbines description

Turbines are rotating devices designed to generate energy, either mechanical or electrical. The gas turbine in a jet aircraft engine is an example of mechanical energy generation. The turbine rotates a compressor that compresses the inlet air. One or more ignition chamber injects and ignites fuel (gas) to heat and expand the air and provide the exhaust thrust necessary to drive the aircraft. In large-scale power plants, the turbine operates a rotating electrical generator to produce electricity for the realm that is serviced by the power plant.

There are several types of turbines, including steam, hydroelectric, solar, wind, and gas turbines. The complexities of the gas turbine machine are of particular interest to the filter and filter medium manufacturer, because of stringent requirements for inlet air cleanliness. According to the US Department of Energy (295) gas turbines basically involve three main sections:

1. The compressor, which draws air into the engine, pressurizes it, and feeds it to the combustion chamber literally at speeds hundreds of miles per hour.
2. The combustion system, typically made up of a ring of fuel injectors that inject a steady stream of fuel (e.g., natural gas) into the combustion chamber where it mixes with the air. The mixture is burned at temperatures of more than 2,000 °F. The combustion produces a high temperature, high pressure gas stream that enters and expands through the turbine section.
3. The turbine is an intricate array of alternate stationary and rotating aerofoil-section blades. As hot combustion air expands through the turbine, it spins the rotating blades. The rotating blades perform a dual function: they drive the compressor to draw more pressurized air into the combustion section, and they spin a generator to produce electricity.

## 8.4.2 Air inlet filtration for gas turbines

A turbine system must contain other components, including an inlet section, an exhaust system, and an exhaust diffuser. Figure 8.21 is a diagram of the inlet air system for a typical turbine-generator layout. The air to the system comes in through the air filter house. In addition to filtration an air inlet system may have provision for noise reduction (silencers) and cooling. The latter is necessary for many installations in hot dry climates such as the Middle East. The cooling makes the air denser. Dense air is more efficient for the turbine process. For installations in cold climates, deicers may be necessary to prevent winter ice condensation.

Turbine rotor assemblies are large, generally 1.5−1.8 m and they spin very fast (several thousand rpm). Even the smallest dust particles can cause microscopic damage and erosion with repeated contact. It is for this reason, that the filtration of the incoming air is so important. The filter houses can be quite large. Note that in Figure 8.22, three filter houses are shown, approximately four stories high. Figure 8.23 shows the types of filters often used in the houses. These are manufactured by TDC Filters in Bolingbrook, Illinois, USA. Note, the two cartridges shown are used in tandem, back to back. Both are pleated medium cartridges, one is conical shaped and the other is a cylinder shape. These are large cartridges, each over two

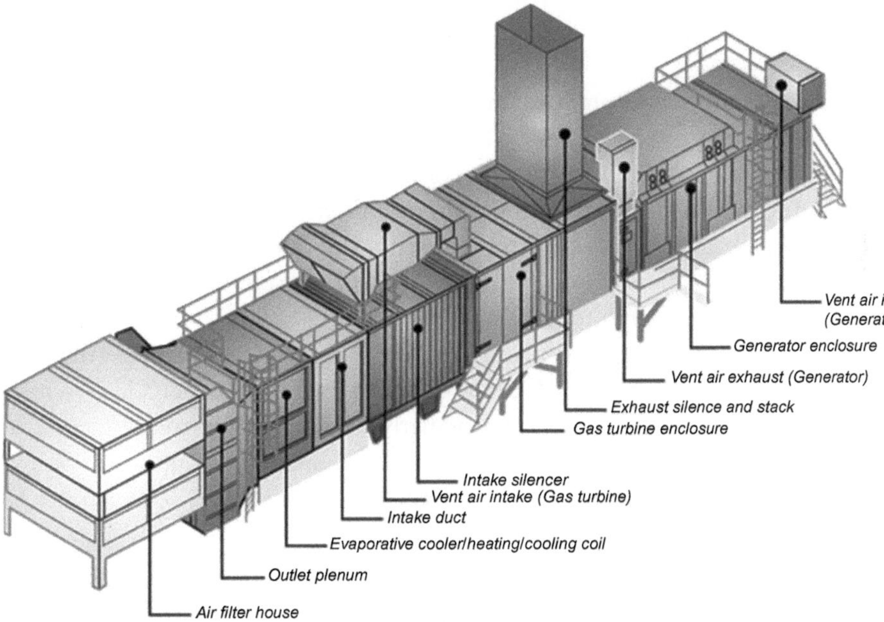


**Figure 8.21** AAF Gas Turbine Total Inlet Air System. *Reproduce with Permission of AAF International, Louisville, Kentucky, USA.*

**Figure 8.22** Filter Houses for a Large Gas Power Plant Installation. *Reproduced with Permission from Donaldson Company, Inc.* 

feet in length. In tandem, they are over four feet. There are hundreds if these type cartridges installed in a typical installation.

The cylinder/cone combination is not the only types of filters installed in gas turbine installations. Figure 8.24 shows the wide variety of filters produced by TDC Filters.

Gas turbine filters are rated in accordance with HVAC and HEPA/ULPA standards discussed above. These include EN779 (Section 8.2.2.1), EN1822 (Section 8.3.5.1), and the ASHRAE 52.2 MERV ratings (Section 8.2.2.2). In addition, MIL STD 282 may be applied to HEPA filters used in gas turbine filters.

There are two primary types of filter houses: static and pulse self-cleaning. Static filter houses are usually panel filters, often pocket filters mounted in filter modules. There are two filter stages: prefilter and final.

Self-cleaning filter houses use both panel filters and circular cartridge filters. AAF International, Louisville, Kentucky, USA produces DuraVee® panel filters which are mini-pleat V-bank filters. They are particularly designed for rotating machinery applications. AAF produces filters for a range of efficiencies from 60% to 95%, based on 3 μm particles. There is also a high capacity version available. The Model DV95 filter element is pictured in Figure 8.25.

Very often the filters are of the conical-cylindrical pair illustrated in Figure 8.23. This type of filter cartridge combination is used in Donaldson's GDX® self-cleaning inlet air filter for gas turbines pictured in Figure 8.26. The GDX® filter house schematic is illustrated in Figure 8.27.

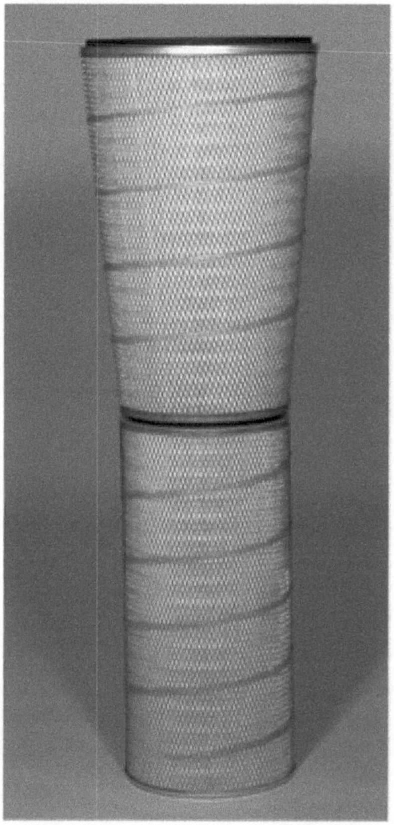

**Figure 8.23** Conical-Cylindrical Filter Pair for Pulse Cleaning Filter Houses. *Reproduced with Permission of Midwesco Filter Resources/TDC Filter, Bolingbrook, Illinois, USA.*

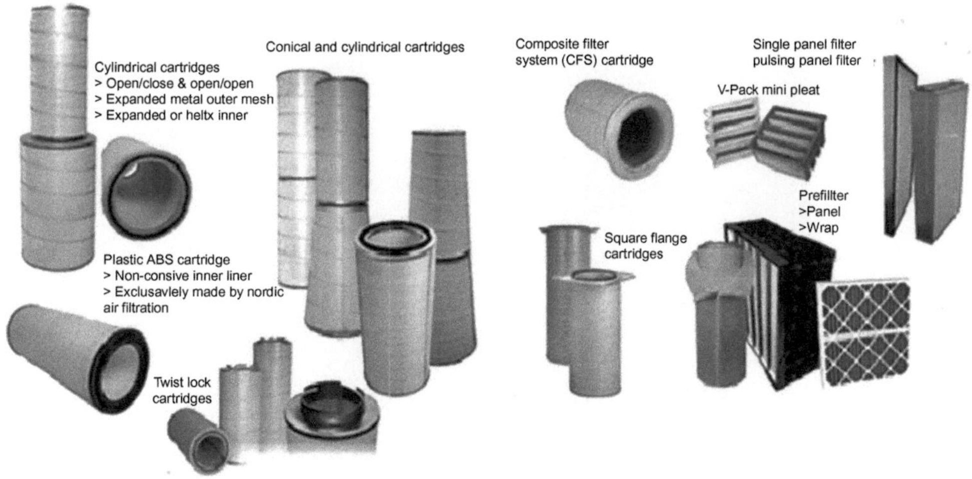


**Figure 8.24** Various Filters for Gas Turbines. *Reproduced with Permission of Midwesco Filter Resources/TDC Filter, Bolingbrook, Illinois, USA.*

**Figure 8.25** DuraVee® Model DV95 Panel Filter by AAF International. *Reproduced with Permission of AAF International, Louisville, Kentucky, USA.*

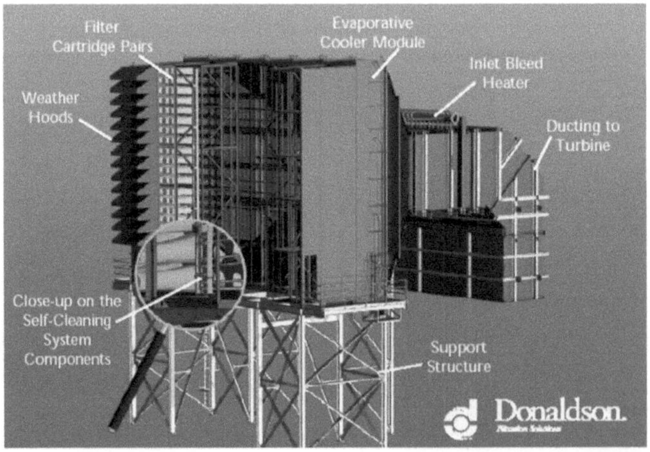


**Figure 8.26** Donaldson GDX® Self-Cleaning Inlet Air Filter System for Gas Turbines. *Reproduced with Permission of Donaldson Co. Inc. © Donaldson Company, Inc. All rights reserved.*

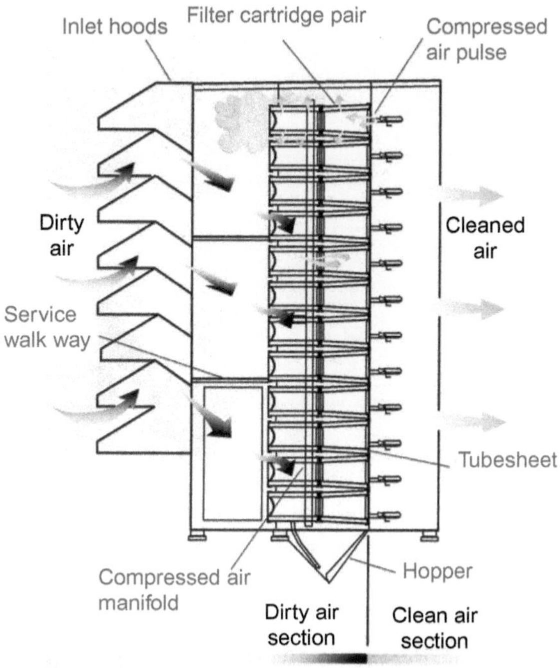


**Figure 8.27** Schematic of Donaldson's GDX® Self-Cleaning Inlet Air Filter System for Gas Turbines. *Reproduced with Permission of Donaldson Co. Inc. © Donaldson Company, Inc. All rights reserved.*

The schematic of the GDX® system as diagrammed in Figure 8.26 includes the pulse cleaning action. Monitoring devices trigger a reverse blast of air when the pressure drop across the filters reaches a certain point. This point can be set to fit the conditions of the environment.

Donaldson Company also produces a TTD® self-cleaning system. Figure 8.28 is that of a TTD® system. The system also utilizes a pulse cleaning action and is referred to as the TTD® huff and puff system. This systems uses vertically mounted cylinder cartridges as illustrated in Figure 8.28.

## 8.4.3 Filter media for gas turbine filtration

Filter media for gas turbine filtration depend on a number of factors including location, environment, desired filtration efficiency, pressure drop, and mechanical properties. Medium types include cellulose, blends of cellulose and polyester, glass microfiber, blends of glass microfibers and synthetic fibers, and composite structures containing a melt-blown, or electrospun layer. High loft mats of glass or synthetic fibers are also available for prefilter applications. They can be used as a prefilter wrap around the final filters. For example, the conical/cylindrical cartridge filter combinations discussed above can be wrapped with a high loft prefilter mat.

**Figure 8.28** Vertical Filter Installation for Donaldson TTD® Vertical Cartridge Filter System. *Reproduced with Permission of Donaldson Co. Inc. © Donaldson Company, Inc. All rights reserved.*

TDC Filters offers several medium choices for their gas turbine intake filters (296). These are tabulated in Table 8.19 below. The TurboWeb® structures are composites with a high efficiency layer: either a membrane or an efficient web structure. There is a lot to be learned about turbine filter selection from the data of Table 8.19. Not only does the filtration properties have to be considered, but also where and how the medium will be used; whether it will be used for pulse or static filtration systems, the type of chemical atmosphere it will be exposed, and to the type of climate it will be exposed. These factors are important in selecting a medium. For example if the medium is used in an offshore sea facility, it will have to be able to salt water and moisture. In a tropical climate it will have to resist extreme humidity.

Donaldson Company offers several medium choices for their gas turbine filters (297).

1. Synthetic—human-made fibers that are sturdy, durable, and moisture resistant.
2. Duratek®—a special blend of human-made fibers and natural (cellulose) fibers that has superior dust holding capacity and moisture resistance.
3. Cellulose—natural fibers that stop a range of particulate particles sizes.
4. Spider-Web®—an electrospun nanofiber layer that catches very fine particulates before they reach the medium substrate. The composite structure can be formed by adding the nanofiber layer (or coating as it is sometimes called) to either synthetic or Duratek®.

AAF offers a broad variety of media for their various rotating machinery filter applications. This includes cellulose–polyester, glass microfiber media that include gradient density versions, high loft melt-blown, and continuous filament glass mats.

**Table 8.19** TDC Gas Turbine Filter Media

| Media | Description | Pulse/ static | Efficiency rating | Chemical resistance to: | Suitable for: |
|---|---|---|---|---|---|
| TurboWeb®-XE Ultra High Efficiency | 100 % synthetic with efficiency layer—e class membrane | Pulse | EN1822: H12 HEPA-99.5% | Moisture | Desert Offshore Arctic Tropical Urban Coastal |
| TurboWeb® UE Ultra High Efficiency | 3 layer, 100% synthetic with efficiency layer—e class membrane | Static | EN1822: H12 HEPA-99.5% | Salt Water | Desert Offshore Arctic Tropical Urban Coastal |
| TurboWeb® Nano High Efficiency | 100% synthetic with efficiency laminate | Pulse | EN779: F9 ASHRAE 52.2: MERV 16 | Moisture Salt Water | Desert Offshore Arctic Tropical Urban Coastal |
| TurboWeb® HE High Efficiency | 100% synthetic with efficiency laminate | Static | EN779: F9 ASHRAE 52.2: MERV 16 | Moisture Salt Water | Desert Offshore Arctic Tropical Urban Coastal |
| 171GT | 100% Synthetic | – | 97% @ 450 Pa | Moisture Chemical corrosives | Desert Offshore Arctic Tropical Urban Coastal |
| 143 GT | Cellulose/Synthetic with nanosurface—melt-blown | Pulse | 98% @ 450 Pa | Moisture | Desert Offshore Arctic Tropical Urban |
| 140 GT | Cellulose with synthetic fibers | – | 96% @ 450 Pa | Recommended for dry climates | Desert Arctic Urban |

Reproduced with permission of Midwesco Filter Resources/TDC Filter, Bolingbrook, Illinois, USA.

## 8.5 RESPIRATORS, GAS MASKS, AND FACEMASKS

In the United States respirators, gas masks and face masks are regulated by 42 CFR (U.S.), Chapter 1, Part 84. Commonly known as Part 84 (298), it is discussed in Section 8.5.1. The regulations are administered by NIOSH (the National Institute for Occupational Safety and Health) and by MSHA (the Mine Safety and Health Administration).

Other U.S Standards include 29 CFR 1910.134 which is administered by OSHA. This standard is discussed in Section 8.5.2.

In Europe these items are regulated by a number of CEN standards. They are discussed in Section 8.5.3.

### 8.5.1 Respirators and gas masks-42 [CFR (U.S.), Chapter 1, Part 84]

Respirators are a form of facemask (and often called facemasks) designed to provide respiratory protection against hostile environments. They come with a face piece that will cover nose, mouth and very often the eyes. Respirators can range from simple fabric filters that cover the nose and mouth and provide home and workplace protection against dust and pollen to complex devices that provide protection against very dangerous dusts (such as asbestos), microbes, fumes, and chemical vapors.

Two types of respirators are air purifying respirators (APRs) and air supplying respirators. The latter, either have a self-contained air supply or are supplied by an external air source.

APRs have three basic components:

1. A face mask.
2. A filter or cartridge filter to remove dust, smaller particles, and mists.
3. A cartridge filter to remove chemical gases and fumes.

Respirators are regulated by 42 CFR (U.S.), Chapter 1, Part 84 (often referred to as Part 84)—Approval of Respiratory Protective Devices. NIOSH (the National Institute for Occupational Safety and Health) and MSHA (the Mine Safety and Health Administration) are the United Stated Government agencies responsible for administering the regulation.

The regulation defines a respirator "as any device designed to provide the wearer with respiratory protection against inhalation from a hazardous atmosphere." It also defines two types of respirators: respirators for escape only and respirators for entry into and escape from. Escape only respirators are respiratory devices providing protection only during escape from hazardous atmospheres. Entry into and escape from respirators are respiratory devices providing protection during entry into and escape from hazardous atmospheres. The regulation identifies the following type of respirators.

#### *8.5.1.1 Self-contained breathing apparatus (SCBA)*

An SCBA is used for entry into and escape from hazardous atmospheres and contains a source of oxygen. There is no filter or canister associated with an SCBA. The source of oxygen can be compressed oxygen, chemical oxygen, or liquid oxygen. There are two types of apparatus; closed-circuit apparatus and open-circuit apparatus. The closed-circuit apparatus provides for rebreathing of the exhaled air by removing carbon dioxide and restoring the air to a suitable oxygen content. An open-circuit apparatus does not recycle the exhaled air, but rather it is vented to the atmosphere and not rebreathed.

#### *8.5.1.2 Gas masks*

Gas masks, also known as an "air-purifying respirators" (APRs), are defined by the regulation as devices designed for use during entry into atmospheres not immediately dangerous to life or health or escape only from hazardous atmospheres containing adequate oxygen to support life. Types of gas masks are:

**a.** Front mounted or back mounted—A gas mask which consists of a full face piece, a breathing tube, a canister at the back or front, a canister harness and associated connections.

**b.** Chin-style gas mask—A gas mask, which consists of a full face piece, a canister which is usually attached to the face piece, and associated connections.

**c.** Escape gas masks—A gas mask designed for use during escape only from hazardous atmospheres, which consists of a face piece or mouthpiece, and associated connections. This type of mask often comes in the form of a half mask, which provides no eye protection.

A gas mask is designed to protect the face and lungs against noxious gases and fumes, chemical agents, and biological substances as in warfare or terror attacks or in certain industrial environments. The gas mask includes a cartridge or canister containing activated carbon and/or chemicals to remove dangerous gas and vapor contaminants. A particulate filter may be attached to the cartridge or canister. Gas masks are also described by the types of gases or vapors against which they are designed to provide protection as indicated by Table 8.20.

The regulation specifies canister bench tests to qualify a gas mask. The test conditions are based on the type of gas mask (front-mounted and back-mounted, chin-style, and escape), and the canister type in accordance with Table 8.20. For each gas mask type and canister type the regulation specifies a challenge gas or vapor, a feed concentration, a flow rate, a specified number of tests, a maximum allowable penetration and a minimum service life. As an example, Table 8.21 in Section 8.5.1.6 below compares specific test requirements for ammonia as the challenge gas. The regulation also specifies particulate tests for those gas masks containing a particulate filter. In this regard, the gas mask becomes like a non-powered air-purifying particulate arrestor and is subject to the testing protocol as discussed in Section 8.5.1.4.

**Table 8.20** Gas Mask Protection Categories

| |
|---|
| Types for front-mouthed or back-mouthed gas mask |
| Acid gas |
| Ammonia |
| Carbon monoxide |
| Organic vapor |
| Other gases and vapor(s) |
| Combinations of two or more of the above |
| Combinations of all of the above |
| Types for chin-style gas mask |
| Acid gas |
| Ammonia |
| Carbon monoxide |
| Organic vapor |
| Other gas(es) and vapor |
| Combinations of two or more of the above |
| Types for escape gas mask |
| Acid gas |
| Ammonia |
| Carbon monoxide |
| Organic vapor |
| Other gas(es) and vapor(s) |
| Combinations of two or more of the above |

Note: The regulation has further restrictions on whether a gas mask can be described as above. The regulation should be consulted for the details of these restrictions.

**Table 8.21** Respirator Requirement for Ammonia

| Type | Test atmosphere concentration (parts per million) | Flow rate (liters per minute) | Number of tests | Maximum allowable penetration (parts per million) | Minimum service life (minutes) |
|---|---|---|---|---|---|
| Front and back-mounted gas mask canisters | 30,000 | 64 as received 32 equilibrated | 3 as received 4 equilibrated | 50 | 12 |
| Chin style gas mask canisters | 5,000 | 64 | 3 | 50 | 12 |
| Escape gas mask canisters | 5,000 | 64 as received 32 equilibrated | 3 as received 4 equilibrated | 50 | 12 |
| Chemical cartridge respirators | 1000 | 64 as received 32 equilibrated | 3 as received 4 equilibrated | 50 | 50 |

#### 8.5.1.3 Supplied-air respirators

Supplied-air respirators (SARs) rely on a source of supplied respirable breathing air. They are designed to provide respiratory protection during entry into and escape from atmospheres not immediately dangerous to life or health. Like the SCBAs discussed in Section 8.5.1.1, the supplied air respirators do not have a filter canister, cartridge, or element. The regulation defines three types of SARs:

Type A—Type "A" SARs provides air through a motor-driven or hand-operated blower. The device is designed to provide for the free entrance of air when the blower is not operating.

Type B—The wearer of a Type "B" SAR draws inspired air by use of his lungs alone.

Type C—Type "C" SARs are airline respirators. They differ from the Type "B" SARs in that they have a detachable coupling, a control valve, an orifice, and a demand valve or pressure demand valve.

All three SAR types above also have an "E" version (Type "AE", Type "BE", and Type "CE"). These versions have protection devices such as a helmet to protect the wearer's head and neck against impact and abrasion from rebounding abrasive material.

#### 8.5.1.4 Non-powered air-purifying particulate arrestors

Non-powered air-purifying particulate arrestors (NPAPPRs) are also referred to as negative pressure particulate respirators. They utilize the wearers negative inhalation to draw ambient air through the air-purifying filter elements (filters) to remove particulates from the ambient air. They are designed for use as respiratory protection against atmospheres with particulate contamination (e.g., dusts, fumes, mists) that are not immediately dangerous to life or health and that contain adequate oxygen to support life. Note that an NPAPPR differs from a gas mask in that:

**a.** Gas masks are designed for atmospheres and hazardous atmospheres where the contaminants are vapors or gases. NPAPPRs are designed for atmospheres with particulate contamination.

**b.** Gas masks are designed for entry and escape only situations, whereas NPAPPRs allow the wearer to stay in the contaminated area for as long as the filter holds out.

The regulation applies efficiency ratings for NPAPPRs and for gas masks with particulate filters based on three classification series; N-series, R-series, and P-series. N-Series filters are restricted to use in those workplaces free of oil aerosols. The R- and P-Series are intended for removal of any particulate that includes oil-based liquid particulates. The difference between R- and P-Series filters is the end point at which minimum filtration efficiency is determined. The manufacturer is required to label certified filters with the appropriate rating. The filters ratings are determined and certified by NIOSH based on their testing:

**a.** N100, R100, and P100 filters shall demonstrate a minimum efficiency level of 99.97%.

**b.** N99, R99, and P99 filters shall demonstrate a minimum efficiency level of 99%.

**c.** N95, R95, and P95 filters shall demonstrate a minimum efficiency level of 95%.

N-Series filters are tested against a sodium chloride or equivalent solid test aerosol. The median particle size is specified at an equivalent diameter of $0.075 \pm 0.020$ μm. The P- and R-Series filters are tested against a liquid test aerosol of dioctyl phthalate (DOP). The median particle size is specified at an equivalent diameter of $0.185 \pm 0.020$ μm. The test is conducted until minimum efficiency is achieved or until $200 \pm 5$ mg of test aerosol has been loaded unto the filter. For P-Series filters, if the efficiency is still decreasing when the $200 \pm 5$ mg challenge point is reached, the test is continued until there is no further decrease in efficiency. The value reported is the minimum efficiency during the course of the test. The test protocol calls for 20 filters to be tested. Each filter shall exceed the minimum efficiency criterion for which the filter is rated.

#### 8.5.1.5 Powered air-purifying respirators

Powered air-purifying respirators (PAPRs) utilize a blower to force air through the respirator filter or canister. Examples of PAPRs are depicted in Figures 8.29 and 8.30.

The testing requirements on PAPRs are regulated by Part 11, and described in Subpart KK of Part 84. Subpart KK limits licensing of PAPR particulate filters to only high efficiency or HEPA filters. These are defined as "Air-purifying respirators, with replaceable filters, designed as respiratory protection against dusts, fumes, and mists

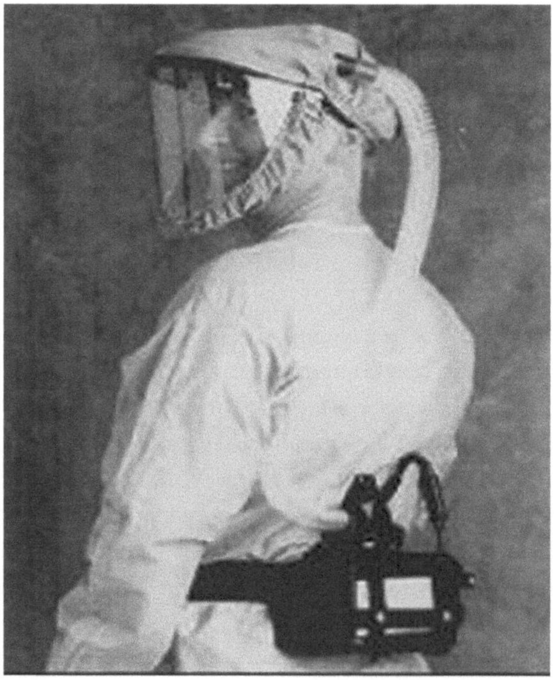

**Figure 8.29** Protective Clothing with PAPR. *Photograph provided by Ernest S. Moyer, NIOSH, Morgantown, West Virginia, USA.*

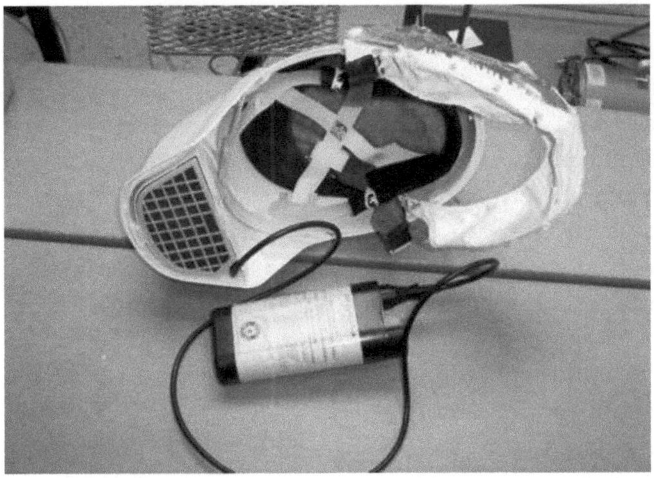

**Figure 8.30** Powered Air-Purifying Respirator Assembly. *Photograph provided by Ernest S. Moyer, NIOSH, Morgantown, West Virginia, USA.*

having an air contamination level less than 0.05 milligram per cubic meter, including but not limited to lithium hydride and beryllium, and against radio nuclides."

PAPRs are subject to either a silica dust test for dust type respirators and a lead fume test for fume type respirators. The test involves a filter test stand. Ground flint is the test contaminant for the silica dust test. The contaminant for the lead fume test is obtained by impinging an oxygen-gas flame on molten lead. The test requirement is that the total amount of unretained test suspension shall not exceed 1.5 mg for the silica test and 1.8 g for the lead fume test.

### *8.5.1.6 Chemical cartridge respirators*

Chemical cartridge respirators are designed for use as respiratory protection during entry into or escape from atmospheres not immediately dangerous to life and health. They are described according to the specific gases or vapors against which they are designed to provide respiratory protection. These are ammonia previously mentioned in Section 8.5.1.2, chlorine, hydrogen chloride, methylamine, organic vapor, sulfur dioxide, and vinyl chloride. Chemical cartridge respirators may have a blower. They differ from gas masks in that they are to be used in less hazardous atmospheres and accordingly should last longer. To illustrate, the author has chosen ammonia as the contaminant type and compared the test requirements in Table 8.21 above.

Specific performance characteristics for surgical and medical facemasks are discussed in Section 8.5.4 below.

### *8.5.1.7 Filter media for respirators*

Glass microfiber media of the ULPA and HEPA (EN 1822) types or of the high efficiency ASHRAE 52-2 or EN779 types are well fit for the performance requirements discussed above. Composite media containing a melt-blown layer, a nanofiber electrospun layer, or a membrane filter layer also find utility in applications where the flammability issue is not critical.

Activated carbon is the most frequently used adsorbent to remove dangerous chemical and fume components from the air. Usually the canisters contain a bed of activated carbon granules or pellets. In some applications, the activated carbon is combined with nonwoven layers to form a composite structure. Other forms of nonwovens containing activated carbon are utilized, such as wet laid webs impregnated with carbon particles and webs coated with activated carbon particles (see Chapter 3, Section 3.10.1).

For severe chemical applications the activated carbon may be treated with metal salts of copper, silver, zinc, and molybdenum and with triethylenediamine (TEDA) for enhanced chem-sorption properties. ASZM-TEDA is the designation for this type of treated activated carbon. Calgon Carbon Corporation of Pittsburgh, Pennsylvania, USA produces this for military gas mask applications.

### 8.5.2 OSHA standard

29 CFR 1910.134 (299), Personal Protective Equipment is the standard called out by OSHA (Occupational Safety & Health Administration) for respirators used in the workplace. Among other things it spells out employer requirements, guidelines and procedures for selecting respirators, employee training, evaluation proceedings. respirator requirements, etc. It does not define filter testing requirements, however, it specifies that filters must be NIOSH certified in accordance with 42 CFR (U.S.), Chapter 1, Part 84 (see Section 8.5).

### 8.5.3 European standards

In the European Union, the use of filtering respiratory protection devices for the protection of your health is regulated by the EC Treaty of June 17, 1987. The minimum requirement for the protection of employees' health must be implemented in national legislation of all EC Member States based on the EC Treaty.

#### *8.5.3.1 The EN 529 standard*

EN 529, is a document which provides employers/users who are required to deal with respiratory protective devices with the information they need to be able to establish and implement an efficient system for respiratory protection management. The standard acts as a guideline document for practical use and contains the basic principles for the selection, use, care, and maintenance of respiratory protective devices. The standard serves as an aid when compiling regulations resulting from national legislation. This standard is another step toward the harmonization of occupational safety and health in Europe.

#### 8.5.3.2 Individual product standards

Following is an overview of the most important European filtering respiratory protection standards. EN standards are drafted and issued by the European Committee for Standardisation (CEN).

1. EN 136: This standard sets the minimum requirements for full face masks. Requirements include temperature stability, impermeability, concentration of carbon monoxide, breathing resistance, speech diaphragms, field of vision.
2. EN 140: Standard providing the minimum requirements for half and quarter masks: This includes inward leakage testing, breathing resistance, flammability, and the mechanical tightness of valves and connections.
3. EN 14387: Standard providing the minimum requirements for gas and combination filters: this primarily covers the classification of filters into various types (e.g., A, B, E, and K) as well as categories (e.g., filter classes 1, 2, and 3).

   Types:
   - Type A: Used to protect against organic gases and vapors with a boiling point of >65 °C
   - Type AX: Used to protect against organic gases and vapors with a boiling point <65 °C, as specified by the manufacturer
   - Type B: Used to protect against inorganic gases and vapors
   - Type E: Used to protect against sulfur dioxide and other acid gases and vapors
   - Type K: Used to protect against ammonia and organic ammonia derivatives
   - Type SX: Used to protect against specially named gases and vapors, as specified by the manufacturer
   - Type Hg-P3: Used to protect against mercury vapors and particles
   - Type CO: Used to protect against carbon monoxide
   - Type NO-P3: Used to protect against nitrate monoxide, nitro gases and vapors and particles

   Filter classes:
   - Class 1: Filters with a low capacity (max. 1000 ppm)
   - Class 2: Filters with a medium capacity (max. 5000 ppm)
   - Class 3: Filters with a high capacity (max. 10000 ppm)
4. EN 143: Standard providing the minimum requirements for particle filters. As is the case with gas filters, particle filters are also subdivided into classes based on their level of performance:
   - P1 filter: Low filter performance (80% efficiency)
   - P2 filter: Medium filter performance (94% efficiency)
   - P3 filter: High filter performance (99.97% efficiency)

5. EN 149: Standard providing the minimum requirements for filtering facepieces for protection against particles. Masks are also subdivided into three filtering face piece (FFP) classes:
   - FFP 1: Low filter performance (80% efficiency)
   - FFP 2: Medium filter performance (94% efficiency)
   - FFP 3: High filter performance (99.97% efficiency)
6. EN 12941: Standard for power-assisted filter systems with helmet or hood. A power-assisted filter system with helmet or hood is classified as follows: TH1, TH2, and TH3.
7. EN 12942: Standard power-assisted full face masks, half masks, or quarter masks. A power-assisted filter system with a full, half, or quarter mask can have one of these classifications: TM1, TM2, and TM3.

#### *8.5.3.3 BS EN 13274*

BS EN 13274 is a supplement to the specific device standards for respiratory protective devices. Test methods are specified for complete or parts of devices. If deviations from the test method given in this standard are necessary, these deviations will be specified in the relevant device standard. Following are the parts of EN13274:

1. BS EN 13274-1:2001—Determination of Inward Leakage and Total Inward Leakage
2. BS EN 13274-2:2002—Practical Performance Tests
3. BE EN 13274-3:2001—Determination of Breathing Resistance
4. BS EN 13274-4:2001—Flame Tests
5. BS EN 13274-5:2001—Climatic Conditions
6. BS EN 13274-6:2001—Determination of Carbon Dioxide Content of Inhalation Air
7. BS EN 13274-7:2008—Determination of Particle Filter Penetration
8. BS EN 13274-8:2002—Determination of Dolomite Dust Clogging.

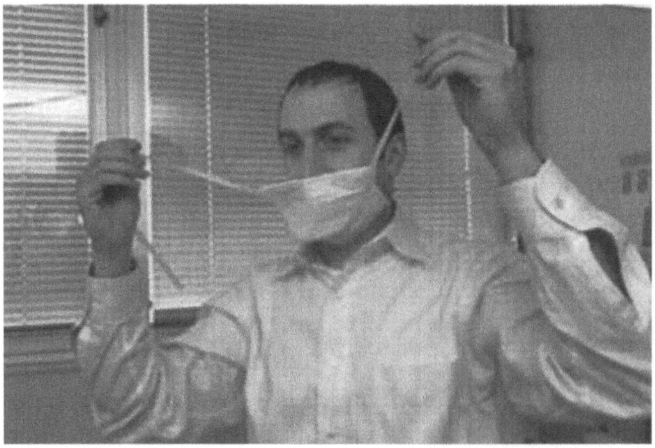

**Figure 8.31** Surgical Face Mask.

### 8.5.4 Surgical and medical face masks

The principal purpose of surgical and healthcare face masks is to protect the wearer against colloidal microorganisms and blood splatters associated with patient care and surgical procedures. Tuneval (300) has suggested that in protecting patients, face masks worn by operating room personnel do very little to reduce bacterial levels in operating rooms. Rawson (301) counters that "the use of surgical masks in protecting the surgical team from contaminated contacts is the principle to consider." The contamination can be in the form of blood, body fluids, secretions, and excretions. Often a shield is worn in conjunction with the facemask to provide eye protection.

Johnson (302) identified five specific performance characteristics for facemasks:

**a.** Bacterial filtration efficiency
**b.** Sub-micron particulate filtration
**c.** Differential pressure (an indication of breathing comfort)
**d.** Fluid penetration resistance
**e.** Resistance to flammability.

Medical face masks generally follow the NIOSH requirements of 42 CFR (US), Chapter 1, Part 84 (181). The efficiency ratings of Section 8.5.1.4 are usually applied. Surgical face masks are medical devices as defined by the United States Food and Drug Administration (FDA). As such they are regulated by the FDA. Other US organizations concerned with standards and guidelines for healthcare face masks are the Centers for Disease Control and Prevention (CDC), Atlanta, Georgia, USA, the American National Standards Institute (ANSI), and the Association of periOperative Registered Nurses (AORN). ASTM International (ASTM) has established standards for filtration including ASTM F2101-07 (bacterial filtration efficiency, BFE), ASTM F2299-03 (initial efficiency by latex spheres), ASTM F1862-13 (resistance of medical face masks to penetration by synthetic blood).

The filter media for surgical masks are usually three layers of three-ply material. This 3 ply material is made up from a melt-blown sandwiched between two non-woven fabrics. The melt-blown material acts as the filter that stops microbes from entering or exiting the mask. Most surgical masks feature pleat/folds allowing the user to expand the mask so it covers the nose and under the chin.

## 8.6 VACUUM CLEANER SYSTEMS

Vacuum cleaners are often classified as bagged or bagless.

Traditional filters for vacuum cleaners are bag filters composed of resin treated filter paper. The filter paper for vacuum cleaner bags is of sufficient strength so as not to burst in its application. The Mullen burst test is commonly used to test for this property. The paper has to be processable in the bag-making equipment. Moisture or

**Figure 8.32** Round HEPA Filter Cartridges for Vacuum Cleaner Systems. *Reproduced with Permission of Tiger-Vac USA Dania, Florida, USA.*

volatiles content is an important property related to the paper's processability. Most vacuum bag media are composed of high quality wood pulp fibers. In some cases where superior strength is required, long vegetable fibers such as Manila hemp (abaca) and/or sisal are utilized. Sometimes, the filter paper is creped to provide more surface area for filtration. Hsieh (303) reported on several significant vacuum bag developments. To name a few, there are "synthetic ultra-fine fiber media, electrets, and composite media." A popular form of composite media developed in recent years is vacuum bag filter paper inner-lined with a melt-blown substrate. In most cases the filter bag is disposable after being filled with dirt. There are some filter bag designs whereby the bag can be emptied and reused. In some cases the bag is washable.

Bagless vacuum cleaners use non-bag filter elements to trap dirt in a dirt cup or chamber that can be emptied and reused. However, the filters often have to be either brushed, washed and dried, or completely replaced.

In many vacuum cleaner applications, the bag or filter element still leaks some dust and allergens that may affect the filter motor and/or leak to the local environment. The motor often requires protection filters. Leakage to the environment may be bothersome to people with allergy problems. As a result many vacuum cleaners are now equipped with post-filters, often a HEPA filter to capture residual traces of dust leaking through the bag.

More elaborate filters are used for centralized and commercial vacuum systems. Figure 8.32 above depicts some of the round HEPA filters used in vacuum systems.

Freudenberg (304) identifies three filter stages for vacuum cleaners. They are tabulated in Table 8.22 below.

Freudenberg also manufactures Separet® filter media designed for vacuum cleaners and room air cleaners. In vacuum cleaners the medium serves as the medium for motor protection filters and exhaust filters. It is an electrostatic medium composed of

**Table 8.22** Filtration Stages in Domestic Vacuum Cleaners

| Filter stage | Aerosol type | Application | Filter media | Remarks |
|---|---|---|---|---|
| **1. Bag** | Coarse dust | Pre-filter | Paper (cellulose) Nonwovens (polymers) | Fine and ultra-fine particles cannot be retained by the paper bag: additional filtration stages are needed. |
| **2. Motor protection filter** | Coarse dust, fine dust | Fine and policing filters | Coarse filter mat, electrets filter or combination of the two | Arrestance of fine particles. In the event of the paper bag being destroyed or incorrectly installed, this prevents the motor being clogged with dust. |
| **3. Exhaust filter** | Ultra-fine dust, micro-organisms, pollens, etc. | Ultra-fine filters (hygiene filters) | Electret filters | Effective final filtration of even ultra-fine particles, including abraded carbon from the motor. |

This combination of filters ensures that 99 % of particles >3 μm and practically 100 % of particles >5 μm are arrested.



a blend of polyester, polypropylene, and polyacrylonitrile fibers. Properties of three grades are listed in Table 8.23 below.

Figure 8.33 is a diagram of a HEPA-ULPA system of the type manufactured by Tiger-Vac USA, Dania, Florida, USA. Note that in this particular design, there is application for five stages of filtration.

This is an example of how a combination of ULPA, HEPA and other types of filters can ensure the trapping of contaminants. There are many other types of configuration, depending on the vacuum cleaner system. This system is a 5-stage filtration system. The 5 stages listed in Figure 8.33 are as follows:

1. Paper filter bag with liner (one micron efficiency) collects the bulk of recovered materials and keeps the other filters clean.
2. Dralon filter with smooth surface prevents caking of fine dust.
3. Polyester filter (one micron efficiency) prolongs the life of the ULPA filter.
4. ULPA filter: Large enough to handle the amount of airflow that the vacuum produces without any significant pressure drop.
5. (Optional 5th stage) Internal HEPA filter can be incorporated in the filtration system for a total of 5 stages. The filter is installed before the motor and is used when recovering toxic or nuisance dusts. It protects the motor from contamination and it allows the motor to be serviced without danger to the operator.

**Table 8.23** Freudenberg Separet® Filter Media for Vacuum Cleaners and Room Air Cleaners

| Property | Units | Separet 402 | Separet 404 | Separet 405 |
|---|---|---|---|---|
| Basis Weight | g/m | Approx. 190 | Approx. 300 | Approx 210 |
| Thickness | mm | Approx. 2.5 | Approx. 3.5 | Approx. 2.4 |
| Color | – | White | White | White |
| Composition | – | Polyester + polypropylene + polyacrylonitrile fibers | Polyester + polypropylene + polyacrylonitrile fibers | Polyester + polypropylene + polyacrylonitrile fibers |
| Availability | – | – Die cuts<br>– Cut Pieces<br>– In widely variable geometries | – Die cuts<br>– Cut Pieces<br>– In widely variable geometries | – Die cuts<br>– Cut Pieces<br>– In widely variable geometries |
| NaCl Penetration @ 8 cm/s | % | ≤6% | ≤1% | ≤0.9% |
| Pressure drop @ 8 cm/s | pa | ≤12 | ≤30 | ≤20 |
| Air permeability | l/s $m^2$ @ 2 bar | ≥1400 | ≥600 | ≥700 |

Reproduced with permission of Freudenberg Filtration Technologies, Weinheim, Germany.

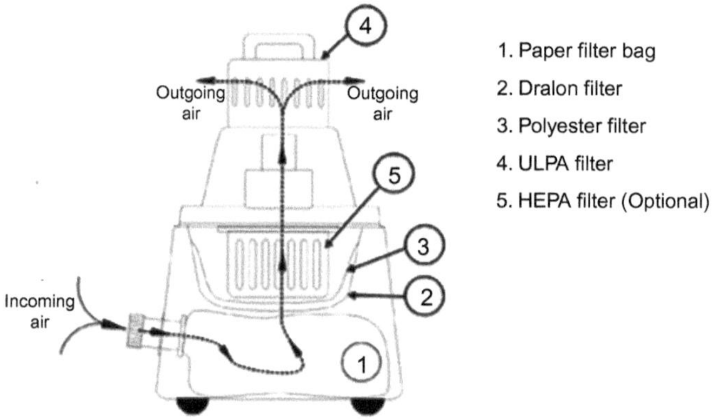


**Figure 8.33** HEPA-ULPA Vacuum System. *Reproduced with Permission of Tiger-Vac USA Dania, Florida, USA.*

ASTM F1977 is a standard test method to determine the initial fractional efficiency of a number of different types of vacuum cleaners including household and commercial canister (tank type), stick, hand held, upright, and utility vacuum cleaner systems. The test is a fractional efficiency test and tests the filter at six discrete particle sizes (0.3, 0.5, 0.7, 1.0, 2.0, and >3 μm). Neutralized potassium chloride (KCl) is the challenge contaminant for this test.

## 8.7 AIR PURIFIERS

Air purifiers are also known as room air cleaners and residential air cleaners. They are usually portable units whose purpose is to provide for clean air in residential environments to include particulate and odor removal. There is increasing customer interest in these units because of concerns about indoor air quality (IAQ). The concerns include odor, gaseous contaminants, microbial activity, allergens, and other particulate irritants such as dust, animal dander, and tobacco smoke. They are perceived as helpful by people who have health problems such as allergies and asthma. Air purifier units come in a number of shapes and sizes. They also utilize a number of different separation technologies. Many do not contain a filter medium but rely on electronic means to separate particles from. Some basic types of air cleaners are mechanical air cleaners, electronic air cleaners, hybrid air cleaners, adsorption cleaners, and ultraviolet cleaners[6].

Mechanical air cleaners contain a medium filter such as a HEPA filter composed of microglass fiber, or an electret type medium. These units are likely to be installed in the ventilation system of a home or as a finishing filter. The filters may be circular cartridge filters, panel filters, or something in between. Usually the medium is in pleated form. The unit may contain an adsorbent such as activated carbon. There is usually an electrostatic charge induced to improve the efficiency of particulate removal. Sometimes the charge is in the form of an electret filter medium. Technostat®, a needle punch felt medium with electret properties, produced by Hollingsworth & Vose Company (305) is one type of medium marketed for air purifier applications. Other forms of filter media include glass microfiber HEPA media, blends of synthetic fiber and glass microfiber, and nanofiber media.

Electronic air cleaners use an electrostatic charge to attract dirt to an electrically charged surface. Filter media is not involved in the separation. These air purifiers use a method called corona discharge to create charged molecules called ions. The corona discharge is a small but intense electrical field. Particles passing through it will pick up an electrical charge. The larger the particle in air the more likely it is to be charged by the corona discharges. Once a particle is charged, it will be attracted to anything with the opposite charge. Paired metal plates within the purifier are charged (one negative, one positive) to attract these particles. Also, particles in the air will be drawn to each other if they have opposite charges. As the particles clump together, they get heavier and eventually they settle out of the air. Very often an electronic air cleaner will contain a prefilter to take out very large particles before the electronic separation. This prefilter may be no more than a tightly woven wire mesh screen

[6] Not included here are ozone generator type of air purifiers. There is some question about the health effects of this type of unit.

**Table 8.24** Measurable CADR Ranges

| Particle type | Measurable CADR rating | Measurable particle size range |
|---|---|---|
| Tobacco smoke | 10 to 450 cfm | 0.1 to 1.0 μm |
| Dust | 10 to 400 cfm | 0.5 to 3.0 μm |
| Pollen | 25 to 450 cfm | 5 to 11 μm |

designed to catch very large particles. It may also be something like the H&V Technostat media mentioned above.

Hybrid air purifiers combine the electronic cleaner mechanism with a high efficiency filter often a HEPA filter. The intent is for the HEPA filter to get the very fine particles such as cigarette smoke.

Adsorption or gas phase cleaners utilize an adsorbent such as activated carbon and/or sodium bicarbonate (baking soda) to remove odors, and undesirable vapors and contaminants from the air. Often this function is combined with a particulate air filter media in hybrid air filters.

Sometimes an ultraviolet light source is incorporated into the cleaner. The purpose is to kill undesirable micro-organisms and germs.

AHAM (the Association for Home Appliance Manufacturers), Washington DC, USA) has developed an ANSI (American National Standards Institute, Washington DC, USA) approved standard for air purifiers, ANSI/AHAM AC-1-2006. The ratings are based on CADR (clean air delivery rate) ratings. CADR is defined as "the measure of the delivery of contaminant free air by a portable household electric room air cleaner, expressed in cubic feet per minute (cfm)" (306). The higher the CADR number better is the purifier performance. The ratings are related to room size and percentage of particles removed (smoke, dust, and pollen). CADR ratings are based on an air purifiers performance in removing tobacco smoke, dust, and pollen. An individual CADR rating is assigned to each of the contaminants. The measurable range of each contaminant is listed in Table 8.24 above.

The test is designed to test any kind of air purifier regardless of separation technology. The unit, to be tested, is installed in a standardized sealed chamber, 1008 $ft^3$ in volume. The chamber is conditioned, then the contaminant is introduced. There is a separate test for each contaminant. The test is started with a natural decay test, that is, the purifier device is not operating. It lasts 10 or 20 minutes depending on the particle type being tested (10 minutes for pollen, 20 minutes for tobacco). The test is repeated with the purifier device turned on. The measure rate of decay is determined and compared to the natural decay rate. The CADR is determined from Eq. 8.6.

$$\text{CADR} = V(k_e - k_n) \tag{8.6}$$

where

CADR is the clean air delivery rating, cfm.

$V$ is the test chamber volume, $ft^3$ (in this case 1008 $ft^3$).

$k_e$ is the total decay (measured) rate, $min^{-1}$.

$k_n$ is the natural decay rate, $min^{-1}$.

AHAM is accredited by ANSI to certify CADR ratings and issue certification labels.

CHAPTER 9

# Engine Filtration (Transportation)

## 9.1 INTRODUCTION

### 9.1.1 Introduction to engine filtration

Engine filtration is often referred to as transportation filtration. It is the term applied to the filtration requirements of engine-powered vehicles and equipments and their fuel supply support systems. This includes automobiles, trucks, farm equipments, off-road construction equipments, aircrafts, railroad trains, boats, and a multitude of specialty engine-operated equipments ranging from mine-drilling equipments to lawn mowers. It also includes the filtration needs of the fuel supply system to keep these engines running. The global market size of engine filtration (estimated to be 150,000−200,000 tons/year in 2016) combined with the complexities of the various filtration mechanisms justifies a complete chapter in this book for the subject. Figure 9.1 illustrates the complexities of engine filtration.

Most of the medium tonnage for engine filtration are in the form of wet laid resin-treated media, however there is a continuing development of dry formed media particularly in air filtration. In addition, nanofiber media are finding more and more utility in both air and liquid filtration applications. The medium is usually used in a pleated form, either as a cartridge or as a panel filter. The filter shape can be quite varied depending on application and location. Often the medium will be corrugated, particularly for lube oil applications. Traditionally, the base sheet for resin-treated media is cellulose, however, there has evolved a growing increase of synthetic fibers. Composites are developing as basic medium in many engine filter applications. Cabin air filtration has emerged as a standard component of almost all automobiles. Much of the filter media being presented for cabin air filtration are in composite form. This includes filter layers coated with melt-blown webs and electrospun nanofibers. Many of the composites include activated carbon either as a composite layer or as an impregnant.

The major volume uses for engine filters are identified as follows:

**a.** Lube oil filtration
**b.** Air intake filtration (AIF)
**c.** Fuel filtration
**d.** Cabin air filtration.

In addition, there are filters for transmission fluids, power steering fluids, brake systems, exhaust systems, crankcase emissions, etc. Computerization of the engine has added filters to protect the electronic components.





*Handbook of Nonwoven Filter Media*
ISBN: 978-0-08-098301-1

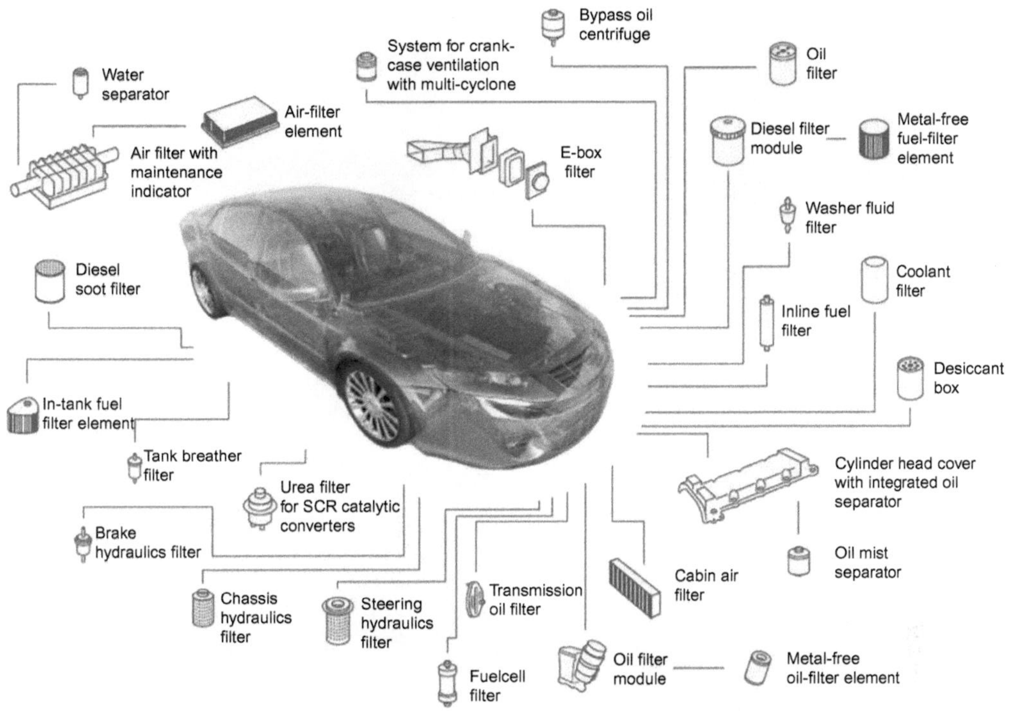


**Figure 9.1** The Filters of an Automobile. *Reproduced with Permission of Freudenberg Filtration Technologies, Weinheim, Germany.*

It is important to mention that the development and growth of hybrid and electric vehicles will revise engine filtration systems: eliminating some filter requirements and creating new ones. For example, it is imagined that 100% electric powered vehicles will no longer need fuel filters.

## 9.1.2 The filters in engine-powered vehicles

In Figure 9.1 above Kreuter (307) identified 25 of the various filters that can be found in an automobile. An engine powered vehicle requires both liquid and air filters. Donaldson Co., Inc. has provided Figure 9.2 which reveals the filters in several types of heavy-duty vehicles.

## 9.1.3 Global Automotive Declarable Substance List (GADSL)

A recent development in the regulation and management of automotive parts, which includes filter media, is the Global Automotive Declarable Substance List (GADSL). The GADSL covers declaration of certain information about substances relevant to parts and materials supplied by the supply chain to automobile manufacturers. GADSL has become significant enough that suppliers are now advertising their

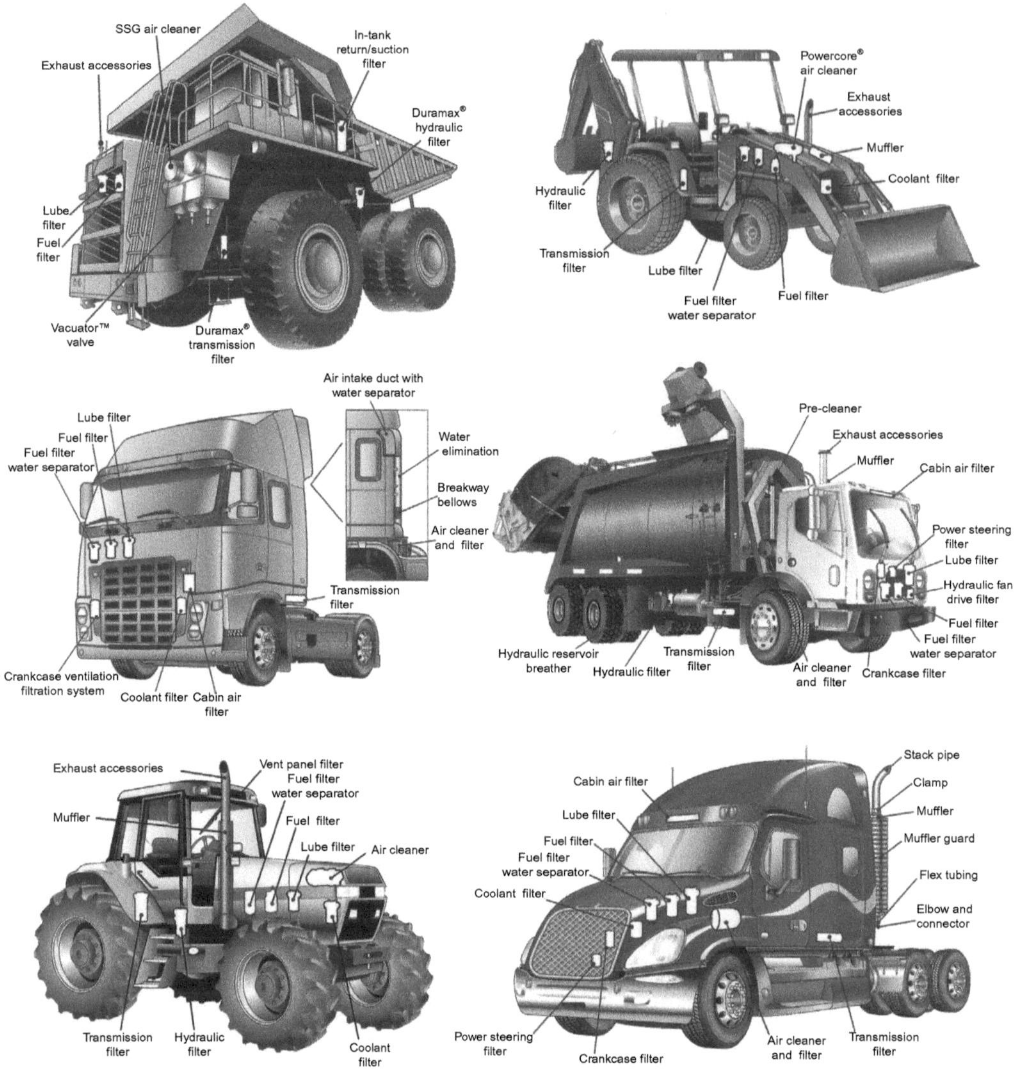


**Figure 9.2** The Filters of Heavy Duty Vehicles. *Reproduced with Permission of Donaldson Company, Inc. © Donaldson Company, Inc. All rights reserved.*

products as being GADSL-compliant; For example H&V (Hollingsworth and Vose Company) advertises its F260 flame-retardant phenolic resin system as being "fully GADSL-compliant." The intent of GADSL is to become the company-specific list for declaration of parts composition within the automotive industry. It provides a definitive list of substances. The scope is to cover declarable substances in the flow of information relevant to parts and materials supplied to the automotive industry; from production to the end-of-the-life phase.

GADSL was first organized in 2005 and is administered by a group of automobile manufacturers, tier suppliers, and material suppliers called the Global Automotive Stakeholder Group (GASG). It represents three regions: Americas, Europe/Africa/Middle East, and Asia/Pacific.

## 9.2 LUBE OIL FILTRATION

Lubrication oil filters play the important role of protecting the engine from wear and damage due to particulate buildup. Particles develop in the engine from a number of sources. This includes fine particles of dust from the outside air that penetrates even the most efficient lube oil filters. There are residual contaminants from the initial production of the engine. The engine operation generates metal fines from wear and soot from fuel combustion. In addition, the oil gets contaminated with acids from the combustion process and diluted with unburned fuels. The oil itself is exposed to an adversarial environment of high temperatures and constant shear. In time, decomposition occurs and the products of decomposition also become a member of the family of oil contaminants.

The purpose of the lubricating oil is to reduce friction in bearings, clearance gaps, and moving parts. It dissipates heat and provides protection against corrosion. Thin films of the oil provide a seal to the combustion chamber. Finally the oil keeps the engine clean by loosening and dispersing impurities and preventing them from forming deposits. If oil is not sufficiently clean, contaminants can cause wear and abrasion of engine parts. This is particularly significant when the oil operates in the narrow gaps between moving parts such as bearings and the annular space between piston and wall.

If the engine oil is not changed at appropriate intervals and/or it is not well filtered, the engine will experience accelerated wear and tear, suffer in performance, and be exposed to an increased risk in damage. Figure 9.3 is a picture of an automotive cylinder and a sample of the wear that can be caused by particle abrasion.

Particle size is significantly important as a factor affecting engine wear. The control of contaminants and their particle size is an important function of the lube oil filter. Staley (308) determined that clearances in diesel and gasoline engines ranged from 2 to 22 μm, suggesting that particles in this size range were lost deterious to engine wear. Staley also demonstrated that increasing the single-pass efficiency of the oil filter will effectively reduce abrasive engine wear. Improved production techniques have reduced lubricating clearances to the point where particle sizes as small as 1 μm are becoming critical, especially when they occur in high concentrations. Durst et al. (309) state that severe engine wear is caused by particles in the range of 8 to 60 μm. They provide Figure 9.4 showing the effect of wear relevance (abrasion) as a function of particle diameter.

There is a large assortment of lube oil filters based on the make, year, and style of the vehicle in which they are installed. Figure 9.5 is an illustration of filters made from various media.

**Figure 9.3** Traces of Wear on a Piston Caused by Particle Abrasion.

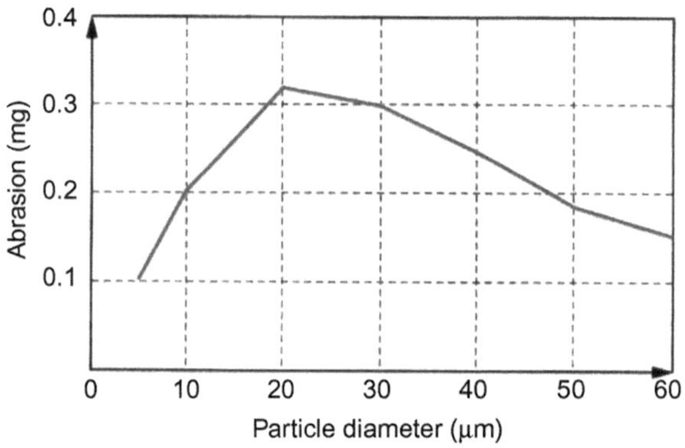


**Figure 9.4** Graph Showing the Effect of Wear Relevance by Particle Size. *Reproduced by Permission of Verlag Moderne Industrie, Germany.*

Knize (310) undertook a methodical study of the various brands of lube oil filters by manufacturer and their internal construction. Figure 9.6 is a sampling from the large selection of filters examined by Knize.

The general operation of a lubricating oil filter system is based on three possible systems:

1. Full flow—all the flow of oil is transported through the filter
2. By-pass—some or all of the oil is diverted through a by-pass line. Since all of the oil cycling through the engine will sooner or later go through the filter, by-pass filtration allows for a more efficient medium. By-pass filtration also protects against catastrophic engine failure due to filter pluggage. If the filter plugs, a differential pressure will direct all of the flow through the by-pass line. The rational is that, even if the oil is not filtered, the engine will continue to function for some period and the detrimental effect to the engine will be gradual.
3. Full flow–bypass combination

**Figure 9.5** Selection of Oil Filers (Metal-Free Elements and Spin-Ons with Paper, Composite, or Synthetic Media). *Reproduced by Permission of Verlag Moderne Industrie, Germany.*

**Figure 9.6** Sampling of Oil Filters Examined by Knize. *Reproduced with Permission of R.W. Knize.*

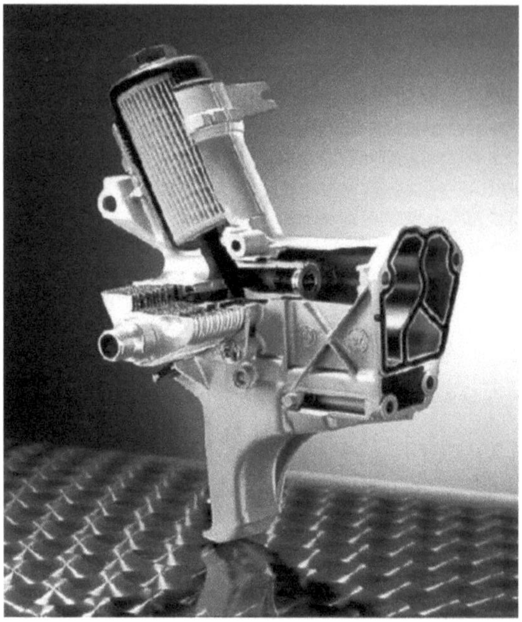

**Figure 9.7** Oil Filter Module. *Reproduced by Permission of Verlag Moderne Industries, Germany (310).*

The types of filters illustrated in Figures 9.5 and 9.6 are of the spin-on type. They consist of pleated media configured into a star-shaped configuration. Pictures of a spin-on type oil filter, showing the inside construction has already been demonstrated in Figures 7.13 and 7.14 of Chapter 7. The spin-on is currently the most common form of lube oil filter used in vehicles. A modern and more expensive option is the oil filter module. A module for a four-cylinder diesel engine is depicted in Figure 9.7 above. The module combines the oil filter with other functions for managing the oil and its flow. This includes the oil cooler, oil pressure and temperature monitors, and control of liquid flows (oil and water). A major advantage of the module is that it takes up less of the critical space in the engine compartment, and yet still provide for easy change of the disposable filter element.

Just as the design of filters for engine oil lubricants in a state of dynamic change, so is the nature of the filter medium. The traditional filter medium for engine oil filtration is a resin-treated paper. The medium is usually treated with a phenol formaldehyde resin to provide for resistance and durability in the hot oil environment. The media is produced as "B-Stage" for the pleating operation and then fully cured for the application. B-stage is the condition where the medium has been dried to a specified volatiles content, but has not been fully cured. It is still soft and pliable for the pleating operation. This is one advantage of the phenolic resin, because once the paper is pleated and formed into a filter element, curing will provide the paper pleats with the stiffness and rigidity to stand up to the adversity of hot oil flow (up to 120 °C).

Most often, resin-treated paper for lube oil filtration is corrugated. Very often, it will be embossed with dimples to provide for pleat separation (see Figure 3.10 of Chapter 3).

It should be noted that one disadvantage of B-stage paper is that it contains up to 10% volatiles that have to be removed in the pleat-curing operation. This results in some undesirable contaminants that have to be environmentally managed from the curing oven. These volatiles include volatile organic compounds (VOCs) and formaldehyde. medium suppliers are therefore offering advance cure and/or precured filter paper to reduce oven volatiles and operating cost. Hollingsworth and Vose Company (H&V) offers an advanced cure resin (ACR) technology (190)[1] that they claim offers the following advantages:

1. Lower running costs with reduced or no oven use
2. Higher throughput and productivity
3. Lower emissions and reduced environmental impact

Table 9.1 illustrates the variety of lube oil filter grades offered by H&V. Note that flat (plain) and corrugated grades are available. Basis weight ranges from 74 to 132 lb per 3000 $ft^2$. Permeabilities range from 26 cfm to as high as 109 cfm; the lower cfm being the more efficient media. Of the grades listed two are for heavy duty applications and the others are for automotive lube oil filtration.

The demand for more modern forms of filter media is motivated by the following concerns:

a. Environmental concerns—Not only do solvent-based phenolics emit VOCs and formaldehyde, but so do water-based resins although to a lesser extent; both undesirable for the air environment and regulated by the environmental protection organizations of many countries. Waste paper treated with phenolic resin is also a disposal concern.
b. Filtration performance—The need for higher quality filtration is demanded by the higher quality of engines being manufactured. Improved filtration efficiency is needed to address the increased concerns of engine performance. High dirt-holding capacity is demanded because the filters are becoming smaller. There is also a desire for longer filter life so as to reduce the disposal quantities of used filters after replacement.
c. Space—Space in the engine compartment is on short supply because of all the modernization occurring in this location of the vehicle. Smaller filters to fit in these spaces demand filter media with improved dirt-holding capacity to maintain filtration quality.

[1] See Chapter 4, Section 4.4.1 for a discussion of H&V resin technollogy.

**Table 9.1** H & V Lube Media Grades

| Grade | Description | Basis weight (lb/3000 sqft) | Max pore (μm) | Fraizer permability (cfm) | Thickness (in) | Corrugation (in) |
|---|---|---|---|---|---|---|
| 1599 VH255 | HD-Full Flow Lube | 122 | 68 | 48 | 0.032 | Plain |
| FL50262 | HD-Full Flow Lube | 131 | 79 | 80 | 0.054 | 0.014 |
| 1808 VH255 | Auto-High Performance Phenolic | 74 | 80 | 104 | 0.024 | Plain |
| 930/19 VH188 | Auto-Flexible Phenolic | 80 | 87 | 109 | 0.027 | 0.011 |
| FL50165 | Auto-Aftermarket Common Lube | 82 | 79 | 62 | 0.034 | 0.015 |
| 1558 VH188 | Auto-Flexible Phenolic | 85 | 73 | 80 | 0.032 | Plain |
| 1819/1 A250/3 | Auto-High Performance Phenolic | 89 | 78 | 94 | 0.028 | Plain |
| 390/1 VH188 | Auto-Flexible Phenolic | 115 | 72 | 47 | 0.035 | 0.011 |
| FL50128 | Auto-High Efficiency Private Brand | 132 | 52 | 26 | 0.047 | 0.013 |

Reproduced with permission of Hollingsworth and Vose Company, Inc. East Walpole, Massachusetts, USA.

**d.** Electric and hybrid vehicles

In response to the challenges, the following media developments are in continuing progress:

**a.** Increased use of synthetic fibers

**b.** The use of dry formed nonwovens

**c.** The advent of composite structures

**d.** The development of advance cure resin systems.

Durst et al. (309) compared a number of different media including a composite structure as well as high synthetic content media. Their results are presented in Figure 9.8. The data suggests that by the appropriate choice of synthetic fibers and/or composite structures, the filtration efficiency performance of a lube oil filter can be greatly improved.

The use of synthetic fibers can improve other properties in addition to filtration efficiency. The presence of synthetic fibers in filter media improves flex resistance. In one of their advertising brochures (311), Ahlstrom, a major manufacturer of filter media for engine filtration, demonstrated that synthetic fiber in one of their oil filter grades gave better wet oil burst strength after prolonged aging in hot oil (see Figure 9.9).

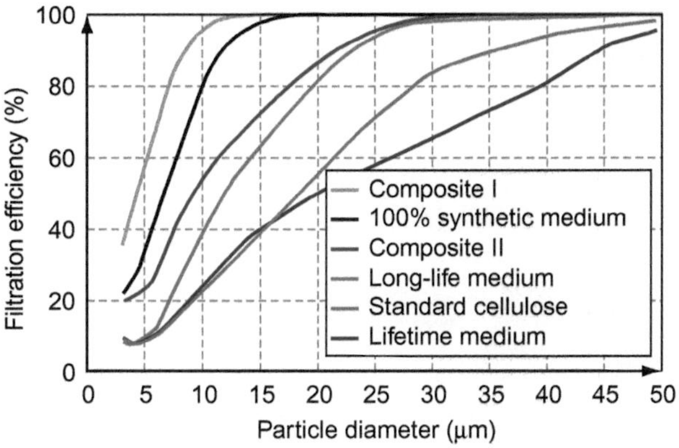


**Figure 9.8** Comparison of Different Filter Media for Lube Oil Filtration efficiency. *Reproduced by Permission of Verlag Moderne Industrie, Germany (p. 54).*

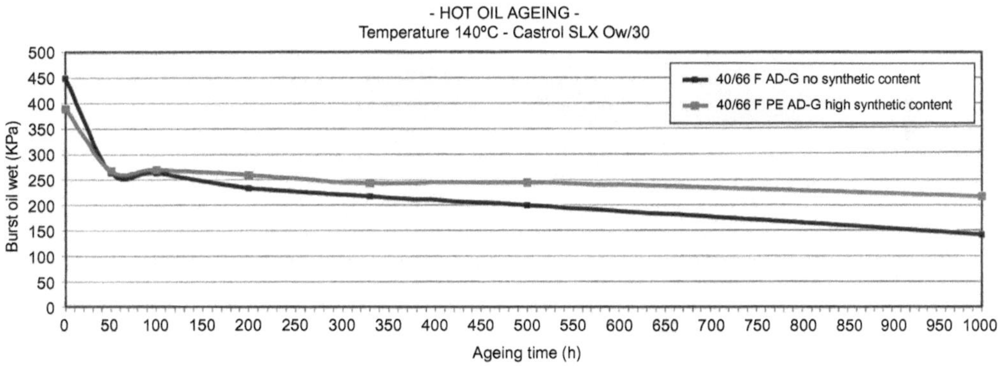


**Figure 9.9** Effect of Hot Oil Aging on the Wet Oil Burst Strength of Lube Oil Filter Media with and without High Synthetic Content. *Reproduced by Permission of Ahlstrom, Madison, Kentucky USA.*

Cummins Filtration (formerly Fleetgard) offers a multilayer synthetic medium labeled as Stratopure™. It contains melt-blown technology and is designed to give a gradient density effect. Cummins (312) report the following benefits:

1. Strength and durability
2. Less compromise of efficiency and life due to multilayering
3. Organic contaminant tolerance
4. Lower filter restriction

Test procedures for lube oil filters is governed by SAE code HS-806/2009 (313). Among other things the code provides test procedures for:

1. Resistance to flow
2. Filter capacity and contaminant removal characteristics of full flow filters
3. Single-pass particle retention capability test

4. Media migration test
5. Ability to meet environmental conditions, which includes water, high oil temperature, and engine oil additives

## 9.3 AIR INTAKE FILTRATION (AIF)

**Figure 9.10** Engine Air Intake Filters by Donaldson. *Reproduced with Permission from Donaldson Company, Inc.*

The purpose of an engine air induction system is to deliver clean air to the throttle body of the engine, while providing engine sound tuning with minimal power loss. Critical to the operation of the system is the air induction filter. The air intake system can be the major source of contaminants entering an engine if its filtration performance is inadequate. Engine wear caused by contaminants is a function of total contaminant mass, chemical composition, and size.

The two major AIF categories for engine-powered vehicles are automotive air intake filtration (AIF) and heavy duty air filtration (HDA). The medium for these two applications are quite different in properties. This is evident in the tabulation of some H&V engine inhale air filter grades in Table 9.2. Listed are three HDA grades (FA6405LW, FA2209, and 1822VH255) and eight AIF grades. The three HDA grades have lower permeabilities and lower maximum pore sizes. This is because HDA media requires more efficiency because heavy duty equipment such as tractors and construction equipment may operate in very dusty environments.

Both phenolic and acrylic resins are used for air intake filters depending on the application. Phenolics are often preferred where heat resistance is an issue. The temperature underneath the engine hood can be as high as 250°F. Flame resistance is

**Table 9.2** H & V Engine Air Intake Grades

| Grade | Description | Basis weight (lbs/3000 $ft^2$) | Max pore (microns) | Frazier permeability (CFM) | Overall caliper (in) | Corrugation (in) |
|---|---|---|---|---|---|---|
| FA6405LW | OE HD Air 99% Initial Eff. | 68 | 39 | 13 | 0.026 | 0.016 |
| FA2209 | Cured Light Duty Truck | 80 | 62 | 35 | 0.030 | 0.018 |
| 1822 VH255 | HD-High Performance Phenolic | 73 | 42 | 20 | 0.019 | 0.011 |
| FA7195 | HD-Pleated Safety Media and Auto Air | 74 | 88 | 70 | 0.028 | 0.011 |
| FA7077E | Auto-Aftermarket | 70 | 80 | 85 | 0.022 | Plain |
| 1833 VH255 | Auto-High Performance Phenolic | 77 | 47 | 29 | 0.020 | 0.011 |
| 930/20 VH134 | Auto-Acrylic, No cure required | 78 | 88 | 110 | 0.023 | 0.011 |
| FA50198SRC | Auto OE Cured, Smolder Resistant | 82 | 65 | 50 | 0.032 | 0.011 |
| 1837/22 VH255 | Auto-High Performance Phenolic | 85 | 61 | 60 | 0.025 | Plain |
| FA6456 | Auto-Private Brand | 86 | 82 | 90 | 0.027 | Plain |
| FA50303 | Auto-OE Air Curing | 93 | 84 | 70 | 0.026 | Plain |

often a required property of intake air filters. Depending on the application, the medium may be corrugated or flat. If the medium is flat, usually it will be emboss dimpled to provide for pleat separation.

### 9.3.1 Automotive air filtration

Compared to HDA applications, media for automotive air filters are more porous and the filters are comparatively small. They can be broken down into light- and medium-duty AIFs. They are designed to fit in the engine compartment of the vehicle which includes automobiles, SUVs, panel trucks, vans, and small commercial trucks. The design attempts to maximize filter life (dirt-holding capacity) and maintain adequate filtration efficiency to protect the engine. The medium grammage or basis weight ranges from as low as 110 $g/m^2$ for light weight panel air filter media to as high as 210 $g/m^2$ for composite dual layer constructions. Permeabilities range from 250–400 $l/dm^2/min$ @12.7 mm. Heat and temperature resistance are important requirements of an automotive air intake filter exposed to the heat of a hot running engine. In many cases the medium is expected to be flame retardant. The filter elements are for one time use and disposed after achieving their useful life. No attempt is made to recover the used elements for reuse.

### 9.3.2 Heavy duty air (HDA) filters

Heavy duty air filters are designed for large trucks, trailer tractors, farm equipment, off-road construction equipment, buses, railroad engines, and military vehicles such as tanks and helicopters. The filters are much larger and usually installed outside the engine compartment. The engines that are designed to protect are high capital items; therefore the filtration efficiency demands for HDA are much greater than for automotive AIF. The grammage is in the range of 90–160 $g/m^2$. Air permeabilities range from 30–90 $l/dm^2/min$ @12.7 mm. HDAs, because of their size and efficiency requirements, can be quite expensive and are designed to be washed and reused. It is estimated that the typical HDA is washed and reused 2–3 (need reference) times before it is discarded.

Bugli (314) has provided Tables 9.3 and 9.4 that compare the difference in requirements for light/medium duty AIF versus heavy duty AIF. Table 9.3 compares service requirements. Note that the HDA AIF uses larger filters and are expected to last up to 500,000 miles (1,100,000 km) whereas, the light/medium duty filers are expected to last at the most to 100,000 miles (220,000 km). The HDA life is optimized based on pressure are changed based on pressure drop. The filters for smaller vehicles tend to be underutilized and changed, based on the vehicle owner's judgment. Also note from the design specifications of Table 9.4 that the HDA filters are expected to have much higher dust capacities and much higher filtration efficiencies.

**Table 9.3** Comparison of Service Conditions for Light/Medium Duty AIF and HDA AIF

| Service conditions | Light/medium duty AIF | Heavy duty AIF |
|---|---|---|
| AIF service criteria | by distance (miles, km.) | by a set restriction rise |
| Measurement method | NA | Pressure/vacuum gage |
| For normal driving conditions | ~30K–45K miles | @ 6.25–7.5 kPa vacuum |
| For severe driving conditions | ~ 5K–10K miles | @ 6.25–7.5 kPa vacuum |
| AIF usage | under utilized | Optimized |
| Real World service intervals - normal conditions | ~30K–100K miles | ~150K–500K miles |
| Typical Customer | over-services every 5K–15K miles | services when required (at design intent) |
| AIF loading | lower than design intent | Optimized |
| AIF efficiency | lower than recommended | Optimized |
| Engine wear | may be higher | significantly lower |
| Cost of ownership | Higher | Lower |
| Value to customer | Lower | Higher |

Reproduced by Permission of the American Filtration and Separations Society.

**Table 9.4** Comparison of Design Specification as for Light/Medium Duty AIF and HDA AIF

| Design specifications | Light/medium duty AIF | Heavy duty AIF |
|---|---|---|
| Dust Capacity | ≥ 100–400 g | >1000–5000 g |
| Initial Efficiency | >97%–99% | >99.5%–99.9%+ |
| Overall Efficiency | >98%–99% | >99.9%–99.99% |
| Filter restriction | ~0.125–0.5 kPa | ~0.125–0.5 kPa |
| AIS System restriction | >2.0–4.0 kPa | >2.0–3.5 kPa |
| Flow rates | 2.5–18 $cm^3$/min | 15–50 $m^3$/min |
| Media face velocities | 20–35 cm/s (paper)<br>25–75 cm/s (felts) | 1.5–5 cm/s (paper) |
| Media area | >0.3–2.0 $m^2$ | >2.5–15.0 + $m^2$ |
| pre-separators | NA | cyclonic/inertia type |
| pre-separator efficiency | NA | ~50%–85% |
| Pre-filters | foam, felts | foam, felts |
| ***Typical AIF sizes: Typical*** | | |
| Panel (L × W × plHt) | 150–400 mm × 125–300 mm × 25–75 mm | 575 mm × 675 mm × 50 mm |
| Cylindrical (OD × OHt × plHt) | 100–300 mm × 150–300 mm × 20–63 mm | 300–650 mm × 700–1000 mm × 25–50 mm |

Reproduced by permission of the American Filtration and Separations Society and Neville Bugli.

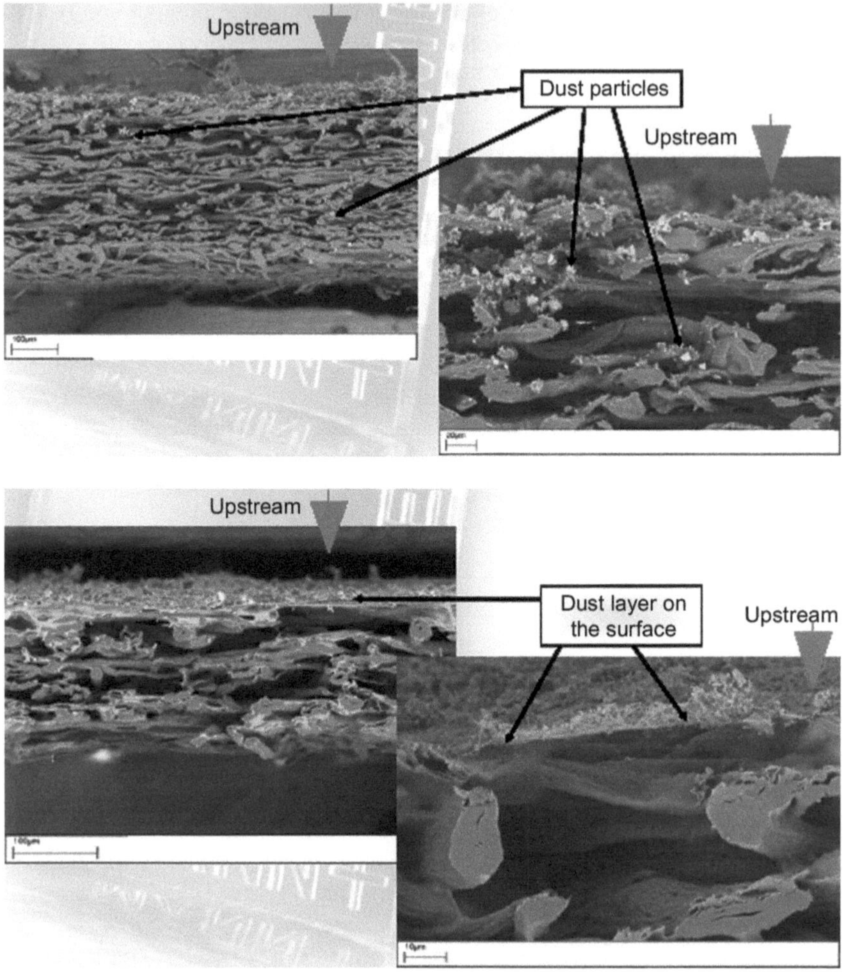


**Figure 9.11** The Depth Penetration of Particles into a Cellulose Filter Medium Without Nanofiber Coating and the Surface Filtration Effect of the Cellulose Medium with Coating. *Reproduced with Permission of Mann + Hummel GmbH, Ludwigsburg, Germany.*

Resin-treated cellulose media from the wet lay process are the most frequently used media for AIF applications. In some applications, such as for military vehicles, the medium may be reinforced with synthetic fibers for high burst strength and greater durability.

Interest has generated for nanofiber coated filter media. Trautmann et al. (315) illustrated the filtration effect in the microphotographs of Figure 9.11. Presumably, the nanofiber coating displayed here is from the electrospun process. The photographs are cross sections of a cellulose medium with and without the nanofiber coating and at two levels of magnification. The photographs were taken after the medium had been tested in accordance with ISO 5011:2000 (since updated to ISO 5011:2014)—the performance test standard for inlet air cleaning equipment for internal combustion engines and compressors. They show that without the nanofiber coating, the dust particles penetrate the medium, a depth filtration effect. On the other hand, the dust particles do not penetrate the nanofiber-coated medium, but rather take over the surface of the nanofiber coating. This is consistent with a study by Graham et al. (29) discussed previously (Chapter 3, Section 3.5.1.1) that indicated that dust particles have a greater affinity for the nanofiber layer in a nanofiber composite.

The air filter industry uses the ISO 5011 testing protocol when testing filtration efficiency and dust capacity. The test protocol calls for the introduction of a measured amount of "test dust" into the air filter at a selected airflow rate. The test is then terminated after the filter reaches a selected level of restriction (terminal test pressure). The test protocol then measures the percentage of dust retained by the filter (efficiency) and the total amount of dust held by the filter capacity.

## 9.4 CABIN AIR FILTRATION

According to Bräunling et al. (316) cabin air filtration began approximately 60 years ago when a cabin air filter was introduced in the 1938 Nash truck. In the 1970s, Ford explored a cabin air filter based on filter paper impregnated with activated carbon. Also in the 1970s, a small Colorado Company called Dickson, Inc. introduced a hat-shaped fluted medium composed of paper impregnated with activated carbon. This filter was inserted into the air intake chamber that was located between the front window and the rear of the engine that was typical of automobiles of that time. Testing revealed that the Ford and Dickson filters were quite effective in protecting the cabin against on the road automotive emissions, local odors from industry and farms (pig farms, for example), and from allergy causing pollens. For various reasons, the developments did not reach the marketplace or were not successful if they did. The concepts lay dormant until the 1980s when Freudenberg introduced the first effective air filter to hold 100% of the pollen, mold spores, and other contaminants found in air (Bräunling et al. (316)). These filters were first produced for Ford, BMW, and Mercedes Benz. In 1994, Freudenberg introduced the first cabin air filters into the North American market. These were in the Ford Contour and Mystique platforms. In more recent times the use of cabin air filters has spread so that almost every automobile manufactured today contains some form of cabin air filter. In addition, the filtration protection is now being applied to heavy duty and off-road equipment. Truck drivers need protection against pollen, allergens, and fumes as much as drivers and passengers of automobiles. Construction sites can

**Table 9.5** Hollingsworth and Vose Technostat® Grades for Cabin Air Filtration

| Grade | Application | Basis weight ($g/m^2$) | Thickness (mils) | Penetration (NaCl @ 95 lpm, 100 $cm^2$, %) | Resistance (@ 95 lpm, 100 $cm^2$, Pascals) |
|---|---|---|---|---|---|
| **T70/15** | Technostat for Cabin Air | 85 | 1.6 | 18.6 | 6.9 |
| **T50/15** | Technostat for Cabin Air | 65 | 1.3 | 31.0 | 5.0 |
| **T30/15** | Technostat for Cabin Air | 45 | 0.9 | 60.0 | 3.5 |
| **T70(+)15** | Technostat PLUS for Cabin Air | 65 | 1.6 | 10.0 | 8.0 |
| **T50(+)15** | Technostat PLUS for Cabin Air | 45 | 1.3 | 16.0 | 5.0 |

Reproduced by permission of Hollingsworth and Vose Co. Inc., East Walpole, Massachusetts, USA.

generate large amounts of dust, pollen, and lint that can make the operating cabins of the construction equipment very uncomfortable, if the cabin air is not filtered. This is particularly true of mining equipment.

The primary function of cabin air filters is to provide for comfort and protection of the passengers by preventing the intake of particulates such as pollens, spores, allergens and other harmful particles. Adsorbent filters also provide protection against odors and harmful gas pollutants. The filter has a second function in that it protects the vehicle HVAC system from buildup of dirt and debris that might impair the performance of these systems. The life span of the unit is thereby increased and the maintenance decreased.

Bräunling et al. (316) identify the following as the types of medium that might be used in cabin air filters

**a.** Electrospun nonwovens for nanofiber filtration
**b.** Triboelectrically charged needle-punched nonwovens
**c.** Split charged fiber media
**d.** Melt-blown media
**e.** Spunbonded nonwovens
**f.** Dry laid webs
**g.** Wet laid webs
**h.** Adsorptive media (activated carbon and activated alumina)
**i.** Composite structures of the above.

Section 3.7.3 in Chapter 3 discusses triboelectric media and mentions the Hollingsworth and Vose Technostat® triboelectric needle punch medium. Several of these grades designed for cabin air filtration are tabulated in Table 9.5.

**Table 9.6** Hollingsworth and Vose Activated Carbon Grades for Cabin Air Filtration

| Grade | Activated carbon (g/m$^2$) | Particle filter | Particle filter type | Main market (OE/AM) |
|---|---|---|---|---|
| 5243 V3 | 400 | Yes | Technostat | OE |
| 5224 V3 | 390 | Yes | Technostat | OE |
| 5236 V3 | 200 | Yes | Technostat | OE |
| 5338 V261 | 400 | Yes | Standard | OE |
| 5321 V272 | 300 | Yes | Standard | AM |
| 5321 V273 | 300 | Yes | Standard | AM |
| 5342 V273 | 340 | Yes | Standard | AM |
| 5234 V2 | 300 | No | – | AM |
| 5236 V2 | 200 | No | – | AM |

Reproduced by permission of Hollingsworth and Vose Co. Inc., East Walpole, Massachusetts, USA.

Hollingsworth and Vose also promote a line of activated carbon media for cabin air filters. This includes activated carbon versions of the Technostat® grades listed in Table 9.5 above and in addition three-layer laminated media, the carbon layer being in the middle. Table 9.6 lists several of these grades including their carbon content.

Ahlstrom has introduced its Trinetex® technology for cabin air particulate filtration. In this unique technology, a three-layer wet laid media is formed, each layer containing a different furnish of synthetic fibers. The three-layer construction provides for a gradient density effect to increase dirt-holding capacity. It combines mechanical and electrostatic filtration to optimize filtration efficiency. Ahlstrom reports that the medium is available with optional flame retardancy and antimicrobial treatments.

Ahlstrom also markets a line of cabin air filter products under the SafeCabin™ trade name. Table 9.7 lists typical properties of several of these grades. There are three categories: regular, carbon and edge.

ISO 11155-2:2009 specifies a test comprising methods for measuring the dynamic gas adsorption of cabin air filters in the passenger compartments of road vehicles. These laboratory test methods are applicable to air filters that improve air quality by reducing concentrations of gaseous, odorous or hazardous components from ambient or recirculated air, or both, in the vehicle cabin. They provide a means of measuring air pressure loss, as well as gas and vapor removal characteristics.

## 9.5 FUEL FILTRATION

### 9.5.1 Engine fuel filtration

Basically, there are two types of fuel systems for vehicle and equipment engines: gasoline and diesel fuel. Diesel fuel is typically used in larger trucks and off-road equipment, but there are a rising number of cars that run on diesel fuel. Diesel fuel is

**Table 9.7** Alhstrom SafeCabin™ Product Portfolio—Technical Data

| Grade | Basis weight | | Air permeability | | Caliper | | Stiffness |
|---|---|---|---|---|---|---|---|
| | gsm | lb/3000ft² | l/m²/S | cfm | μm | Mils | mg |
| SafeCabin™80 | 80 | 50 | 2100 | 263 | 500 | 20 | 700 |
| SafeCabin™80 HC | 80 | 50 | 4800 | 600 | 760 | 30 | 1000 |
| SafeCabin™ 100 | 100 | 60 | 2000 | 250 | 500 | 20 | 1300 |
| SafeCabin™ 100 FR | 100 | 60 | 2700 | 338 | 700 | 28 | 1000 |
| SafeCabin™ 100 Nano | 100 | 60 | 1200 | 150 | 500 | 20 | 1300 |
| SafeCabin™120 | 120 | 70 | 1600 | 200 | 720 | 24 | 1500 |
| SafeCabin™120 Charge | 120 | 70 | 1500 | 188 | 610 | 24 | 580 |
| SafeCabin™ Carbon 520 Charge | 520 | 330 | 800 | 100 | 2100 | 83 | 6600 |
| SafeCabin™ Carbon 170 | 170 | 100 | 2700 | 338 | 1150 | 45 | 2200 |
| SafeCabin™ Carbon 300 | 300 | 180 | 1400 | 175 | 1700 | 67 | 6300 |
| SafeCabin™ Carbon 500 | 500 | 310 | 1300 | 163 | 2050 | 81 | 6500 |
| SafeCabin™ Edge 200 | 200 | 120 | 800 | 100 | 1070 | 42 | 7900 |
| SafeCabin™ Edge 250 | 250 | 150 | 660 | 83 | 1270 | 50 | 13900 |

Data to be considered as first parameter indicators.

more popular in Europe than the United States, since Europe has less strict emissions standards that allow them to have diesel vehicles, whereas in the United States, not all diesel cars are allowed.

There may be as many as five filters for fuel filtration in a diesel fuel-powered automobile. These include an in-tank fuel filter, a diesel particulate filter, a tank ventilation filter, a fuel filter module with a metal free fuel filter element, and an in-line fuel filter. A range of filtration media are available for these filters depending on performance required and geographical considerations. Petri et al. (317) has identified the following as being part of that range.

- 100% cellulose
  - Cellulose—single layer media
  - Cellulose—single layer gradient density
- Cellulose-polyester blends
- Cellulose/polyester single layer media
- Cellulose-microfiber glass blends
- Cellulose/microfiber glass-single layer media
- Cellulose/microfiber glass—gradient density/dual-phase
- Melt-blown composites
  - Cellulose/melt-blown
  - Cellulose—microfiber glass/melt-blown
  - Cellulose (gradient density)/melt-blown.

In more recent times nanofiber composites are being developed for more stringent fuel filter applications.

The level of removal required to protect modern diesel engines depends on the application. For less demanding, light duty applications beta ratios as low as 64 may be adequate at 4 μm(c). For more severe duty applications beta ratios well in excess of 200 at 4 μm(c) may be necessary.

The important test method for fuel filtration is ISO 16889:2008 (see Chapter 6, Section 6.14.1.5), a multi-pass efficiency test. The test stand is capable of testing both flat sheet media or filter elements and cartridges.

## 9.5.2 Fuel transportation systems

There are a number of distribution systems that distribute oil and fuel to various locations. The transportation system includes ocean tankers, barges, rail, trucks, and pipelines. A pipeline system might direct fuel from a refinery to a terminal to a local terminal to the point of use. For example, the point of use for aviation fuel would be an airport. The airport would have a storage facility to receive the fuel and capabilities to deliver the fuel to the aircraft. Chevron Products Corporation has issued a company publication (Hemighaus et al. (318)) that adequately describes the pipeline fuel distribution system for aviation fuel. Chevron particularly notes the problems of contamination that may be inherent with pipeline fuel. These contaminants include particulate matter and moisture. Fuel surfactants are interference because they inhibit the ability of water particles to coalesce in a fuel filter/separator system. Micro-organisms such as bacteria and fungi may have the ability to flourish in jet fuel. Their biogrowth may result in solids that are very effective in plugging fuel filters.

Chevron discusses the cleaning techniques for contaminated fuel. This includes filters to remove particulate matter and salt driers to remove moisture. Chevron (318) states that "The filter/separator is the workhorse of the processes used to remove free water from aviation fuels." A schematic of a fuel filter/separator is illustrated in Figure 9.12. A manufacturer of this type of equipment is Velcon Filters of Colorado Springs, Colorado, USA. The equipment consists of two filter elements: an inlet coalescing element that coalesces minute droplets of water into drops that are large enough to be gravity separated downstream of the coalescing element; and a separator element, containing a very hydrophobic medium, that rejects any water particles that might be entrained in the exiting fuel stream. The rejected water gravity settles in the sump at bottom of the housing.

The typical coalescing media for the inlet element is usually a resin-treated wet laid medium composed of a blend of cellulose fibers and glass microfibers. Sometimes synthetic fibers such as polyester are added for greater flex resistance. Coalescing

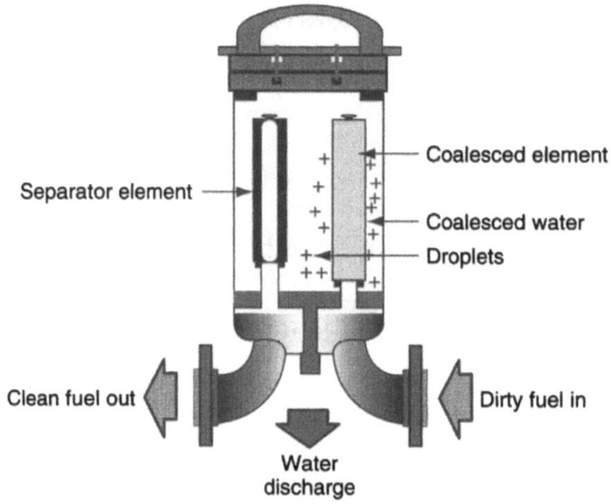


**Figure 9.12** Schematic of a Fuel/Filter Separator.

efficiency is governed by the content and diameter of the glass microfibers. The surface energy of the medium is important. What is desired is a fairly hydrophilic fiber surface that attracts water droplets and holds on to them as they grow in size by being struck by additional droplets. Eventually the coalesced water particles will be of sufficient size to be blown off the fibers by the wind of the fluid stream passing through the medium. Downstream of the medium, the water droplets are of sufficient size to gravity settle in the settling sump. Media treated with phenolic resin generally have the right surface properties for good coalescing and separation efficiencies. On the other hand medium with low surface energy coatings such as fluorocarbons or silicones will generally have poor coalescing and separation efficiencies.

The exit element or separator needs to be very hydrophobic. It must reject any small water particles that are entrained in the exiting fuel stream. The medium for the exit elements will be coated or treated with a very low energy additive such as a fluorocarbon or silicone.

Standards for fuel filter/separators are set-up by a joint committee of the American Petroleum Institute (API), Washington, DC, USA and the Energy Institute (EI), London, England. The Energy Institute was set up in 2003 as a result of a merger between the Institute of Petroleum (IP) and the Institute of Energy (InstE). Standards were given an API/IP designation. This designation still exists even though the Institute of Petroleum (IP) no longer exists by that name. For example one of the more important specifications concerning fuel filter/separators is API/IP 1581 5th Edition. This test procedure specifies a multipass filter test. Other API/EI standards include: API/IP 1582 1st Edition (similarity for AP/IP 1581), IP 1583 5th edition (aviation fuel monitors), API/IP 1590 2nd edition (microfilters), and API/IP 1599 1st edition (aviation fuel dirt defense). A recognized laboratory for testing fuel filter/

separators and other aviation fuel filtration devices is the Southwest Research Institute (SwRI), San Antonio, Texas, USA.

Filter monitors or fuse filters are a type of filter that uses a super absorbent polymer to absorb moisture from a fuel stream. If there is an over abundance of moisture or water in the fuel the polymer will swell and expand to shut down a pipeline fuel stream and protect the system to wherever that fuel is flowing This is well discussed by Bresee (319). Fuse monitors are usually installed at the point where fuel is loaded into the fuel tanks of an aircraft. The polymer medium is a highly absorbent material, such as carboxy methyl cellulose (CMC) that can absorb several times its weight in water, and of the type that is used in highly absorbent hygienic pads. Fuse filters are regulated by API/IP Specification 1583.

# CHAPTER 10

# Federal Law and Regulatory Activity That May Relate to Nonwoven Filter Media

Nonwoven filter media are affected by Federal Laws and regulatory activities. The world is going green and the main concerns are environmental control and public safety. Sustainability is now a significant point of discussion in business negotiations. New law regulations and standards will affect how the medium is made, how it is used, how it is processed, how it is disposed, what the raw materials may be, how it is packaged and labeled, how it is specified, and how it affects the environment. Nonwoven filter media may be affected not only by local regulation of specific nations, but also by global organizations such as the United Nations. Global standards and controls developing that will have an effect on the manufacture, design, and application of filters and filtration media.

It is important to note that the emergence of green standards do not only provide additional challenges to the production of nonwoven filter media, but also provide great opportunity for new technologies, new products, and new markets for nonwoven filter media. Smart filtration companies will enhance their business by taking advantage of "green projects" and becoming the technology leaders of these programs.

The United States Environmental Protection Agency (EPA) is the most prominent government organization involved in the management, administration, implementation, and enforcement of most laws that have environmental concerns.

Following are some of the more significant laws, executive orders, regulations, and guidelines of the United States Government and the global community that may affect nonwoven filter media and filters made from such media.

## 10.1 SUSTAINABILITY

Sustainability is the endurance of systems and processes. The organizing principle for sustainability is sustainable development, which includes the four interconnected domains: ecology, economics, politics, and culture.

Sustainability is concerned with the preservation of our natural resources and biological systems. Some of the major purposes are to minimize environmental impact and to reduce carbon foot print.[1] Many organizations are producing annual

[1] Carbon footprint is the amount of carbon dioxide and other carbon compounds emitted due to the consumption of fossil fuels by a particular person, group, etc.



*Handbook of Nonwoven Filter Media*
ISBN: 978-0-08-098301-1

sustainability reports; an organizational report that gives information about economic, environmental, social and governance performance that relates to their businesses and what they are doing about it. In dealing with sustainability a manufacturer will have to deal with the following issues:

- Biodegradability
- Recyclability
- Ozone safe
- Free of ----
- Carbon offsets
- Made with renewable materials and energy.

The United States EPA administers the sustainability programs in the United States. They state that (320) "EPA efforts in the area of sustainability practices and approaches include labeling green products and promoting green chemistry and engineering, managing materials rather than creating waste, using green infrastructure to manage storm water runoff, and supporting the sustainable design of communities."

## 10.2 CLEAN AIR ACT (CAA), TITLE 42 U.S.C., CHAPTER 85

The Clean Air Act (CAA) (321) is the law that defines EPA's responsibilities for protecting and improving the nation's air quality and the stratospheric ozone layer. Originally passed in 1970, the law was amended by Congress in 1990 with the Clean Air Act Amendments of 1990. Legislation passed since then has made several minor changes. The law was incorporated into the United States Code as Title 42, Chapter 85. The House of Representatives maintains a current version of the U.S. Code, which includes Clean Air Act changes enacted since 1990.

CAA regulates air emissions from stationary and mobile sources. Among other things, this law authorizes EPA to establish National Ambient Air Quality Standards (NAAQS) to protect public health and public welfare and to regulate emissions of hazardous air pollutants. EPA is required to set NAAQS standards for the following six common air pollutants identified as criteria pollutants: ozone, particulate matter, carbon monoxide, nitrogen oxides, sulfur dioxide, and lead.

Section 112 of the Clean Air Act addresses emissions of hazardous air pollutants. CAA Section 112 requires issuance of technology-based standards for major sources and certain area sources. "Major sources" are defined as a stationary source or group of stationary sources that emit or have the potential to emit 10 tons/year or more of a hazardous air pollutant or 25 tons/year or more of a combination of hazardous air pollutants. An "area source" is any stationary source that is not a major source (for example, a dry cleaner).

For major sources, Section 112 requires that EPA establish emission standards that require the maximum degree of reduction in emissions of hazardous air pollutants.

These emission standards are commonly referred to as "maximum achievable control technology" or "MACT" standards.

The new law includes a list of 189 toxic air pollutants whose emissions must be reduced. EPA must publish a list of source categories that emit certain levels of these pollutants. The list of source categories includes: (1) major sources emitting 10 tons/year of any one, or 25 tons/year of any combination of those pollutants; and, (2) area sources. Included in this list is formaldehyde, a known carcinogen. In addition to CAA, formaldehyde is also regulated by the Occupational Safety and Health Act (OSHA) (322) (See Section 10.5) and by the Toxic Substance Control Act (TSCA) (See Section 10.6).

## 10.2.1 Volatile organic compound emissions

The Clean Air Act regulates the emission of volatile organic compounds (VOCs) and hazardous air pollutants (HAP). VOCs are organic chemicals that have a significant vapor pressure at ordinary, room-temperature conditions. If they are allowed to build up in the air, they can harm human health or damage the environment. A significant concern is the reactivity with $NO_x$ compounds to form ground-level ozone, a major constituent of smog. VOCs come from many different classes of organic chemicals, including aliphatic, aromatic and halogenated hydrocarbons, aldehydes, ketones, esters, ethers, alcohols, acids, and amines. VOC emissions are a regulated concern in many industrial operations concerned with the production of filter media. This includes raw materials, processes, product, and customer operations. Even during transportation, VOC content material must be contained and/or packaged so as not to emit emissions during shipment. Note that formaldehyde is a VOC and water is not.

### *10.2.1.1 Sources of VOC emissions*

In the processing of filter media, VOCs most often result from solution-based polymers. This includes solvent-based resins such as phenol formaldehyde that are used as a resin binder for wet lay filter media. Solvent-based polymer solutions used for the spinning of fibers and webs also emit VOCs that have to be controlled. This includes the type of solvents that are used in the flash spun and electrospun processes as well as those used in the spinning of filament fibers such as Rayon. Sources of VOC emissions include coating and saturating lines, fiber and web spinning operations, storage tanks containing solvent or solvent-based materials, drum storage, and fuel storage.

### *10.2.1.2 Control of VOC emissions*

Vatavuk et al. (323) described the various control technologies for VOCs. They can be divided into two general groups: solvent recovery and oxidation. Oxidation can be further divided into two forms: thermal oxidation and catalytic oxidation.

#### 10.2.1.2.1 Solvent recovery

Solvent recovery systems usually involve an adsorbent such as activated carbon, activated alumina, silica gel, or molecular sieves to recover solvent VOCs. The VOCs are desorbed from the adsorbent material with superheated steam. The mixture of steam and VOC vapors are then condensed to liquid form. The VOC solvent is then separated for recovery and recycle. If the solvent is methanol, then the condensate is a miscible mixture of water and methanol and distillation is required for the separation. In such cases the cost of the recovered methanol may be more than the cost of virgin methanol purchased in the marketplace.

#### 10.2.1.2.2 Thermal oxidation

Whereas, solvent recovery attempts to recover and recycle VOC material, thermal oxidation is a process that oxidizes the VOCs to nothing but water and carbon dioxide. The oxidation process may generate heat and energy that might either be recycled within the process and/or be utilized elsewhere in the facility. Figure 10.1 is a general presentation of the thermal oxidation process. The waste stream from the emission source contains the VOCs and HAPs. The heart of the thermal incinerator is a nozzle-stabilized flame maintained by a combination of auxiliary fuel, waste gas compounds, and supplemental air added when necessary. Upon passing through the flame, the waste gas is heated to its ignition temperature (usually 1200–1800°F). The mixture ignites and continues to react as it flows through the combustion chamber. Residence time in the combustion chamber ranges from less than 1 to 2 s. The hot discharge from the combustion chamber is used to preheat the incoming waste stream from the emission source. This is done in a heat exchanger called the waste gas preheater. Additional energy from the exhaust stream may be recovered in another heat exchanger called the secondary energy recovery.

There are four general forms of thermal oxidation: direct fire thermal oxidation, recuperative thermal oxidation, regenerative thermal oxidation, and catalytic oxidation.

**10.2.1.2.2.1 Direct fired thermal oxidation** A direct fired thermal oxidizer is sometimes referred to as an afterburner, enclosed flare, or a vapor combustor. Usually there is no heat recovery. It is used in VOC stream, where the VOC content is high

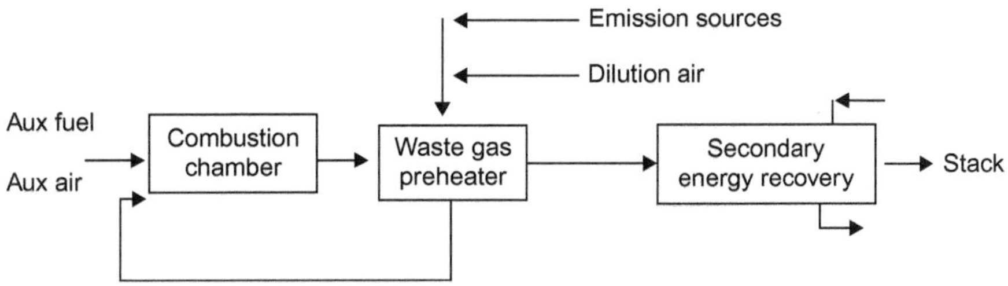


**Figure 10.1** Thermal Incinerator—General Case.

enough to support combustion without additional fuel. There is no waste gas preheater. The waste stream from the emission source exhaust is introduced directly into the combustion chamber. The oxidizer uses a specially designed burner to raise the temperature of the waste air stream to the predetermined combustion temperature. The clean, heated exhaust gas is then discharged to the atmosphere. It is possible to send the gas through an optional heat recovery system.

**10.2.1.2.2.2 Recuperative thermal oxidation** The recuperative thermal oxidizer is designed for streams with low VOC concentrations, and is based on volume of airflow, organic vapor concentrations, and desired destruction efficiency. The operation pretty much follows the process as described for Figure 10.1 above. It includes waste gas preheating and secondary energy recovery. The thermal oxidation temperature generally ranges from 1,200 °F to 1,800 °F (650 °C to 1,000 °C). The specified residence time is 0.5–2.0 seconds. The contaminant-free air is exhausted into the atmosphere. The waste gas heater minimizes the system's fuel consumption with the system being self-sustaining at moderate lower explosive limits.

**10.2.1.2.2.3 Regenerative thermal oxidation (RTO)** The regenerative thermal oxidizer (RTO) is best for VOC streams with very low VOC concentrations. It utilizes ceramic media to recover most of the heat generated during emission combustion. It is capable of near fuel free operation. RTO destroys HAPs, VOCs, and odorous emissions that are often discharged from industrial or manufacturing processes.

Referring to Figure 10.2, waste gas, from an emission source, is either pushed or pulled into the inlet manifold of the ceramic oxidizer via a system fan. Poppet control

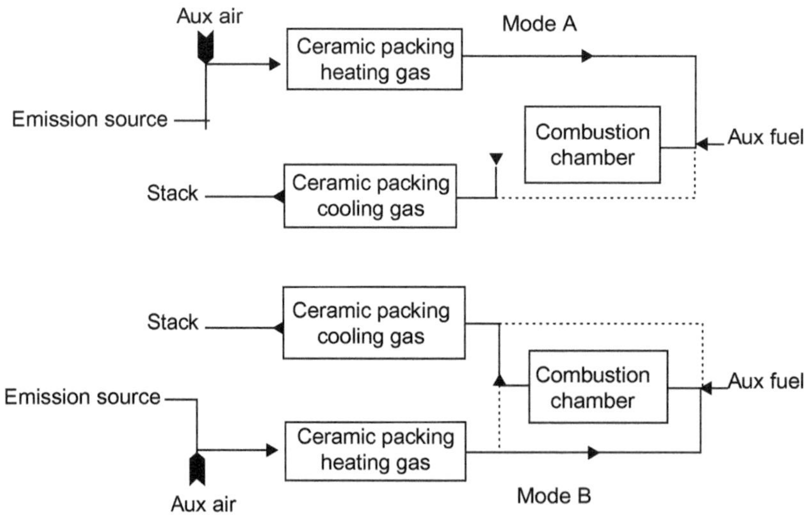


**Figure 10.2** Regenerative Thermal Oxidizer.

valves then direct this gas into energy recovery chambers where it is preheated. The process gas and contaminants are progressively heated in the ceramic medium beds as they move toward the combustion chamber.

Once oxidized in the combustion chamber, the hot purified air releases thermal energy as it passes through the medium bed in the outlet flow direction. The outlet bed is heated and the gas is cooled so that the outlet gas temperature is only slightly higher than the process inlet temperature. Poppet control valves alternate the airflow direction into the medium beds from Mode A to Mode B and vice versa, to maximize energy recovery within the oxidizer. The high energy recovery within these oxidizers reduces the auxiliary fuel requirement and saves operating cost.

#### 10.2.1.2.3 Catalytic oxidation

Catalytic incinerators employ a bed of active material (catalyst) that facilitates the overall combustion reaction. The catalyst has the effect of increasing the reaction rate, thereby enabling conversion at lower reaction temperatures than in thermal incinerator units. Nevertheless, the waste stream must still be preheated to a temperature sufficiently high (usually from 300 to 900 °F) to initiate the oxidation reactions. The waste stream is preheated either directly in a preheater combustion chamber or indirectly by heat exchange with the incinerator's effluent or other process heat or both (Figure 10.3). The preheated gas stream is then passed over the catalyst bed. The chemical reaction (combustion) between the oxygen in the gas stream and the gaseous pollutants takes place at the catalyst surface. Catalytic incineration can, in principle, be used to destroy essentially any oxidizable compound in an air stream. However, caution must be taken because some compounds have a poisoning effect on the catalyst.

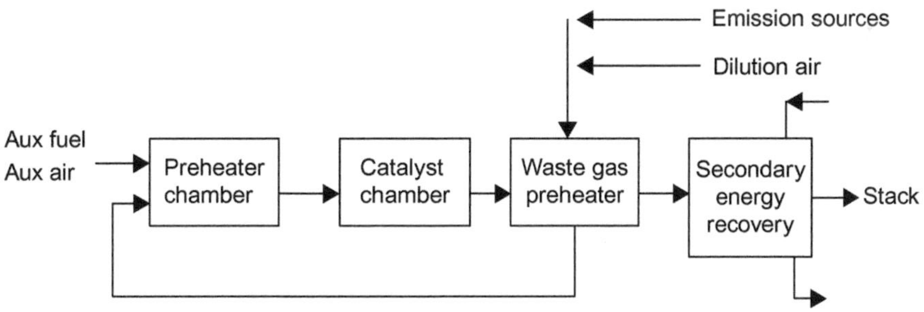


**Figure 10.3** Catalytic Oxidizer.

### *10.2.1.3 Calculation of VOC emissions*

For nonwoven medium manufacturers, the guidelines for calculating VOC emissions are a bit tricky. The original guidelines are based on coating operations and the basic unit is the lb of VOC per unit gallon of coating (SOLIDS). The coater operator is able to measure the thickness of his coating and is able to calcultate the gallons of coating solids applied, by knowing the area of applied coating. Then knowing the pounds of organic solvent associated with each gallon of coating he can calculate the total VOC emissions. The nonwoven manufacturer who uses solvent-based materials would be better served if the units were the other way, that is, gallon of VOC per lb of coating. For example, if the process is a solvent-based resin saturation applied to a nonwoven substrate, it is very difficult to determine a resin thickness and applied area of resin coating in the resin impregnated fibers. Suppliers to coating operations will often supply data based on lb VOC per gallon coating solids on their product data sheets or Material Safety Data Sheets (MSDS), or at least the density of the coating solids. For solvent-based resins such as phenolics, the solid volume or density is not easy to get at.

One method allowed or even recommended by several state authorities is to determine the volume of VOC solvent in a known volume of resin solution (as applied). After subtracting the VOC volume from the total volume, what is left over is the solid volume. To do this, one needs to know the volume of resin solution as applied, its solids content, and the solution density. These are often routine numbers used in a resin saturating operation. To illustrate, consider a hypothetical operation that impregnates a wet lay medium with a methanol solvent based phenolic resin. Assume that the resin solution is prepared as 1,000 gallon batches at 30% solids and has a density of 7.6 lb/gal. The 1000 gallon batch then weighs 7,600 lb. Based on 30% solids, the batch contains 2280 lb of solid resin and 5320 lb of methanol solvent. The methanol solvent at 6.75 lb/galon (obtained from the literature or Google) amounts to 788 gallon of VOC in the1000 gallon resin batch. Subtracting the 788 gallon from 1000 gallon batch leaves 212 gallon of resin solids. Dividing this into the 5320 lb of methanol solids gives 25.1 lb of VOC/gallon resin solids. Note that some of the volatile material may be non-VOC (such as water) and this should be corrected for in the calculation.

The state authorities may have some maximum limit on the allowable emissions expressed as lb VOC per gallon solids. From this the desired efficiency of the VOC control unit can be calculated. It should be noted that the EPA delegates its authority for administering and enforcing state programs to state governments. Any calculation scheme for VOC emissions should be approved by the authorized state agency as part of the permit application.

## 10.3 FOOD AND DRUG ADMINISTRATION MODERNIZATION ACT (FDAMA) OF 1997

Filters in contact with aqueous and fatty foods are regulated by FDAMA. The following paragraphs or sections may apply:

- Generally recognized as safe—GRAS, 21 CFR Part 182
- Indirect food additives—21 CFR Part 177

- Resin bonded filters—21 CFR 177.2260
- Medical devices—21 CFR Subchapter H.

For filters the ingredients must be GRAS. 21 CFR Part 182 lists additives that are directly added to food. 21 CFR Part 177 lists additives that are in contact with food such as by processing, manufacture, or as a packaging material. The substance must be shown to be "generally recognized" as safe under the conditions of its intended use. The proponent of the exemption has the burden of proving that the use of the substance is "generally recognized" as safe. Exemptions to this are food additive items that were regularly used prior to 1958 and substances that were generally recognized as safe by a pertinent body of scientists and experts.

21CFR 177.2260 lists items generally recognized as safe for filter materials. If the substance is not listed in 21CFR 177.2260 then supporting evidence and data must be provided to be approved for 177.2260. Based on the way 177.2260 is written, one would reasonably assume that if a substance were listed under Part 182 or Part 177, it would be automatically safe for filter materials. This is not the case. If you are a maker of filters or filter media for a food application and a substance is not listed in 177.2260, do not use it until it is listed or you take the necessary steps to have it listed.

Filters and filter media used as medical devices or as components in a medical device are regulated by 21 CFR Subpart H.

## 10.4 CLEAN WATER ACT (CWA) 33 U.S.C. PARA 1251, ETC.—EPA

The Clean Water Act (CWA) (324) establishes the basic structure for regulating discharges of pollutants into the waters of the United States and regulating quality standards for surface waters. The basis of the CWA was enacted in 1948 and was called the Federal Water Pollution Control Act. The Act was significantly reorganized and expanded in 1972. "Clean Water Act" became the Act's common name with amendments in 1972. It is of concern to nonwoven filter medium operations that discharge large amounts of water. Examples are wet lay operations and hydroentanglement facilities. In addition, nonwoven filter media manufacturers who discharge waste chemicals (such as waste resin) into the discharge waters of the system are carefully regulated by the CWA.

The US EPA is responsible for implementing pollution control programs such as setting wastewater standards for industry. It also sets water quality standards for all contaminants in surface waters.

The heart of the act is EPA's National Pollutant Discharge Elimination System (NPDES) permit program which controls discharges into government waters. Under this program industrial, municipal, and other facilities must obtain permits if their discharges go directly to surface waters. Point sources are discrete conveyances such as pipes or man-made ditches. In most cases, the NPDES permit program is administered by authorized states. The realm of the act also defines pretreatment facilities.

Most discharges are directed through Publicly Owned Treatment Works (POTWs) to remove undesirable contaminants prior to discharge to surface waters. EPA provides standards for the quality of water discharged from the POTWs as well as standards on water that can be accepted by the POTWs from industrial users.

The EPA and some authorized states also issue NPDES storm water permits. This controls water discharges that come from excessive rain storms and melting snow that travel over ground and possibly picks up undesirable contaminants.

## 10.5 OCCUPATIONAL SAFETY AND HEALTH ACT (OSHA) 29 U.S.C. 655

Occupational Safety and Health Act (OSHA) (325) was passed into law in 1970. An important result of this law was the creation of the the United States Occupational Safety and Health Administration (OSHA). The purpose is to assure safe and healthful working conditions for working men and women by setting and enforcing standards and by providing training, outreach, education and assistance. Nonwoven filter media manufacturers must pay attention to the standards and regulations of OSHA. In recent years at least one nonwoven manufacturer failed to abide by OSHA requirements and was severely fined for failing to do so.

Chemicals are the source of the polymers, fibers, resins, and additives used in nonwoven fitter media. Nonwoven manufacturers are very much affected by this act. It requires them to keep extensive records not only on the materials they buy from their suppliers, but also on the roll and sheet products they ship to their customers. They must maintain safety data sheet files on all chemical materials they purchase and provide safety data sheets to their customers on all the materials they ship to them. This includes experimental and trial materials. Hazard warning labels must be applied to all shipped materials that have anything hazardous about them. Proper labeling procedures must be followed.

### 10.5.1 Hazard Communication Standards Code (HCS)

In order to ensure chemical safety in the workplace, OSHA insists that information about the identities and hazards of the chemicals must be available and understandable to workers. OSHA's Hazard Communication Standard (HCS) (326), often referred to as HazCom requires the development and dissemination of such information. This standard was being incorporated into the Globally Harmonized System of Classification and Labeling of Chemicals (GHS) as this book was written (see Section 10.1.4.1.2 below). Chemical manufacturers and importers are required to evaluate the hazards of the chemicals they produce or import, and prepare labels and MSDS, now called safety data sheets to convey the hazard information to downstream customers. Employers with hazardous chemicals in their workplaces must have labels and MSDS for their exposed workers, and train them to handle the chemicals appropriately.

### 10.5.2 Globally Harmonized System of Classification and Labeling of Chemicals (GHS)

In 2003, the United Nations (UN) adopted the GHS. The GHS (327) includes criteria for the classification of health, physical, and environmental hazards, as well as specifying what information should be included on labels of hazardous chemicals as well as safety data sheets. The United States was an active participant in the development of the GHS, and is a member of the UN bodies established to maintain and coordinate implementation of the system. The GHS was adopted by the United States in 2012. OSHA's HCS is being modified to conform to GHS. Completion is scheduled for 2016 (328).

The purpose of GHS is to provide a globally harmonized system of standards for classifying chemicals, and labels for these chemicals. It includes guidelines for safety data sheets. GHS guidelines applies to pure substances, their dilute solutions, and mixtures. Hazard classification includes risk assessment based on the following simple formula:

$$\text{Risk} = \text{hazard} \times \text{exposure}$$

Hazard classification involves the following three steps:

1. Identification of relevant data regarding the hazards of a substance or mixture.
2. Subsequent review of those data to ascertain the hazards associated with the substance or mixture.
3. A decision on whether the substance or mixture will be classified as a hazardous substance or mixture and the degree of hazard, where appropriate, by comparison of the data with agreed hazard classification criteria.

GHS spells out three hazard classifications for chemicals: physical, health, and environmental. The subcategories are as follows:

1. Physical
   - Explosives
   - Flammable gases
   - Flammable aerosols
   - Oxidizing gases
   - Gases under pressure
   - Flammable liquids
   - Flammable solids
   - Self-reactive substances
   - Pyrophoric liquids
   - Pyrophoric solids

- Self-heating substances
- Substances which, in contact with water emit flammable gases
- Oxidizing liquids
- Oxidizing solids
- Organic peroxides
- Corrosive to metals

2. Health
   - Acute toxicity
   - Skin corrosion/irritation
   - Serious eye damage/eye irritation
   - Respiratory or skin sensitization
   - Germ cell mutagenicity
   - Carcinogenicity
   - Reproductive toxicology
   - Target organ systemic toxicity—single exposure
   - Target organ systemic toxicity—repeated exposure
   - Aspiration toxicity
3. Environmental
   - Hazardous to the aquatic environment
     - Acute aquatic toxicity
     - Chronic aquatic toxicity
       - Bioaccumulation potential
       - Rapid degradability

#### *10.5.2.1 Safety data sheets (SDS)*

As this book was being written, the term Material Safety Data Sheet (MSDS) is being replaced with the term Safety Data Sheet (SDS). SDS are also referred to as Product Safety Data Sheets (PSDS). According to OSHA (329) the SDS contain 16 sections. Sections 1 through 8 contain general information about the chemical, identification, hazards, composition, safe handling practices, and emergency control measures (e.g., fire fighting). This information should be helpful to those that need to get the information quickly. Sections 9 through 11 and 16 contain other technical and scientific information, such as physical and chemical properties, stability and reactivity information, toxicological information, exposure control information, and other information including the date of preparation or last revision. The SDS must also state that no applicable information was found when the preparer does not find relevant information for any required element.

#### *10.5.2.2 Labeling*

OSHA's most recent update (330) requires implementation of GHS harmonized labels as of June 15, 2015. The required label will have the following elements:

- Pictograms
- A signal word
- Hazard and precautionary statements
- The product identifier
- Supplier identification.

A sample revised HCS label, identifying the required label elements, is shown as Figure 10.4 below. Supplemental information can also be provided on the label as needed.

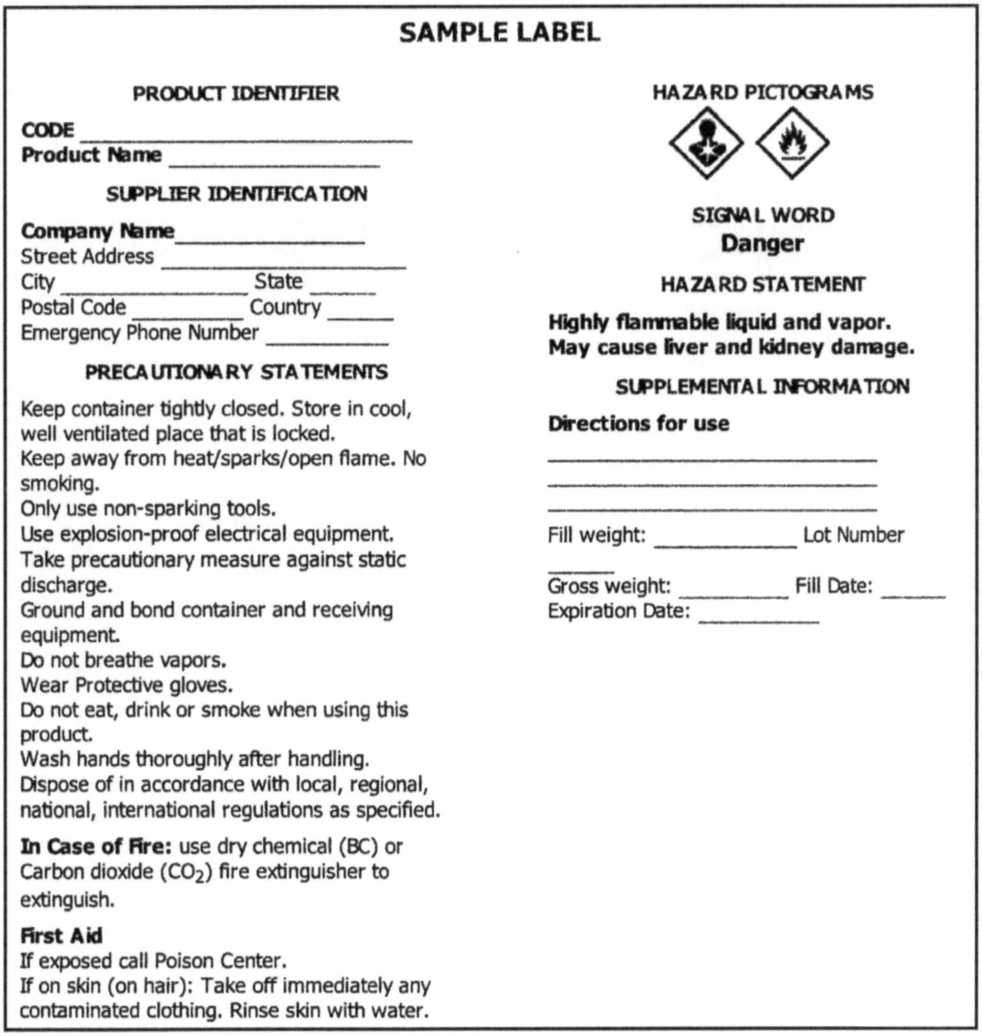

**SAMPLE LABEL**

**PRODUCT IDENTIFIER**

**CODE** ____________________

**Product Name** ______________

**SUPPLIER IDENTIFICATION**

**Company Name** ______________
Street Address ______________
City ____________ State _____
Postal Code ________ Country _____
Emergency Phone Number ________

**PRECAUTIONARY STATEMENTS**

Keep container tightly closed. Store in cool, well ventilated place that is locked.
Keep away from heat/sparks/open flame. No smoking.
Only use non-sparking tools.
Use explosion-proof electrical equipment.
Take precautionary measure against static discharge.
Ground and bond container and receiving equipment.
Do not breathe vapors.
Wear Protective gloves.
Do not eat, drink or smoke when using this product.
Wash hands thoroughly after handling.
Dispose of in accordance with local, regional, national, international regulations as specified.

**In Case of Fire:** use dry chemical (BC) or Carbon dioxide ($CO_2$) fire extinguisher to extinguish.

**First Aid**
If exposed call Poison Center.
If on skin (on hair): Take off immediately any contaminated clothing. Rinse skin with water.

**HAZARD PICTOGRAMS**

**SIGNAL WORD**
**Danger**

**HAZARD STATEMENT**

**Highly flammable liquid and vapor.**
**May cause liver and kidney damage.**

**SUPPLEMENTAL INFORMATION**

**Directions for use**
____________________
____________________
____________________

Fill weight: ________ Lot Number ____
Gross weight: ________ Fill Date: _____
Expiration Date: ________

**Figure 10.4** OSHA Sample Label.

#### 10.5.2.2.1 49 CFR 171, Section 171.8 HAZMAT

Pictograms are hazard symbols used to quickly identify the hazardous nature of the material in the package.

Table 10.1 is a tabulation of OSHA authorized pictograms. It is required by OSHA that the diamond border, around the pictogram symbol, be in red. Unfortunately this book is published in black and white, and the red diamond border cannot be demonstrated. Although listed by GHS, the environmental pictogram is not mandatory by OSHA, however, some companies may desire to use it on an optional basis. OSHA pictograms do not replace the diamond shaped labels required by the

**Table 10.1** OSHA Authorized Hazard Pictograms

| **Health Hazard** | **Flame** | **Exclamation Mark** |
|---|---|---|
|  |  |  |
| • Carcinogen<br>• Mutagenicity<br>• Reproductive Toxicity<br>• Respiratory Sensitizer<br>• Target Organ Toxicity<br>• Aspiration Toxicity | • Flammables<br>• Pyrophorics<br>• Self-Heating<br>• Emits Flammable Gas<br>• Self-Reactives<br>• Organic Peroxides | • Irritant (skin and eye)<br>• Skin Sensitizer<br>• Acute Toxicity (harmful)<br>• Narcotic Effects<br>• Respiratory Tract Irritant<br>• Hazardous to Ozone Layer (Non Mandatory) |
| **Gas Cylinder**  | **Corrosion**  | **Exploding Bomb**  |
| • Gases under Pressure | • Skin Corrosion/ burns<br>• Eye Damage<br>• Corrosive to Metals | • Explosives<br>• Self-Reactives<br>• Organic Peroxides |
| **Flame over Circle**  | **Environment (Non Mandatory)**  | **Skull and Crossbones**  |
| • Oxidizers | • Aquatic Toxicity | • Acute Toxicity (fatal or toxic) |

United States Department of Transportation (DOT). The DOT labels must be on a shipped container and must meet requirements set forth in 49 CFR 172 Subpart E.

OSHA provides guidelines for workplace labels, however, employers have the flexibility to create their own.

Labeling for consumer products are regulated by the Federal Hazardous Substances Act (16 CFR Chapter II, Subchapter C). This act is administered by the U.S. Consumer Products Safety Commission (CPSC).

### 10.5.3 Formaldehyde labeling 29 CFR 1910.1048 (also identified as a VOC)

Formaldehyde is an annoying chemical that is used and/or emitted by many producers of nonwoven filter media, particularly wet lay producers. It is often an ingredient of the resins used to bond wet lay products. OSHA requires manufacturers of products that contain or release formaldehyde to include information about formaldehyde and its hazards on the label and in the SDS. Formaldehyde must be listed if it is in the product at 0.1% or more (as a gas or in solution) or if the product releases formaldehyde above 0.1 parts of formaldehyde per million parts (ppm) of air. Employers that directly import products from other countries have the same responsibilities as a manufacturer under HCS, that is, they must determine the hazards of the product and develop labels and MSDSs that communicate the hazards to users.

Test methods for formaldehyde content are discussed in Chapter 6, Section 6.3. If a product contains 0.1% or more formaldehyde or can release formaldehyde into the air above 0.1 ppm, then the product label must include the following information, as required by OSHA's formaldehyde standard, 29 CFR 1910.1048(m)(3):

- A statement that the product has formaldehyde in it;
- The name and address of the manufacturer, importer, or other company responsible for the product;
- A statement that the employer and MSDSs can readily give health hazard information.

Additionally, if the product can release formaldehyde into the air above 0.5 ppm, the label must also have the following information:

- A list of all product health and safety hazards;
- The phrase "Potential Cancer Hazard."

The SDS must include the following information, as required by OSHA's Formaldehyde Standard, 29 CFR 1910.1048(m)(4) and OSHA's Hazard Communication Standard, 29 CFR 1910.1200(g):

- Name of the product used on its label.
- Name (including common names) of all hazardous chemicals in the product at more than 1% (or more than 0.1% for carcinogens such as formaldehyde), or that could be released into the air above the limits set by OSHA or the American

Conference of Governmental Industrial Hygienists (ACGIH), or could be a health risk to employees.
- Health and safety information for each listed hazardous chemical—including physical properties and health hazards.
- Common ways that people are exposed to the product and its hazardous ingredients.
- Exposure limits (the limit of how much can be in the air).
- Whether the chemical is listed as a carcinogen by OSHA, the International Agency for Research on Cancer (IARC) or the National Toxicology Program (NTP).
- How to safely store and use the product, including what protective equipment to wear and what to do in an emergency.
- The name, address, and telephone number of the company or person responsible for preparing the SDS, and the date it was manufactured.

Formaldehyde is one of five substances CPSC has identified as a strong sensitizer—a substance that can cause hypersensitivity through recurring contact—under the Federal Hazardous Substances Act (16 CFR Chapter II, Subchapter C). Therefore, formaldehyde and any products containing one percent of formaldehyde or more (10,000 parts per million) are required to bear a warning label. The 1 percent refers to the concentration of formaldehyde in solutions and products, not in the air.

## 10.6 TOXIC SUBSTANCE CONTROL ACT (TSCA) 15 U.S.C. §2601 ET SEQ. (1976)

The Toxic Substances Control Act of 1976 (331) provides EPA with authority to require reporting, record-keeping and testing requirements, and restrictions relating to chemical substances and/or mixtures. Certain substances are generally excluded from TSCA, including, among others, food, drugs, cosmetics and pesticides. The act requires:
- Under Section 5, premanufacture notification for "new chemical substances" before manufacture. Section 5 of TSCA requires anyone who plans to manufacture (including import) a new chemical substance for a non-exempt commercial purpose to provide EPA with notice before initiating the activity. This premanufacture notice, or PMN, must be submitted at least 90 days prior to the manufacture of the chemical.
- Under Section 4, testing of chemicals by manufacturers, importers, and processors where risks or exposures of concern are found.
- Significant New Use Rules (SNURS) Section 5(a) of TSCA authorizing EPA to designate by rule certain uses of a chemical to be "significant new uses" and require any company to notify EPA at least 90 days prior to producing, importing or processing such a chemical for such a use.
- Maintenance of the TSCA Inventory, under Section 8, which contains more than 83,000 chemicals. As new chemicals are commercially manufactured or imported, they are placed on the list.

- That those importing or exporting chemicals, under Sections 12(b) and 13, comply with certification reporting and/or other requirements.
- Under Section 8, reporting and record-keeping by persons who manufacture, import, process, and/or distribute chemical substances in commerce.
- Under Section 8(e), that any person who manufactures (including imports), processes, or distributes in commerce a chemical substance or mixture and who obtains information which reasonably supports the conclusion that such substance or mixture presents a substantial risk of injury to health or the environment to immediately inform EPA, except where EPA has been adequately informed of such information. EPA screens all TSCA b Section 8(e) submissions as well as voluntary "For Your Information" (FYI) submissions. The latter are not required by law, but are submitted by industry and public interest groups for a variety of reasons.

When the TSCA was put into place, all existing chemicals were considered to be safe for use and subsequently grandfathered in. This action contrasts the Registration, Evaluation and Authorization of Chemicals (REACh) legislation of the European Union. There are three main objectives to assess and regulate new commercial chemicals: before their entrance into the market, to regulate chemicals (which were already existing in 1976) that posed an "unreasonable risk to health or to the environment," and to regulate their distribution and use.

Contrary to what the name implies, TSCA does not separate chemicals into categories of toxic and non-toxic. Rather it prohibits the manufacture or importation of chemicals that are not on the TSCA Inventory (or subject to one of many exemptions). Chemicals that are listed on the TSCA Inventory are referred to as "existing chemicals," while chemicals not listed are referred to as new chemicals. The TSCA defines 'chemical substance' as "any organic or inorganic substance of a particular molecular identity, including any combination of these substances occurring in whole or in part as a result of a chemical reaction or occurring in nature, and any element or uncombined radical." Generally, manufacturers must submit premanufacturing notification to the U.S. Environmental Protection Agency (EPA) prior to manufacturing (or importing) new chemicals for commercial purposes.

As this book was being written EPA was pursuing a comprehensive regulatory approach under TSCA. This four-pronged approach includes: Premanufacture notifications; a Significant New Use Rule; an information gathering rule; and a test rule. A major concern is the many nanoscale materials that are regarded as "chemical substances" under the Toxic Substances Control Act (TSCA). One substance of specific concern is carbon nanotubes. The purpose is to insure that nanoscale materials are manufactured and used in a manner that protects against unreasonable risks to human health and the environment. Nanoscale materials include many of the nanofiber type materials being developed for nonwoven filter media.

A major problem is to distinguish the properties of a nanoscale material from a macro material of the same chemical structure. The nano structure may have significantly different properties than its macro structure including health and toxicity effects. Because of this and other complications EPA is developing a SNUR under section 5(a)(2) of TSCA to ensure that nanoscale materials receive appropriate regulatory review.

## 10.7 RESOURCE CONSERVATION AND RECOVERY ACT (RCRA) 42 U.S.C. §6901 ET SEQ

RCRA regulates land-based disposal of waste (and focuses on hazardous waste). It has the goal of reducing waste and encouraging recycling. It is not a ban on land-based disposal, but rather a regulation, thereof, which uses "manifests" and the "cradle-to-grave" tracking system. All hazardous waste must obtain an identification number, and be accompanied by a "manifest" which tracks the waste. Each time the waste changes hands, a copy is sent back, ensuring that everyone along the chain is informed, and preventing unidentified wastes from arriving at disposal facilities. It is a part of Environmental Law and administered by the EPA.

RCRA includes a Congressional mandate directing EPA to develop a comprehensive set of regulations to implement the law. The hazardous waste program, under RCRA Subtitle C (40 CFR Part 260), establishes a system for controlling hazardous waste from the time it is generated until its ultimate disposal—in effect, from "cradle to grave."

RCRA is limited to solid waste, however, solid waste is very broadly defined. It includes solids, sludge, liquid, semisolids, or contained gaseous material. In defining waste, the EPA was focused on actual waste, not materials that were still part of the manufacturing process. For example, wet lay operations generate a lot of in-house "broke"; wet lay fiber material that is recovered and recycled back into the fiber process stream. This is not considered as waste by the EPA definition. RCRA Section 261.2(c) designates as solid wastes certain materials that are recycled in particular manners (i.e., used in a manner constituting disposal, burned for energy recovery, reclaimed, and speculatively accumulated). Used or residual waste-like materials that can potentially be solid and hazardous wastes are divided into five groups of secondary materials: spent materials, sludges, by-products, commercial chemical products, and scrap metal. Whether or not the material is a solid waste depends on the type of material being recycled and the type of recycle. One must refer to RCRA Section 261.2 (332) for guidance.

### 10.7.1 Hazardous waste

Hazardous waste is a serious concern for nonwoven filter medium manufacturers. There is a broad range of chemical types used in the production of filter media and for one reason or another; some of these may be classified as hazardous waste. The problem this creates is that the chemical waste can no longer be disposed of by hauling to a landfill or nearby incinerator. Rather it must be shipped to an authorized hazardous waste site for disposal. To illustrate, as this book was being written, environmental groups were petitioning the EPA to declare poly vinyl chloride (PVC) polymer as a hazardous waste. If the petition is granted then such things as PVC piping will no longer be allowed in landfills and incinerators. The concern is that PVC exudes toxic gases when burned and may emit slow amounts of toxic vapors when it degrades in a landfill. The medium manufacturer has to be aware of this possibility, if PVC is present in his raw materials or product, as a fiber, resin, or additive. There are other materials that the medium manufacturer has to consider as a possible hazardous waste. Examples are solvents, resins, additives, flame retardants, formaldehyde containing compounds, and waste cleansers from clean-up operations. Following are some sources of hazardous waste:

- Off-quality materials.
- Spill clean-ups.
- Unused residues from production operations.
- Clean-up chemicals – particularly if they are a solvent, caustic, or acid.
- Residues after shutdown of process runs or operations.
- Left over inventory from unused or discontinued raw materials, intermediates, or product.
- Disposal of laboratory chemicals.
- Chemicals from non-production operations—examples: transformer fluids, coolant from air conditioning units, wastes fuels, machine shop chemicals, etc.
- Contaminated items such as gloves, aprons, and clothing that were contaminated with a hazardous material.

There are four steps to be taken to determine if a material is a hazardous waste:

1. Is the material a solid waste?—This is discussed in Section 10.1.6.1 above.
2. Is the waste excluded from regulation?—This is done by determining whether a solid or hazardous waste exclusion in RCRA §261.4 applies to the given material.
3. Is the waste a listed hazardous waste in 40 CFR Part 261, Subpart D?—The next step is determining whether the waste is a listed hazardous waste. The three categories of listed hazardous wastes are the F, K, and P and U listed wastes:
   - The F list designates as hazardous particular wastes from certain common industrial or manufacturing processes.
   - The K list designates as hazardous particular wastestreams from certain specific industries. Does the waste match one of the specific K-list waste descriptions?
   - The P list and the U list are similar in that both list as hazardous; pure or commercial grade formulations of certain specific unused chemicals. P listed wastes are acutely hazardous and U listed wastes are toxic unless otherwise specified as hazardous for another reason (e.g., hazardous for ignitability, corrosivity, and reactivity).
   - One should refer to Subpart D for other criteria related to the listings.

**4.** Does the waste exhibit a characteristic in Part 261, Subpart C?—There are four principal characteristics used to identify a material as a hazardous waste:
   - Ignitable
   - Corrosive
   - Reactive
   - Toxic.

One should refer to Part 261, Subpart D to determine if his waste exhibits any of the characteristics. Usually knowledge of the process and/or EPA prescribed testing may be necessary.

## 10.7.2 Disposal of hazardous waste

There are three parties involved in hazardous waste management: generators, transporters, and disposal operators.

### *10.7.2.1 Generators*

Generators are the people or facilities that create or generate the hazardous waste. They provide the containers, arrange for storage, labeling and permitting. Once a generator determines or declares that a waste is hazardous, he has 90 days to dispose of it (up to 270 days for small business). The generator arranges for approved containers and labeling of the waste. He also arranges for transportation and disposal. The generator must be specifically careful that the transporters and disposal sites are properly permitted by the EPA and operate in accordance with proper and regulated hazardous waste management practice. There have been a number of instances where a disposal operator has been shut down and no longer able to operate. The generator may still be responsible for any of his hazardous waste still on the site including subsequent transportation and disposal at another facility. Worse if the site is ordered to be cleaned up, the generator may have to pay for his share of the clean-up costs, in accordance with "Superfund" (Comprehensive Environmental Response, Compensation, and Liability Act (CERCLA)) and these costs may be very formidable.

### *10.7.2.2 Transporters*

Hazardous waste must be packaged and transported in accordance with strict procedures and guidelines, while disposal must be coordinated and managed in properly permitted Treatment, Storage and Disposal Facilities (TSDF). Special arrangements have to be made if the material is to be imported or exported.

### *10.7.2.3 Disposal operators*

Disposal operators are more correctly identified as TSDFs. TSDFs manage hazardous wastes under RCRA Subtitle C and generally must have a permit in order to operate. They are the facilities that ultimately treat/dispose of or recycle the hazardous waste.

While, most facilities have RCRA permits, some continue to operate under what are called "interim status." Interim status requirements appear in 40 CFR Part 265.

The permitting requirements for TSDFs appear in 40 CFR Parts 264 and 270. TSDFs manage (treat, store, or dispose) hazardous wastes in units that may include: container storage areas, tanks, surface impoundments, waste piles, land treatment units, landfills, incinerators, containment buildings, and/or drip pads. The unit-specific permitting and operational requirements are described in further detail in 40 CFR Part 264, Subparts J through DD.

#### *10.7.2.4 Cradle to grave requirements*

RCRA Subtitle C, directs EPA to establish controls on the management of hazardous wastes from their point of generation, through their transportation and treatment, storage and/or disposal. Because RCRA requires controls on hazardous waste, generators, transporters, and treatment, storage and disposal facilities, the overall regulatory framework has become known as the "cradle-to-grave" system. The program exacts stringent recordkeeping and reporting requirements on generators, transporters, and operators of TSDFs handling hazardous waste.

#### *10.7.2.5 Containers*

The issue of containers for hazardous waste is important enough that the EPA has issued a "Handboook for Hazardous Waste Containers" (333). The Handbook specifies that the waste must be stored in containers that meet the definition of a portable container (i.e. 55 gallon drums). Mounted tanks, surface impoundments, and waste piles are not considered as containers.

In choosing a container one must make sure that the waste is compatible with the container. If the generator is mixing different wastes in the same container, this is permissible, but he must make sure the components are compatible with each other as well as with the container. The size of the container is also a consideration. Of course the container must be properly labelled and contain the proper hazard warning signs.

One problem with raw materials shipped to the generator is that they come in containers that have to be disposed or recycled in some manner. If the contents are a hazardous material then the container has to be properly cleaned otherwise the container and its residual contents are hazardous waste and have to be disposed of that way.

Following is the EPA definition of empty containers:

- No more than 1 in of residue can remain on the bottom of the container or the inner liner.
- No more than 3% (by weight) of the container's total capacity (less than or equal to 119 gallons) can remain in the container; or no more than 0.3% (by weight) of the container's total capacity (greater than 119 gallons) can remain in the container.

- For acute wastes:
  - the container is triple rinsed with a solvent capable of removing the acute waste, or
  - the container has been cleaned by another method that has been shown in the scientific literature, or by tests conducted by the generator, to achieve equivalent removal, or
  - the inner liner that prevented contact between the acute waste and the container is removed.

The US DOT has a different definition: A container that once held a hazardous material is considered empty when it has been sufficiently cleaned of residue and purged of vapor. According to 49 CFR 171.8, the DOT includes hazardous wastes in the definition of a hazardous material. The difference may cause problems, For example a container containing less than one inch of residue may satisfy the EPA requirement, but not the DOT definition.

## 10.8 FEDERAL HAZMAT LAW 49 U.S.C. SECTION 5101 ET SEQ

The Federal Hazmat Law is designed to protect against the risks to life, property, and the environment which are inherent in the transportation of hazardous materials in intrastate, interstate, and foreign commerce. It is administered by the US DOT. It is the basic statute regulating hazardous materials transportation in the United States. Section 5103 provides that the Secretary of Transportation shall designate material (including an explosive, radioactive material, infectious substance, flammable or combustible liquid, solid or gas, toxic, oxidizing, or corrosive material, and compressed gas) or a group or class of material as hazardous when the Secretary determines that transporting the material in commerce in a particular amount and form may pose an unreasonable risk to health and safety or property. The Secretary of Transportation shall also issue regulations for the safe transportation, including security, of hazardous material in intrastate, interstate, and foreign commerce.

# CHAPTER 11

# Standards for Nonwoven Filter Media

## 11.1 STANDARDS AND TRADE ORGANIZATIONS THAT HAVE STANDARDS, REGISTRATIONS, CERTIFICATIONS, AND/OR OTHER INFORMATION RELATED TO NONWOVEN FILTER MEDIA

Listed in Section 11.2 below are several organizations whose activities include developing, coordinating, promulgating, revising, amending, reissuing, interpreting, certifying, regulating, or otherwise maintaining standards that are related to filtration and filter media. The list is by no means complete. It is an attempt to identify at least some of the organizations that have standards related to filter media and their applications. In almost all the cases, the main concern of the organization is a broader objective that may include filters or filter media. For example, organizations such as Technical Association of the Pulp and Paper Industry (TAPPI), Technical Association of the Australian and New Zealand Pulp and Paper Industry (APPITA), and Scandinavian Pulp, Paper and Board Testing Committee (SCAN) set standards for pulp and paper testing, some of which may be applied to filter media. An example is tensile strength. The test for this paper property is also used to determine the tensile strength of filter media. European Disposables and Nonwovens Association (EDANA) and Association of the Nonwovens Fabrics Industry (INDA) are interested in nonwoven standards, many of which are pertinent to filter media. Organizations that set standards for a particular industry may include standards for the filters that are part of the products of that industry. For example, The American Society of Heating Refrigerating and Air Conditioning Engineers (ASHRAE) sets standards for heating, ventilation, and air conditioning (HVAC), The Society of Automotive Engineers (SAE) sets standards for automotive products. In both cases, associated filters and filter systems are included in the standards and test methods of these organizations.

Organizations such as Association Francaise de Normalisation (AFNOR, France), British Standards Institution (BSI, England), and Deutsches Institut für Normung (DIN, Germany) set standards for a particular country in Europe. National organizations must stop work on a standard if it is going to become a European standard (Comité de Européen Normalisation (CEN)) or an international or global standard (International Organization for Standards (ISO)). A national standards organization whose standard has been established by ISO may have its standard recognized by having the ISO designation included with the standard number. For example DIN 10121-2 (concerning air cleaning media) became DIN EN ISO 10121-2 when this standard became a European standard (CEN) and an international standard (ISO).

*Handbook of Nonwoven Filter Media*
ISBN: 978-0-08-098301-1

American-based standards organizations such as ASTM International (ASTM) may have their standards certified as national standards by the American National Standards Institute (ANSI). The letters ANSI will then appear with the standards designation. For example, the ASHRAE Standard 52.2-2012 is listed as ANSI/ASHRAE Standard 52.2-2012, thus identifying it as a national standard.

Some organizations do not by themselves develop standards; however, they are entrusted with the responsibility of administering, regulating, certifying, and/or monitoring certain standards.

There is an ongoing global effort to harmonize standards and alleviate conflicts between the laws of different nations. Harmonized standards are standards of the same subject approved by different standardizing organizations to establish interchangeability of products, process and services, mutual understanding of test results, or information provided according to these standards. Section 11.4 below includes a listing (Table 11.2) of the harmonized standards developed jointly be INDA and EDANA.

Government or federal bodies set standards on products based on environmental and safety concerns. For example, the United States Environmental Protection Agency (USEPA) places limits on volatile organic compound (VOC) emissions that apply to filter media as well as non-filter related products.

There is an ongoing global effort to harmonize standards and alleviate conflicts between the laws of different nations. Harmonized standards are standards on the same subject approved by different standardizing organizations to establish interchangeability of products, process, and services, or mutual understanding of test results, or information provided according to these standards. Section 11.4 below includes a listing (Table 11.2) of the harmonized standards developed jointly be INDA and EDANA.

## 11.2 STANDARDS ORGANIZATIONS

**AATCC** American Association of Textile Chemists and Colorists, One Davis Drive, P.O. Box 12215. Research Triangle Park, NC 27709-2215, USA. Telephone: +1 (919)-549-8141; fax: + 1 (919)-549-8933; website: www.aatcc.org/

**ABINT** Brazilian Association of Nonwovens and Technical Industries, Avenida Brigadeiro Faria Lima, 1478-3 floor. cj 312. São Paulo/SP, Brazil. Telephone: 11-3032-3015; email: abint@abint.org.br; website: http://www.abint.org.br/

**AFSS** American Filtration and Separations Society, 618 Church Street, Suite 220, Nashville, TN 37219, USA. Telephone +1 (615) 250-7792; fax +1 (615) 254-7047; website: http://www.afssociety.org; email: afs@afssociety.org

**AFNOR** Association Francaise de Normalisation, 11 rue Francis de Pressensé, 93571 La Plaine Sai. nt-Denis, Cedex, France. Telephone: +33 (0)1 41 62 80 00; fax: 33(0)149 1790 00; website: www.afnor.fr!portail.asp

**AHAM** Association for Home Appliance Manufacturers, 1111, 19th Street, NW, Suite 402, Washington, DC 20036, USA. Telephone: +1 (202)-872-5955; website: www.aham.org; email: infoaham.org

**AIA** Aerospace Industries of America. 1000. Wilson Boulevard, Suite 1700, Arlington, VA 22209-3928, USA. Telephone: +1 (703) 358 1000; website: www.aia-aerospace.org/

**ANNA** All Nippon Nonwovens Association, Mengyokaikan Honkan 4F 5-8, Bingo-Machi 2-Chome, Chuo-Ku, Osaka, 541-0051, Japan. Telephone: (81)6-6233-0842; fax: (81)6-6233-0843; E-mail: annasibu@jasmine.ocn.ne.jp, website: http://www.anna.gr.jp/

**ANSI** American National Standards Institute, 1899 L Street, NW, (between 18th and 19th Streets), 11th floor, Washington, DC 20036, USA. Telephone: F 1 (202) 293 8020; fax: + 1 (202) 298 9287; website: www.ansi.org/ 1899 L Street, NW, 11th Floor

**ANFA** Asia Nonwovens Fabrics Association—See liaison offices in Asia for contact information: ANNA (Japan), CNTA (China), KNIC (South Korea) TNFIA (Taiwan).

**APCER** Associação Portuguesa de Certificação Building Services Exponor, 2nd Floor, Av. Dr. Antonio Macedo, Leca da Palmeira 4450-617, Portugal. Telephone: +351 229 993 600; fax: +351 229 993 601; email: info@apcer.pt, website: http://www.apcer.pt/

**ASHRAE** American Society of Heating, Refrigerating and Air-Conditioning Engineers. Inc., 1791 Tullie Circle. N.E., Atlanta, GA 30329, USA. Telephone: +1 (404) 636-8400; fax: +1 (404) 321-5478; website: www.ashrae.org/

**ASME** ASME International. Two Park Avenue, New York, NY 10016-5990, USA. Telephone: +1 (800) 843 2763 (US/Canada), +001-800-843-2763 (Mexico), (973) 882 1170 (outside North America); fax: (973) 882-1717; website: www.asne.org/, email: CustomerCare@asme.org

**ASTM** ASTM International, 100 Barr Harbor Drive, West Conshohocken, PA 19428-2959, USA. Telephone: +1 (610) 832-9585; fax: +1 (610) 832-9555; website: www.astm.org/cgi-bin/SoftCart, exe.Yindex.shtml?E-mystore

**ASQ** American Society for Quality, P.O. Box 3005, Milwaukee, WI 53201-3005, USA. Telephone: 800 248 1946; website: http://asq.org/index.aspx, email: help@asq.org

**APPITA** Technical Association of the Australian and New Zealand Pulp and Paper Industry Australia: 47/255, Drummond Street, Carlton, VIC 3053, Australia. Telephone: +61 3 9347 2377; fax: +613 3 9348 1206; New Zealand: P.O. Box 6042, Whakarewarewa, Rotorua, New Zealand. Telephone: + 64 7 350 2252; fax: +64 7 350 2232; website: www.appita.com.au

**BSI** BSI British Standards, BSI Group Press Office, British Standards House, 389 Chiswick High Road, London W4 4AL, UK. Telephone: +44 (0) 208 996 9001; fax: +44 (0) 208 996 7001; website: www.bsi-global.comfindex.xalter; email: info@bsi-global.com

**CEN** CEN-CENELEC Comité Européen de Normalisation (French)—European Committee for Standardization (English), Management Center, Avenue Mannix 17, B-1000, Brussels, Belgium. Telephone:32 2 550 08 11; fax: 32 2 550 08 19; website: https://www.cen.eu/Pages/default.aspx

**CFR** U.S. Code of Federal Regulations. U.S. Government Printing Office 732 North Capitol Street, NW, Washington, DC 20401-0001, USA. Telephone: 202 512 1800 (Washington area) and (866) 312 1800 (toll-free); fax: +1 (202) 312 2104; email: ContactCenter@gpo.gov; website: http://www.gpo.gov/fdsys/browse/collectionCfr.action?collectionCode = CFR

**CINITA** China Nonwovens and Industrial Textiles Association website: 124 East Chang'an St., Beijing, PRC. Telelephone: 0086 108 522 9421; fax: 0086 108 522 8425; email: cnita2001@sina.org; website: http://www.cnita.org.cn/en/

**CNTA** China Nonwovens Technical Association, 1525 Zhong Shan Xi Lu, Tech & Trading Plaza, Rm 118-120, Shanghai, PRC 200235, China. Telephone: 86-21-6464 2530; fax: 86-21-6481 2993; email: info@cnta.org; website; http://www.chinanonwovens.com/EN/

**DIN** German Institute for Standardization. Deutsches Institut für Normung e. V, Am DIN-Platz, Burggrafenstraße 6, 10787 Berlin, Germany. Telephone: 49 30 2601-0; fax: 49 30 2601-1231; website: http://www.din.de/cmd?level = tpl-bereich&menuid = 47566&languageid = en&cmsareaid = 47566

**DOE** United States Department of Energy, 1000 Independence Ave, SW, Washington, DC 20585, USA. Telelphone: 202-586-5000; website: http://www.energy.gov/

**EDANA** European Disposables and Nonwovens Association, Avenue Herrmann Debroux, 46-B-1160, Brussels, Belgium. Telephone: 32 2 734 9310; fax: 32 2 733 3318; email: info@edana.org; website: www.edana.org

**EI** Energy Institute, 61 New Cavendish Street, London W1G 7AR, UK. Telephone: +44 (0) 20 7467 7100; fax: +44 (0) 20 7255 1472; email: info@energyinst.org, website: www.energyinst.org/home

**EPA** United States Environmental Protection Agency, 1200 Pennsylvania Avenue, N.W., Washington, DC 20460, USA. Telephone: 1 202-272-0167; website: www.epa.gov/

**Eureovent Association** European Association of Air Handling & Refrigeration Equipment Manufacturers, Diamant Building, Bd. A. Reyers 80, B-1030 Brussels, Belgium. Telephone: + 32 (2) 706 79 62; fax: +32 (2) 706 79 66; website http://www.eurovent-association.eu/

**Eurovent Certification** 48-50 rue de la victoire, 75009 Paris, France. Telephone: +33 (0)1 7544 7171; website: http://www.eurovent-certification.com/

**FAA** Federal Aviation Administration, 800 Independence Avenue, SW, Washington, DC 20591, USA. Telephone: 866 835 5322, website: http://www.faa.gov/

**FDA** United States Food and Drug Administration. 10903 New Hampshire Avenue, Silver Spring, MD 20993, USA. Telephone: 1 888-463-6332; website: http://www.fda.gov/default.htm

**FMC** Filter Manufacturers Council, P.O. Box 13968, Research Triangle Park, NC 27709-3966 USA. Telephone: 919 406 8817; fax: 919 406 1306; website: www.filtercouncil.org

**GOST R** Russia: Federal Agency on Technical Regulating and Metrology, Leninsky Prospekt 9, RU-Moscow, V-49, GSP-1, 119991. Telephone: +7 495 236 40 44 / +7 499 236 40 44; fax: +7 495 236 62 31 / +7 499 236 62 31; email: iso@gost.ru; website: http://www.gost.ru/

**IEC** International Electrotechnical Commission, IEC Central Office, 3 rue de Varembé, P.O. Box 131, 1211 Geneva 20, Switzerland. Telephone: + 41 22 919 02 11; website: http://www.iec.ch/

**IEEE** Institute of Electrical and Electronics Engineers, 3 Park Avenue, 17th Floor New York, NY 10016-5997, USA. Telephone: +1 212 419 7900, +1 732 981 0060 (Standards); fax: +1 212 752 4929; website: http://www.ieee.org/index.html

**IEST** The Institute of Environmental Sciences and Technology, 2340 S Arlington Heights Rd, Arlington Hts, IL 60005, USA. Telephone: 847 981 0100; fax: 847 981 4130; email: information@-iest.org; website: http://www.iest.org/

**INDA** Association of the Nonwovens Fabrics Industry, 1200 Crescent Green, Suite 100, P.O. Box 1288 Cary, NC 27511, USA. Telephone: +1 (919) 233-1210; fax: +1 (919) 233-1282; website: www.inda.org/

**ISO** International Organization for Standardization, ISO Central Secretariat, ch. de la Voie-Creuse CP 56, CH1211, Geneva 20, Switzerland. Telephone: 41 22 749 01 11; fax: 41 22 733 34 30; email: central@iso.org; website: http://www.iso.org/iso/home/about.htm

**JSA** Japanese Standards Association, Mita Bldg., Minato-ku, Tokyo 108-0073, Japan. Telephone: +81-3-4231-8503; fax: +81-3-4231-8650; website: http://www.jsa.or.jp/

**KNIC** Korean Nonwoven Industry Cooperative. Telephone: (02) 365-2332-4; fax: (02) 393-5098; website: www.nonwoven.or.kr/

**MIL STD** U.S. Military Standards, Superintendent of Documents, U.S. Government Printing Office, 732 North Capitol Street, NW, Washington, DC 20401-0001, USA. Telephone: 202 512 1800; fax: 202 512 2104; email: ContactCenter@gpo.gov; website http://www.gpo.gov/about/bookstore.htm

**MVSS** Motor Vehicle Safety Standards — NHTSA (National Highway Traffic Safety Administration) 1200 New Jersey Avenue, SE, West Building, Washington, DC 20590, USA. Telephone: 1-888-327-4236; fax: 1-800-424-9153; website: http://www.nhtsa.gov/cars/rules/standards/

**NAFA** National Air Filtration Association, 291 Independence Blvd #515, Virginia Beach, VA 23462, USA. Telephone: +1 (757) 313 7400; fax: +1(757) 313 7401; email: nafa@nafahq.org; website: www.nafahq.org

**NAS** National Aerospace Standards, AIA Aerospace Industries Association, 1000 Wilson Blvd, Suite 1700, Arlington, VA 22209, USA. Telephone: 703 358 1000; website: www.aia-aerospace.org

**NFPA** National Fire Protection Association, 1 Battervmarch Park, Quincy, MA 02169-7471, USA. Telephone: US (617) 770 3000; fax: 800 593-6372; website: www.nfpa.org/about-nfpa/contact-us

**NFPA** National Fluid Power Association, 6737 W. Washington Street, Suite 2350 Milwaukee, WI 53214, USA. Telephone: +1 (414) 778 3344; fax: +1 (414) 778 3361; email: nfpa@nfpa.com; website: http://www.nfpa.com/

**NSF** NSF International, World Headquarters: P.O. Box 130140, 789 N. Dixboro Road, Ann Arbor, MI 48105, USA. Telephone: (734) 769 8010; fax: (734) 769 0109; email: info@nsf.org; website: http://www.nsf.org/

**NIOSH** National Institute of Occupation Safety and Health (U.S.) — part of the Centers for Disease Control and Prevention (CDC) in the Department of Health and Human Services — Centers for Disease Control and Prevention, 1600 Clifton Rd., Atlanta, GA 30333, USA. Telephone: (800) 232 4636; fax: (888) 232-6348; website: http://www.cdc.gov/niosh/

**NIST** National Institute of Standards and Technology; NIST. 100 Bureau Drive Stop 1070, Gaithersburg, MD 20899-1070, USA. Telephone: + 1 (301) 975 6478; website: www.nist.gov/; email: inquiries@nist.gov

**PAPTAC** Pulp and Paper Association of Canada, 740 Notre Dame Street, Montreal, Quebec, H3C 3X6 Canada. Telephone: 514 392 0265; Fax: 514 392 0369; email: tech@paptac.ca; website: www.paptac.ca

**SA** Standards Australia, Level 10, The Exchange Centre, 20 Bridge Street, Sydney NSW 2001, or GPO Box 476 Sydney NSW 2001. Telephone: + 61 2 9237 6171; website: http://www.standards.org.au/

**SAE** SAE International, SAE World Headquarters, 400 Commonwealth Drive, Warrendale, PA 15096-000, USA. Telephone: +1 (724) 776 4841; fax: +1 (724) 776 0790; website: http://www.sae.org/

**SCAN-Test** Scandinavian Pulp. Paper and Board Testing Committee, Box 3604. SEl 14 86. Stockholm, Sweden. Telephone: +46 8 676 71 30; fax: +46 8 24 7277; website: wwwstfi-packforsk.se/ uploadi2 839 /nie-04 1 0.pdl

**TAPPI** Technical Association of the Pulp and Paper Industry (U.S.), 15 Technology Parkway South, Suite 115, Peachtree Corners, GA 30092, USA. Telephone: + 1 (800) 332 8686 (US), (800) 446 9431 (Canada), (770) 446 1400 (Worldwide); fax: (770) 446 6947; email: memberconnection@tappi.org; website: http://www.tappi.org/

**TNFIA** Taiwan Nonwoven Fabrics Industry Association, 5 Fl. No.30, Nan King West Road, Taipei, Taiwan 10352. Telephone: +886-2-25580516; fax: + 886-2-25527330; e-mail: tnfiaroc@ms24.hinet.net; Website: http://www.nonwoven.org.tw

**UL** Underwriters Laboratories Inc.., 333 Pfingsten Road. Northbrook, IL 60062-2096, USA. Telephone: +1 (847) 272 8800; fax: +1 (847) 272 8129; email: cec.us@us.ul.com cec.us@us.ul.com cec.us@us.ul.com; website: http://ul.com/

**VDI** Verein Deutscher Ingenieure, Abteilung VDI-Richtlinien, Postfach 10 11 39. 40002 Dusseldort, Germany. Telephone: +49 211 62 14-2 37 and +49 211 62 14-3 39; fax: +49 211 62 14-i 74; website: www.vdi.de/; email: rili@vdi.de

## 11.3 STANDARDS RELATED TO NONWOVEN FILTER MEDIA

**Table 11.1** Standards Related to Nonwoven Filter Media

| Standard Org. | Number | Title |
|---|---|---|
| AATCC | AATCC Test Method 127-2003 | Water Resistance: Hydrostatic Pressure Test |
| AATCC | AATCC Test Method 22-2010 | Water Repellency: Spray Test |
| AATCC | AATCC-112-2008 | Formaldehyde Release from Fabric, Determination of Sealed Jar Method |
| AATCC | AATCC-33-1962 | Test for the Flammability of Clothing Textiles |
| AFNOR | NF Q03-057 (1973) | Determination of Bursting Strength of Paper and Paperboard Wet |
| AFNOR | NF EN 12083 (1998). | Respiratory Protective Devices-Filters with Breathing Hoses, Particulate Filters, Gas Filters and Combined Filters (Non-Mask Mounted Filters)-Requirements, Testing, Marking |
| AFNOR | NF EN 15798 (2010) | Products Used for Treatment of Swimming Pool Water-Filter Media |
| AFNOR | NF EN ISO 1924-2 (2009) | Paper and Board-Determination of Tensile Properties-Part 2: Method Constant Rate of Expansion of Elongation (20 mm/min) |
| AFNOR | NF EN ISO 1974 (2012) | Paper-Determination of Tear Resistance-Elmendorf Method |
| AFNOR | NF EN ISO 2758 (2004) | Paper-Determination of Bursting Strength |
| AFNOR | NF EN ISO 536 November 2012 | Paper and Board-Determination of Grammage |
| AFNOR | NF EN ISO 9073 (2008) | Textiles-Test Methods for Nonwovens-Part 15: Determination of Air Permeability |
| AFNOR | NF EN ISO 9073 (2009) | Textiles-Test Methods for Nonwovens-Part 16: Determination of Resistance to Penetration by Water (Hydrostatic Pressure) |
| AFNOR | NF EN ISO 9073-4 (1997) | Textiles-Test Methods for Nonwovens-Part 4: Determination of Tear Resistance |
| AFNOR | NF P90-319 (2011) | Private Pools for Family Use-Groups and Filtration Systems-Method for Measuring the Filtration Efficiency, Capacity Retention, and Strength |
| AFNOR | NF Q03-056 (1973) | Determination of the Resistance to Tensile Strength of Paper and Paperboard |
| AFNOR | NF X44-013 (1983) | Separators Ventilation-Test Method Filters the Aerosol of Sodium Chloride by Flame Photometry (HEPA Test) |
| AHAM | ANSI/AHAM AC-1 2006 | Method for Measuring the Performance of Portable Household Electric Room Air Cleaners |
| ANSI | ANSI K13.1-1973 | American National Standard for Identification of Air-Purifying Respirator Canisters and Cartridges |

| | | |
|---|---|---|
| APPITA | APPITA P431 rp-73 | Stiffness of Paper and Paperboard |
| ASHRAE | ANSI/ASHRAE Standard 52.2-2012 | Method of Testing General Ventilation Air-Cleaning Devices for Removal Efficiency by Particle Size |
| ASHRAE | ASHRAE Standard 145.1-2008R | Laboratory Test Method for Assessing the Performance of Gas-Phase Air-Cleaning Systems: Loose Granular Media |
| ASME | ANSI/ASME N509-2002 | Nuclear Power Plant Air Cleaning Units and Components |
| ASME | ASME 510-2007 | Testing of Nuclear Air-Cleaning Systems |
| ASME | ASME AG-1:2012, Section FC | Code On Nuclear Air and Gas Treatment—HEPA Filters |
| ASME | ASME AG-1:2012, Section FK | Code On Nuclear Air And Gas Treatment—Special HEPA Filters |
| ASQ | ANSI/ASQ Z1.9-2008 (2013) | Sampling Procedures and Tables for Inspection by Variables for Percent Nonconforming |
| ASTM | ASTM 119-14 | Standard Test Methods for Fire Tests of Building Construction and Materials |
| ASTM | ASTM D1230-10 (2010) | Standard Test Method for Flammability of Apparel Textiles |
| ASTM | ASTM D1729-96 (2009) | Standard Practice for Visual Appraisal of Colors and Color Differences of Diffusely-Illuminated Opaque Materials |
| ASTM | ASTM D2068-14 (2014) | Standard Test Method for Determining Filter Blocking Tendency |
| ASTM | ASTM D2261-13 (2013) | Standard Test Method for Tearing Strength of Fabric by the Tongue (Single Rip) Procedure (Constant-Rate-of-Extension Tensile Testing Machine) |
| ASTM | ASTM D2863-13 (2013) | Standard Test Method for Measuring the Minimum Oxygen Concentration to Support Candle-Like Combustion of Plastics (Oxygen Index) |
| ASTM | ASTM D4632 / D4632M-08 (2013)e1 | Standard Test Method for Grab Breaking Load and Elongation of Geotextiles |
| ASTM | ASTM D6830-02 (2008) | Standard test Method for Characterizing the Pressure Drop and Filtration Performance of Cleanable Filter Media |
| ASTM | ASTM D7224-13 | Standard Test Method for Determining Water Separation Characteristics of Kerosine-Type Aviation Turbine Fuels Containing Additives by Portable Separometer |
| ASTM | ASTM D737-04(2012) | Standard Test Method for Air Permeability of Textile Fabric |
| ASTM | ASTM D774/D774M-97(2007) | Standard Test Method for Bursting Strength of Paper |
| ASTM | ASTM E128 -1989 | Test Method for Maximum Pore Diameter and Permeability of Rigid Porous Filters for Laboratory Use |

*(Continued)*

**Table 11.1** (Continued)

| Standard Org. | Number | Title |
|---|---|---|
| ASTM | ASTM E1545-11 (2011) | Test Method for Glass Transition Temperature by Thermomechanical Analysis |
| ASTM | ASTM E1824-13 | Standard Test Method for Assignment of a Glass Transition Temperature Using Thermomechanical Analysis Under Tension |
| ASTM | ASTM E84-14 | Standard Test Method for Surface Burning Characteristics of Building Materials |
| ASTM | ASTM F1862-13 | Standard Test Method for Resistance of Medical Face Masks to Penetration by Synthetic Blood (Horizontal Projection of Fixed Volume at a Known Velocity) |
| ASTM | ASTM F1977−04 (2010) | Standard Test Method for Determining Initial, Fractional, Filtration Efficiency of a Vacuum Cleaner System |
| ASTM | ASTM F2101-07 | Standard Test Method for Evaluating the Bacterial Filtration Efficiency (BFE) of Medical Face Mask Materials, Using a Biological Aerosol of Staphylococcus aureus |
| ASTM | ASTM F2299-03 (2010) | Standard Test Method for Determining the Initial Efficiency of Materials Used in Medical Face Masks to Penetration by Particulates Using Latex Spheres |
| ASTM | ASTM F316−03 (2011) | Pore Size Characteristics of Membrane Filters by Bubble Point and Mean Flow Pore Testing |
| ASTM | ASTM F778-88(2001) | Standard Methods for Gas Flow Resistance of Filtration Media |
| BSI | BS 3137:1972 | Methods for Determining the Bursting Strength of Paper and Board |
| BSI | BS 3928:1969 | Method for Sodium Flame Test for Air Filters (Other Than for Air Supply to I.C. Engines and Compressors) |
| BSI | BS 4415-1:1992 | Determination of the Tensile Properties of Paper and Board−Constant Loading (Withdrawn in 2012) |
| BSI | BS 5852:2006 | Fire Test to Upholstered Seating |
| BSI | BS EN 13274-1:2001 | Respiratory Protective Devices. Methods of Test -Determination of Inward Leakage and Total Inward Leakage |
| BSI | BS EN 13274-2:2001 | Respiratory Protective Devices. Methods of Test-Practical Performance Tests |
| BSI | BS EN 13274-3:2001 | Respiratory Protective Devices. Methods of Test—Determination of Breathing Resistance |
| BSI | BS EN 13274-4:2001 | Respiratory Protective Devices. Methods of Test—Flame Tests |

| | | |
|---|---|---|
| BSI | BS EN 13274-5:2001 | Respiratory Protective Devices. Methods of Test—Climatic Conditions |
| BSI | BS EN 13274-6:2001 | Respiratory Protective Devices. Methods of Test—Determination of Carbon Dioxide Content of Inhalation Air |
| BSI | BS EN 13274-7:2008 | Respiratory Protective Devices. Methods of Test—Determination of Particle Filter Penetration |
| BSI | BS EN 13274-8:2002 | Respiratory Protective Devices. Methods of Test- Determination of Dolomite Dust Clogging |
| BSI | BS EN 21974:1994 | Method for the Determination of the Internal Tearing Resistance of Paper |
| BSI | BS EN ISO 974:2012 | Determination of Tearing Resistance. Elmendorf Method |
| BSI | BS EN ISO 9863-1:2005 | Geosynthetics. Determination of Thickness at Specified Pressures. Single Layers |
| BSI | BS ISO 2493-1:2010 | Paper and Board. Determination of Bending Resistance, Constant Rate of Deflection |
| BSI | BS ISO 4589-2:1999 | Plastics. Determination of Burning Behaviour by Oxygen Index. Ambient-Temperature Test |
| CEN | EN 149:2001 | Disposable Particulate Respirator |
| CEN | EN 1822:2009 | High Efficiency Air Filters (EPA, HEPA, and ULPA) |
| CEN | EN 779:2012 | Particulate Air Filters for General Ventilation—Determination of the Filtration Performance |
| CEN | EN13779:2007 | Ventilation for Non-Residential Buildings-Performance Requirements for Ventilation and Room-Conditioning Systems |
| CFR | 16 CFR Part 1610 | Standard for the Flammability of Clothing Textiles |
| CFR | 16 CFR Part 1633 | Standard for the Flammability (Open Flame) of Mattress Sets |
| CFR | 21 CFR 177 SUBCHAPTER B | Subpart B—Substances for Use as Basic Components of Single and Repeated Use Food Contact Surfaces |
| CFR | 21 CFR 177 SUBCHAPTER B Sec. 177.2260 | Filters, Resin-bonded |
| CFR | 29 CFR 1910.1048 (2012) | Formaldehyde |
| CFR | 29 CFR 1910.134 | Personal Protective Equipment |
| CFR | 29 CFR 1926.150 | Safety and Health Regulations for Construction, Fire Protection and Prevention |
| CFR | 42 CFR (U.S.), Chapter 1, Part 84 | Approval of Respiratory Protective Devices |

(*Continued*)

**Table 11.1** (Continued)

| Standard Org. | Number | Title |
|---|---|---|
| DIN | DIN 71460-1:2006-4 | Air Filters for Passenger Compartments; Test for Gaseous Filtration |
| DIN | DIN EN ISO 10121-2: 2013-08 | Test Methods for Assessing the Performance of Gas-Phase Air Cleaning Media and Devices for General Ventilation-Part 2: Gas-Phase Air Cleaning Devices |
| DIN | DIN EN ISO 23328-1:2008 | Breathing System Filters for Anesthetic and Respiratory Use-Part 1: Salt Test Method to Assess Filtration Performance |
| DOE | DOE-STD-3020-2005 | Specification for HEPA Filters Used by DOE Contractors |
| EI | EI 1581 5th Edition (2003) | Specifications and Qualification Procedures for Aviation Jet Fuel Filter/Separators |
| EI | EI 1582 1st Edition (2001) | Specification for Similarity for API/IP 1581 Aviation Jet Fuel Filter/Separators |
| EI | EI 1583 5th edition, (2006) | Laboratory Tests and Minimum Performance Levels for Aviation Fuel Filter Monitors |
| EI | EI 1590 2nd edition, (2002) | Specifications and Qualification Procedures for Aviation Fuel Microfilters |
| EI | EI 1599 1st edition, (2007) | Laboratory Tests and Minimum Performance Levels for Aviation Fuel Dirt Defence Filters |
| Eurovent | Guideline 4/11 (2014) | The Energy Efficiency Classification of Air Filters for General Ventilation Purposes |
| FAA | AC 25.853-1 | Flammability Requirements for Aircraft Seat Cushions |
| IEC | IEC 60312-1:2010 | Vacuum Cleaners for Household Use-Part 1: Dry Vacuum Cleaners—Methods for Measuring the Performance |
| IEEE | IEEE 634-2005 | IEEE Standard Cable—Penetration Fire Stop Qualification Test |
| IEST | IEST-RP-CC001.5 (2009) | HEPA and ULPA Filters |
| IEST | IEST-RP-CC0021.3 (1995) | Testing HEPA and ULPA Filter Media |
| IEST | IEST-RP-CC0034.3 (2009) | HEPA and ULPA Filter Leak Tests |
| IEST | IEST-RP-CC007.2 (2007) | Testing ULPA Filters |
| ISO | ISO 11155-2:2009 | Road Vehicles–Air Filters for Passenger Compartments—Part 2: Test for Gaseous Filtration |
| ISO | ISO 11171:2010 | Hydraulic fluid power—Calibration of Automatic Particle Counters for Liquids |
| ISO | ISO 12103-1:1997 | Road Vehicles Test–Dust for Filter Evaluation—Part 1: Arizona Road Dust |
| ISO | ISO 13357-2:2005 | Determination of the Filterability of Lubricating Oils—Part 2: Procedure for Dry Oils |

| | | |
|---|---|---|
| ISO | ISO 14644-1:1999 | Part 1: Classification of Air Cleanliness by Particle Concentration |
| ISO | ISO 15373:2001 | (Method A): Plastics—Polymer dispersions—Determination of Free Formaldehyde |
| ISO | ISO 15373:2001 | (Method B): Plastics—Polymer dispersions—Determination of free formaldehyde |
| ISO | ISO 16889:2008 | Hydraulic Fluid Power–Filters—Multi-pass Method for Evaluating Filtration Performance of a Filter Element |
| ISO | ISO 1924-2:2008 | Determination of the Tensile Properties of Paper and Board–Constant Loading |
| ISO | ISO 19438:2003 | Diesel Fuel and Petrol Filters for Internal Combustion Engines—Filtration Efficiency Using Particle Counting and Contaminant Retention Capacity |
| ISO | ISO 1974:2012 | Elmendorf Tear Tester |
| ISO | ISO 2493-2:2011 | Paper and Board—Determination of bending resistance—Part 2: Taber-type Tester |
| ISO | ISO 2758:2001 | Paper—Determination of Bursting Strength |
| ISO | ISO 2758:2003 | Paper—Determination of Bursting Strength |
| ISO | ISO 29461-1:2013 | Air Intake Filter Systems for Rotary Machinery—Test Methods—Part 1: Static Filter Elements |
| ISO | ISO 4003:1977 | Permeable Sintered Metal Materials—Determination of Bubble Test Pore Size |
| ISO | ISO 4020:2001 | Road Vehicles—Fuel Filters for Diesel Engines |
| ISO | ISO 4402:1991 | Hydraulic Fluid Power—Calibration of Automatic-count Instruments for Particles Suspended in Liquids—Method Using Classified AC Fine Test Dust Contaminant |
| ISO | ISO 4405:1991 | Hydraulic Fluid Power—Fluid Contamination—Determination of Particulate Contamination by the Gravimetric Method |
| ISO | ISO 4406:1999 | Hydraulic Fluid Power—Fluids—Method for Coding the Level of Contamination by Solid Particles |
| ISO | ISO 4548-12:2000 | Methods of Test for Full-Flow Lubricating Oil Filters for Internal Combustion Engines—Part 12: Filtration Efficiency Using Particle Counting, and Contaminant Retention Capacity |
| ISO | ISO 4548-12:2000 | Methods of Test for Full-Flow Lubricating Oil Filters for Internal Combustion Engines—Part 12: Filtration Efficiency Using Particle Counting, and Contaminant Retention Capacity |
| ISO | ISO 4572-81 | Hydraulic Fluid Power-Filters—Multi-pass Method for Evaluating Filtration Performance—Discontinued |

(*Continued*)

**Table 11.1** (Continued)

| Standard Org. | Number | Title |
|---|---|---|
| ISO | ISO 4589-2:1996 | Plastics—Determination of Burning Behaviour by Oxygen Index—Part 2: Ambient-Temperature Test |
| ISO | ISO 4589-2:2006 | Burning Test for Oxygen Index |
| ISO | ISO 5011:2014 | Inlet Air Cleaning Equipment for Internal Combustion Engines and Compressors—Performance Testing |
| ISO | ISO 5635:1978 | Paper—Measurement of Dimensional Change After Immersion in Water |
| ISO | ISO 5636-1:1984(en) | Paper and Board—Determination of Air Permeance (Medium Range)—Part 1: General Method |
| ISO | ISO 5636-2:1984 | Paper and Board—Determination of Air Permeance (Medium Range)—Part 2: Schopper Method |
| ISO | ISO 5636-3:2013 | Paper and board—Determination of Air Permeance (Medium Range)—Part 3: Bendtsen Method |
| ISO | ISO 5636-4:1986 | Paper and Board—Determination of Air Permeance (Medium Range)—Part 4: Sheffield Method |
| ISO | ISO 5636-5:2013 | Paper and Board—Determination of Air Permeance and Air Resistance (Medium Range)—Part 5: Gurley Method |
| ISO | ISO 6944-2:2009 | Fire Containment—Elements of Building Construction—Part 2: Kitchen Extract Ducts |
| ISO | ISO 789-8 | Agricultural Tractors—Test Procedures—Part 8: Engine Air Cleaner |
| ISO | ISO 8226-1:1994 | Paper and Board—Measurement of Hygroexpansivity—Part 1: Hygroexpansivity up to a Maximum Relative Humidity of 68 % |
| ISO | ISO 8226-2:1990 | Paper and Board—Measurement of Hygroexpansivity—Part 2: Hygroexpansivity up to a Maximum Relative Humidity of 86% |
| ISO | ISO 9073-1:1989 | Textiles—Test Methods for Nonwovens—Part 1: Determination of Mass per Unit Area |
| ISO | ISO 9073-15:2007 | Textiles—Test Methods for Nonwovens—Part 15: Evaluation of Air Permeability |
| ISO | ISO 9073-16:2007 | Textiles—Test Methods for Nonwovens—Part 16: Evaluation of Water Resistance (Hydrostatic Pressure Test) |

| | | |
|---|---|---|
| ISO | ISO 9073-17:2008 | Textiles—Test Methods for Nonwovens—Part 17: Evaluation of Water Penetration (Spray Impact) |
| ISO | ISO 9073-18:2007 | Textiles—Test Methods for Nonwovens—Part 18: Determination of Breaking Length and Elongation of Nonwoven Materials Using the Grab Tensile Test |
| ISO | ISO 9073-2:1995 | Textiles—Test Methods for Nonwovens—Part 2: Determination of Thickness-Test Method C |
| ISO | ISO 9073-3:1989 | Textiles—Test Methods for Nonwovens—Part 3: Determination of Tensile Strength and Elongation |
| ISO | ISO 9092:2011 | Textiles-Nonwovens—Definition |
| ISO | ISO 9865:1991 | Textiles—Determination of Water Repellency of Fabrics by the Bundesmann Rain Shower Test |
| ISO | ISO/DIS 14644-1 (2010) | Part 1: Classification of Air Cleanliness by Particle Concentration |
| ISO | ISO/DIS 14644-12 (under development) | Cleanrooms and Associated Controlled Environments—Part 12: Classification of Air Cleanliness by Nanoscale Particle Concentration |
| ISO | ISO/DIS 14644-2 (2010) | Part 2: Specifications for Monitoring and Periodic Testing to Prove Continued Compliance with ISO 14644-1 |
| ISO | ISO/TS 13353:2002 | Diesel Fuel and Petrol Filters for Internal Combustion Engines—Initial Efficiency by Particle Counting Revised by: ISO 19438:2003 |
| ISO | ISO/TS 16332:2006 | Diesel Engines—Fuel Filters—Method for Evaluating Fuel/Water Separation Efficiency |
| JSA | JIS D 1611-1:2003 | Automotive Parts—Lubricating Oil Filters for Internal Combustion Engines—Part 1: General Test Methods |
| JSA | JIS L 1041:2011 | Test Methods for Resin Finished Textile |
| JSA | JIS P 8125:2000 | Testing Method for Stiffness of Paperboard by Bending Load |
| JSA | JIS P 8223:2006 | Pulps—Laboratory Sheets—Determination of Physical Properties |
| MIL-STD | MIL-H-5606 9/10/94 | Hydraulic Fluid |
| MIL-STD | MIL-STD-105D, 5/10/89 | Sampling Procedures and Tables for Inspection by Attributes [S/S BY MIL-STD-1916 AND ANSI/ASQ Z1.4] |
| MIL-STD | MIL-STD-1916, 4/1/1996 | Department of Defense Test Method Standard DOD Preferred Methods for Acceptance of Product |

(*Continued*)

**Table 11.1** (Continued)

| Standard Org. | Number | Title |
|---|---|---|
| MIL-STD | MIL-STD-282A(1) 9/24/12 | Filter Units, Protective Clothing, Gas-Mask Components, and Related Products: Performance Test Methods |
| MIL-STD | MIL-STD-414 6/11/1957 | Sampling Procedures and Tables for Inspection by Variables for Percent Defective |
| MIL-STD | MIL-STD-810G | Environmental Engineering Considerations and Laboratory Tests |
| MVVS | MVVS-302 | Flammability of Interior Materials |
| NAS | NAS 1638 | Cleanliness Requirements of Parts Used in Hydraulic Systems |
| NFPA | NFPA 255 | Standard Method of Test of Surface Burning Characteristics of Building Materials |
| NFPA | NFPA 90A | Installation of Air Conditioning and Ventilating Systems (Other Than Residence Type) |
| NFPA | NFPA 90B | Installation of Warm Air Heating and Air Conditioning Systems (Residence Type) |
| NFPA | NFPA-701 (2010) | Fire Tests for Flame Resistant Textiles and Films |
| NIOISH | TEB-APR-STP-0057 | Determination of Particulate Filter Efficiency for N100 Series Filters Against Solid Particulates for Non-Powered, Air-Purifying Respirators Standard Testing Procedure (STP) |
| NIOSH | 42 CFR 84 (US) | Chapter 1, Part 84—Approval of Respiratory Protective Devices |
| NIOSH | TEB-APR-STP-0051 (2012) | Determination of Particulate Filter Efficiency for P100 Series Filters Against Liquid Particulates for Non-Powered, Air-Purifying Respirators Standard Testing Procedure (STP) |
| NSF | NSF 50 | Standard for Pools and Spas |
| PAPTAC | D.4 (2013) | Thickness and Apparent Density of Paper and Paperboard |
| PAPTAC | D.6H (1984) | Tensile Breaking Strength of Pulp and Paperboard |
| PAPTAC | D.8 (205) | Bursting Strength of Paper |
| PAPTAC | D.9 (1993) | Internal Tearing Resistance of Paper, Paperboard and Pulp Handsheet |
| SA | AS 1324.1-2001 | Air Filters for Use in General Ventilation and Air Conditioning-Application, Performance and Construction |
| SA | AS 4260-1997 | High Efficiency Particulate Air (HEPA) Filters—Classification, Construction, and Performance |
| SAE | SAE HS-806/2009 | SAE Oil Filter Test Procedure |

| | | |
|---|---|---|
| SAE | SAE J 1124:2012-05-29 | Glossary of Terms Related to Fluid Filters and Filter Testing |
| SAE | SAE J1839-201010 | Coarse Droplet Water/Fuel Separation Test Procedure |
| SAE | SAE J905-200909 | Fuel Filter Test Methods |
| SAE | SAEJ1488-201010 | Emulsified Water/Fuel Separation Test Procedure |
| SCAN | SCAN P16 | Tensile Strength and Stretch of Paper and Paperboard Determined with a Pendulum Tester |
| SCAN | SCAN P24:68 | Paper Burst Strength |
| SCAN | SCAN P29:95 | Bending Resistance |
| SCAN | SCAN-P 80:98 | Internal Bond Test (Z Directional Tensile Test) |
| TAPPI | | |
| TAPPI | T 558 om-10 | Surface Wettability and Absorbency of Sheeted Materials Using an Automated Contact Angle Tester |
| TAPPI | T 833 wd-01 | Test for Interfiber Bond Using the Internal Bond Tester |
| TAPPI | TAPPI T 234 om-84 | Coarseness of Pulp Fibers |
| TAPPI | TAPPI T 400 sp-11 | Sampling and Accepting a Single Lot of Paper, Paperboard, Containerboard, or Related Product |
| TAPPI | TAPPI T 402 sp-13 | Standard Conditioning and Testing Atmospheres for Paper, Board, Pulp Handsheets, and Related Products |
| TAPPI | TAPPI T 403 om-10 | Bursting Strength of Paper |
| TAPPI | TAPPI T 411 om-10 | Thickness (Caliper) of Paper and Paperboard and Combined Board |
| TAPPI | TAPPI T 413 om-11 | Ash in Wood, Pulp, Paper and Paperboard: Combustion at 900 Degrees C (Used to Determine Organic Content of Glass Microfiber HEPA and ULPA Filter Media) |
| TAPPI | TAPPI T 414 om-12 | Internal Tearing Resistance of Paper (Elmendorf-type Method) |
| TAPPI | TAPPI T 456 om-11 | Tensile Breaking Strength of Water-Saturated Paper and Paperboard (Wet Tensile Strength) |
| TAPPI | TAPPI T 460 om-11 | Air Resistance of Paper (Gurley Method) |
| TAPPI | TAPPI T 461 om-09 | Flammability or Treated Paper and Paperboard |
| TAPPI | TAPPI T 489 om-13 | Stiffness of Paper and Paperboard (Taber Stiffness Tester) |
| TAPPI | TAPPI T 494 om-13 | Tensile Properties of Paper and Paperboard (Using Constant Rate of Elongation Apparatus) |

(*Continued*)

**Table 11.1** (Continued)

| Standard Org. | Number | Title |
|---|---|---|
| TAPPI | TAPPI T 541 om-10 | Internal Bond Strength of Paperboard (Z-Direction Tensile) |
| TAPPI | TAPPI T 543 om-11 | Bending Resistance of Paper (Gurley-type Tester) |
| TAPPI | TAPPI T 550 om-13 | Determination of Equilibrium Moisture in Pulp, Paper, and Paperboard for Chemical Analysis |
| TAPPI | TAPPI T1215 sp-12 | The Determination of Instrumental Color Differences |
| TAPPI | TAPPI T401 om-08 | Fiber Analysis of Paper and Paperboard |
| TAPPI | TAPPI T410 om-13 | Grammage of Paper and Paperboard (Weight per Unit Area) |
| TAPPI | TAPPI/ANSI T 524 om-13 | Color of Paper and Paperboard (45/0, C/2) |
| TAPPI | TAPPI/ANSI T 527 om-13 | Color of Paper and Paperboard (d/0, C/2) |
| UL | UL 586-90 | Standard for Safety High Efficiency, Particulate Air Filter Units |
| UL | UL 72 | Tests for Fire Resistance of Record Protection Equipment |
| UL | UL 900 | Test Performance of Air Filter Units |
| UL | UL 94 | Standard for Safety of Flammability of Plastic Materials for Parts in Devices and Appliances |
| UL | UL1479 | Fire Tests of Through-Penetration Firestops |
| VD! | VDI 3926, Sheet 1: 1994-12 | Testing of Filter Media for Cleanable Filters |
| VDI | VDI 3926, Part 2, 1994 | Testing of Filter Media for Cleanable Filters Under Operational Conditions |

## 11.4 INDA–EDANA HARMONIZED TEST METHODS

INDA and EDANA have responded to industry needs by publishing the 2012 edition of their Harmonized Test Methods Manual for the nonwoven and related industries. The methods are listed in the Table of Contents of Table 11.2 below. Featured is a new, intuitive numbering system. It has been reviewed and approved by INDA and EDANA's joint task force of members. In this edition, several testing areas have been expanded. This includes permeability, stiffness, superabsorbent materials and other useful methods. In addition, new methods in filtration, hygiene, absorption, retention and physical characteristics have been added.

## 11.5 ISO MANAGEMENT STANDARDS

When ISO management standards were first introduced, it was viewed as a prestige tool. A company achieved ISO 9000 status because it looked good on their resume and made a good impression on their customers. The company more or less devised its own quality system and the major requirement was "do what you say and say what you do." Times have changed. ISO 9000 certification is no longer a prestige venture, but a requirement in certain markets. For example, filter manufacturers vying for part supplier status for most automobile manufacturers must be ISO/TS16949 certified in order to be considered (see Section 11.5.3). ISO/TS16949 is the automotive version of ISO 9001:2008. One of the requirements on part supplier is that their suppliers be TS 16949 certified. Of course, this includes manufacturers of the filter medium used in the filters. The requirements not only tell the supplier what he must do, but also how to do it. In short, many companies either have to be certified and in compliance with the pertinent ISO Management Standards are out of business. Following is a brief discussion in what is involved in the more prominent ISO standards.

**Table 11.2** Harmonized Test Methods Nonwovens and Related Industries

| Description | Method number | Equivalent methods | ISO reference | Page number |
|---|---|---|---|---|
| ***(1) Guidance Documents*** | | | | |
| Glossary of Terms | WSP 001.0.R3 (12) | Original Document | | 1.1 |
| How to Write a Test Method | WSP 002.0.R3 (12) | Original Document | | 1.59 |
| Sample and Laboratory Conditioning | WSP 003.0.R3 (12) | ERT 60.2 (99) | | 1.76 |
| Worldwide Associations | WSP 004.0.R2 (12) | ERT Useful Addresses | | 1.77 |
| Sampling | WSP 005.0.R3 (12) | ERT 130.2 (99) | | 1.84 |
| List of Equipment Vendors | WSP 006.0.R2 (12) | IST Useful Vendor's List | | 1.87 |
| Guidance to Evaluating Nonwoven Fabrics | WSP 007.0.R2 (12) | INDA Guidance Document | | 1.94 |
| Guidance to Evaluating Nonwoven Felts | WSP 008.0.R2 (12) | INDA Guidance Document | | 1.100 |
| ***(2) Absorption*** | | | | |
| Nonwoven Absorption | WSP 010.1.R3 (12) | ERT 10.4 (02) | 9073 – 6:2000 | 2.1 |
| Rate of Sorption of Wiping Materials | WSP 010.2.R3 (12) | IST 10.2 (01) | | 2.11 |
| Demand Absorbency | WSP 010.3.R3 (12) | ERT 230.1 (02) | 9073 – 12:2002 | 2.16 |
| Evaluation of Oil and Fatty Liquid Absorption | WSP 010.4.R3 (12) | Original Document | | 2.28 |
| ***(3) Abrasion Resistance*** | | | | |
| Inflated Diaphragm | WSP 020.1.R3 (12) | Modified IST 20.1 (01) | | 3.1 |
| Flexing and Abrasion | WSP 020.2.R3 (12) | Modified IST 20.2 (01) | | 3.9 |
| Double Rotary Platform (Taber) | WSP 020.4.R3 (12) | Modified IST 20.4 (01) | | 3.20 |
| Modified Mortindale | WSP 020.5.R3 (12) | Modified IST 20.5 (01) | | 3.29 |
| ***(4) Bursting Strength*** | | | | |
| Diaphragm Burst | WSP 030.1.R3 (12) | Modified IST 30.1 (01) | | 4.1 |
| Burst | WSP 030.2.R3 (12) | ERT 80.4 (02) | | 4.7 |
| ***(5) Electrostatic Properties*** | | | | |
| Surface Resistivity | WSP 040.1.R3 (12) | IST 40.1 (01) | | 5.1 |
| Decay | WSP 040.2.R3 (12) | IST 40.2 (01) | | 5.8 |

| | | | | |
|---|---|---|---|---|
| ***(6) Optical Properties*** | | | | |
| Opacity (INDA) | WSP 060.1.R3 (12) | IST 60.1 (01) | | 6.1 |
| Brightness (INDA) | WSP 060.2.R3 (12) | IST 60.2 (01) | | 6.6 |
| Brightness (EDANA) | WSP 060.3.R3 (12) | ERT 100.1 (78) | | 6.11 |
| Opacity (EDANA) | WSP 060.4.R3 (12) | ERT 110.1 (78) | | 6.16 |
| ***(7) Permeability*** | | | | |
| Air Permeability | WSP 070.1.R3 (12) | Modified IST 70.1 (01) | 9073 – 15:2007 | 7.1 |
| Liquid Strike-Through | WSP 070.3.R3 (12) | ERT 150.5 (02) | 9073 – 8:1995 | 7.7 |
| Water Vapor Transmission Rate Mocon/ INDA | WSP 070.4.R3 (12) | IST 70.4 (01) | | 7.15 |
| Water Vapor Transmission Rate Mocon/ EDANA, Part 1 | WSP 070.5.R3 (12) | Original Method | | 7.27 |
| Water Vapor Transmission Rate Lyssy/ EDANA, Part 2 | WSP 070.6.R3 (12) | Original Method | | 7.39 |
| Repeated Liquid Strike-Through Time | WSP 070.7.R4 (12) | ERT 153.0 (02) | 9073 – 13:2006 | 7.45 |
| Wetback After Repeated Strike-Through | WSP 070.8.R4 (12) | ERT 154.0 (02) | 9073 – 14:2006 | 7.57 |
| Rate of Acquisition and Re-Wet | WSP 070.9.R1 (12) | New Method | | 7.68 |
| Centrifugal Liquid Retention (Dry Weight vs. Wet Spun Weight) | WSP 070.10.R1 (12) | New Method | | 7.75 |
| ***(8) Repellency*** | | | | |
| Surface Wetting Spray | WSP 080.1.R4 (12) | IST 80.1 (01) | | 8.1 |
| Penetration by Water (Rain Test) | WSP 080.2.R3 (12) | IST 80.2 (01) | | 8.7 |
| Penetration by Water (Spray Impact) | WSP 080.3.R4 (12) | Modified IST 80.2 (01) | 9073 – 17:2008 | 8.14 |
| Penetration by Saline Solution (Automated Mason Jar) | WSP 080.5.R4 (12) | IST 80.5 (01) | | 8.23 |
| Water Resistance (Hydrostatic Pressure) | WSP080.6.R4 (12) | IST 80.6 (01)/IST 80.4 (01) | 9073 – 16:2007 | 8.29 |
| Penetration by Oil (Hydrocarbon Resistance) | WSP080.7.R4 (12) | IST 80.7 (01) | | 8.38 |
| Alcohol Repellency | WSP080.8.R4 (12) | IST 80.8 (01) | | 8.43 |
| Run-Off | WSP080.9.R4 (12) | ERT 152.2 (02) | 9073 – 11:2002 | 8.49 |

*(Continued)*

**Table 11.2** (Continued)

| Description | Method number | Equivalent methods | ISO reference | Page number |
|---|---|---|---|---|
| Coverstock Wetback | WSP080.10.R3 (12) | ERT 151.3 (02) | | 8.63 |
| Wet Barrier Mason Jar | WSP080.11.R4 (12) | ERT 170.1 (02) | | 8.73 |
| ***(9) Stiffness*** | | | | |
| Cantilever Bending (INDA) | WSP 090.1.R4 (12) | Modified IST 90.1 (01) | 9073 – 7:1995 | 9.1 |
| Stiffness Using the Gurley | WSP 090.2.R4 (12) | Modified IST 90.2 (01) | | 9.9 |
| Handle-O-Meter | WSP 090.3.R4 (12) | Modified IST 90.3 (01) | 9073 – 9:1995 | 9.15 |
| Cusick Drape | WSP 090.4.R4 (12) | ERT 90.4 (99) | | 9.21 |
| Bending Length (EDANA) | WSP 090.5.R4 (12) | ERT 50.5 (99) | | 9.28 |
| Drapeability | WSP 090.6.R4 (12) | New Document | | 9.36 |
| ***(10) Tear Strength*** | | | | |
| Elmendorf | WSP 100.1.R3 (12) | Modified IST 100.1 (01) | | 10.1 |
| Trapezoid Tear | WSP 100.2.R3 (12) | IST 110.2 (01)/ERT 70.4 (99) | 9073 – 4:2008 | 10.11 |
| Tongue Tear | WSP 100.3.R3 (12) | Modified IST 100.3 (01) | | 10.23 |
| ***(11) Tensile*** | | | | |
| Grab Tensile | WSP 110.1.R4 (12) | Modified IST 110.1 (01) | 9073 – 18:2007 | 11.1 |
| Bond Strength | WSP 110.3.R4 (12) | | | 11.11 |
| Strip Tensile | WSP 110.4.R4 (12) | IST 110.5 (02)/ERT 20.2 (89) | 9073 – 3:1989 | 11.16 |
| Ball Burst | WSP 110.5.R4 (12) | Original Method | ISO/DIS 9073 – 5 | 11.26 |
| ***(12) Thickness*** | | | | |
| Thickness (INDA) | WSP 120.1.R4 (12) | Modified IST 120.1 (01) | | 12.1 |
| Thickness of Highloft | WSP 120.2.R4 (12) | Modified IST 120.2 (01) | | 12.10 |
| Compression and Recovery, Highloft | WSP 120.3.R4 (12) | IST 120.3 (01) | | 12.17 |
| Compression and Recovery, Highloft Using Weights and Plates | WSP 120.4.R4 (12) | IST 120.4 (01) | | 12.22 |
| Compression and Recovery, Highloft High Temperature/High Humidity | WSP 120.5.R4 (12) | IST 120.5 (01) | | 12.29 |
| Thickness (EDANA) | WSP 120.6.R4 (12) | ERT 30.5 (99) | 9073 – 2:1995 | 12.35 |

| | | | | |
|---|---|---|---|---|
| ***(13) Weight*** | | | | |
| Mass per Unit Area | WSP 130.1.R4 (12) | ERT 40.3 (90) | 9073 – 1:1989 | 13.1 |
| ***(14) Binder/Appearance/Dry Cleaning*** | | | | |
| Binder Distribution/Penetration | WSP 150.1.R4 (12) | IST 50.1 (01) | | 14.1 |
| Appearance and Integrity of Highloft Batting | WSP 150.2.R4 (12) | IST 50.2 (01) | | 14.7 |
| ***(15) Linting*** | | | | |
| Particulate Shedding, Dry | WSP 160.1.R4 (12) | IST 160.1 (01)/ERT 220.1(02) | 9073 – 10:2003 | 15.1 |
| Particulate Shedding, Wet | WSP 160.2.R4 (12) | IST 160.2 (01) | | 15.12 |
| Fibrous Debris from Nonwovens | WSP 160.3.R4 (12) | IST 160.3 (01) | | 15.20 |
| Fibrous Debris from Hydrophobic Nonwovens | WSP 160.4.R3 (12) | IST 160.4 (01) | | 15.29 |
| ***(16) Geotextiles*** | | | | |
| Conditioning | WSP 003.0.R3 (12) | ERT 60.2 (99)/ISO 554 | | 1.76 |
| Sampling | WSP 005.0.R3 (12) | ERT 130.2 (89) | | 1.84 |
| Bursting Strength | WSP 030.1.R3 (12) | Modified IST 30.1 (01) | | 4.1 |
| Air permeability | WSP 070.1.R3 (12) | Modified IST 70.1 (01) | | 7.1 |
| Cantilever Stiffness | WSP 090.1.R4 (12) | Modified IST 90.1 (01) | | 9.1 |
| Trapezoid Tear | WSP 100.2.R3 (12) | Modified IST 100.3 (01) | | 10.11 |
| Force and Elongation Grab | WSP 110.1.R4 (12) | Modified IST 110.1 (01) | | 11.1 |
| Breaking Force Strip Test | WSP 110.4.R4 (12) | IST 110.5 (02)/ERT 20.2 (89) | | 11.16 |
| Thickness | WSP 120.1.R4 (12) | Modified IST 120.1 (01) | | 12.1 |
| Thickness of Highloft | WSP 120.2.R4 (12) | Modified IST 120.2 (01) | | 12.10 |
| Mass per Unit Area | WSP 130.1.R4 (12) | ERT 40.3 (90) | | 13.1 |

*(Continued)*

**Table 11.2** (Continued)

| Description | Method number | Equivalent methods | ISO reference | Page number |
|---|---|---|---|---|
| ***(17) Superabsorbent Materials*** | | | | |
| pH of Polyacrylate (PA) Powders | WSP 200.2.R3 (12) | ERT 400.2 (02)/IST 200.1 (02) | 17190 – 1:2001 | 17.1 |
| Residual Monomers | WSP 210.2.R3 (12) | ERT 410.2 (02)/IST 210.2 (02) | 17190 – 2:2001 | 17.7 |
| Particle Size Distribution | WSP 220.2.R3 (12) | ERT 420.2 (02)/IST 220.0 (02) | 17190 – 3:2001 | 17.18 |
| Mass Loss Upon Heating | WSP 230.2.R3 (12) | ERT 430.2 (02)/IST 230.0 (02) | 17190 – 4:2001 | 17.26 |
| Free Swell Capacity in Saline, Gravimetric Determination | WSP 240.2.R3 (12) | ERT 440.2 (02)/IST 240.2 (02) | 17190 – 5:2001 | 17.32 |
| Fluid Retention Capacity in Saline, After Centrifugation | WSP 241.2.R3 (12) | ERT 441.2 (02)/IST 241.2 (02) | 17190 – 6:2001 | 17.41 |
| Absorption Under Pressure, Gravimetric Determination | WSP 242.2.R3 (12) | ERT 442.2 (02)/IST 241.2 (02) | 17190 – 7:2001 | 17.52 |
| Absorption Under Pressure | WSP 243.2.R1 (12) | New Method | | 17.63 |
| Flow-rate & Density, Gravimetric Determination | WSP 251.0.R1 (12) | New Method | | 17.74 |
| Extractable | WSP 270.2.R3 (12) | ERT 470.2 (02)/IST 270.2 (02) | 17190 – 10:2001 | 17.85 |
| ***(18) Bacterial*** | | | | |
| Filtration Efficiency | WSP 300.0.R4 (12) | ERT 180.0 (89) | | 18.1 |
| Dry Bacterial Penetration | WSP 301.0.R4 (12) | ERT 190.1 (02) | | 18.12 |
| Wet Bacterial Penetration | WSP 302.0.R4 (12) | ERT 200.1 (02) | | 18.20 |
| ***(19) Formaldehyde*** | | | | |
| Water Extraction Method I | WSP 310.1.R4 (12) | ERT 210.1 (99) | | 19.1 |
| Stressed Extraction Method II | WSP 311.1.R4 (12) | ERT 211.1 (99) | | 19.11 |

| | | | | |
|---|---|---|---|---|
| Free Formaldehyde Determination HPLC, Method III | WSP 312.0.R4 (12) | ERT 212.0 (96) | | 19.13 |
| Free Formaldehyde in Processing Method IV | WSP 313.1.R4 (12) | ERT 213.0 (99) | | 19.23 |
| ***(20) Absorbent Hygiene Products*** | | | | |
| Tampons Absorbency-Syngina Method | WSP 350.1.R3 (12) | ERT 350.0 (02) | | 20.1 |
| Ethanol–Extractable Organotin 1 | WSP 351.0.R4 (12) | ERT 360.0 (02) | | 20.9 |
| Acetone Extractable Finish | WSP 353.0.R2 (12) | New Method | | 20.34 |
| Absorption Before Leakage Using an Adult Mannequin | WSP 354 0.R2 (12) | Original Method | | 20.43 |
| ***(21) Useful Methods*** | | | | |
| Surface Linting | WSP 400.0.R2 (12) | Original Method | | 21.1 |
| Lamination Strength | WSP 401.0.R2 (12) | Original Method | | 21.10 |
| Cup Crush | WSP 402.0.R2 (12) | Original Method | | 21.16 |
| Thermal Barrier Performance | WSP 403.0.R2 (12) | Original Method | | 21.25 |
| Homogenization of Absorbent Hygiene Products | WSP 404.0.R1 (12) | New Method | | 21.38 |
| Respirable Particles | WSP 405.0.R1 (12) | New Method | | 21.46 |
| Dust in Collection, Sodium Atomic Absorption/Emission Spectrometry | WSP 406.0.R1 (12) | New Method | | 21.53 |
| Fiber Orientation Distribution | WSP 407.0.R1 (12) | Was WSP 60.5 | | |

### 11.5.1 ISO 9000

ISO 9000 is a series of documents that define requirements for the Quality Management System Standards. ISO 9000:2005[1] is the most significant of these standards. It is a guidance standard and not a certification standard. For certification one shoud refer to ISO 9001:2008 (Section 11.5.2) or ISO/TS 16949 (Section 11.5.3). ISO 9000:2005 explains the fundamentals of quality management systems, such as ISO 9001, and vocabulary used in the ISO 9000 Standards. It describes the following:

- Describes the fundamentals of the QMS in the ISO 9000 family.
- Defines the related terms in the ISO 9000 QMS.
- Is used as guidance when implementing ISO 9000 Standards.
- Does NOT have any requirements, so an organization does NOT certify to ISO 9001.
- It is scheduled to be updated in the next few years.

The ISO 9000 family addresses various aspects of quality management and contains some of ISO's best known standards. The standards provide guidance and tools for companies and organizations who want to ensure that their products and services consistently meet customer's requirements, and that quality is consistently improved.

### 11.5.2 ISO 9001:2008

An organization may be certified to be in compliance with ISO 9001:2008. ISO 9001:2008 specifies requirements for a quality management system where an organization needs to demonstrate its ability to consistently provide product that meets customer and applicable statutory and regulatory requirements, and aims to enhance customer satisfaction through the effective application of the system, including processes for continual improvement of the system and the assurance of conformity to customer and applicable statutory and regulatory requirements. All requirements of ISO 9001:2008 are generic and are intended to be applicable to all organizations, regardless of type, size and product provided. Where any requirement(s) of ISO 9001:2008 cannot be applied due to the nature of an organization and its product, this can be considered for exclusion.

### 11.5.3 TS16949: automotive version of ISO 9001:2008

ISO/TS 16949 is an ISO technical specification which aligns existing US, German, French, and Italian automotive quality system standards within the global automotive industry. It specifies the quality system requirements for the design/development, production, installation, and servicing of automotive-related products.

[1] The last four digits of an ISO standard is the year it was initiated or revised. For example ISO 9000:2008 was revised in 2008.

ISO/TS 16949 has been developed to encourage improvement in both the supply chain and the certification process. In fact, for the majority of leading vehicle manufacturers, certification to this specification is a mandatory requirement for doing business. ISO/TS 16949 is relevant to all types of automotive supply companies, from small manufacturers to multisite, multinational organizations located anywhere in the world. However, it is only applicable to sites where production or service parts are manufactured for the original equipment market. Organizations wishing to break into the automotive market must wait until they are on an automotive customer's potential supplier list before they can progress with certification to this specification. Once a supplier is on the customer's potential supplier list and once the supplier becomes ISO/TS certified he still has to have his part approved in accordance with customer requirements and in compliance with the ISO/TS 16949 protocol. For the part approval process, the following documentation may be required:

Part Submission Warrant Approval (PPA)
Process Flow Chart
Process Failure Mode and Effect Analysis (PFMEA)
Control Plan
Production Part Approval—Dimensional Results (PPA)
Gauge R&R—Repeatability and Reproducibility Analysis Report.

### 11.5.4 ISO 14000 environmental management

The ISO 14000 family addresses various aspects of environmental management. It provides practical tools for companies and organizations looking to identify and control their environmental impact and constantly improve their environmental performance. ISO 14001:2004 and ISO 14004:2004 focus on environmental management systems. The other standards in the family focus on specific environmental aspects such as life cycle analysis, communication and auditing. It is expected that with the current flow of legislation and regulation, it will eventually become necessary for all companies to become ISO 14000 compliant and certified in order to do business.

### 11.5.5 OHSAS 18000 occupational safety and health management

OHSAS 18000 is an international occupational health and safety management system specification. It comprises two parts, 18001 and 18002, and embraces BS8800 and a number of other publications. The OHSAS specification is applicable to any organization that wishes to:

- Establish an OH&S management system to eliminate or minimize risk to employees and other interested parties who may be exposed to OH&S risks associated with its activities.
- Assure itself of its conformance with its stated OH&S policy.

- Demonstrate such conformance to others.
- Implement, maintain and continually improve an OH&S management system.
- Make a self-determination and declaration of conformance with this OHSAS specification.
- Seek certification/registration of its OH&S management system by an external organization.

# GLOSSARY

**Abaca** Also known as manila hemp. A long natural fiber that is produced from the Abaca tree which is related to the banana tree (without the bananas). Abaca is used in media such as teabag, vacuum cleaner bag, and other wet lay products.

**Absorption** Involves the taking of molecules from one substance directly into another substance. It is contrasted with adsorption in which the molecules adhere only to the surface of the second substance. The process may be physical or chemical. Physical absorption involves such factors as solubility and vapor pressure relationships. Chemical absorption involves chemical reactions between the absorbed substance and the absorbing medium.

**Acrylic resin** Any of numerous thermoplastic or thermosetting polymers or copolymers of acrylic acid, methacrylic acid, esters of these acids, or acrylonitrile used to produce latex resins and other polymeric forms such as paints, synthetic rubbers, and lightweight plastics.

**Advance cure** The state of a resin in a filter medium that has been partially precured or partially cured prior to shipment to a customer. The degree of precuring is controlled to the customer's requirements.

**Additive** Chemicals and/or substances used to modify the properties of nonwoven webs, polymers, and /or fibers. Types of additives include reagents, fillers, viscosity modifiers, dyes and pigments, flame retardants, anti-microbial agents, water repellents, and wet strength agents.

**Adsorption** A separation process by which molecular fluid contaminants are attracted to the surface of the separation medium. Three forms of adsorption are

1. Physical adsorption: A molecular species is attracted to a surface and held in place by weak forces such as van der Waals forces.
2. Chemical adsorption: A molecular species is attracted to a surface and reacts with that surface or something on that surface to form a chemical bond which holds it in place. Typically considered irreversible at constant conditions.
3. Catalysis: A molecular species is attracted to the surface which encourages a chemical reaction. The reaction products move away leaving the surface unchanged. Typically the reaction is considered irreversible.

**Aerosol** A system of colloidal particles dispersed in a gas, smoke, or fog.

**Air permeability or permeance** The property of a medium that relates to the passage of air when a pressure difference exists across the flow boundaries of the medium.

**Air purifier** A device which aims to free air from contaminants. Various techniques are used to accomplish this which includes mechanical particle filtration, adsorbents such as activated carbon or alumina, and electrostatic charge.

**Air purifying respirators (APRs)** A respirator containing the following components: a face mask, a filter or cartridge filter to remove dust, smaller particles, and mists, and a cartridge filter to remove chemical gases and fumes

**Alpha Value** See Figure of merit (FOM)

**Anisotropic** A feature of most nonwoven filter media is that they have different properties in different directions. This is a result of fiber orientation when the medium was made. The properties are usually referred to as machine direction (MD) properties, cross direction (CD), and Z-direction properties.

**Anti-microbial** An additive intended to inhibit microbes (bacteria, fungus, mildew) from propagating on the surface and/or interior fiber surface of a filter medium.

**Arrestance** A measure of an air filtration device's ability to remove dust from the air. It is part of the test procedures in EN779 and ASHRAE 52.2. In the ASHRAE 52.2 air duct test procedure, the equation is:

$$A = 100[1 - W_u/W_d]$$

where $A$, is the percentage of test dust, $W_d$, that is captured by the test filter. It is determined by measuring the weight of test dust fed, $W_d$, and the weight of test dust, $W_u$, captured by the final filter downstream of the test filter.

**Ash content** The noncombustible content of a filter medium (see loss on ignition).

**Aspect Ratio** The length to diameter ratio of a short cut, staple, or pulp fiber.

**Asymmetric** Membranes that have changing or different structures in the direction of their thicknesses. Also known as anisotropic membranes.

**Atactic** Random substituent groups along the chain of a vinyl polymer.

**Barrier**
1. Relates to the ability of a filter medium to prevent the penetration of particles in a fluid stream that passes through the medium. May be considered an another term for filtration efficiency.
2. A type of medium that is used to contain or prevent the passage of oil from oil spills in water.

**Bacterial filtration efficiency (BFE)** A test use to determine the filtration efficiency of a filter material such as face mask by using an aerosol of bacteria organisms. The test procedure is in accordance with ASTM F2101-1.

**Basalt** A material produced from volcanic rock that is used to produce inorganic fibers used in some filter medium applications where high temperature resistance is required.

**Base sheet** Wet lay media that are either not yet resin treated downstream of the forming operation or will not be resin treated for its application. See Raw stock.

**Basis weight** The weight of a unit area of medium. It is synonymous with the term grammage; however, it is more often used with nonmetric units such as $lb/3{,}000\ ft^2$ or $oz/yd^2$.

**Bast fibers** Vegetable fibers obtained from the inner bark or phloem of a woody plant.

**Beater** An oval shaped vessel used for the preparation of laboratory pulp fiber slurries for handsheet making. The slurry or stock circulates the oval path of the beater and in each cycle passes past a beater roll that is rotating at a high speed. The beater roll contains protruding bars that rub the fibers against a flat plate and causes separation and dispersion of the individual fibers. Fibrillation and possible cutting of the fibers also occurs, the amount depends on the beating time and pressure. It is possible that large-scale beaters are still used in production operations; however, they are quite obsolete when compared to modern day pulpers.

**Beater addition (beater ad)** The process of adding a non-fibrous additive to the stock slurry in the wet end of a wet lay operation. The beater additive, often called a beater ad, can be a bonding latex, a wet strength agent, a powdered filler such as diamataceous earth or activated carbon, or other agents designed to provide some specific property or process capability to the web.

**Beta ($\beta$) ratio** The efficiency term for multi-pass filter testing. It is the number of particles upstream of the filter divided by the number of particles downstream of the filter. The particles are grouped by particle size so that there is a $\beta$ ratio for each particle size or each group of particle sizes.

**BET (Brunauer, Emmet, Teller) Theory** Theory based on the attraction of an inert gas to the surface of the substrate being tested. A technology for measuring the specific surface area and pore structure of a filter medium.

**Biharmonic equation** A fourth-order partial differential equation which arises in elasticity theory, often expressed as $\nabla^4\psi = 0$ where $\psi$ is the streaming function.

**Biofiber** Fiber sourced from natural sources such as trees, vegetable, or animals (e.g., sheep wool) or synthesized from natural raw materials (e.g., rayon)

**Boehmite** An aluminum oxide hydroxide ($\gamma$−AlOOH) particle, containing extra hydroxyl groups on its surface. It is a component of the aluminum ore, bauxite. It exhibits a cationic charge that makes it useful for electrokinetic media.

**Bone Dry** The basis weight or grammage of the web after all volatiles have been removed, often determined by oven heating the web at a specified temperature for a specified length of time (usually 10 min @ 150°C).

**B-stage** The state of a medium, treated with a thermosetting resin, which has been dried to a specific volatile content, but has not yet been cured to the final end-use state. The term is most often applied to resin-treated filter media that will be subject to a pleating operation and cured after pleating.

**Bubble Point** The term often used to express the maximum pore size of a filter medium. See Maximum pore size.

**CADR Value** Clean air delivery rate—a rating for air purifiers or room air cleaners as determined by ANSI/AHAM AC-1 which is an ANSI approved standard developed by AHAM

**Calender** A set of at least two highly polished nip rolls that apply a very high pressure to a web in order to give it a smooth surface treatment.

**Candle filter** A form of cylindrical cartridge filter that is mounted vertically.

**Capacity layer** A bulky porous layer in a composite structure where a downstream layer is a highly efficient filter medium such as a membrane or a nanofiber web. The capacity layer upstream to flow traps larger particles and provides dirt holding capacity whereas the efficiency layer only has to deal with the fine particles that penetrate the capacity layer.

**Carding** The web forming operation in a dry laid process. The purpose of a carding machine is to break down entangled fibers fed to it by a feed shute and form them into a coherent nonwoven web. It does this by passing the fibers between two surfaces, (often roll surfaces) containing metallic coating (highly engineered sawtooth barbed wire). The surfaces move at different speeds so that the barbs perform a roughing action on the fibers. The action separates and individualizes fibers, breaks down tufts of fiber, orients the fibers, separates tufts for further processing and transfers the web from one operation to the next. The metallic coated surfaces are usually rolls. Sometimes flat surfaces are used as the second surface.

**Cartridge filter** That part of a filtration system that is removable from the main body of the system for disposal and replacement, or for cleansing and reuse. Generally associated as being cylindrically shaped, although not necessarily so

**Cast membranes** A membrane formed by casting a polymer into porous film. There are basically three types of casting procedures:

1. Air casting: The solution polymer is cast unto a cold chilling roll to form a partially solidified film which is then processed to a finished membrane.
2. Solvent or emission casting: A dissolved polymer solution is cast on a moving roll surface and then submerged into a nondissolving liquid that kicks out the polymer from the solution. Pores are formed as the dissolving solvent tries to escape from the polymer mass.
3. Thermal casting: Heat and subsequent cooling is used to precipitate the polymer from the polymer solution and form it into a porous membrane form.

**Catalytic media** A separation medium that contains a catalytic agent or substance to remove or break down undesirable contaminants in a fluid.

**Chopped fiber** See Short cut fibers

**Chopped strand** Rovings, usually glass, that have been chopped to shorter lengths for nonwoven manufacture.

**Cleanroom** A room in which the concentration of airborne particles is controlled and which is constructed and used in a manner to minimize the introduction, generation, and retention of particles inside the room, and in which the relevant parameters (e.g., temperature, humidity and pressure) are controlled as necessary.

**Coarseness, fiber** The weight of fiber wall material in a specified fiber length, usually expressed in units of mg/100 m fiber.

**Coat hanger type distributor** A type of polymer feed distributer in the die block of meltspun processes. It is referred to as a coat hanger because of the parabolic shape of the wall around the distributer chamber.

**Coating** **1.** The application of a nanofiber or microfiber web to a substrate.

**2.** The application of a polymer (resin, adhesive, hotmelt, etc.) to a substrate so as to cover the surface of the substrate

A coater is a device that applies the coating to the substrate.

**Coir** Coconut fiber

**Collector** An advancing porous or foraminous device on a nonwoven former that collects the fibers and helps form them into a web, Often the device is a foraminous belt, fabric, or screen. In a wet lay process, the Fourdrinier wire would be considered as the collector. In many cases, the collector is in the form of a belt traversing between rollers that provide guidance and tension. Sometimes the collector is the foraminous surface of a rotary drum. In composite structures one of the layers may serve as collector for the other layer(s)

**Colloid** A substance made up of a system of particles with linear dimensions in the range of .001 to 0.5 μm dispersed in a continuous gas, liquid, or solid medium whose properties depend on large surface area. The particles can be large molecules like proteins or solid, liquid, or gaseous aggregates and they remain dispersed indefinitely.

**Color space** A mathematical model that maps the colors that can be reproduced by a device to a standard color model.

**Comber noils** The short fibers removed by a combing operation. Usually refers to cotton.

**Composite structure** Two or more web technology layers that are combined to optimize the characteristics of a filter medium. Composites provide the opportunity to bring more than one separation technology into one medium.

**Computational fluid dynamics (CFD)** The application of numerical methods to solve the general field equations of mass, momentum, energy, and species balances. It can include multi-physics mechanisms (electromagnetics, radiant heat transfer, photonics and others). CFD is used in many industries such as automotive and aerospace to predict performance of designs of vehicles and vehicle components. CFD can also be used to predict the performance of filter media.

**Condensing roll** Rolls in the carding process that works with doffer rolls to randomize the fiber structure and put more fibers in the cross direction.

**Coniferous** Relating to or part of trees or shrubs bearing cones and evergreen leaves such as pines, spruces, and firs, and to plants and shrubs bearing foliage throughout the year. Coniferous trees are the source of that type of wood pulp known as softwood.

**Consistency** The papermakers term for the solids or bone dry fiber content of a wet end stock or slurry.

**Copolymer** A polymer composed of more than one monomer, usually two.

**Corona discharge** An electrical discharge brought on by the ionization of a fluid surrounding a conductor.

**Corrugation** A sine wave-like filter medium configuration in the cross direction of the medium (usually wet laid) obtained by embossing the medium between two meshing corrugating rolls that have a sign wave configuration to their surfaces.

**Couching (pronoonced "kooching")** The process of transferring a wet laid fabric or paper from the forming fabric or Fourdrinier wire to another surface.

**Creep** Also known as "cold flow," the change in dimensions, with time, of a medium that is under constant load.

**Creeping Flow** A region of very slow flow, and of very low Reynolds number, $Re << 1$.

**Creping** A process which continuously applies crass direction wrinkles to a filter medium to increase its surface area and its extensibility.

**Crimp** "The property of a fiber that exhibits bends, waves, or curls. It can be natural such as the waviness of wool fibers as they naturally grow on sheep or it can be mechanically" induced in the manufacturing process.

**Cross direction (CD)** The in-plane direction across a medium crosswise to the flow direction of the machine on which the medium was made.

**Cross layering** A process step for producing thicker and heavier carded webs by lapping them back and forth unto a conveyor moving at a right angle to the cross-layering motion.

**Curl** A property of fiber that is defined as the amount of curvature over any specified length of fiber.

**Curtain coater** A type of applicator of bonding resin or coating that applies the resin or coating as a falling film from an inclined flat surface.

**Decigrex** Papermakers units for coarseness, the weight in milligrams of 100 m of fiber or the weight in decigrams of 10 km of fiber (17).

**Decitex (dtex)** The weight of fiber in grams of 10,000 m of fiber. See Tex.

**Deckle** A term for wet lay machines that indicates the width of the web that can be formed on the machine.

**Degree of polymerization (dp)** Describes the molecular size of a polymer. The number of repeat units in the polymer structure.

**Demister** A coalescing-type filter designed to remove moisture and/or organic liquid particles from gas streams.

**Denier** The weight of fiber in grams of 9,000 m of fiber

**Die** The part of a spinnerette that contains small holes through which molten or solution polymer passes through to form fibers.

**Die block** The part of a melt spinning (or solution spinning) machine that distributes the polymer flow and contains the spinnerette from which fibers are formed.

**Die cutting** A machine operation that cuts a medium into filter sheets or filter pads of various desired shapes. It is like a cookie cutter. A die-cutting machine uses steel rule die shapes to cut through a wide range of filter materials quickly and easily.

**Diffusion coefficient (particle)** $D_{AB}$ Relates the movement of a particle A in a binary mixture of A and B.

**Dimensional stability** The ability of a medium to maintain its structural dimensions when exposed to changes in humidity and moisture content, temperature, the application of load and/or the passage of time.

**Dioctyl phthalate (DOP)** An oily type of fluid used to generate liquid test aerosols in many high-efficiency air filtration test procedures.

**Dirt holding capacity** A measure of filter life as determined by a prescribed filter test. It is the amount of test dust or dirt a filter will hold under flow conditions until the filter reaches a predetermined pressure drop. At this point the test is terminated and the dirt or dust on the filter is measured to determine its dirt holding capacity.

**Disintegrator** A laboratory device used to disperse pulp fibers in water for handsheet making, The disintegrating action is similar to that of a Waring blender.

**Dispersion** A system of dispersed particles suspended in a solid, liquid, or gas.

**Distributor** That part of a die block that controls and balances the polymer flow.

**Doffer rolls** Rolls in a carding machine that are used to transfer the web from the master cylinder to a moving bed.

**Drum dryers** Heated rotating cylinders used to apply heat to a nonwoven fabric and/or paper for purposes of drying or other purposes. The heat is applied to the fabric as it contacts the surface of the cylinder as it passes over and under the rotating dryer. Heat sources are most often from steam in the hollow interior of the cylindrical drum. Hot oil and electricity are also used as heat sources.

**Dry Smoke** A form of industrial fume usually consisting of solid particles in air such as that emitted by welding operations or certain types of machining operations. Contrast to wet smoke below.

**Dye** A natural or synthetic substance applied to a web or to the fibers of a web to add a color to the web or to change its color. The dye is usually applied as a solution. It differs from a pigment in that a pigment is not a solution but rather a dispersion of particles.

**Earlywood** See Springwood

**Efficiency** See Filtration efficiency

**Efficiency Particulate Air (EPA)** Refers to a class of high efficiency filters identified by EN 1822:2009 as having sub-HEPA filter capabilities and designated as E10, E11, and E12. Filtration efficiencies for this group range from >85% to <99.95%.

**Electret** Materials that have an electrostatic charge or exhibit an external electric field in the absence of an applied field.

**Electrodialysis** An electronic separation process whereby ions are removed from a polar liquid by passing a current through that liquid and forcing dissolved salt ions to migrate through cationic and anionic membranes. Most often used for the desalinization of brackish water.

**Electrokinetic** Electrical charges in moving fluids such as water. It concerns particle motion which is the direct result of applied electric fields. It relates to the property (often referred to as zeta potential) of a filter medium that has electrically charged sites, usually positive (cationic), to enhance its filtration performance in liquid filtration applications.

**Electrospun** A process that creates nano- or micro-sized fiber through an electrically charged jet of polymer solution or polymer melt.

**Electrostatic** Refers to a filter with static electric charge, usually opposite to the electrical charge of the particles it needs to attract. Particles drawn to the fiber surface are trapped by their electrostatic attraction to the fibers.

**Elmendorf tear test** A tear test that involves a swinging pendulum. The specimen is slit so that one side of the slit is attached to a stationary clamp and the other side to the moving pendulum. When the pendulum is released, it continues the tear along the slit. The force to propagate the tear is then determined.

**Elongation** Tensile property that has several possible definitions or understandings in connection with the forced stretching or expansion of a strip of nonwoven or paper. It is generally a measure of the change in length of a specimen when a stress is applied. It can be used as an alternate or synonym for strain. Some definitions are as follows:

1. The change in length ($\Delta L$) of a nonwoven specimen, in mm or inches when a stress is applied.
2. A synonym for strain or percent of original length that a nonwoven specimen is stretched when a stress is applied.
3. The fraction of original length a nonwoven specimen is stretched when a stress is applied.
4. The tensile strain at maximum tensile strength before rupture. This is more or less synonymous with the definition of stretch below.
5. % Elongation—A synonym for stretch as defined in TAPPI 494, which is the maximum *strain* before rupture when a stress is applied.

**Embossing** A three-dimensional pattern pressed into a filter medium as a result of pressing the medium between two engraved hard surfaces. The surfaces are usually two embossing rolls engraved with the desired pattern. The embossing is applied by passing the medium under pressure and heat through the nip of the embossing rolls. This is a technique used for point thermobonding of nonwoven webs. It is also a process for putting dimples or pleat separators into a filter medium that is to be pleated. Corrugation is a form of embossing.

**Emulsion** Colloidal suspension of liquid particles in another liquid.

**Epitropic fibers** Fibers with an altered surface property, e.g., electrically conducting, abrasive, etc. Electrically conducting forms usually contain imbedded carbon particles.

**Extruder** A part of a melt spun line that contains a rotating screw shaft to force polymer through several heating zones. The polymer is in a molten state by the time it reaches the filter screen on the exit side of the extruder.

**Face mask** A filter designed to cover nose and mouth and protect the wearer from irritating, toxic, or infectious contaminants in the air. There are different kinds of facemasks. For example, a dusk mask usually has one supporting strap and is designed to protect the wearer from typical atmospheric contaminants such as pollen, dust, and other large irritating particles in the air. A surgical face mask is a more efficient filter designed to protect medical workers from infectious diseases and microorganisms. A respirator is a device designed to protect the wearer from inhaling dangerous substances such as toxic chemicals and infectious particles. Respirators and some surgical face masks are subject to a NIOSH rating

**Fancy roll** A roll associated with a garnett or carding machine that is used to produce higher loft mats.

**Felt** A fabric of matted, compressed animal fibers, such as wool or fur, sometimes mixed with vegetable and/or synthetic fibers. Wet milled felts are bonded by the application of moisture and physical pressure in a procedure referred as fulling.

**Felt side** Felt side and wire side are terms usually applied to wet laid media. The felt side of the medium is that away from or not in contact with the Foudrinier wire when the sheet is formed. The wire side is the side in contact with the wire. Generally, for nonwoven filter media the felt side is less dense and less smooth then the wire side.

**Fiber Coarseness** The papermaker's designation for fiber density. It is the weight in milligrams of 100 m of fiber or the weight in decigrams of 10 km of fiber (17). See Decigrex.

**Fibril** Sometimes referred to as microfibrils, fibrils are bundles of cellulose molecules composing the S2 layer of a tracheid or pulp fiber.

**Fibril angle** The orientation of fibrils in a pulp fiber relative to the long axis of the fiber.

**Fibrillation** The shearing of wet wood pulp fibers to loosen the micro-fibrils from the fiber surface and fiber wall.

**Field Emission Scanning Electron Microscopy (FESEM)** A scanning electron microscopy technique that provides topographical and elemental information at magnifications of 10 × to 300,000 ×, with virtually unlimited depth of field. Compared with conventional scanning electron microscopy (SEM), field emission SEM (FESEM) produces clearer, less electrostatically distorted images with spatial resolution down to 1.5 nanometers—three to six times better.

**Figure of merit (FOM)** Also known as the alpha value and sometimes the gamma value, FOM is a way of measuring the relative performance of HEPA and ULPA media and filters.

$$\mathrm{FOM} = ((-\log 10(\mathrm{Penetration}/100)) \times 100)/\mathrm{Resistance}$$

Where Penetration is in percentage and Resistance is in mm $H_2O$

**Filament** A very slender or natural synthetic fiber. Continuous filaments are very long or never ending fibers before they have been cut or chopped

**Film** **1.** A thin sheet or strip of flexible material, such as a cellulose derivative or a thermoplastic polymer.
**2.** A continuous polymer coating on a substrate or fiber.

**Filter cloth** A loose term used for a fabric used as a medium for filtration. Fabrics used for filter press operations are often referred to as filter cloths. They are also used for air filtration applications such as bag filters. Filter cloths may be a woven, nonwoven, or felted material. They are also composed of a variety of polymer fibers depending on the application.

**Filter element** That part of the filter that fits inside the housing and contains the medium that does the actual filtering.

**Filter house** The air inlet structure of a gas turbine-generator where the filters are located.

**Filter life** The amount of time a filter will last till it is no longer functioning in its specified manner or until failure occurs. Dirt holding capacity may be one measure of filter life, however, other factors such as degradation due to mechanical agitation, chemical exposure, and climate and environmental influences may all affect filter life and be a cause of failure.

**Filtration** A mechanism or process for separating substances or contaminants from a fluid by passing the fluid through a porous medium designed to trap the contaminant, but let the main body of fluid flow through. Filtration may be used to separate contaminants from a fluid or recover value-added materials, such as minerals, chemicals, or foodstuffs in a process operation.

**Filtration efficiency** The percentage or fraction of a contaminant removed by a filter. It is usually based on weight of the contaminant; however it can be based on other factors such as particle count, volume, cross-sectional area of the particles, etc. Whatever the basis the formula is,

$$e = 100\left[\frac{M_u - M_d}{M_u}\right]$$

Where

$e$ is the filtration efficiency, %.

$M_u$ is the upstream measurement of contaminant.

$M_d$ is the downstream measurement of contaminant.

**Finish** **1.** A substance or mixture of substances added to textile materials to impart desired properties.

**2.** A process, physical or chemical, applied to textile materials to produce a desired effect.

**3.** Properties such as smoothness, drape, luster, water repellency, flame retardancy, or crease resistance which is produced by 1 and/or 2 above.

**4.** The state of a textile material as it leaves a process.

**Flash spinning** The process by which polymer fibers or webs are spun from a polymer solution in contrast to spunbonded webs that are spun from a molten polymer. The polyethylene spunbond known as Tyvek® is a flash spun web.

**Flame resistance** The characteristic of a fabric to resist ignition and to self-extinguish if ignited.

**Flame retardant** A chemical substance used to impart flame resistance to a nonwoven fabric. It is not part of the basic web chemistry.

**Fleece** A fabric with a soft pile such as that might come out of a carding process prior to any bonding activity.

**Fluff pulp** A chemical, mechanical, or combination chemical–mechanical pulp, usually bleached, used as an absorbent medium in disposable diapers, bedpads, and hygienic personal products. This type of pulp imparts bulk to filter media and is a very important fiber raw material for wet laid cellulosic filter media.

**Flux** The rate of flow of a property per unit area. The property can be volume flow, mass flow, particle flow, heat flow, diffusion, radiation, energy, and momentum. For filtration purposes, the most frequent form is volume flow through a filter medium and is measured in $m^3/m^2 \cdot s$

**Fluid** Any liquid or gas or generally any material that cannot sustain a tangential, or shearing, force when at rest and that undergoes a continuous change in shape when subjected to such a stress.

**Foam bonding** A method of applying a latex binder to a nonwoven medium by applying it as a foam or froth.

**Foraminous** A material such as metal screen, netting, woven fabric, etc that has openings, holes, punctures, etc penetrating through its entire depth.

**Fractional efficiency** A filtration efficiency system that rates filter efficiency over a range of particle sizes or particle size groups. It is also used to rate different filter efficiencies for the same particle size.

**Freeness** A wet lay process fiber property. It is the rate at which water drains from a stock or slurry suspension of wet lay fibers or pulps. Although freeness is an important pulp property for the manufacture of paper and paperboard, it is not commonly used in wet laid filter medium products. For filter media, air permeability is a better indicator of the "tightness" or "openness" of the fibers.

**Fulling** The process of cleansing, shrinking, and thickening a wet milled felt with heat, pressure, and moisture.

**Furnish** Papermaker's term—the recipe of fibers and other ingredients that are slurried in water to produce the medium of the wet lay process.

**Gamma number** See Figure of Merit

**Garnett** A machine very similar to a card in forming dry laid webs. In general, garnetts differ from cards in that they form more randomized webs. They also produce higher loft webs for filter applications.

**Gas Mask** A masklike device containing or attached to a component that filters the air inhaled by the wearer for protecting the face and lungs against noxious fumes, as in warfare or in certain industrial processes.

**Ginning** The process by which cottonseed is separated from the fiber.

**Glass transition temperature (T*g*)** The temperature at which a polymer transforms from a hard glassy state to a soft rubbery state. The polymer loses its hardness or brittleness, becomes more flexible, and takes on rubbery or leathery properties. At this transition temperature, noticeable changes in the specific volume, thermal conductivity, refractive index, stiffness, heat content, and dielectric loss are apparent.

**Glazing** The process by which a web (usually a felt or needle felt) passes through a set of heated, high temperature, nip rolls that are rotating at slightly different surface speeds. The effect is to "iron" into the web a glazed glasslike surface

**Gradient density media** A type of nonwoven filter medium that has a variation in porosity and density through the depth of the medium. The usual purpose is to provide a filter medium that will remove the largest particles at or near its upstream surface, and more efficiently remove smaller and smaller particles, as the fluid flow transports through the depth of the medium. Generally this is accomplished using more than one ply to the medium; however there are other techniques for doing this.

**Graft polymer** A polymer containing a main chain or backbone that has attached to its side chains containing atoms or groups different from those in the main chain.

**Grammage** The weight of a unit area of medium. It is synonymous with the term basis weight, however, it is more often used with metric measures such as $g/m^2$.

**Gravure press** A resin applicator that contains a Gravure cylinder which is engraved with small recessed cells (or dots) that act as tiny wells. Their depth and size determines the amount of resin that gets transferred to the substrate.

**Gurley unit** The unit of measurement for the Gurley stiffness tester. One Gurley unit equals one milligram force (mgf) equivalent to 102 millinewtons (mN). 1/0.009807 (=124) millinewtons (mN).

**Hagotan** A motorized, tapered spindle used for stripping the fiber layer of the abaca plant.

**Handsheet** A sheet of wet lay fiber formed by draining the slurry through sheet mold which contains a wire or forming fabric through which the slurry drains. The fiber is filtered on the wire to form the handsheet. The handsheet is then dried on a thermal device for subsequent testing.

**Headbox** That part of a wet lay or paper machine where the stock or fiber slurry is distributed to a forming fabric or collector (often called a wire or forming fabric). The headbox is designed to provide a continuous and uniform flow to the wire or forming fabric.

**High efficiency air (HEA)** Description of air filters not only in the HEPA and ULPA range but also in the sub-HEPA range. Includes all the H and U class filters of EN1822 and filters with MERV ratings of MERV 15 and higher. Generally efficiencies of 85% or better are considered as HEA filters

**High loft** A low density fiber medium characterized by a high ratio of thickness to weight per unit area. Often referred to as a high loft mat or batting. High loft nonwovens are generally defined as having a solidity of less than 10% and a thickness of at least 3 mm (.12 in)

**High volume Instrument (HVI)** A test instrument designrd for the classification or classing of cotton. It is based on three prime parameters: micronaire, length, and strength. These and other properties are measured by the HVI.

**Homopolymer** A polymer consisting of a single species of monomer.

**Horizon Light** A type of light source used in color booths for comparing colors. It simulates early morning sunrise and/or late afternoon sunset.

**Hurd** An impurity in hemp fiber.

**Hydroentanglement** A process that uses very fine high velocity jets of water to entangle the fibers in a nonwoven web—sew spunlace

**Hydrolysis** A reaction involving the breaking of a bond in a molecule using water. The reaction mainly occurs between an ion and water molecules and often changes the pH of a solution.

**Hydrophilic** A material that has a strong affinity for water

**Hydrophobic** A material that has little or no affinity for water.

**Hygroexpansivity** The change in dimensions of a medium due to changes in humidity and associated moisture content.

**Inclined wire** A type of wet lay headbox, specifically designed for filter media and wet lay nonwovens, that forms the wet lay web unto a moving forming fabric or wire that is inclined as it passes through the forming area in the pond of the headbox. The inclined configuration allows for high levels of water to dilute the stock slurry and allow for forming of webs from long fibers.

**Internal bond strength** Generally refers to the Z-direction tensile strength or the delamination strength of a medium.

**Ion exchange** An interchange of ions present on an insoluble solid with ions of like charge present in a solution surrounding the solid.

**Islands in the sea** A form of bicomponent (or possibly multi-component) fiber that contains one polymer as fibril type threads that run the length of the fiber and the other component as a polymer mass that surrounds and embeds the fibril threads. The crossection of the fiber reveals the fibril threads as islands surrounded by the sea on the embedding polymer.

**Isotactic** All the substituent groups are the same on every repeat unit of a vinyl polymer chain

**Isotropic**
1. A nonwoven medium that has identical properties in all directions. Such a medium would be unusual because most nonwovens are inherently anisotropic from the manufacturing process.
2. Microporous membranes that have the same chemical and physical structure throughout their thicknesses.

**Kinkiness** The property of a fiber that is clfosely twisted of tightly curled.

**Knudsen number** The Knudsen number is a dimensionless number that provides guidance as to what analytical approach should be used in analyzing fluid dynamics problems. For example, for air flow past small diameter fibers such as nanofibers, the Knudsen number can be used to determine if there is slip flow past the fiber surface and if so become part of the equation used for the calculation. The Knudsen number for flow past a fiber is calculated as:

$$Kn = \frac{2\lambda}{d_f}$$

Where

$Kn$ is the Knudsen number

$\lambda$ is the mean free path of gas molecules around the fiber

$d_f$ is the fiber diameter.

**Lamella** Lamella is the singular form, lamellae is the plural form. A thin plate-like structure, often one amongst many lamellae very close to one another, with open space between. Among other things, they appear in stretched microporous membranes where high permeability is important.

**Latewood** See summerwood.

**Latex** An emulsion in water of finely divided particles of synthetic rubber or plastic.

**L/D Ratio** For a fiber, the ratio of its length to diameter.

**Likerin Roll** The metallic coated roll in a card or Garnett that accepts lap or batting that is condensed from fibers fed from a chute or hopper, opens them into small tufts, and feeds them to the master cylinder.

**Limiting oxygen index (LOI)** The minimum oxygen content in air that will sustain combustion of a material.

**Linear density** The density of a fiber in terms of its mass per specified length. Fiber coarseness, denier, decitex, and decigram are examples of linear density.

**Linters** The fine, silky fibers which remain adhered to the seeds of the cotton plant after ginning, including parts of the longer textile fibers, or "lint", as well as coarse, short fuzz fibers.

**Loss on ignition** The combustible part of a filter medium that is burned off when exposed to a high temperature in a muffler furnace. The noncombustible residue, often called the ash content is a measure of inorganic content. It is a way of determining fiber content in a filter medium composed of inorganic fibers such as glass microfibers.

**Lumen** The central canal of a wood pulp or vegetable fiber.

**Machine direction** The in-plane direction of a medium parallel to the flow direction of the machine on which the medium was made.

**Main cylinder** Often referred to as the master cylinder. The main roll in a card or Garnett. Its function is to transfer the web from one point in the process to the next. It also provides the surface on which other rolls such as stripper rolls and worker rolls can do their work.

**Maximum pore size** A test on a filter medium that relates to its largest pore size. Assuming cylindrical pores through the medium, the maximum pore size (also known as the "bubble point") is determined at the pressure drop at which the first air bubble is observed emerging from the medium wetted with a layer of liquid of known surface tension. The maximum flow pore size is calculated as follows:

$$D_{\rm m} = \frac{4^{\gamma} \cos \theta}{\Delta P}$$

where

$D_{\rm m}$ is the maximum pore size, μm.

$\gamma$ is the surface tension of the liquid layer wetting the medium, mN/m.

$\theta$ is the contact angle between the pore surface and the liquid layer, rad.

$\Delta P$ is the pressure drop at which the first air bubble is observed to emerge from the medium, kPa.

**Mean flow pore size (MFP)** A test on a filter medium that relates to its pore size. Assuming cylindrical pores through the medium the mean flow pore size is determined at the pressure drop at which the flow through a medium wetted with a liquid of known surface tension is 50% of the flow through a dry medium at that same pressure drop. The mean flow pore size is calculated form the following equation:

$$D_{\rm m} = \frac{4^{\gamma} \cos \theta}{\Delta P}$$

where:

$D_{\rm m}$ is the mean flow pore size, μm.

$\gamma$ is the surface tension of the wetting liquid, mN/m.

$\theta$ is the contact angle between the pore surface and the wetting liquid, rad.

$\Delta P$ is the pressure drop where the flow through the wetted medium is 50% of the flow through the dry medium, kPa.

**Medium** A porous structure of the type that can be used in filtration. The plural form is media.

**Medium migration** The release into the fluid stream of "fines" or small fibers and particles in the medium structure that were there as a result of the manufacturing process.

**Melt-blown** A process for forming nonwoven webs by forcing molten thermoplastic polymer such as polypropylene through a multi-hole spinnerette and collecting the formed fibers on a moving collector to form the web. A converging airflow acts on the molten fibers as they leave the spinnerette to attenuate them and randomize their direction, so as to form a random nonwoven fiber structure on the collector.

**Meltspun** A generic term referring to nonwoven webs such as spunbond and melt-blown that are spun from a molten polymer.

**Membrane filtration** A type of filtration or separation where the medium is a porous membrane structure and the separation consists of separating micro- or nano-sized particles. There are four primary types of membrane filtration:

1. Microfiltration
2. Ultrafiltration
3. Nanofiltration
4. Reverse osmosis

**MERV Rating** Minimum Efficiency Reporting Value: an efficiency rating for HVAC filters determined by the testing methodology and criteria of ASHAE 52.2-2013

**Metallic Clothing** A coating of highly engineered spirally wound saw-tooth wire used on the rolls in dry forming operations.

**Microfiber** Generally defined as a fiber whose diameter is 10 μm or less.

**Microfibrils** See fibril. A very fine fibril, or fiber-like strand that forms the surface of a cellulose fiber such as wood. The angle it makes with the axial direction of the wood fiber greatly determines the properties of that fiber. Exposure to high shearing force or refining can fray the microfibrils from the wood fiber surface and provide for increased bonding performance of the fiber. Microfibrils diameters are estimated to be as small as of 3.5 nm and as such can be considered as a nanofiber.

**Micron** Synonym for micrometer—the length measurement that is one-millionth of a meter or .000039 in. In filter media it is usually the measurements for fiber diameter and pore size.

**Micronaire** A measurement of the fiber fineness or linear density of cotton expressed as μg/in. It is converted to denier by dividing by 2.82

**Microporous** Describes a thin, porous film or hollow fiber having pores ranging from 0.01 to 10 μm

**Minimum efficiency (ME)** In accordance with EN779:2012, ME is defined as defined as the lowest value of three different tests for 0.4 μm particles: initial efficiency, efficiency throughout the test's loading procedure, and efficiency after discharging.

**Mini-pleat** A panel filter construction utilizing 7/8−1¼ in deep pleats with very narrow air spaces (1/8 in) in between, making it possible to pack more filter paper onto the standard frame sizes than can be done with conventional deep-pleat. corrugated separators

**Modacrylic** Fiber whose acrylonitrile content is 35-85%.

**Monofil** A single strand of man-made fiber.

**Most penetrating particle size (MPPS)** The size of the particles that achieve maximum penetration of the filter medium. Particles that are smaller or larger than the most penetrating size exhibit a lower rate of penetration. The reduced penetration of the smaller particles is due to diffusion mechanisms, while for the large particles it is due to interception and inertial effects. The most penetrating size is a function of the structure of the filter medium, the velocity of the airflow through the filter, and the physical and chemical nature of the particles. Efficiency ratings for EN 1822 (HEPA and ULPA media) are based on MPPS.

**Motes** Small, immature seeds with attached fiber that are removed at a different state of the cotton gin stand than the mature seeds. The cotton fiber can be removed from the motes using a delinting machine. This fiber is gin mote fiber and is used in nonwoven products

**Mullen burst** A strength test for paper and nonwovens, wherein the medium specimen is placed between annular clamps and subject to an increasing pressure from a rubber diaphragm pressed against it by a hydraulic pressure. The pressure is increase at a controlled rate till the specimen ruptures. The burst strength is the pressure at rupture.

**Multiplex structures** Also known as hybrid structures. A form of composite structure that is the result of combining two or more primary and/or secondary structures, at least one of which is a recognized flexible, fibrous structure to obtain a unified/integrated structure.

**Nanofibers** Term generally applied to fibers with diameters less than 0.5 $\mu$m. Typical nanofibers have diameters between 50 and 300 nanometers. They cannot be seen without visual amplification. Other terms often applied to very fine fibers are micro-denier, sub-micron, and superfine.

**Nanofiber membranes** A nanofiber web or a web with a nanofiber coating that has membrane-like filtration properties. The nanofiber part of the structure may be melt-blown, electrospun, or centrifugal spun. Such filter media are limited to microfiltration and ultrafiltration.

**Neat Resin** The condition and properties of a resin as produced by the manufacturer and as received by the customer or user. The resin may be further modified and diluted by the user in his formulating process.

**Needlefelt** A nonwoven structure resulting from the mechanical bonding of a nonwoven web or batt by needle punching.

**Needle punch** The process by which needle felts are produced by using vertical barbed needles to puncture the web (usually carded), entangle the fibers, and consolidate the structure. Sometimes referred to as needling.

**Neps** Small tangled knots of cotton fiber caused be mechanical processing.

**NIOSH rating** A rating system for respirators requiring NIOSH certification.

**N** A form of respirator that utilizes the wearers negative inhalation to draw ambient air through the air-purifying filter element (filter) to remove particulates from the ambient air. It is designed for use as respiratory protection against atmospheres with particulate contamination (e.g. dusts, fumes, mists) that are not immediately dangerous to life or health and that contain adequate oxygen to support life. An NPAPPR differs from a gas mask in that:

- **a.** Gas masks are designed for atmospheres and hazardous atmospheres where the contaminants are vapors or gases. NPAPPRS are designed for atmospheres with particulate contamination.
- **b.** Gas masks are designed for entry and escape only situations, whereas NPAPPRS allow the wearer to stay in the contaminated area for as long as the filter holds out.

**Novolak** A form of phenol formaldehyde resin that is the result of an acid catalyzed reaction of phenol and formaldehyde.

**Oblate ellipsoid** See prolate ellipsoid.

**Packing density** See solidity.

**Padder** A device used to apply latex resin to a nonwoven web by using an applicator roll rotating through a resin bath and applying the resin film to the web that is in surface contact with the applicator roll. Often the applicator action is followed by nip roll squeezing to remove excess resin.

**PAN based carbon fiber** Carbon fiber whose source is the alteration of organic fibers such as rayon, acrylics, etc.

**Part 11** Refers to 42 CFR (U.S.), Part 11

**Part 84** Refers to 42 CFR (U.S.), Chapter 1, Part 84

**Particle Filtration Efficiency** A test to determine filtration efficiency of surgical face mask materials by using a challenge aerosol of latex particles, usually 0.1 μm in diameter.

**Permeate** That part of a feed stream in a membrane separation operation that passes through the membrane

**Pervaporation** A membrane separation process whereby a volatile component is separated from a liquid mixture (the retentate) by selectively absorbing the volatile component (the permeate) into the membrane structure, allowing it to penetrate the thickness of the membrane and desorb from the other side of the membrane by a volatilization step such as a vacuum.

**Phenolic resin** A resin used as a bonding agent for wet laid filter media. It is the result of a reaction between phenol and some form of aldehydes. Phenol formaldehyde resin is, by far, the most commonly used form of phenolic resin

**Pigment** A substance used as a coloring agent, usually an insoluble powder, to be mixed with water, oil, or another base to produce paint and similar products

**Pitch based carbon fiber** Carbon fiber whose source is the remains of petroleum or tar distillation.

**Pleat pack** The pleated filter medium pack after it has come out of the pleating machine and cut to size, but before it is assembled into a panel or cartridge type filter element.

**Pocket filter** A filter containing multiple pockets to offer extended surface area and increased dirt holding capacity,

**Polar fluid** A fluid where the constituent molecules have a polarization such as a fluid of molecules that have a magnetic spin moment, or something like $H_2O$ where each individual molecule has a nonzero electric dipole.

**Pond regulator** Also known as a boat slice and as a back wall. A device in a Rotoforner® or inclined wire type of wet lay machine that controls the flow of water and stock over the forming fabric or wire at the point where fibers are deposited on the fabric and where the medium is formed. The amount of flow through the pond regulator and the flow relationship to the speed of the forming fabric has an important effect on the properties, formation, and fiber orientation of the medium.

**Pore** The channel of open space in a filter medium that transverses from one surface of the medium to the other surface. In theoretical development the channels are idealized as a multitude of perfect cylinders whose pore size is the diameter of the cylinder. The size of the pore diameters are determined by test procedures that measure the air pressure to displace a liquid of known surface tension from the medium.

**Porometry** Methods to determine pore size distribution data on media that are wetted with a liquid wetting agent. Extrusion flow porometry utilizes air to force the liquid from the pores of the medium. The pore size distribution is calculated form the flow—pressure relationship. Capillary flow porometry uses a non-reacting gas to displace liquid from the pores.

**Porosimetry** There are two forms of porosimetry: liquid intrusion porosimetry and liquid extrusion porosimetry. Liquid intrusion forces a liquid such as mercury into the porous medium. By measuring the volume of liquid forced into the pores and the pressure; pore volume and diameter can be calculated. Liquid extrusion works in the opposite manner by using a gas to force the liquid from the pores. Again pore volume and diameter can be determined.

**Porosity** The volume fraction in a filter medium that is nothing, that is, void or empty space. The porosity is one minus the solidity of the medium. (see solidity)

**Pot life** A measure of how long a resin formulation will last after it has been prepared.

**Potential flow** An idealized theoretical fluid flow in which there is no viscosity.

**Powered Air-Purifying Respirators (PAPR)** A respirator that utilizes a blower to force air through the respirator filter or canister. Licensing of PAPR particulate filters is limited to only high efficiency or HEPA filters, They are defined as "air-purifying respirators", with replaceable filters, that provide respiratory protection against dusts, fumes, and mists.

**Precured** A filter medium containing a resin that has been cured prior to being shipped to the customer.

**Presser foot** That part of a thickness gauge that rests on top of a medium when the thickness measurement is made. It is the distance between the presser foot and the base on which the medium rests that determined the caliper or thickness

**Prolate ellipsoid** Obtained by rotating an ellipse around its long axis so that the three dimensional shape is something like an egg. An oblate ellipsoid is obtained by rotating an ellipse around its short axis. Then the three-dimensional shape is something like a disc or pancake.

**Pulp** A mass of wood or vegetable material, composed of cellulose that has been reduced to a fibrous form.

**Pulper** A machine, usually a tank with suitable agitation, designed to break up, defiber, and disperse pulps and other fibers into a suitable slurry for the wet lay process. It is also used for blending other materials into the slurry.

**Pulse jet cleaner** A used filter cleaning mechanism that applies a high pressure reverse air flow to remove the cake from the filter and enabling its reuse. The reverse flow is usually from a high pressure pulse jet or fan pulse

**Pycnometer** A laboratory device use for measuring the density of a specimen.

**Qualitative filter papers** Laboratory filter paper for use in qualitative analytical techniques designed for identifying materials.

**Quantitative filter papers** Laboratory filter paper designed for use in analytical techniques intended to quantify the composition of materials. Purity and composition of the filter paper are of critical importance.

**Randomizing roll** A high-speed roll inserted between the master cylinder and the doffing roll. The randomizing roll doffs the web from the master cylinder and by centrifugal force transfers the fibers to the doffing roll. Random webs with MD:CD ratios of 4:1 are obtainable with this arrangement

**Rapidity** A measure of the liquid permeability of a filter sheet. Usually used for industrial filter paper. The rapidity is measure by the flow of water in a 100 cm tube through the specimen clamped at the bottom of the tube. The water height is maintained at 100 cm. throughout the test. Flow through the medium is the reported in ml/min.

**Raw Stock** A term applied to the form of base sheet that will later be resin treated either in-line with the forming process or treated off-line after the base sheet is produced.

**Reel** The machine that is that part of a web making process that winds up the finished web into jumbo rolls which are then subject to further processing.

**Reentrainment** The release of particles, trapped in the medium, back into the fluid stream. The release can be caused by a number of factors, such as high fluid velocity, rising pressure drop. mechanical vibrations, flow fluctuations, etc.

**Refining** A paper making or wet lay process step that mechanically treats pulp fibers, in aqueous suspension, by shearing them between two surfaces, at least one of which contains metal bars, the two surfaces not moving at the same speed. The effect is to hydrate, fibrillate, and/or cut the fibers to shorter lengths. Refining generally results in wet laid webs that are stronger and tighter (lower permeability).

**Resin** The general term for a chemical binder. A resin is a polymeric material that bonds the fiber cross-ections in a nonwoven web. It includes polymeric additives such as crosslinking agents that support the resin system. A resin may be a thermobonding resin or it may be thermosetting. It may be water based or solvent based. Water-based forms include latex emulsions, dispersions, and solutions. Solvent based resins are usually solutions of the resin dissolved in organic solvent.

**Resin penetration** The process by which a liquid resin applied to a nonwoven filter medium transports from the surface of the medium to the center or interior of the medium. The goal of a fully penetrated or saturated medium is to achieve uniform resin content throughout the depth of the medium. If the center of the medium has not been penetrated, "soft centers" may result, that is the medium is unbonded at its center and can easily be delaminated into two layers.

**Resin treated filter paper** Filter paper treated with a polymeric binder or resin to provide strength and structural properties.

**Resole** A form of phenol formaldehyde resin that is the result of a base catalyzed reaction of phenol and formaldehyde.

**Respirator** Any device designed to provide the wearer with respiratory protection against inhalation from a hazardous atmosphere. *A respirator for escape* only means respiratory devices providing protection only during escape from hazardous atmospheres. *Respirators for entry into and escape* from means respiratory devices providing protection during entry into and escape from hazardous atmospheres.

**Respiratory protective devices** See Respirator

**Retentate** The main part of a feed stream to a pervaporation unit.

**Retting** A natural, enzymatic process lasting approximately two weeks. The process breaks down the woody cores holding the fiber bundles together. The fibers are then easily separated. Hemp, flax, and coir are among several vegetable fibers processed by this technique.

**Reverse air cleaner** A means of cleaning a filter by reversing the fluid flow through the medium so as to dislodge the cake, dirt, and/or dust from the medium. The reverse flow is relatively low pressure compared to pulse jet cleaners.

**Reverse osmosis** The passage of a solvent (e.g., water), under high pressure, through a semi-permeable membrane from a solution of high concentration to a solution of lower concentration. The intent is to further dilute the contaminant in the solution of low concentration such as in water desalinization. The high pressure is required to reverse the normal osmotic pressure inherent in the process. Typical pressures are. 30–60 bar (3,000–6,000 kPa).

**Rewinder** A converting machine that accepts a jumbo roll from a web making process or a web converting operation and rewinds the web so that it can be trimmed and slit into finished rolls either for shipment to a customer or for further downstream processing. A rewinder that does both trimming and slitting is often referred to as a slitter/rewinder

**Reynolds Number (R*e*)** A dimensionless number that describes the nature of a fluid flow. In filter medium it is determined by a characteristic distance of the medium such as pore size or fiber diameter, multiplied by the flow velocity through the medium times the density of the fluid all divided by the fluid viscosity.

**Rotoformer®** A type of wet lay headbox specifically designed for filter media and wet lay nonwovens that contains a rotating drum whose surface is a forming fabric or wire. The wet lay medium is formed on the fabric of the rotating drum and transferred from the drum to downstream processing.

**Rovings** A collection of bundles of continuous filaments in untwisted strands.

**Saturator** A device that applies a resin to a substrate in such a way that the resin is uniformly penetrates the thickness of the substrate and is fully distributed throughout the thickness. Saturation is the way of applying resin binders to nonwoven webs.

**Scrim** A lightweight porous fabric, often woven, sometimes nonwoven, used as a support for nonwoven fabrics such as needle felts.

**Secant modulus** The slope of a line drawn from the origin of the stress-strain to a point of interest on the stress-strain curve. Usually this point of interest is maximum stress where the tensile strength is determined. In this case the secant modulus is the tensile strength divided by the stretch.

**Sheet mold** A device that is used to form laboratory sheets of paper or wet lay web by screening a slurry of the fiber so that the water passes vertically downward through the screen. The fiber is retained uniformly across the surface of the screen (often called a wire) to form a uniform wet sheet. The wet lay web so formed is then separated from the screen and dried.

**Shelf life** The length of time that foods, beverages, pharmaceutical drugs, chemicals, and other perishable items are given before they are considered unsuitable for sale, use, or consumption.

**Short cut fibers** Filament fibers that have been cut to shorter lengths. Also known as chopped fibers. Lengths generally range from 3 to 40 mm.

**Single fiber efficiency** The theoretical development of filtration efficiency based on the filtration efficiency of a single fiber in a fluid stream. The efficiency is based on the probability as a particle in that steam striking the fiber. Interception, diffusion, impaction, and combinations thereof are all considered.

**Singeing** An operation that burns off protruding surface fibers from a felt or needle felt.

**Size press** A device that applies liquid resins, chemicals, and/or additives to a wet lay web by passing the web through a set of rolls that either apply the liquid material as a coating or as a saturant. The rolls may apply the coating or saturant as a film by contacting the web with the coated surface of the rolls, or the rolls may be used to squeeze out excess coating and/or saturant already applied to the web from a size press pan, a shower, a spray, or other applicator device. The size press is the paper maker's version of the "padder" used to apply liquid additives to textile materials

**Slice** In a traditional type of wet lay machine or paper machine, the slice is the long horizontal gap across the width of the machine through which the diluted stock flows from the pond of the headbox to the forming area over forming fabric or Fourdrinier wire. The slice gap is adjustable and ranges from one to eight cm. It fixes the jet to wire ratio of the water flow over the wire. Thus in turn determines the MD−CD fiber orientation of the wet laid web.

**Slip** The slurry of clay that is poured into a designed mold to form ceramic materials. More widely known for its use in making ceramic pottery and other cast ceramic pieces it is also the source for sintered inorganic membrane media.

**Slip flow** The theoretical case where the fluid flow velocity at a nanofiber surface "slips" and is not zero. For larger fibers, theory assumes no slippage and the fluid velocity at the fiber surface is zero. The degree of slip is related to the Knudsen number.

**Slitter** A form of rewinder that rewinds and slits rolls into smaller sizes for customer requirements or for further downstream processing. A slitter differs from rewinder in that it is designed for special roll sizes that cannot be handled on a rewinder or that is awkward for a rewinder. An example is very narrow rows finished rolls that are less than 25 mm in width. The slitter usually is smaller than and often accepts split rolls from a rewinder for further slitting to smaller roll sizes.

**Sol** Colloidal dispersions of solid inorganic particles or polymers in a continuous liquid medium.

**Solidity** The volume of fibers or solid material per unit volume of filter medium. Synonyms are packing density and volume fraction. The solidity is equal to one minus the porosity.

**Solution resin** A resin system, either water based or solvent based, where the polymer is dissolved in the solvent to the point where particle size is on the order of molecular dimensions

**Solvent** An organic or non-aqueous liquid that is used for resin solutions.

**Solvent based resin** A resin that is solubilized in an organic solvent, such as methanol, toluene, methylene chloride, et al, and applied either as a coating or saturant to a nonwoven web. Phenolic resin dissolve in methanol or isopropanol is a large volume example of a solvent based resin.

**Sorption** A physical and chemical process by which one substance becomes attached to another. It is the common term used for both absorption and adsorption.

**Specific gravity** The ratio of the density of a substance to the density (mass of the same unit volume) of a reference substance. Most often the reference substance is water at 40°C. For gases the reference substance is usually air at 21°C in both cases the pressure is specified at 1 at (101.325 kPa). Apparent specific gravity is the ratio of the weight of a volume of the substance to the weight of an equal volume of the reference substance. It should be noted that specific gravity is not the same as density. Density is the weight of a specified volume of the substance. If the specified volume is 1 cc and the weight is in gm, then the density and specific gravity with water as the reference will be very close. On the other hand if the density is expressed in other units such as lb/gal then the numbers will be quite different.

**Spinnerette** That part of a die block assembly for a spun web process that contains the little holes through which the molten or solution polymer is spun into fiber. It is a single block of metal having several thousand drilled orifices or holes.

**Spin pump** Also known as a gear pump or a metering pump. A positive displacement pump for feeding molt polymer into the die assembly of a spunbond or melt-blown process.

**Springwood, (early wood)** That part of a tree that grows in the spring. Usually characterized by thin wall structure and small lumen.

**Spunbonded** A nonwoven forming process by which filaments are extruded from a spinnerette, drawn, and laid on a moving collector. Air turbulence acting on the fibers as they are drawn to the collector disorients the fibers and causes a random fiber orientation to the formed web.

**Spunlace** An accepted generic term for hydroentanglement—see hydroentanglement.

**Standards organization** An organization whose activities include developing, coordinating, promulgating, revising, amending, reissuing, interpreting, certifying, regulating, or otherwise maintaining standards that are related to filtration and filter media.

**Staple fibers** Non-filament fiber such as cotton or filament fibers cut to fiber lengths of 25–200 mm. staple fibers differ from short cut or chopped fibers by the way they are cut and because they are longer. They are usually used for dry formed carding operations.

**Stereoregularity** The degree to which successive configurations of a polymer in the space along the chain follow a simple rule. Also called tacticity.

**Stock** The aqueous slurry of fibers and ingredients that are fed to the headbox of the wet lay machine. "Thick stock" refers to the stock as prepared in a pulper and processed in the wet end. "Thin stock" refers to the slurry after it has been diluted for sheet forming in the headbox

**Stokes Number, *St*** Dimensionless group expressing the ratio of kinetic energy of a moving particle divided by the viscous drag forces acting on it. The stokes number is defined by:

$$St = \frac{d_p \rho_p v_p}{9 \mu R}$$

Where

St is the Stokes number.

$d_p$ is the particle diameter.

$p_p$ is the density.

$v_p$ is the particle density.

$\mu$ is the air or fluid viscosity

$R$ is the viscous drag distance

**Strain** Tensile property—The increase in length of a nonwoven specimen as stress is applied. Strain is measured as a percent of the original length before the load was applied.

**Stream function** In fluid mechanics, a mathematical concept which defines a flow field of two space coordinates, x and y, resulting in a useful stream function, $\psi$(x,y). This s function is not only mathematically useful, but also has physical meaning. Lines of constant $\psi$ are streamlines of the flow, i.e. they are everywhere parallel to the local velocity vector. No flow can exist normal to a streamline, thus, selected $\psi$ lines can be interpreted as solid boundaries of the flow. $\psi$ is also quantitatively useful for any two points in the plane flow field, the difference in their stream function values represents the volume flow between the points. The function is defined for incompressible (divergence-free) flows in two dimensions—as well as in three dimensions with asymmetry. The flow velocity components can be expressed as the derivatives of the scalar stream function. The stream function can be used to plot streamlines, which represent the trajectories of particles in a steady flow. In filtration celll model theory it is used to represent flow pasts a single fiber.

**Stress** Tensile property—The load applied to the specimen, ever increasing, as the tension increases. It is measured in force per unit width of specimen

**Stretch** Tensile property—The strain on a nonwoven specimen at the point of rupture. Like strain, it is measured as percent of original specimen length. It is often referred to as % elongation (see strain and elongation).

**Stripper roll** In the carding process stripper rolls are used to transfer the web from one roll to another. They have the wire points in the same direction as the master cylinder.

**Summerwood** That part of a tree that grows in the summer. Usually characterized by thick wall structure and small lumen

**Super absorbent polymers (SAP)** Polymer structures that absorb many times their weight in water.

**Suspension** A state in which particles are mixed with a fluid but are not dissolved. A mechanical suspension is one in which the particles are kept dispersed by agitation. A colloidal suspension is one in which the particles are kept dispersed by the molecular motion in the surrounding medium.

**Sustainability** Development that meets the needs of the present without compromising the ability of future generations to meet their own needs.

**Sustainable manufacturing** For the purposes of Commerce's Sustainable Manufacturing Initiative, sustainable manufacturing is defined as the creation of manufactured products that use processes that minimize negative environmental impacts, conserve energy and natural resources, are safe for employees, communities, and consumers and are economically sound.

**Swollen fiber** The fiber cell structure in a porous filter medium that is not available to permeable fluid flow.

**Syndiotactic** The substituent groups alternate on the repeat units of a vinyl polymer chain.

**Taylor Cones** Cone shaped obtrusions that emerge from the surface of a polymer solution under high electrical charge. When the electrical charge voltage attains a certain high level, Taylor Cones will erupt and emit a fine fiber like mass. This is the method of nanofiber production in electrospinning,

**Tear Strength** The force to begin or continue a tear in a fabric under specified conditions.

**Tenacity** The tensile stress on a fiber when expressed as a force per unit linear density of the unstrained specimen (expressed as $g_f$/den, N/tex, or cN/dtex). The breaking tenacity is the tenacity at the point of rupture.

**Tensile Energy Absorption (TEA)** Tensile property—An indication of the energy or work to break a tensile strip of nonwoven medium. This property is related to the toughness of the medium. In the constant rate of elongation tensile test, it is the integrated area under the stress–strain curve up to the point of rupture.

**Terpolymer** A polymer consisting of three different monomers

**Tex** The weight in grams of 1,000 m of fiber. decitex or dtex is the weight in grams of 10,000 m of fiber.

**Thermal oxidation** A way of controlling volatile organic compounds (VOCs) by oxidizing the vapors at very high temperatures (incineration).

**Thermobonding** A way of bonding nonwoven materials by using a low melting thermoplastic polymer. The polymer is melted in process and fuses with neighboring fibers so as to form thermal bonds between the fibers. The thermal bonding polymer may be in the form of a powder, a fiber, a low melting bead applied to the web, or a bicomponent fiber where one component is the low melting polymer.

**Tongue tear test** A tear test where a slit is cut down the middle of a 3 × 8 in (7.6 × 20.3 cm) specimen, The specimen is tested on a CRE (constant rate of elongation) tensile test machine by fastening one side of the slit to an upper clamp. The other side is turned down and fastened to a lower clamp. when the clamp movement is turned on, the tongue tear strength is the force necessary to propagate the tear from the slit.

**Tortuosity factor** A term used to define the Kozeny constant in the Kozeny-Carman equation. It defines the tortuosity of a porous medium as the length of the tortuous pore paths through the medium divided by the thickness of the medium.

**Tracheid** The long tapering cells of wood that become the fibers.

**Track edge membranes** Isotropic porous filter membranes where the pores are "drilled" by a radiation beam and are uniform is size.

**Trapezoid tear test** A trapezoid shaped test specimen is tested. The specimen is clamped in a constant-rate-of-extension (GRE) tester so that the diagonal sides of the specimen are in the clamp fixture. The pull is in-plane and normal to the direction of the slit cut at the top of the specimen. It tears the specimen by pulling it apart at the slit.

**Triboelectric** An electrostatic charge imparted by rubbing or intimate contact of two dissimilar dielectric materials. The kinetic energy dissipated from the rubbing action induces a charge transfer.

**Tristimulus** Relates to values giving the amounts of the three colored lights red, green, and blue that when combined additively, produce a match for the color being considered.

**Turbine** A rotating device designed to generate energy from an outside source of another energy. The energy generated may be mechanical energy or electrical energy. The turbine acts by the action of a very high velocity fluid flowing past the vanes of the turbine. The fluid can be gas, water, steam, or wind. There has to be a driving force to force the fluid past the vanes. In the case of water, hydroelectric turbines are driven by the gravity flow of water from a waterfall or dam that is channeled through the turbines. Gas turbines are driven by injected fuel ignition that heats and expands compressed air to force its flow past the turbine.

**Ultimate** The fusiform (taper ended) cells that make up a sisal fiber or filament. Also applies to other plant fibers such as flax.

**Viral Filtration Efficiency (VFE)** A test for microbial filtration efficiency using Bacterlophage OX174 as the challenge aerosol. The bacteriophage is one of the smallest of viral organisms, having a diameter in the range of .025–027 μm. The testing is very similar to the BFE test.

**Volatile Organic Compounds (VOC)** Organic chemicals that have a high vapor pressure at ordinary, room-temperature conditions. Does not include water.

**Volume fraction** See solidity.

**V-pack** Panel filters organized into a "V" array of subcomponent filters. The subcomponents are usually panels of mini-pleat media. By organizing the panels into the V-pack design, greater surface area of filter medium can be incorporated into the panel volume. This allows for higher flow volume, lower pressure drop, and greater filter life (dust holding capacity).

**Water based resin** A resin form where the polymer is contained in a water medium. Thus includes solution resins, emulsions, dispersions, lattices, and any other polymer form contained in a water medium.

**Water repellency** The ability of a substrate to resist wetting by water.

**Water repellent** The property of a material that has a finish that resists, but is not impervious to water.

**Water resistance** The ability of a material to retain its strength, stiffness, and dimensional stability after being wetted out with water.

**Water tolerance (or water solubility)** The maximum amount of water, in weight percent which, when added to a phenolic resin solution, before the resin kicks out enough for the resulting mixture to become turbid.

**Wet milling** An ancient process that is still used today for forming felt materials from wool and other fibers. The fibers are soaked with soapy water and then subject to repeated agitation and compression until the fibres hook together into a single piece of fabric. The fabric may then be cleansed and densified by fulling (see fulling)

**Wet smoke** A form of industrial fume that is an aerosol of liquid particles, such as oil mists, Contrast to dry smoke above.

**Wire** A fabric or woven screen upon which paper or wet laid products is formed during the wet lay process. Often called a Fourdrinier wire.

**Wire side** See Felt side.

**Worker roll** In the carding process, worker rolls act in conjunction with the master cylinder to do the fundamental carding action. Carding takes place between the point of the master cylinder and the worker roll

**X-ray microtomography** A radiation technique that can reproduce graphic three-dimensional images of small sections of nonwoven and paper structures.

**Z-direction** The direction in a medium that is perpendicular or normal to the surfaces

**Zeta potential** The electric potential across diffuse layers of ions surrounding a charged colloidal particle.

# NOMENCLATURE

## ENGLISH ALPHABET SYMBOLS

| | |
|---|---|
| $a$ | 1. The long axis of a prolate ellipsoid, m. $b$ is the short radius, m<br>2. The radius of the fiber in cylinder or cell models, m |
| $A$ | 1. Filter area. $m^2$<br>2. Tensile test—the area under the load-elongation curve, J<br>3. ASHRAE arrestance, ASHRAE 52-2, dimensionless, % |
| $A_{avg}$ | ASHRAE incremental arrestance, ASHRAE 52-2, dimensionless |
| $A_f$ | LRV equation: concentration of microorganiama in the feed, microorganisms/unit |
| $\acute{A}_i$ | ASHRAE average arrestance, ASHRAE 52-2, dimensionless |
| $a/b$ | The length/diameter ratio of a prolate ellipsoid, dimensionless |
| $b$ | 1. Diameter of a prolate ellipsoid at its center or thickest part, m<br>2. The radius of the outer cylinder in cylinder or cell models, m |
| $B$ | 1. Langmuir correction factor for determining the resistance of fibers transverse to the flow from the resistance of fibers parallel to the flow, dimensionless<br>2. Bechold correction factor or capillarity constant for tortuosity in the bubble point test. It has the value of 0.715 in ASTM F316, dimensionless |
| $B_f$ | LRV equation: the concentration of microorganisms in the filtrate, microorganisms/unit |
| $BL$ | Breaking length, m |
| $c$ | 1. The mass concentration of fiber per unit volume of medium, $kg/m^3$<br>2. Solid volume fraction (Chase) or solidity $\chi$ |
| $c_p$ | Particle concentration of an air stream approaching a filter element, $particles/m^3$ |
| $C$ | 1. A constant in the solution to the biharmonic equation<br>2. Brunauer, Emmet, Teller (BET) equation—a constant relating the adsorption energy of a gas to a solid substrate as compared to the liquefaction energy of the gas, dimensionless |
| **CADR** | Clean Air Delivery Rate – AHEM, $ft^3/min$ |
| $C_d$ | The drag coefficient, dimensionless |
| $C_{d\chi}$ | Drag coefficient of a cylinder moving in a tank—White's equation, dimensionless |
| $C''_D$ | Pavet correction to single fiber efficiency models, dimensionless |
| $C_0$ **and** $C_R$ | Correction terms applied by Rubow to the Lee and Liu single fiber efficiency models, dimensionless |
| $d$ | 1. The maximum pore diameter in the bubble point test, μm<br>2. The pore diameter in Kelvin's law, m |
| $d_b$ | Inter-fiber distance, m |
| $d_f$ | Mean fiber diameter, m |
| $d_{fe}$ | Effective fiber diameter, m |
| $(d_f)_s$ | Surface average fiber diameter, m |
| $d_p$ | Particle diameter, m |

**$\bar{d}$** The mean fiber diameter in the Piekaar and Clarenburg structure model, m

**$2\bar{d}$** Thickness of a layer in the Piekaar and Clarenburg structure model equal to twice the mean fiber diameter, m

**$dp/dz$** Pressure gradient across the z-direction or thickness direction of a medium, Pa/m

**$D$**
1. The particle diffusion coefficient. $m^2/s$
2. The diameter of a fiber in the aspect ratio $L/D$, m
3. A constant in the solution to the biharmonic equation

**$D_c$** The inside diameter of the cylinderin the Hagen–Poiselle equation, m

**$D_f$** Fiber diameter, m

**$D_h$** Hydraulic diameter, m

**$DHC$** Total dust holding capacity, ASHRAE 52-2, gm

**$DHC_i$** Incremental dust holding capacity, ASHRAE 52-2, gm

**$\mathscr{D}_{AB}$** Particle diffusion coefficient describing the motion of particles A. in a fluid B, $m^2/s$

**$e$**
1. $2e$ – The distance between layers in layered fiber structures, m
2. Liquid filtration efficiency by gravimetric or number means. %

**$E$**
1. Single fiber efficiency, dimensionless
2. Young's modulus, $N/m^2$

**$E_D$** Single fiber efficiency by diffusion, dimensionless

**$E'_D$** Single fiber efficiency by diffusion—correction by Payet, dimensionless

**$E_{DR}$** Combination term for E efficiency due to diffusion and interception in the Kuwabara flow field, dimensionless

**$E_I$** Single fiber efficiency by inertia, dimensionless

**$E_R$** Single fiber efficiency by interception, dimensionless

**$E_{RI}$** Combined single fiber efficiency term from $E_R$ and $E_I$—Stechkina et al. model, dimensionless

**$E'$** A constant in the solution to the biharmonic equation

**$F$**
1. A constant in the solution to the biharmonic equation
2. The drag force in the drag equation, N

**$F_L$** The force to move a prolate ellipsoid with its long axis parallel to the flow, N or $mg \cdot kg/s^2$

**$F_T$** The force to move a prolate ellipsoid with its long axis transverse to the flow, N or $mg \cdot kg/s^2$

**$F_T/F_L$** Force ratio acting on a prolate ellipsoid, dimensionless

**$g$** The gravitational constant (981 $cm/s^2$)

**$g$** The acceleration of gravity, $m/s^2$

**$G$** The grammage of a web, $g/m^2$

**$h$**
1. Height of the liquid reservoir in the bubble point test, cm
2. Filter thickness (Chase) m/s

**$k$** The permeability constant of Darcy's law, $m^2$

**$k_e$** The total decay rate (measured) – CADR equation, $min^{-1}$

**$k'$ and $k''$** Constants of White's drag equation, dimensionless

**$k_4$ and $k_5$** Constants of Chen's drag model equation, dimensionless

**$k_n$** The natural decay rate – CADR equation, $min^{-1}$

**$k_0$** A factor of the Kozeny constant that depends on the shape and sire of the cross-sectional areas of the pores, $\overline{L}$

**$K$** The Kozeny constant

| | |
|---|---|
| ***Ku*** | Kuwabara hydrodynamic factor |
| ***L*** | Length of a fiber in the Piekaar and Clarenburg structure model, m |
| ***L*** | **1.** The length of a fiber ïn the aspect ratio $L/D$, m<br>**2.** $2L$ – The distance between parallel fibers within a layer of layered fiber structures, m<br>**3.** The thickness of a filter fabric, pad, or medium, m<br>**4.** The length of the cylinder in the Hagen—Poiseuille equation, m<br>**5.** The original length of a specimen in the tensile test, mm |
| $L_e$ | The length of a tortuous pore path through a porous medium, m |
| $L_f$ | **1.** The length of fiber in the unit thickness h of a filter pad, m<br>**2.** Fiber length in a medium, m |
| $\overline{L}$, $\overline{M}$, & $\overline{N}$ | $\overline{L}$ is a unit of distance in the Kahn offset screen model, m. $\overline{L} \times \overline{L}$ is an area of a layer in the offset screen model composed of $\overline{N}$ equally spaced fibers in the first layer and $\overline{M}$ equally spaced fibers in the second layer and perpendicular to the fibers in the first layer. $\overline{N}$ and $\overline{M}$ are dimensionless |
| $L/D$ | The length to diameter or aspect ratio of a fiber, dimensionless $\overline{M}\,\overline{N}$ |
| $L_e/D$ | Tortuositv factor, dimensionless |
| ***LRV*** | Log reduction value, dimensionless |
| $\Delta L$ | **1.** An end correction for the fluid entering the cylinder in the Langmuir cylinder model, m<br>**2.** The change in length of the specimen in the tensile test, mm |
| $\Delta L/L$ | Strain |
| $m$ | Brunauer. Emmet. Teller (BET)—the mass of the specimen being tested, g |
| $m_h$ | The hydraulic radius in the Piekaar and Clarenburg structure model, m |
| $m_{hg}$ | Geometric mean radius based on a log-normal hydraulic radius distribution in the Piekaar and Clarenburg structure model, m |
| $M_d$ | The downstream mass of a contaminant in liquid filtration, kg |
| $M_h$ | The maximal hydraulic radius in the Piekaar and Clarenburg structure model, m |
| $M_s$ | The medium grammage, g/m$^2$ |
| $M_U$ | The upstream mass of a contaminant in liquid filtration, kg |
| $n$ | **1.** The concentration of aerosol particles leavng the filter—single fiber theory, partides/m$^3$<br>**2.** The total number of line interceptions per unit of surface area in the Piekaar and Clarenburg structure model, number/m$^2$<br>**3.** The mean number o! vertices in a polygon—Piekaar and Clarenburg structure model, number |
| $n_o$ | The concentration of aerosol particles entering a filter—single fiber theory, particles/m$^3$ |
| $n'_p$ | Corrected total number of pores in a slice of thickness $2\overline{d}$ and a unit square $l^2$ – Piekaar and Clarenburg structure model, number/m$^2$ |
| $n_v$ | Number of polygons sharing one vertex—Piekaar arid Clarenburg structure model, number |
| $N$ | Number of Lines per unit surface area Piekaar and Clarenburg structure model, number/m$^2$ |
| $N_{d,D}$ | The downstream particle count for particles of diameter $d$ or greater in a liquid filtration system, number |
| $N_{d,U}$ | The upstream particle count for particles of diameter $d$ or greater in a liquid filtration system, number |
| $N_f$ | The number of fibers per unit area of medium, number/in$^2$ |
| $N_{p'}$ | The number of pores per unit area of medium, number/m$^2$ |

| | |
|---|---|
| $p$ | 1. Pressure, Pa<br>2. In the bubble point test, the pressure at which the first bubble is observed, Pa |
| $p_a$ | Average pressure $(Po—P_L)/2$ in the Langmuir cylinder model, Pa |
| $p_0$ | Brunauer. Emmet, Telier (BET) equation and Kelvin's law—equilibrium vapor pressure, Pa |
| $p_v$ | 1. Brunauer. Emmet, Telier (BET) equation – gas vapor pressure, Pa<br>2. Kelvin's law—the vapor pressure at which condensation occurs, Pa |
| $P$ | 1. The probability that two lines intersect—Piekaar and Clarenburg structure model, dimensionless<br>2. The penetration of particles through a filter medium, percent or fraction<br>3. Penetration in the single fiber efficiency model, dimensionless |
| $P_L$ | The pressure at the exit side of the cylinder in the Hagen–Poiseuille equation, Pa |
| $P_0$ | The pressure at the inlet side of the cylinder in the Hagen–Poiseuille equation, Pa |
| $\Delta P$ | Pressure drop across the thickness of a fabric or medium, Pa |
| $\overline{\Delta p}$ | The average pressure drop for the Eurovent 4/11 energy consumption equation, Pa |
| $\Delta P_s$ | Pressure drop with slip flow, Pa |
| $\Delta P_{sn}$ | Pressure drop with no slip flow, Pa |
| $q_f$ | Quality factor for air filtration performance. Also known as the alpha ($\alpha$) or gamma number ($\gamma$) |
| $q_v$ | The air flow in the Eurovent 4/11 energy consumption equation, $m^3/s$ |
| $Q$ | Volumetric flow rate, $m^3/s$ |
| $r$ | Radial component in a polar coordinate system, m |
| $r_c$ | The inside radius of the cylinder in the Hagen–Poiseuille equation, m |
| $r_f$ | Fiber radius, m |
| $r_h$ | Hydraulic radius, m |
| $R$ | 1. The resistance to flow in the Langmuir cylinder model, $Pa/s/m^3$<br>2. The viscous drag distance in the Stokes Number, m<br>3. The interception parameter. $d_p/d_f$ in the Lee and Liu model, dimensionless<br>4. Grammage in the breaking length equation, $g/m^2$ |
| $R'$ | Basis weight in breaking length equation, lb/3,000 $ft^2$ |
| $R_0$ | The Hagen–Poiseuille resistance of the flow within a cylinder, $Pa \cdot s/m^3$ |
| $R_e$ | The Reynolds number, dimensionless |
| $R_{pa}$ | Anisotnopy parameter in Kahn's "offset screen model." $R_{pa} = \overline{M}/\overline{N}$, dimensionless – (see $\overline{L}$, $\overline{M}$, & $\overline{N}$ above) |
| $s$ & $t_2$ | Factors in the Piekaar and Clarenburg structure model equation for hydraulic radius, dimensionless |
| $\bar{s}$ | Mean surface area of a polygon in the Piekaar and Clarenhurg structure model, $m^2$ |
| $\bar{s}'$ | Corrected mean surface area of a polygon in the Piekaar and Clarenburg structure model, $m^2$ |
| $\bar{s}_h$ | Mean hydraulic surface area of the polygons in the Piekaar and Clarenburg structure model, $m^2$ |
| $\bar{s}'_h$ | Term which corrects for the mean line width or fiber diameter in the Piekaar and Clarenburg structure model, $m^2$ |
| $S$ | 1. Flow exposed surface of fibers per unit mass of medium, commonly called the specific surface area, $m^2/kg$<br>2. Pore surface area as determined by intrusion methods, $m^2/g$ |
| $S_i$ | Tensile stiffness, N/m S |

| | |
|---|---|
| $S_o$ | The effective surface area per unit volume of solid material in a medium, $m^2/m^3$ |
| $S_g$ | Gurley stiffness, mgf (Gurley unit) |
| $S_t$ | Taber stiffness, gm-cm |
| $t$ | 1. Time variable, s<br>2. The thickness of a web or medium, μm or m<br>3. The operation time for he Eurovent 4/11 energy consumption equation, h |
| $T$ | 1. Absolute temperature, °K<br>2. Tensile strength, kN/m |
| $T'$ | Tensile strength in the breaking length equation, lbf/in |
| $TEA$ | Tensile energy absorption as determined from the tensile test, $J/m^2$ |
| $T_g$ | Glass transition temperature, °C |
| $TI$ | Tensile index, m |
| $T_m$ | Melting point, °C |
| TU | Triboelectric unit, dimensionless |
| $u$ | 1. The velocity of the cyìnder in the Happel theory, m/s<br>2. The velocity of an object moving through a fluid in drag theory, m/s |
| $u'$ | The average velocity through a cylinder in the Hagen–Poiseuille equation, m/s |
| $U$ | Face velocity of a gas at the filter surface (Chase) m/s |
| $V$ | 1. Pore volume, $m^3$<br>2. Test chamber volume – CADR equation, $ft^3$ |
| $\overline{V}$ | The molar volume of the condensed vapor in Kelvin's law, $m^3/mol$ |
| $W$ | 1. Brunauer. Emmet, Teller (BET) equation – the amount of adsorbed gas, mol<br>2. Wïdth of a specimen in the tensile test, mm<br>3. The Eurovent 4/11 energy consumption as determined from filter testing in accordance with EN779, kWh<br>4. ASHRAE 52-2: the total dust fed during the life of the test, gm |
| $W_m$ | Brunauer. Emmet. Teller EETi Equation – the amount of gas to form a monolayer, mol |
| $W_d$ | The weight of synthetic test dust fed to the test filter – ASHRAE 52-2, gm |
| $W_i$ | Incremental amount of dust fed – ASHRAE 52-2, gm |
| $W_u$ | The weight of synthetic test dust captured on the final filter – ASHRAE 52-2, gm |
| $x$ | 1. Partice size of packed beds, m<br>2. The distance along an axis in the Langmuir cylinder model, m<br>3. The direction of fluid flow normal to the surfaces of a filter in the single fiber efficiency model, m |
| y | The width of a stream some distance upstream of a fiber where all theparticles in this stream are the ones that will strike the fiber—single fiber efficiency model, m |

## GREEK ALPHABET SYMBOLS

| | |
|---|---|
| $\alpha$ | 1. The volume of swollen fiber per unit mass of fiber in a medium, $m^3/kg$<br>2. Brunauer. Emmet. Teller (BET) – cross-sectional area of an adsorbed gas molecule, $cm^2$<br>3. $\alpha$ value: an air filtration quality factor for evaluating HEPA media. See $q_f$. |
| $\beta$ | The conversion factor from effective polygon area to real polygon area in the Piekaar and Clarenburg structure model, dimensionless |

| | |
|---|---|
| $\beta_w$ | Bulk of a web or medium. $cm^3/g$ |
| **$\beta$-ratio** | The ratio of upstream particles to downstream particles in a liquid filtration system, dimensionless |
| $\gamma$ | **1.** Filtration index in single fiber efficiency, $\gamma = EL_f d_f$, $m^{-1}$<br>**2.** $\gamma$ value: quality factor for filtration performance − see $q_f$ |
| $\gamma_L$ | The surface tension at a liquid surface, $mJ/m^2$ |
| $\gamma_{SG}$ | The interfacial tension at a solid−gas interface, $mJ/m^2$ |
| $\gamma_{SL}$ | The interfacial tension at a liquid-solid interface, $mJ/m^2$ |
| $\gamma_{r\theta}$ | Shearing strain, dimensionless |
| $\Delta f$ | The difference between the two force level in the determination of tensile stiffness, N |
| $\nabla$ | Operator, $m^{-1}$ |
| $\nabla \cdot \rho v$ | Vector operator indicating the divergence of the mass flux $\rho v$, $kg/m^3/s$ |
| $\nabla \cdot \tau$ | Viscous force on an element per unit volume, $kg \cdot m^3/s^2$ |
| $\varepsilon$ | Total void volume or porosity of a porous structure expressed as a decimal fraction of the total media volume, $m^3/m^3$ |
| $\varepsilon_0$ | The permeable or extra-fiber porosity of a porous medium expressed as a decimal fraction of the total medium volume, $m^3/m^3$ |
| $\varepsilon_{eff}$ | Effective flow porosity, the volume fraction of pore space open to flow, $m^3/m^3$ |
| $\varepsilon_{rel}$ | The fraction of total pore space open to flow, $m^3/m^3$ |
| $\varepsilon_{rr}$ | Normal strain, m/m |
| $\eta$ | The fan efficiency for the Eurovent 4/11 energy consumption equation, % |
| $\theta$ | **1.** Angular component in a polar coordinate system, rad<br>**2.** Bubble point test—the contact angle between the reservoir liquid and the pore wall, (°)<br>**3.** Wetting angle in Kelvins law, (°)<br>**4.** The contact angle between a liquid drop and the solid surface it is in contact withs |
| $\eta$ | Absolute viscosity, $N \cdot s/m^2$ |
| $\lambda$ | The mean free path of gas molecules, m |
| $\mu$ | Fluid viscosity, kg/m/s or Pa · s |
| $\mu_p$ | Particle mobility, s/kg |
| $\nu$ | **1.** The velocity through a tortuous pore, m/s<br>**2.** Mean velocity inside a filter medium, m/s |
| $\nu_f$ | Superficial velocity of the cylinder in the Happel theory, m/s |
| $\nu_\theta$ | Velocity component in the angular direction, m/s |
| $v_r$ | Velocity component in the radial direction, m/s |
| $\delta x$ | Length of a filter element in a fibrous filter of unit cross-section − Davies theory, m |
| $\rho$ | Density, $kg/m^3$ or $g/cm^3$ |
| $\rho_f$ | Density of the fiber in a medium, $g/cm^3$ |
| $\rho_p$ | The particle density in the Stokes number, $kg/m^3$ |
| $\rho_L$ | Density of the reservoir fluid at 20°C used in the bubble point test, $g/cm^3$ |
| $\rho_m$ | Density of the materials (fibers, resins, etc.) in a nonwoven medium, $g/cm^3$ |
| $\rho_w$ | The density of a web or medium, $g/cm^3$ |
| $\rho Dv/Dt$ | Rate of momentum accumulation per unit volume, $kg/m^2$ |
| $\sigma$ | **1.** Surfacetension.at 20°C, N/rn<br>**2.** The superficial tension in Kelvin's Law, N/m |
| $\sigma_{hg}$ | The standard deviation radius based on a log-normal hydraulic radius distribution in the Piekaar and Clarenburg structure model, m |

| | |
|---|---|
| $\sigma_{rr}$ | Normal stress, Pa |
| $\sigma_{r\theta}$ | Shearing stress, Pa |
| $\tau$ | The shear stress tensor, $kg/m/s^2$ |
| $\upsilon$ | Velocity vector, m/s |
| $\upsilon_0$ | Face velocity in a filter, m/s |
| $\upsilon_0$ | Superficial velocity of Darcy's law. m/s |
| $\upsilon_p$ | The particle velocity in the Stokes number, m/s |
| $\varphi$ | A reciprocal function in Langmuir cylinder model |
| $\chi$ | Solidty or packing density—solid volume of a porous structure expressed as a decimal fraction of the total medium volume, $m^3/m^3$ |
| $\chi_0$ | The apparent solidity of a porous web that would include voids or cells within the surface of the filter media fibers or particles as part of the solid volume expressed as a decimal fraction of the total medium volume, $m^3/m^3$ |
| $\psi$ | Stream function, $m^2/s$ |

## NUMBERS AND CONSTANTS

**Ck** Cunningham slip factor, dimensionless:

$$C_k = 1 + \left(\frac{\lambda}{d_p}\right)\left(2.492 + 0.84e^{\left(\frac{-0.435d_p}{\lambda}\right)}\right)$$

**$k_b$** Boltzman constant, $1.3805 \times 10^{(-23)}$ J/K

**Kn** Knudsen number, $Kn = 2\lambda/d_f$

**Ku** Kuwabara hydrodynamic factor,

$$Ku = -\frac{1}{2}\ln\chi - 3/4 + \chi - \frac{\chi^2}{4}$$

**$q_F$ or $\gamma$** Gamma number relating penetration to pressure drop, $Pa^{-1}$

$$\gamma = q_F = [\ln(1/P)]/(\Delta p)$$

**N** Avagadro number, $6.02 \times 10^{23}$ molecules or atoms/mol

**Pe** Peclet Number,
$Pe = \frac{1 \times 10^6 \upsilon_0 d_f}{D}$ where $D$ is the particle diffusion coefficient

**R** The universal gas law constant, 8.31 J/mol/°K

**Re** The Reynolds number

1. $Re = \rho v x/\mu$ for packed beds, Wakeman and Tarleton (32)
2. $Re = v d_f/\mu$ for fibrous media, Davies (33)
3. $Re = u d_f \rho/\mu$ for an isolated fiber, Lamb (56)

**St** Stokes number,

$$St = \frac{d_p \rho_p v_p}{9\mu R}$$

# BIBLIOGRAPHY


[1] INDA - EDANA. Leading Applications by Major Regions. INDA-EDANA. Worldwide Outlook for the Nonwovens Industry. INDA - EDANA, (2012).

[2] B. Kalil, Filtration, in: B. Kalil (Ed.), North American Nonwivens Industry Outlook, 2013–2018, INDA, Cary, NC, 2015.

[3] B. Kalil, Section II - macro drivers and assumptions, D – Energy and the manufacturing environment, North American Nonwove Industry Ouylook, 2013–2018, INDA, Cary, NC, 2015.

[4] J. Elaftherakis, Optimizing the Design Process, Filtration Technology - One Day Seminar, Fluid Technologies, Inc., 1999.

[5] I. Butler, Nonwoven Fabrics Handbook, INDA, Cary, NC, 1999.

[6] Encyclopedia Britanica. Encyclopedia Britannica, Inc. Encyclopedia Britannica, (1977).

[7] R. Krcma, Manual of Nonwovens, Textile Trade Press, 1971.

[8] ASTM. D1117, D13, ASTM Standards, ASTM, 1989.

[9] Hoechst Celanese Corp, Man-Made Fiber and Textile Dictionary, Hoechst Celanese Corp, Charlotte, NC, 1988.

[10] W.E. Houfek, Chapter 2 - nonwoven terminology, in: A.E. Turback (Ed.), Nonwovens: Theory, Process, Performancce, and Testing, TAPPI Press, Atlanta, GA, 1993.

[11] American Paper Institute, Inc, The Dictionary of Paper, fourth ed., American Paper Institute Inc., 1989.

[12] EDANA, What are Nonwovens. edana.org. [Online] [Cited: April 30, 2012.]. Available from: <http://www.edana.org/content/default.asp?PageID=33>.

[13] International Organization for Standardization, ISO 9092:2011, Definition for the Term: Nonwoven, ISO, Geneva, Switzerland, 2011.

[14] S.K. Batra, S.P. Hersh, R.L. Barker, B.S. Buchanan, T.W. Gupta, T.W. George, et al., A New system for classifying nonwovens, in: A.F. Turbak, T.L. Vigo (Eds.), Nonwovens, An Advanced Tutorial, TAPPI Press, 1989, pp. 9–22.

[15] S.M. Hansen, Chapter 5: nonwovens: theory, process, performance, and testing, in: A.F. Turbak (Ed.), Nonwoven Engineering Pricipals, TAPPI Press, Atlanta, GA, 1993, p. 55.

[16] S.K. Batra Ph.D., B. Pourdeyhimi Ph.D., Introduction to Nonwovens Technology, DEStech Publications, Inc., 2011.

[17] R.J. Wakeman, Filtration Dictionary and Glossary, The Filtration Society (UK), 1985.

[18] K. Sutherland, D.B. Purchas, Elsevier 2002. Handbook of Filter Media, second ed., Elsevier, 2002.

[19] Wikepedia, Sorption. Wikepedia, the Free Encyclopedia. [Online] Wikepedia. [Cited: May 12, 2012.] Available from: <en.wikipedia.org/wiki/Sorption>.

[20] T.C. Savage, R.C. Carlozzi, O. Petzoldt, Design and Testing of Antistatic Filters for Industrial Air Filtration. Powder & Bulk Solids Conference and Exhibition, (May 2001). Available from: <https://www.gore.com/MungoBlobs/581/280/article_design_test_antistatic_filters.pdf>.

[21] Zeta Plus® A Series Media, 3m.com. [Online] Cuno Incorporated, (1990). Available from: <http://multimedia.3m.com/mws/mediawebserver?mwsId = SSSSSu7zK1fslxtUnx2B48_Uev7qe17zHvTSevTSeSSSSSS>.

[22] Ahlstrom Corporation, Ahlstrom Disruptor®. Ahlstrom. [Online] Ahlstrom Corporation, (2013). Available from: <http://www.ahlstrom.com/en/products/enduseApplication/foodAndRetail/waterFiltration/Pages/AhlstromDisruptor.aspx>.

[23] R.W. Murphey, Composite Filter Media. Personnel communication with Hutten I.M. (Ed.), unpublished, (2004).

[24] L. Crook, Dry laid systems, Chapter 7, in: A.F. Turbak (Ed.), Nonwovens: Theory, Process, Performance, and Testing, TAPPI Press, 1993.

[25] S.R. Malkan, L.C. Wadsworth, Polymer laid systems, Chapter 8. Turbak A.F. (Ed.), TAPPI Press, Atlanta, Georgia, USA, (1993).

[26] A. Formhals, Process and Apparatus for Preparing Artificial Threads. U.S. Patent 1975504 October 5, (1934).
[27] Y.M. Kim, K.R. Ahn, R. Kyoung, Y.B. Sung, R.S. Jang, Manufacturing device and the method of preparing nanofibers via electro-blown spinning process. U.S. 7618579 November 17, (2009). Signed to Dupont.
[28] S. Frisk, Nanofiber Processes and Products: Recent Developments. American Filtration Society - Fall 2013 Meeting, (2013).
[29] K. Graham, M. Ouyang, T. Raether, T. Grafe, M. Bruce, P. Knauf, Polymeric Nanofibers in Air Filtration Applications. AFS,15th Annual Tech. Conf. & Expo, (2002).
[30] O. Lastow, A. Podgorski, Chapter 3 - single fiber collection efficiency, in: K. Spurny (Ed.), Advances in Aerosol Filtration, Lewis Publishers, 1998.
[31] I.M. Hutten, "Nonwoven Filter Media" INDA Training Course, INDA, Cary, NC, 2012.
[32] CEN, EN1822-5:2009, High efficiency air filters (EPA, HEPA and ULPA) Part 5: Determining the efficiency of filter elements. Brussels Belgium : CEN European Committee for Standardization, (2009).
[33] S.H. Goldsmith, High Efficiency Air Filter Media - Synthetic and Glass Media. INDA - Filtration 2004, (December, 2004).
[34] K. Spurny, Advances in aerosol filtration at the end of the 20th century, Chapter 2, in: K. Spurny (Ed.), Advances in Aerosol Filtration, Lewis Publishers, 1998.
[35] E.A. Ramskill, W.L. Anderson, The inertial mechanism in the mechanical filtration of aerosols, J. Colloid Sci. 6 (1951) 416–428.
[36] C.N. Davies, Air Filtration, Acadamic Press, 1973.
[37] Wikipedia, Elecrodialysis. wikipedia.org. [Online] (February 2014).
[38] KMX Corporation, Pervaporation Membranes. kmxcorp.com. [Online] KMX Corporation, (2014). Available from: <http://www.kmxcorp.com/water_chemical_purification.php?division=Technologies&area=Pervaporation%20Membranes&page=Introduction>.
[39] R.J. Wakeman, E.S. Tarleton, Filtration: equipment selection, Modeling and Process Simulation, Elsevier, 1999.
[40] J. Happel, Viscous flow in multiparticle systems: slow motion of fluid relative to beds of spherical particles, AICHe J. 4 (1958) 197–201.
[41] J. Happel, H. Brenner, Low Reynolds Number Hydrodrodynamics, Prentice-Hall, 1965.
[42] J.W. Schaeffer, An Ivestigation of Pressure Drop across Fibrous Filter Media. Ph.D. Thesis, University of Minnesota, (1976).
[43] R.B. Bird, W.E. Stewart, E.N. Lightfoot, Transport Phenomena, second ed., J. Wiley & Sons, Inc., 2002.
[44] J. Kozeny, Uber kappillare Leitubg des Wassers im Boden, Sitzungsber. Akad. Wiss. 136 (1927) 271–306.
[45] P.C. Carman, Fundamental principles of industrial filtration, Trans. IChemE. 16 (1938).
[46] A. Jena, K. Gupta, Advanced Technology for Evaluation of Pore Structure Characteristics of Filter Media to Optimize Their Design and Performance. pmiapp.com. [Online] (2002). Available from: <http://www.pmiapp.com/publications/index.html>.
[47] J.D. Lindsay, The Relative Flow Porosity of Paper in Liquid Flows. Tappi Nonwovens Conference Proceedings, 1993.
[48] C.N. Davies, The Separation of Airborne Dust and Particles, Proceedings of the Institution of Mechanical Engineers, vol. B1, (1952), pp. 185–198.
[49] E. Albrecht, Theoretische Untersuchungen ubfer die Ablagerung Von Staun aus der loft and ihre Anwendung auf die Theorie de Staubfilter, Phys. Z. 32 (1931) 48.
[50] W. Sell, Staubabscheidung an einfachen Korpern and Luftfiltern, VDI Ber. 247 (1931) 1–14.
[51] A. Kaufmann, Diz Faserstoffe fur Atemschuhfilter. Z. Verein Deutshes. Ing., 80, p 593.
[52] I. Langmuir, OSRD Report 865. U.S. Government Report, (1942), September 4.
[53] J. Happel, Viscous flow relative to arrays of cylinders, AIChE J. 5 (1965) 174–177.
[54] S. Kuwabara, The forces experienced by randomly distributed parallel circular cylinders or spheres in viscous flow at small reynolds numbers, Phy. Soc. Japan, J. Phys. Soc. 14 (1959) 527–532.
[55] S.W. Yuan, Fondations of Fluid Dynamics, Prentice Hall, 1967.

[56] A.A. Kirsch, N.A. Fuchs, The fluid flow in a system of parallel cylinders perpendicular to the flow direction at small reynolds numbers, J. Phys. Soc. Jpn. 22 (1967) 1251–1255.
[57] R.C. Brown, Airflow through filters-beyond single fiber theory, in: K. Spurny (Ed.), Advances in Aerosol Filtration, Lewis Publishers, 1998, pp. 153–172.
[58] J. Pich, Theory of aerosol filtration by fibrous and membrane filtes, in: C.N. Davies (Ed.), Aerosol Science, Acadamic Press, 1966, pp. 223–285.
[59] J. Pich, Pressure drop of fibrous filters at small Knudsen numbers, Ann. Occup. Hyg. 9 (1966) 23–27.
[60] T. Grafe, K. Graham, Nanofiber Webs from Electrospinning Nonwovens in Filtration. Stutgard, Germany: Fifth International Conference, (2003).
[61] G. Chase, What is Slip-flow and How Does it Affect Filter Performance? Nanofiltration Filtration & Separations Society. [Online] Available from: <http://www.afssociety.org/nanofiltration/84-what-is-slip-flow-and-how-does-it-affect-filter-performance>.
[62] Myagi, Viscous flow at low reynolds numbers past an infinite row of equal circular cylinders, J. Phys. Soc. Jpn. 13 (1958) 493–496.
[63] A.A. Kirsch, N.A. Fuchs, Studies on fibrous aerosol filters, II. Pressure drops in systems of parallel cylinders, Ann. Occup. Hyg. 10 (1967) 23–30.
[64] A.A. Kirsch, N.A. Fuchs, Studies on fibrous aerosol filters, III. Diffusional deposition of aerosols in fibrous filters, Ann. Occup. Hyg. 11 (1968) 299–304.
[65] N.A. Fuchs, A.A. Kirsch, I.B. Stechkina, A Contribution to the Theory of Fibrous Aerosol Filters. Faraday Division of the Chemical Society Symposium, No.7, Fogs and Smokes, (1973).
[66] A.S. Sangani, A. Acrivos, Slow flow past periodic arrays of cylinders with application to heat transfer., Int. J. Multiphas. Flow 8 (1982) 193–206.
[67] A.M. Khan, Historical review of filtration theories in fibrous media, AFSS Adv. Filtr. Sep. Technol. 9 (1995).
[68] B. Fardi, B.Y.H. Liu, Flow field and pressure drop of filters with rectangular fibers, Aerosol Sci. Technol. 17 (1992).
[69] R.C. Brown, A many-fibered model of airflow through a fibrous media, J. Aerosol Sci. 15 (5) (1984) 583–593.
[70] R.C. Brown, Theory of airflow through filters modeled as arrays of parallel fibers, Chem. Eng. Sci. 48 (20) (1993).
[71] M.L. Hildyard, D.B. Ingham, P.J. Heggs, M.A. Kalmanson, Integral equation solution of viscous flow through a fibrous filter, in: C.A. Maier, G. Brebbia (Eds.), Boundry Elements VII, Spring-Verlag, 1985.
[72] H. Lamb, Hydrodynamics, sixth ed., Cambridge University Press, 1932. p. 77.
[73] C.Y. Chen, Filtration of aerosols by fibrous media, Chem. Rev. 55 (3) (1955) 595–623.
[74] White, The drag of cylinders in fluids at slow speeds, Proc. Roy. Soc. London A 186 (1946) 472.
[75] A.M. Kahn, Offset Screen and Pseudo-Diffusion Models for Prediction of Filtration Phenomena in Real Fibrous Media. Ph.D. Dissertation. University of Rhode Island, Mechanical Engineering Department, (1990).
[76] D. Bouland, A. Renoux, Stationary and nonstationary filtration of liquid aerosols by fibrous filters-Chapter 4, in: K. Spurny (Ed.), Advances in Aerosol Filtration, Lewis Publishers, 1998, pp. 53–83.
[77] W.J. Kowalski, W.P. Bahnfleth, T.S. Whitman, Filtration of airborne microorganisms: modeling and prediction, ASHRAE (1999) 1–23.
[78] R.H. Perry, D.W. Green, Perry's Chemical Engineers' Handbook, seventh ed., McGraw-Hill, 1997.
[79] I.B. Stechkina, N.A. Fuchs, Studies in fibrous aerosol filters. I. Calculation of diffusional deposition of aerosols on fibrous filters, Ann. Occup. Hyg. 9 (1966).
[80] K.W. Lee, B.Y.H. Liu, Experimental study of aerosol filtration by fibrous filters, Aerosol Sci. Technol. 1 (1981).
[81] K.W. Lee, B.Y.H. Liu, Theoretical study of aerosol filtration by fibrous filters, Aerosol Sci. Technol. 1 (1982).
[82] B.Y.H. Liu, K.L. Rubow, Efficiency, pressure drop. and figure of merit of high efficiency fibrous and membrane filter media. Proceedings of the World Filtration Congress, (1990). p. 112.

[83] S. Payet, These de l'Universite Paris XII. Repport CEA-R-5589. Universite Paris, (1991).
[84] S. Payet, D. Bouland, G. Madelaine, A. Renoux, Penetration and pressure drop of a HEPA filter during loading with submicron liquid particles, J. Aerosol Sci. 23 (1992).
[85] J.B. Stechkina, A.A. Kirsch, N.A. Fuchs, Studies on fibrous aerosol filters. IV. Calculation of aerosol deposition in model filters in the range of maximum penetration, Ann. Occup. Hyg. 12 (1) (1969) 1–8.
[86] H.W. Piekaar, L.A. Clarenburg, Aerosol filters - pore size distribution in fibrous filters, Chemical Engineering Science, Pergamon Press, vol. 22. pp. 1399–1408.
[87] R.E. Miles, Proc. Natl. Acad. Sci. 52 (1964) 901.
[88] P. Kulkani, Mixed Hyduophilic/Hydrpphobic Fiber Midia for Liquid-Liquid Coalescence, University of Akron, 2011.
[89] CFD Review - Serving the CFD Community with News/Articles/Discussion. cfdreview.com. [Online] CFD Review, (December 2013). Available from: <http://www.cfdreview.com/>.
[90] Geo Dict. Geo Dict. geodict.de. [Online] Geo Dict, (2013). Available from: <http://www.geodict.de/>.
[91] R. Holdich, Fundamentals of Particle Technology, Midland Information Technology and Publishing, Shepshed, Leicestershire, U.K., 2002.
[92] C. Crowe, M. Summerfield, Y. Tsugi, Multiphase Flows with Droplets and Particles, CRC Press, Boca Raton, Florida, USA, 1998.
[93] A.G. Konstandopoulos, M. Kostoglou, N. Vlachos, E. Kladopoulou, Advances in the science and technology of diesel particulate filter simulation, Adv. Chem. Eng. Vols 33 (2008) 213–275.
[94] F. Cortés-Juan, B. Balannec, T. Renouard, CFD-assisted design improvement of a bench-scale nanofiltrarion cell, Sep. Purif. Technol. 82 (2011) 177–184.
[95] R.C. Brown, Air Filtration, Pergamon Press, Oxford, England, 1993.
[96] M. Knudsen, The Kinetic Theory of Gases, third ed., Melthuen, London, England, 1950.
[97] Kennard, Kinetic Theory of Gases, McGraw Hill, New York City, USA, 1938.
[98] S.A. Hosseinui, H.V. Tafreshi, Modeling permeability of 3-D nanofiber media in slip flow regime, Chem. Eng. Sci. 65 (2010) 2249*–22254.
[99] G. Jackson, D. James, The permeability of fibrous porous media, Can. J. Chem. Eng. 64 (1986) 364.
[100] L. Spielman, S.L. Goren, Model for predicting pressure drop and filtration efficiency in fibrous media, Environ. Sci. Technol. 2 (1968) 279.
[101] B. Ogorodnikov, Pressure-drop across FP fiber filters under gas slip-flow and in transition regime, Colloid J. USSR 38 (1976) 168–172.
[102] H.V. Tafreshi, M.S.A. Rahman, S. Jaganathan, Q. Wang, B. Pourdeyhimi, Analytic expressions for predicting permeability of bimodal fibrous porous media, Chem. Eng. Sci. 64 (2009) 1154.
[103] R.C. Brown, A. Thorpe, Glass-fiber filters with bimodal fiber size distributions 118 (1–2) (2001) 3–9.
[104] D.S. Clague, R.J. Phillips, A numerical calculation of the hydraulic permeability of three dimensional disordered fibrous media, Phys. Fluids 9 (6) (1997) 1562–1572.
[105] M. Rabai, M. Prat, P. Meireles, P. Schmitz, R. Baclet, A semi-analytical model for gas flow in pleated filters, Chem. Eng. Sci. 65 (2010) 2835–2846.
[106] A. Dittler, G. Kasper, Simulation of operational behavior of patchily regenerated, rigid gas cleaning filter media, Chem. Eng. Process. 38 (1999) 321–327.
[107] W. Tanthapanichakoon, T. Charinpanikul, W. Jintaworn, J. Laksmeearunotai, M. Amomkitbamrung, T. Fukui, et al., CFD investigation of high-temperature gas filtration using twin ceramic filters, 180 (2008) 245–252.
[108] A. Skyt, Advance products from narrow airlaid machines, Nonwovens World 13 (2004) 13.
[109] Kamath et al. present in text. Please provide reference list details.
[110] C.F. White, Hydroentanglement Technology Applied to Wet Formed and Other Precursor Webs. TAPPI Nonwoven Conference Proceedings, (1990), 177–187.
[111] J. Rupp, Spunlaced or hydroentangled nonwovens, part three. Available from: <http://www.textileworld.com/Issues/2008/July-August/Nonwovens-Technical_Textiles/Spunlaced_Or_Hydroentangled_Nonwovens> Textile World, Billian Publishing Inc., 2008.

[112] Voith Paperand and Trützschler Nonwovens. Of Wet Laying and Spunlacing. truetzschler-nonwovens.de. [Online] Voith.com, April 8, (2013). Available from: <http://www.voith.com/en/press/press-releases-99_43759.html>.

[113] SCS Global Sevices. Freudenberg Nonwovens Achieves Recycled Content Certification for Lutradur® Polyester Spunlaid Products. scsglobalservices.com. [Online] SCS Global Sevices, (April, 2014). Available from: <http://www.scsglobalservices.com/freudenberg-nonwovens-achieves-recycled-content-certification-for-lutradurr-polyester-spunlaid>.

[114] S. Ramkumar, V. Singh, Nanofiber-New Developments, Nonwoven Industry, April 7, 2011.

[115] B.W. Davis, K. Gramley, Moving Beyond Electrospinning: Melt Nanofiber Production. textile-world.com. [Online] (March/April 2014). Available from: <http://www.textileworld.com/Issues/2013/March-April/Nonwovens-Technical_Textiles/Moving_Beyond_Electrospinning-Melt_Nanofiber_Production>.

[116] J.E. Armantrout, A.B. Bryner, C.B. Spiers, -DuPont. Electroblowing Fiber Spinning Process. US7582247 B2 USA, (August 17, 2005). Grant.

[117] J. Wertz, I. Schneiders, Filtration Media: Advantages of Nanofiber Coating Technology. filtsep.com. [Online] (September 7, 2009). Available from: <http://www.filtsep.com/view/3826/filtration-media-advantages-of-nanofibre-coating-technology/>.

[118] D.L. Myers, B.D. Arnold, Electret media for HVAC filtration applications, Int. Nonwovens J. (Winter, 2003) 43–45.

[119] J. van Turnhout, Method for the Manufacture of an Electret Fibrous Filter. 3998916 USA, (December 21, 1976).

[120] B.D. Arnold, Filtration Medium with Enhanced Particle Holding Characteristics. 6815383 USA, (November 11, 2004).

[121] P.P. Tsai, L.C. Wadsworth, Methods and Apparatus for the Electrostatic Charging of a Web or Film. 5401446 (November 28, 1995).

[122] R.C. Brown, Nature, stability, and effectiveness of electric charge in filters, (Chapter 11), in: K. Spurny (Ed.), Advances in Aerosol Filtration, Lewis Publishers, 1998.

[123] N.L. Lifshutz, The Significance of Triboelectric Difference on Filtration Performance in Webs of Blended Fiber. AFSS 14th Annual Technical Conference, (May 2001).

[124] Hollingsworth and Vose Air Filtration Ltd. Technostat™ Electrostatically Charged Filter Media. Hollingsworth and Vose Air Filtration Ltd. Hollingsworth and Vose Co., Inc.

[125] E.S. Carter, A.E. Ortega, Description of Static – Dissipating Properties of Nystat™. Email from Ortega A.E. to Hutten I.M. Cerex Advanced fabrics, (2014).

[126] R.J. Brock, G.H. Meitner, Nonwoven Thermoplastic Fabric. US Patent 4041203 United States, (August 9, 1977). Assigned to Kimberly Clark.

[127] R. Chen, B. Martin, Innovative Coalescing Media SFM and Its Filtration Applications. Boca Raton, FL: AFFS - AFS 2012 Annual Conference, (2012).

[128] M. Middlebrooks, Filtration 101: Air Filtration of Gaseous Contaminants. INDA - Filtration 2004, (2004).

[129] Permatron Inc. Pleated Activated Carbon Filters for Odor, Fume and Gas Removal. permatron.com. [Online] Permatron Inc., (2010). Available from: <http://www.permatron.com/products/odors—fume-removing-filters/commercial-filters-dp2-ac-series/activated-carbon-pleated-filters.aspx>.

[130] D.L. Price, R.B. Simmons, D.G. Ahearn, Antimicrobial preservatives in air filter media - assessment of efficacy and significance of use, AFSS Adv. Filtr. Sep. Technol. 11 (1997).

[131] VOC and Odor Removal (Cabin Air Filters): Future Technologies – PCRA. pall.com. [Online] Pall Corporation, (2014). Available from: <http://www.pall.com/main/aerospace-defense-marine/voc-and-odor-removal-cabin-air-filters.page#future>.

[132] P. Chuaybamroong, R. Chotigawin, S. Supothing, P. Sribenjalux, C. Larpkiattaworn, Y. Wu, Efficacy of Photocatalytic HEPA Filter on Microorganism Removal, Indoor Air 20 (2010) 246–254.

[133] DuPont™. Hybrid Membrane Technology (HMT). dupont.com. [Online] DuPont™, (2009). Available from: <http://www2.dupont.com/Separation_Solutions/en_US/tech_info/hmt/hmt.html>.

[134] J.C. Crittendon, R.R. Trussell, K.J. Hand, K.J. Howe, G. Tchobanoglous, Water Treatment: Principals and Design, second ed., John Wiley & Sons, 2005.

[135] R.G. Guillen, E.M.V. Hoek, Development and Testing of "Smart" Nanofiltration Membranes. [Final Project Report]. Naval Water Research Institute, (2010). Available from: <http://www.nwri-usa.org/pdfs/NWRIReportonSmartNFMembranes07.2010.pdf>.

[136] G.L. Wilkes, Preparation of Microporous Membranes Through A Uniaxial Stretching Method. che.vt.ed. [Online] Virginia Tech University, (2000).

[137] A.E. Ortega, J.T. Walker, W.W. Whitfield, J. Bostwick, Materials and Methods for Removing Oil fronm Bodies of Water. US 2011/0272361 United States of America, November 10, (2011).

[138] J.K. Stille, Introduction to Polymer Chemistry, John Wiley and Sons Inc., 1967.

[139] MiniFIBERS Inc. Comparison of Fiber Properties. Available from: <http://www.minifibers.com>. [Online] MiniFIBERS Inc. Available from: <http://www.minifibers.com/techdata/TD-60.pdf>.

[140] C. Peart, F. Ludwig, The effect of synthetic fiber diameter on number of pores, pore size and efficiency, AFSS Adv. Filt. Sep. Technol. 14 (272) (2000).

[141] R.M. Broughton Jr., P.H. Brady, Structure and Properties of Paper - An Introductory Text, Auburn University, 1984.

[142] MiniFIBERS Inc. wide variety of precision cut fibers, synthetic pulps and papers, and specialty yarns. Available from: <http://www.minifibers.com>. [Online] MiniFIBERS Inc. Available from: <http://www.minifibers.com/products.htm>.

[143] G.A. Smook, Handbook for Pulp and Paper Technologists, second ed., Angus Wilde Publications, 1997.

[144] J. Boyer, Cotton in Nonwovens. TIP 1205-2 Technical Information Sheet. TAPPI, 1995.

[145] J. Dugan, Critical Factors in Engineering Segmented Bicomponent Fibers for Specific End Uses. Fiber Innovation Technology, Inc. - Presented at INDATEC 1999.

[146] Fiber Innovation Technologies. FIT Publications - 4DG™ Fibers. fitfibers.com. [Online] January (2000). Available from: <http://www.fitfibers.com/publications.htm>.

[147] R.R. Bresee, General effects of aging on textiles, J. Am. Inst. Conserv. 25 (1) (1986).

[148] R.A. Parhman, H.M. Kaustinen, Papermaking Materials - An Atlas of Electron Micrographs, Institute of Paper Science and Technology, 1974.

[149] J.S. Han, Fiber Property Comparison, TAPPI North American Nonwood Fiber Symposium. TAPPI Press, (1988), pp. 77–85.

[150] Cotton Inc. Cotton Fiber Technical Guide. cottoninc.com. [Online] Cotton Inc. Available from: <http://www.cottoninc.com/product/NonWovens/Nonwoven-Technical-Guide/>.

[151] Abaca. pulpspecialties.com. [Online] Isarog Pulp & Paper Co., Inc., 2007. [Cited: August 25, 2013.] Available from: <http://www.pulpspecialties.com/index-1.html>.

[152] J.M. Martinez, J. Gras, R. Ripol, J. Rio, Nonwoody Raw Fibers Quality: Influence on the Pulp Properties. TAPPI Press: TAPPI North American Nonwood Fiber Symposium, 1998, Atlanta, Georgia USA.

[153] Gesamtverband der Deutschen Versicherungswirtschaft e. V. (GDV). Esparto. TIS. [Online] © Gesamtverband der Deutschen Versicherungswirtschaft e. V. (GDV), (2002–2004). Available from: <http://www.tis-gdv.de/tis_e/ware/fasern/esparto/esparto.htm>.

[154] A. Slater, Tencel® Fibres in Filtration Applications. Email attachment from A. Slater to I.M. Hutten 9/20/2013. Lenzing Fibers A.G., (2013).

[155] Wikipedia. Polylactic acid. Wikepedia, the Free Encyclopedia. [Online] Wikepedia, (2012). Available from: <http://en.wikipedia.org/wiki/Polylactic_acid>.

[156] J.S. Dugan, Novel Properties of PLA Fibers. INDA - INTC 2000.

[157] O. Martin, L. Avérous, Poly(lactic acid): Plasticization and Properties of Biodegradable Multiphase Systems, Polymer 42 (14) (2001) 6237–6262.

[158] TAPPI. TIP 1205-1, Acrylic Fibers for Nonwovens. TAPPI Technical Information Sheet (TIP. TAPPI, (1994).

[159] A.G. Swicofil, Polyester (PET). swicofil.com. [Online] Swicofil AG, (2013). Available from: <http://www.swicofil.com/polyester.html>.

[160] www.fibre2fashion.com. Soaring Cotton Prices Benefits Polyester Fiber. fibre2fashion.com. [Online] Fibre2fashion.com, (2005). Available from: <http://www.fibre2fashion.com/industry-article/pdffiles/9/822.pdf>.
[161] ICIS, Weak Demand is Keeping Polyester Fiber Pricing Sluggish. icis.com. [Online] ICIS, October 28, (2011). Available from: <http://www.icis.com/Articles/2011/10/31/9503668/weak+demand+is+keeping+polyester+fiber+pricing+sluggish.html>.
[162] Fibersource, Nylon Fiber. fibersource.com. [Online] Fibersource. Available from: <http://www.fibersource.com/f-tutor/nylon.htm>.
[163] A.E. Ortega, Privarte Email: A. Ortega to I. Hutten. Albert Ortega, Cerex Advanced Products, Inc., (September 17, 2014).
[164] Evonik Fibers, P-84 - Engineering Materials. evonik.com. [Online] Evonik Industries, 2012. Available from: <http://www.p84.com/product/p84/en/application-areas/engineering-material/pages/default.aspx>.
[165] W.C. Smith, High Performance and High Temperature Resistant Fibers - Emphasis on Portective Clothing. intexa.com. [Online] PCC'99 Industrial Textile Associates, (September 9, 1999). Available from: <http://www.intexa.com/downloads/hightemp.pdf>.
[166] NF New Fibres, Inorganic Fibers. NF New Fibres. [Online] [Cited: October 2, 2013.] Available from: <http://technica.net//NF/NF2/efibreinorganiche.htm>.
[167] N.B. Scheffel, Recent developments in microglass media, AFSS Adv. Filt. Sep. Tech. 9 (1995).
[168] Lauscha Fiber International, Glass Micrifiber for Air and Liquid Filtration. unifrax.com. [Online] Lauscha Fiber International, (July 2013). Available from: <http://www.unifrax.com/web/Audit.nsf/ByUNID/86F293A69DE1C1A785257DA2000D130A/$File/Glass%20Microfiber.pdf>.
[169] Lauscha Fiber International, Glass Micrfiber. Technical Data Sheet. Lauscha Fiber International, (2003).
[170] R.D. Nixdorf, High Temperature Ceramic Fiber Nonwoven Filter Media Applied in Self Cleaning Filter Systems. AFSS 18th Annual Conference, (2005).
[171] Wikepedia, Carbon Nanofiber. en.wikipedia.org. [Online] Wikepedia, The Free Encyclopedia, (March 10, 2014). [Cited: March 14, 2014.] Available from: <https://en.wikipedia.org/wiki/Carbon_nanofiber>.
[172] WAL SA, Carbon Nanotube Filtration Media (WalSafeWater WMF-30). walsafewater.com. [Online] Technologies WAL S.A., (2014). Available from: <http://www.walsafewater.com/en/technology/carbon-nanotube-filtration-media.html>.
[173] Mott Corporation, Fiber Metal Gas Filtration. mottcorp.com. [Online] Mott Corporation. Available from: <http://www.mottcorp.com/resource/pdf/PSFIBERfinal.pdf>.
[174] P. Hulme, Invitation to discuss Sintered Metal fiber filtration media at Filtration 2013. [E-mail 10/14/2013]. Bekaert, (2013).
[175] A.N. Lisakovsky, Y.L. Tsybula, A.A. Medvedyev, Yarns of Basalt Continuous Fibers. The Fiber Society, New Frontiers in Fiber Science, Spring Conference, (2001).
[176] Heat Shield™, affco.com. [Online] American Felt and Filter Company (AFFCO). [Cited: October 9, 2013.] Available from: <http://www.affco.com/heatshield>.
[177] R. Komlenic, Ahlstrom Disruptor®. ahlstrom.com. [Online] Ahlstrom Filtration LLC, (2013). Available from: <http://www.ahlstrom.com/en/products/enduseApplication/foodAndRetail/waterFiltration/Pages/AhlstromDisruptor.aspx>.
[178] J. Dugan, E.Homonoff, Synthetic Split Microfiber for Filtration. Filtration (2000).
[179] H. Schreuder-Gibson, P. Gibson, Use of Electrospun Nanofibers for Aerosol Filtration in Textile Structures. ASC.com. [Online] U.S. Army Soldier Systems Center AMSSB-RMS-Ms (N), Natick, Massachusetts, USA, (2004). Available from: <http://www.asc2004.com/23rdASC/oral_-summaries/L/LO-05.PDF.2004>.
[180] S.-S. Choi, S.G. Lee, S.S. Im, S.H. Kim, Y.L. Joo, Silica nanofibers from electrospinning/Sol-gel process, Journal of Material Science Letters, Kluwar Acadamic Publishers, Korea, vol. 22, 2003.

[181] Z.-M. Huang, Y.-Z. Zhangb, M. Kotakic, S. Ramakrishnab, A review on polymer nanofibers by electrospinning and their applications in nanocomposites, Compos. Sci. Technol. 63 (2003).

[182] E.C. Homonoff, R.E. Evans, C.D. Weaver, Nanofibrillated Cellolose Fibers: Opening New Markets to Nanofiber Usage. St.Louis, Mo. TAPPI Nanofiber Conference, (2008).

[183] M de la. Giroday, Paper and Fibre Research Institute holds nanocellulose party/seminar. FrogHeart. [Online] FrogHeart Communications, (2012). Available from: <http://www.frogheart.ca/?tag=cellulose-nanowhiskers>.

[184] S.Gangwal, M. Wright, XanoShear™: A Novel, Scalable Technology Platform for Producing Polymeric Nanofibers. Advance preview provided by Xanofli Inc. Xanofi Inc., (2013).

[185] J.E. Hagewood, Spinning of submicron diameter fibers, AFSS Adv. Filtr. Sep. Technol. 14 (2000).

[186] J. Hagewood, Multi-Component Fiber Technology for Filtration and Other Applications. AFSS Ninth World Filtration Congress, (April, 2004).

[187] Intelligensys Ltd. Dry Add for Phenol Formaldihyde Resins. Intelligensys. [Online] Intelligensys Ltd, (2012). Available from: <http://www.intelligensys.co.uk/phenapps.htm>.

[188] R.C. Bell, M.G. Cove, Impregnated Papers for Automotive Filtration. Filtration Separation Magazine. (September/October, 1978).

[189] R. Dupré, Low Emission and Faster Curing Phenolic Resins for Liquid Filtration. INTC 2004 Conference, (2004).

[190] Hollingsworth and Vose Company, Resin Technology for Engine and Industrial Filtration. hollingsworth-vose.com. [Online] Hollingsworth and Vose Company, (October 2012). Available from: http://www.hollingsworth-vose.com/Documents/Product%20Literature-Filtration/ResinTechnologyInsert-EIF.pdf.

[191] R.W. Murphy, R.W. Ravel, Advance Cure Resin Systems for Oil Filtration. INTC 2004 Conference 2004, (2004).

[192] R.C. Bell, J.J. Latimer, C.E. Closthe. A New Approach to High Performance Glass Mats. Proceedings 1985 Nonwovens Symposium, TAPPI Press, (1985), pp. 33–38.

[193] M. Mlynar, ECO Product Line, Rohm and Haas Company, 2005.

[194] W.E. deVry, Chapter 5: Latex bonding chemistry and processes, in: T.L. Vigo, A.F. Turbak (Eds.), Nonwovens - An Advanced Tutorial, TAPPI Press, 1989.

[195] P.H. Brouwer, J. Baas, T. Wielema, Anionic Wet End Starch. TAPPI Paper Summit 2002, (2002).

[196] Calgon Carbon Corporation, Product Bulletin: ASZM-TEDA Impregnated Carbon. calgoncarbon.com. [Online] Calgon Carbon Corporation, (2012). Available from: <http://www.calgoncarbon.com/media/images/site_library/4_ASZM_TEDA_1pg_0917.pdf>.

[197] International, AAF, American Air Filters AmAir®/C, AmAir®/C+SAAFO xi™, AmAir®/C+SAAFO xi™, AmAir®/CP - Disposable Filters for Economic, Effective, Long-Lasting Gas-Phase and Particulate Filtration. aafintl.com. [Online] AAF International, JHune (2013). Available from: <http://www.aafintl.com/products/~/media/product%20media/Gas%20Phase/Gas%20Phase%20Brochures/amair%20c%20afp%201%20247%20pdf.ashx>.

[198] EFRA, Flame Retardants - Integral to Fire Safety. efra.com. [Online] EFRA (Euriopean Flame Retardants Associstion), (2005). Available from: <http://www.cefic-efra.com/index.php?option=com_content&view=article&id=6&Itemid=11>.

[199] TECHNICAL BULLETIN 117-2013: Requirements, Test Procedure and Apparatus for Testing the Smolder Resistance of Materials Used in Upholstered Furniture. Department of Consumer Affairs, State of California, USA. State of California USA, (January 2013).

[200] U.S. Department of Energy, Nuclear Air Cleaning Handbook, Chapter 3: Filters for the Nuclear Industry. energy.gov, DOE-HDBK-1169-2003. [Online] U.S. Department of Energy, (2003). Available from: <http://energy.gov/sites/prod/files/2013/09/f3/bibliography3_1.pdf>.

[201] SpecialChem, Plastics and Additives Formulation. specialchem4polymers.com. [Online] SpecialChem S.A., (2013). Available from: <http://www.specialchem4polymers.com/product-directory/biocides-antimicrobials-antibacterial-agents_0_100982/index.aspx?did=0>.

[202] S.W. Foss, D. Keser, A. Tefft, R.V. Sawvell, S.R. Brown, G. Goodwin, et al., Anti-microbial Fiber and Fibrous Products. US 6723428 USA, April 20, 2004. Foss Mfg. Co. Hampton New hampshire.

[203] Foss Manufacturing Co. Filtration. fosshield.com. [Online] Foss Manufacturing Co. Hampton, New Hampshire, (2013). [Cited: November 29, 2013.] Available from: <http://www.fosshield.com/where-it-is-used/filtration/>.

[204] I.L.J. Pittman, L. Irish, Optimizing airlaid wipe substrates through thermal bonding, Nonwovens World, MTS Publications, 2004.

[205] Homanoff, (Ed.), Personal Conversation. (March 5, 2004).

[206] J. Hansen, Personal Conversation. (March 17, 2004).

[207] J. Israel, J. Kiobasa, P. Hao, New Filtration Media Promising Twice the Life and Lower Emissions - Dura-LifeTM. AFSS 18th Annual Conference, (2005).

[208] J.E. Williamson, Chapter 6 - wet laid systems, in: A.F. Turbak (Ed.), Nonwovens: Theory, Process, Performance, and Testing, TAPPI Press, 1993.

[209] E. Schmalz, H. Fuchs, M. Brodtka, J. Schreiber, Innovative products from hydroentangled-stitch-bonded nonwoven composites for technical applications. Technica.net. [Online] (2001). Available from: <http://www.technica.net/NT/NT3/StitchBondedHydroentangled.htm>.

[210] E. Schmalz, Stitch-Bonded Nonwovenswith High Volume for the Separation of Coarse and Fine Components from Gaseous Media. TU Chemnitz, Dissertation. [Online] (July 2003). Available from: <http://archiv.tu-chemnitz.de/pub/2003/0114>.

[211] O. Jirsak, L.C. Wadsworth, Nonwoven Textiles, Carolina Press, 1999.

[212] E. Rodriguez, Principles of Polymer Systems, Routledge/Taylor and Francis Group, LLC, 1982.

[213] R.R. Bresee, U.A. Qureshi, M.C. Pelham, Influence of Processing Conditions on Melt-blown Web Structure. Toronto, Canada : INDA Intenational Nonwovens Conference (INTC), (2004).

[214] Elmarco s.r.o. NanospiderTM technology parameters. elmarco.com. [Online] Elmarco s.r.o., Czech Republic. Available from: <http://www.elmarco.com/upload/soubory/obsah/63-1-nanospidertm-technology-parameters.pdf>.

[215] DuPont. DuPont™ Hybrid Membrane Technology. dupont.com. [Online] (2007). Available from: <http://www2.dupont.com/Separation_Solutions/en_US/assets/downloads/DHM702_HMTSheetRev_me05.pdf>.

[216] K. Graham, H. Schreuder-Gibson, M. Gogins, Incorporation of Electrospun Nanofibers into Functional Structures, INTC, 2003.

[217] A. Lang, E. Schmalz, A New Generation in Filter Media. INDA Filtration 2004, (2004).

[218] C.P. Wood, Personal Communication. (September 2004).

[219] JCEM, Innovation. JCEM engineering and manufacturing. [Online] JCEM GmbH, (2013). Available from: <http://www.jcem.ch/>.

[220] Polyester Membrane Filter, sterlitech.com. [Online] Sterlitech Corporation. [Cited: January 26, 2013.] Available from: <http://www.sterlitech.com/membrane-disc-filters/polyester-pete-membranes/polyester-membrane-filter-pet0225100.html>.

[221] EMD Millipore Corporation, Membrane resources for researchers and decision makers. Millipore.com. [Online] EMD Millipore Corporation, January 12, (2013). [Cited: January 12, 2013.] Available from: <http://www.millipore.com/membrane/flx4/filter_manufacture_hm&tab1=3#tab1=3>.

[222] K. Li, Separation Membranes for Separation and Reaction, John Wiley & Sons Inc., West Sussex, England, 2007.

[223] M. Khalilia, S. Sabbaghia, H. Daneshmanda, M.M. Zerafatb, Synthesis and Characterization of $\alpha$-alumina membrane support and effects of binders, Int. J. Chem. Environ. Eng. 3 (2012).

[224] S. Petrik, M. Maly, K. Rubacek, J. Marcak, D. Stranska, J. Duchoslav, et al., Electrospun Nanofiber - The Tiny Layers that Add Great Value to Nonwovens, Edana - International Nonwovens Symposium, Stockholm, Sweden, 2009.

[225] M. Fretz, Personal Email: Mark Fretz to Irwin Hutten, Wacker Chemical Corporation, 2013.

[226] United States Government Accountability Office, GAO-10-875 - Formaldehyde in Textiles. United States Government Accountability Office, (August 2010).

[227] Company, Frazier Instrument, Conversion of Standard Units used by Frazier Permeability Tester. frazierinstrument.com. [Online] Frazier Instrument Company, (2003). Available from: <http://www.frazierinstrument.com/p-perm-faq.htm>.

[228] R. Singh, The Saturated Permeability of Composite Pulp Fiber and Filler Mats, ProQuest, Ann Arbor, Michigan USA, 2007.

[229] D19.08, ASTM Subcommittee: Standard Test Methods for Pore Size Characteristics of Membrane Filters by Bubble Point and Mean Flow Pore Test. ASTM F316 - 03 (2011): Standard Test Methods for Pore Size Characteristics of Membrane Filters by Bubble Point and Mean Flow Pore Test. ASTM International, (2011).

[230] A. Jena, K. Gupta, Characterization of Pore Structure of Filtration Media. pmiapp.com. [Online] (2002). Available from: <http://www.pmiapp.com/publications/docs/Characterization_of_pore_2002.pdf>.

[231] R.R. Bresee, Personal Conversations and Communications - Image Analysis. Dr. Randall Bresee, (2004).

[232] X. Huang, R.R. Bresee, Characterization of Nonwovens Web Structure Using Image Analysis Technology, Part I: Pore Analysis in Thin webs, vol. 5, INDA, 1993, pp. 14321.

[233] T.J. Ptak, D. Cady, Filter media pore size analysis, AFSS Adv. Filtr. Sep. Technol. 9 (1995) 37–45.

[234] S.S. Sreedhara, N.R. Tata, A novel method for measurement of porosity in nanofiber mat using pycnometer in filtration, J. Eng. Fibers Fabrics 8 (4) (2013).

[235] J.F. Bloch, Paper Structure Characteristics Using X-Ray Microtomography. AFSS - Ninth World Filtration Congress, (April, 2004).

[236] X. Thibakt, P. Cloetens, J.F. Bloch, F. Peyrkin, Nonwoven Characterization Using X-Ray Synchrotron Radiation microtomography. AFSS - Ninth World Filtration Congress, (April, 2004).

[237] C.L. Lin, J.D. Miller, April 14–17, 1999. Network Analysis of Filter Cake Pore Structure, 1st World Congress on Industrial Process Tomography, Greater Manchester, England, 1999.

[238] S. Brunauer, P.H. Emmett, E. Teller, Adsorption of gases in multimolecular layers, J. Am. Chem. Soc. 602 (1938) (1938) 309.

[239] Porosity Characterization Chart, pmiapp.com. [Online] Porous Materials Inc., (2013). Available from: <http://www.pmiapp.com/products/docs/Product_Explorer.pdf>.

[240] TAPPI T 494 0m-13: Tensile Breaking Properties of Paper and Paperboard (Using Constant Rate of Elongation Apparatus). TAPPI, (2013).

[241] ASME (American Society of Mechanical Engineers). HEPA Filters. ASME AG-1 Section FC. ASME, (2003).

[242] C. Habeger, D. Coffin, The role of stress concentrations in accelerated creep and sorption-induced physical aging, J. Pulp. Pap. Sci. 26 (2000).

[243] L. Pulkikinen, J. Fiskari, V. Alopaeus, The effect of sample size and shape on the hygroexpansion coefficient - A study with advanced methods of for hygroexpansion measurement, Taapsa J. (2009).

[244] Hygroexpansion - measure it accurately. innventia.com. [Online] Innventia AB, (2013). Available from: <http://www.innventia.com/en/Our-Expertise/Papermaking/Paper-mechanics-/Hygroexpansion—measure-it-accurately-/>.

[245] Paper Stability and Creeping, techpap.com. [Online] Techpap SAS, (2013). Available from: <http://www.techpap.com/paper-stability-and-creeping,lab-device,34.html>.

[246] P.A. Mango, Choosing the right Binder for Flame Retardant Nonwovens. TAPPI Nowovens Conference, (April, 1986).

[247] GTI Graphic Technology, Inc. gti True Light True Color. gtilite.com. [Online] GTI Graphic Technology, Inc, (2014). Available from: <http://gtilite.com/index.html>.

[248] TAPPI, T 1215 sp-12: The determination of instrumental color differences, TAPPI Standards, Technical Information Papers, and Useful Methods Online, TAPPI, Atlanta, Georgia, 2013.

[249] Filter Manufacturers Council, Technical Service Bulletin 89-5R3: The Micron Rating for Media in Fluid Filters. Filter Manufacturers Council, Revised (2005).

[250] ASHRAE, Method of Testing General Ventilation Air-Cleaning Devices for Removal Efficiency by Particle Size. ANSI/ASHRAE 52.2-2012. ASHRAE.

[251] CEN, Particulate Air Filters for General Ventilation - Determination of Filtration Performance. EN 779:2012. Comité de Européen Normalisation (CEN), (2012).

[252] (CEN, Comité de Européen Normalisation. EN 1822 "High Efficiency Air Filters (EPA, HEPA and ULPA)". CEN, (2010).

[253] Lydall Filtration/Separation Inc. Lydall Performance Materials. lydallfiltration.com. [Online] (2014). Available from: <http://www.lydallfiltration.com/app/air/comm.shtml>.

[254] N.L. Lifshutz, Multipass Filter Testing, Beta Ratio and Beta Rating. INDA Filtration 2004, (December 2004).

[255] AE-5A Aerospace Fuel, Inerting and Lubrication Sys Committee. AIR5455 Impact of Changes in Test Dust Contaminants and Particle Counter Calibration on Laboratory Filter Element Performance and Fluid Cleanliness Classes). SAE International, (January 3, 2012).

[256] Powder Technology Inc, Filtration Standards & Specifications. powdertechnologyinc.com. [Online] Powder Technology Inc., (2013). [Cited: March 11, 2014.] Available from: <http://www.powdertechnologyinc.com/test-dust-contaminants/filtration-standards-specifications/>.

[257] J. Warren, H. Marin, I. Marshall. The Evolution of a Spunbonded Nonwoven for Filtation Applications. DuPont Technical Paper, (2001).

[258] E. Mayer, J. Warren, Tyvek® SoloFlo™ Filtration Medium to Replace Tyvek® T-980. AFFS 11th Annual Conference and Exposition, (1998).

[259] DuPont™. dupont.com. DuPont™ Flashspun HDPE Filtration Media. [Online] DuPont™, (2007). Available from: <http://www2.dupont.com/Separation_Solutions/en_US/assets/downloads/K17043HDPE.pdf>.

[260] M. Wang, M. Korzeb, R. Sale, M. Entezarian, Single-Use Depth-Filtration System - Tutorial. No. 7, GEN Genetic Engineering and Biotechnology News, (April, 2012), vol. 32. Available from: <http://www.genengnews.com/gen-articles/single-use-depth-filtration-system/4058/>.

[261] C. Roman, Management of Metal Removal Fluid Environment - Disposable Media. awareservices.com. [Online] Organization Resources Counselors, Inc., 1999. <http://aware-services.com/orc/disposab.htm>.

[262] E. Mayer, Heavy Metal Removal eigth Dupont/Oberlin Microfiltration Technology, Fluid/Part. Separ. J. 6 (1) (1993) 20–26.

[263] Schroeder Industries, Excellement®. schroederindustries.com. [Online] Schroeder Industries, (2012). Available from: <http://www.schroederindustries.com/Products/ElementTechnology/ElementMedia/Excellement/default.aspx>.

[264] J.G. Kaukopaasi, N.B. Shah, Multi-layered Sheet - The Future Media for Filtration Applications. Tappi 1991 Nonwovens Conference, (1991) pp. 137–154.

[265] Bensch, L. Impact of Changes to ISO Standards on Contamination Control Programs. particle.com. [Online] Pall Corporation. <http://www.particle.com/whitepapers_hiac/ISO%20Changes.pdf>.

[266] Schroeder Industries LLC. Filter Systems - Product Catalogue. [Online] Schroeder Industries LLC, (2008). Available from: <http://www.schroederindustries.com/sites/IntranetContent/Departments/Marketing/Literature_Anon/L-2681C_FilterSystemsCatalogWeb.pdf>.

[267] M. Day, What Happened to NAS 1638? machinerylubrication.com. [Online] MKachinery Lubrication. [Cited: April 26, 3014.] Available from: <http://www.machinerylubrication.com/Read/409/nas-1638>.

[268] Docstoc. 1 2, Flat sheet tangential flow module The flat sheet tangential. docstoc.com. [Online] Docstoc, (2013). Available from: <http://www.docstoc.com/docs/99895133/1-2-Flat-sheet-tangential-flow-module-The-flat-sheet-tangential>.

[269] K. Sutherland, Filters and Filtration Handbook, second ed., Elsevier Ltd., 2008.

[270] Z.F. Cui, Y. Jiang, R.W. Field, Fundamentals of Pressure Driven Membrane Separation Processes, in: Z.F. Muralidhara, H.S. Cui (Eds.) Membrane Technology, Elsevier, 2010.

[271] A.F. Cui, H.S. Muralidhara, A practical guide to membrane technology and applications in food and bioprocessing. Elsevier Ltd. Membrane Technology, 2010.

[272] National Air Filtration Association, NAFA Guide to Air Filtration, fourth ed., National AIr Filtration Association, 2007.

[273] T.C. Dickenson, Filters and Filtration Handbook, fourth ed., Elsevier, 1997.

[274] W.L. Gore, Asociates, GORE® Membrane Technology versus Non-Membrane Media. gore.com. [Online] (2014). Available from: <http://www.gore.com/en_xx/products/filtration/baghouse/membrane_vs_non_membrane.html?isAjax = true>.

[275] T. Godbey, Ensuring effective dust collection in challenging environments, Int. Filtr. News 32 (2) (2013).

[276] J. Wertz, I. Schneiders, NANO WEB - Advantages of a New and Advanced Nanofiber Coating Technology for Filtration Media Compared to the Electrospinning Process. hollingsworth-vose.com. [Online] (November 2008). Available from: <http://www.hollingsworth-vose.com/Documents/White%20Papers/0821053%20NANOWEB%20WhitePaper.pdf>.

[277] AFFCO, Heat Shield™. affco.com. [Online] (2014). Available from: <http://www.affco.com/heatshield>.

[278] S. Fulko, Dura-LifeTM: A New Filter Ensures Twice the Filter Life-Time. donaldson.com. [Online] (November 4, 2004). Available from: <http://.emea.donaldson.com/about/news/033672.html>.

[279] AAF International, OptiFlo WFE 150 and Optiflo WFE 300 Carbo. aafgb.com. [Online] (February 2004). Available from: <http://www.aafgb.com/apc/pdf/wfe150c-300c.pdf>.

[280] K.E. Kinzer, R.C. Moreno, Performance characteristics of residential in-duct air cleaning devices, AFSS Adv. Filtr. Technol. 11 (1997).

[281] R. Martin, Nonwoven fabrics: mechano-electret filter media provides for IAQ advantages, Int. Filtr. News 33 (5) (2014).

[282] Erovent, EUROVENT 4/11 - ENERGY EFFICIENCY. eurovent-association.eu. [Online] (2012). Available from: <http://www.eurovent-association.eu/fic_bdd/magazine_en_fichier/Eurovent%20REC04-11%20Energy%20Efficiency%20Classification%20Filters,%20version%202011.01_13089042561.pdf>.

[283] AAF International, EN779:2012 and Energy Classification. slideshare.net. [Online] (2014). Available from: <http://www.slideshare.net/aafeurope/en779>.

[284] V.K. LaMer, D. Sinclair, OSRD 1668: Portable Optical Instrument for the Measurement of the Particle Size in Smokes, The OWL, an Improved Homogeneneous Aerosol Generator. The United States Government of Technical Services, (1943).

[285] H. Gilbert, High-Efficiency Particulate Air Filter Units, Inspection, Handling. Installation. United States Atomic Energy Commission, (1961). AEC Report TID-7023, NTIS.

[286] U.S. Government, MIL-F-51068: Filter, Particulates, High-Eficiency, Fire Resistant. United States. US Military Standard.

[287] United States. MIL-F-51079: Filter Medium, Fire Resistant, High Efficiency. United States Government. Military Standard.

[288] J. Larzelere, New and Novel Technologies in Particulate Air Filtration. jocotas.natick.army.mil. [Online] Naval Surface Warfare Center, Dahlgren Division, (2002). Available from: <http://jocotas.natick.army.mil/ColPro_Papers/Larzelere.pdf>.

[289] W. ter Kuile, R. Doig, Circular HEPA Filters for Use in Nuclear Containment and Ventilation Sytems. U.S. DOE/NRC, 25th DOE/NRC Air Cleaning and Treatment Conference, (1997).

[290] A. Boni, M. Clark, Realizing Higher Performance, Improved Processability, and Lower Costs with Next-Generation HEPA/ULPA Filter Media. hollingsworth-vose.com. [Online] Hollingsworth and Vose Co., Inc. <http://www.hollingsworth-vose.com/Documents/White%20Papers/0821026%20PerFormWhitePaperFinal.pdf>.

[291] ASME, Section FC: HEPA Filters. ASME AG-1, Code on Nuclear Air and Gas Treatment. (2012).

[292] ASME, Section FK: Radial Flow HEPA Filters. ASME AG-1, Code on Nuclear Air and Gas Treatment. ASME, (2012).

[293] W. Wang, P. Winters, Comparing test standards for HEPA and ULPA Filters 10–15 Perf. Rev. J. Control. Environ. Test. Assoc. (CETA) (2005).

[294] International Organization for Standardization. ISO 14644 Cleanrooms and Associated Controlled Environments – Part 1: Classification of Air Ceanliness. International Organization for Standardization, (1999).

[295] U.S. Department of Energy (DOE), How Gas Turbines Work. fe.doe.gov. [Online] U.S. Department of Energy. Available from: <http://www.fe.doe.gov./programs/powersystems/turbines/turbines_howitworks.html>.

[296] TDC, nordic-air-filtration.co. [Online] Available from: <http://www.nordic-air-filtration.com/en/filtermedia/gas-turbine-filter-media>.

[297] Donaldson Company, Inc. Self-Cleaning TTD Air Filter for Operating Airflows up to 88,000 cfm/2492 m3/min. Available from: <http://donaldsongts.com/>. [Online] Donaldson Company, Inc., (2002). Available from: <http://donaldsongts.com/donaldsongts-literatura/TTDsistema-donaldson.pdf>.

[298] United States Department of Labor, Part 84: Approval of Respiratory Protective Devices. 42 CFR. OSHA, (2013).

[299] Labor, United States Department of Personal Protective equipment. 29 CFR 1910.134. United States Department of Labor.

[300] T.G. Turnivall, Postoperative wound infections and surgical face masks: a controlled study, World J. Surg. 15 (1991) 383–388.

[301] D. Rawson, The basics of surgical face mask selection, Infec. Contr. Today (2003).

[302] R.A. Johnson, Nonwoven facemasks for healthcare applications, Techn. Text. Technol. (2003).

[303] K.C. Hsieh, Performance Comparison of Vacuum Cleaner Filters Made from Different Types of Filter Media. AFSS 14th Annual Technical Conference and Exosition, (May 2001).

[304] Freudenberg Filtration Technologies, Separet Filters in vacuum cleaners: freudenberg-filter.com. [Online] Freudenberg Filtration Technologies, (2013). Available from: <http://www.freudenberg-filter.com/fileadmin/templates/EN_downloads/Staubsaugerprospekt_E_2-09.pdf>.

[305] Hollingsworth and Vose Company. Appliance. hollingsworth-vose.com. [Online] (2012). Available from: <http://www.hollingsworth-vose.com/en/Products/Filtration-Media/Air-Filtration1/Appliance-Filters/>.

[306] R. Saar Test Method for Rating Performance of Portable Room Air Cleaners. Association for Home Appliance Manufacurers, (2006).

[307] A. Kreuter, Trends in Automotive – Implications for Filtration, INDA Filtration, Philadelphia, USA, 2012.

[308] D.R. Staley, Paper 881825: Correleting Lube Oil Filtration Efficiencies with Engine Wear. SAE Truck and Bus Meeting and Exposition, (1988).

[309] M. Durst, G-M. Klein, M. Moser, Automotive Filtration. Verlag Modern Industries, Technical collaboration with Mann and Hummel GMBH, (2002).

[310] R.W. Knize, Engine Oil Filter Study. Available from: <http://minimopar.knizefamily.net/oilfilterstudy.html> - no long active. [Online] R.W. KNIZE>, (2005).

[311] Ahlstrom, Engine Filtration Oil Synthetic Media - advertising brochure. Ahlstrom.

[312] Cummins Filtration, StrataPore™ Filtration Media. cummins filtration. [Online] Cummins Filtration, (2014). Available from: <http://www.cumminsfiltration.com/html/en/innovation/technology/stratapore.html>.

[313] SAE International, SAE HS-806/2009. SAE Oil Filter Test Procedure. SAE International, (2009).

[314] N. Bugli, Service life requirements for engine air induction filters, AFSS Adv. Filtr. Sep. Technol. 12 (1998) 38–49.

[315] P. Trautman, M. Durst, A. Pelz, N. Moser, High Performance Nanofiber Coated Filter Media for Engine Intake Air Filtration. AFSS 18th Annual Conference, (April, 2005).

[316] V. Bräunling, A. Yung, D. Miller, J. Kalets, A brief overview of cabin air filtration, AFSS Adv. Filtr. Sep.Technol. 14 (2000).

[317] H.R. Petri, T. Kessinger, N. Lifshutz, Comparisons of High Performance Filter Media. AFSS18th Annual Conference, (April, 2005).

[318] G. Hemighaus, T. Boval, J. Bacha, F. Barnes, M. Franklin, L. Gibbs, et al., Aviation Fuels Technical Review. cgabusinessdesk.com. [Online] Chevron Corporation, 2006. Available from: <https://www.cgabusinessdesk.com/document/aviation_tech_review.pdf>.

[319] G.B. Bessee, Keeping jet fuel clean and dry. Technol. Today. (Summer 2011). Available from: <http://www.swri.org/3pubs/ttoday/Summer11/PDFs/JetFuel.pdf>.

[320] EPA, Sustainability. epa.gov. [Online] United States Environmental Protection Agency. Available from: <http://www.epa.gov/sustainability/>.

[321] United States Environmental Agency (EPA), Clean Air Act (CAA),Title 42 U.S.C., Chapter 85. United States Environmental Agency (EPA, 1970, amended (1990).
[322] Agency, United States Environmental Protection. Occupational Safety and Health Act (OSHA), 29 CFR Part 1910. United States Government, (1970).
[323] W. Vatavuk, D.R. van der Varrt, J.J. Spivey, EPA/452/B-02-001, Section 3.2, VOC Destruction Controls. Available from: <http://cfpub.epa.gov/EPA>: Air Pollution Control Cost Manual, Section 3.2, Chapter 2 - Incinerators. [Online] United States Environmental Protection Agency, September (2000). Available from: <http://www.epa.gov/ttn/catc/dir1/cs3-2ch2.pdf>.
[324] Agency, United States Environmental Protection. Federal Water Pollution Control Act aka Clean Water Act (CWA). 33 U.S.C. 1251 et seq. (1972).
[325] OSHA, Occupational Safety and Health Act (OSHA) 29 USC 655. (1970): United States Government.
[326] OSHA, Hazard Communication Standard (HCS). 29 CFR 1910.1200(g). Occupational Safety and Health Administration (OSHA), (2012).
[327] United Nations, Globally Harmonized System of Classification and labeling of Chemicals, 4th Revised Edition. United Nations, (2011). Vol. ST/SG/AC.10/30/Rev.4.
[328] OSHA, Modification of the Hazard Communication Standard (HCS) to conform with the United Nations' (UN) Globally Harmonized System of Classification and Labeling of Chemicals (GHS). Hazard Communication. United States Department of Labor, (2014). Available from: <https://www.osha.gov/dsg/hazcom/hazcom-faq.html>.
[329] OSHA, OSHA Brief, Hazard Communication Standard: Safety Data Sheets. Hazard Communication Standard (HCS) (29 CFR 1910.1200(g)). United States Department of Labor - OSHA, (2012).
[330] OSHA, Hazard Communication Standard Labels. OSHA Quick Card. United States Depsrtment of Labor, (2014). Available from: <https://www.osha.gov/Publications/HazComm_QuickCard_Labels.html>.
[331] United States Environmental Protection Agency (EPA), Toxic Substances Control Act. 15 U.S.C., Para 2601 etc. United States Environmental Protection Agency (EPA), (1976).
[332] United States Govenment, RCRA Section 261.2 - Definition of Solid Waste. CFR Title 40 Chapter I Subchapter I Part 261 Subpart A Section 261.2. United States Government.
[333] A.T. Kearney, Inc. Handbook for Hazardous Waste Containers. Best Management Practices. Environmental Protection Agency (EPA) Region 6, (1997). Available from: <http://www.epa.gov/region6/6en/h/handbk4.pdf>.
[334] K.W. Lee, B.Y.H. Liu.
[335] C E. Richard, H. Viazmensky, J E. Williamson, Water entanglement process and product. US5009747A USA, April 23, (1991). Grant.
[336] (GDV), Gesamtverband der Deutschen Versicherungswirtschaft e. V. Coconut. TIS. [Online] Gesamtverband der Deutschen Versicherungswirtschaft e. V. (GDV).
[337] Porous metal solutions (Table p. 2). mottcorp.com. [Online] Mott Corporation. Available from: <http://www.mottcorp.com/resource/pdf/pmover.pdf>.
[338] P. Cusack, P. Patel, Environmentally-Friendly Inorganic Fire Retardants, ITRI Limited, 2011. Available from: <https://www.itri.co.uk/index.php?option=com_mtree&task=att_download&link_id=49894&cf_id=24>.
[339] H. Bechold, Durchlassugkeit von Ultrafiltern, Zeitschrift fuer Physik und Chemie 64 (1908) 328.
[340] R. Vijayakumar, IEST-RP-CC991.4: HEPA and ULPA Filters, Clean Room Mag. 20 (2) (2006).
[341] Encyclo Online Encyclopedia, Look Up: Count Median Diameter. encyclo.co.uk. [Online] (2012). Available from: <http://www.encyclo.co.uk/define/count%20median%20diameter>.
[342] Nuclear Ventilation High Efficiency Articulate Air. Available from: <http://www.eh.doe.gov/techstds/standard/hdbk1169/appendixc.pdf.2003>.
[343] American Bar Association, Section of Environment, Energy, and Resources. Regulation of Nanoscale Materials under the Toxic Substances Control Act. American Bar Association, (June 2006). Available from: <http://www.americanbar.org/content/dam/aba/migrated/environ/nanotech/pdf/TSCA.authcheckdam.pdf>.

# INDEX

## G

## I



## J



## K

## L



## M

## Q



## R

## S

FSC
www.fsc.org
MIX
Papier | Fördert
gute Waldnutzung
FSC® C083411

Kolibri 360 GmbH
Ferdinand-Jühlke-Straße 7
99095 Erfurt, Deutschland
produktsicherheit@kolibri360.de